Handbook of Experimental Pharmacology

Volume 169

Anxiety and Anxiolytic Drugs

Contributors

A. Bilkei-Gorzo, E.B. Binder, D.S. Charney, F. Crestani, R.J. Daher, W. Danysz, C.H. Duman, R.S. Duman, F. Holsboer, M.E. Keck, R. Landgraf, K.P. Lesch, R. Lieb, K.-M. Lin, M.T. Lin, A.C.E. Linthorst, N.C.P. Low, H. Möhler, M.B. Müller, K.R. Merikangas, J.R. Nash, A. Neumeister, D.J. Nutt, F. Ohl, C.G. Parsons, U. Rudolph, A. Ströhle, C.W. Turck, K. Vogt, C.T. Wotjak, R. Yehuda, W. Zieglgänsberger, A. Zimmer

Editors

Florian Holsboer and Andreas Ströhle

Professor Dr. Dr. Florian Holsboer
Max-Planck-Institut für Psychiatrie
Kraepelinstr. 10
80804 München, Germany
e-mail: holsboer@mpipsykl.mpg.de

Priv.-Doz. Dr. Andreas Ströhle
Klinik für Psychiatrie und Psychotherapie
Charité Campus Mitte
Charité – Universitätsmedizin Berlin
Schuhmannstr. 20/21
10117 Berlin, Germany
e-mail: andreas.stroehle@charite.de

With 139 Figures and 30 Tables

ISSN 0171-2004

ISBN-10 3-540-22568-4 Springer Berlin Heidelberg New York

ISBN-13 978-3-540-22568-3 Springer Berlin Heidelberg New York

Library of Congress Control Number: 2004115902

Springer is a part of Springer Science + Business Media
springeronline.com

Printed in Germany

Editor: Dr. P. Roos
Desk Editor: S. Dathe
Cover design: *design&production* GmbH, Heidelberg, Germany
Typesetting and production: LE-TEX Jelonek, Schmidt & Vöckler GbR, Leipzig, Germany
Printed on acid-free paper 27/3151-YL - 5 4 3 2 1 0

Preface

Research on anxiety and anxiety disorders is undergoing a paradigmatic transformation as disparate areas of psychiatric nosology, epidemiology, pharmacology and cognitive neuroscience converge towards an integrated understanding of the pathophysiology of these disorders.

In the last century, the basic treatment indicated for patients with anxiety disorders was to employ psychotherapy to facilitate changes in behaviour and develop ways of coping with stressful life events. A wide spectrum of somatic treatments from catharsis and emetics to opium and strengthening tonics, from atropine and digitalis to potassium bromide and chloral hydrate, from benzodiazepines to antidepressants came to be used as well. Systematic studies of antidepressants revealed that these drugs have antipanic properties independent of their antidepressive effects. This finding stirred a new classification of anxiety disorders, which is reflected in the current classification systems, such as the international classification of diseases (ICD) published by the World Health Organization (WHO). Anxiety has evolved as a defensive mechanism disposing the individual to recognize changes. As a warning signal, anxiety has life-saving qualities, and a species without appropriate anxiety would not survive. While normal anxiety is beneficial to co-ordinate response patterns in a threatening situation, pathological anxiety has many facets that can burden an individual substantially and warrants therapeutic intervention. The new classification of anxiety disorders encouraged basic and clinical research on the pathophysiology and treatment of pathological anxiety. Using newly developed methods and techniques, we are now beginning to understand the molecular mechanisms of anxiety, anxiety disorders and their treatment. In parallel, new drug targets have been generated and the first clinical studies with new compounds have been started.

In the first chapter, C.T. Wotjak describes the results of studies on the cellular basis of learning and memory together with a description of the methods that led to these discoveries. Aversively motivated learning and memory enable us to recognize and to appropriately respond to potentially dangerous situations. These abilities, which ensured the survival of humans and animals throughout evolution, bear the risk of pathological alteration that might be directly linked to distinct human anxiety disorders, such as phobias or post-traumatic stress disorder.

A detailed overview of animal models for anxiety-related behaviour is presented by F. Ohl. These models are indispensable tools to unravel the neurobiological mechanisms underlying normal anxiety as well as its pathological variations. The main concepts in generating animals models for anxiety, i.e. selective breeding, experience-related models, genetically engineered mice, and phenotype-driven approaches, are described and the potential opportunities and caveats of current models as well as the emerging possibilities offered by gene technology are discussed.

Although current views emphasize the joint influence of genes and environmental sources during early brain development, the physiological complexities of multiple gene and environment interactions as well as cross-talk between minor gene variants in the developmental neurobiology of fear and anxiety remain poorly understood. Focusing on the hypothalamic–pituitary–adrenocortical system, substance P and the serotonergic system, three chapters describe the impact of mutagenesis and knockout techniques on our current understanding of anxiety-related behaviour. K.P. Lesch reviews findings showing that variations in genes coding for proteins that control serotonin (5-HT) system development and plasticity establish 5-HT neuron identity and modulate 5-HT receptor-mediated signal transduction, and cellular pathways have been implicated in the genetics of anxiety and related disorders. In particular, pertinent approaches regarding phenotypic changes in mice bearing inactivation mutations of 5-HT receptors, 5-HT transporter, monoamine oxidase A and other genes related to 5-HT signalling are discussed. M.E. Keck and M.B. Müller describe how neuroendocrine and behavioural phenotypes of anxiety disorders are at least in part mediated via modulation of corticotropin-releasing-hormone (CRH) and vasopressin (AVP) neurocircuitry and that normalization of an altered neurotransmission after treatment may lead to restoration of disease-related alterations. A. Bilkei-Gorzo and A. Zimmer show that anxiety and depression-related phenotypes are profoundly affected by the tachykinin system.

The genetic epidemiology of anxiety disorders is reviewed by K.R. Merikangas and N.C.P. Low. They conclude that better comprehension of the phenomenology of the specific anxiety disorders and their overlap should guide the development of the next phase of diagnostic categories. In light of the rapidly accumulating information on genetic variations associated with anxiety disorders, we can expect that based on these genetic data new drugs will emerge not only for better treatment of the clinical conditions but also for preventing their onset.

The interactions between CRH and 5-HT and the implications for the aetiology and treatment of anxiety disorders are reviewed by A.C.E. Linthorst. A. Neumeister, R.J. Daher and D.S. Charney focus on the central role of noradrenergic neurotransmission for fear, anxiety and consequently the development and treatment of anxiety disorders. H. Möhler, K. Vogt, F. Crestani and U. Rudolph review the pathophysiology and pharmacology of the γ-aminobutyric acid $(GABA)_A$ receptors. The diversity of the $GABA_A$ receptors as described in the past decade is the basis for novel subtype selective benzodiazepine site ligands with hypnotic, anxiolytic, anticonvulsive or memory-enhancing activity.

The physiology and pathology of excitatory amino acid neurotransmission is described by C.G. Parsons, W. Danysz and W. Zieglgänsberger. At present, there seems to be a consensus that competitive AMPA and *N*-methyl-D-aspartate (NMDA) receptor antagonists have a low chance of finding therapeutic applications. Antagonists showing moderate affinity and satisfactory selectivity for certain NMDA receptor subtypes seem to have a more favourable profile.

C.H. and R.S. Duman focus on signal transduction and neural plasticity in the neurobiology and therapy of anxiety. The challenge of identifying intracellular signalling pathways and related molecular and structural changes that are critical to the aetiology and treatment of anxiety disorders will further confirm the importance of mechanisms of neuronal plasticity in functional outcome and improve treatment strategies.

Anxiety modulation by neuropeptides is described by R. Landgraf. Particularly due to their high number and diversity, the dynamics of their central release and the multiple and variable modes of interneuronal communication they are involved in, neuropeptides play a major role in the regulation of anxiety-related behaviour. Despite the immense progress in the field of neuropeptides and anxiety, we are far from mimicking these processes simply by administering synthetic agonists or selectively attenuating the pathology by administration of receptor antagonists. The only exception seems the development of antagonists blocking the effects of CRH. One of the CRH receptor antagonists has been probed in a clinical study with promising results. From the clinical perspective, R. Yehuda describes the neuroendocrine aspects of post-traumatic stress disorder (PTSD). The observations in PTSD are part of a growing body of neuroendocrine data providing evidence of insufficient glucocorticoid signalling in stress-related psychiatric disorders.

The clinical presentation of anxiety disorders according to the fourth edition of the *Diagnostic and Statistical Manual of Mental Disorders* is summarized by R. Lieb. In addition, selected aspects (prevalence, correlates, risk factors and comorbidity) of epidemiological knowledge on anxiety disorders are presented.

Pharmacogenetics is a field of research increasing our knowledge on the use of psychotropic drugs in different ethnic patient populations. K.-M. and M.T. Lin's chapter on transcultural issues summarizes current knowledge on the metabolism of anxiolytic agents with emphasis on pharmacogenetics and ethnic variations in drug responses.

Challenge studies in anxiety disorders are highlighted by M.E. Keck and A. Ströhle. The heterogeneity of agents capable of producing panic attacks in susceptible patients and the inconsistency of autonomic responses during a panic attack has led to the assumption that panic originates in an abnormally sensitive fear network, which includes the prefrontal cortex, insula, thalamus, amygdala and amygdalar projections to the brainstem and hypothalamus. The differences in sensitivity to certain panicogens, therefore, might be fruitful in serving as biological markers of subtypes of panic disorders and should be a major focus of research, as the identification of reliable endophenotypes is currently one of the major rate-limiting steps in psychiatric genetic studies.

The current state on the pharmacotherapy of anxiety disorders is summarized by J.R. Nash and D.J. Nutt. The recent shift in clinical practice towards the use of antidepressants, particularly SSRIs, for the first-line treatment of anxiety disorders is supported by research evidence from randomized controlled trials. It is only in recent years that drugs acting via GABA neurotransmission have been supplanted as first-line treatments, and new drugs in this class with improved tolerability compared to the benzodiazepines are likely to be marketed in the near future.

New developments in the pharmacological treatment of anxiety disorders are summarized by A. Ströhle. Further characterization of pathophysiological processes including evolving techniques of genomics and proteomics will generate new drug targets. Drug development design will generate new pharmacological substances with specific action at specific neurotransmitter and neuropeptide receptors or their reuptake and metabolism. New anxiolytic drugs may target receptor systems which only recently have been linked to anxiety-related behaviour. Combining psychopharmacological and psychotherapeutical interventions is a further field where benefits for the treatment of anxiety disorders could be achieved. Although the road of drug development is arduous, improvements in the pharmacological treatment of anxiety disorders are expected for the near future.

Pharmacogenetic strategies in anxiety disorders are described by E.B. Binder and F. Holsboer. This field holds great promise for the treatment of anxiety disorders, and in the future psychiatrists may be able to base the decision regarding the type and dose of a described drug on more objective parameters than only the diagnostic attributions used so far. This will limit adverse drug reactions and could reduce time to response, resulting in a more individualized pharmacotherapy.

Introducing proteomics, C.W. Turck shows that the comprehensive analysis of the protein complement of the genome of an organism is becoming an increasingly important discipline for the identification of disease targets. The effects of drug treatment and metabolism can now be studied on the protein level in a comprehensive manner.

We thank all the contributing authors for their excellent manuscripts. We thank K. Starke, who has initiated this volume and the Springer-Verlag team, especially S. Dathe, for the smooth co-operation. With this volume of the

Handbook of Experimental Pharmacology, we are happy to present an overview on the current state of basic and clinical research on "Anxiety and Anxiolytic Drugs".

Munich and Berlin, March 2005 F. Holsboer, A. Ströhle

List of Contents

List of Contributors

(Addresses stated at the beginning of respective chapters)

HEP (2005) 169:1–34

Learning and Memory

C. T. Wotjak

Research Group Neuronal Plasticity/Mouse Behaviour, Max-Planck-Institute of Psychiatry, Kraepelinstr. 2-10, 80804 Munich, Germany
wotjak@mpipsykl.mpg.de

Abstract Learning and memory processes are thought to underlie a variety of human psychiatric disorders, including generalised anxiety disorder and post-traumatic stress disorder. Basic research performed in laboratory animals may help to elucidate the aetiology of the respective diseases. This chapter gives a short introduction into theoretical and practical aspects of animal experiments aimed at investigating acquisition, consolidation and extinction of aversive memories. It describes the behavioural paradigms most commonly used as well as neuroanatomical, cellular and molecular correlates of aversive memories. Finally, it discusses clinical implications of the results obtained in animal experiments in respect to the development of novel pharmacotherapeutic strategies for the treatment of human patients.

Keywords Learning · Memory · Fear · Anxiety · Conditioning · Reconsolidation · Extinction · Sensitisation · PTSD

1 Introduction

One common characteristic of animals throughout the animal kingdom is their ability to adapt to suddenly changed environmental conditions. If these adaptations rest on modifications of the nervous system and become evident at the behavioural level, we say that the animals have learned. Whereas learning describes the process of adaptation, memory refers to the state and persistence of the adaptive behavioural changes. A typical learning curve consists of several subsequent phases. Memory acquisition is the phase of acute interaction of an organism with its changing environment that is characterised by admission of sensory information. During memory consolidation, the acquired information is further processed within the brain, leading to transient or lasting changes in interneuronal communication, i.e. to formation of memory engrams. Animals show memory retention as long as they are principally able to retain the adaptive changes of their behavioural performance. This, however, does not mean that they are always able to retrieve/recall the memory, i.e. to 'translate' the altered interneuronal communication into the adaptive behavioural changes. Most memories dissipate with time, due to reversal of the original changes in interneuronal communication or to additional modification of the respective neuronal circuits by new learning processes that disturb retrieval of the original memory. Memories can also be actively extinguished by training. The resulting state of reduced memory performance is called retention of memory extinction and is thought to rest on the formation of new memories that counteract the retrieval of the original one. The existence of various phases of memory processing underscores the importance of clearly describing which stage of the learning curve is targeted by a pharmacological study.

With the advent of molecular biology, refinements of neurophysiological tools and selection of suitable animal models, it became more and more feasible to search for the cellular basis of memory. This chapter will briefly summarise the results of this search together with a description of the methods that led to the discoveries. In this context, I will largely concentrate on aversively motivated learning and memory that enable us to recognise and to appropriately respond to potentially dangerous situations. These abilities, which ensured the survival of man and animal throughout evolution, bear the risk of pathological alteration that might be directly linked to distinct human anxiety disorder, such as phobias or post-traumatic stress disorder. In the beginning of this chapter, I will briefly introduce behavioural paradigms and animal models that turned out to be useful for the study of aversive memories, followed by a short description of neurological substrates and cellular mechanisms that are involved in the respective memory processes. I will end with a discussion of how closely animal experiments resemble the situation in human beings.

This chapter has been written for pharmacologists and physicians interested in the ways of studying the involvement of learning and memory in the aetiology of pathological anxiety. Hopefully, it will provide a guideline for better understanding the rationales and experimental strategies of the respective animal experiments and stimulate the searching for novel therapeutic targets. At the same time, it should sharpen the attention for potential caveats of various methodological approaches (cf. Wotjak 2004). It is not my intention to give an exhaustive review on the formation and extinction of aversive memories and the clinical impact of these processes. Readers interested in more detailed information will be referred to more specialised publications.

2 Behavioural Paradigms for Studying Aversive Memories

Motivation is essential for both memory acquisition/consolidation and memory retrieval. Principally, animals are motivated to approach rewarding and to avoid aversive situations. This has been used for the development of behavioural paradigms that help to study memory processing of appetitive (rewarding) and aversive (punishing) events. Most of these paradigms are of associative nature and can be assigned either to classical or to instrumental conditioning. However, aversive memories might be formed also in non-associative manner, e.g. by sensitisation (for a detailed description of theoretical and practical aspects of the learning paradigms see Mackintosh 1974; Dickinson 1980; Dudai 1989; Eichenbaum and Cohen 2001).

For human beings, two different memory categories have been introduced. According to Schacter, implicit (or unconscious/unaware) memory is revealed when previous experiences facilitate performance on a task that does not require conscious or intentional recollection of those experiences. Explicit memory, in turn, is revealed when the performance of a task requires conscious recollection of previous experiences. These are descriptive concepts that are primarily concerned with a person's psychological experience at the time of memory retrieval. Accordingly, the concepts of implicit and explicit memory neither refer to nor imply the existence of two independent or separate memory systems (Schacter 1987). As these two memory categories cannot be easily applied to the situation in animals, they will not be further considered in this chapter.

2.1 Classical Conditioning

Classical ('Pavlovian') conditioning is a process whereby a subject learns the associative relationships between discrete elemental or configural stimuli, with one stimulus being initially 'neutral' (or innocuous) to the animals (conditioned stimulus, CS) and the other (unconditioned stimulus, UCS) being able to evoke an unconditioned response. A distinct CS (designated CS+) comes to gain control over eliciting a conditioned response if the probability of a UCS occurrence in combination with the CS exceeds that of its unsignalled occurrence. If the two probabilities are equal, the CS has apparently no predictive value, in which case the lack of predictability itself is learned ('learned irrelevance'). If the probability of a UCS occurrence alone exceeds that of its combination with the CS, the CS (conditioned inhibitor, CS−) predicts the omission of the UCS. In latent inhibition studies, the CS will be presented several times before its pairing with a rewarding or aversive stimulus, with the consequence that animals will show a diminished conditioned response to it. Explicit unpairing of CS and UCS is often used as a control for the specificity of learning-induced changes in interneuronal communication, as both paired and unpaired protocols share similarities in number and intensity of CS/UCS presentation and differ solely in the temporal relationship between the two stimuli. However, this kind of control could be inappropriate, as unpairing induces a learning process as well, in that animals will regard the CS as a conditioned inhibitor.

The conditioned response elicited by a CS+ might be similar to the unconditioned response to the UCS. However, it seems to be more appropriate to assume that the conditioned response is elicited by the anticipation of the UCS rather than necessarily consisting of any component of the unconditioned response (Fanselow 1994; Gray and McNaughton 2000). The nature of the conditioned response depends on the UCS and the behavioural repertoire of a distinct species and cannot be controlled by the experimenter.

Fear conditioning and eyelid conditioning are the most frequently used paradigms of aversive classical conditioning. In these tasks, a tone or light (CS+) will be associated with a mild electric shock (UCS) applied either to the feet (fear conditioning) or to the eye (eyelid conditioning). As a consequence of this pairing, the CS+ will elicit a fear reaction that can be measured at the behavioural level as freezing (immobility except for breathing-related movements), fear-potentiated startle (potentiation of a normal startle reaction to a loud tone during presentation of the CS+, typically a light signal), conditioned suppression of an operant behaviour (e.g. lever pressing for food or water) or reflexive closure of the eye (eyelid or eye-blink conditioning). The conditioned emotional response becomes manifest also at the hormonal (increased secretion of stress hormones) and autonomic (e.g. tachycardia, galvanic skin response, rise in blood pressure) levels (Davis 2000). The majority of studies analyse the animals' freezing response to the CS. Other than measurements of startle responses or eyelid closures, this analysis does not require a sophisticated technical apparatus and enables behavioural observations in free, non-restrained animals. In any case, detailed knowledge about the neural basis of a selected behavioural response turns out to be essential for correct interpretation of the data in respect to the strength and persistence of the aversive memory. For instance, electrical stimulation of the sensory pathway that relays information of the tone signal to the lateral amygdala triggers a sequence of different fear reactions ranging from increased vigilance via freezing to escape behaviours, depending on the intensity of stimulation (Lamprea et al. 2002). Under these circumstances, an extraordinarily strong association between tone and shock can result in panic-like behaviour rather than a pronounced freezing response. Uncritical reduction of a complex behavioural phenotype to a single behavioural parameter could, therefore, easily lead to false-negative or false-positive findings.

Fear conditioning depends on the temporal overlap of CS and UCS (contiguity). In a common conditioning protocol, the onset of the tone precedes the shock by several seconds and co-terminates with it (delay conditioning). In other cases (trace conditioning) there is an interval between the end of the tone and application of the footshock that can last from milliseconds to several seconds. On longer intervals, the CS will usually not be associated with the UCS anymore. The situation is different for conditioned taste aversion (see also Sect. 2.2), for which CS and UCS presentation can be separated by several hours (for review, see Welzl et al. 2001). In general, contingency of a particular CS (i.e. its ability to predict the occurrence of the UCS) seems to be more critical for memory acquisition than contiguity (Mackintosh 1983; Rescorla 1988).

Animals form associations not only between a discrete elemental or unimodal CS (i.e. tone, light or odour) and the UCS, but also between the more complex test situation (configural or polymodal CS) and the shock. Configural CS are composed of a complex 'meshwork' of different unimodal CS (such as the shape, structure, material and smell of the conditioning environment), of the handling procedure and of information about the inner state of the animals. Memory of elemental CS is called cued memory; memory of the configural CS, contextual memory. In a common auditory fear-conditioning task, the appearance of the tone is temporally connected tightly with the presentation of foot shock (foreground conditioning), whereas the conditioning context is more latent (background conditioning).

2.2 Instrumental Conditioning

Instrumental ('Thorndikian', operant) conditioning is the process whereby the animals acquire new behavioural patterns that enable them to alter the frequency of their exposure to stimulus events. Whereas in classical conditioning subjects learn about relations between signal and significant events such as food or danger (stimulus–stimulus association), in instrumental conditioning they learn about relations between their behaviour and those significant events (response–stimulus learning). In instrumental conditioning, the experimenter controls the occurrence of the stimuli. The animals, by contrast, have more control over the occurrence of the response. If the occurrence of the stimulus is completely independent of the occurrence of responding, animals either do not change their baseline response rate (in case of a rewarding stimulus: learned irrelevance) or develop a special type of learned irrelevance called 'learned helplessness' (in case of aversive stimuli). If the probability that a response is followed by a rewarding stimulus (also called a positive reinforcer) is above chance levels, animals will show the behavioural response more frequently than during baseline conditions in order to maximise reward. Contrarily, if the response is followed by a punishment (also called a negative reinforcer), animals will reduce their response below baseline performance in order to minimise punishment. The latter situation is typical for passive (or inhibitory) avoidance paradigms in which animals are in an approach–avoidance conflict that performance of a natural behavioural response would lead to a punishment. Animals can avoid this punishment only if they remain passive. Typical examples for passive avoidance paradigms are step-down avoidance, step-through avoidance and conditioned taste aversion for rats and mice as well as bead pecking for chicks. In the step-down task, animals will be placed on a neutral platform that is localised on a metal grid. Animals receive a mild electric footshock as soon as they leave the platform. In the step-through task, animals will be placed onto the brightly illuminated floor of a test box that consists of a lit and a dark compartment connected by a sliding door. Because of their innate aversion to brightly lit environments, the animals will 'escape' to the dark compartment where they will receive a footshock. Memory performance is generally assessed by measuring the time until animals step down from the platform or leave the lit compartment on re-exposure to the respective test situation. In conditioned taste aversion (which contains aspects of both classical and instrumental conditioning), thirsty animals will be exposed to a fluid of novel taste (commonly a sucrose solution), followed by an injection of lithium chloride that causes nausea and discomfort. Animals will avoid consuming this fluid in the future, which is taken as a measure of the aversive memory. With bead-pecking avoidance, a similar task has been established for chicks. In this task, coloured beads are coated with a distasteful chemical compound and exposed to day-old chicks. Chicks that peck such beads show a disgust reaction and will avoid a similarly coloured but dry food in the future.

If in instrumental conditioning a response is not followed by a punishment, but its absence, animals will increase this response in order to minimise punishment. In active avoidance tasks, for instance, animals have to show a distinct behavioural response in order to avoid a punishment. Typical examples would be shuttle-box experiments and jump-up avoidance. A shuttle box consists of two compartments that are connected by a sliding door. The punishment will be signalled by either a tone or a light stimulus (CS). Animals have to leave the compartment in which the CS was presented within a selected amount of time, after which the CS would be followed by a footshock. In the pole-jump test, the occurrence of the footshock will be signalled by a tone or light stimulus as well. Animals can avoid the punishment if they jump onto a vertical wooden rod.

Memory performance will be assessed by the number of anticipatory responses (i.e. escape reactions during CS presentation). Whereas fear conditioning and passive avoidance tasks can be acquired within a single trial, active avoidance learning usually requires more intensive training. In case of repeated training, the distribution of learning events into several sessions (spaced learning) results in stronger memories than equivalent amounts of training crammed into a single session (massed learning).

2.3 Sensitisation

Both classical and instrumental conditioning are based on associative learning processes. However, animals might show an intensified or reduced behavioural response following non-associative learning as well. During sensitisation, a stressful, aversive event (e.g. footshock) leads to an unspecific increase in the sensitivity/reactivity to distinct sensory stimuli (Rosen and Schulkin 1998; Stam et al. 2000). The resulting aversive memory intensifies the animals' innate defence reaction. This definition implies that animals only become more sensitive to sensory stimuli that are generally able to elicit defensive reactions.

3 Animal Models

The majority of cellular signalling cascades involved in memory processing have been described in invertebrates, namely in the giant marine snail *Aplysia californica* (Abel and Kandel 1998; Kandel 2001) and the fruit fly *Drosophila melanogaster* (Dubnau and Tully 1998). Moreover, basic principles of memory consolidation have been discovered in chicks (Rose and Stewart 1999; Rose 2000). Nevertheless, this chapter will largely concentrate on rats and mice, which are the preferred experimental subjects for the study of cellular correlates of aversive learning and memory in mammals and most closely resemble neural processes of human being (Denny and Justice 2000; Bucan and Abel 2002).

Today, there are hundreds of different rat and mouse lines available. Newcomers to the field of animal experimentation might wonder what species and strains to use for a selected experimental question. Rats have the clear advantage over mice in that they are bigger (in particular when it comes to the stereotaxic targeting of small brain structures), less impulsive and superior to mice pertaining to the complexity of their behavioural 'repertoire' (Whishaw et al. 2001). Mice, in contrast, are the preferred subjects of geneticist. Their genome has been sequenced, and genetical tools for specific and sophisticated manipulations of the genome have been established exclusively for this species.

Moreover, their housing is less space- and cost-intensive, which predestines them for large mutagenesis screens, selective breeding and quantitative trait loci studies. Although rats are principally indispensable for behavioural experiments, mice will clearly dominate the experimental analysis of learning and memory for the next decade.

Before selecting one of the different mous and rat strains available from commercial suppliers for a given experiment, one has to carefully consider the rationale of the planned study (Andrews 1996; Crawley et al. 1997; Owen et al. 1997). Mice from C57BL/6 strains, for instance, are good learners in a variety of memory tasks, including amygdala- and hippocampus-dependent conditioning. DBA/2 mice, by contrast, are poor learners in hippocampus-dependent paradigms, including contextual fear conditioning (Paylor et al. 1994; Gerlai 1998). The selection of C57BL/6 or DBA/2 strains would therefore depend on whether one expects impairment or amelioration of memory performance after a certain pharmacological treatment.

With the advent of modern mouse genetics, mice could be generated that bear either a transgene, which will be expressed under control of a specific promoter (transgenic mice), a specific point mutation in a given protein, a null mutation of a gene (conventional 'knock-outs') or ablation of a gene in temporally and locally restricted manner (conditional 'knock-outs') (Picciotto and Wickman 1998). The most advanced generation of mutant mice (inducible 'knock-outs') allows the timed inactivation of a given gene by pharmacological means (Mayford and Kandel 1999). The latter animals turn out to be extremely useful for analysis of the involvement of the respective gene product in different phases of the learning curve (e.g. Shimizu et al. 2000; Genoux et al. 2002; Kida et al. 2002). Unlike conventional and conditional 'knock-outs', these animals do not bear the risk that alterations in their memory performance are due to developmental defects or compensatory processes (Gingrich and Hen 2000; Gross et al. 2002). Strictly taken, studies performed in conventional 'knock-outs' investigate the animals' ability to cope with the life-long and ubiquitous ablation of a given gene product. Quite often, this ability depends on the genetic background of the animals, indicating that the mutation targeted a 'specific gene ensemble' rather than a single gene (Routtenberg 2002). In any case, it is strongly recommended to validate major findings obtained with mutant mice by comparing them with intact control animals using pharmacological means.

Embryonic stem cells for the generation of 'knock-out' mice have been available from a small subset of inbred strains only (i.e. 129 strains), which, ironically enough, turned out to be poor learners in a variety of learning and memory tasks (Montkowski et al. 1997; Cook et al. 2002). Experimental success, thus, largely depends on an optimal breeding strategy (Wolfer et al. 2002), which, however, requires the co-ordination between molecular biologists and behavioural scientist at early stages of mouse generation. As soon as two different mouse strains are mixed, we face the problem of genetic background, in particular for the generation of null mutants by homologous recombination

(Gerlai 1996, 2001). In such cases, animals have to be crossed with a pure inbred strain (commonly C57BL/6J lines) for at least 6–8 generations (backcrossing). For correct interpretation of data obtained in genetically modified mice it is indispensable to follow the strict rules of strain nomenclature (Wotjak 2003).

The breeding scheme of 'knock-out' mice defines not only the genetic background, but also the availability of control animals. Control animals (wild-type) bear no mutation on their chromosomes. Their behaviour will be compared with that of mice with a mutation on each allele (homozygous) or one allele only (heterozygous). For behavioural experiments, animals of all three genotypes should derive from the same heterozygous breeding pairs. As littermates, they share a similar life history with respect to maternal care and stress. Homozygous breeding pairs should be avoided, as the mutation might affect maternal behaviour, which, in turn, has strong influences on stress susceptibility and memory capabilities of the offspring (Meaney 2001). However, even for heterozygous breeding pairs, there is still a risk of observing false-positive differences between homozygous null-mutants and their wild-type littermate controls. The offspring are not passively nursed by their mothers, but interact with them and compete among each other for resources. As mutants are commonly weaker than their wild-type littermates, this situation might be of disadvantage for them and lead to long-lasting changes in their emotionality.

Housing conditions are another important factor that influences memory performance (Würbel 2001). In standard laboratory conditions, rats and mice are housed under sensory deprivation in an extremely impoverished environment. It is, therefore, not surprising that animals that grew up in an enriched environment are better learners (van Praag et al. 2000). However, enrichment might eventually 'overwrite' effects of a mutation (Rampon et al. 2000). Furthermore, it might increase the variability among the experimental subjects, with negative consequences for the statistical interpretation of the data.

4 Neurological Substrates of Aversive Memories

For a long time, scientists had been sceptical that memory could ever be assigned to specific brain regions. However, as distinct mental functions such as movement co-ordination, perception, attention and language could be localised to different regions, it turned out that memory processes also critically depend on selected brain structures (Milner et al. 1998). As learned behaviour can be regarded as a refinement and further development of intrinsic (innate or inherited) behaviour (Vanderwolf and Cain 1994), learning-induced alterations in interneuronal communication primarily occur in those brain circuits that are involved in expression of the respective behavioural response. These brain circuits might differ in a number of brain structures, depending on the characteristics of the stimuli processed (e.g. olfactory vs auditory fear conditioning vs conditioned taste aversion). Nevertheless, there are a few brain structures that seem to be of general importance for most types of aversive learning. These structures include the amygdala (crucial for consolidation and, possibly, also storage of aversive memories; McGaugh 2000; LeDoux 2000) and the hippocampus (critical for learning and memory tasks in which discontiguous items must be associated, in terms of their temporal or spatial positioning; Wallenstein et al. 1998). Before I come to a short description of anatomical and functional features of these brain structures, I will briefly consider the methodological approaches that have led to the characterisation of such structures' importance for aversive memories.

4.1 How to Find a Candidate Brain Structure

First indicators for an involvement of a given brain structure in learning and memory processes are local changes in neuronal activity. These changes can be measured during different phases of the learning curve. In vivo methods [such as functional magnetic resonance imaging (fMRI), microdialysis procedures or electrophysiological recordings of electroencephalograms (EEG), sensory evoked field potentials and single unit activity] enable the monitoring of neuronal activity in conscious animals during memory performance. With these techniques, dynamic changes in neuronal activity can be observed over several learning phases within the same experimental subject. A disadvantage is, however, that most of these methods are cost intensive, invasive, show relatively poor temporal and spatial resolution (e.g. fMRI, EEG) and allow the simultaneous monitoring of a relatively small number of brain structures only.

In vitro (in situ) methods monitor neuronal activity off-line. For this purpose, animals have to be killed at a given time point of the learning curve, and markers of neuronal activity (e.g. expression patterns of immediate early genes or local accumulation of specific metabolic markers such as 2-deoxyglucose) (Sharp et al. 1993; Herdegen and Leah 1998; Sokoloff 2000) are visualised in the dissected brain. In vitro methods are less cost-intensive and less technically demanding than in vivo approaches. They allow, furthermore, analysis of a high number of brain structures. Elaborate statistical tools enable the characterisation of functionally relevant neuronal circuits (e.g. McIntosh and Gonzalez-Lima 1994). A disadvantage is, however, that these methods provide only a snapshot of neuronal activity during a distinct phase of the learning curve.

4.2 How to Prove a Causal Involvement in Learning Processes

Changed neuronal activity during a distinct learning phase provides at best an indirect hint for a critical involvement of this brain structure in memory processes. Lesion studies are, therefore, indispensable as a proof of causalities. We distinguish between permanent and transient lesions. Permanent lesions can be achieved by electrocoagulation, aspiration, knife cuts or, preferably, local administration of excitotoxins (Jarrad 2001, 2002). Permanent lesions have generally the disadvantage that they cannot be confined to distinct phases of the learning curve. Transient lesions, by contrast, can be achieved by cooling of distinct brain structures or local administration of anaesthetics, tetrodotoxin (which blocks the propagation of action potentials) and muscimol [an agonist of the γ-aminobutyric acid $(GABA)_A$ receptor]. Transient lesions allow for the dissection of the role of a brain structure for a given learning phase and provide information not only on 'where' but also 'when' and for 'how long' these processes take place, thus adding the chronological dimension to the topographical one (Ambrogi Lorenzini et al. 1999). The causal involvement of a distinct brain structure in learning and memory can, furthermore, be assessed by local pharmacological treatments. Compared to lesioning, this approach has not only the advantage that it is not destructive, but it is also informative as to the mechanisms of memory processing (Izquierdo and Medina 1998; McGaugh and Izquierdo 2000).

Brain structures can be lesioned either before (anterograde) or after (retrograde) a distinct phase of the learning curve. Notably, for anterograde lesioning there is the risk that animals bypass the lesioned brain structure and still show relatively normal memory performance, although the brain structure would have been involved in intact animals. In any case, care has to be taken that lesions or pharmacological treatments do not interfere with general locomotion, motivation or processing of sensory information, but specifically with memory processes.

4.3 Candidate Brain Structures

A variety of brain structures seem to be essential for aversive memories. In the following, I will briefly introduce the hippocampus and amygdala involvement, without disregarding the importance, for instance, of the cerebellum for eyeblink conditioning (Thompson et al. 1997, 2000; Medina et al. 2002) and the insular cortex for conditioned taste aversion (e.g. Berman and Dudai 2001).

4.3.1 Hippocampus

The hippocampus received its name from the similarity of the human hippocampus to the tail of a seahorse (Latin name, hippocampus). In mice and rats, however, there is little resemblance to a seahorse. In fact, in these species the hippocampus has a rather 'banana-like' shape in its rostro-caudal extension. Morphologically and functionally, scientists differentiate between the dorsal pole (also septal pole because of its close connections with the septum) and the ventral pole of the hippocampus. The hippocampus contains several anatomically and functionally well-defined cell fields (CA1 to CA4, named after *Cornu ammonis*, a snail that stimulated the morphologists' imagination in a similar manner as the seahorse did when it came to the description of the human hippocampus). Together with the entorhinal cortex, the dentate gyrus and the subiculum, the hippocampus comprises the hippocampal formation (Amaral and Witter 1989).

Inputs to the hippocampus are spread to the different cell fields primarily by the famous trisynaptic pathway (Amaral and Witter 1989). According to this simplified circuit, the entorhinal cortex projects to the dentate gyrus via the perforant path, the dentate gyrus to CA3 region via the mossy fibres and the CA3 region to the CA1 region via the Schaffer collateral. This trisynaptic pathway turned out to represent an excellent model system for studying cellular processes of synaptic plasticity. The fact that this pathway remains intact in coronal sections of the rat and mouse brain and that the subfields can be easily visualised opened the avenue for studying synaptic plasticity under in vitro conditions.

Plenty of evidence suggests an essential role for the hippocampus in the formation and extinction of aversive memories, in particular in passive avoidance learning (Izquierdo and Medina 1997). The hippocampus is not essential for acquisition and recall of cued fear memories in delay fear conditioning (Kim and Fanselow 1992; Phillips and LeDoux 1992). In contrast, it plays an important role for the processing of aversive memories following trace fear conditioning (Berger and Thompson 1976; McEchron et al. 1998) and background contextual conditioning tasks (Maren and Holt 2000; Sanders et al. 2003; but see also Gewirtz et al. 2000). As background contextual conditioning occurs in parallel to the acquisition of cued fear memories, analysis of drug effects on each of the two components enables the dissection of hippocampus-dependent from hippocampus-independent memory processes and unspecific effects of a pharmacological treatment or mutation (e.g. on locomotion, emotionality, general sensitivity to sensory inputs).

4.3.2 Amygdala

The amygdala is the most prominent brain structure pertaining to the generation of negative emotions, including fear and anxiety (LeDoux 2000; Adolphs 2002; Dolan 2002). It has been named after its structural similarities in humans with an almond (Latin name, amygdala). The amygdala is a heterogeneous collection of interconnected nuclei in the depth of the temporal lobe that differ morphologically and functionally. A detailed description of its complex structural organisation and functions is given elsewhere (Swanson and Petrovich 1998; Pitkänen 2000). In brief, the amygdala contains cortical and striatal components. The cortical components (i.e. the basolateral amygdala complex that combines lateral, basolateral and basal amygdala) seem to be essential for both cued and contextual fear conditioning, in particular for the association between CS and UCS. Efferents of the basolateral amygdala to extra-amygdaloid brain structures are thought to regulate active responses to potentially dangerous stimuli or situations (Killcross et al. 1997). The striatal components comprise the medial and central nuclei, the latter of which receives inputs from the basolateral amygdala complex and orchestrates the defensive reactions of the animals to the aversive stimulus events. The role of the basolateral amygdala complex as a place where aversive memories are not only acquired, but also consolidated and stored, is disputed (Cahill et al. 1999; Fanselow and LeDoux 1999). According to the consolidation hypothesis (McGaugh 2000), the basolateral amygdala complex facilitates the consolidation of aversive memories in other brain structures, but does not serve as a storage site itself. In any case, the amygdala (in particular the basolateral amygdala complex) is essential for memory acquisition and consolidation in passive avoidance tasks and fear conditioning (delay and trace cued conditioning, contextual conditioning) (LeDoux 2000; Maren 2001).

Sensory information about auditory stimuli reach the lateral amygdala via two different pathways, either directly from thalamic relay structures, such as the medial geniculate nucleus, or from cortical structures (auditory cortex). Information transfer via the thalamus is faster but less precise when it comes to the exact recognition of the sensory stimulus. In contrast, information processed by cortical structures is precise but needs longer to reach the amygdala (LeDoux 1998, 2000). An ultimate explanation of this parallel processing of sensory stimuli might be that it seems less devastating for an animal to react immediately to a potentially harmful stimulus with a false alarm, than to react to it too late (LeDoux 1996).

5
Cellular Mechanisms Underlying Aversive Learning and Memory

More than a century ago, Ramón y Cajal and Tanzi postulated that cellular mechanisms of learning and memory include both the formation of new synaptic connections and the restructuring of the existing ones to make the interneuronal communication more efficacious (for review and references see Geinisman 2000). Another 50 years later, Donald Hebb formulated his famous principles of memory encoding, stating that "When an axon of cell A is near enough to excite a cell B and repeatedly or persistently takes part in firing it, some growth process or metabolic change takes place in one or both cells such that A's efficiency, as one of the cells firing B, is increased" (Hebb 1949; for review see Sejnowski 1999). Today, it is generally accepted that learning leads to transient or permanent modifications in interneuronal communication via morphological or functional changes of synaptic contacts (Milner et al. 1998;

Woolf 1998; Geinisman 2000). A variety of cellular models of learning and memory have been established, including long-term potentiation (LTP), long-term depression (LTD; for comprehensive review see Martin et al. 2000) and kindling (Adamec and Young 2000; Hannesson and Corcoran 2000). LTP and kindling are induced by repetitive high-frequency stimulation of discrete brain areas or specific pathways and characterised by long-lasting hyperexcitability to single electrical pulses. LTD, in turn, is induced by low-frequency stimulation and stands for decreases in neural excitability. Despite the ongoing debate about their physiological significance (e.g. Hölscher 1997; McEachern and Shaw 1999), LTP and LTP have a high number of cellular processes in common with learning and memory (Martin et al. 2000; Blair et al. 2001; Braunewell and Manahan-Vaughan 2001; Goosens and Maren 2002).

5.1 Memory Acquisition

There is good evidence that both associative and non-associative learning lead to a strengthening of synaptic contacts. For instance, in auditory cued fear conditioning, coincident depolarisation of principal (i.e. glutamatergic pyramidal) neurones within the lateral amygdala by a tone and a footshock results in a potentiation of those synapses, which relay the auditory information to that brain structure (Rogan et al. 1997; Collins and Paré 2000; Tang et al. 2001). This potentiation becomes evident by an increase in evoked field potentials compared to baseline responses. However, as field potentials integrate over the activity of multiple neurones and might even be volume conducted from other brain structures, care has to be taken that the changes in interneuronal communication indeed originate from the lateral amygdala. Studies verified the lateral amygdala as the place of learning-induced changes in synaptic transmission by local infusion of anaesthetics (Tang et al. 2003) as well as by single-unit (e.g. Collins and Paré 2000) and intracellular recordings (Rosenkranz and Grace 2002). The potentiation of synaptic transmission may affect each of the two principal inputs to the lateral amygdala, the thalamic (Rogan et al. 1997) and the cortical pathways (Tsvetkov et al. 2002). The similarities of memory acquisition and LTP induction include cooperativity (a neurone must reach a threshold of depolarisation before learning-induced or LTP-induced synaptic changes can occur) and associativity (pairing stimulation of a weak pathway with stimulation of a strong pathway results in facilitated synaptic transmission in both pathways). Both memory acquisition and LTP induction depend on a special form of ionotropic glutamate receptors [*N*-methyl-D-aspartate (NMDA) receptors], protein kinases, voltage-gated calcium channels and protein synthesis (e.g. Schafe et al. 2001; Blair et al. 2001). Recently, with the gastrin-releasing peptide, a transmitter could be described that is specifically expressed in the brain circuit responsible for fear conditioning and involved in both induction of LTP in the cortical afferents to the lateral amygdala and auditory-cued fear conditioning (Shumyatsky et al. 2002). Other studies revealed that learning-induced changes in synapses of the cortical projections are under negative control of the medial prefrontal cortex (Grace and Rosenkranz 2002). In this context, dopamine seems to play a crucial role for memory acquisition, as it overrides the inhibitory input from the medial prefrontal cortex and potentiates the cortical input.

The amygdala is not the only brain structure where learning causes a rebuilding of synaptic contacts during aversive learning. Sensitisation by application of an electrical footshock, for instance, affects synaptic transmission in the septal-hippocampal system (Thomas 1988; Garcia 2002). Most prominent, however, are the refinements of receptive fields in the auditory cortex. On more intensive training protocols than the few tone-shock parings usually applied in fear conditioning paradigms, neurones of the auditory cortex become sensitive to the frequency of the tone used during conditioning (Weinberger 1998; Edeline 1999). Importantly, these changes are not restricted to classical conditioning, but are also evident following active avoidance learning. They include modifications that allow gerbils (a species with excellent hearing capabilities) to form categories about special features of more complex tone signals (Ohl et al. 2001).

Auditory-cued fear conditioning leads to a potentiation of field potentials not only within the lateral amygdala but also within the hippocampus (Doyere et al. 1993; Tang et al. 2003). On first glance, this observation has been astonishing, as the hippocampus is not essential for acquisition and recall of aversive memory to the tone during delay conditioning (Kim and Fanselow 1992; Phillips and LeDoux 1992). However, a recent publication provides the first evidence for its physiological relevance (Moita et al. 2003) that might only become evident in more complex test situations (Doyere et al. 1993).

Only a few studies used in vivo electrophysiological recordings for contextual conditioning or passive avoidance learning (e.g. Sacchetti et al. 2001). In contrast, cellular mechanisms underlying memory acquisition in these tasks were extensively studied by pharmacological and genetical means (Izquierdo and Medina 1997; Schafe et al. 2001; Silva 2003). Similarities between induction of hippocampal LTP and memory acquisition are evident. However, a causal relationship between these two processes still remains to be shown (Gerlai 2002).

With the technical progress being made in molecular biology, it has become possible to screen for genes that might be critically involved in acquisition (and consolidation) of aversive memories. These techniques include mutagenesis screens and quantitative trait loci studies. In large mutagenesis screens, breeding pairs are treated, for instance, with the highly mutagenic compound *N*-ethyl-*N*-nitrosourea (ENU) (Anderson 2000; Brown and Balling 2001). Offspring of these breeding pairs are tested for their memory capabilities. The quantitative trait loci approach, in contrast, is based on the different behavioural performance of genetically heterogeneous mice. In a typical experimental situation, animals of two different inbred strains are crossed, with the consequence that the F1 generation shares 50% homology of their genome with each of its parental strains. F1 animals are then crossed with mice from their parental strains, with the consequence that animals of the F2 generation are genetically and behaviourally heterogeneous due to homologue recombination during meiosis (crossing-over). Animals of the F2 generation are ranked according to their behavioural performance in the learning task. For the upper and the lower 10%, the contribution of genes from the two parental strains to the behavioural phenotype are estimated using polymorphism markers (Wehner et al. 2001).

5.2
Memory Consolidation

More than a century ago, Müller and Pilzecker proposed the perseveration–consolidation hypothesis, according to which new memories initially persist in a fragile state and consolidate over time to reach a state in which they are insensitive to disruption (Lechner et al. 1999 and references therein). About 50 years later, Gerad and Hebb independently from each other came up with the dual-trace theories of memory, suggesting that short-term and long-term memories are sequentially linked, and stabilisation of reverberating neural activity (underlying short-term memory, lasting for seconds to hours) produces long-term memory (lasting for hours to months) (McGaugh 2000 and references therein). Later on, however, it could be demonstrated that drugs might selectively block either short-term or long-term memory, indicating that these two processes occur independently and in parallel (McGaugh 2000; Izquierdo et al. 2002).

Memory consolidation becomes manifest in morphological and functional changes of synaptic contacts. The underlying cellular mechanisms have been studied by pharmacological manipulation, activity monitoring and genetic approaches (for review see Martin et al. 2000; Kandel 2001; Silva 2003). In this way, two different groups of agents could be characterised: permissive agents and instructive agents. Permissive agents may 'arouse' brain structures. They are necessary, since they aid the instructive agents, but are not sufficient for memory storage. Instructive agents, by contrast, directly modify synaptic strength (Shobe 2002), for instance by directly altering transmitter release, by receptor sensitisation/desensitisation and by structural rearrangements. Whereas permissive agents are rather ubiquitously distributed throughout a neurone following memory acquisition, instructive agents are confined to those synaptic terminals that undergo functional and/or morphological changes during learning. The mechanisms underlying this synapse specificity still remain elusive. However, with the synaptic tag hypothesis there is a promising concept for future investigations (Frey and Morris 1998). According to this hypothesis, consolidation at local sites represents a dual process: memory acquisition induces both cell-wide expression of macromolecules and the formation of local postsynaptic 'tags' that 'hijack' only the macromolecules to those synapses that are involved in the memory engram.

Changes of neural processes initiated by memory acquisition follow different time courses for different brain structures. For instance, transient inactivation of the basolateral amygdala interfered with consolidation of cued and contextual memory for 48 h, whereas the perirhinal cortex was sensitive to retrograde amnesia for 192 h (Sacchetti et al. 1999). Long-term consolidation of contextual fear (Shimizu et al. 2000) and spatial memory (Riedel et al. 1999) requires recurrent activation of hippocampal ionotropic glutamate receptors for about 1 week following conditioning, which led to the synaptic re-entry reinforcement hypothesis of memory consolidation (Wittenberg and Tsien 2002). Obviously, it seems to be an evolutionary advantage to delay memory consolidation until the significance of an experience could be evaluated. In fact, events that precede or follow memory acquisition are able to interrupt the consolidation process by proactive and retroactive inhibition (Xu et al. 1998; Izquierdo et al. 1999), possibly via LTD-like mechanisms (Manahan-Vaughan and Braunewell 1999). Accordingly, there seem to be similar critical phases for consolidation of LTP, during which a depotentiation is possible (Huang and Hsu 2001; Lin et al. 2003a).

Consolidation processes for short-term and long-term memories are distinguished by their dependency on de novo protein synthesis (Davis and Squire 1984; Matthies 1989; Izquierdo et al. 2002). Blockade of transcription or translation by drug infusion into lateral amygdala or hippocampus revealed that the consolidation of long-term but not short-term memories for cued and contextual fear conditioning as well as passive avoidance learning required protein synthesis (Schafe and LeDoux 2000; Kida et al. 2002; Muller Igaz et al. 2002). Interestingly, there seem to be at least two different waves of protein synthesis necessary for memory consolidation, with peaks between 0–1 h and 3–6 h after conditioning (Bourtchouladze et al. 1998; Muller Igaz et al. 2002), corresponding to those described for bead pecking in chicks (Freeman et al. 1995). Protein synthesis will be initiated via a cascade of second messenger systems and protein kinases that, in turn, activate transcription factors [such as the cAMP-responsive element binding protein (CREB)] and finally transcription (Milner et al. 1998; Clayton 2000; Kandel 2001). The activity of this consolidation cascade is negatively controlled at various levels, for instance by protein phosphatases and repressors of transcriptional activity (Abel and Kandel 1998; Cardin and Abel 1999; Kandel 2001; Genoux et al. 2002).

The characterisation of memory-related genes and proteins belongs to the hot spots of current memory research (D'Agata and Cavallaro 2002). Respective studies in the field of aversive memories employ different molecular biological methods including in situ hybridisation (Ressler et al. 2002), differential display (Huang et al. 1998), subtractive hybridisation (Stork et al. 2001) and DNA microarrays (Kida et al. 2002). Most critical for the correct interpretation of the data are the selections of appropriate controls, as unspecific changes in gene expression may occur already from handling of the animals and exposure to the test context. Another critical point is the dissection of biological material. So far, the methods employed, in particular for analysis of the proteome, require relatively high amounts of protein, with the consequence that whole brain structures have to be analysed. However, only a small subset of neurones or even synapses of a given neurone might be involved in a distinct memory process, and neurones are highly heterogeneous in their gene expression profiles (Kamme et al. 2003; Levsky and Singer 2003). Consequently, the signal-to-noise ratio would be very small on average over whole brain structures, making subtle changes in gene expression, expected for learning events, hard to detect (Geschwind 2000).

The role of the amygdala for consolidation of aversive memories is generally accepted. However, as stated before, it is still debated as to whether or not the amygdala is also the storage site for the consolidated memories. Some authors suggest that the amygdala is essential for memory consolidation and storage in other brain structures only. According to their hypothesis, punishment used for aversively motivated learning activates the two major hormonal stress systems of the organism, the hypothalamic–pituitary–adrenal axis (with corticotropin and corticosterone/cortisol) and the sympatho-adrenergic system (with noradrenaline and adrenaline). Both stress systems seem to funnel into the same regulatory system at the level of the amygdala. The resulting potentiation of the local effects of noradrenaline leads to an activation of efferent projections that are known to modulate plastic changes, for instance within the hippocampus (McGaugh and Roozendaal 2002).

Interestingly, the role of the hippocampus as a storage site of aversive memories is temporally limited. Contextual fear memories are susceptible to lesions of the hippocampus only for 3–4 weeks after memory acquisition (Kim and Fanselow 1992). During this time, they become finally consolidated in neocortical structures in a process that is called systems reconsolidation. Memories that have become independent of the hippocampus with time are referred to as 'remote' memories. Cellular correlates of systems reconsolidation are fairly unknown. It is conceivable that these processes involve reiteration of learning-induced changes in neuronal activity during wakening and sleep (for reviews see Sejnowski and Destexhe 2000; Sutherland and McNaughton 2000; Graves et al. 2001; Paré et al. 2002) as well as principles of homeostatic plasticity (Turrigiano and Nelson 2000).

Another striking characteristic of aversive memories is their ability to reconsolidate on reactivation. By definition, memories should be insensitive to disruption, for instance by electroconvulsive shocks or drugs, once they have been consolidated. This is, in fact, the case as long as the treatments do not coincide with memory recall. Reactivation of a memory, however, makes it 'labile' again because of reconsolidation processes (Sara 2000; Nader 2003). Reconsolidation resembles consolidation in that similar cascades of molecular events seem to be activated, including phosphorylation of the transcription factor CREB (Hall et al. 2001a; Kida et al. 2002), expression of immediate early genes (Hall et al. 2001a,b) and protein synthesis (Nader et al. 2000). However, reconsolidation occurs faster and is more sensitive to amnesic challenge than initial memory consolidation (Nader 2003). Even remote aversive memories return to a labile hippocampus-dependent state on reactivation. Reconsolidation processes initiated in this way are sensitive to the memory-disrupting effects of protein synthesis blockade again (Debiec et al. 2002). However, not all forms of remote aversive memories seem to undergo reconsolidation in a labile state (Milekic and Alberini 2002), an observation that deserves further investigation. Nevertheless, retrieval-induced reconsolidation might have evolved as a useful mechanism for dynamically integrating new information into pre-existing memory engrams.

5.3 Memory Retrieval

Memory retrieval is the only direct measure of memory in animal experiments. However, in the absence of controls that closely match the conditions for performance, it is difficult to make inferences about the role of a neurobiological process in retrieval. Depending on the circumstances, memory retrieval might lead to memory reconsolidation or memory extinction (Nader 2003). It is, therefore, not surprising that it shares cellular mechanisms, such as dependency on protein kinases and activation of immediate early genes (Hall et al. 2001a,b; Szapiro et al. 2002), with each of the two other processes. Interestingly, a recent study (Murchison et al. 2004) ascribes an important role in retrieval of aversive memories to noradrenaline, thereby challenging concepts about its primary involvement in consolidation processes (McGaugh and Roozendaal 2002).

5.4 Memory Retention

Memory retention and memory decay describe the same phenomenon but from two different perspectives: either we emphasise that memories are relatively stable over time (memory retention) or that they dissipate with time (decay, temporal degradation). So far, little is known about the cellular correlates of memory retention/decay. If memory acquisition and consolidation indeed lead to long-term changes in synaptic contacts and interneuronal communication, the question remains as to how these changes are maintained over long period of times, despite the regular protein turnover and hormonally mediated structural reorganisation of dendritic arbours. Recent data suggest that these processes include subunit-specific dynamic changes in the expression of distinct ionotropic glutamate receptors in the postsynaptic membrane (for review see Malinow and Malenka 2002). Other authors postulate 'mnemogenic' chemical reactions as the basis of memory retention, including phosphorylation/autophosphorylation and conformational changes (Roberson and Sweatt 1999). Failures of these processes or counterregulatory mechanisms (e.g. dephosphorylation) would, consequently, lead to memory decay (e.g. Genoux et al. 2002).

5.5 Memory Decay and Extinction

Everyone experiences how memories may dissipate with time. Responsible for this phenomenon might be spontaneous forgetting, the suppression of memory retrieval and memory extinction. Spontaneous forgetting describes the loss of learned performances that is often observed when time elapses between memory acquisition and memory retrieval. It results from different processes (Bouton et al. 1999). First, memory traces might dissipate over time. Second, information might increasingly interfere with retrieval of the original memory in a proactive or retroactive manner. Third, memory retrieval might be disturbed transiently or permanently (e.g. following head injury or stroke). Fourth, a recent study suggests that neurogenesis in the dentate gyrus might play a role in the clearance or destabilisation of outdated hippocampal memory traces after systems reconsolidation, thereby saving the hippocampus from overload (Feng et al. 2001). Furthermore, on confrontation with reminders of unpleasant or traumatic events, we often try to refocus attention and ignore the unwanted memory. The ability to suppress memory retrieval is accomplished, in part, by executive-control mechanisms as a special case of response-override situations (Levy and Anderson 2002).

It is a matter of debate whether memories can be erased at all (Jacobs and Nadel 1985). However, there is some evidence from animal experiments that processes that lead to learning or LTP-induced changes in interneuronal communication can be reversed by specific opponents of the consolidation cascade, as for instance phosphatases (Genoux et al. 2002; Lin et al. 2003a,b). The decay of both LTP and memory seems to depend on NMDA receptors. However, there are conflicting data as to whether recurrent activation of NMDA receptors promotes decay or long-term consolidation processes (Shimizu et al. 2000; Villarreal et al. 2002).

Whereas memory decay describes the actual loss of memory, the term extinction stands for an active learning process that suppresses rather than erases the original memory (Bouton and Swartzentruber 1991; Myers and Davis 2002). Extinction requires memory retrieval in the absence of positive or negative reinforcement. The conclusion that extinction can be regarded as an active process bases on the following observations: First, extinction retention dissipates over time thus resulting in re-occurrence of the extinguished conditioned response (spontaneous recovery). Second, the extinguished memory performance reappears in a context different from that used for extinction training (renewal). Third, presentation of the UCS following extinction training re-activates the extinguished CS–UCS association (memory reinstatement).

Extinction seems to engage the same brain structures as memory acquisition and consolidation (Myers and Davis 2002; Schwaerzel et al. 2002). A variety of neurotransmitter and second messenger systems could be characterised that contribute to extinction of aversive memories, including glutamate (via NMDA receptors), GABA, dopamine, noradrenaline (via β-adrenergic receptors), selected forms of voltage-gated calcium channels, protein kinases, phosphatases and endocannabinoids (for comprehensive review see Myers and Davis 2002). If extinction training indeed represents an active learning process during which the animals learn that the CS does not predict the occurrence of the UCS anymore, it would be interesting to analyse what transmitter systems are of particular importance for which phase of the extinction learning curve. So far, however, animals have mostly been treated before extinction training, thus leaving open whether the pharmacological intervention interfered with the initiation or consolidation of extinction.

As mentioned before, retrieval renders the consolidated memories labile again. It seems to depend on the test situation, whether this labile state is followed by reconsolidation or extinction of the aversive memory. Blockade of reconsolidation processes (e.g. by interrupting protein synthesis within the lateral amygdala), for instance, has led to extinction of the fear responses to the CS in an auditory fear-conditioning paradigm (Nader et al. 2000).

Several studies suggest that the prefrontal cortex plays an important role in retention of extinction by reducing amygdala-dependent fear reactions. On inappropriate functioning, animals show abnormal perseveration of fear responses (Garcia 2002). However, data about the involvement of the prefrontal cortex in fear conditioning are relatively inconsistent (Morgan and LeDoux 1995; Gewirtz et al. 1997), which might be related to the anatomical and functional heterogeneity of this brain area. In fact, lesions that spared with the caudal infralimbic nucleus a prominent part of the ventromedial prefrontal cortex had no effect on extinction (Gewirtz et al. 1997). By contrast, lesions that included this brain structure resulted in impaired extinction consolidation and retention (Quirk et al. 2000). Furthermore, infralimbic neurones showed an increased activity to the CS (i.e. tone) following extinction training that was negatively correlated with the animals' freezing reaction. Importantly, stimulation of infralimbic neurones by electrical impulses that resembled extinction-induced changes in neuronal activity simulated extinction memory in the absence of extinction training (Milad and Quirk 2002).

The medial division of the prefrontal cortex seems to contribute to memory extinction as well. It is under negative control of the basolateral amygdala (Pérez-Jaranay and Vives 1991; Garcia et al. 1999), most likely via activation of the mesocortical dopamine system (Davis et al. 1994; Morrow et al. 1999; Tzschentke 2001). After complete extinction of the fear response, which coincides with a return to baseline of learning-induced potentiation of neural activity within the basolateral amygdala complex (Rogan et al. 1997; Tang et al. 2001), the medial prefrontal cortex will be activated and gain inhibitory control over the amygdala (Garcia 2002). The latter process likely involves the activation of inhibitory interneurones within the basolateral amygdala (Grace and Rosenkranz 2002).

The septal-hippocampal system is another prominent brain circuit involved in extinction of aversive memories. It is the major component of the behavioural inhibitory system of the brain and essential for the suppression of aversive emotional states (Thomas 1988; Gray and McNaughton 2000; Garcia 2002). For instance, long-term changes in septal-hippocampal efficacy turned out to be critical for the inhibition of behavioural despair, including the expression of freezing behaviour (Garcia 2002).

It is of importance to note that cellular mechanisms of extinction might vary for classical and instrumental conditioning paradigms. In classical conditioning tasks, the experimenter can control the rate of extinction by repeated presentation of the CS that cannot be avoided by the animals. In commonly used protocols of instrumental conditioning, by contrast, the animal decides when to start extinction, for instance by stepping down from a platform (step-down avoidance) or starting to consume the sucrose solution again (conditioned taste aversion). These procedural differences might well explain the inconsistencies in molecular correlates of extinction that have been reported for the two conditioning tasks (Myers and Davis 2002). Future studies on extinction of avoidance learning and conditioned taste aversion should, therefore, consider withholding the reinforcers of avoidance through blocking avoidance (response prevention) and prolonging the exposure to the CS/aversive situation (flooding, Baum 1973).

Extinction can be induced either by a few long-lasting presentations of the CS (massive extinction training) or by a series of short-lasting CS (graded extinction training). It critically depends on the training protocol, as the circumstances of memory retrieval seem to define whether the re-activated memory undergoes reconsolidation or extinction. Short-lasting presentations of the CS may cause a flashback of the aversive memory and, together with the aroused state of the animals, reconsolidation. Presentation of longer-lasting CS, in contrast, often goes along with an acute within-session extinction of the fear response that might be essential for consolidation and retention of extinction (Nader 2003). Moreover, the strength of the originally formed aversive memory determines whether memory retrieval leads to reconsolidation (strong memories) or extinction (weak memories) (Eisenberg et al. 2003). This has to be taken into account if extinction of aversive memories should be modulated by pharmacological means.

6
Clinical Implications

Most basic researchers and clinicians believe that detailed knowledge about cellular mechanisms underlying the formation and extinction of aversive memories will lead to the development of novel therapeutic strategies for the treatment of human anxiety disorders. Our current knowledge about these mechanisms largely arises from results obtained in animal experiments. However, the transferability of such experimental data to the human situation depends on the validity of the experimental models chosen. Numerous sets of criteria have been developed for evaluating experimental models (McKinney and Bunney 1969; Willner 1984; Newport et al. 2002). Human psychiatric disorders may develop as a consequence of genetic and developmental predisposition that affect sensitivity to life stress and the initiation of pathological processes. Consequently, it appears rather unlikely that comprehensive animal models can be developed that accurately reflect the human situation (Shekhar et al. 2001). These limitations apply also for fear conditioning and sensitisation (e.g. Yehuda and Antelman 1993) that are, nevertheless, frequently discussed as experimental models for pathological anxiety (i.e. phobias, generalised anxiety disorders, panic disorder and post-traumatic stress disorder) (Marks and Tobena 1990; Charney and Deutch 1996; Rosen and Schulkin 1998; Bouton et al. 2001; Garcia 2002). Rosen and Schulkin (1998) suggest that pathological anxiety evolves directly from normal fear responses. The pathology of these anxiety disorders would include hyperexcitability in the amygdala and the bed nucleus of the stria terminalis, caused by a process of neural sensitisation or kindling in which psychosocial stressors initiate changes in the fear circuits that lead to enhanced perception and response to subsequent threat and danger. On the other hand, one of the key functions of the amygdala might be the potentiation of vigilance by lowering neuronal thresholds in sensory systems. As a consequence, pathological anxiety may not be a disorder of fear, but a disorder of vigilance (Davis and Wahlen 2001). Both concepts are based on hyperexcitability of the amygdala, which can be experimentally induced by conditioning and sensitisation as well as LTP and kindling protocols. Hence, the respective animal and cellular models seem to be at least analogous (if not homologous) to the situation in patients and might, therefore, guide our search for novel therapeutic strategies for the treatment of pathological anxiety.

Data of animal experiments discussed in this chapter suggest a variety of potential pharmacological targets for the treatment of pathological anxiety (Fig. 1). As the occurrence of traumatic events is usually unpredictable, it seems more promising to interfere with consolidation than with acquisition processes. In this context, the sympatho-adrenergic and the hypothalamic–pituitary–adrenal system are of particular interest. Both noradrenaline and corticosterone/cortisol are known to facilitate memory consolidation, in par-

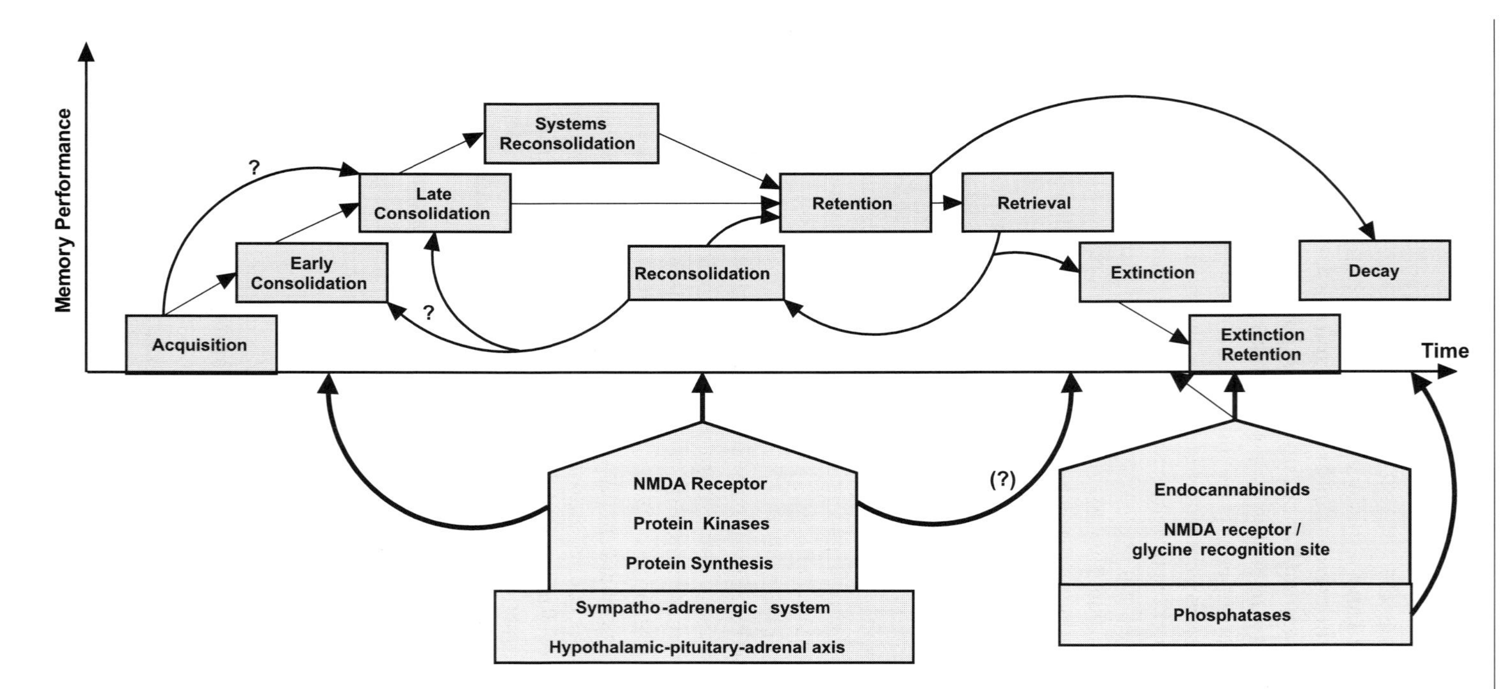

Fig. 1 Potential therapeutic targets for the treatment of pathological forms of aversive memories. The *upper panel* illustrates a typical learning curve. Subsequential phases are connected by *arrows*; still-uncertain interrelations are indicated by *question marks*. The *lower panel* shows some of the most promising targets for the pharmacotherapy of aversive memories. Consolidation, reconsolidation and extinction share a variety of cellular processes and, consequently, a common set of potential therapeutic targets. Other therapeutic targets are rather specific for memory extinction and decay. This schematic illustrates that pharmacotherapy has to be carefully adjusted to the different phases of the learning curve

ticular for aversive events (Cahill and McGaugh 1998; Korte 2001; McGaugh and Roozendaal 2002).

Treatments targeting these stress hormone systems would be restricted to a relatively narrow time window around the learning event. Quite often, however, patients might consult a physician long after the traumatic incident, when aversive memories have already been consolidated. Under these circumstances, the reconsolidation hypothesis gains particular importance, according to which (aversive) memories return to a labile state on retrieval. It appears to be possible to 'erase' aversive memories, if memory recall would be combined with a pharmacological treatment or electroconvulsive therapy (Sara 2000; Nader et al. 2000; Davis and Myers 2002). Depending on the protocol used, retrieval might initiate memory extinction rather than memory reconsolidation. This observation has already been used in the clinical praxis in form of exposure therapy (Marks and Tobena 1990), for instance for the treatment of post-traumatic stress disorder (Ballenger et al. 2000; Foa 2000; Rothbaum and Schwartz 2002). Exposure therapies are laborious for both patients and therapists. Data from animal experiments suggest that such therapies could become more efficient if they would be combined with pharmacotherapy (Myers and Davis 2002; Davis and Myers 2002) targeting, for instance, the glycine recognition site of the NMDA receptor (Walker et al. 2002), the endocannabinoid system of the brain (Marsicano et al. 2002) or protein kinases (Lu et al. 2001; Cohen 2002). Another potential target would be the GABAergic system, which, however, seems to be involved primarily in the expression of extinction that has already been acquired and not the extinction process per se (Davis and Myers 2002). Experiments performed with D-cycloserine, an agonist of the glycine recognition site of the NMDA receptor, demonstrate how efficiently animal studies on fear extinction (Walker et al. 2002; Ledgerwood et al. 2004; Richardson et al. 2004) can be transferred into the pharmacotherapy of human anxiety disorders (Ressler et al. 2004). Certainly, with the implementation of genomics, proteomics and pharmacogenomics in animal experiments on aversive memories, many novel therapeutical targets will be discovered for the benefit of patients. Exciting times!

References

Abel T, Kandel E (1998) Positive and negative regulatory mechanisms that mediate long-term memory storage. Brain Res Brain Res Rev 26:360–378

Adamec R, Young B (2000) Neuroplasticity in specific limbic system circuits may mediate specific kindling induced changes in animal affect-implications for understanding anxiety associated with epilepsy. Neurosci Biobehav Rev 24:705–723

Adolphs R (2002) Neural systems for recognizing emotion. Curr Opin Neurobiol 12:169–177

Amaral DG, Witter MP (1989) The three-dimensional organization of the hippocampal formation: a review of anatomical data. Neuroscience 31:571–591

Ambrogi Lorenzini CG, Baldi E, Bucherelli C, Sacchetti B, Tassoni G (1999) Neural topography and chronology of memory consolidation: a review of functional inactivation findings. Neurobiol Learn Mem 71:1–18
Anderson KV (2000) Finding the genes that direct mammalian development: ENU mutagenesis in the mouse. Trends Genet 16:99–102
Andrews JS (1996) Possible confounding influence of strain, age and gender on cognitive performance in rats. Brain Res Cogn Brain Res 3:251–267
Ballenger JC, Davidson JR, Lecrubier Y, Nutt DJ, Foa EB, Kessler RC, McFarlane AC, Shalev AY (2000) Consensus statement on posttraumatic stress disorder from the International Consensus Group on Depression and Anxiety. J Clin Psychiatry 61 Suppl 5:60–66
Baum M (1973) Extinction of avoidance in rats: the effects of chlorpromazine and methylphenidate administered in conjunction with flooding response (prevention). Behav Res Ther 11:165–169
Berger TW, Alger B, Thompson RF (1976) Neuronal substrate of classical conditioning in the hippocampus. Science 192:483–485
Berman DE, Dudai Y (2001) Memory extinction, learning anew, and learning the new: dissociations in the molecular machinery of learning in cortex. Science 291:2417–2419
Blair HT, Schafe GE, Bauer EP, Rodrigues SM, LeDoux JE (2001) Synaptic plasticity in the lateral amygdala: a cellular hypothesis of fear conditioning. Learn Mem 8:229–242
Bourtchouladze R, Abel T, Berman N, Gordon R, Lapidus K, Kandel ER (1998) Different training procedures recruit either one or two critical periods for contextual memory consolidation, each of which requires protein synthesis and PKA. Learn Mem 5:365–374
Bouton ME, Swartzentruber D (1991) Sources of relapse after extinction in Pavlovian and instrumental conditioning. Clin Psychol Rev 11:123–140
Bouton ME, Nelson JB, Rosas JM (1999) Stimulus generalization, context change, and forgetting. Psychol Bull 125:171–186
Bouton ME, Mineka S, Barlow DH (2001) A modern learning theory perspective on the etiology of panic disorder. Psychol Rev 108:4–32
Braunewell KH, Manahan-Vaughan D (2001) Long-term depression: a cellular basis for learning? Rev Neurosci 12:121–140
Brown SD, Balling R (2001) Systematic approaches to mouse mutagenesis. Curr Opin Genet Dev 11:268–273
Bucan M, Abel T (2002) The mouse: genetics meets behaviour. Nat Rev Genet 3:114–123
Cahill L, McGaugh JL (1998) Mechanisms of emotional arousal and lasting declarative memory. Trends Neurosci 21:294–299
Cahill L, Weinberger NM, Roozendaal B, McGaugh JL (1999) Is the amygdala a locus of "conditioned fear"? Some questions and caveats. Neuron 23:227–228
Cardin JA, Abel T (1999) Memory suppressor genes: enhancing the relationship between synaptic plasticity and memory storage. J Neurosci Res 58:10–23
Charney DS, Deutch A (1996) A functional neuroanatomy of anxiety and fear: implications for the pathophysiology and treatment of anxiety disorders. Crit Rev Neurobiol 10:419–446
Clayton DF (2000) The genomic action potential. Neurobiol Learn Mem 74:185–216
Cohen P (2002) Protein kinases—the major drug targets of the twenty-first century? Nat Rev Drug Discov 1:309–315
Collins DR, Pare D (2000) Differential fear conditioning induces reciprocal changes in the sensory responses of lateral amygdala neurons to the CS(+) and CS(−). Learn Mem 7:97–103

Cook MN, Bolivar VJ, McFadyen MP, Flaherty L (2002) Behavioral differences among 129 substrains: implications for knockout and transgenic mice. Behav Neurosci 116:600–611
Crawley JN, Belknap JK, Collins A, Crabbe JC, Frankel W, Henderson N, Hitzemann RJ, Maxson SC, Miner LL, Silva AJ, Wehner JM, Wynshaw-Boris A, Paylor R (1997) Behavioral phenotypes of inbred mouse strains: implications and recommendations for molecular studies. Psychopharmacology (Berl) 132:107–124
D'Agata V, Cavallaro S (2002) Gene expression profiles—a new dynamic and functional dimension to the exploration of learning and memory. Rev Neurosci 13:209–219
Davis HP, Squire LR (1984) Protein synthesis and memory: a review. Psychol Bull 96:518–559
Davis M (2000) The role of the amygdala in conditioned and unconditioned fear and anxiety. In: Aggleton JP (ed) The amygdala. A functional analysis. Oxford University Press, New York, p 213–287
Davis M, Myers KM (2002) The role of glutamate and gamma-aminobutyric acid in fear extinction: clinical implications for exposure therapy. Biol Psychiatry 52:998–1007
Davis M, Whalen PJ (2001) The amygdala: vigilance and emotion. Mol Psychiatry 6:13–34
Davis M, Hitchcock JM, Bowers MB, Berridge CW, Melia KR, Roth RH (1994) Stress-induced activation of prefrontal cortex dopamine turnover: blockade by lesions of the amygdala. Brain Res 664:207–210
Debiec J, LeDoux JE, Nader K (2002) Cellular and systems reconsolidation in the hippocampus. Neuron 36:527–538
Denny P, Justice MJ (2000) Mouse as the measure of man? Trends Genet 16:283–287
Dickinson A (1980) Contemporary animal learning theory. Cambridge University Press, Cambridge
Dolan RJ (2002) Emotion, cognition, and behavior. Science 298:1191–1194
Doyere V, Burette F, Negro CR, Laroche S (1993) Long-term potentiation of hippocampal afferents and efferents to prefrontal cortex: implications for associative learning. Neuropsychologia 31:1031–1053
Dubnau J, Tully T (1998) Gene discovery in Drosophila: new insights for learning and memory. Annu Rev Neurosci 21:407–444
Dudai Y (1989) The neurobiology of memory—concepts, findings, trends. Oxford University Press, New York
Edeline JM (1999) Learning-induced physiological plasticity in the thalamo-cortical sensory systems: a critical evaluation of receptive field plasticity, map changes and their potential mechanisms. Prog Neurobiol 57:165–224
Eichenbaum H, Cohen NJ (2001) From conditioning to conscious recollection. Memory systems of the brain. Oxford University Press, New York
Eisenberg M, Kobilo T, Berman DE, Dudai Y (2003) Stability of retrieved memory: inverse correlation with trace dominance. Science 301:1102–1104
Fanselow MS (1994) Neural organization of the defensive behavior system responsible for fear. Psychon Bull Rev 1:429–438
Fanselow MS, LeDoux JE (1999) Why we think plasticity underlying Pavlovian fear conditioning occurs in the basolateral amygdala. Neuron 23:229–232
Feng R, Rampon C, Tang YP, Shrom D, Jin J, Kyin M, Sopher B, Miller MW, Ware CB, Martin GM, Kim SH, Langdon RB, Sisodia SS, Tsien JZ (2001) Deficient neurogenesis in forebrain-specific presenilin-1 knockout mice is associated with reduced clearance of hippocampal memory traces. Neuron 32:911–926
Foa EB (2000) Psychosocial treatment of posttraumatic stress disorder. J Clin Psychiatry 61[Suppl]5:43–48

Freeman FM, Rose SP, Scholey AB (1995) Two time windows of anisomycin-induced amnesia for passive avoidance training in the day-old chick. Neurobiol Learn Mem 63:291–295
Frey U, Morris RG (1998) Synaptic tagging: implications for late maintenance of hippocampal long-term potentiation. Trends Neurosci 21:181–188
Garcia R (2002) Stress, synaptic plasticity, and psychopathology. Rev Neurosci 13:195–208
Garcia R, Vouimba RM, Baudry M, Thompson RF (1999) The amygdala modulates prefrontal cortex activity relative to conditioned fear. Nature 402:294–296
Geinisman Y (2000) Structural synaptic modifications associated with hippocampal LTP and behavioral learning. Cereb Cortex 10:952–962
Genoux D, Haditsch U, Knobloch M, Michalon A, Storm D, Mansuy IM (2002) Protein phosphatase 1 is a molecular constraint on learning and memory. Nature 418:970–975
Gerlai R (1996) Gene-targeting studies of mammalian behavior: is it the mutation or the background genotype? [see comments] [published erratum appears in Trends Neurosci 1996 Jul;19(7):271]. Trends Neurosci 19:177–181
Gerlai R (1998) Contextual learning and cue association in fear conditioning in mice: a strain comparison and a lesion study. Behav Brain Res 95:191–203
Gerlai R (2001) Gene targeting: technical confounds and potential solutions in behavioral brain research. Behav Brain Res 125:13–21
Gerlai R (2002) Hippocampal LTP and memory in mouse strains: is there evidence for a causal relationship? Hippocampus 12:657–666
Geschwind DH (2000) Mice, microarrays, and the genetic diversity of the brain. Proc Natl Acad Sci U S A 97:10676–10678
Gewirtz JC, Falls WA, Davis M (1997) Normal conditioned inhibition and extinction of freezing and fear-potentiated startle following electrolytic lesions of medical prefrontal cortex in rats. Behav Neurosci 111:712–726
Gewirtz JC, McNish KA, Davis M (2000) Is the hippocampus necessary for contextual fear conditioning? Behav Brain Res 110:83–95
Gingrich JA, Hen R (2000) The broken mouse: the role of development, plasticity and environment in the interpretation of phenotypic changes in knockout mice. Curr Opin Neurobiol 10:146–152
Goosens KA, Maren S (2002) Long-term potentiation as a substrate for memory: evidence from studies of amygdaloid plasticity and Pavlovian fear conditioning. Hippocampus 12:592–599
Grace AA, Rosenkranz JA (2002) Regulation of conditioned responses of basolateral amygdala neurons. Physiol Behav 77:489–493
Graves L, Pack A, Abel T (2001) Sleep and memory: a molecular perspective. Trends Neurosci 24:237–243
Gray JA, McNaughton N (2000) The neuropsychology of anxiety. An enquiry into the functions of the septo-hippocampal system, 2nd edn. Oxford Psychology Series No. 33. Oxford University Press, New York
Gross C, Zhuang X, Stark K, Ramboz S, Oosting R, Kirby L, Santarelli L, Beck S, Hen R (2002) Serotonin1A receptor acts during development to establish normal anxiety-like behaviour in the adult. Nature 416:396–400
Hall J, Thomas KL, Everitt BJ (2001a) Cellular imaging of zif268 expression in the hippocampus and amygdala during contextual and cued fear memory retrieval: selective activation of hippocampal CA1 neurons during the recall of contextual memories. J Neurosci 21:2186–2193
Hall J, Thomas KL, Everitt BJ (2001b) Fear memory retrieval induces CREB phosphorylation and Fos expression within the amygdala. Eur J Neurosci 13:1453–1458

Hannesson DK, Corcoran ME (2000) The mnemonic effects of kindling. Neurosci Biobehav Rev 24:725–751

Hebb DO (1949) Organization of behavior: a neuropsychological theory. John Wiley and Sons, New York, p 62

Herdegen T, Leah JD (1998) Inducible and constitutive transcription factors in the mammalian nervous system: control of gene expression by Jun, Fos and Krox, and CREB/ATF proteins. Brain Res Brain Res Rev 28:370–490

Hölscher C (1997) Long-term potentiation: a good model for learning and memory? Prog Neuropsychopharmacol Biol Psychiatry 21:47–68

Huang AM, Wang HL, Tang YP, Lee EH (1998) Expression of integrin-associated protein gene associated with memory formation in rats. J Neurosci 18:4305–4313

Huang CC, Hsu KS (2001) Progress in understanding the factors regulating reversibility of long-term potentiation. Rev Neurosci 12:51–68

Izquierdo I, Medina JH (1997) Memory formation: the sequence of biochemical events in the hippocampus and its connection to activity in other brain structures. Neurobiol Learn Mem 68:285–316

Izquierdo I, Medina JH (1998) On brain lesions, the milkman and Sigmunda. Trends Neurosci 21:423–426

Izquierdo I, Schroder N, Netto CA, Medina JH (1999) Novelty causes time-dependent retrograde amnesia for one-trial avoidance in rats through NMDA receptor- and CaMKII-dependent mechanisms in the hippocampus. Eur J Neurosci 11:3323–3328

Izquierdo LA, Barros DM, Vianna MR, Coitinho A, deDavid e Silva T, Choi H, Moletta B, Medina JH, Izquierdo I (2002) Molecular pharmacological dissection of short- and long-term memory. Cell Mol Neurobiol 22:269–287

Jacobs WJ, Nadel L (1985) Stress-induced recovery of fears and phobias. Psychol Rev 92:512–531

Jarrard LE (2001) Retrograde amnesia and consolidation: anatomical and lesion considerations. Hippocampus 11:43–49

Jarrard LE (2002) Use of excitotoxins to lesion the hippocampus: update. Hippocampus 12:405–414

Kamme F, Salunga R, Yu J, Tran DT, Zhu J, Luo L, Bittner A, Guo HQ, Miller N, Wan J, Erlander M (2003) Single-cell microarray analysis in hippocampus CA1: demonstration and validation of cellular heterogeneity. J Neurosci 23:3607–3615

Kandel ER (2001) The molecular biology of memory storage: a dialogue between genes and synapses. Science 294:1030–1038

Kida S, Josselyn SA, de Ortiz SP, Kogan JH, Chevere I, Masushige S, Silva AJ (2002) CREB required for the stability of new and reactivated fear memories. Nat Neurosci 5:348–355

Killcross S, Robbins TW, Everitt BJ (1997) Different types of fear-conditioned behaviour mediated by separate nuclei within amygdala. Nature 388:377–380

Kim JJ, Fanselow MS (1992) Modality-specific retrograde amnesia of fear. Science 256: 675–677

Korte SM (2001) Corticosteroids in relation to fear, anxiety and psychopathology. Neurosci Biobehav Rev 25:117–142

Lamprea MR, Cardenas FP, Vianna DM, Castilho VM, Cruz-Morales SE, Brandao ML (2002) The distribution of fos immunoreactivity in rat brain following freezing and escape responses elicited by electrical stimulation of the inferior colliculus. Brain Res 950:186–194

Lechner HA, Squire LR, Byrne JH (1999) 100 years of consolidation—remembering Muller and Pilzecker. Learn Mem 6:77–87

Ledgerwood L, Richardson R, Cranney J (2004) d-Cycloserine and the facilitation of extinction of conditioned fear: consequences for reinstatement. Behav Neurosci 118:505–513

LeDoux J (1996) The emotional brain. The mysterious underpinnings of emotional life. Simon and Schuster, New York, pp 175–190
LeDoux J (1998) Fear and the brain: where have we been, and where are we going? [see comments]. Biol Psychiatry 44:1229–1238
LeDoux JE (2000) Emotion circuits in the brain. Annu Rev Neurosci 23:155–184
Levsky JM, Singer RH (2003) Gene expression and the myth of the average cell. Trends Cell Biol 13:4–6
Levy BJ, Anderson MC (2002) Inhibitory processes and the control of memory retrieval. Trends Cogn Sci 6:299–305
Lin CH, Lee CC, Gean PW (2003a) Involvement of a calcineurin cascade in amygdala depotentiation and quenching of fear memory. Mol Pharmacol 63:44–52
Lin CH, Yeh SH, Leu TH, Chang WC, Wang ST, Gean PW (2003b) Identification of calcineurin as a key signal in the extinction of fear memory. J Neurosci 23:1574–1579
Lu KT, Walker DL, Davis M (2001) Mitogen-activated protein kinase cascade in the basolateral nucleus of amygdala is involved in extinction of fear-potentiated startle. J Neurosci 21:RC162
Mackintosh NJ (1974) The psychology of animal learning. Academic Press, London
Mackintosh NJ (1983) Conditioning and associative learning. Oxford University Press, New York
Malinow R, Malenka RC (2002) AMPA receptor trafficking and synaptic plasticity. Annu Rev Neurosci 25:103–126
Manahan-Vaughan D, Braunewell KH (1999) Novelty acquisition is associated with induction of hippocampal long-term depression. Proc Natl Acad Sci U S A 96:8739–8744
Maren S (2001) Neurobiology of Pavlovian fear conditioning. Annu Rev Neurosci 24:897–931
Maren S, Holt W (2000) The hippocampus and contextual memory retrieval in Pavlovian conditioning. Behav Brain Res 110:97–108
Marks I, Tobena A (1990) Learning and unlearning fear: a clinical and evolutionary perspective. Neurosci Biobehav Rev 14:365–384
Marsicano G, Wotjak CT, Azad SC, Bisogno T, Rammes G, Cascio MG, Hermann H, Tang J, Hofmann C, Zieglgansberger W, Di Marzo V, Lutz B (2002) The endogenous cannabinoid system controls extinction of aversive memories. Nature 418:530–534
Martin SJ, Grimwood PD, Morris RG (2000) Synaptic plasticity and memory: an evaluation of the hypothesis. Annu Rev Neurosci 23:649–711
Matthies H (1989) In search of cellular mechanisms of memory. Prog Neurobiol 32:277–349
Mayford M, Kandel ER (1999) Genetic approaches to memory storage. Trends Genet 15:463–470
McEachern JC, Shaw CA (1999) The plasticity-pathology continuum: defining a role for the LTP phenomenon. J Neurosci Res 58:42–61
McEchron MD, Bouwmeester H, Tseng W, Weiss C, Disterhoft JF (1998) Hippocampectomy disrupts auditory trace fear conditioning and contextual fear conditioning in the rat. Hippocampus 8:638–646
McGaugh JL (2000) Memory—a century of consolidation. Science 287:248–251
McGaugh JL, Izquierdo I (2000) The contribution of pharmacology to research on the mechanisms of memory formation. Trends Pharmacol Sci 21:208–210
McGaugh JL, Roozendaal B (2002) Role of adrenal stress hormones in forming lasting memories in the brain. Curr Opin Neurobiol 12:205–210
McIntosh AR, Gonzalez-Lima F (1994) Structural equation modeling and its application to network analysis in functional brain imaging. Hum Brain Mapp 2:2–22
McKinney WT Jr, Bunney WE Jr (1969) Animal model of depression. I. Review of evidence: implications for research. Arch Gen Psychiatry 21:240–248

Meaney MJ (2001) Maternal care, gene expression, and the transmission of individual differences in stress reactivity across generations. Annu Rev Neurosci 24:1161–1192
Medina JF, Christopher RJ, Mauk MD, LeDoux JE (2002) Parallels between cerebellum- and amygdala-dependent conditioning. Nat Rev Neurosci 3:122–131
Milad MR, Quirk GJ (2002) Neurons in medial prefrontal cortex signal memory for fear extinction. Nature 420:70–74
Milekic MH, Alberini CM (2002) Temporally graded requirement for protein synthesis following memory reactivation. Neuron 36:521–525
Milner B, Squire LR, Kandel ER (1998) Cognitive neuroscience and the study of memory. Neuron 20:445–468
Moita MA, Rosis S, Zhou Y, LeDoux JE, Blair HT (2003) Hippocampal place cells acquire location-specific responses to the conditioned stimulus during auditory fear conditioning. Neuron 37:485–497
Montkowski A, Poettig M, Mederer A, Holsboer F (1997) Behavioural performance in three substrains of mouse strain 129. Brain Res 762:12–18
Morgan MA, LeDoux JE (1995) Differential contribution of dorsal and ventral medial prefrontal cortex to the acquisition and extinction of conditioned fear in rats. Behav Neurosci 109:681–688
Morrow BA, Elsworth JD, Rasmusson AM, Roth RH (1999) The role of mesoprefrontal dopamine neurons in the acquisition and expression of conditioned fear in the rat. Neuroscience 92:553–564
Muller Igaz L, Vianna MR, Medina JH, Izquierdo I (2002) Two time periods of hippocampal mRNA synthesis are required for memory consolidation of fear-motivated learning. J Neurosci 22:6781–6789
Murchison CF, Zhang XY, Zhang WP, Ouyang M, Lee A, Thomas SA (2004) A distinct role for norepinephrine in memory retrieval. Cell 117:131–143
Myers KM, Davis M (2002) Behavioral and neural analysis of extinction. Neuron 36:567–584
Nader K (2003) Memory traces unbound. Trends Neurosci 26:65–72
Nader K, Schafe GE, Le Doux JE (2000) Fear memories require protein synthesis in the amygdala for reconsolidation after retrieval. Nature 406:722–726
Newport DJ, Stowe ZN, Nemeroff CB (2002) Parental depression: animal models of an adverse life event. Am J Psychiatry 159:1265–1283
Ohl FW, Scheich H, Freeman WJ (2001) Change in pattern of ongoing cortical activity with auditory category learning. Nature 412:733–736
Owen EH, Logue SF, Rasmussen DL, Wehner JM (1997) Assessment of learning by the Morris water task and fear conditioning in inbred mouse strains and F1 hybrids: implications of genetic background for single gene mutations and quantitative trait loci analyses. Neuroscience 80:1087–1099
Pare D, Collins DR, Pelletier JG (2002) Amygdala oscillations and the consolidation of emotional memories. Trends Cogn Sci 6:306–314
Paylor R, Tracy R, Wehner J, Rudy JW (1994) DBA/2 and C57BL/6 mice differ in contextual fear but not auditory fear conditioning. Behav Neurosci 108:810–817
Pérez-Jaranay JM, Vives F (1991) Electrophysiological study of the response of medial prefrontal cortex neurons to stimulation of the basolateral nucleus of the amygdala in the rat. Brain Res 564:97–101
Phillips RG, LeDoux JE (1992) Differential contribution of amygdala and hippocampus to cued and contextual fear conditioning. Behav Neurosci 106:274–285
Picciotto MR, Wickman K (1998) Using knockout and transgenic mice to study neurophysiology and behavior. Physiol Rev 78:1131–1163

Pitkänen A (2000) Connectivity of the rat amygdaloid complex. In: Aggleton JP (ed) The amygdala. A functional analysis. Oxford University Press, New York, p 31–115
Quirk GJ, Russo GK, Barron JL, Lebron K (2000) The role of ventromedial prefrontal cortex in the recovery of extinguished fear. J Neurosci 20:6225–6231
Rampon C, Tang YP, Goodhouse J, Shimizu E, Kyin M, Tsien JZ (2000) Enrichment induces structural changes and recovery from nonspatial memory deficits in CA1 NMDAR1-knockout mice. Nat Neurosci 3:238–244
Rescorla RA (1988) Pavlovian conditioning: it's not what you think it is. Am Psychol 43:151–160
Ressler KJ, Paschall G, Zhou XL, Davis M (2002) Regulation of synaptic plasticity genes during consolidation of fear conditioning. J Neurosci 22:7892–7902
Ressler KJ, Rothbaum BO, Tannenbaum L, Anderson P, Graap K, Zimand E, Hodges L, Davis M (2004) Cognitive enhancers as adjuncts to psychotherapy: use of D-cycloserine in phobic individuals to facilitate extinction of fear. Arch Gen Psychiatry 61:1136–1144
Richardson R, Ledgerwood L, Cranney J (2004) Facilitation of fear extinction by d-cycloserine: theoretical and clinical implications. Learn Mem 11:510–516
Riedel G, Micheau J, Lam AG, Roloff E, Martin SJ, Bridge H, Hoz L, Poeschel B, McCulloch J, Morris RG (1999) Reversible neural inactivation reveals hippocampal participation in several memory processes. Nat Neurosci 2:898–905
Roberson ED, Sweatt JD (1999) A biochemical blueprint for long-term memory. Learn Mem 6:381–388
Rogan MT, Staubli UV, LeDoux JE (1997) Fear conditioning induces associative long-term potentiation in the amygdala [see comments] [published erratum appears in Nature 1998 Feb 19;391(6669):818]. Nature 390:604–607
Rose SP (2000) God's organism? The chick as a model system for memory studies. Learn Mem 7:1–17
Rose SP, Stewart MG (1999) Cellular correlates of stages of memory formation in the chick following passive avoidance training. Behav Brain Res 98:237–243
Rosen JB, Schulkin J (1998) From normal fear to pathological anxiety. Psychol Rev 105:325–350
Rosenkranz JA, Grace AA (2002) Dopamine-mediated modulation of odour-evoked amygdala potentials during pavlovian conditioning. Nature 417:282–287
Rothbaum BO, Schwartz AC (2002) Exposure therapy for posttraumatic stress disorder. Am J Psychother 56:59–75
Routtenberg A (2002) Targeting the "species gene ensemble". Hippocampus 12:105–108
Sacchetti B, Lorenzini CA, Baldi E, Tassoni G, Bucherelli C (1999) Auditory thalamus, dorsal hippocampus, basolateral amygdala, and perirhinal cortex role in the consolidation of conditioned freezing to context and to acoustic conditioned stimulus in the rat. J Neurosci 19:9570–9578
Sacchetti B, Lorenzini CA, Baldi E, Bucherelli C, Roberto M, Tassoni G, Brunelli M (2001) Long-lasting hippocampal potentiation and contextual memory consolidation. Eur J Neurosci 13:2291–2298
Sanders MJ, Wiltgen BJ, Fanselow MS (2003) The place of the hippocampus in fear conditioning. Eur J Pharmacol 463:217–223
Sara SJ (2000) Retrieval and reconsolidation: toward a neurobiology of remembering. Learn Mem 7:73–84
Schacter DL (1987) Implicit memory: history and current status. J Exp Psychol 13:501–518
Schafe GE, LeDoux JE (2000) Memory consolidation of auditory Pavlovian fear conditioning requires protein synthesis and protein kinase A in the amygdala. J Neurosci 20:RC96
Schafe GE, Nader K, Blair HT, LeDoux JE (2001) Memory consolidation of Pavlovian fear conditioning: a cellular and molecular perspective. Trends Neurosci 24:540–546

Schwaerzel M, Heisenberg M, Zars T (2002) Extinction antagonizes olfactory memory at the subcellular level. Neuron 35:951–960
Sejnowski TJ (1999) The book of Hebb. Neuron 24:773–776
Sejnowski TJ, Destexhe A (2000) Why do we sleep? Brain Res 886:208–223
Sharp FR, Sagar SM, Swanson RA (1993) Metabolic mapping with cellular resolution: c-fos vs 2-deoxyglucose. Crit Rev Neurobiol 7:205–228
Shekhar A, McCann UD, Meaney MJ, Blanchard DC, Davis M, Frey KA, Liberzon I, Overall KL, Shear MK, Tecott LH, Winsky L (2001) Summary of a National Institute of Mental Health workshop: developing animal models of anxiety disorders. Psychopharmacology (Berl) 157:327–339
Shimizu E, Tang YP, Rampon C, Tsien JZ (2000) NMDA receptor-dependent synaptic reinforcement as a crucial process for memory consolidation. Science 290:1170–1174
Shobe J (2002) The role of PKA, CaMKII, and PKC in avoidance conditioning: permissive or instructive? Neurobiol Learn Mem 77:291–312
Shumyatsky GP, Tsvetkov E, Malleret G, Vronskaya S, Hatton M, Hampton L, Battey JF, Dulac C, Kandel ER, Bolshakov VY (2002) Identification of a signaling network in lateral nucleus of amygdala important for inhibiting memory specifically related to learned fear. Cell 111:905–918
Silva AJ (2003) Molecular and cellular cognitive studies of the role of synaptic plasticity in memory. J Neurobiol 54:224–237
Sokoloff L (2000) In vivo veritas: probing brain function through the use of quantitative in vivo biochemical techniques. Annu Rev Physiol 62:1–24
Stam R, Bruijnzeel AW, Wiegant VM (2000) Long-lasting stress sensitisation. Eur J Pharmacol 405:217–224
Stork O, Stork S, Pape HC, Obata K (2001) Identification of genes expressed in the amygdala during the formation of fear memory. Learn Mem 8:209–219
Sutherland GR, McNaughton B (2000) Memory trace reactivation in hippocampal and neocortical neuronal ensembles. Curr Opin Neurobiol 10:180–186
Swanson LW, Petrovich GD (1998) What is the amygdala? Trends Neurosci 21:323–331
Szapiro G, Galante JM, Barros DM, Levi dS, Vianna MR, Izquierdo LA, Izquierdo I, Medina JH (2002) Molecular mechanisms of memory retrieval. Neurochem Res 27:1491–1498
Tang J, Wotjak CT, Wagner S, Williams G, Schachner M, Dityatev A (2001) Potentiated amygdaloid auditory-evoked potentials and freezing behavior after fear conditioning in mice. Brain Res 919:232–241
Tang J, Wagner S, Schachner M, Dityatev A, Wotjak CT (2003) Potentiation of amygdaloid and hippocampal auditory evoked potentials in a discriminatory fear-conditioning task in mice as a function of tone pattern and context. Eur J Neurosci 18:639–650
Thomas E (1988) Forebrain mechanisms in the relief of fear: the role of the lateral septum. Psychobiology 16:36–44
Thompson RF, Bao S, Chen L, Cipriano BD, Grethe JS, Kim JJ, Thompson JK, Tracy JA, Weninger MS, Krupa DJ (1997) Associative learning. Int Rev Neurobiol 41:151–189
Thompson RF, Swain R, Clark R, Shinkman P (2000) Intracerebellar conditioning—Brogden and Gantt revisited. Behav Brain Res 110:3–11
Tsvetkov E, Carlezon WA, Benes FM, Kandel ER, Bolshakov VY (2002) Fear conditioning occludes LTP-induced presynaptic enhancement of synaptic transmission in the cortical pathway to the lateral amygdala. Neuron 34:289–300
Turrigiano GG, Nelson SB (2000) Hebb and homeostasis in neuronal plasticity. Curr Opin Neurobiol 10:358–364
Tzschentke TM (2001) Pharmacology and behavioral pharmacology of the mesocortical dopamine system. Prog Neurobiol 63:241–320

van Praag H, Kempermann G, Gage FH (2000) Neural consequences of environmental enrichment. Nat Rev Neurosci 1:191–198

Vanderwolf CH, Cain DP (1994) The behavioral neurobiology of learning and memory: a conceptual reorientation. Brain Res Brain Res Rev 19:264–297

Villarreal DM, Do V, Haddad E, Derrick BE (2002) NMDA receptor antagonists sustain LTP and spatial memory: active processes mediate LTP decay. Nat Neurosci 5:48–52

Walker DL, Ressler KJ, Lu KT, Davis M (2002) Facilitation of conditioned fear extinction by systemic administration or intra-amygdala infusions of d-cycloserine as assessed with fear-potentiated startle in rats. J Neurosci 22:2343–2351

Wallenstein GV, Eichenbaum H, Hasselmo ME (1998) The hippocampus as an associator of discontiguous events. Trends Neurosci 21:317–323

Wehner JM, Radcliffe RA, Bowers BJ (2001) Quantitative genetics and mouse behavior. Annu Rev Neurosci 24:845–867

Weinberger NM (1998) Physiological memory in primary auditory cortex: characteristics and mechanisms. Neurobiol Learn Mem 70:226–251

Welzl H, D'Adamo P, Lipp HP (2001) Conditioned taste aversion as a learning and memory paradigm. Behav Brain Res 125:205–213

Whishaw IQ, Metz GA, Kolb B, Pellis SM (2001) Accelerated nervous system development contributes to behavioral efficiency in the laboratory mouse: a behavioral review and theoretical proposal. Dev Psychobiol 39:151–170

Willner P (1984) The validity of animal models of depression. Psychopharmacology (Berl) 83:1–16

Wittenberg GM, Tsien JZ (2002) An emerging molecular and cellular framework for memory processing by the hippocampus. Trends Neurosci 25:501–505

Wolfer DP, Crusio WE, Lipp HP (2002) Knockout mice: simple solutions to the problems of genetic background and flanking genes. Trends Neurosci 25:336–340

Woolf NJ (1998) A structural basis for memory storage in mammals. Prog Neurobiol 55:59–77

Wotjak CT (2003) C57BLack/BOX? The importance of exact mouse strain nomenclature. Trends Genet 19:183–184

Wotjak CT (2004) Of mice and men: potentials and caveats of behavioural experiments in mice. BIF Futura 19:158–169

Würbel H (2001) Ideal homes? Housing effects on rodent brain and behaviour. Trends Neurosci 24:207–211

Xu L, Anwyl R, Rowan MJ (1998) Spatial exploration induces a persistent reversal of long-term potentiation in rat hippocampus [see comments]. Nature 394:891–894

Yehuda R, Antelman SM (1993) Criteria for rationally evaluating animal models of post-traumatic stress disorder. Biol Psychiatry 33:479–486

HEP (2005) 169:35–69

Animal Models of Anxiety

F. Ohl

Laboratory Animal Science, University Utrecht, PO Box 80166, 3508 TD Utrecht, The Netherlands
ohl@mpipsykl.mpg.de

Abstract Animal models for anxiety-related behavior are based on the assumption that anxiety in animals is comparable to anxiety in humans. Being anxious is an adaptive response to an unfamiliar environment, especially when confronted with danger or threat. However, pathological variants of anxiety can strongly impede the daily life of those affected. To unravel neurobiological mechanisms underlying normal anxiety as well as its pathological variations, animal models are indispensable tools. What are the characteristics of an ideal animal model? First, it should display reduced anxiety when treated with anxiolytics (predictive validity). Second, the behavioral response of an animal model to a threatening stimulus should be comparable to the response known for humans (face validity). And third, the mechanisms underlying anxiety as well as the psychological causes should be identical (construct validity). Meeting these three requirements is difficult for any animal model. Since both the physiological and the behavioral response to aversive (threatening) stimuli are similar in humans and animals, it can be assumed that animal models can serve at least two distinct purposes: as (1) behavioral tests to screen for potential anxiolytic and antidepressant effects of new drugs and (2) tools to investigate specific pathogenetic aspects of cardinal symptoms of anxiety disorders. The examples presented in this chapter have been selected to illustrate the potential as well as the caveats of current models and the emerging possibilities offered by gene technology. The main concepts in generating animal models for anxiety—that is, selective breeding of rat lines, experience-related models, genetically engineered mice, and phenotype-driven approaches—are concisely introduced and discussed. Independent of the animal model used, one major challenge remains, which is to reliably identify animal behavioral characteristics. Therefore, a description of behavioral expressions of anxiety in rodents as well as tests assays to measure anxiety-related behavior in these animals is also included in this chapter.

Keywords Anxiety · Animal model · Behavioral phenotyping · Ethological testing

1 Introduction

Anxiety is an essential emotion that is highly conserved during evolution. In interaction with cognitive parameters, anxiety regulates behavior in humans and other animals. The assessment of anxiety-related behavior in animal models is based on the assumption that anxiety in animals is comparable to anxiety in humans. As a matter of fact, it cannot be proved that rodents, the prime species in basic research, experience anxiety in the same way as human beings. However, it is undisputed that distinct behavioral and physiological patterns in rodents indicate anxiety, i.e., behavioral and peripheral changes presumed to accompany high sympathetic nervous activity (Hall 1936). From this, an analogy, if not a homology, between anxiety in humans and rodents may be assumed.

In principle, being anxious is an adaptive reaction when confronted with danger or threat. Behavioral and physiological responses accompanying anxiety prepare an individual to react appropriately to such situations, i.e., by displaying defense behavior such as flight or fight. Thus, anxiety enables an individual to escape from dangerous situations and to avoid them in the first place (Livesey 1986). However, pathological variants of anxiety can strongly impede the daily life of those affected. Anxiety disorders are reported to be the most prevalent psychiatric diseases. Additionally, a large proportion of psychiatric disorders comprise anxiety and depression, and these two conditions demonstrate a considerable overlap of clinical symptoms and pathophysiological processes. Therefore, some animal models with features of depression will be discussed as well.

The fact that the pharmacological treatment of choice has remained more or less the same for several decades (see chapter by Binder and Holsboer, this volume) emphasizes the urgent need for novel treatment strategies. To reach this goal, the neurobiological mechanisms underlying normal anxiety and its pathological variations have to be assessed. The limitations of research in humans, however, demand appropriate animal models. For decades, rodents have proved to be excellent animal models for the advancement of medicine as a whole. Nowadays rodents are becoming increasingly important for research in psychiatry (Geyer and Markou 1995).

The ideal animal model for any human clinical condition must fulfill three criteria (McKinney and Bunney 1969): (1) pharmacological treatments known to be effective in patients should induce comparable effects in the animal model (predictive validity); (2) the responses or symptoms observed in patients should be the same in the animal model (face validity); (3) the underlying rationale should be the same in both humans and animal models (construct validity). In other words, the ideal animal model for anxiety has to respond to treatment with anxiolytics such as benzodiazepines with reduced anxiety; it has to display defense behavior when confronted with a threatening stimulus; the mechanisms underlying anxiety as well as the psychological causes must be identical.

Meeting all three validity criteria is difficult for an animal model of anxiety as there are many heterogeneous forms of pathological anxiety. According to the Diagnostic and Statistical Manual of Mental Disorders (DSM-IV 1994), pathological anxiety is classified in five types: obsessive-compulsive disorder, phobias, panic disorder, post-traumatic stress disorder, and generalized anxiety disorder. Some of these pathological types of anxiety are hard to model in animals given the very human characteristics of their cardinal features, e.g., fear of dying in patients suffering from panic attacks; re-experiencing traumatic events in post-traumatic stress disorder. These symptoms are defined by a subjective verbal report, something that can never be modeled in an animal. Other aspects, such as the physiological and the behavioral response to aversive stimuli, are similar in humans and animals, allowing animal models to be used for at least two distinct purposes: (1) as behavioral tests to screen for potential anxiolytic and antidepressant properties of new drugs and (2) tools to investigate specific pathogenetic aspects of cardinal symptoms of anxiety disorders as well as other psychiatric diseases.

In literature, the term "animal model of anxiety" is used for animals altered in their anxiety-related behavior (Flint et al. 1995; Henn et al. 1993; Liebsch et al. 1998) as well as for test assays conceptualized to assess anxiety-related behavior in animals (Menard and Treit 1999; Rodgers et al. 1997). In the following, examples of both will be given, but the term "animal model" will be restricted to the former while the latter will be summarized under the heading "tests for anxiety."

2 Tests for Anxiety

Various test paradigms have been developed to assess behavioral parameters indicating anxiety in rodents. In the following, some well-established and available tests for anxiety will be described. These tests and also other test paradigms of unconditioned and conditioned anxiety, not explicitly mentioned here, are valuable tools in determining the implication of genetic factors in the whole complexity of behavior, and specifically in identifying the profile of anxiety-related behavior in rodents. Moreover, they are known to be extremely useful in behaviorally phenotyping drugs that potentially affect distinct aspects of anxiety. However, it should be taken into account that behavioral expressions represent a combination of behavioral dimensions influenced by genetic as well as environmental factors. The results of behavioral tests might be strongly influenced by testing conditions and the test procedure used. Therefore, it is essential to carefully define these factors when testing for anxiety in animals. Before describing how to test anxiety in rodents, some general points should be taken into consideration.

First, anxiety is not a unitary phenomenon, as it includes innate (trait) anxiety, which is considered to be an enduring feature of an individual, and situation-evoked or experience-related (state) anxiety. Since tests for anxiety in rodents are always restricted to the evaluation of situation-evoked behavior, it might be difficult to investigate trait anxiety in animals. However, the two phenomena are not separable from each other, as individuals with a high trait-anxiety often will show an increased tendency to also display high state anxiety. Thus, the term "anxiety" will be used without an a priori assumption of trait or state anxiety.

Second, modeling anxiety in animals is critically dependent on the test systems used. As standard behavioral test assays for anxiety were developed for and validated by classical anxiolytics, they might be of limited use towards discovering a new treatment strategy. The design of novel animal models and tests requires a different approach, namely the implementation of the ethological relevance of behavior. The minimum requirement for such a test in rodents must allow the animal to display its natural anxiety-related behaviors. It is also important to take into consideration behavioral dimensions related to anxiety, such as, for example, exploration, locomotor activity, or cognitive processes, as these dimensions are potentially confounding factors when assessing anxiety.

Third, during recent years, a tendency towards automation of such tests has established itself in order to standardize measurements and to avoid subjective interpretation of the animals' behavior. As only very few behavioral patterns can be recorded automatically, this increase in standardization causes a lack of sensitivity, because more subtle behavioral alterations of drug effects, such as changes in risk assessment behavior, remain undiscovered. Moreover, yet-unknown behavioral characteristics, which may occur in genetically modified animals, might not be detected by automatic recordings, and alterations in exploratory strategies are likely to produce wrong results when analyzed by use of predefined parameters (Ohl et al. 2001b). Thus, more complex, observation-based analyses should be performed, with respect to the rich behavioral repertoire displayed by rodents (Belzung 1999; Rodgers et al. 1997).

2.1 Behavioral Expression of Anxiety

2.1.1 Avoidance of Unprotected Areas as Index for Anxiety

Independent of the test set-up, there are species-specific behavioral expressions that are related to anxiety in rodents. For example, it is well known from both field studies and laboratory observations that rodents tend to avoid the unprotected area of a novel environment when first entering it (Barnett 1963; Belzung and LePape 1994; Treit and Fundytus 1989). In an experimental set-up, usually represented by a defined area, rodents will typically start to explore the environment along the walls while avoiding the open, i.e., unprotected, area (Fig. 1). The aversive character of an area can be modulated by illumination levels, with a brightly lit area being more aversive for a rat or mouse, thus producing a more pronounced avoidance behavior than a dark area. Another

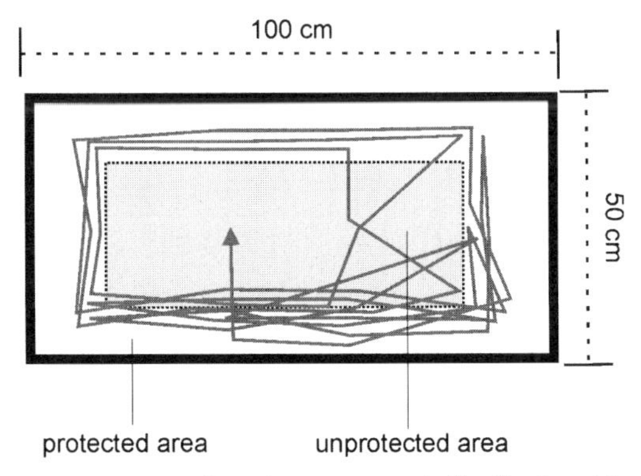

Fig. 1 Exploration pattern in a novel environment typically displayed by rodents. Placed in one corner of the environment, a rodent will first explore the protected area along the walls (thigmotaxis) before entering the unprotected area

way to increase a rodent's aversion against an unprotected area can be achieved by elevating it and by enabling the animal to see the edge. In fact, the expression of avoidance behavior depends on the visual capabilities of the animal and can further be influenced by its locomotor activity, motivational factors, and also by its exploration strategy (see Sect. 2.1.4). In general, a large body of literature reports avoidance behavior in rodents is sensitive to compounds with anxiolytic activity in humans (Belzung and Berton 1997; Chaouloff et al. 1997; Martin 1998).

2.1.2 Exploration: The Counterpart of Anxiety

Being confronted with novelty, behavior in rodents is determined by the conflict between the drive to explore the unknown area/object and the motivation to avoid potential danger. Exploration behavior summarizes a broad spectrum of behavioral patterns such as risk assessment behaviors, walking, rearing, climbing, sniffing, and manipulating objects (Barnett 1963; Kelley 1993; Sheldon 1968). It is suggested that exploration is gradually inhibited by anxiety, and,

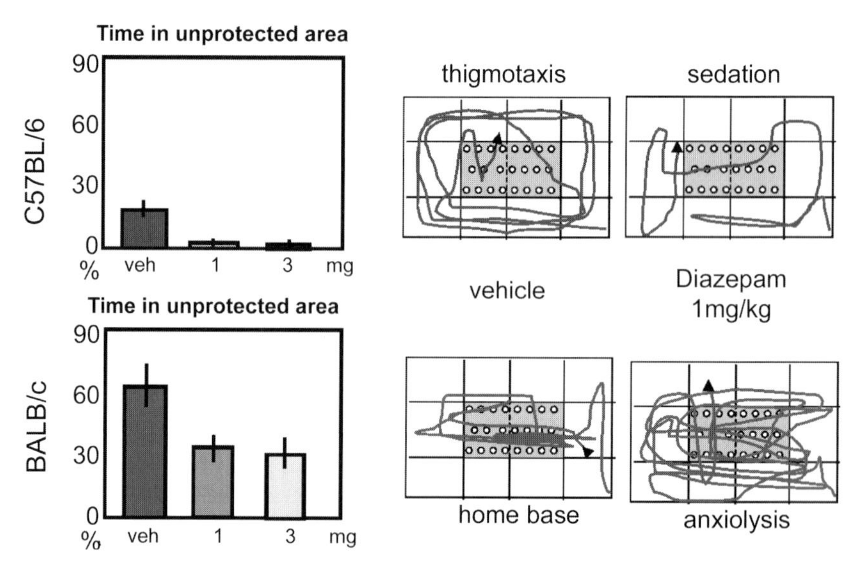

Fig. 2 Different exploration strategies in mice. While C57BL/6 mice explore a novel environment along the walls (thigmotaxis), BALB/c mice build a home base by first exploring the area close to the starting point. Referring to the standard measure for anxiety, i.e., the time spent in the unprotected area, BALB/c mice seem to be less anxious than C57BL/6. Moreover, treatment with diazepam decreases the time spent in the unprotected area in both mouse strains, suggesting an anxiogenic effect. However, looking at the exploration patterns, it is obvious that diazepam is sedative in C57BL/6 mice while anxiolysis is indicated in BALB/c mice by disintegration of the home base. (Modified from Ohl et al. 2001c)

therefore, might represent an indirect measurement of anxiety (Crawley and Goodwin 1980; Handley and Mithani 1984; Pellow et al. 1985). The inhibition of exploration behavior can be reversed by anxiolytic compounds (Belzung and Berton 1997; Griebel et al. 1993; Rodgers et al. 1992) but primary alterations in exploratory motivation may confound measures of anxiety (Belzung 1999), which has to be taken into account when behaviorally phenotyping rodents.

This is nicely demonstrated by the following example: As described above, a mouse or rat will typically start to walk around the walls of an unknown area (thigmotaxis). This is one strategy but, notably, different exploration strategies exist in rodents (Golani et al. 1999; Ohl et al. 2001b). Using a "home base" building strategy, some inbred mouse strains first explore the close surroundings of their starting point instead of walking around the area. Consequently, standard parameters that use avoidance behavior will not detect anxiety, while alterations in the exploratory strategy will indicate anxiolytic effects (Fig. 2).

2.1.3 Risk Assessment: A Sensitive Indicator for Anxiolytic Activity?

When confronted with a threatening stimulus, rodents display species-specific behavioral patterns, such as stretched-attend posture and directed sniffing, which are categorized as risk assessment behavior (Blanchard and Blanchard 1989; Cruz et al. 1994; Rodgers et al. 1997). The biological function of these behaviors is to gather information regarding the potential threat by cautiously approaching the threatening stimulus or by scanning the surroundings. Risk assessment behavior is thought to be a defense behavior (Blanchard et al. 1993), thus being indicative of anxiety. It is of note that factor analyses on complex ethological measures found risk assessment behavior to represent a behavioral dimension independent from avoidance behavior

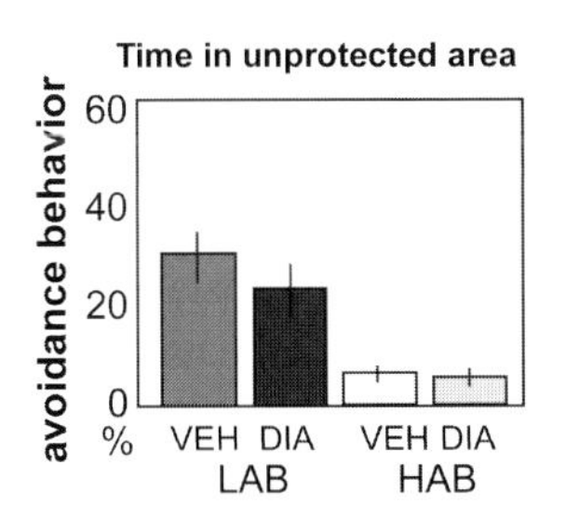

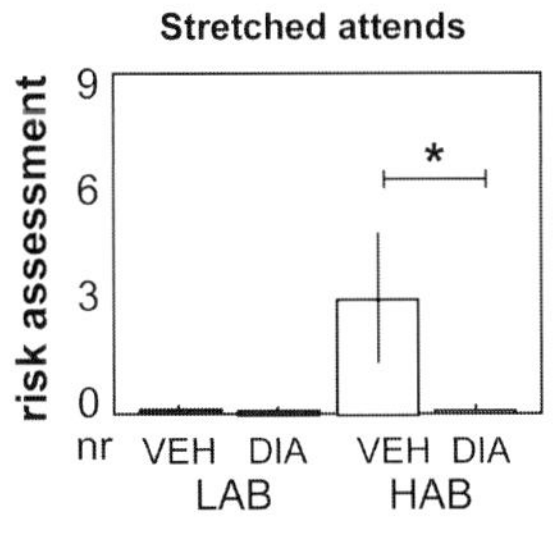

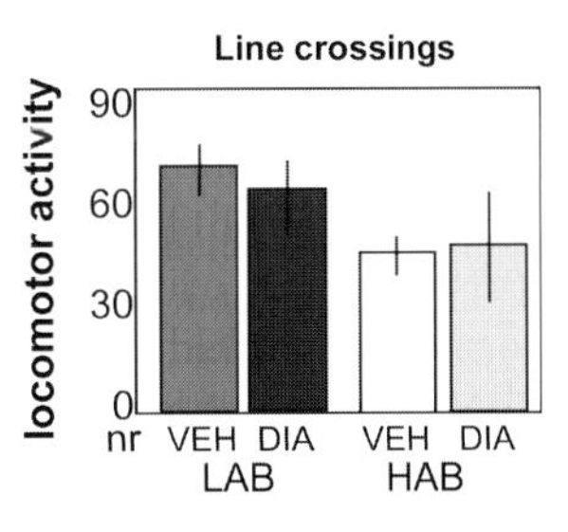

Fig. 3 Behavior of high (*HAB*) and low (*LAB*) anxiety rats (see Sect. 2.1.2) after treatment with either vehicle or diazepam (1 mg/kg). While avoidance behavior towards an unprotected area remains unaffected, the number of stretched-attend postures is significantly reduced, indicating a high sensitivity of risk assessment behavior for anxiety-modulating drug effects. The lack of effects on locomotor activity shows that treatment with this relatively low dose of diazepam was not sedative

(Cruz et al. 1994; Ohl et al. 2001c; Rodgers and Johnson 1995). These behavioral patterns of risk assessment appear to be even more sensitive for anxiety-modulating drug effects than standard measures of avoidance behavior (Rodgers and Cole 1994; Shephard et al. 1994) (Fig. 3), which might be due to the fact that risk assessment behaviors still are displayed when the animal has already overcome its avoidance of, for example, an unprotected area. Thus, risk assessment behavior represents the longest lasting expression of anxiety in rodents.

2.1.4
Flight-Related Behavior: A Dissociation Between Panic and Generalized Anxiety?

Within the group of anxiety disorders, panic disorder shows an increasing prevalence (Lecrubier and Ustun 1998), thus giving rise to an urgent need for causal treatment strategies. As panic disorder is classified by the DSM-IV system by the presence of symptoms that mostly are impossible to model in animals, such as fear of dying or paresthesias, the assessment of the reliability of tests for panic behavior predominantly is based on pharmacological validation. Specific tests based on conditioned behavior, such as the conditioned suppression of drinking, have been shown to be indicative for antipanic effects of drugs (Commissaris et al. 1990; Ellis et al. 1990; Fontana et al. 1998, 1989). Although not revealing the expected effects of panicogenic compounds, the conditioned suppression of drinking is considered to represent a reliable test for panic disorder (Blanchard et al. 2001a). Behavioral assays such as the suppressed drinking test are based on the conflict between approach and avoidance tendencies in an animal with avoidance behavior possibly substituting "flight" behavior in this specific set-up. Taking into account that patients experiencing a panic attack frequently report an urge to flee, panic disorder was hypothesized to represent a defense-related human psychopathology (Ashcroft et al. 1993). In contrast, other authors explicitly state that the precursor of pathological anxiety lies in the exaggeration of normal fear, not in altered defense behavior (Rosen and Schulkin 1998).

Extensive studies have been performed, especially by Caroline and Robert Blanchard and colleagues, (1) to evaluate whether human and rodent defensive behaviors in response to threat show parallels (Blanchard et al. 2001b) and (2) to characterize the predictive validity of specific defense patterns in rodents for anxiety-modulating compounds (for review Blanchard et al. 1997). The Mouse Defense Test Battery (Blanchard et al. 2003) revealed that distinct defense patterns such as defensive threat or risk assessment are sensitive to drugs known to be effective in the treatment of generalized anxiety disorder, while flight responses are modulated by panicolytic or panicogenic drugs (Blanchard et al. 2001a; Graeff 2002). Thus, rodent flight behavior appears to be a useful tool to investigate mechanisms of panic disorder.

2.1.5 Food Intake Inhibition

Food intake is considered a reliable indicator for the anxiolytic properties of drugs. Rodents usually are reluctant to eat unknown food (Boissier et al. 1976; Soubrie et al. 1975). When both familiar and unknown food are presented, rodents will typically show a longer latency to the first intake of unknown food compared to the intake of familiar food. Anxiolytic drugs not only reverse this food intake inhibition (Fletcher and Davies 1990; Hodges et al. 1981) but also result in an increased consumption of food (Britton and Britton 1981).

2.1.6 Cognition: A Primary Feature of Pathological Anxiety?

In recent years a fundamental relation between anxiety and cognitive processes has been demonstrated (Belzung and Beuzen 1995; McNaughton 1997). It has been argued that cognitive alterations may be the primary presenting feature of pathological anxiety (Hindmarch 1998). Gray (1990) already suggested that anxiety emerges when there is a mismatch between the information perceived by an individual and the information already stored. McNaughton (1997) hypothesized that generalized anxiety disorder could be the consequence of a purely cognitive dysfunction that results in inappropriate emotional responses. On the other hand, there is extensive evidence that emotional arousal modulates both affective memory and declarative memory (i.e., factual knowledge) for emotional events (Cahill and McGaugh 1998; McGaugh et al. 1996). Still, little is known about the interaction between emotionality and non-emotional cognitive processes.

Using animals to characterize the interaction of emotion and cognition, one should most carefully select appropriate animal models and test paradigms. In studies assessing cognitive processes, the experimental conditions often interfere with the performance of experimental animals, since stressors such as novel environment or food deprivation are part of the experimental setting (Conrad et al. 1996; Hodges 1996; Beck and Luine 1999; Croiset et al. 2000). Comparable problems occur when pharmacological treatments are used: Benzodiazepines (see also the chapter by Duman and Duman, this volume), for example, are well known to act both as an anxiolytic and amnestic (Belzung and Beuzen 1995).

Recently, evidence was provided that in a rat model of innate high or low emotionality (see Sect. 3.1.1) the degree of anxiety is differentially associated with enhanced performance of distinct informational processes (Ohl et al. 2002). It was hypothesized that the increased anxiety-related behavior may be due to these differences in cognitive processing. Concerning the analysis of the interaction between anxiety and cognition, inbred mouse strains (see Sect. 3.1.2, such as C57BL/6 (BL6) and DBA/2 (DBA) mice, are also of interest.

These mouse strains do not only differ in distinct cognitive abilities (for review: Rossi-Arnaud and Ammassari-Teule 1998; Cabib et al. 2002; Parmigiani et al. 1999), but also in their anxiety-related behavior (Griebel et al. 1997; Trullas and Skolnick 1993; Crawley et al. 1997). Specifically, DBA/2 mice are considered to represent a genetic animal model of hippocampal dysfunction due to their deficits in spatial memory performance (Thinus-Blanc et al. 1996), while being indistinguishable from C57BL/6 mice in visually cued tasks (Ammassari-Teule et al. 1999). As to their anxiety-related behavior, DBA/2 mice are reported to display high anxiety compared to C57BL/6 mice only in some tests of unconditioned anxiety (Griebel et al. 1997; Trullas and Skolnick 1993; Rogers et al. 1999). A recent study using these two inbred mouse strains (Ohl et al. 2003) indicated an interaction between behavioral and cognitive characteristics comparable to that found in high- and low-anxiety rats: High anxiety again was paralleled by an increased cognitive performance. It was, therefore, hypothesized that the expression of high anxiety might represent a behavioral inhibition that is cognitively driven. Consequently, cognitive processes should be carefully controlled, especially when characterizing animal models for anxiety.

2.2 Tests for Unconditioned Avoidance Behavior

Among the most frequently used paradigms are tests for unconditioned anxiety that are thought to be indicative for human generalized anxiety symptoms (Crawley 1999). In these tests, rodents usually are confronted with a novel environment or stimulus, and behavioral patterns related to anxiety (see Sect. 2.1.1) are measured. In the following, the most commonly used tests for unconditioned behavior will be briefly described.

2.2.1 Open Field

As a test for unconditioned anxiety, the open field was first described by Hall (1936), who evaluated the behavior of rats in a circular, brightly lit area surrounded by a wall. Meanwhile, different types of open fields have been designed, varying in size, shape (from circular to square), illumination (from dimly lit to bright illumination), and enrichment (by offering objects or food) (Fig. 4). Also, the testing procedure differs widely: Testing duration ranges from 2 min to several hours (Golani et al. 1999) with the most frequent duration of 5 or 10 min. The standard procedure implies a forced confrontation with the open field: The animal is placed either in the center or in the periphery of the area. In a specific variation, also referred to as free exploration test, an open field is connected to the home cage of the animal, which is then permitted free access to the novel environment (Kopp et al. 1997). The avoidance behavior towards the unprotected area is the indicator for anxiety. Investigations of behavioral

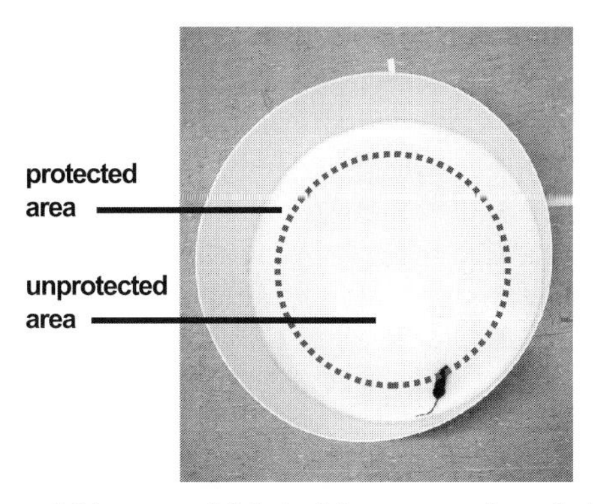

Fig. 4 Example of an open field set-up which, in this case, consists of white polyvinyl chloride (*PVC*) and is evenly lit

patterns related to anxiety are sometimes used to gain further information (Clement et al. 1997; Crabbe 1986; Flint et al. 1995). Although evidence exists that the open field may be useful in detecting genetic or pharmacological effects of anxiety (Treit and Fundytus 1989; Prut and Belzung 2003), some studies also report a lack of sensitivity for anxiety-modulations (Saudou et al. 1994).

2.2.2
Elevated Plus Maze

Probably the most frequently used test for unconditioned anxiety is the elevated plus maze (EPM), which was first introduced by File and coworkers (Pellow et al. 1985). The test consists of an elevated, plus-sign-shaped runway with two opposing arms being closed by walls and the other two arms being open, i.e., unprotected (Fig. 5). The animal usually is placed in the center of the EPM,

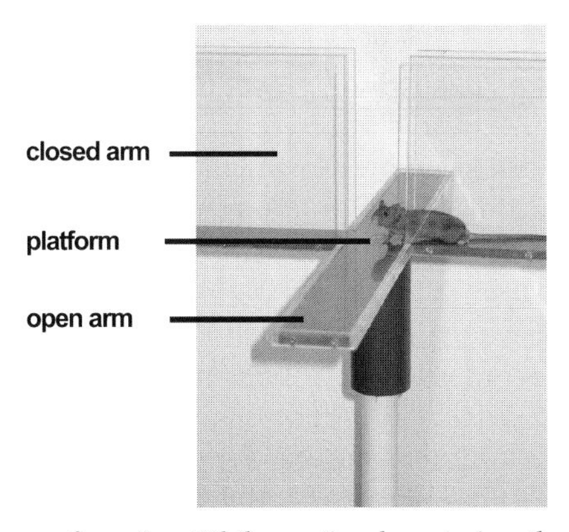

Fig. 5 Elevated plus maze for mice. While cautiously entering the platform, the mouse displays risk assessment behavior

where the four arms cross each other, facing a closed arm. The EPM is based on the observation that rodents tend to avoid elevated areas (Montgomery 1958), which, in the case of the EPM, usually are brightly lit in addition. Following the concept of avoidance behavior in rodents (see Sect. 2.1.1), avoidance of the open arms is interpreted as anxiety (Lister 1990; Pellow et al. 1985; Rodgers et al. 1997). Moreover, it has been argued that the EPM also allows investigators to control for locomotor activity, thus representing a reliable test for anxiety-modulating properties of pharmacological compounds (Belzung and Griebel 2001; Hogg 1996; Rodgers et al. 1992). The reliability and sensitivity of this test is increased by using more detailed approaches to analyze rodent behavior on the EPM, such as including, for example, risk assessment behavior (Cruz et al. 1994; Griebel et al. 1993; Rodgers and Johnson 1995; Rodgers et al. 1997; Weiss et al. 1998).

2.2.3 Dark/Light Box

Making use of the conflict between a rodent's motivation to explore a novel environment and its avoidance of brightly lit areas, the dark/light box (also referred to as light/dark box or black–white box) comprises one aversive (i.e., light) and one less aversive (i.e., dark) compartment (Hascoet et al. 2001). In the original set-up, the dark compartment is smaller than the lit one, and both compartments are separated by a partition containing an opening (Crawley and Goodwin 1980; Crawley 1999). Later on, modifications were made in that the two compartments were equal in size and connected by a tunnel (Belzung and Berton 1997). Transitions between the compartments and time spent exploring each are interpreted as indicators of anxiety and are sensitive to anxiety-affecting drugs (Belzung 1999; Chaouloff et al. 1997; Crawley 1999). As a limitation, it has been reported that dark/light transitions are likely to be confounded by alterations in general activity (Bourin and Hascoet et al. 2003). It was suggested that the behavioral expression of decreased anxiety in the dark/light box might be determined by genetically based spontaneous exploration (Crawley 1999). Again, more complex behavioral analyses can increase the reliability of measuring anxiety (Rodgers et al. 1992).

2.2.4 The Forced Swim Test

Behavioral alterations induced by stressful experiences frequently include increased anxiety-related behavior. Accordingly, anxiety has been hypothesized to play a role in stress-coping behavior (Ferre et al. 1994). The most regularly used tests for stress-coping behavior is the so-called forced swim test, which has been developed by Porsolt et al. (1977) as a behavioral paradigm to identify compounds with antidepressant efficacy in humans: Mice or rats are forced to swim in a glass cylinder filled with water, preventing the animal from escaping. The behavior of the animal in this situation is classified in three behavioral patterns: struggling (interpreted as escape behavior), swimming (mostly interpreted somehow neutrally), and floating (interpreted as despair). During the first exposure, the animal is thought to learn that it cannot escape the situation and, consequently, in subsequent tests the time spent immobile is increased. More recently, immobility has been discussed as a successful strategy that conserves energy and allows the animal to float for prolonged periods of time, thereby improving its chances of survival (West 1990). Others interpret floating as adaptive disengagement from the persistent stress and the immobile posture alternates with active escape as part of a search-waiting coping strategy (Thierry et al. 1984).

Profound evidence exists that the forced swim test and modified versions (Cryan et al. 2002) have a good predictive validity, especially for antidepressant drugs (Porsolt et al. 1978). It is, however, an ongoing matter of discussion whether or not the observed increase in escape-oriented behavior may rather be secondary to changes in cognitive performance (Montkowski et al. 1995; De Pablo et al. 1989) or to anxiety-related behavior (Ferre et al. 1994). This discussion will continue. The argument that genetic factors contribute to the behavioral performance of rodents in this test—some strains of mice do not decrease but rather increase immobility after antidepressant treatment (Lucki et al. 2001)—is attracting increasing attention.

2.2.5
The Modified Hole Board: Assessing Dimensions of Behavior

Most of the test procedures described above are predictive of only a small spectrum of behavioral patterns. Consequently, additional tests are recommended to control for possible confounding factors, e.g., locomotor activity (Escorihuela et al. 1999; Ohl et al. 2001c). However, testing the same animal in a multiple test battery is likely to induce interferences between distinct tests (Belzung and LePape 1994). Studies based on behavioral tests, which are focused on a more detailed ethological analysis of experimental animals in a single complex paradigm, may overcome these disadvantages (Cruz et al. 1994; Lister 1990; Rodgers et al. 1997; Wilson 2000). The modified hole board (mHB, Fig. 6) test is based on the concept that the rich behavioral repertoire of rodents can only be displayed in an adequate, i.e., rich, testing environment. The test essentially comprises the characteristics of a hole board (File and Wardill 1975; Lister 1990) and an open field test. In the mHB set-up, a hole board—with all holes covered by a movable lid—is placed in the middle of a box, thus representing the central area of an open field (Fig. 4). The experimental box is enlarged by an additional compartment where the group mates of the experimental animal are placed during the test period, being separated from the test area by a transparent partition. In both rats and mice it was

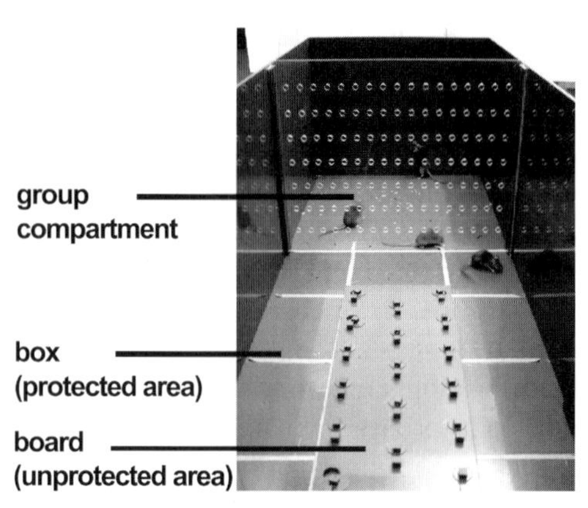

Fig. 6 The modified hole board. In the center of the experimental box, i.e., in the unprotected area, a hole board is placed. The box is enlarged by an additional compartment, where the group-mates of the experimental animal are placed during testing

demonstrated that the mHB enables the investigator to detect alterations in a wide range of behaviors, including anxiety-related behavior, risk assessment, exploration strategies, locomotor activity, arousal, social affinity, and cognition (Ohl et al. 2001b,c, 2002). Although validation in terms of sensitivity and specificity for pharmacological effects has yet to be extensively worked out, there is good evidence that anxiolytic effects may be dissociated from sedative effects and general alterations in exploratory behavior by use of the mHB (Ohl et al. 2001b). In contrast to other behavioral tests, the mHB, due to the presence of the group mates, avoids isolation stress in experimental animals during testing, a factor that is well known to affect behavioral performance and especially anxiety-related behavior (Ohl et al. 2001a). In general, test assays such as the mHB enable the animal to display a complex behavioral repertoire that, in combination with careful analysis by a trained observer, offers the opportunity to also discover subtle and unexpected behavioral effects of novel treatment strategies.

2.3
Tests for Conditioned Anxiety

Another behavioral approach used to assess aspects of anxiety in animals relies on conflict paradigms in combination with punishment, mostly induced by electric foot shock. Due to ethical and also ethological considerations, paradigms based on electric shock are less often used than tests for unconditioned anxiety. However, it has been hypothesized that behavioral expressions displayed in tests for unconditioned and conditioned anxiety may reflect profoundly different aspects of anxiety (File 1995; Griebel 1996; Millan and Brocco 2003). Thus, shock paradigms are quite frequently included in behavioral phenotyping procedure used to characterize novel animal models for anxiety or anxiety-modulating compounds.

The Vogel conflict test (Vogel et al. 1971) has been modified from the Geller–Seifter test (Geller and Seifter 1960) in that the initially used food reward was replaced by a water reward. The test is based on an operant conditioning procedure, where water-deprived animals are given access to a water bottle during the test situation, but randomly a lick is accompanied by electric shock (see also Sect. 2.1.3). Anxiolytic effects are indicated by modulation of the shock-induced suppression of licking (Kopp et al. 1999; Stefanski et al. 1993).

Referred to as a conditioned fear paradigm, the fear potentiated startle response was first described by Brown et al. (1951). In the original test, an acoustic stimulus is presented in the presence of a conditioned stimulus that has previously been paired with an aversive, unconditioned stimulus. The amplitude of the acoustic startle response is thought to indicate the degree of conditioned anxiety, which can be reduced by anxiolytic drugs (Davis et al. 1993; Hijzen et al. 1995).

3 Animal Models

Comprehensive studies based on rodent models of anxiety have not only underlined that anxiety in itself represents a complex behavioral system but also that it is determined by both genetic and environmental factors as well as by the interaction between both. The examples used in this section have been selected to illustrate both the potential and the caveats of current models and the emerging possibilities offered by gene technology. These examples are thought to be representative of the different concepts followed in generating animal models.

3.1 Genetically Based Animal Models of Anxiety

3.1.1 Selectively Bred Rat Lines

Animal models based on genetic selection are intended to model the genetically based susceptibility known to be one risk factor for the development of anxiety disorders. Animal models of high innate anxiety, gained by selective breeding, are valuable tools since these animals do not exhibit pathological anxiety due to stress exposure (see also Sect. 3.2); instead, the anxiety is the result of an enduring feature of a strain or an individual, probably involving multiple genetic and environmental factors. Studies on selective breeding of rats began already some 80 years ago, when Edward C. Tolman selected rats for their learning capacities (1924). Several years later, Calvin Hall initiated the first rat breeding lines selected for emotionality (1938). In the following section, some still-existing and available selectively bred rat lines for emotionality will be introduced.

3.1.1.1
The Maudsley Strains

Selected for differences in their defecation rate when exposed to the open-field test (see Sect. 2.2.1), the Maudsley reactive and non-reactive strains represent the eldest, still-available selection strain (Broadhurst 1960). By some authors, these rat strains are reported to represent a model for high and low emotionality, respectively (Berrettini et al. 1994; Ahmed et al. 2002). Other investigators found the Maudsley strains to differ only in some tasks reflecting emotionality (Overstreet et al. 1992). Questioning the face validity of the Maudsley strains as an animal model for anxiety, no strain differences were found in terms of endocrine response to a variety of stimulating conditions such as forced swimming or foot shock (Blizard and Adams 2002). Notably, alterations in the activation of the central noradrenergic system were found in response to stress, being correlated to the strain differences in open field defecation rates (Blizard et al. 1982; McQuade and Stanford 2001). Surprisingly, in these studies non-reactive rats had a greater concentration of noradrenaline than reactive rats (Blizard et al. 1982), indicating a lower emotionality of the latter strain. From these results it was hypothesized that the higher release of noradrenaline during exposure to the open field may result in an inhibition of colonic motility and, thus, a lower defecation rate. In summary, while the Maudsley strains undoubtedly are of high use to study, for example, environmental influences on distinct behavioral characteristics, they are unlikely to represent an animal model for anxiety in general.

3.1.1.2
Roman High- and Low-Avoidance Rats

Already several decades ago, two lines were selected by Broadhurst and Bignami (1965) from Wistar rats, which showed either good or bad performance in a two-way active avoidance task. These rat lines were called Roman high- (RHA) and low- (RLA) avoidance rats and have been extensively evaluated in terms of behavioral and neurocrine/neurochemical parameters. Corresponding to the fact that performance of active avoidance tasks strongly depends on emotional factors (Driscoll and Bättig 1982; Escorihuela et al. 1999; Fernandez-Teruel et al. 2002), RLA rats have been shown to be highly emotional and more passive in their coping strategy compared to RHA rats (Ferre et al. 1994; Gentsch et al. 1982). Moreover, the endocrine response to stressful situations is increased in RHA rats compared to their RLA counterparts (Steimer et al. 1998; Walker et al. 1989). These differences between RHA and RLA rats are hypothesized to have a genetic basis (Castanon et al. 1994, 1995), determining a specific sensitivity to environmental factors (Steimer et al. 1998). Notably, differences in a variety of additional behavioral and cognitive processes are described for the two rat lines, such as reduced locomotor activity and exploration (Corda et al. 1997; Giorgo et al. 1997), as well as better spatial learning and memory performance (Escorihuela et al. 1995) in highly emotional RLA rats. Thus, although representing an interesting animal model for a complex behavioral trait, including altered emotional reactivity, the face value of RLA rats as a model for anxiety or mood disorders is difficult to assess.

3.1.1.3
The Flinders Sensitive Rats

At Flinders University in Australia, two other lines of rats were developed in the late 1970s (Overstreet et al. 1979) by selective breeding for differences in the hypothermic response to the cholinesterase inhibitor diisopropylfluorophosphate. This approach aimed at reversing the order of questions asked by typical selective breeding programs; it chose a physiological response to specific pharmacological agents as the selection criterion. Only afterwards did behavioral alterations enter the evaluation in order to identify possible anxiety-related or depressive-like characteristics. From a number of various findings it was concluded that the Flinders sensitive line, being hypersensitive to cholinergic agonists, may rather represent an animal model of depression more than anxiety disorder, because these rats exhibit several symptom patterns of depression, such as reduced locomotor activity, reduced body weight, increased rapid eye movement (REM) sleep, and cognitive (learning) deficits (Overstreet 1993). However, evidence that cholinergic hypersensitivity might be the leading cause of depression is limited, and this hypothesis is further to be questioned by the therapeutic efficacy of new antidepressants lacking anticholinergic properties. Notably, Flinders sensitive line rats show exaggerated immobility in the forced swim test (see Sect. 2.2.4) and this behavior returns to normal after treatment with antidepressants, including those without anticholinergic effects (Overstreet et al. 1995). This in turn is of interest in connection with recently reported changes in serotonergic activity in these animals (Zangen et al. 1997).

3.1.1.4
Hyperanxious Rats

One recent selectively bred rat line is the rat model of high and low anxiety-related behavior, developed by Landgraf and colleagues (Liebsch et al. 1998). Wistar rats displaying either extremely high (HAB) or low (LAB) anxiety-related behavior on the elevated plus maze (see Sect. 2.2.2) were selected and bred over a decade (Liebsch et al. 1998). The resulting two rat lines distinctly differ in their inborn emotionality, which has been shown in detailed ethological studies (Henniger et al. 2000; Ohl et al. 2001b,c). Moreover, rats with high or low levels of anxiety-related behavior show marked differences in their stress-coping strategies and their neuroendocrine susceptibility to stressors in that rats with high levels of anxiety-related behavior displayed an enhanced stress hormone release when stressed (Landgraf et al. 1999). This may be interpreted as a dysregulation of the stress hormone system, the hypothalamus–pituitary–adrenocortical (HPA) system, which also occurs in the majority of patients suffering from depression (Holsboer 2000). The predictive validity of this rat model was shown in a variety of preclinical studies (Liebsch et al. 1998; Keck et al. 2001a,b, 2002, 2003; Landgraf and Wigger 2002). It may also be assumed that this model possesses construct validity regarding anxiety-related disorders and maybe regarding depression, although the effectiveness of classical antidepressants has to be studied further. Recently, the first molecular genetic approaches were performed in high and low anxiety rats, revealing single nucleotide polymorphisms in the vasopressin promoter area of these strains (Landgraf and Wigger 2002). These results underline the potential value of selectively bred rat lines for the identification of genetic determinants underlying anxiety.

3.1.2
Inbred Mouse Strains

During the past 90 years, starting from when the first inbred mouse strains were being described, this species has become an organism of choice for modeling human diseases. More than 450 inbred strains of mice are available now, representing a variety of genotypes but also phenotypes. Thus, in the search for animals models of anxiety disorders, inbred mouse strains are an interesting resource.

As an example, the two inbred mouse strains C57BL/6 and BALB/c have been shown to differ in their behavior in several tests of unconditioned anxiety (Belzung et al. 2000; Griebel et al. 2000; Kopp et al. 1999; Rogers et al. 1999; Trullas and Skolnick 1993). Thus, these mouse strains may represent a possible parallel to the high and low anxiety rats (see Sect. 3.1.1). However, perusing literature, the findings seem to be contradictory. Although differences in terms of anxiety are consistently reported for these two strains, the rank order of their level of anxiety varies between different tests and studies. While C57BL/6 mice were found to be more anxious than BALB/c in the elevated plus maze, opposite results were found in the open field in the same study (Avgustinovich et al. 2000). Rogers et al. (1999) also reported that BALB/c displayed the lowest level of anxiety compared to five other mouse strains including C57BL/6 in the elevated plus maze. In the more complex modified hole board test the two strains fail to demonstrate profound differences in terms of classical parameters of anxiety-related behavior, i.e., no avoidance at all of an unprotected area

(Ohl et al. 2001a). Still, BALB/c mice displayed significantly more risk assessment behavior and showed a higher arousal than C57BL/6 mice. Interestingly, the paths taken by the two strains show that those parameters indicating avoidance behavior are without relevance because of differences in the exploratory strategy of these strains. C57BL/6 mice perform a pronounced thigmotaxis by first walking along the wall, and only later do they enter the unprotected area; in contrast, BALB/c mice rarely display any avoidance behavior towards the unprotected area but tend to build a pronounced home base (Golani et al. 1999) near the starting point, which may be either within the protected or unprotected area (see Fig. 3). When using anxiety tests based on the avoidance of aversive areas, the BALB/c strain, consequently, is of limited use as animal model for increased anxiety.

DBA/2 mice have also been suggested to represent an opposite extreme in terms of emotionality in comparison to the C57BL/6 strain (Griebel et al. 2000; Trullas and Skolnick 1993; Crawley et al. 1997; Rogers et al. 1999). And again, studies on the anxiety-related behavior in these mouse strains showed contradictory results. The findings ranged from DBA/2 mice being less anxious (elevated plus maze: Trullas and Skolnick 1993), to both strains being indistinguishable (elevated plus maze: Griebel et al. 1997), and to C57BL/6 mice being less anxious (elevated plus maze: Rogers et al. 1999). The same holds true for different tests of unconditioned behavior, such as the dark/light box

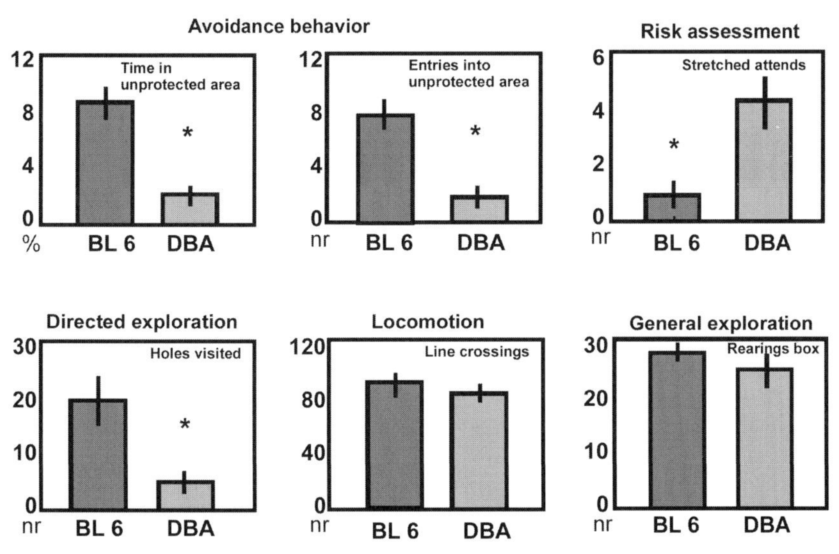

Fig. 7 Results from behavioral testing in the modified hole board (see Sect. 2.2.5). While DBA/2 mice (*DBA*) display a more pronounced avoidance behavior towards an unprotected area as well as more risk assessment behavior than C57BL/6 (*BL6*) mice, no differences in term of general exploration and locomotion are found in these strains. Therefore, these strains can be considered to represent an interesting model for high and low anxiety

(Crawley et al. 1997; Griebel et al. 1997) and for other behavioral dimensions, such as locomotor activity (Griebel 1997; Rogers et al. 1999). In the modified hole board test, DBA/2 mice proved to be more anxious than C57BL/6 mice by performing a pronounced avoidance behavior towards the unprotected area and an increased risk assessment behavior. Importantly, the two strains did not differ in terms of locomotor activity or general exploration, suggesting that differences between DBA/2 and C57BL/6 in anxiety-related behavior are not secondary to characteristics in general activity (Fig. 7). These findings strongly point towards the idea that C57BL/6 and DBA/2 mice represent low and high anxiety strains, respectively.

These examples underline that in the search for animal models of anxiety disorders it is not sufficient to screen for anxiety-related behavioral characteristics. On the contrary, it is of fundamental importance to phenotype extensively and carefully each potential animal model, even the well-established inbred mouse strains.

3.1.3 Genetically Engineered Mice

The traditional routes to animal models of mood disorders have recently been expanded by the possibility of studying mice with behavioral changes that are not experience-related, but are secondary to the insertion of a transgene or to a targeted disruption of a single gene, thus potentially allowing consideration of behavioral effects that are mediated by distinct genes (Jaenisch 1998; Gerlai 1996; Holmes 2001; Holsboer 1997; Picciotto 1999; Wurst et al. 1995). In recent years, several methods have emerged to alter the mouse genome in a direct or "reverse" genetic manner, presenting a new molecular approach to behavior. Reverse genetics refers to a set of techniques such as transgenesis and gene targeting in which a single cloned gene is used to generate a line of mice with an alteration specifically in that gene. Genetically engineered mice were not originally generated to produce animal models with face validity for psychiatric disorders, but rather to delineate the role of a specific gene product for the behavioral phenotype (Müller and Keck 2002).

For more than 40 years, the most frequently used anxiolytic compounds have been benzodiazepines. Benzodiazepine agonists such as diazepam act via modulation of γ-aminobutyric acidergic (GABAergic) transmission at $GABA_A$ receptors (see chapter by Duman and Duman, this volume), which consist of about 20 subunits. Several mutant mice with deletions of different subunits have been engineered (Holmes 2001) and most of these mutants perform altered anxiety-related behavior, thus emphasizing the intimate link between benzodiazepine receptors and anxiety.

Another major focus of interest for the investigation of anxiety disorders is the monoamine neurotransmitter serotonin (5-HT; see also chapter by Möhler et al., this volume) because of reduced levels of 5-HT receptors found in patients suffering from anxiety and mood disorders (Toth 2003) and mostly because of the effectiveness of 5-HT re-uptake inhibitors in the treatment of anxiety disorders (Argyropoulus et al. 2000). Consequently, mouse models with genetically modified functions of the 5-HT system have been engineered. Nowadays, several 5-HT receptor subtypes are known, and knocking out one of these subtypes produces quite specific behavioral phenotypes. While the 5-HT_{2C} receptor has been shown to regulate feeding behavior (Tecott et al. 1995), the 5-HT_{1B} knock-out mouse shows an altered emotional learning (Dulawa et al. 1997). More recently, the 5-HT_{1A} receptor has been knocked out by three independent research groups each of them using a different genetic background mouse strain and, notably, all three knock-out lines displayed increased anxiety-related behavior under stressful conditions compared to the respective background strain (Gingrich and Hen 2001; Olivier et al. 2001).

Taking into account the central role of the HPA-axis for the regulation of anxiety, a variety of genetically altered mice has been developed, aimed at targeting the hormonal stress system (Müller and Keck 2002; Sillaber et al. 2002; Stenzel-Poore et al. 1992; Timpl et al. 1998). These models are described extensively in the chapter by Keck and Müller and will, therefore, be omitted here.

3.1.4
ENU-Mutagenized Mice: The "Phenotype-Driven" Approach

Although a large number of knock-out mutants will be engineered in the future, genetic analysis requires the availability of multiple alleles of the same gene or of different genes involved in the pathogenesis of the same disease, including hypomorphs, alleles of different strength, and gain of function alleles. Such alleles can be obtained after treatment with chemical mutagens, such as ethyl-nitrosourea (ENU), which is suggested to be the most powerful mutagen in mice. In contrast to, for example, radiation, ENU primarily induces point mutations (Balling 2001). The injection of ENU in a male mouse mutagenizes premeiotic stem cells, leading to a large number of F1 animals carrying different mutations.

Complementary to the "gene-driven" analysis of gene function, "phenotype-driven" approaches can be performed and may be equally important. Large-scale mutagenesis represents a possible way to create animal models that approximate the underlying genetic etiology. The success of this approach was demonstrated by the genetic and molecular dissection of the pathways that set-up the Drosophila body pattern (Lee et al. 1989). In the meantime, mutagenesis screens have also been carried out in mice (Balling 2001; Brown and Balling 2001; Brown and Hardisty 2003; Hrabe de Angelis et al. 2000; Nolan et al. 2000).

Using a large-scale screening procedure for behavioral phenotyping in mice, the first behavioral screenings in ENU-mutagenesis mice demonstrated that it is possible to reliably identify and dissociate mutants with alterations in

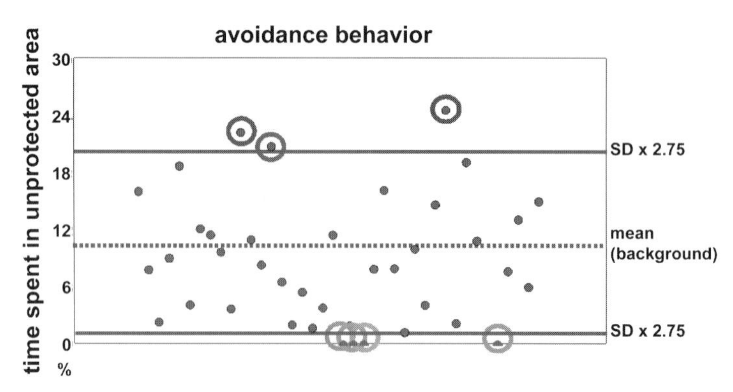

Fig. 8. Avoidance behavior towards an unprotected area displayed by the offspring of mice treated with the mutagen ethyl-nitrosourea (ENU). The behavior of the offspring is compared to a mean value calculated from the genetic background strain. Mice differing more than 2.75 SDs from the mean value (indicated by *circles*) are considered to be potential mutants and, after confirmation of the phenotype, are selected as founders for a ENU-mutant line

distinct behavioral dimensions. For example, ENU-mice have been found that performed an increased or decreased anxiety-related behavior (Fig. 8), while being inconspicuous regarding other behavioral patterns. A well-known example of this technique was given by the identification of the Clock gene based on studies performed in the mid-1990s (Vitaterna et al. 1994; Antoch et al. 1997). These findings gained fundamental insights into the molecular mechanisms underlying mammalian circadian rhythms, thus emphasizing the power of the phenotype-driven approach.

3.2 Experience-Related Models of Altered Emotionality

There exists a broad spectrum of anxiety disorders. These disorders are known to have a high comorbidity between each other. In addition, mood disorders (especially) are often accompanied by symptoms of altered anxiety (Holsboer 1999; Nesse 1999; Ramos and Mormede 1998; Reul and Holsboer 2002). Anxiety disorders and depression have been classified as separate types of disorders for decades. This view is under discussion now (Nemeroff 2002), because the efficacy of major psychotropic drugs is known for the treatment of both anxiety and mood disorders. Moreover, anxiety and depression share at some points a common pathophysiology (Holsboer 1995) and there is increasing evidence that both disorders share a common genetic background (Kendler 2002). Therefore, animal models conceptualized to elucidate mechanisms underlying depression often show altered anxiety-related behavior (Henn et al. 1993; Müller and Keck 2002), thus making them of high use to assist us in gaining new insights into the genetics and neurobiology of anxiety.

3.2.1 Early Life Stressors

Adversity in early life has been recognized as a fundamental factor determining susceptibility to psychiatric disorders in adulthood. To model this condition, investigators have administered stressors, such as maternal deprivation to newborn rats or mice and followed them throughout adulthood. These animals were frequently found to display increased anxiety-related behavior and also to be more vulnerable to stress (Dirks et al. 2002; Schmidt et al. 2002). Separation models are also used in nonhuman primates, species that, due to the evolutionary proximity to humans, seem to be particularly suited to provide insights into the biobehavioral underpinnings of emotionality. Infant monkeys respond to maternal separation with agitation, sleep disturbances, and altered emotionality that is interpreted as "despair" (Hinde et al. 1978; McKinney and Bunney 1969). Studies by many laboratories have also suggested that "depressive" responses during "despair" can be predicted by the amount of stress hormones released immediately following separation. The link between the stress hormone system and anxiety and depression-like behavior (see also Merikangas and Low, this volume) has been studied extensively (Coplan et al. 1993; Kalin et al. 1989). Infant monkeys with elevated levels of central nervous corticotropin-releasing hormone (CRH) developed a phenotype similar to the "behavioral despair" produced by maternal separation. This interesting finding is seen as further support for the causal role of exaggerated production and release of CRH, which is thought not only to activate the stress hormone system but also to coordinate the behavioral and vegetative responses to stress. According to the neuroendocrine hypothesis, if the balance between stress-related elevation of CRH and corticosteroid-induced suppression is severely disturbed, anxiety is increased and depression develops (Holsboer 2000). If this animal model is applicable to humans, the conclusion can be drawn that early stressors such as neglect or abuse lead to persistent elevations of CRH, rendering an individual vulnerable to depression, anxiety, or both.

3.2.2 Chronic Mild Stress

The chronic mild stress (CMS) model is based on the finding that rodents are highly vulnerable to variable, i.e., unpredictable stress. After exposure to changing stressors, such as changing housing conditions, loud noise, or constant bright light, for 2 to 3 weeks, rats show a number of long-lasting behavioral changes that are similar to depressive symptoms. These include not only changes in psychomotor behavior, as evidenced by reduced open field activity, but also increased anxiety, and a reduced sensitivity to rewards (interpreted as anhedonia), such as a decrease in the consumption of sucrose solution in comparison to tap water (Willner et al. 1992), both being interpreted as core features of depression. The finding of reduced reward-sensitivity was supported by studies using intracranial self-stimulation (Moreau et al. 1992), which showed that CMS causes an increase in the threshold current required to perform intracranial self-stimulation. Place conditioning studies showed that CMS also attenuates the ability to associate rewards with a specific environment, suggesting that CMS causes a generalized decrease in sensitivity to rewards and reflects anhedonia (for review see Willner 1997). Recently, the group led by Willner could show changes in REM sleep, including a reduced latency to the onset of the first REM period (Cheeta et al. 1997) after 21 days of CMS, a phenomenon that is common in depression. Notably, the specific stress schedule used as well as the choice of strain might be very critical, as some groups report problems of replication (Nielsen et al. 2000). Nevertheless, the predictive validity of the CMS paradigm can be stated as an established fact because long-term treatment with various antidepressants reverses sucrose intake to initial levels.

3.2.3
Learned Helplessness

Unpredictability also is a central feature in the concept of learned helplessness. This concept, using uncontrollable shock, was introduced by Overmier and Seligman (1967) and is based on the observation that animals exposed to an invariable stressor such as electric foot shock, which, due to the experimental set-up, is uncontrollable in nature, developed behavioral deficits. As first shown by Weiss (1968), rats exposed to uncontrollable shock showed significant weight loss due to decreased food and water intake. Moreover, these animals spent more time immobile in the forced swim test, and they revealed altered sleep patterns as well as a weakened response to previously rewarding brain stimulation, i.e., anhedonia (Henn et al. 1985; Weiss 1991). Importantly, these changes are not seen in animals that receive the same shocks but can exert control over their duration.

Seligman and Beagley (1975) linked the behavioral consequences of uncontrollable shock in rats to the clinical condition of depression, and because patients with depression also have feelings of helplessness, the term "learned helplessness" was coined for this response in animals. Other investigators have argued that it is extremely difficult, if not impossible, to prove the existence of feeling of helplessness as a cognitive response to external events in rats (Weiss 1980; Anisman et al. 1991). A more pragmatic interpretation is that the depression-like symptoms resulting form uncontrollable shock are stress-induced and the observed behavioral changes, including deficits in learning ability, are secondary to increased anxiety. It also remains unclear whether "learned helplessness" is a behavioral process characterizing depressed people (Gilbert and Allan 1998). Only a fraction of healthy normal humans develop symptoms of depression when exposed to uncontrollable stress, which also applies to healthy rats, as only 10%–15% of them develop this syndrome under these circumstances.

4
General Conclusions

In looking at the different animal models for anxiety, two questions are of central importance: (1) Do these animal models display predictive, face, and construct validity? (2) Are there reliable test assays to evaluate anxiety in animals? From the examples given above, which are not to be considered as being exhaustive but representative, the reply may be Yes. However, there also are some caveats that should be taken into account when working with animal models and test assays for anxiety.

To investigate mechanisms underlying pathological variants of anxiety, selectively bred rat lines have been demonstrated to be of high importance. Following this concept for several decades now, scientists were able to fundamentally increase the knowledge about neuronal and hormonal circuits involved in the regulation of distinct behavioral characteristics. Moreover, selectively bred rat lines have been proved to offer a good predictive validity in screening for potential anxiolytic and also antidepressant pharmaceutical drugs. Although some promising results have been obtained in the search of genetic determinants underlying altered anxiety in such rat lines, one should be aware of the fact that the selection of complex behavioral traits includes the possibility to accidentally co-select factors independent from anxiety.

In general, the same holds true for inbred mouse strains, which represent the prime organism of choice for modeling human disease. A variety of mice strains are available and some of them have been shown to display extremely different, probably genetically controlled, anxiety-related behavior. Thus, it has been hypothesized that, for example, the measurement of global gene expression profiles by use of DNA micro-arrays in combination with quantitative trait locus identification may be a promising strategy to identify genes underlying human diseases. However, it is of note that tremendous differences regarding the behavioral phenotype of inbred mouse strains occasionally are reported, pointing towards the need for standardized phenotyping assays and procedures.

This latter topic has been under discussion among behavioral pharmacologists already for quite some time, but during recent years it has become a focus of attention in the whole field of biomedical research, for example because of difficulties in reproducing behavioral phenotypes in knock-out mice. One strategy in developing more standardized behavioral tests is to automate established test set-ups. As already pointed out in Sect. 3.1.1, such automation risks causing a lack of sensitivity, since subtle behavioral alterations may remain undetected, and yet-unknown behavioral patterns are likely to go unrecognized. Another approach is to establish home cage phenotyping assays that may become of high use especially for the dissociation of behavioral traits and states. Independent from the test assay used, behavioral parameters and their analysis remain most critical. Profound evidence exists that complex analyses based on ethological observation are necessary to realize a high reliability and sensitivity with behavioral assays.

References

Ahmed FP, McLaughlin DP, Stanford SC, Stamford JA (2002) Maudsley reactive and non-reactive (MNRA) rats display behavioral contrasts on exposure to an open field, the elevated plus maze or the dark-light shuttle box. Abstract, FENS, Paris, France

Ammassari-Teule A, Milhaud JM, Passino E, Restivo L, Lassalle JM (1999) Defective processing of contextual information may be involved in the poor performance of DBA/2 mice in spatial tasks. Behav Genet 29:283–289

Anisman H, Zalcman S, Shanks N, Zacharko RM (1991) Multisystem regulation of performance deficits induced by stressors: an animal model of depression. In: Boulton AA, Baker GB, Martin-Iverson MT (eds) Animal models in psychiatry, vol 2. Humana Press, Clifton, pp 1–59

Antoch MP, Song EJ, Chang AM, Vitaterna MH, Zhao Y, Wilsbacher LD, Sangoram AM, King DP, Pinto LH, Takahashi JS (1997) Functional identification of the mouse circadian Clock gene by transgenic BAC rescue. Cell 89:655–667

Argyropoulus SV, Sandford JJ, Nutt DJ (2000) The psychobiology of anxiolytic drug. Part 2. Pharmacological treatments of anxiety. Pharmacol Ther 88:213–227

Ashcroft GW, Walker LG, Lyle A (1993) A psychobiological model for panic: including models for the mechanisms involved in the regulation of mood and anxiety and implications for behavioral and pharmacological therapies. In: Leonard BE, Butler J, O'Rourke D, Fahy TJ (eds) Psychopharmacology of panic. Oxford University Press, Oxford, pp 131–143

Avgustinovich DF, Lipina TV, Bondar NP, Alekseyenko OV, Kudriavtseva NN (2000) Features of the genetically defined anxiety in mice. Behav Genet 30:101–109

Balling R (2001) ENU mutagenesis: analyzing gene function in mice. Annu Rev Genomics Hum Genet 2:463–492

Barnett SA (1963) The rat: a study in behavior. Metheun, London

Beck KD, Luine VN (1999) Food deprivation modulates chronic stress on object recognition in male rats: role of monoamines and amino acids. Brain Res 830:56–71

Belzung C (1999) Measuring rodent exploratory behavior. In: Crusio WE, Gerlai RT (eds) Handbook of molecular genetic techniques for brain and behavior research (techniques in the behavioral and neural sciences). Elsevier, Amsterdam, pp 738–749

Belzung C, Berton F (1997) Further pharmacological validation of the BALB/c neophobia in the free exploratory paradigm as an animal model of anxiety. Behav Pharmacol 8:541–548

Belzung C, Beuzen A (1995) Link between emotional memory and anxiety states: a study by principal component analysis. Physiol Behav 58:111–118

Belzung C, Griebel G (2001) Measurement of normal and pathological anxiety-like behaviour in mice: a review. Behav Brain Res 125:141–149

Belzung C, Le Pape G (1994) Comparison of different behavioral test situations used in psychopharmacology for measurements of anxiety. Physiol Behav 56:623–628

Belzung C, Le Guisquet AM, Crestani F (2000) Flumazenil induces benzodiazepine partial agonist-like effects in BALB/c but not C57BL/6 mice. Psychopharmacology (Berl) 148:24–32

Berrettini WH, Harris N, Ferraro TN, Vogel WH (1994) Maudsley reactive and non-reactive rats differ in exploratory behavior but not learning. Psychiatr Genet 4:91–94

Bizard DA, Altman HJ, Freedman LS (1982) The peripheral sympathetic nervous system in rat strains selectively bred for differences in response to stress. Behav Neural Biol 34:319–325

Blanchard DC, Griebel G, Blanchard RJ (2001a) Mouse defensive behavior: pharmacological and behavioral assays for anxiety and panic. Neurosci Biobehav Rev 25:205–218

Blanchard DC, Hynd AL, Minke KA, Minemoto T, Blanchard RJ (2001b) Human defense behaviors to threat scenarios show parallels to fear- and anxiety-related defense patterns of non-human mammals. Neurosci Biobehav Rev 25:761–770

Blanchard DC, Griebel G, Blanchard RJ (2003) The mouse defense test battery: pharmacological and behavioral assays for anxiety and panic. Eur J Pharmacol 463:97–116

Blanchard RJ, Blanchard DC (1989) Anti-predator defense behaviors in a visible burrow system. J Comp Psychol 103:70–82

Blanchard RJ, Yudko EB, Rodgers RJ, Blanchard DC (1993) Defense system psychopharmacology: an ethological approach to the pharmacology of fear and anxiety. Behav Brain Res 58:155–165

Blanchard RJ, Griebel G, Henrie JA, Blanchard DC (1997) Differentiation of anxiolytic and panicolytic drugs by effects on rat and mouse defense test batteries. Neurosci Biobehav Rev 21:783–789

Blizard DA, Adams N (2002) The Maudsley reactive and nonreactive strains: a new perspective. Behav Genet 32:277–299

Boissier JR, Simon P, Soubrie P (1976) New approaches to the study of anxiety and anxiolytic drugs in animals. In: Airaksinen M (ed) CNS and behavioral pharmacology. Pergamon Press, New York, pp 213–222

Bourin M, Hascoet M (2003) The mouse light/dark box test. Eur J Pharmacol 463:55–65

Britton DR, Britton KT (1981) A sensitive open field measure of anxiolytic drug activity. Pharmacol Biochem Behav 15:577–582

Broadhurst PL (1960) Experiments in psychogenetics: applications of biometrical genetics to inheritance of behavior. In: Eysenck HJ (ed) Experiments in personality: psychogenetics and psychopharmacology, vol. 1. Routledge and Kegan Paul, London, pp 1–102

Broadhurst PL, Bignami G (1965) Correlative effects of psychogenetic selection: a study of the Roman high and low avoidance strains of rats. Behav Res Ther 2:273–280

Brown JS, Kalish HI, Farber IE (1951) Conditioned fear as revealed by magnitude of startle response to an auditory stimulus. J Exp Psychol 1:317–328

Brown SDM, Balling R (2001) Systematic approaches to mouse mutagenesis. Curr Opin Genet Dev 11:268–273

Brown SDM, Hardisty RE (2003) Mutagenesis strategies for identifying novel loci associated with disease phenotypes. Semin Cell Dev Biol 14:19–24

Cabib S, Puglisi-Allegra S, Ventura R (2002) The contribution of comparative studies in inbred strains of mice to the understanding of the hyperactive phenotype. Behav Brain Res 130:103–109

Cahill J, McGaugh JL (1998) Mechanism of emotional arousal and lasting memory. Trends Neurosci 21:294–299

Castanon N, Dulluc J, LeMoal M, Mormede P (1994) Maturation of the behavioral and neurocrine differences between the Roman rat lines. Physiol Behav 55:775–782

Castanon N, Perez-Diaz F, Mormede P (1995) Genetic analysis of the relationship between behavioral and neurocrine traits in Roman high and low avoidance rat lines. Behav Genet 25:371–383

Chaouloff F, Durand M, Mormede P (1997) Anxiety and anxiety-related effects of diazepam and chlordiazepoxide in the rat light/dark and dark/light test. Behav Brain Res 85:27–35
Cheeta S, Ruigt G, van Proosdij J, Willner P (1997) Changes in sleep architecture following chronic mild stress. Biol Psychiatry 41:419–427
Clement Y, Proeschel MF, Bondoux D, Girard F, Launay JM, Chapouthier G (1997) Genetic factors regulate processes related to anxiety in mice. Brain Res 752:127–135
Commissaris RL, Ellis DM, Hill TJ, Schefke DM, Becker CA, Fontana DJ (1990) Chronic antidepressant and clonidine treatment effects on conflict behavior in the rat. Horm Behav 37:167–176
Conrad CD, Galea LAM, Kuroda Y, McEwen B (1996) Chronic stress impairs rat spatial memory on the y maze, and this effect is blocked by tianeptine pretreatment. Behav Neurosci 110:1321–1334
Coplan JD, Andrewa MW, Rosenblum LA, Owens MJ, Freidman S, Gorman JM, Nemeroff CB (1996) Persistent elevations of cerebrospinal fluid concentrations of corticotropin-releasing factor in adult nonhuman primates exposed to early-life stressors: implications for the pathophysiology of mood and anxiety disorders. Proc Natl Acad Sci USA 93:1619–1623
Corda MG, Lecca D, Piras G, Di Chiara G, Giorgo O (1997) Biochemical parameters of dopaminergic and GABAergic neurotransmission in the CNS of Roman high-avoidance and Roman low-avoidance rats. Behav Genet 27:527–536
Crabbe JC (1986) Genetic differences in locomotor activity in mice. Pharmacol Biochem Behav 25:289–292
Crawley JN (1999) Evaluating anxiety in rodents. In: Crusio WE, Gerlai RT (eds) Handbook of molecular genetic techniques for brain and behavior research (techniques in the behavioral and neural sciences). Elsevier, Amsterdam, pp 667–673
Crawley JN, Belknap JK, Collins A, Crabbe JC, Frankel W, Henderson N, Hitzemann RJ, Maxson SC, Miner LJ, Silva AJ, Wehner JM, Wynshaw-Boris A, Paylor R (1997) Behavioral phenotypes of inbred mouse strains: implications and recommendations for molecular studies. Psychopharmacology (Berl) 132:107–124
Crawley LN, Goodwin FK (1980) Preliminary report of a simple animal behavior model for the anxiolytic effects of benzodiazepines. Pharmacol Biochem Behav 13:167–170
Croiset G, Nijsen MJMA, Kamphuis PJGH (2000) Role of corticotropin-releasing factor, vasopressin and the autonomic nervous system in learning and memory. Eur J Pharmacol 405:225–234
Cruz APM, Frei F, Graeff FG (1994) Ethopharmacological analysis of rat behavior on the elevated plus-maze. Pharmacol Biochem Behav 49:171–176
Cryan JF, Markou A, Lucki I (2002) Assessing antidepressant activity in rodents: recent developments and future needs. Trends Pharmacol Sci 23:238–245
Davis M, Falls WA, Campeau S, Kim M (1993) Fear-potentiated startle: a neural and pharmacological analysis. Behav Brain Res 58:175–198
De Pablo JM, Parra A, Segovia S, Guillarmón A (1989) Learned immobility explains the behavior of rats in the forced swimming test. Physiol Behav 46:229–237
Dirks A, Fish EW, Kikusui T, van der Gugten J, Groenink L, Olivier B, Miczek KA (2002) Effects of corticotropin-releasing hormone on distress vocalizations and locomotion in maternally separated mouse pups. Pharmacol Biochem Behav 72:993–999
Driscoll P, Bättig K (1982) Behavioral, emotional and neurochemical profiles of rats selected for extreme differences in active, two-way avoidance performance. In: Lieblich I (ed) Genetics of the brain. Elsevier, Amsterdam, pp 95–123
DSM-IV (1994) American Psychiatric Association diagnostic and statistical manual of mental disorders. American Psychiatric Press, Washington

Dulawa SC, Hen R, Scearce-Levie K, Geyer M (1997) Serotonin 1B receptor modulation of startle reactivity, habituation, and prepulse inhibition in wildtype and serotonin 1B knock-out mice. Neuroreport 132:125–134

Ellis DM, Fontana DJ, McCloskey TC, Commissaris RL (1990) Chronic anxiolytic treatment effects on conflict behavior in the rat. Pharmacol Biochem Behav 37:177–186

Escorihuela RM, Tobena A, Driscoll P, Fernandez-Teruel A (1995) Effects of training, early handling, and perinatal flumazenil on shuttle box acquisition in Roman low-avoidance rats: towards overcoming a genetic deficit. Neurosci Biobehav Rev 19:353–367

Escorihuela RM, Fernandez-Teruel A, Gil L, Aguilar R, Tobena A, Driscoll P (1999) Inbred Roman high- and low-avoidance rats: differences in anxiety, novelty-seeking, and shuttlebox behavior. Physiol Behav 67:19–26

Fernandez-Teruel A, Driscoll P, Gil L, Tobena A, Escorihuela RM (2002) Enduring effects of environmental enrichment on novelty seeking, saccharin and ethanol intake in two rat lines (RHA/Verh and RLA/Verh) differing in incentive-seeking behavior. Pharmacol Biochem Behav 73:225–231

Ferre P, Fernandez-Teruel, Escorihuela RM, Driscoll P, Gercia E, Zapata A, Tabena A (1994) Struggling and flumazenil effects in the swimming test are related to the level of anxiety in mice. Neuropsychobiology 29:23–27

Ferre P, Fernandez-Teruel, Escorihuela RM, Driscoll P, Corda MG, Giorgo O, Tabena A (1995) Behavior of the Roman/Verh high- and low-avoidance rat lines in anxiety-tests: relationship with defecation and self-grooming. Physiol Behav 58:1209–1213

File SE (1995) Animal models of different anxiety states. In: Biggio G, Sanna E, Costa E (eds) GABAa receptors and anxiety: from neurobiology to treatment. Raven Press, New York, pp 93–113

File SE, Wardill AG (1975) Validity of head-dipping as a measure of exploration in a modified hole-board. Psychopharmacologia 44:53–59

Fletcher PJ, Davies M (1990) Effects of 8-OH-DPAT, buspirone and ICS 205–930 on feeding in a novel environment: comparison with chlordiazepoxide and FG 1742. Psychopharmacology (Berl) 102:301–308

Flint J, Corley R, De Fries JC, Fulker DW, Gray JA, Miller S, Collins AC (1995) A simple genetic basis for a complex psychological trait in laboratory mice. Science 269:1432–1435

Fontana DJ, Commissaris RL (1988) Effects of acute and chronic imipramine administration on conflict behavior in the rat: a potential 'animal model' for the study of panic disorder? Psychopharmacology (Berl) 95:147–150

Fontana DJ, Carbary TJ, Commissaris RL (1989) Effects of acute and chronic antipanic drug administration on conflict behavior in the rat. Psychopharmacology (Berl) 98:157–162

Geller I, Seifter J (1960) The effects of meptobamate, barbiturates, d-amphetamine and promazine on experimentally induced conflict in the rat. Psychopharmacology (Berl) 1:482–492

Gentsch C, Lichtsteiner M, Driscoll P, Feer H (1982) Differential hormonal and physiological responses to stress in Roman high- and low-avoidance rats. Physiol Behav 28:259–263

Gerlai R (1996) Gene-targeting studies of mammalian behavior: is it the mutation or the background genotype? Trends Neurosci 19:177–181

Geyer MA, Markou A (1995) Animal models of psychiatric disorders. In: Bloom FE, Kupfer DJ (eds) Psychopharmacology: the fourth generation of progress. Raven Press, New York, pp 787–798

Gilbert P, Allan S (1998) The role of defeat and entrapment (arrested flight) in depression: an exploration of an evolutionary view. Psychol Med 28:585–598

Gingrich JA, Hen R (2001) Dissecting the role of the serotonin system in neuropsychiatric disorders using knockout mice. Psychopharmacology (Berl) 155:1–10

Giorgo O, Corda MG, Carboni G, Frau V, Valentini V, Di Chiara G (1997) Effects of cocaine and morphine in rats from two psychogenetically selected rat lines: a behavioral and brain dialysis study. Behav Genet 27:537–546

Golani I, Kafkafi N, Drai D (1999) Phenotyping stereotypic behavior: collective variables, range of variation and predictability. Appl Anim Behav Sci 65:191–220

Graeff FG (2002) On serotonin and experimental anxiety. Psychopharmacology (Berl) 163:467–476

Gray TS (1990) Amygdaloid CRF pathways: role in autonomic, neuroendocrine, and behavioral responses ot stress. In: De Souza EB, Nemeroff CB (eds) Annals of the New York Academy of Sciences. Corticotropin-releasing factor and cytokines: Proceedings of the Hans Selye Symposium on Neuroendocrinology and Stress, vol 697. New York Academy of Sciences, New York, pp 53–60

Griebel G (1996) Variability in the effects of 5-HT related compounds in experimental models of anxiety: evidence for multiple mechanisms of 5-HT in anxiety or never ending story? Pol J Pharmacol 48:129–136

Griebel G, Moreau J-L, Jenck F, Martin JR, Misslin R (1993) Some critical determinants of the behavior of rats in the elevated plus-maze. Behav Process 29:37–48

Griebel G, Sanger DJ, Perrault G (1997) Genetic differences in the mouse defense test battery. Aggress Behav 23:10–31

Griebel G, Belzung C, Perrault G, Sanger DJ (2000) Differences in anxiety-related behaviours and in sensitivity to diazepam in inbred and outbred strains in mice. Psychopharmacology (Berl) 148:164–170

Hall CS (1936) Emotional behavior in the rat. III. The relationship between emotionality and ambulatory activity. J Comp Physiol Psychol 22:345–352

Hall CS (1938) The inheritance of emotionality. Am Sci 26:17–27

Handley SL, Mithani S (1984) Effects of alpha-adrenoceptor agonists and antagonists in a maze exploration model for "fear-motivated" behavior. Naunyn Schmiedebergs Arch Pharmacol 327:1–5

Hascoet M, Bourin M, Dhonnchadha BAN (2001) The mouse light-dark paradigm: a review. Prog Neuropsychopharmacol Biol Psychiatry 25:141–166

Henn FA, Johnson J, Edwards E, Anderson D (1985) Melancholia in rodents: neurobiology and pharmacology. Psychopharmacol Bull 21:443–446

Henn FA, Edwards E, Muneyyirci J (1993) Animal models in depression. Clin Neurosci 1:152–156

Henniger MSH, Ohl F, Hölter SM, Weißenbacher P, Toschi N, Lörscher P, Wigger A, Spanagel R, Landgraf R (2000) Unconditioned anxiety and social behaviour in two rat lines selectively bred for high and low anxiety-related behaviour. Behav Brain Res 111:153–163

Hijzen TH, Houtzager SW, Joordens RJ, Olivier B, Slangen JL (1995) Predictive validity of the potentiated startle response as a behavioral model for anxiolytic drugs. Psychopharmacology (Berl) 118:150–154

Hinde RA, Leighton-Shapiro ME, McGinnis L (1978) Effects of various types of separation experience on rhesus monkeys 5 months later. J Child Psychol Psychiatry 19:199–211

Hindmarch I (1998) Cognition and anxiety: the cognitive effects of anti-anxiety medication. Acta Psychiatr Scand 98:89–94

Hodges H (1996) Maze procedures: the radial-arm and water maze compared. Cogn Brain Res 3:167–181

Hodges HM, Green SE, Crewes H, Mathers I (1981) Effects of chronic chlordiazepoxide treatment on novel and familiar food preference in rats. Psychopharmacology (Berl) 75:311–314

Hogg S (1996) A review of the validity and variability of the elevated plus maze as an animal model of anxiety. Pharmacol Biochem Behav 54:21–30

Holmes A (2001) Targeted gene mutation approaches to the study of anxiety-like behavior in mice. Neurosci Biobehav Rev 25:261–273

Holsboer F (1995) Neuroendocrinology of mood disorders. In: Bloom FE, Kupfer DJ (eds) Psychopharmacology: the fourth generation of progress. Raven Press, New York, pp 957–969

Holsboer F (1997) Transgenic mouse models: new tools for psychiatric research. Neuroscientist 3:328–336

Holsboer F (1999) The rationale for corticotropin-releasing hormone receptor (CRH-R) antagonists to treat depression and anxiety. J Psychiatr Res 33:181–214

Holsboer F (2000) The corticosteroid receptor hypothesis of depression. Neuropsychopharmacology 23:477–501

Holsboer F, Lauer CJ, Schreiber W, Krieg JC (1995) Altered hypothalamic-pituitary-adrenocortical regulation in healthy subjects at high familial risk for depression. Neuroendocrinology 62:659–664

Hrabé de Angelis M, Flaswinkel H, Fuchs H, Rathkolb B, Soewarto D, Marschall S, Heffner S, Pargent W, Wuensch K, Jung M, Reis A, Richter T, Alessandrini F, Jakob T, Fuchs E, Kolb H, Kremmer E, Schaeble K, Rollinski B, Roscher A, Peters C, Meitinger T, Strom T, Steckler T, Holsboer F, Klopstock T, Gekeler T, Schindewolf C, Jung T, Avraham K, Behrendt H, Ring J, Zimmer A, Schughart K, Pfeffer K, Wolf E, Balling R (2000) Genome-wide, large-scale production of mutant mice by ENU mutagenesis. Nat Genet 25:444–447

Jaenisch R (1998) Transgenic animals. Science 240:1468–1472

Kalin NH, Shelton SE, Barksdale CM (1989) Behavioral and physiologic effects of CRH administration to infant primates undergoing maternal separation. Neuropsychopharmacology 2:97–104

Keck ME, Welt T, Post A, Müller MB, Toschi N, Wigger A, Landgraf R, Holsboer F, Engelmann M (2001a) Neuroendocrine and behavioral effects of repetitive transcranial magnetic stimulation in a psychopathological animal model are suggestive of antidepressant-like effects. Neuropsychopharmacology 24:337–349

Keck ME, Welt T, Wigger A, Renner U, Engelmann M, Holsboer F, Landgraf R (2001b) The anxiolytic effect of the CRH1 receptor antagonist R121919 depends on innate emotionality in rats. Eur J Neurosci 13:373–380

Keck ME, Wigger A, Welt T, Müller MB, Gesing A, Reul JMHM, Holsboer F, Landgraf R, Neumann ID (2002) Vasopressin mediates the response of the combined dexamethasone/CRH test in hyper-anxious rats: implications for pathogenesis of affective disorders. Neuropsychopharmacology 26:94–105

Keck ME, Welt T, Müller MB, Uhr M, Ohl F, Wigger A, Toschi N, Holsboer F, Landgraf R (2003) Reduction of hypothalamic vasopressinergic hyperdrive contributes to clinically relevant behavioral and neuroendocrine effects of the antidepressant paroxetine in a psychopathological rat model. Neuropsychopharmacology 28:235–243

Kelley AE (1993) Locomotor activity and exploration. In: Sahgal A (ed) Behavioural neuroscience: a practical approach. Oxford University Press, Oxford, pp 1–21

Kendler KS (2002) Psychiatric genetics: an intellectual journey. The Salmon Lecture. Clin Neurosci Res 2:110–119

Kopp C, Vogel E, Misslin R (1999) Comparative study of emotional behavior in three inbred strains of mice. Behav Processes 47:161–174
Landgraf R, Wigger A (2002) HAB and LAB rats as a psychological animal model of extremes in innate anxiety. Behav Genet 32:301–314
Landgraf R, Wigger A, Holsboer F, Neumann ID (1999) Hyper-reactive hypothalamo-pituitary adrenocortical axis in rats bred for high anxiety-related behaviour. J Neuroendocrinol 11:405–407
Lecrubier Y, Ustun TB (1998) Panic and depression: a worldwide primary care perspective. Int Clin Psychopharmacol 13:S7–S11
Lee WR, Arbour P, Fossett NG, Kilroy G, Mahmoud J, McDaniel ML, Tucker A (1989) Sequence-analysis of X-ray and ENU induced mutations at the ADH locus in Drosophila-melanogaster. Genetics 122:S38
Liebsch G, Linthorst ACE, Neumann ID, Reul JMHM, Holsboer F, Landgraf R (1998) Behavioral, physiological, and neuroendocrine stress responses and differential sensitivity to diazepam in two Wistar rat lines selectively bred for high and low anxiety-related behavior. Neuropsychopharmacology 19:381–396
Lister RG (1990) Ethologically based animal models of anxiety disorders. Pharmacol Ther 46:321–340
Livesey PJ (1986) Learning and emotion: a biological synthesis. Lawrence Erlbaum, Hillsdale
Lucki I, Dalvi A, Mayorga AJ (2001) Sensitivity to the effects of pharmacologically selective antidepressants in different strains of mice. Psychopharmacology (Berl) 155:315–322
Martin P (1998) Animal models sensitive to anti-anxiety agents. Acta Psychiatr Scand Suppl 393:74–80
McGaugh JL, Cahill L, Roozendaal B (1996) Involvement of the amygdala in memory storage: interaction with other brain systems. Proc Natl Acad Sci USA 93:13508–13514
McKinney WT Jr, Bunney WE Jr (1969) Animal model of depression: I. Review of evidence: implications for research. Arch Gen Psychiatry 21:240–248
McNaughton N (1997) Cognitive dysfunction resulting from hippocampal hyperactivity—a possible cause of anxiety disorder? Pharmacol Biochem Behav 56:603–611
McQuade R, Stanford SC (2001) Differences in central noradrenergic and behavioral responses of Maudsley non-reactive and Maudsley reactive inbred rats on exposure to an aversive novel environment. J Neurochem 76:21–28
Menard J, Treit D (1999) Effects of centrally administered anxiolytic compounds in animal models of anxiety. Neurosci Biobehav Rev 23:591–631
Millan MJ, Brocco M (2003) The Vogel conflict test: procedural aspects, gamma-aminobutyric acid, glutamate and monoamines. Eur J Pharmacol 463:67–96
Montgomery KC (1958) The relation between fear induced by novel stimulation and exploratory behavior. J Comp Physiol Psychol 48:254–260
Montkowski A, Pöttig M, Mederer A, Holsboer F (1997) Behavioral performance in three substrains of mouse strain 129. Brain Res 762:12–18
Moreau JL, Jenck F, Martin JR, Mortas P, Haefely WE (1992) Antidepressant treatment prevents chronic unpredictable mild stress-induced anhedonia as assessed by ventral tegmentum self-stimulation behavior in rats. Eur Neuropsychopharmacol 2:43–49
Müller MB, Keck ME (2002) Genetically engineered mice for studies of stress-related clinical conditions. J Psychiatr Res 36:53–76
Nemeroff CB (2002) Comorbidity of mood and anxiety disorders: the rule, not the exception? Am J Psychiatry 159:3–4
Nesse RM (1999) Proximate and evolutionary studies of anxiety, stress and depression: synergy of interface. Neurosci Biobehav Rev 23:895–903

Nielsen CK, Arnt J, Sanchez C (2000) Intracranial self-stimulation and sucrose intake differ as hedonic measures following chronic mild stress: interstrain and interindividual differences. Behav Brain Res 107:21–33

Nolan PM, Peters J, Strivens M, Rogers D, Hagan J, Spurr N, Gray IC, Vizor L, Brooker D, Whitehill E, Washbourne R, Hough T, Greenaway S, Hewitt M, Liu X, McCormack S, Pickford K, Selley R, Wells C, Tymowska-Lalanne Z, Roby P, Glenister P, Thornton C, Thaung C, Stevenson JA, Arkell R, Mburu P, Hardisty R, Kiernan A, Erven A, Steel KP, Voegeling S, Guenet JL, Nickols C, Sadri R, Nasse M, Isaacs A, Davies K, Browne M, Fisher EM, Martin J, Rastan S, Brown SD, Hunter J (2000) A systematic, genome-wide, phenotype-driven mutagenesis programme for gene function studies in the mouse. Nat Genet 25:440–443

Ohl F, Holsboer F, Landgraf R (2001a) The modified hole board as differential screen for behavior in rodents. Behav Res Methods Instrum Comput 33:392–397

Ohl F, Sillaber I, Binder E, Keck ME, Holsboer F (2001b) Differential analysis of basal behavior and diazepam-induced alterations in C57BL/6 and BALB/c mice using the modified hole board. J Psychiatr Res 35:147–154

Ohl F, Toschi N, Wigger A, Henniger MSH, Landgraf R (2001c) Dimensions of emotionality in an animal model of inborn hyperanxiety. Behav Neurosci 115:429–436

Ohl F, Roedel A, Storch C, Holsboer F, Landgraf R (2002) Cognitive performance in rats differing in their inborn anxiety. Behav Neurosci 116:464–471

Ohl F, Roedel A, Binder E, Holsboer F (2003) Impact of high and low anxiety on cognitive performance in a modified hole board test in inbred mice strains C57BL/6 and DBA/2. Eur J Neurosci 17:128–136

Olivier B, Pattij T, Wood SJ, Oosting R, Sarnyai Z, Toth M (2001) The 5-HT1A receptor knockout mouse and anxiety. Behav Pharmacol 12:439–450

Overmier JB, Seligman MEP (1967) Effects of inescapable shock upon subsequent escape and avoidance learning. J Comp Physiol Psychol 63:28–33

Overstreet DH (1993) The Flinders sensitive line rats: a genetic animal model of depression. Neurosci Biobehav Rev 17:51–68

Overstreet DH, Russel RW, Helps SC, Messenger M (1979) Selective breeding for sensitivity to the anticholinesterase, DFP. Psychopharmacology (Berl) 65:15–20

Overstreet DH, Rezvani AH, Janowsky DS (1992) Maudsley reactive and nonreactive rats differ only in some tasks reflecting emotionality. Physiol Behav 52:149–152

Overstreet DH, Pucilowski O, Rezvani AH, Janowsky DS (1995) Administration of antidepressants, diazepam and psychomotor stimulants further confirms the utility of Flinders sensitive line rats as an animal model of depression. Psychopharmacology (Berl) 121:27–37

Parmigiani S, Palanza P, Rodgers J, Ferrari PF (1999) Selection, evolution of behavior and animal models in behavioral neuroscience. Neurosci Biobehav Rev 23:957–970

Pellow S, Chopin P, File SE, Briley M (1985) Validation of open:closed arm entries in an elevated plus-maze as a measure of anxiety in the rat. J Neurosci Methods 14:149–167

Picciotto MR (1999) Knock-out mouse models used to study neurobiological systems. Crit Rev Neurobiol 13:103–149

Porsolt RC, Chermat R, Lenegre A, Avril I, Janvier S, Steru L (1987) Use of the automated tail suspension test for the primary screening of psychotropic agents. Arch Int Pharmacodyn Ther 288:11–30

Porsolt RD, Bertin A, Jalfre M (1977) Behavioral despair in mice: a primary screening test for antidepressants. Arch Int Pharmacodyn Ther 229:327–336

Prut L, Belzung C (2003) The open field as a paradigm to measure the effects of drugs on anxiety-like behaviors: a review. Eur J Pharmacol 463:3–33

Ramos A, Mormede P (1998) Stress and emotionality: a multidimensional and genetic approach. Neurosci Biobehav Rev 22:33–57
Reul JMHM, Holsboer F (2002) Corticotropin-releasing factor receptors 1 and 2 in anxiety and depression. Curr Opin Pharmacol 2:23–33
Rodgers RJ, Cole JC (1994) The elevated plus-maze: pharmacology, methodology and ethology. In: Cooper SJ, Hendrie CA (eds) Ethology and psychopharmacology. John Wiley and Sons, Chichester, pp 9–44
Rodgers RJ, Johnson NJT (1995) Factor analysis of spatiotemporal and ethological measures in the murine elevated plus-maze test of anxiety. Pharmacol Biochem Behav 52:297–303
Rodgers RJ, Cole JC, Cobain MR, Daly P, Doran PJ, Eells JR, Wallis P (1992) Anxiogenic-like effects of fluprazine and eltoprazine in the mouse elevated plus maze: profile comparison with 8-OH-DPAT, TFMPP and mCPP. Behav Pharmacol 3:621–624
Rodgers RJ, Cao BJ, Dalvi A, Holmes A (1997) Animal models of anxiety: an ethological perspective. Braz J Med Biol Res 30:289–304
Rogers DC, Jones DNC, Nelson PR, Jones CM, Quilter CA, Robinson TL, Hagan JJ (1999) Use of SHIRPA and discriminant analysis to characterise marked differences in the behavioural phenotype of six inbred mouse strains. Behav Brain Res 105:207–217
Rosen JB, Schulkin J (1998) From normal fear to pathological anxiety. Psychol Rev 105:325–350
Rossi-Arnaud C, Ammassari-Teule M (1998) What do comparative studies of inbred mice add to current investigations of the neural basis of spatial behaviours? Exp Brain Res 123:36–44
Saudou F, Ait Amara D, Dierich A, Lemeur M, Ramboz S, Segu L, Buhot MC, Hen R (1994) Enhanced aggressive behavior in mice lacking 5-HT1b receptor. Science 265:1875–1878
Schmidt M, Oitzl MS, Levine S, de Kloet ER (2002) The HPA system during the postnatal development of CD1 mice and the effects of maternal deprivation. Brain Res Dev Brain Res 139:39–49
Seligman MEP, Beagley G (1975) Learned helplessness in the rat. J Comp Physiol Psychol 88:534–541
Sheldon MH (1968) Exploratory behaviour: the inadequacy of activity measures. Psychol Sci 11:38
Shephard JK, Grewal SS, Fletcher A, Bill DJ, Dourish CT (1994) Behavioral and pharmacological characterization of the elevated "zero-maze" as an animal model of anxiety. J Psychopharmacol 116:56–64
Sillaber I, Rammes G, Zimmermann S, Mahal B, Zieglgänsberger W, Wurst W, Holsboer F, Spanagel R (2002) Enhanced and delayed stress-induced alcohol drinking in mice lacking a functional CRHR 1 receptor. Science 296:931–933
Soubrie P, Kulkarni S, Simon P, Boissier JR (1975) Effets des anxiolytiques sur la prose de nourriture des rats et des souris placés en situation nouvelle ou familière. Psychopharmacologia 45:203–210
Stefanski R, Paleijko W, Bidzinski A, Kostowski W, Plaznik A (1993) Serotonergic innervation of the hippocampus and nucleus-accumbens septi and the anxiolytic-like action of the 5-HT3 receptor antagonist. Neuropharmacology 32:987–993
Steimer T, Escorihuela RM, Fernandez-Teruel A, Driscoll P (1998) Long-term behavioral and neurocrine changes in Roman high-(RHA/Verh) and low-(RLA/Verh) avoidance rats following neonatal handling. Int J Dev Neurosci 16:165–174
Stenzel-Poore P, Heinrichs SC, Rivest S, Koob GF, Vale WW (1994) Overproduction of corticotropin-releasing factor in transgenic mice: a genetic model of anxiogenic behavior. J Neurosci 14:2579–2584

Anxiety in rodents is defined as a high level of avoidance of novel and unfamiliar environment and increased fear reaction (Finn et al. 2003; Weiss et al. 2000). Other components such as autonomic activation, increased stress reactivity, and neuroendocrine abnormalities are an integral part of anxiety responses. The design of an anxiety-like phenotype in mice partially or completely lacking a gene of interest during all stages of development (constitutive knockout) or in a spatio-temporal context (conditional knockout) is among the prime strategies directed at elucidating the role of genetic factors in fear and anxiety. In many cases targeted inactivation of key players of the serotonergic gene pathway have been able to confirm what has already been anticipated based on pharmacological studies with serotonin system selective compounds. In other instances, studies in knockout (KO) mice have changed views of the relevance of serotonin (5-hydroxytryptamine, 5-HT) homeostasis in brain development and plasticity as well as processes underlying emotional behavior.

Molecules suspected to mediate emotionality are commonly derived from hypothesized pathogenetic mechanisms of an anxiety disorder or from observations of therapeutic response. Above all, molecular components of neural circuits mediating the effects of the 5-HT system, and which appear to be involved in anxiety-like behavior of KO mice, are being identified at a steady rate, thus leading to new candidate genes of presumed pathophysiological pathways for genetic evaluation in humans. Here, I describe fundamental aspects of the genetics of anxiety-related traits and emotional responses. A thorough appraisal of behavioral and physiological consequences in mice with genetic manipulation of the serotonergic pathway is also provided. Finally, I will emphasize conceptual issues in the search for candidate genes for anxiety and for the development of mouse models of anxiety disorders.

2 Anxiety-Like Behavior in Knockout Mice

Recent advances in gene targeting (constitutive or conditional KO/knockin techniques) are increasingly impacting our understanding of the neurobiological basis of anxiety- and depression-related behavior in mice (Lesch 2001a). However, the majority of neural substrates and circuitries that regulate emotional processes or cause anxiety disorders remain remarkably elusive. Among the reasons for the lack of progress are several conceptual deficiencies regarding the psychobiology of fear and anxiety, which make it difficult to develop and validate reliable models. The clinical presentation of anxiety disorders and the lack of consensus on clinical phenotypes or categories further complicates the development of mouse models for specific anxiety disorders. In addition, human anxiety disorders encompass not only the behavioral trait of inappropriate fear but also the cognitive response towards this disposition.

This response, however, is substantially modulated by environmental factors including cultural determinants such as rearing and education as well as sociocultural and socioeconomic context. Investigations on the neurobiological basis of anxiety disorders therefore rely on the accurate dissection of behavioral dysfunctioning from other factors. The dilemma that no single paradigm mimics the diagnostic entities or treatment response of anxiety disorders may reflect the fact that current classification systems are not based on the neurobiology of disease, rather than being the result of the failure to develop valid mouse models.

Various approaches have been employed to detect and quantify "anxiety-like" behaviors in mice, and the majority postulates that aversive stimuli, such as novelty or potentially harmful environments, induce a central state of fear and defensive reactions, which can be assessed and quantified through physiological and behavioral paradigms (Crawley 1999; Crawley and Paylor 1997). When rodents are introduced into a novel environment, they tend to move around the perimeter of the environment ("open field"). They stop occasionally and rear up, sniffing the walls and the floor. They initially spend very little time in the open center of the area. If they have a choice, they will spend more time in a dark than in a brightly lit area ("light-dark box"). Rodents will also spend more time in a small, elevated area enclosed by walls than in an elevated area without walls ("elevated plus maze"). When they move from one delimited area into another, they often engage in a type of stretching-out behavior. Anxiety-like behavior often appears to contrast with exploratory behavior, indicating that avoidance and curiosity or novelty seeking are biologically related and share common physiological mechanisms.

While substantial similarities between human and murine avoidance, defense, aggression, or escape response exist, it remains obscure whether mice also experience subjective anxiety and associated cognitive processes similar to humans or whether defense responses or aggressiveness represent pathological expression of anxiety in humans. In general, pathological anxiety may reflect an inappropriate activation of a normally adaptive, evolutionarily conserved defense reaction. It should therefore be practicable to elucidate both physiological and pathological anxiety by studying avoidant and defensive behavior in mice using a broad range of anxiety models to ensure comprehensive characterization of the behavioral phenotype.

3
Functional Neuroanatomy of Emotionality: Focus on the Serotonin System

A neural circuit composed of several regions of the prefrontal cortex, amygdala, hippocampus, medial preoptic area, hypothalamus, anterior cingulate cortex, insular cortex, ventral striatum, and other interconnected structures has been implicated in emotion regulation including the associated affective phenomena of fear and anxiety (Gorman et al. 2000). Fear and anxiety-related circuits involve pathways transmitting information to and from the amygdala to various neural networks that control the expression of avoidant, defensive, or aggressive reactions, including behavioral, autonomic, and stress hormone responses. While pathways from the thalamus and cortex (sensory and prefrontal) project to the amygdala, inputs are processed within intra-amygdaloid circuitries and outputs are directed to the hippocampus, brain stem, hypothalamus, and other regions.

Perception of danger or threat are transmitted to the lateral nucleus of the amygdala, which projects to the basal nuclei where information regarding the social context derived from orbitofrontal projections is integrated with the perceptual information. Behavioral responses can then be initiated via activation of projections from the basal nuclei to various association cortices, while physiological responses can be produced via projections from the basal nuclei to the central nucleus and then to the hypothalamus and brainstem. Excessive or insufficient activation of the amygdaloid complex leads to either disproportionate negative emotionality or impaired sensitivity to social signals. The orbitofrontal cortex, through its connections with other domains of the prefrontal cortex and with the amygdala, plays a critical role in limiting emotional outbursts, and the anterior cingulate cortex recruits other neural systems during arousal and other emotions. Although the brain systems mediating anxiety-related responses appear to be fairly constant among mammals, several details of the regulatory pathways are species specific. While genetic and environmental factors contribute to the structure and function of this circuitry, the amygdala-associated neural network is critical to processes of learning to associate stimuli with events that are either punishing or rewarding.

In humans, non-human primates, rodents, and other mammals, preclinical and clinical studies have accumulated substantial evidence that serotonergic signaling is a major modulator of emotional behavior including fear and anxiety, as well as aggression, and it integrates complex brain functions such as cognition, sensory processing, and motor activity. This diversity of these functions is because 5-HT orchestrates the activity and interaction of several other neurotransmitter systems. The central 5-HT system, which originates in the midbrain and brainstem raphe complex, is widely distributed throughout the brain, and its chemical messenger is viewed as a master control neurotransmitter within this highly elaborate system of neural communication mediated by 14+ pre- and postsynaptic receptor subtypes with a multitude of isoforms (e.g., functionally relevant splice variants) and subunits. The prefrontal cortex receives major serotonergic input, which appears dysfunctional in individuals who are emotionally unstable and stress reactive. Individuals vulnerable to faulty regulation of emotionality are therefore at risk for anxiety-related disorders.

The level of 5-HT in the synaptic (and extrasynaptic) space is restricted by the synchronized action of at least three components. Firing of raphe 5-HT neurons is controlled by 5-HT$_{1A}$ autoreceptors located in the somatodendritic section of neurons. Release of 5-HT at the terminal fields is regulated by the 5-HT$_{1B}$ receptor. Once released, 5-HT is taken up by the 5-HT transporter located at the terminals (as well as the somatodendritic fraction) of 5-HT neurons, where monoamine oxidase A eventually metabolizes it. The action of 5-HT as a messenger is tightly regulated by its synthesizing and metabolizing enzymes, and the 5-HT transporter. It is therefore likely that modification or removal of one of these components affects extracellular levels of 5-HT.

Serotonergic raphe neurons diffusely project to all brain regions implicated in anxiety-related behavior, while neurons in anxiety-mediating areas are rich in both 5-HT$_1$ and 5-HT$_2$ receptor subtypes. In addition to its role as a neurotransmitter, 5-HT is, via its receptors, an important regulator of morphogenetic activities during early brain development as well as during adult neurogenesis and plasticity, including cell proliferation, migration, differentiation, and synaptogenesis (Azmitia and Whitaker-Azmitia 1997; Di Pino et al. 2004; Gaspar et al. 2003; Lauder 1993).

3.1 Serotonin Receptors

Some 14 different 5-HT receptors confer the effects of 5-HT upon neuronal or other cells. Pharmacological classification based on studies of ligand binding to receptor subtypes and of signal transduction pathway responses to agonists/antagonists were traditionally employed to delineate four 5-HT receptor subfamilies, 5-HT$_{1–4}$. Gene identification efforts have eventually not only validated this classification but also uncovered the existence of several novel 5-HT receptor subtypes (5-HT$_{1E/F}$, 5-HT$_{3A/B}$, 5-HT$_{5A/B}$, 5-HT$_6$, and 5-HT$_7$) (Barnes and Sharp 1999; Hoyer and Martin 1997). In 5-HT$_{2–7}$ receptor genes, the coding region is interrupted by introns, whereas the genes for 5-HT$_{1A–F}$ receptors contain no intronic sequences. The 5-HT$_{2B}$, 5-HT$_4$, 5-HT$_6$, and 5-HT$_7$ receptors are alternatively spliced and RNA editing of the 5-HT$_{2C}$ receptor subtype in the second intracellular loop has been reported to confer differential receptor properties. A genetic 5-HT$_{1A}$ receptor variant, Gly22Ser, shows differences in agonist-induced downregulation compared with the wildtype 5-HT$_{1A}$ receptor allele, thus increasing the complexity of naturally occurring 5-HT receptor structural variants. The challenge now is to identify the physiological impact of these gene products, establish their specific functionality with respect to distinct neurocircuits of emotion regulation, design selective agonists/antagonists, and determine potential therapeutic application of these novel compounds.

The molecular characterization of different 5-HT receptor subtypes has simplified the elucidation of gene transcription, mRNA processing, and translation as well as intracellular trafficking and posttranslational modification relevant to synaptic and postreceptor signaling. Transcriptional control regions have been cloned for several 5-HT receptor subtypes and functional promoter mapping data are available for the 5-HT$_{1A}$, 5-HT$_{2A}$, 5-HT$_{2C}$, and 5-HT$_3$ receptor genes. The analysis of genomic regulatory regions of 5-HT receptor gene transcription and the modeling of variable 5-HT receptor gene function in genetically modified mice (constitutive and conditional knockout/in) provides critical knowledge regarding the respective role of these receptors in neurodevelopment, synaptic plasticity, and behavior (Bonasera and Tecott 2000).

3.1.1
Serotonin Receptor 1A

The 5-HT$_{1A}$ receptor subtype has long been implicated in the pathophysiology of anxiety and depression; its role as a molecular target of anxiolytic and antidepressant drugs is well established (Griebel 1995; Griebel et al. 2000; Olivier et al. 1999). Patients with panic disorder and depression display an attenuation of 5-HT$_{1A}$ receptor-mediated hypothermic and neuroendocrine responses, reflecting a reduced responsivity of both pre- and postsynaptic 5-HT$_{1A}$ receptors (Lesch et al. 1990b; Lesch et al. 1992). Likewise, a decrease in 5-HT$_{1A}$ ligand binding has been shown in postmortem brain of depressed suicide victims (Cheetham et al. 1990) as well as in forebrain areas such as the medial temporal lobe and in the raphe of depressed patients elicited by positron emission tomography (PET) (Drevets et al. 1999; Sargent et al. 2000). Both glucocorticoid administration and chronic stress, a pathogenetic factor in affective disorders, have also been demonstrated to result in downregulation of 5-HT$_{1A}$ receptors in the hippocampus in animals (Flugge 1995; Lopez et al. 1998; Wissink et al. 2000). While deficits in hippocampal 5-HT$_{1A}$ receptor function may contribute to the cognitive abnormalities associated with affective disorders, recent work suggests that activation of this receptor stimulates neurogenesis in the dentate gyrus of the hippocampus. By using both a mouse model with a targeted ablation of the 5-HT$_{1A}$ receptor and radiological methods, Santarelli and coworkers (2003) have provided persuasive evidence that 5-HT$_{1A}$-activated hippocampal neurogenesis is essentially required for the behavioral effects of long-term antidepressant treatment with 5 HT reuptake inhibitors.

Intriguingly, downregulation and hyporesponsivity of 5-HT$_{1A}$ receptors in patients with major depression are not reversed by antidepressant drug treatment (Lesch et al. 1990a; Lesch et al. 1991; Sargent et al. 2000), raising the possibility that low receptor function is a trait feature and therefore a pathogenetic mechanism of the disease. In line with this notion, evidence is accumulating that a polymorphism in the transcriptional control region of the 5-HT$_{1A}$ receptor gene (HTR1A) resulting in allelic variation of 5-HT$_{1A}$ receptor expression, is associated with personality traits of negative emotionality including anxiety and depression (neuroticism and harm avoidance) (Strobel et al. 2003) as well as major depressive disorder, suicidality, and panic disorder (Lemonde et al. 2003; Rothe et al. 2004).

5-HT_{1A} receptors operate both as somatodendritic autoreceptors and as postsynaptic receptors. Somatodendritic 5-HT_{1A} autoreceptors are predominantly located on 5-HT neurons and dendrites in the brainstem raphe complex and their activation by 5-HT or 5-HT_{1A} agonists decreases the firing rate of serotonergic neurons and subsequently reduces the synthesis, turnover, and release of 5-HT from nerve terminals in projection areas. Postsynaptic 5-HT_{1A} receptors are widely distributed in forebrain regions that receive serotonergic input, notably in the cortex, hippocampus, septum, amygdala, and hypothalamus. Their activation results in membrane hyperpolarization and decreased neuronal excitability. Hippocampal heteroreceptors mediate neuronal inhibition by coupling to G protein-gated potassium channel subunit 2 (GIRK2) potassium channels. Physiological responses depend upon the function of the target cells (e.g., hypothermia, activation of the hypothalamic pituitary adrenocortical system) (Hamon et al. 1990). Moreover, 5-HT_{1A} receptor expression is modulated by steroid hormones and 5-HT_{1A}-mediated signaling is an important regulator of gene expression through its coupling to G proteins that inhibit adenylyl cyclase and through modulation of GIRK2 channels.

The effects of 5-HT_{1A} receptor-selective agents, such as the agonist 8-hydroxy-2-(di-*n*-propylamino) tetralin (8-OH-DPAT, and the partial agonists ipsapirone and gepirone, have been extensively studied in rodents (De Vry 1995). Both agonists and partial agonists induce a dose-dependent anxiolytic effect which correlates with the inhibition of serotonergic neuron firing, decrease of 5-HT release as well as the reduction of 5-HT signaling at postsynaptic target receptors. Blockade of the negative feedback by selective 5-HT_{1A} receptor antagonists, such as WAY 100635, increases firing of the serotonergic neurons but exerts no effect on 5-HT neurotransmission or behavior (Olivier and Miczek 1999), while the combination with selective 5-HT reuptake inhibitors augments the increases in 5-HT levels in terminal regions.

The converging lines of evidence that receptor deficiency or dysfunction is involved in mood and anxiety disorders encouraged investigators to genetically manipulate the 5-HT_{1A} receptor in mice (Table 1) (Heisler et al. 1998; Parks et al. 1998; Ramboz et al. 1998). Mice with a targeted inactivation of the 5-HT_{1A} receptor show a complete lack of ligand binding to brain 5-HT_{1A} receptors in null-mutant (−/−) mice, with intermediate binding in the heterozygote (+/−) mice. Importantly, a similar behavioral phenotype characterized by increased anxiety-related behavior and stress reactivity in several avoidance and behavioral despair paradigms was observed in three different KO mouse strains (Lesch and Mössner 1999).

3.1.1.1
Anxiety-Related Behavior

5-HT_{1A} receptor KO mice consistently display a spontaneous phenotype that is associated with a gender-modulated and gene-dose dependent increase

Table 1 Mice with inactivation of serotonergic genes displaying an anxiety-like or related behavioral phenotype

Knockout	Effect of anxiety-related behavior	Additional behavioral phenotypes, special features	Authors
Serotonergic neuron phenotype-specific genes			
5-HT receptor 1A	↑	Consistent in different genetic backgrounds	Heisler et al. 1998 Ramboz et al. 1998 Parks et al. 1998 Sibille et al. 1999
5-HT receptor 1B	↓	Aggression ↑	Brunner et al. 1999
5-HT transporter	↑	Stress reactivity Aggression ↓ ↑	Holmes et al. 2003 Holmes et al. 2002
TPH 1	–	Normal 5-HT in brain	Walther et al. 2003
TPH 2	**n.d.**		
MAOA	↑ (?)	Aggression ↑ Stress reactivity ↑	Cases et al. 1995 Seif and De Maeyer 1999
Signal transduction			
AC VIII	↑		Schaefer et al. 2000
CamKII	↓	Offensive aggression ↑	Chen et al. 1994
GIRK2	↓	Hyperactivity	Blednov et al. 2001
nNOS	↓ (?)	Impulsivity and aggression ↑	Chiavegatto et al. 2001
Developmental factors			
NCAM	↑	5-HT_{1A} response ↑ GIRK2 ↑	Delling et al. 2002 Stork et al. 1999
Pet1	↑	Aggression ↑	Hendricks et al. 2003
BDNF	↑*	*Conditional knockout Constitutive knockout +/–, Aggression ↑ –/–, Not viable	Rios et al. 2001 Lyons et al. 1999
Other 5-HT related systems			
tPA	↑	Gap43, a growth-associated protein of growing 5-HT axons ↓	Pawlak et al. 2002
Neurokinin 1 receptor	↑	Serotonergic function ↑	Santarelli et al. 2001

AC VIII, adenylyl cyclase type VIII; BDNF, brain-derived neurotrophic factor; CamKII, calcium-calmodulin kinase II; GIRK2, G protein-activated inward rectifying potassium 2; MAOA, monoamine oxidase A; n.d., not determined; NCAM, neural cell adhesion molecule; nNOS, neuronal nitric oxide synthase; Pet1, ETS domain transcription factor; tPA, serine protease tissue-plasminogen activator (tPA). ↑/↓, Increase/decrease in anxiety-related behavior. –, No effect.

of anxiety-related behaviors (Heisler et al. 1998; Parks et al. 1998; Ramboz et al. 1998). With the exception of an enhanced sensitivity of terminal 5-HT_{1B} receptors, no major neuroadaptational changes were detected. Worthy of note is that this behavioral phenotype was observed in animals in which the mutation was bred into mice of Swiss–Webster (SW), C57BL/6J, and 129/SV backgrounds, substantiating the assumption that this behavior is an authentic consequence of reduced or absent 5-HT_{1A} receptors. While all investigators used open field exploratory behavior as a model for assessing anxiety, two groups confirmed that *5-HT_{1A}* KO mice had increased anxiety by using other models, the elevated zero maze or elevated plus maze test (Heisler et al. 1998; Ramboz et al. 1998). These ethologically based conflict models test fear and anxiety-related behaviors based on the natural tendencies of rodents to prefer enclosed, dark spaces versus their interest in exploring novel environments.

Activation of presynaptic 5-HT_{1A} receptors provides the brain with an autoinhibitory feedback system controlling 5-HT neurotransmission. Thus, enhanced anxiety-related behavior most likely represents a consequence of increased terminal 5-HT availability resulting from the lack or reduction in presynaptic somatodendritic 5-HT_{1A} autoreceptor negative feedback function (Lesch and Mössner 1999). Although extracellular 5-HT concentrations and 5-HT turnover appear to be unchanged in the brain of *5-HT_{1A}* KO mice on the SW and 129/SV backgrounds, indirect evidence for increased presynaptic serotonergic activity resulting in elevated synaptic 5-HT concentrations is provided by the compensatory upregulation of terminal 5-HT release-inhibiting 5-HT_{1B} receptors (Olivier et al. 2001; Toth 2003). In contrast to *5-HT_{1A}* KO mice with a SW or 129/SV background, extracellular 5-HT concentrations were significantly elevated in mutant C57BL/6 mice in the frontal cortex and hippocampus (Parsons et al. 2001). This may reflect a lack of compensatory changes in 5-HT_{1B} receptor and is consistent with findings that C57BL/6 mice are more aggressive and susceptible to drugs of abuse than many other strains.

Several studies addressed electrophysiological properties of both presynaptic serotonergic neurons and postsynaptic hippocampal neurons in 5-HT_{1A} receptor-deficient mice. A robust increase in the mean firing rate in dorsal raphe neurons was also reported, although a considerable number of neurons was firing in the normal range and 5-HT release was not altered (Richer et al. 2002). Moreover, mutant mice showed an absence of paired-pulse inhibition in the CA1 region and lack of paired-pulse facilitation in the dentate gyrus, suggesting altered hippocampal excitability and impaired plasticity of the hippocampal network with consequence for cognition, learning, and memory (Sibille et al. 2000).

This mechanism is also consistent with models of fear and anxiety that are primarily based upon pharmacologically derived data. The cumulative reduction in serotonergic impulse flow to septohippocampal and other limbic and cortical areas involved in the control of anxiety is believed to explain the anxiolytic effects of ligands with selective affinity for the 5-HT_{1A} receptor in some animal models of anxiety-related behavior. This notion is based, in part, on evidence that 5-HT_{1A} agonists (e.g., 8-OH-DPAT) and antagonists (e.g., WAY 100635) have anxiolytic or anxiogenic effects, respectively. However, to complicate matters further, 8-OH-DPAT has anxiolytic effects when injected in the raphe nucleus, whereas it is anxiogenic when applied to the hippocampus. Thus, stimulation of postsynaptic 5-HT_{1A} receptors has been proposed to elicit anxiogenic effects, while activation of 5-HT_{1A} autoreceptors is thought to induce anxiolytic effects via suppression of serotonergic neuronal firing resulting in attenuated 5-HT release in limbic terminal fields.

Since the 5-HT_{1A} receptor is expressed in different brain subsystems, it is of interest to clarify whether pre- or postsynaptic receptors are required to maintain normal expression of anxiety-related behavior in mice. With an elegant conditional rescue approach, Gross et al. (2002) illustrated that expression of the 5-HT_{1A} receptor in the hippocampus and cortex but not in the raphe nuclei is required to rescue the behavioral phenotype in KO mice. The findings indicate that deletion of the 5-HT_{1A} receptor in mice, specifically in forebrain structures, results in a robust anxiety-related phenotype and that this phenotype in 5-HT_{1A} KO mice is caused by the absence of the receptor during a critical period of postnatal development, whereas inactivation of 5-HT_{1A} in adulthood does not affect anxiety. Even more importantly, the findings further support the notion of a central role for 5-HT in the early development of neurocircuits mediating emotion (Di Pino et al. 2004; Lesch 2003). Although there is converging evidence that the 5-HT_{1A} receptor mediates anxiety-related behavior, the neurodevelopmental mechanism that renders 5-HT_{1A} receptor-deficient mice more anxious is highly complex and remains to be elucidated in detail.

While increased 5-HT availability and activation of other serotonergic receptor subtypes that have been shown to mediate anxiety (e.g., 5-HT_{2C} receptor) may contribute to increased anxiety in rodent models, multiple downstream neurotransmitter pathways or neurocircuits, including γ-aminobutyric acidergic (GABAergic), noradrenergic, glutamatergic, and peptidergic transmission, as suggested by overexpression or targeted inactivation of critical genes within these systems (Lesch 2001a), have been implicated as participating in the processing of this complex behavioral trait. Since avoidance induced by conflict and fear is only one dimension of anxiety-related responses, other components, including autonomic systems activation, responsiveness to stress, 5-HT dynamics, and neuronal excitability in limbic circuitries, appear to be involved in fear and anxiety.

3.1.1.2
Stress Reactivity

As a facet of anxiety-like behavior, 5-HT_{1A} receptor KO mice show genotype-dependent and background strain-unrelated increase in stress reactivity in two paradigms of behavioral despair, the forced swim and tail suspension tests

(Heisler et al. 1998; Parks et al. 1998; Ramboz et al. 1998). The autonomic manifestation of anxiety and stress responsiveness in a novel environment or when exposed to other stressors (increased heart rate and body temperature as well as attenuated release of corticosterone) is also a characteristic of 5-HT_{1A} receptor KO mice (Groenink et al. 2003). The reduced immobility in stress/antidepressant test models is either due to an increased serotonergic tone resulting from the compromised 5-HT_{1A} autoreceptor-dependent negative feedback regulation or enhanced dopamine and norepinephrine function because it is reversed by pretreatment with α-methyl-*para*-tyrosine, but not by *para*-chlorophenylalanine (Mayorga et al. 2001).

Although the behavior of 5-HT_{1A} receptor-deficient mice in various stress-related paradigms is more consistent with increased emotionality, their behavior essentially corresponds with the performance of rodents treated with antidepressants. The role of 5-HT_{1A} receptors in the therapeutic action of antidepressant drugs has attracted extraordinary interest; there is substantial conflicting evidence, however, regarding the involvement of other serotonergic receptor subtypes and neurotransmitter systems or neurocircuits that interact with 5-HT neurotransmission. Electrophysiological studies in rats indicate that each class of antidepressant enhances 5-HT neurotransmission via differential adaptive changes in the 5-HT_{1A} receptor-modulated negative feedback regulation that eventually leads to an overall increase of terminal 5-HT (for review see Blier and de Montigny 1998), and desensitization of 5-HT_{1A} responsivity following antidepressant treatment has been demonstrated in rodents (Le Poul et al. 1995; Li et al. 1993) and humans (Berlin et al. 1998; Sargent et al. 1997; Lesch et al. 1991; Lerer et al. 1999). While the neuroadaptive mechanism of the antidepressant action of tricyclics or selective 5-HT reuptake inhibitors is exceedingly complex, as the onset of clinical improvement commonly takes 2–3 weeks or more after initiation of antidepressant drug administration, progressive functional desensitization of pre- and postsynaptic serotonergic receptors, including 5-HT_{1A}, 5-HT_{1B}, and 5-HT_{2A}, that is set off by blockade of the 5-HT transporter, has been implicated in these delayed therapeutic effects. In conclusion, the phenotypic similarity between anxiety-related behavior and stress reactivity in humans and 5-HT_{1A} receptor KO mice powerfully validates the practicability of KO animal models.

3.1.2
Serotonin Receptor 1B

The 5-HT_{1B} receptor was the first subtype to have its gene inactivated by classical homologous recombination (Saudou et al. 1994). 5-HT_{1B} receptors are expressed in the basal ganglia, central gray, lateral septum, hippocampus, amygdala, and raphe nuclei. They are located predominantly at presynaptic terminals inhibiting 5-HT release or, as heteroreceptors, modulating the release of other neurotransmitters. Selective agonists and antagonists for 5-HT_{1B}

receptors are largely lacking, but indirect pharmacological evidence suggests that 5-HT_{1B} activation influences food intake, sexual activity, locomotion, and emotionality including, particularly, impulsivity and aggression.

Generation of mice with a targeted disruption of the 5-HT_{1B} gene facilitated investigation of the concept of 5-HT-related impulsivity in the context of aggressive behavior (Stark and Hen 1999). Two of the behaviors, locomotion and aggression, each postulated to be modulated by 5-HT_{1B} receptors, were analyzed. Wildtype and null-mutant (5-$HT_{1B}{}^{-/-}$) mice were found to display similar levels of locomotor activity in an open field. However, 5-HT_{1B} receptor KO mice show adaptation in 5-HT_{2C} receptor-mediated functions with smaller reductions in food intake and locomotor activity in response to administration of 5-HT_{2C} receptor agonists. Impulsivity and aggression-related behavior of male 5-$HT_{1B}{}^{-/-}$ mice were assessed by isolation and subsequent exposure to a non-isolated male wildtype intruder mouse. The latency and number of attacks displayed by the KO mice were used as indices of aggression. The 5-HT_{1B} KO mice, when compared with wildtype mice, showed more rapid, more intense, and more frequent attacks. Lactating female 5-$HT_{1B}{}^{-/-}$ mice also attack unfamiliar male mice more rapidly and violently. In addition to increased aggression, KO mice acquire cocaine self-administration faster and ingest more ethanol than controls, indicating that the 5-HT_{1B} receptor not only modulates motor impulsivity and aggression but also addictive behavior (Brunner et al. 1999).

These results further support the notion that distinct receptor subtypes modulate different dimensions of behavior that may be either synergistic or antagonistic. Opposite to 5-HT_{1A} receptor-deficient mice, 5-HT_{1B} KOs are more reactive and more aggressive but show dramatically less anxiety-related behavior than control mice, although both 5-HT_{1A} and 5-HT_{1B} receptors control the tone of the serotonergic system and mediate some of the postsynaptic 5-HT effects (Zhuang et al. 1999). The regional variation of 5-HT receptor expression and the complex autoregulatory processes of 5-HT function that are operational in different brain areas may lead to a plausible hypothesis to explain an inconsistency that is more apparent than real. Assessment of 5-HT_{1A} receptor expression in male mice selected for high and low offensive aggression showed that high-aggressive mice are characterized by a short attack latency, decreased plasma corticosterone concentration, and increased levels of 5-HT_{1A} mRNA in the dorsal hippocampus (dentate gyrus and CA1) compared to low-aggressive mice that had long attack latency and high plasma corticosterone levels (Korte et al. 1996). Correspondingly, increased postsynaptic 5-HT_{1A} receptor radioligand binding was also found in the hippocampal CA1 subdivision, dentate gyrus, lateral septum, and frontal cortex, whereas no difference in ligand binding was found for the 5-HT_{1A} autoreceptor on cell bodies in the dorsal raphe nucleus. These results suggest that high offensive aggression is associated with reduced (circadian peak) plasma corticosterone and increased postsynaptic 5-HT_{1A} receptor availability in limbic and cortical regions.

3.1.3
Other Serotonin Receptors

Prior to the generation of 5-HT_{2C} receptor-deficient mice, studies with nonselective agonists had suggested potential roles for this receptor in the serotonergic regulation of feeding and anxiety. Consistent with the pharmacological evidence, 5-HT_{2C} mutant mice display hyperphagia-evoked weight gain but also infrequent and sporadic spontaneous seizures, suggesting a globally enhanced neuronal network excitability. Behavioral analysis of 5-HT_{2C} KO mice revealed abnormal performance in a spatial learning task and altered exploratory behavior associated with altered long-term potentiation restricted to the dentate gyrus perforant path synapse (Heisler and Tecott 1999). However, abnormalities of hippocampal function-dependent cognitive function were subtle and did not generalize to contextual fear conditioning.

Studies in mice with a targeted inactivation of other 5-HT receptor subtypes, such as the 5-HT_{5A} and 5-HT_7, or a transgenic line that overexpresses 5-HT_3, demonstrate that these receptors modulate the activity of neural circuits involved specifically in exploratory and reward-related behavior. When exposed to novel environments, KO mice lacking the 5-HT_{5A} exhibit increased exploratory activity and an attenuated stimulatory effect of lysergic acid diethylamide (LSD) on exploratory activity but no change in anxiety-related behavior (Grailhe et al. 1999), whereas 5-HT_7 KO mice do not express any overt behavioral phenotype at all (Hedlund et al. 2003).

Since all 5-HT receptor subtypes are remarkably similar in their ligand-binding domains, it has been difficult to design pharmacological compounds that can specifically interact with the other subtypes. The present challenge therefore is to further characterize the physiological and behavioral relevance of the remaining 5-HT receptor gene products as well as to generate and analyze KO mice for each remaining subtype (Compan et al. 2004; Fiorica-Howells et al. 2002; Nebigil et al. 2001). The new insights into neural plasticity and complexity of gene regulation in 5-HT subsystems will eventually provide the means for novel approaches of studying 5-HT receptor subtype-related behaviors at the molecular level.

3.2
Serotonin Synthesizing and Metabolizing Enzymes

3.2.1
Tryptophan Hydroxylases

The first step of 5-HT biosynthesis is catalyzed by the rate-limiting enzyme tryptophan hydroxylase (TPH). Two isoforms, TPH1 and TPH2, have been identified in the periphery and in 5-HT neurons, respectively. Both isoforms are members of the aromatic amino acid hydroxylase gene family, together with phenylalanine (PAH) and tyrosine hydroxylases (TH). The human TPH1 gene located on chromosome *11p15.1*, spans a region of 30 kb, contains at least 11 exons, and an unusual splicing complexity in the 5′-untranslated region (5′-UTR) resulting in at least four *TPH1* mRNA species transcribed from a single transcriptional start site (Boularand et al. 1995). The murine *TPH1* has been mapped to chromosome *7* at 23.5 cM. The human TPH2 gene was found on chromosome *12p21.1*, covers a region of more than 120 kb, and contains 11 exons, and a single *TPH2* mRNA species is transcribed from a unique transcriptional start site (Walther et al. 2003). The murine *TPH2* is located on chromosome *10*. The deduced amino acid sequence of *TPH2* shows 84%, and 86% identity to *TPH1* sequences of man, and mouse, respectively. Mice with an inactivation of the TPH1 gene lack 5-HT in the periphery, whereas 5-HT concentrations in the serotonergic projection region of the brain are in the normal range. Moreover, 5-HT-related avoidance behavior as assessed by the elevated plus maze and hole board tests was not different in *TPH1* KO mice, indicating that the behavioral effects of 5-HT in the brain are uncoupled and thus independent from 5-HT and its metabolites in peripheral tissues.

This finding is critically relevant, since correlations between peripheral levels of 5-HT metabolites and 5-HT function in the brain of patients suffering from psychiatric disorders have extensively been studied. A possible role of TPH1 gene variations in antisocial personality disorder, alcoholism, bipolar disorder, major depression, and associated suicidality has been shown in some but not all studies (Bellivier et al. 1998; Furlong et al. 1998; Manuck et al. 1999; New et al. 1998). Nielson et al. (1994, 1998) reported that the *TPH* A779C polymorphism influences 5-hydroxyindoleacetic acid concentrations (5-HIAA), the major metabolite of 5-HT, in cerebrospinal fluid (CSF), and may predispose to suicidality, a pathophysiological mechanism that may involve impaired impulse control. This finding was subsequently replicated by the several other groups suggesting that functional variant(s) in or close to the TPH1 gene may predispose individuals to affective disorders and suicidality.

In the face of preferential expression of *TPH2* in the brain, the identification of a gene encoding a neuron-specific TPH isoform, and unaltered anxiety-like behavior in *TPH1* KO mice, this conclusion calls for reconsideration. Among the questions that need to be answered are whether *TPH1* expression is restricted to the early stages of brain development and whether peripheral 5-HT impacts on the development of limbic circuits setting the stage of emotional behavior and thus influencing susceptibility to anxiety, depression, and suicidality in adulthood. In anticipation of a mouse model with a targeted disruption of the TPH2 gene, clarification of the effect of brain 5-HT deficiency on anxiety-related and aggressive behavior seems to be imminent.

3.2.2
Monoamine Oxidase A

Monoamine oxidase A (MAOA) oxidizes 5-HT, norepinephrine as well as dopamine, and is expressed in a cell type-specific manner. Abnormalities in

MAOA activity have been implicated in a wide range of psychiatric disorders. Deficiency in MAOA enzyme activity due to a hemizygous chain termination mutation of the MAOA gene has recently been shown to be associated with impulsive aggression and hypersexual behavior in affected males from a single extended pedigree (Brunner syndrome) (Brunner et al. 1993).

Mice with a targeted disruption of the MAOA gene have markedly higher brain 5-HT concentrations and exhibit aggressive behavior in adult males as assessed by the resident-intruder paradigm as well as violent courtship (Seif and De Maeyer 1999). Moreover, *MAOA* KO mice display reduced activity, possibly reflecting increased anxiety in the open field and enhance stress reactivity in the forced swim test. Morphological analyses of brain structures where 5-HT has been suggested to act as a differentiation signal in development revealed a detrimental effect of *MAOA* inactivation on the formation and plasticity of cortical and subcortical structures. Investigations of 5-HT participation in neocortical development and plasticity have been concentrated on the rodent somatosensory cortex (SSC), due to its one-to-one correspondence between each whisker and its cortical barrel-like projection area. The processes underlying patterning of projections in the SSC have been intensively studied with a widely held view that the formation of somatotopic maps does not depend on neural activity. The timing of serotonergic innervation coincides with pronounced growth of the cortex, the period when incoming axons begin to establish synaptic interactions with target neurons and to elaborate a profuse branching pattern.

Interestingly, the brains of *MAOA* KO mice show a lack of these characteristic barrel-like clustering of layer IV neurons in S1, despite relatively preserved trigeminal and thalamic patterns (Cases et al. 1995; Di Pino et al. 2004). Thalamo-cortical afferents display a decrease in branching and excessive tangential distributions, suggesting a deficiency of terminal retraction (Rebsam et al. 2002). Other abnormalities include abnormal segregation of contralateral and ipsilateral retinogeniculate projections (Upton et al. 1999), and aberrant maturation of the brainstem respiratory network (Burnet et al. 2001). The excess 5-HT is likely responsible for these alterations, since barrel formation is restored by the 5-HT synthesis inhibitor *p*-chlorophenylalanine (pCPA), which also restores normal development of retinogeniculate projections and the brainstem respiratory network as well as aggression-related behavior.

Additional evidence for a role of 5-HT in the development of neonatal rodent SSC derives from the transient barrel-like distribution of 5-HT, 5-HT$_{1B}$, and 5-HT$_{2A}$ receptors, and of the 5-HT transporter (Lebrand et al. 1996; Mansour-Robaey et al. 1998).The transient barrel-like 5-HT pattern visualized in layer IV of the SSC of neonatal rodents stems from 5-HT uptake and vesicular storage in thalamocortical neurons, transiently expressing at this developmental stage both 5-HT transporter and the vesicular monoamine transporter (VMAT2) despite their later glutamatergic phenotype (Lebrand et al. 1996).

3.3 Transporter

High-affinity 5-HT transport into the presynaptic neuron is mediated by a single protein, the 5-HT transporter (5-HTT, SERT), which is regarded as initial sites of action of antidepressant drugs and several neurotoxic compounds. Tricyclic antidepressants, such as prototypical imipramine, and the selective 5-HT uptake inhibitors, paroxetine, citalopram, and sertraline, occupy several pharmacologically distinct sites overlapping at least partially the substrate binding site and are widely used in the treatment of depression, anxiety, and impulse control disorders, as well as substance abuse including alcoholism.

While in adult brain 5-HTT expression appears to be restricted to raphe neurons, it has been detected in the sensory areas of the cortex and thalamus during perinatal development (Lesch and Murphy 2003). Cloning of *5-HTT* has identified a protein with 12 transmembrane domains (TMDs) and studies using site-directed mutagenesis and deletion mutants indicate that distinct amino acid residues participate in substrate translocation and competitive antagonist binding. 5-HTT function is acutely modulated by posttranslational modification. Moreover, several intracellular signal transduction pathways converge on the transcriptional apparatus of the 5-HTT gene regulating its expression. A polymorphism in the transcriptional control region of the human 5-HTT gene (*SLC6A4*) that results in allelic variation in functional 5-HTT expression is associated with anxiety, depression, and aggression-related personality traits (Lesch 2003; Lesch et al. 1996). In addition to the exploration of the impact of allelic variation in 5-HTT expression on anxiety, depression, and aggression-related personality traits, a role of the regulatory and structural 5-HTT gene variation has been suggested in a variety of diseases such as depression, bipolar disorder, anxiety disorders, eating disorders, substance abuse, autism, schizophrenia, and neurodegenerative disorders (Lesch and Murphy 2003).

3.3.1 Anxiety-Related Responses

The converging evidence that 5-HTT deficiency plays a role in anxiety and related disorders lead to the generation of mice with a targeted inactivation of the 5-HTT gene (*Slc6a4*). Behavior of the *5-HTT* KO mice was tested in a variety of conditions evaluating fear, avoidance, conflict, stress responsiveness, status of the neuroendocrine system, and effects of various pharmacological agents on the behavior. In particular, anxiety-related behaviors were characterized using a battery of tests including open field, elevated plus maze, and light-dark box. In these tests both male and female *5-HTT* KO mice show consistently increased anxiety-like behavior and inhibited exploratory locomotion. The selective 5-HT_{1A} receptor antagonist WAY 100635 produced an anxiolytic effect in the elevated plus maze in *5-HTT* KO mice, suggesting that the abnormalities in anxiety-like and exploratory behavior is mediated by the 5-HT_{1A} receptor (Holmes et al. 2003). Unlike heterozygous *5-HT_{1A}*$^{+/-}$ mice, *5-HTT*$^{+/-}$ mice, in which transporter binding sites are reduced by approximately 50%, were similar to controls on most measures of anxiety-like behavior. However, changes in exploratory behavior in *5-HTT*$^{+/-}$ mice were limited to specific measures under baseline conditions, but extended to additional measures under more stressful test conditions. This observation is in accordance with reduced aggressive behavior in *5-HTT*$^{+/-}$ mice that is limited to specific measures and test conditions. While male *5-HTT*$^{-/-}$ mice are slower to attack the intruder and attacked with less frequency than control littermates, heterozygous *5-HTT*$^{+/-}$ mice were as quick to attack, but made fewer overall attacks, as compared to controls. Aggression increased with repeated exposure to an intruder in *5-HTT*$^{+/-}$ and control mice, but not in *5-HTT*$^{-/-}$ mice. These subtle behavioral alterations in *5-HTT*$^{+/-}$ mice are contrasted by robust perturbations in serotonergic homeostasis that are intermediate between −/− mice and controls in a gene-dose dependent manner including elevated extracellular 5-HT, decreased 5-HT neuron firing in the dorsal raphe, and reduced 5-HT_{1A} receptor expression and function (Gobbi et al. 2001; Li et al. 1999, 2000). The evidence that serotonergic dysfunction in *5-HTT*$^{+/-}$ mice may manifest and become noticeable as behavioral abnormalities only under challenging environmental conditions strongly support the disposition-stress model of affective and anxiety disorders (Murphy et al. 2003).

3.3.2 Neuroadaptive Changes

Analogous to *5-HT_{1A}* KO mice, the neural mechanisms underlying increased anxiety-related behavior and reduced exploratory locomotion in mice with a disruption of the 5-HTT gene may relate to excess serotonergic neurotransmission which is expected to cause enhanced activation of postsynaptic 5-HT receptors. Both in vivo microdialysis in striatum and in vivo chronoamperometry in hippocampus revealed that 5-HTT null-mutant mice exhibit an approximately fivefold increase in extracellular concentrations of 5-HT and an absence of transporter-mediated clearance, although brain tissue 5-HT concentrations are markedly reduced by 40%–60% (Bengel et al. 1998).

Excess of extracellular 5-HT activates the negative autoinhibitory feedback and reduces cellular 5-HT availability by stimulating 5-HT_{1A} receptors, which results in their desensitization and downregulation in the midbrain raphe complex and, to a lesser extent, in hypothalamus, septum, and amygdala but not in the frontal and hippocampus (Li et al. 2000). Although postsynaptic 5-HT_{1A} receptors appear to be unchanged in frontal cortex and hippocampus, indirect evidence for decreased presynaptic serotonergic activity but reduced 5-HT clearance resulting in elevated synaptic 5-HT concentrations is provided by compensatory alterations in 5-HT synthesis and turnover, downregulation of terminal 5-HT release-inhibiting 5-HT_{1B} receptors (Fabre et al. 2000).

Therefore, a partial downregulation of postsynaptic 5-HT_{1A} receptors in some forebrain regions but a several-fold increase in extracellular concentrations of 5-HT in 5-HTT null-mutant mice could still cause excess net activation of postsynaptic 5-HT_{1A} receptors, resulting in increased anxiety-like behavior and its reversal by WAY 100635 (Holmes et al. 2003). However, administration of WAY 100635 antagonizes not only postsynaptic 5-HT_{1A} receptors in forebrain regions but also acts at somatodendritic autoreceptors in the raphe nuclei, and electrophysiological studies show that WAY 100635 causes a reversal of markedly reduced spontaneous firing rates of 5-HT neurons in the dorsal raphe nucleus of *5-HTT*$^{-/-}$ mice, indicating that the net effect of WAY 100635 on serotonergic neurotransmission in *5-HTT* KO mice may be more complex than anticipated (Gobbi et al. 2001).

Taken together, these findings add to an emerging picture of abnormalities in 5-HTT null mutants across a range of behavioral, neuroendocrine, and physiological parameters associated with emotional disorders, including marked increases in adrenocorticotropin (ACTH) concentrations in responses to stress (Li et al. 1999), increased sensitivity to drugs of abuse such as cocaine (Sora et al. 2001; Sora et al. 1998), altered gastrointestinal motility (Chen et al. 2001), and disturbed rapid eye movement (REM) sleep (Wisor et al. 2003). Finally, given the absence of the 5-HTT throughout ontogeny, *5-HTT* KO mice also provide a research tool for studying the potential for neurodevelopment abnormalities affecting anxiety-like behavior.

3.3.3
Development of the Somatosensory Cortex

Analogous to *MAOA* KO mice, inactivation of the 5-HTT gene profoundly disturbs formation of the SSC with altered cytoarchitecture of cortical layer IV, the layer that contains synapses between thalamocortical terminals and their postsynaptic target neurons (Persico et al. 2001). Brains of *5-HTT* KO mice display no or only very few barrels. Cell bodies as well as terminals, typically more dense in barrel septa, appear homogeneously distributed in layer IV of adult *5-HTT* KO brains. Injections of a 5-HT synthesis inhibitor within a narrow time window of 2 days postnatally completely rescued formation of SSC barrel fields. Of note, heterozygous KO mice develop all SSC barrel fields, but frequently present irregularly shaped barrels and less defined cell gradients between septa and barrel hollows. These findings demonstrate that excessive concentrations of extracellular 5-HT are deleterious to SSC development and suggest that transient 5-HTT expression in thalamocortical neurons is responsible for barrel patterns in neonatal rodents, and its permissive action is required for normal barrel pattern formation, presumably by maintaining extracellular 5-HT concentrations below a critical threshold. Because normal synaptic density in SSC layer IV of *5-HTT* KO mice was shown, it is more likely that 5-HT affects SSC cytoarchitecture by promoting dendritic growth toward the barrel hollows as well as by modulating cytokinetic movements of cortical granule cells, similar to concentration-dependent 5-HT modulation of cell migration described in other tissues. Since the gene-dose dependent reduction in 5-HTT availability in heterozygous KO mice—which leads to a modest delay in 5-HT uptake but distinctive irregularities in barrel and septum shape—is similar to those reported in humans carrying low activity allele of the 5-HTT-LPR (gene-linked polymorphic region), it may be speculated that allelic variation in 5-HTT function also affects the human brain during development with due consequences for disease liability and therapeutic response.

These findings demonstrate that excessive amounts of extracellular 5-HT are detrimental to SSC development and suggest that transient 5-HTT expression and its permissive action is required for barrel pattern formation, presumably by maintaining extracellular 5-HT concentrations below a critical threshold. Two key players of serotonergic neurotransmission appear to mediate the deleterious effects of excess 5-HT: the 5-HTT and the 5-HT_{1B} receptor. Both molecules are expressed in primary sensory thalamic nuclei during the period when the segregation of thalamocortical projections occurs (Bennett-Clarke et al. 1996; Hansson et al. 1998). 5-HT is internalized via 5-HTT in thalamic neurons and is detectable in axon terminals (Cases et al. 1998; Lebrand et al. 1996). The presence of the VMAT2 within the same neurons allows internalized 5-HT to be stored in vesicles and used as a cotransmitter of glutamate. Lack of 5-HT degradation in *MAOA* KO mice as well as severe impairment of 5-HT clearance in mice with an inactivation of *5-HTT* results in an accumulation of 5-HT and overstimulation of 5-HT receptors all along thalamic neurons (Cases et al. 1998). Since 5-HT_{1B} receptors are known to inhibit the release of glutamate in the thalamocortical somatosensory pathway, excessive activation of 5-HT_{1B} receptors could prevent activity-dependent processes involved in the patterning of afferents and barrel structures. This hypothesis is supported by a recent study using a strategy of combined KO of MAOA, 5-HTT, and 5-HT_{1B} receptor genes. While only partial disruption of the patterning of somatosensory thalamocortical projections was observed in *5-HTT* KO, *MAOA–5-HTT* double KO (DKO) mice showed that 5-HT accumulation in the extracellular space causes total disruption of the patterning of these projections (Salichon et al. 2001). Moreover, the removal of 5-HT_{1B} receptors in *MAOA* and *5-HTT* KO as well as in *MAOA–5-HTT* DKO mice allows a normal segregation of the somatosensory projections as well as retinal axons in the lateral geniculate nucleus (Upton et al. 2002). These findings point to an essential role of the 5-HT_{1B} receptor in mediating the deleterious effects of excess 5-HT in the somatosensory system.

The effect of elevated extracellular 5-HT concentration on the modulation of programmed cell death during neural development was also investigated in early postnatal brains of *5-HTT* KO mice. 5-HTT gene inactivations leads to a reduced number of apoptotic cells in striatum, thalamus, hypothalamus, cerebral cortex, and hippocampus on postnatal day 1 (P1) with differences displaying an increasing fronto-caudal gradient and regional specificity (Persico et al. 2003). These findings underscore the role of 5-HT in the regulation of programmed cell death during brain development, and suggest that pharmacological enhancement of serotoninergic neurotransmission may minimize pathological apoptosis.

The evidence that changes in 5-HT system homeostasis exert long-term effects on cortical development and adult brain plasticity may be an important step forward in establishing the psychobiological groundwork for a neurodevelopmental hypothesis of negative emotionality, aggressiveness, and violence (Lesch 2003). Although there is converging evidence that serotonergic dysfunction contributes to anxiety-related behavior, the precise mechanism that renders 5-HTT-deficient mice more anxious and stress responsive remains to be elucidated. While increased 5-HT availability and activation of other serotonergic receptor subtypes that have been shown to mediate anxiety (e.g., 5-HT_{2C} receptor) may contribute to increased anxiety in the rodent model, multiple downstream cellular pathways or neurocircuits, including noradrenergic, GABAergic, glutamatergic, and peptidergic transmission—as suggested by overexpression or targeted inactivation of critical genes within these systems—have been implicated to participate in the processing of this complex behavioral trait. Recent work has therefore been focused on a large number of genes that have known relevance in the neurocircuitries of fear and anxiety, although the KO of some genes that appear not directly involved in anxiety may also lead to an anxiety-related phenotype.

3.4 Signal Transduction and Cellular Pathways

As the next dimension of complexity, signaling through 5-HT receptors involves different transduction pathways, and each receptor subtype modulates distinct, though frequently interacting, second and third messenger systems and multiple effectors.

3.4.1 Adenylyl Cyclase Type VIII

Stress results in alterations in behavior and physiology that can be either adaptive or maladaptive. Although mice deficient in the calcium-stimulated adenylyl cyclase type VIII (AC8) exhibit indices of anxiety comparable with that of wildtype mice at baseline, AC8 KO mice do not show normal increases in behavioral features of anxiety when subjected to repeated stress such as repetitive or post-restraint stress testing in the elevated plus maze test (Schaefer et al. 2000). Although these findings suggest a role for AC8 in the modulation of anxiety, the mechanism by which AC8 deficiency results in impaired stress-induced anxiety may be complex, involving impaired long-term depression (LTD) in the CA1 region of the hippocampus and failure to activate CRE-binding protein (CREB) in the CA1 region after restraint stress. Interestingly, it was recently reported that *CREB1* polymorphisms predispose to depressive disorder in a gender-specific manner, further strengthening the assumption that this pathway is involved in emotion regulation (Zubenko et al. 2003).

3.4.2
Calcium-Calmodulin Kinase II

The gene of the effector enzyme calcium-calmodulin kinase II (CaMKII), which participates in some intracellular responses to 5-HT receptor activation, has also been implicated in aggressive behavior by gene disruption (Chen et al. 1994). While $CaMKII^{-/-}$ mutants showed global behavioral impairment, male mice heterozygous for the inactivated *CaMKII* had a greater tendency to fight with each other when housed together. To be more specific, they showed enhanced offensive aggression, normal defensive aggression, and decreased fear-related responses.

3.4.3
G Protein-Activated Inward Rectifying Potassium Channel 2

The G protein-activated inward rectifying potassium (GIRK) channels regulate synaptic transmission and neuronal firing rates. The GIRK1–4 subunits exhibit unique but overlapping tissue localization patterns and contain cytoplasmic amino and carboxyl termini, two transmembrane domains, and a hydrophobic pore region similar to other potassium-selective channels. Evidence for homo- and heteromultimerization of GIRK subunits has been derived from heterologous expression and biochemical studies (Wischmeyer et al. 1997). They are regulated by neurotransmitters and hormones through G protein-coupled receptors, including muscarinic M2, dopamine D1–3, α_2-adrenoreceptor, 5-HT_{1A}, adenosine A_1, $GABA_B$, μ-, δ-, and κ-opioid, and somatostatin receptors. The GIRK2 (Kir 3.2) channel is abundantly expressed in the mammalian CNS (Karschin et al. 1996) and co-localization with dopamine receptors in the mesolimbic system and 5-HT_{1A} receptors in serotonergic raphe neurons suggests a role in modulation of motor activity and anxiety-like behavior (Luscher et al. 1997). Indeed, GIRK2-deficient mice show evidence of hyperactivity and reduced anxiety-like behavior with initially higher motor activity and slower habituation in a novel situation, increased levels of spontaneous locomotor activity during dark phase, and impaired habituation in the open-field test (Blednov et al. 2001; Blednov et al. 2002). After habituation, GIRK2 KO mice showed enhanced motor activity, which is modulated by D1 agonists and antagonists. Interestingly, increased expression and function of 5-HT_{1A} receptor-stimulated GIRK2 channels in mice with disruption of neural cell adhesion molecule (NCAM) gene may be causal for a lower excitability of target neurons for serotonergic fibers in the limbic system resulting in altered anxiety and aggression-related behavior (Delling et al. 2002) (also see Sect. 3.5.3).

3.4.4 Neuronal Nitric Oxide Synthase

The discovery of a considerable number of hyperaggressive mutant strains in the course of gene KO experiments highlights the extraordinary diversity of genes involved in the genetic influence on emotionality. Interestingly, genetic support for a role of 5-HT in anxiety and aggression also derives from mice lacking specific genes—such as the neuronal nitric oxide synthase (nNOS)—that either directly or indirectly affect 5-HT turnover or 5-HT receptor sensitivity. Male $nNOS^{-/-}$ mice and wildtype mice in which nNOS is pharmacologically suppressed are highly aggressive (Chiavegatto et al. 2001). Excessive aggressiveness and impulsiveness of *nNOS* KO mice depend on the presence of testosterone but seem to be caused by a selective decrease in 5-HT turnover and deficient 5-HT_{1A} and 5-HT_{1B} receptor function in brain regions regulating emotion. These findings indicate an interaction of nNOS and the 5-HT system mediated through 5-HT_{1A} and 5-HT_{1B} receptors, but the specific molecular mechanisms in anxiety and aggression remain to be clarified in more detail (Chiavegatto and Nelson 2003). Increased aggression and hypersexuality in male *nNOS* KO mice largely resemble the behavioral phenotype of Brunner syndrome caused by MAOA gene disruption (also see Sect. 3.2.1), suggesting common downstream pathways. Interestingly, the impact of nNOS inactivation is gender-specific, as female KO mice display reduced aggression during lactation (Gammie and Nelson 1999). Whether this is also due to an interaction with the serotonergic system should be the scope of future investigations.

3.5 Gene–Environment Interaction at the Neurodevelopmental Interface of Anxiety

At the core of the gene-versus-environment debate, the relative influences of adverse experiences early in life on susceptibility to behavioral and psychiatric disorder is still a matter of intense debate. Investigations in rats have shown that maternal behavior has long-lasting consequences on fear-related behavior of the offspring. Maternal separation for several hours a day during the early postnatal period results in increased anxiety-like behaviors as well as increased stress responsivity in adult animals (Kalinichev et al. 2002). Similarly, pups that are raised by mothers that display low licking-and-grooming behavior show higher levels of anxiety-like behavior than pups raised by high licking-and-grooming mothers, and cross-fostering studies show that these influences are primarily environmental (Caldji et al. 1998; Liu et al. 2000). Cross-fostering offspring of low licking-and-grooming mothers to high licking-and-grooming mothers is able to impart low anxiety-like behavior to the offspring, whereas the converse does not influence this behavior. Offspring of high licking-and-grooming mothers raised by low licking-and-grooming mothers do not show high anxiety-like behavior, suggesting that specific genes inherited by the high licking-and-grooming offspring protect them from the effects of low licking-and-grooming mothering. Furthermore, Francis et al. (1999) have shown that the effect of high licking-and-grooming can be passed from one generation to the next. Females raised by high licking-and-grooming mothers themselves become high licking-and-grooming mothers and go on to produce low anxiety offspring regardless of whether their biological mother showed low or high licking-and-grooming. This epigenetic inheritance of anxiety-like behavior underscores the power that environmental influences can exert to persistently remodel circuits in the brain during early development.

Studies using mice of defined genetic backgrounds have also begun to shed light on the molecular mechanisms of specific gene–environment interactions. Anisman et al. (1998) found that mice of the low licking-and-grooming Balb/c inbred strain cross-fostered at birth to the high licking-and-grooming C57BL/6 inbred strain display improvements in a hippocampal-dependent memory task. Because the reverse cross-fostering, where C57BL/6 pups are raised by Balb/c mothers, does not alter the behavior of C57BL/6 mice, it appears that the C57BL/6 genetic background protects the pups from the effects of a Balb/c maternal environment. However, by transplanting C57BL/6 embryos into Balb/c foster mothers shortly after conception, Francis et al. (2003) were able to show that a combined prenatal and postnatal Balb/c maternal environment is sufficient to confer Balb/c behavior on C57BL/6 offspring, demonstrating that intra- and extra-uterine maternal signals are likely to synergistically induce long-term plasticity changes in anxiety- and depression-related neurocircuits.

Since the genetic basis of present-day temperamental and behavioral traits is already laid out in many mammalian species including mice and may reflect selective forces among our remote ancestors, research efforts have recently been focused on nonhuman primates and humans (Barr et al. 2003a,b; Bennett et al. 2002; Caspi et al. 2003; Champoux et al. 2002; Newman et al. 2003).

3.5.1 Developmental Specification Serotonergic Neurons: Setting the Stage

The comparatively small number of serotonergic neurons (~20,000 and ~300,000 in rodents and humans, respectively) are primarily located in the raphe nuclei, on the midline of the rhombencephalon, and in the reticular formation. Although these neurons are clustered in caudal and rostral divisions of the B1–B9 cell groups, the extensive collateralization of their terminals densely innervate all regions of the CNS. While serotonergic neurons are generated during early embryonic development (E10–E12), launch synthesis of 5-HT shortly after that, and extend their axonal tracts to the forebrain and spinal cord, the maturational process shaping the networks is only completed during postnatal development.

Despite the widespread importance of the central serotonergic neurotransmitter system, knowledge of the molecular mechanisms regulating the development of 5-HT neurons is still limited. The specification, differentiation, diversification, phenotype maintenance, and survival of neurons comprising the raphe serotonergic system require a considerable number of transcription factors, other morphogenetic regulators of gene expression, neurotrophins, and growth factors, as well as 5-HT itself to work in concert or in cascade (Fig. 1).

Induction of the floor plate at the ventral midline of the neural tube and definition of the region in which the progenitor cells of 5-HT neurons will be formed are among the initial events in the establishment of dorsoventral polarity in the vertebrate brain. Several secreted positional markers, including the fibroblast growth factors (Fgf4 and 8) and Sonic hedgehog (Shh) synergistically control serotonergic (and dopaminergic) cell fate in the anterior neural plate (Hynes et al. 2000; Ye et al. 1998). Generation of 5-HT neurons in the neural tube depends on the action of the notochord and floor plate-derived Shh, as elicited by constitutive activation of the Shh, Smoothened, in transgenic mice resulting in a dislocation of serotonergic neurons. Fgf4 and Fgf8, expressed in

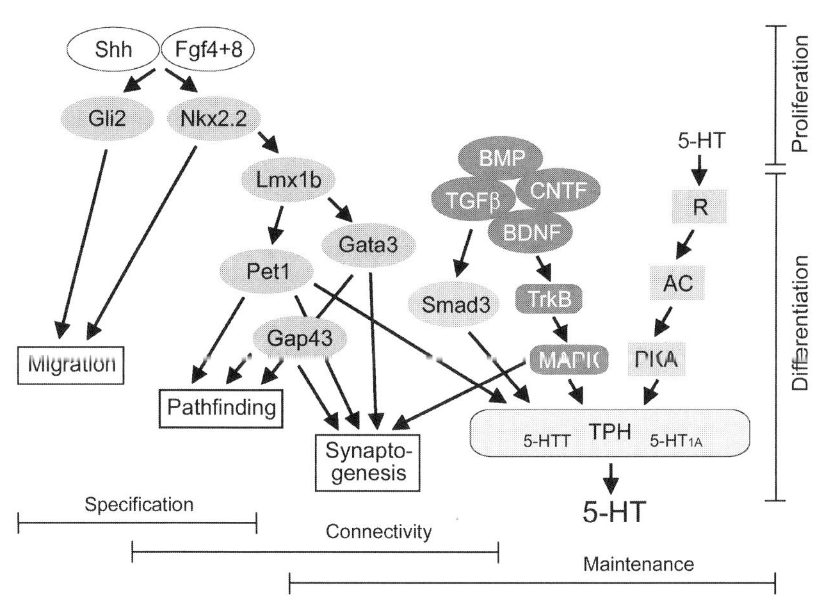

Fig. 1 Genetic pathways in the development of the raphe serotonin (*5-HT*) system (modified from Lesch 2001b). *Shh* and *Fgf4+8* are intrinsic signaling molecules, *Nkx2.2*, *Gli2*, *Lmx1b*, *Pet1*, and *Gata3* are transcription factors. *Gap43*, *Smad3*, *BMP*, *TGFβ*, *CNTF*, and *BDNF* are neurotrophins and other growth factors. *TrkB*, neurotrophin receptor; *MAPK*, MAP kinase; *AC*, adenylyl cyclase; *PKA*, protein kinase A; *TPH*, tryptophan hydroxylase 2; *R*, receptor

the primitive streak and isthmus region, respectively, have been suggested to participate in the formation of an induction and organizing center that specifies the location and identity of rostral 5-HT neurons. The concerted action of Shh with other as-yet-unidentified signaling molecules, but not Fgf8, induces the generation of the caudal cluster of 5-HT neurons.

In addition, the genetic cascade of several transcription factors participates in the development of 5-HT neurons. The homeobox gene *Nkx2.2* and the zinc-finger transcription factor Gli2 are two downstream targets of Shh that operate during early stages of neurogenesis at the boundary between the midbrain and hindbrain (Brodski et al. 2003). KO of *Nkx2.2* results in the absence of some serotonergic neurons in the hindbrain (Briscoe et al. 1999), whereas elimination of Gli2 results in a partial loss and abnormal location of remaining 5-HT neurons in the ventral midline (Matise et al. 1998). Even within the relatively circumscribed serotonergic raphe complex, gene expression in discrete subsystems appears to be differentially controlled by transcriptional regulators. The transcription factor Gata3 is expressed broadly during embryogenesis, including in many but not all 5-HT raphe neurons (van Doorninck et al. 1999). Gata3 seems to play a critical role in the development of serotonergic neurons of the caudal raphe nuclei and thus in locomotor performance (Matise et al. 1998).

Following definition of neuronal precursors' position, several other transcription factors turn out to be instrumental in establishing the serotonergic phenotype, reflected by the expression of genes representing the synthetic and metabolic machinery for 5-HT (e.g., TPH, MAO), receptor-mediated signaling (e.g., 5-HT_{1A} receptor), or uptake-facilitated clearance (e.g., 5-HTT). Transcription factors that are expressed in by now postmitotic cells and that induce expression of these 5-HT markers encompass the Lim homeodomain and ETS domain transcription factor, Lmx1b and Pet1, respectively (Ding et al. 2003; Hendricks et al. 1999). By coupling Nkx2.2-mediated early specification with Pet1-induced terminal differentiation, Lmx1b acts as a critical mediator and thus represents a major determinant in the gene expression cascade resulting in the phenotypic determination of all 5-HT neurons in the CNS.

3.5.2
Transcription Factor Pet1

While the transcription factors Nkx2.2 and Gli2 are also required for induction of floor plate and adjacent cells including both serotonergic and dopaminergic neurons throughout the midbrain, hindbrain, and spinal cord (van Doorninck et al. 1999), expression of Pet1 is restricted to the rostrocaudal extent of hindbrain raphe nuclei and closely associated with developing serotonergic neurons in the raphe nuclei (Hendricks et al. 1999; Pfaar et al. 2002) (Fig. 1). Pet1 is therefore likely to be distinct from other factors because its expression pattern suggests that it performs a strictly serotonergic-specific function in the brain. Moreover, consensus Pet1 binding motifs are present in the transcriptional regulatory regions of both the human and murine 5-HT$_{1A}$ receptor, 5-HTT, tryptophan hydroxylases (Tph1 and 2), and aromatic L-amino acid decarboxylase (Aaad) genes whose expression profile is characteristic of the serotonergic neuron phenotype, i.e., 5-HT synthesis, release, uptake, and metabolism.

In the rat dorsal and median raphe, 5-HT neurons begin to appear at approximately E11 and peak at E13–E14. In these nuclei, it is thought that serotonergic neuron precursors begin to produce 5-HT near the time of their last cell division. The detection of Pet1 as early as E12.5 in the rostral cluster suggests that it is expressed in 5-HT neuron precursors during their terminal differentiation, consistent with its expression before the appearance of 5-HT. Taken together, these findings identify Pet1 as a critical regulator of serotonergic system specification.

While nearly all serotonergic neurons fail to differentiate in mice lacking Pet1, the remaining exhibit deficient expression of genes required for 5-HT synthesis, uptake, and vesicular storage (Hendricks et al. 2003). In target fields including cortex and hippocampus, 5-HT-specific fibers as well as 5-HT and 5-HIAA concentrations were also dramatically reduced in *Pet1* KO mice, whereas no major cytoarchitectural abnormalities in nuclear groups of several brain regions were detected. Interestingly, Pet1-deficient mice show evidence for increased anxiety-like behavior in the elevated plus maze test and enhanced aggressiveness in the resident-intruder test as a consequence of disrupted 5-HT system development. These findings further support the notion that Pet1 may represent the terminal differentiation factor that establishes the final identity of 5-HT neurons. Finally, the Pet1-dependent transcriptional program appears to couple 5-HT neuron differentiation during brain development to serotonergic modulation of behavior related to anxiety and aggression in adulthood.

Beyond the point of transcription initiation and neurotrophin action, the role of messenger RNA elongation and other mechanisms of neural gene regulation are increasingly attracting systematic scrutiny. Foggy, a phosphorylation-dependent, dual regulator of transcript elongation, affects development of 5-HT-containing (and dopamine-containing) neurons in the zebrafish (Guo et al. 2000). In the fruit fly, the homeobox genes *engrailed* (*en*) and *islet*, the zinc finger transcription factor huckebein (hkb), and eagle (eg), which codes for an orphan nuclear receptor with homology to the steroid receptor family, are required for the specification of 5-HT neurons (Goridis and Brunet 1999; Lundell and Hirsh 1998). The human orthologs of *foggy* and *fagle* remain to be identified and their role in anxiety and emotionality to be determined.

3.5.3 Brain-Derived Neurotrophic Factor

An extended assembly of other neurotrophins and growth factors also modulates the phenotype of 5-HT neurons. These include family members of the neurotrophins, such as transforming growth factor-β (TGF-β), bone morphogenetic protein (BMP), and neurokines (ciliary neurotrophic factor, CNTF) (Galter and Unsicker 2000a,b). 5-HT itself regulates the serotonergic phenotype of neurons by sequential activation of the 5-HT_{1A} receptor, brain-derived neurotrophic factor (BDNF), and its receptor TrkB, as well as a wide spectrum of signal transduction pathways. In particular, transcriptional regulation appears to be dependent on stimulation of the adenylyl cyclase/protein kinase A signaling pathway mediated by a family of cyclic AMP (cAMP)-responsive nuclear factors, including CREB, CREM, and ATF-1 (Herdegen and Leah 1998). These factors contain the basic domain/leucine zipper motifs and bind as dimers to cAMP-responsive elements (CREs). Galter and Unsicker (2000a,b) have therefore proposed the neurotrophin receptor TrkB as the master control protein that integrates a diverse array of signals that elicit and maintain serotonergic differentiation and survival.

BDNF is involved in a variety of trophic and modulatory effects that include a critical role in the development and plasticity of dopaminergic, serotonergic, and other neurons (Bonhoeffer 1996; Schuman 1999). Specifically, BDNF enhances differentiation of 5-HT neurons during embryonic development and prevents neurotoxin-induced serotonergic denervation in adult brain (Frechilla et al. 2000; Galter and Unsicker 2000a,b). Furthermore, human fetal mesencephalic cultured cells treated with BDNF exhibit greater neuronal survival and increased tissue 5-HT concentrations (Spenger et al. 1995). BDNF treatment of E14 rat embryos induced a twofold increase in the number of raphe 5-HT neurons and produced a marked extension and ramification of their neurites with greater expression of 5-HTT, 5-HT_{1A}, and 5-HT_{1B} receptors (Galter and Unsicker 2000a; Zhou and Iacovitti 2000). Reduced expression of BDNF modifies synaptic plasticity resulting in specific alterations in spatial learning and memory processes, emotionality, and motor activity in KO mice (Carter et al. 2002; Kernie et al. 2000; Minichiello et al. 1999), whereas targeted inactivation of the BDNF receptor, TrkB, leads to neuronal loss and cortical degenerative changes (Vitalis et al. 2002). In addition, BDNF mediates the effects of repeated stress exposure and long-term antidepressant treatment on neurogenesis and neuronal survival in the hippocampus (D'Sa and Duman 2002). These findings converge with reduced hippocampal plasticity, as reflected by a reduced hippocampal volume; and hippocampus-related memory deficiency plays an critical role in the pathophysiology of emotional and stress-related disorders (Duman 2002).

Mice completely lacking BDNF have reduced sensory neuron survival, other neuronal deficits, and are viable only a few weeks (Ernfors et al. 1995). Heterozygote $BDNF^{+/-}$ mice exhibit gene dose-dependent reductions in BDNF expression in forebrain, hippocampus, and some hypothalamic nuclei (Kernie et al. 2000; MacQueen et al. 2001) as well as decreased striatal dopamine content, decreased potassium-elicited dopamine release (Dluzen et al. 2002), and some evidence of decreased concentrations of forebrain 5-HT concentrations and fiber densities at 18 months of age (Lyons et al. 1999). Furthermore, learning deficits and hyperactivity was revealed in *BDNF*$^{+/-}$ mice (Kernie et al. 2000). They also develop intermale aggressiveness in the resident-intruder test (Lyons et al. 1999), but do not show increased anxiety in the elevated plus maze, nor differences in the antidepressant-sensitive forced swim test (MacQueen et al. 2001). However, conditional deletion of the BDNF gene in the postnatal brain leads to increased anxiety-like behavior in the light-dark box, deficits in context-dependent learning in a fear conditioning paradigm, and hyperactivity (Rios et al. 2001). Both conditional and constitutive *BDNF* KO mice also exhibit obesity, with hyperphagia, elevated serum glucose, insulin and leptin levels, and elevated cell fat content (Kernie et al. 2000; Rios et al. 2001).

Possible gene-interactive alterations in 5-HT function and BDNF expression was recently evaluated in mice with a combined manipulation of the genes for 5-HTT and BDNF. Male but not female *5-HTT*$^{-/-}$–*BDNF*$^{+/-}$ DKO mice showed further decreases in brain 5-HT concentrations as well as further increases in anxiety-like behavior and stress reactivity compared to *5-HTT*$^{-/-}$–*BDNF*$^{+/+}$ controls (Murphy et al. 2003). These findings support the notion of critical role of gene–gene interaction in brain plasticity related to anxiety and related disorders.

3.5.4 Neural Cell Adhesion Molecule

NCAM plays a critical role during brain development and in adult plasticity. In particular, NCAM is involved in neuronal migration, neurite outgrowth, synaptic plasticity, and emotional behavior (Schachner 1997). NCAM-deficient mice display both elevated anxiety and aggression levels (Stork et al. 1999). Although 5-HT$_{1A}$ binding as well as brain 5-HT and 5-HIAA tissue concentrations were unaltered, lower doses of 5-HT$_{1A}$ agonists are necessary to reduce anxiety and aggressiveness in the *NCAM*$^{-/-}$ mice, suggesting a functional change in the 5-HT$_{1A}$ receptor (Stork et al. 1999). Interestingly, the expression of one of the effectors of the 5-HT$_{1A}$ receptor, the G protein-activated inward rectifying potassium channel 2 (GIRK2) is greatly upregulated in *NCAM* KO mice, thus identifying disrupted 5-HT$_{1A}$-activated cellular pathways as an additional cause for their anxiety- and aggression-related behavior (Delling et al. 2002) (also see Sect. 3.4.3). Taken together, these findings indicate an involvement of NCAM impacting on 5-HT system function through the 5-HT$_{1A}$ receptors and its effector, but the specific molecular mechanisms in emotional behavior remain to be elucidated in more detail.

3.6 Other Serotonin-Related Systems

3.6.1 Tissue-Plasminogen Activator and Growth-Associated Protein 43

Adaptive responses to stressful events comprise physiological processes and behavior aimed at sustaining homeostasis, while severe stress may modify this response and lead to exaggerated fear reaction and persisting anxiety and depression. At the center of the functional neuroanatomy of the stress circuit are the amygdala and the hippocampus, which both exhibit dendritic remodeling following repeated inescapable stress. Although several key players of the stress circuit have been characterized, the mechanism that underlies stress-induced neural plasticity leading to anxiety and associated cognitive impairment remains to be elucidated. Recently, Pawlak and coworkers (2003) identified acute restraint stress-induced upregulation of the serine protease tissue-plasminogen activator (tPA) in the amygdala as a critical mechanism in stress-related neural remodeling that is either adaptive and directed toward attenuation of the deleterious impact of stress on the brain or is reflecting the interference with protective mechanisms. Targeted disruption of the tPA gene in mice resulted in attenuated anxiety-like behavior and maladaptive endocrine response as well as compromised neural plasticity. These findings suggest that tPA represents a signal to the postsynaptic machinery to phosphorylate extracellular signal-regulated kinase 1/2 (ERK1/2), a trigger for postsynaptic plasticity-related events. Furthermore, axonal remodeling reflected by decreases in expression of Gap43, a growth-associated protein expressed in growing 5-HT axons and thus a marker of presynaptic plasticity, in the amygdala but not in hippocampus. Interestingly, a gene dose-dependent failure of 5-HT axons to innervate selected forebrain regions including the somatosensory cortex and hippocampus but not the amygdala was revealed in *Gap43* KO mice, suggesting Gap43 as a key regulator in normal pathfinding and arborization of 5-HT axons during early brain development (Donovan et al. 2002; Maier et al. 1999; McIlvain et al. 2003). However, it remains to be elucidated whether Gap43 participates also in adult plasticity of the hippocampus or amygdala, and whether *Gap43* KO mice exhibit changes in anxiety-related or other emotional behaviors.

3.6.2 Neurokinin 1 Receptor

Although substance P (SP) and its receptor, neurokinin 1 receptor (NK1R), have been implicated in the control of mood, anxiety, and stress, the efficacy of NK1R antagonists as both antidepressants and anxiolytics has been matter of considerable debate (Lesch 2001a). Santarelli and associates (2001) have recently made a strong argument for a critical role of the SP/NK1R system the modulation of anxiety-related behaviors in mice. Targeted inactivation of the NK1R produced a phenotype that is associated with an increase of fear and anxiety in the elevated plus maze, novelty suppressed feeding, and maternal separation paradigms. Results derived from pharmacologic, immunohistochemical, autoradiographic, endocrine, and electrophysiological studies convincingly identify the 5-HT system as a important participant in anxiety-related responses to NK1R KO, while an association of the NK1R with noradrenergic neurons seems to mediate this behavioral phenotype. Thus, NK1R antagonists may exert their anxiolytic effect by modulating the activity of noradrenergic neurons, which in turn modulate serotonergic function.

4 Clinical Implications and Outlook

Mutant mice of genes controlling 5-HT-system development and plasticity, of genes establishing 5-HT neuron identity, and of genes modulating 5-HT receptor-mediated signal transduction and cellular pathways provide practical models to study how genomic variation in these genes modulate human emotional behavior (Lesch 2001a). Allelic variation in the expression of human genes and function of their respective protein products, which are determinants of the serotonergic neuron phenotype, have been implicated in anxiety-related traits, such as $5\text{-}HT_{1A}$ and *5-HTT* (Reif and Lesch 2003; Rothe et al. 2004; Strobel et al. 2003), and aggressive behavior, such as $5\text{-}HT_{1B}$ and *MAOA* (Lappalainen et al. 1998; Lesch and Merschdorf 2000; Samochowiec et al. 1999) (Table 2). For instance, consistent with the finding that 5-HTT gene inactivation in mice leads to increased anxiety-like and reduced aggressive behavior, the allelic variation of the human *5-HTT* polymorphism is associated with increased trait anxiety/stress reactivity and neural responses to fear (Hariri et al. 2002; Lesch 1996, 2003). Moreover, genetically driven variation in 5-HTT function also modifies the risk for anxiety and depressive disorders and their response to treatment (Caspi et al. 2003; Collier et al. 1996; Lesch 2001b; Lesch and Mössner 1998). Another example is a repeat polymorphism in the promoter region of the human and nonhuman primate MAOA gene that differentially modulates gene transcription (Deckert et al. 1999; Newman et al. 2003). Allelic variation in MAOA gene expression and enzyme activity is not only associated with increased aggressiveness and violence but also with panic disorder. Complex traits such as anxiety, depression, and aggression are most likely to be generated by a complex interaction of environmental and experiential factors with a number of genes and their products, as has been documented extensively for *5-HTT* (Barr et al. 2003a,b; Bennett et al. 2002; Caspi et al. 2003; Champoux et al. 2002; Lesch et al. 2002). Even pivotal regulatory proteins of neurocircuits have only a modest impact, while noise from epigenetic mechanisms obstructs identification of relevant gene variants.

Table 2 Functional serotonergic gene variations associated with behavioral phenotype, psychopathology, and psychiatric disorders

Gene	Functional variation	Behavioral traits	Psychopathology/disorders
5-HT receptors			
1A	*C-1019G*	Anxiety, depression	Depression, suicidality, panic disorder
1B	*G861C* (in linkage disequilibrium with promoter haplotype: *T-261G*, *A-161T*, and *-182INS/DEL-181*)	Impulsivity, aggression	Alcoholism
2C	*Cys23Ser HTR2C-LPR*	Feeding, learning, and memory, anxiety?	Hallucinatory psychosis, eating disorders
5-HTT	*5HTT-LPR*	Anxiety, depression, stress reactivity, aggression	Depression/suicidality, alcoholism, OCD, autism, ADHD, eating disorders
MAOA	*MAOA-LPR*	Aggression, anxiety antisocial behavior	Alcoholism, panic disorder, antisocial personality disorder

ADHD, attention-deficit/hyperactivity disorder; OCD, obsessive–compulsive disorder.

Although current methods for the detection of gene–environment interaction in behavioral genetics are largely indirect, the most pertinent consequence of gene identification for behavioral traits may be that it will provide the tools required to systematically elucidate the effects of gene–environment interaction.

Finally, future benefits will stem from the potential development of strategies involving spatio-temporally specific conditional knockouts with and gene transfer technology that could facilitate novel drug design (Lesch 2001a). Paralleling the resolution of gene–gene and gene–environment interactions and as the dogma fades that neurons are highly vulnerable and their capacity for regeneration, reproducibility, and plasticity is limited, it is being realized that advanced gene transfer strategies may eventually be applicable to complex behavioral disorders.

References

Anisman H, Zaharia MD, Meaney MJ, Merali Z (1998) Do early-life events permanently alter behavioral and hormonal responses to stressors? Int J Dev Neurosci 16:149–164

Azmitia EC, Whitaker-Azmitia PM (1997) Development and adult plasticity of serotonergic neurons and their target cells. In: Baumgarten HG, Göthert M (eds) Serotonergic neurons and 5-HT receptors in the CNS. Springer, Berlin, Heidelberg, New York, pp 1–39

Barnes NM, Sharp T (1999) A review of central 5-HT receptors and their function. Neuropharmacology 38:1083–1152
Barr CS, Newman TK, Becker ML, Champoux M, Lesch KP, Suomi SJ, Goldman D, Higley JD (2003a) Serotonin transporter gene variation is associated with alcohol sensitivity in rhesus macaques exposed to early-life stress. Alcohol Clin Exp Res 27:812–817
Barr CS, Newman TK, Becker ML, Parker CC, Champoux M, Lesch KP, Goldman D, Suomi SJ, Higley JD (2003b) The utility of the non-human primate; model for studying gene by environment interactions in behavioral research. Genes Brain Behav 2:336–340
Bellivier F, Leboyer M, Courtet P, Buresi C, Beaufils B, Samolyk D, Allilaire JF, Feingold J, Mallet J, Malafosse A (1998) Association between the tryptophan hydroxylase gene and manic-depressive illness [see comments]. Arch Gen Psychiatry 55:33–37
Bengel D, Murphy DL, Andrews AM, Wichems CH, Feltner D, Heils A, Mossner R, Westphal H, Lesch KP (1998) Altered brain serotonin homeostasis and locomotor insensitivity to 3, 4-methylenedioxymethamphetamine ("Ecstasy") in serotonin transporter-deficient mice. Mol Pharmacol 53:649–655
Bennett AJ, Lesch KP, Heils A, Long JC, Lorenz JG, Shoaf SE, Champoux M, Suomi SJ, Linnoila MV, Higley JD (2002) Early experience and serotonin transporter gene variation interact to influence primate CNS function. Mol Psychiatry 7:118–122
Bennett-Clarke CA, Chiaia NL, Rhoades RW (1996) Thalamocortical afferents in rat transiently express high-affinity serotonin uptake sites. Brain Res 733:301–360
Blednov YA, Stoffel M, Chang SR, Harris RA (2001) GIRK2 deficient mice. Evidence for hyperactivity and reduced anxiety. Physiol Behav 74:109–117
Blednov YA, Stoffel M, Cooper R, Wallace D, Mane N, Harris RA (2002) Hyperactivity and dopamine D1 receptor activation in mice lacking girk2 channels. Psychopharmacology (Berl) 159:370–378
Blier P, de Montigny C (1998) Possible serotonergic mechanisms underlying the antidepressant and anti-obsessive-compulsive disorder responses. Biol Psychiatry 44:313–323
Bonasera SJ, Tecott LH (2000) Mouse models of serotonin receptor function: toward a genetic dissection of serotonin systems. Pharmacol Ther 88:133–142
Bonhoeffer T (1996) Neurotrophins and activity-dependent development of the neocortex. Curr Opin Neurobiol 6:119–126
Boularand S, Darmon MC, Mallet J (1995) The human tryptophan hydroxylase gene. An unusual splicing complexity in the 5'-untranslated region. J Biol Chem 270:3748–3756
Briscoe J, Sussel L, Serup P, Hartigan-O'Connor D, Jessell TM, Rubenstein JL, Ericson J (1999) Homeobox gene Nkx2.2 and specification of neuronal identity by graded Sonic hedgehog signalling. Nature 398:622–627
Brodski C, Weisenhorn DM, Signore M, Sillaber I, Oesterheld M, Broccoli V, Acampora D, Simeone A, Wurst W (2003) Location and size of dopaminergic and serotonergic cell populations are controlled by the position of the midbrain-hindbrain organizer. J Neurosci 23:4199–4207
Brunner D, Buhot MC, Hen R, Hofer M (1999) Anxiety, motor activation, and maternal-infant interactions in 5HT1B knockout mice. Behav Neurosci 113:587–601
Brunner HG, Nelen M, Breakefield XO, Ropers HH, van Oost BA (1993) Abnormal behavior associated with a point mutation in the structural gene for monoamine oxidase A. Science 262:578–580
Burnet H, Bevengut M, Chakri F, Bou-Flores C, Coulon P, Gaytan S, Pasaro R, Hilaire G (2001) Altered respiratory activity and respiratory regulations in adult monoamine oxidase A-deficient mice. J Neurosci 21:5212–5221

Caldji C, Tannenbaum B, Sharma S, Francis D, Plotsky PM, Meaney MJ (1998) Maternal care during infancy regulates the development of neural systems mediating the expression of fearfulness in the rat. Proc Natl Acad Sci U S A 95:5335–5340

Carter AR, Chen C, Schwartz PM, Segal RA (2002) Brain-derived neurotrophic factor modulates cerebellar plasticity and synaptic ultrastructure. J Neurosci 22:1316–1327

Cases O, Seif I, Grimsby J, Gaspar P, Chen K, Pournin S, Muller U, Aguet M, Babinet C, Shih JC, et al (1995) Aggressive behavior and altered amounts of brain serotonin and norepinephrine in mice lacking MAOA [see comments]. Science 268:1763–1766

Cases O, Lebrand C, Giros B, Vitalis T, De Maeyer E, Caron MG, Price DJ, Gaspar P, Seif I (1998) Plasma membrane transporters of serotonin, dopamine, and norepinephrine mediate serotonin accumulation in atypical locations in the developing brain of monoamine oxidase A knock-outs. J Neurosci 18:6914–6927

Caspi A, Sugden K, Moffitt TE, Taylor A, Craig IW, Harrington H, McClay J, Mill J, Martin J, Braithwaite A, Poulton R (2003) Influence of life stress on depression: moderation by a polymorphism in the 5-HTT gene. Science 301:386–389

Champoux M, Bennett A, Shannon C, Higley JD, Lesch KP, Suomi SJ (2002) Serotonin transporter gene polymorphism, differential early rearing, and behavior in rhesus monkey neonates. Mol Psychiatry 7:1058–1063

Cheetham SC, Crompton MR, Katona CL, Horton RW (1990) Brain 5-HT1 binding sites in depressed suicides. Psychopharmacology (Berl) 102:544–548

Chen C, Rainnie DG, Greene RW, Tonegawa S (1994) Abnormal fear response and aggressive behavior in mutant mice deficient for alpha-calcium-calmodulin kinase II. Science 266:291–294

Chen JJ, Li Z, Pan H, Murphy DL, Tamir H, Koepsell H, Gershon MD (2001) Maintenance of serotonin in the intestinal mucosa and ganglia of mice that lack the high-affinity serotonin transporter: Abnormal intestinal motility and the expression of cation transporters. J Neurosci 21:6348–6361

Chiavegatto S, Nelson RJ (2003) Interaction of nitric oxide and serotonin in aggressive behavior. Horm Behav 44:233–241

Chiavegatto S, Dawson VL, Mamounas LA, Koliatsos VE, Dawson TM, Nelson RJ (2001) Brain serotonin dysfunction accounts for aggression in male mice lacking neuronal nitric oxide synthase. Proc Natl Acad Sci U S A 98:1277–1281

Collier DA, Stober G, Li T, Heils A, Catalano M, Di Bella D, Arranz MJ, Murray RM, Vallada HP, Bengel D, Muller CR, Roberts GW, Smeraldi E, Kirov G, Sham P, Lesch KP (1996) A novel functional polymorphism within the promoter of the serotonin transporter gene: possible role in susceptibility to affective disorders. Mol Psychiatry 1:453–460

Compan V, Zhou M, Grailhe R, Gazzara RA, Martin R, Gingrich J, Dumuis A, Brunner D, Bockaert J, Hen R (2004) Attenuated response to stress and novelty and hypersensitivity to seizures in 5-HT4 receptor knock-out mice. J Neurosci 24:412–419

Crawley JN (1999) Behavioral phenotyping of transgenic and knockout mice: experimental design and evaluation of general health, sensory functions, motor abilities, and specific behavioral tests. Brain Res 835:18–26

Crawley JN, Paylor R (1997) A proposed test battery and constellations of specific behavioral paradigms to investigate the behavioral phenotypes of transgenic and knockout mice. Horm Behav 31:197–211

D'Sa C, Duman RS (2002) Antidepressants and neuroplasticity. Bipolar Disord 4:183–194

De Vry J (1995) 5-HT1A receptor agonists: recent developments and controversial issues. Psychopharmacology (Berl) 121:1–26

Deckert J, Catalano M, Syagailo YV, Bosi M, Okladnova O, Di Bella D, Nothen MM, Maffei P, Franke P, Fritze J, Maier W, Propping P, Beckmann H, Bellodi L, Lesch KP (1999) Excess of high activity monoamine oxidase A gene promoter alleles in female patients with panic disorder. Hum Mol Genet 8:621–624

Delling M, Wischmeyer E, Dityatev A, Sytnyk V, Veh RW, Karschin A, Schachner M (2002) The neural cell adhesion molecule regulates cell-surface delivery of G-protein-activated inwardly rectifying potassium channels via lipid rafts. J Neurosci 22:7154–7164

Di Pino G, Mössner R, Lesch KP, Lauder JM, Persico AM (2004) Serotonin roles in neurodevelopment: more than just neural transmission. Curr Neuropharmacol 2:403–417

Ding YQ, Marklund U, Yuan W, Yin J, Wegman L, Ericson J, Deneris E, Johnson RL, Chen ZF (2003) Lmx1b is essential for the development of serotonergic neurons. Nat Neurosci 6:933–938

Dluzen DE, Anderson LI, McDermott JL, Kucera J, Walro JM (2002) Striatal dopamine output is compromised within +/– BDNF mice. Synapse 43:112–117

Donovan SL, Mamounas LA, Andrews AM, Blue ME, McCasland JS (2002) GAP-43 is critical for normal development of the serotonergic innervation in forebrain. J Neurosci 22:3543–3552

Drevets WC, Frank E, Price JC, Kupfer DJ, Holt D, Greer PJ, Huang Y, Gautier C, Mathis C (1999) PET imaging of serotonin 1A receptor binding in depression. Biol Psychiatry 46:1375–1387

Duman RS (2002) Synaptic plasticity and mood disorders. Mol Psychiatry 7 Suppl 1:S29–S34

Ernfors P, Kucera J, Lee KF, Loring J, Jaenisch R (1995) Studies on the physiological role of brain-derived neurotrophic factor and neurotrophin-3 in knockout mice. Int J Dev Biol 39:799–807

Fabre V, Rioux A, Lesch KP, Murphy DL, Lanfumey L, Hamon M, Marres MP (2000) Altered expression and coupling of the serotonin 5-HT1A and 5-HT1B receptors in knock-out mice lacking the 5-HT transporter. Eur J Neurosci 12:2299–2310

Finn DA, Rutledge-Gorman MT, Crabbe JC (2003) Genetic animal models of anxiety. Neurogenetics 4:109–135

Fiorica-Howells E, Hen R, Gingrich J, Li Z, Gershon MD (2002) 5-HT(2A) receptors: location and functional analysis in intestines of wild-type and 5-HT(2A) knockout mice. Am J Physiol Gastrointest Liver Physiol 282:G877–G893

Flugge G (1995) Dynamics of central nervous 5-HT1A-receptors under psychosocial stress. J Neurosci 15:7132–7140

Francis D, Diorio J, Liu D, Meaney MJ (1999) Nongenomic transmission across generations of maternal behavior and stress responses in the rat. Science 286:1155–1158

Francis DD, Szegda K, Campbell G, Martin WD, Insel TR (2003) Epigenetic sources of behavioral differences in mice. Nat Neurosci 6:445–446

Frechilla D, Insausti R, Ruiz-Golvano P, Garcia-Osta A, Rubio MP, Almendral JM, Del Rio J (2000) Implanted BDNF-producing fibroblasts prevent neurotoxin-induced serotonergic denervation in the rat striatum. Brain Res Mol Brain Res 76:306–314

Furlong RA, Ho L, Rubinsztein JS, Walsh C, Paykel ES, Rubinsztein DC (1998) No association of the tryptophan hydroxylase gene with bipolar affective disorder, unipolar affective disorder, or suicidal behaviour in major affective disorder. Am J Med Genet 81:245–247

Galter D, Unsicker K (2000a) Brain-derived neurotrophic factor and trkB are essential for cAMP-mediated induction of the serotonergic neuronal phenotype. J Neurosci Res 61:295–301

Galter D, Unsicker K (2000b) Sequential activation of the 5-HT1(A) serotonin receptor and TrkB induces the serotonergic neuronal phenotype. Mol Cell Neurosci 15:446–455

Gammie SC, Nelson RJ (1999) Maternal aggression is reduced in neuronal nitric oxide synthase-deficient mice. J Neurosci 19:8027–8035
Gaspar P, Cases O, Maroteaux L (2003) The developmental role of serotonin: news from mouse molecular genetics. Nat Rev Neurosci 4:1002–1012
Gobbi G, Murphy DL, Lesch K, Blier P (2001) Modifications of the serotonergic system in mice lacking serotonin transporters: an in vivo electrophysiological study. J Pharmacol Exp Ther 296:987–995
Goridis C, Brunet JF (1999) Transcriptional control of neurotransmitter phenotype. Curr Opin Neurobiol 9:47–53
Gorman JM, Kent JM, Sullivan GM, Coplan JD (2000) Neuroanatomical hypothesis of panic disorder, revised. Am J Psychiatry 157:493–505
Grailhe R, Waeber C, Dulawa SC, Hornung JP, Zhuang X, Brunner D, Geyer MA, Hen R (1999) Increased exploratory activity and altered response to LSD in mice lacking the 5-HT(5A) receptor. Neuron 22:581–591
Griebel G (1995) 5-Hydroxytryptamine-interacting drugs in animal models of anxiety disorders: more than 30 years of research. Pharmacol Ther 65:319–395
Griebel G, Belzung C, Perrault G, Sanger DJ (2000) Differences in anxiety-related behaviours and in sensitivity to diazepam in inbred and outbred strains of mice. Psychopharmacology (Berl) 148:164–170
Groenink L, Pattij T, De Jongh R, Van der Gugten J, Oosting RS, Dirks A, Olivier B (2003) 5-HT(1A) receptor knockout mice and mice overexpressing corticotropin-releasing hormone in models of anxiety. Eur J Pharmacol 463:185–197
Gross C, Zhuang X, Stark K, Ramboz S, Oosting R, Kirby L, Santarelli L, Beck S, Hen R (2002) Serotonin1A receptor acts during development to establish normal anxiety-like behaviour in the adult. Nature 416:396–400
Guo S, Yamaguchi Y, Schilbach S, Wada T, Lee J, Goddard A, French D, Handa H, Rosenthal A (2000) A regulator of transcriptional elongation controls vertebrate neuronal development. Nature 408:366–369
Hansson SR, Mezey E, Hoffman BJ (1998) Serotonin transporter messenger RNA in the developing rat brain: early expression in serotonergic neurons and transient expression in non-serotonergic neurons. Neuroscience 83:1185–1201
Hariri AR, Mattay VS, Tessitore A, Kolachana B, Fera F, Goldman D, Egan MF, Weinberger DR (2002) Serotonin transporter genetic variation and the response of the human amygdala. Science 297:400–403
Hedlund PB, Danielson PE, Thomas EA, Slanina K, Carson MJ, Sutcliffe JG (2003) No hypothermic response to serotonin in 5-HT7 receptor knockout mice. Proc Natl Acad Sci U S A 100:1375–1380
Heisler LK, Tecott LH (1999) Knockout corner—neurobehavioural consequences of a serotonin 5-HT(2C) receptor gene mutation. Int J Neuropsychopharmacol 2:67–69
Heisler LK, Chu HM, Brennan TJ, Danao JA, Bajwa P, Parsons LH, Tecott LH (1998) Elevated anxiety and antidepressant-like responses in serotonin 5-HT1A receptor mutant mice [see comments]. Proc Natl Acad Sci U S A 95:15049–15054
Hendricks T, Francis N, Fyodorov D, Deneris ES (1999) The ETS domain factor Pet-1 is an early and precise marker of central serotonin neurons and interacts with a conserved element in serotonergic genes. J Neurosci 19:10348–10356
Hendricks TJ, Fyodorov DV, Wegman LJ, Lelutiu NB, Pehek EA, Yamamoto B, Silver J, Weeber EJ, Sweatt JD, Deneris ES (2003) Pet-1 ETS gene plays a critical role in 5-HT neuron development and is required for normal anxiety-like and aggressive behavior. Neuron 37:233–247

Herdegen T, Leah JD (1998) Inducible and constitutive transcription factors in the mammalian nervous system: control of gene expression by Jun, Fos and Krox, and CREB/ATF proteins. Brain Res Brain Res Rev 28:370–490
Holmes A, Lit Q, Murphy DL, Gold E, Crawley JN (2003) Abnormal anxiety-related behavior in serotonin transporter null mutant mice: the influence of genetic background. Genes Brain Behav 2:365–380
Hoyer D, Martin G (1997) 5-HT receptor classification and nomenclature: towards a harmonization with the human genome. Neuropharmacology 36:419–428
Hynes M, Ye W, Wang K, Stone D, Murone M, Sauvage F, Rosenthal A (2000) The seven-transmembrane receptor smoothened cell-autonomously induces multiple ventral cell types. Nat Neurosci 3:41–46
Kalinichev M, Easterling KW, Holtzman SG (2002) Early neonatal experience of Long-Evans rats results in long-lasting changes in reactivity to a novel environment and morphine-induced sensitization and tolerance. Neuropsychopharmacology 27:518–533
Karschin C, Dissmann E, Stuhmer W, Karschin A (1996) IRK(1–3) and GIRK(1–4) inwardly rectifying K+ channel mRNAs are differentially expressed in the adult rat brain. J Neurosci 16:3559–3570
Kernie SG, Liebl DJ, Parada LF (2000) BDNF regulates eating behavior and locomotor activity in mice. Embo J 19:1290–1300
Korte SM, Meijer OC, de Kloet ER, Buwalda B, Keijser J, Sluyter F, van Oortmerssen G, Bohus B (1996) Enhanced 5-HT1A receptor expression in forebrain regions of aggressive house mice. Brain Res 736:338–343
Lappalainen J, Long JC, Eggert M, Ozaki N, Robin RW, Brown GL, Naukkarinen H, Virkkunen M, Linnoila M, Goldman D (1998) Linkage of antisocial alcoholism to the serotonin 5-HT1B receptor gene in 2 populations. Arch Gen Psychiatry 55:989–994
Lauder JM (1993) Neurotransmitters as growth regulatory signals: role of receptors and second messengers. Trends Neurosci 16:233–240
Lebrand C, Cases O, Adelbrecht C, Doye A, Alvarez C, Elmestikawy S, Seif I, Gaspar P (1996) Transient uptake and storage of serotonin in developing thalamic neurons. Neuron 17:823–835
Lemonde S, Turecki G, Bakish D, Du L, Hrdina PD, Bown CD, Sequeira A, Kushwaha N, Morris SJ, Basak A, Ou XM, Albert PR (2003) Impaired repression at a 5-hydroxytryptamine 1A receptor gene polymorphism associated with major depression and suicide. J Neurosci 23:8788–8799
Lesch KP (2001a) Mouse anxiety: the power of knockout. Pharmacogenomics J 1:187–192
Lesch KP (2001b) Variation of serotonergic gene expression: neurodevelopment and the complexity of response to psychopharmacologic drugs. Eur Neuropsychopharmacol 11:457–474
Lesch KP (2003) Neuroticism and serotonin: a developmental genetic perspective. In: Plomin R, DeFries J, Craig I, McGuffin P (eds) Behavioral genetics in the postgenomic era. American Psychiatric Press, Washington, pp 389–423
Lesch KP, Merschdorf U (2000) Impulsivity, aggression, and serotonin: a molecular psychobiological perspective. Behav Sci Law 18:581–604
Lesch KP, Mössner R (1998) Genetically driven variation in serotonin uptake: is there a link to affective spectrum, neurodevelopmental, and neurodegenerative disorders? Biol Psychiatry 44:179–192
Lesch KP, Mössner R (1999) 5-HT1A receptor inactivation: anxiety or depression as a murine experience. Int J Neuropsychopharmacol 2:327–331

Lesch KP, Murphy DL (2003) Molecular genetics of transporters for norepinephrine, dopamine, and serotonin in behavioral traits and complex diseases. In: Broeer S, Wagner CA (eds) Membrane transport diseases: molecular basis of inherited transport defects. Kluwer Academic/Plenum, New York, pp 349–364

Lesch KP, Disselkamp-Tietze J, Schmidtke A (1990a) 5-HT1A receptor function in depression: effect of chronic amitriptyline treatment. J Neural Transm Gen Sect 80:157–161

Lesch KP, Mayer S, Disselkamp-Tietze J, Hoh A, Wiesmann M, Osterheider M, Schulte HM (1990b) 5-HT1A receptor responsivity in unipolar depression. Evaluation of ipsapirone-induced ACTH and cortisol secretion in patients and controls. Biol Psychiatry 28:620–628

Lesch KP, Hoh A, Schulte HM, Osterheider M, Muller T (1991) Long-term fluoxetine treatment decreases 5-HT1A receptor responsivity in obsessive-compulsive disorder. Psychopharmacology (Berl) 105:415–420

Lesch KP, Wiesmann M, Hoh A, Muller T, Disselkamp-Tietze J, Osterheider M, Schulte HM (1992) 5-HT1A receptor-effector system responsivity in panic disorder. Psychopharmacology (Berl) 106:111–117

Lesch KP, Bengel D, Heils A, Sabol SZ, Greenberg BD, Petri S, Benjamin J, Müller CR, Hamer DH, Murphy DL (1996) Association of anxiety-related traits with a polymorphism in the serotonin transporter gene regulatory region. Science 274:1527–1531

Lesch KP, Greenberg BD, Higley JD, Murphy DL (2002) Serotonin transporter, personality, and behavior: toward dissection of gene-gene and gene-environment interaction. In: Benjamin J, Ebstein R, Belmaker RH (eds) Molecular genetics and the human personality. American Psychiatric Press, Washington, pp 109–135

Li Q, Wichems C, Heils A, Van De Kar LD, Lesch KP, Murphy DL (1999) Reduction of 5-hydroxytryptamine (5-HT)(1A)-mediated temperature and neuroendocrine responses and 5-HT(1A) binding sites in 5-HT transporter knockout mice. J Pharmacol Exp Ther 291:999–1007

Li Q, Wichems C, Heils A, Lesch KP, Murphy DL (2000) Reduction in the density and expression, but not G-protein coupling, of serotonin receptors (5-HT1A) in 5-HT transporter knock-out mice: gender and brain region differences. J Neurosci 20:7888–7895

Liu D, Diorio J, Day JC, Francis DD, Meaney MJ (2000) Maternal care, hippocampal synaptogenesis and cognitive development in rats. Nat Neurosci 3:799–806

Lopez JF, Chalmers DT, Little KY, Watson SJ (1998) A.E. Bennett Research Award. Regulation of serotonin1A, glucocorticoid, and mineralocorticoid receptor in rat and human hippocampus: implications for the neurobiology of depression. Biol Psychiatry 43:547–573

Lundell MJ, Hirsh J (1998) eagle is required for the specification of serotonin neurons and other neuroblast 7–3 progeny in the Drosophila CNS. Development 125:463–472

Luscher C, Jan LY, Stoffel M, Malenka RC, Nicoll RA (1997) G protein-coupled inwardly rectifying K+ channels (GIRKs) mediate postsynaptic but not presynaptic transmitter actions in hippocampal neurons. Neuron 19:687–695

Lyons WE, Mamounas LA, Ricaurte GA, Coppola V, Reid SW, Bora SH, Wihler C, Koliatsos VE, Tessarollo L (1999) Brain-derived neurotrophic factor-deficient mice develop aggressiveness and hyperphagia in conjunction with brain serotonergic abnormalities. Proc Natl Acad Sci U S A 96:15239–15244

MacQueen GM, Ramakrishnan K, Croll SD, Siuciak JA, Yu G, Young LT, Fahnestock M (2001) Performance of heterozygous brain-derived neurotrophic factor knockout mice on behavioral analogues of anxiety, nociception, and depression. Behav Neurosci 115:1145–1153

Maier DL, Mani S, Donovan SL, Soppet D, Tessarollo L, McCasland JS, Meiri KF (1999) Disrupted cortical map and absence of cortical barrels in growth-associated protein (GAP)-43 knockout mice. Proc Natl Acad Sci U S A 96:9397–9402

Mansour-Robaey S, Mechawar N, Radja F, Beaulieu C, Descarries L (1998) Quantified distribution of serotonin transporter and receptors during the postnatal development of the rat barrel field cortex. Brain Res Dev Brain Res 107:159–163

Manuck SB, Flory JD, Ferrell RE, Dent KM, Mann JJ, Muldoon MF (1999) Aggression and anger-related traits associated with a polymorphism of the tryptophan hydroxylase gene. Biol Psychiatry 45:603–614

Matise MP, Epstein DJ, Park HL, Platt KA, Joyner AL (1998) Gli2 is required for induction of floor plate and adjacent cells, but not most ventral neurons in the mouse central nervous system. Development 125:2759–2770

Mayorga AJ, Dalvi A, Page ME, Zimov-Levinson S, Hen R, Lucki I (2001) Antidepressant-like behavioral effects in 5-hydroxytryptamine(1A) and 5-hydroxytryptamine(1B) receptor mutant mice. J Pharmacol Exp Ther 298:1101–1107

McIlvain VA, Robertson DR, Maimone MM, McCasland JS (2003) Abnormal thalamocortical pathfinding and terminal arbors lead to enlarged barrels in neonatal GAP-43 heterozygous mice. J Comp Neurol 462:252–264

Minichiello L, Korte M, Wolfer D, Kuhn R, Unsicker K, Cestari V, Rossi-Arnaud C, Lipp HP, Bonhoeffer T, Klein R (1999) Essential role for TrkB receptors in hippocampus-mediated learning. Neuron 24:401–414

Murphy DL, Uhl GR, Holmes A, Ren-Patterson R, Hall FS, Sora I, Detera-Wadleigh S, Lesch KP (2003) Experimental gene interaction studies with SERT mutant mice as models for human polygenic and epistatic traits and disorders. Genes Brain Behav 2:350–364

Nebigil CG, Hickel P, Messaddeq N, Vonesch JL, Douchet MP, Monassier L, Gyorgy K, Matz R, Andriantsitohaina R, Manivet P, Launay JM, Maroteaux L (2001) Ablation of serotonin 5-HT(2B) receptors in mice leads to abnormal cardiac structure and function. Circulation 103:2973–2979

New AS, Gelernter J, Yovell Y, Trestman RL, Nielsen DA, Silverman J, Mitropoulou V, Siever LJ (1998) Tryptophan hydroxylase genotype is associated with impulsive-aggression measures: a preliminary study. Am J Med Genet 81:13–17

Newman TK, Syagailo Y, Barr CS, Wendland J, Champoux M, Graessle M, Suomi SJ, D HJ, P LK (2003) Monoamine oxidase A gene promoter polymorphism and infant rearing experience interact to influence aggression and injuries in rhesus monkeys. Biol Psychiatry 57:167–172

Nielsen DA, Goldman D, Virkkunen M, Tokola R, Rawlings R, Linnoila M (1994) Suicidality and 5-hydroxyindoleacetic acid concentration associated with a tryptophan hydroxylase polymorphism. Arch Gen Psychiatry 51:34–38

Nielsen DA, Virkkunen M, Lappalainen J, Eggert M, Brown GL, Long JC, Goldman D, Linnoila M (1998) A tryptophan hydroxylase gene marker for suicidality and alcoholism. Arch Gen Psychiatry 55:593–602

Ninan PT (1999) The functional anatomy, neurochemistry, and pharmacology of anxiety. J Clin Psychiatry 60 Suppl 22:12–17

Olivier B, Miczek KA (1999) Fear and anxiety: mechanisms, models and molecules. In: Dodman N, Shuster I (eds) Psychopharmacology of animal behavior disorders. Blackwell, London, pp 105–121

Olivier B, Soudijn W, van Wijngaarden I (1999) The 5-HT1A receptor and its ligands: structure and function. Prog Drug Res 52:103–165

Olivier B, Pattij T, Wood SJ, Oosting R, Sarnyai Z, Toth M (2001) The 5-HT(1A) receptor knockout mouse and anxiety. Behav Pharmacol 12:439–450

Parks CL, Robinson PS, Sibille E, Shenk T, Toth M (1998) Increased anxiety of mice lacking the serotonin1A receptor. Proc Natl Acad Sci U S A 95:10734–10739
Parsons LH, Kerr TM, Tecott LH (2001) 5-HT(1A) receptor mutant mice exhibit enhanced tonic, stress-induced and fluoxetine-induced serotonergic neurotransmission. J Neurochem 77:607–617
Pawlak R, Magarinos AM, Melchor J, McEwen B, Strickland S (2003) Tissue plasminogen activator in the amygdala is critical for stress-induced anxiety-like behavior. Nat Neurosci 6:168–174
Persico AM, Revay RS, Mössner R, Conciatori M, Marino R, Baldi A, Cabib S, Pascucci T, Sora I, Uhl GR, Murphy DL, Lesch KP, Keller F (2001) Barrel pattern formation in somatosensory cortical layer IV requires serotonin uptake by thalamocortical endings, while vesicular monoamine release is necessary for development of supragranular layers. J Neurosci 21:6862–6873
Persico AM, Baldi A, Dell'Acqua ML, Moessner R, Murphy DL, Lesch KP, Keller F (2003) Reduced programmed cell death in brains of serotonin transporter knockout mice. Neuroreport 14:341–344
Pfaar H, von Holst A, Vogt Weisenhorn DM, Brodski C, Guimera J, Wurst W (2002) mPet-1, a mouse ETS-domain transcription factor, is expressed in central serotonergic neurons. Dev Genes Evol 212:43–46
Ramboz S, Oosting R, Amara DA, Kung HF, Blier P, Mendelsohn M, Mann JJ, Brunner D, Hen R (1998) Serotonin receptor 1A knockout: an animal model of anxiety-related disorder. Proc Natl Acad Sci U S A 95:14476–14481
Rebsam A, Seif I, Gaspar P (2002) Refinement of thalamocortical arbors and emergence of barrel domains in the primary somatosensory cortex: a study of normal and monoamine oxidase a knock-out mice. J Neurosci 22:8541–8552
Reif A, Lesch KP (2003) Toward a molecular architecture of personality. Behav Brain Res 139:1–20
Richer M, Hen R, Blier P (2002) Modification of serotonin neuron properties in mice lacking 5-HT1A receptors. Eur J Pharmacol 435:195–203
Rios M, Fan G, Fekete C, Kelly J, Bates B, Kuehn R, Lechan RM, Jaenisch R (2001) Conditional deletion of brain-derived neurotrophic factor in the postnatal brain leads to obesity and hyperactivity. Mol Endocrinol 15:1748–1757
Rothe C, Gutknecht L, Freitag CM, Tauber R, Franke P, Fritze J, Wagner G, Peikert G, Wenda B, Sand P, Jacob C, Rietschel M, Nöthen MM, Garritsen H, Fimmers R, Deckert J, Lesch KP (2004) Association of a functional 1019C>G 5-HT1A receptor gene polymorphism with panic disorder with agoraphobia. Int J Neuropsychopharmacol 7:189–192
Salichon N, Gaspar P, Upton AL, Picaud S, Hanoun N, Hamon M, De Maeyer EE, Murphy DL, Mossner R, Lesch KP, Hen R, Seif I (2001) Excessive activation of serotonin (5-HT) 1B receptors disrupts the formation of sensory maps in monoamine oxidase A and 5-HT transporter knock-out mice. J Neurosci 21:884–896
Samochowiec J, Lesch KP, Rottmann M, Smolka M, Syagailo YV, Okladnova O, Rommelspacher H, Winterer G, Schmidt LG, Sander T (1999) Association of a regulatory polymorphism in the promoter region of the monoamine oxidase A gene with antisocial alcoholism. Psychiatry Res 86:67–72
Santarelli L, Gobbi G, Debs PC, Sibille EL, Blier P, Hen R, Heath MJS (2001) Genetic and pharmacological disruption of neurokinin 1 receptor function decreases anxiety-related behaviors and increases serotonergic function. Proc Natl Acad Sci U S A 98:1912–1917

Sargent PA, Kjaer KH, Bench CJ, Rabiner EA, Messa C, Meyer J, Gunn RN, Grasby PM, Cowen PJ (2000) Brain serotonin1A receptor binding measured by positron emission tomography with [11C]WAY-100635: effects of depression and antidepressant treatment. Arch Gen Psychiatry 57:174–180

Saudou F, Amara DA, Dierich A, LeMeur M, Ramboz S, Segu L, Buhot MC, Hen R (1994) Enhanced aggressive behavior in mice lacking 5-HT1B receptor. Science 265:1875–1878

Schachner M (1997) Neural recognition molecules and synaptic plasticity. Curr Opin Cell Biol 9:627–634

Schaefer ML, Wong ST, Wozniak DF, Muglia LM, Liauw JA, Zhuo M, Nardi A, Hartman RE, Vogt SK, Luedke CE, Storm DR, Muglia LJ (2000) Altered stress-induced anxiety in adenylyl cyclase type VIII-deficient mice. J Neurosci 20:4809–4820

Schuman EM (1999) Neurotrophin regulation of synaptic transmission. Curr Opin Neurobiol 9:105–109

Seif I, De Maeyer E (1999) Knockout corner—knockout mice for monoamine oxidase A. Int J Neuropsychopharmacol 2:241–243

Sibille E, Pavlides C, Benke D, Toth M (2000) Genetic inactivation of the Serotonin(1A) receptor in mice results in downregulation of major GABA(A) receptor alpha subunits, reduction of GABA(A) receptor binding, and benzodiazepine-resistant anxiety. J Neurosci 20:2758–2765

Sora I, Wichems C, Takahashi N, Li XF, Zeng Z, Revay R, Lesch KP, Murphy DL, Uhl GR (1998) Cocaine reward models: conditioned place preference can be established in dopamine- and in serotonin-transporter knockout mice. Proc Natl Acad Sci U S A 95:7699–7704

Sora I, Hall FS, Andrews AM, Itokawa M, Li XF, Wei HB, Wichems C, Lesch KP, Murphy DL, Uhl GR (2001) Molecular mechanisms of cocaine reward: combined dopamine and serotonin transporter knockouts eliminate cocaine place preference. Proc Natl Acad Sci U S A 98:5300–5305

Spenger C, Hyman C, Studer L, Egli M, Evtouchenko L, Jackson C, Dahl-Jorgensen A, Lindsay RM, Seiler RW (1995) Effects of BDNF on dopaminergic, serotonergic, and GABAergic neurons in cultures of human fetal ventral mesencephalon. Exp Neurol 133:50–63

Stark KL, Hen R (1999) Knockout corner—5-HT1B receptor knockout mice: a review. Int J Neuropsychopharmacol 2:145–150

Stork O, Welzl H, Wotjak CT, Hoyer D, Delling M, Cremer H, Schachner M (1999) Anxiety and increased 5-HT1A receptor response in NCAM null mutant mice. J Neurobiol 40:343–355

Strobel A, Gutknecht L, Zheng Y, Reif A, Brocke B, Lesch KP (2003) Allelic variation of serotonin receptor 1A function is associated with anxiety- and depression-related traits. J Neural Transm 110:1445–1453

Toth M (2003) 5-HT(1A) receptor knockout mouse as a genetic model of anxiety. Eur J Pharmacol 463:177–184

Upton AL, Salichon N, Lebrand C, Ravary A, Blakely R, Seif I, Gaspar P (1999) Excess of serotonin (5-HT) alters the segregation of ipsilateral and contralateral retinal projections in monoamine oxidase A knock-out mice: possible role of 5-HT uptake in retinal ganglion cells during development. J Neurosci 19:7007–7024

Upton AL, Ravary A, Salichon N, Moessner R, Lesch KP, Hen R, Seif I, Gaspar P (2002) Lack of 5-HT(1B) receptor and of serotonin transporter have different effects on the segregation of retinal axons in the lateral geniculate nucleus compared to the superior colliculus. Neuroscience 111:597–610

van Doorninck JH, van Der Wees J, Karis A, Goedknegt E, Engel JD, Coesmans M, Rutteman M, Grosveld F, De Zeeuw CI (1999) GATA-3 is involved in the development of serotonergic neurons in the caudal raphe nuclei. J Neurosci 19:RC12

Vitalis T, Cases O, Gillies K, Hanoun N, Hamon M, Seif I, Gaspar P, Kind P, Price DJ (2002) Interactions between TrkB signaling and serotonin excess in the developing murine somatosensory cortex: a role in tangential and radial organization of thalamocortical axons. J Neurosci 22:4987–5000

Walther DJ, Peter JU, Bashammakh S, Hortnagl H, Voits M, Fink H, Bader M (2003) Synthesis of serotonin by a second tryptophan hydroxylase isoform. Science 299:76

Weiss SM, Lightowler S, Stanhope KJ, Kennett GA, Dourish CT (2000) Measurement of anxiety in transgenic mice. Rev Neurosci 11:59–74

Wischmeyer E, Doring F, Spauschus A, Thomzig A, Veh R, Karschin A (1997) Subunit interactions in the assembly of neuronal Kir3.0 inwardly rectifying K+ channels. Mol Cell Neurosci 9:194–206

Wisor JP, Wurts SW, Hall FS, Lesch KP, Murphy DL, Uhl GR, Edgar DM (2003) Altered rapid eye movement sleep timing in serotonin transporter knockout mice. Neuroreport 14:233–238

Wissink S, Meijer O, Pearce D, van Der Burg B, van Der Saag PT (2000) Regulation of the rat serotonin-1A receptor gene by corticosteroids. J Biol Chem 275:1321–1326

Ye W, Shimamura K, Rubenstein JL, Hynes MA, Rosenthal A (1998) FGF and Shh signals control dopaminergic and serotonergic cell fate in the anterior neural plate. Cell 93:755–766

Zhou J, Iacovitti L (2000) Mechanisms governing the differentiation of a serotonergic phenotype in culture. Brain Res 877:37–46

Zhuang X, Gross C, Santarelli L, Compan V, Trillat AC, Hen R (1999) Altered emotional states in knockout mice lacking 5-HT1A or 5-HT1B receptors. Neuropsychopharmacology 21:52S–60S

Zubenko GS, Hughes HB, Stiffler JS, Brechbiel A, Zubenko WN, Maher BS, Marazita ML (2003) Sequence variations in CREB1 cosegregate with depressive disorders in women. Mol Psychiatry 8:611–618

HEP (2005) 169:113–141

Mutagenesis and Knockout Models: Hypothalamic–Pituitary–Adrenocortical System

M. E. Keck (✉) · M. B. Müller

Max Planck Institute of Psychiatry, Kraepelinstrasse 2-10, 80804 Munich, Germany
keck@mpipsykl.mpg.de

Abstract Hyperactivity of central neuropeptidergic circuits such as the corticotropin-releasing hormone (CRH) and vasopressin (AVP) neuronal systems is thought to play a causal role in the etiology and symptomatology of anxiety disorders. Indeed, there is increasing evidence from basic science that chronic stress-induced perturbation of CRH and AVP neurocircuitries may contribute to abnormal neuronal communication in conditions of pathological anxiety. Anxiety disorders aggregate in families, and accumulating evidence supports the notion that the major source of familial risk is genetic. In this context, refined molecular technologies and the creation of genetically engineered mice have allowed us to specifically target individual genes involved in the regulation of the elements of the CRH (e.g., CRH peptides, CRH-related peptides, their receptors, binding protein). During the past few years, studies performed in such mice have complemented and extended our knowledge. The cumulative evidence makes a strong case implicating dysfunction of CRH-related systems in the pathogenesis of anxiety disorders and depression and leads us beyond the monoaminergic synapse in search of eagerly anticipated strategies to discover and develop better therapies.

Keywords CRH · CRF · Depression · Anxiety · CRH receptor antagonist · R121919 · NBI 30775 · CRH receptor type 1 · CRH receptor type 2 · Transgenic mice · Conditional knockout

1
Introduction—Understanding Endocrine-Behavior Interactions: Lessons from Mutant Mice

Anxiety disorders are common, and lifetime prevalence for the group of disorders is estimated to be as high as 25% (Kessler et al. 1994). These disorders display a substantial lifetime and episode comorbidity between each other and between other psychiatric conditions, particularly mood disorders (Hettema et al. 2001). With respect to the neuroendocrine phenotype, increased concentrations of corticotropin-releasing hormone (CRH) in the cerebrospinal fluid have been reported in stress-related clinical conditions (Holsboer 1999, 2003). In major depression, the combined dexamethasone (DEX)/CRH test, in which DEX-pretreated subjects receive a single dose of CRH, has proved to be the most sensitive tool for the detection of altered hypothalamic–pituitary–adrenocortical (HPA) regulation. Depending on age and gender, up to 90% of patients with depression show this neuroendocrine phenomenon (Heuser et al. 1994). Panic disorder patients do not show any relevant alterations in their basal pituitary–adrenocortical hormone secretion pattern (Holsboer 1999). Studies using the DEX/CRH test, however, support the hypothesis that HPA system functioning is altered in these patients and that this dysregulation is directly involved in the pathogenesis of the disorder (Schreiber et al. 1996). Accordingly, hyperactivity of central neuropeptidergic circuits such as the CRH and vasopressin (AVP) neuronal systems is thought to play a causal role in the etiology and symptomatology of anxiety disorders (Hökfelt et al. 2000; Holsboer 2000). Indeed, there is increasing evidence from basic science that chronic stress-induced perturbation of CRH and AVP neurocircuitries may contribute to abnormal neuronal communication in conditions of pathological anxiety (e.g., Antoni 1993; Griebel et al. 2002; Keck et al. 2002, 2003b; Müller et al. 2002). Anxiety disorders aggregate in families, and accumulating evidence supports the notion that the major source of familial risk is genetic (Hettema et al. 2001). In this context, genetically engineered mice with a specific deletion of targeted genes (e.g., "conventional" and "conditional" knockouts) provide a novel and useful tool to study the endogenous mechanisms underlying aberrant anxiety-related behavior. In recent years, refined molecular technologies have allowed the targeting of individual genes involved in HPA system regulation. The generation of "conventional" knockout mice allows for deleting a gene of interest in every cell of the body. Equally important for the studies of gene function in mice is the use of tissue-specific regulatory systems that allow gene inactivation to be restricted to specific tissues and, in some cases, to specific time points during development ("conditional" knockout). These gene-targeting methods have become valuable tools for dissecting the functions of individual components of complex biological systems (e.g., Müller and Keck 2002).

2 The Stress Hormone (Hypothalamic–Pituitary–Adrenocortical) System

Every disturbance of the body, either real or imagined, either physical or psychological, evokes a stress response. This stress response involves a large number of mechanisms and processes that, altogether, serve to restrain the body's defense reactions to stress, so as to restore homeostasis and to facilitate adaptation. CRH is the primary hypothalamic hypophysiotropic factor that regulates both basal and stress-induced release of pituitary corticotropin (ACTH) and is the major constituent of the HPA system (Vale et al. 1981). At the pituitary level, the effects of CRH are amplified by AVP, which, after prolonged stress, is increasingly co-expressed and co-secreted from hypothalamic CRH neurons (Antoni et al. 1993; Keck et al. 2000, 2002). CRH triggers the immediate release of ACTH from the anterior pituitary, subsequently leading to release of glucocorticoid hormones (GC, cortisol in humans and corticosterone in rats and mice) from the adrenal cortex (Fig. 1).

GC, in turn, exert a very sensitive negative feedback on the HPA system at the level of the paraventricular nucleus of the hypothalamus (PVN) and the anterior pituitary, and also at the level of the hippocampus, which projects to the bed nucleus of the stria terminalis, the latter which sends off projections to the PVN. In concert with other components of the stress hormone system, the action of corticosterone displays two modes of operation (for review see De Kloet et al. 1998). In the first "proactive" mode, GC maintain basal activity of the HPA system and control the sensitivity or threshold of the system's response to stress. GC promote coordination of circadian events, such as the

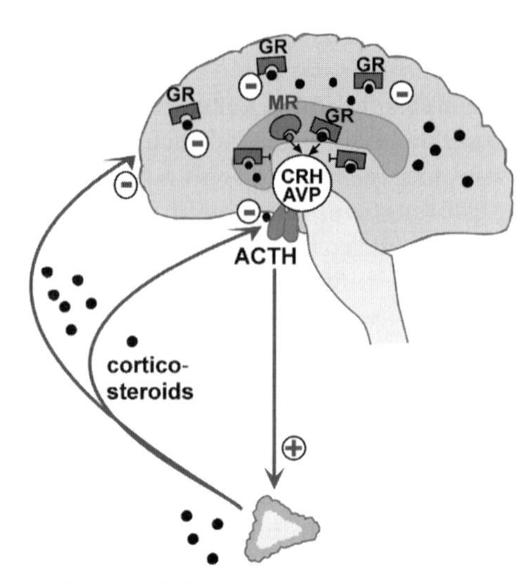

Fig. 1 The regulation of the hypothalamic–pituitary–adrenocortical (HPA) system under basal, physiological conditions. Hypothalamic corticotropin-releasing hormone (*CRH*) and vasopressin (*AVP*) trigger the release of corticotropin (*ACTH*) from the anterior pituitary. ACTH, in turn, stimulates secretion of glucocorticoid hormones (e.g., cortisol in humans, corticosterone in rats and mice) from the adrenal cortex. The increase in glucocorticoid levels suppresses hypothalamic CRH and AVP expression via negative feedback through hippocampal and hypothalamic corticosteroid receptors (glucocorticoid and mineralocorticoid receptors, *GR* and *MR*). CRH/AVP central nervous system pathways regulating neuroendocrine function as reflected by HPA measurements (i.e., plasma ACTH and cortisol) are independent from those circuitries modulating behavior

sleep/wake cycle and food intake, and are involved in processes underlying selective attention, integration of sensory information, and response selection. In the second "reactive" mode, GC feedback helps to terminate stress-induced HPA system activation. GC facilitate an animal's ability to cope with, adapt to, and recover from stress (Korte 2001).

2.1 Corticosteroid Effects Are Mediated via Two Receptor Subtypes

GC exert their regulatory effects on the HPA system via two types of corticosteroid receptors: the glucocorticoid receptor (GR) and the mineralocorticoid receptor (MR) (Reul and De Kloet 1985). GRs occur everywhere in the brain but are most abundant in hypothalamic CRH neurons and pituitary corticotropes. MRs, in contrast, are highly expressed in the hippocampus and, at lower expression levels, in hypothalamic sites involved in the regulation of salt appetite and autonomic outflow. The MR binds GC with a tenfold higher affinity than does the GR (Reul and De Kloet 1985). These findings on corticosteroid receptor diversity led to the working hypothesis that the tonic influences of corticosterone are exerted via hippocampal MRs, while the additional occupancy of GRs with higher levels of corticosterone mediates feedback actions aimed to restore disturbances in homeostasis. The progressive activation of MRs with a low concentration of corticosterone and additional activation when steroid levels rise can cause profound changes in neuronal integrity and neuronal function (Joels and De Kloet 1992) associated with changes in neuroendocrine regulation (Jacobson and Sapolsky 1991) and behavior (Oitzl and De Kloet 1992). Thus, the balance in MR- and GR-mediated effects exerted by corticosterone seems to be critical for homeostatic control (for review: De Kloet et al. 1998). Recently, a new mechanism of crosstalk between the CRH neuropeptidergic systems and hippocampal MRs was described: acute stressors act via a CRH receptor-mediated action to cause an elevation in MR levels in the hippocampus, which is associated with an augmented MR-mediated inhibition of HPA activity (Gesing et al. 2001). Thus, CRH receptors are involved in strengthening an important control instrument of the HPA system. It is still unclear, however, which CRH receptor subtype mediates this phenomenon.

Activation of GRs at the level of the PVN reduces CRH and AVP activity (Erkut et al. 1998). This negative feedback is a fundamental way in which the HPA system is restrained during stress and activity, and this restraint of HPA activation by glucocorticoids is rapid and profound. In contrast, induction of CRH expression by increasing glucocorticoid levels has been described to occur at the level of the central amygdala and the bed nucleus of the stria terminalis (BNST) (for review: Schulkin et al. 1998; Watts 1996). The latter is derived embryologically from the amygdala, and plays a fundamental role in the regulation of the HPA system during stress. This dual mechanism of glucocorticoid action on the central nervous system suggests that corticosterone appears capable of interacting with at least two different neuronal mechanisms to regulate CRH gene transcription. One is seen in paraventricular neurosecretory neurons, where increasing corticosteroid concentrations reduce CRH mRNA level. The other mechanism, seen in neurons in the central nucleus of the amygdala and the lateral BNST, acts to increase the CRH mRNA level. This "paradoxical" elevation of CRH gene expression by glucocorticoids in the central nucleus of the amygdala and lateral BNST may underlie a number of functional as well as pathological emotional states in which elevated circulating levels of glucocorticoids are accompanied by increased anxiety (for review: Holsboer 2000).

In summary, stress initially activates the hypothalamic CRH and AVP system, resulting in the hypersecretion of glucocorticoids from the adrenal gland. In addition, the psychological component of the stressor stimulates the amygdaloid CRH system. Chronic stressful life events could result in a loss of capacity of CRH or CRH-related peptides to upregulate hippocampal MR levels, leading to a loosening of the tonic inhibitory influence on parvocellular neurons in the PVN. Consequently, levels of CRH and AVP will increase in these neurons, providing an enhanced drive on HPA activity. Subsequently, the elevated circulating glucocorticoid levels will raise CRH expression in the central nucleus of the amygdala, resulting in an enhanced stimulatory influence on the PVN. In addition, in the chronic phase of stress, downregulation of GR in the PVN and other brain structures such as the locus coeruleus fails to restrain hyperfunction of the HPA system, and persistent activation of the HPA axis further upregulates the amygdaloid CRH system. In this manner, a feed-forward loop develops, accelerating the establishment of a state of sustained HPA hyperactivity (Reul and Holsboer 2002). Thus, the hypothalamic and the amygdaloid CRH systems cooperatively constitute stress-responsive, anxiety-producing neurocircuitry during chronic stress, which is responsible for the clinical manifestation of stress-associated disorders. Similarly, central AVP neurocircuits are well known to be upregulated under conditions of long-term activation of the HPA system, such as innate anxiety and chronic stress in rats (Keck et al. 2002, 2003b; de Goeij et al. 1992), human and rodent aging (Lucassen et al. 1993; Keck et al. 2000), and human depression (Purba et al. 1996).

2.2 The Dual Action of CRH: Activator of the HPA System and Neurotransmitter

Parvocellular neurons of the hypothalamic PVN are the major source of CRH within the central nervous system. These parvocellular neurons project via the external zone of the median eminence to the anterior pituitary where CRH is released into hypophyseal portal blood vessels to activate the HPA system by triggering ACTH release from pituitary corticotropes through activation of CRH 1 receptors (CRHR1). Furthermore, CRH acts as a neurotransmitter in several brain areas. High densities of CRH-like immunoreactivity have been observed throughout the neocortex (particularly in the prefrontal and cingulate cortices), the central nucleus of the amygdala (Van Bockstaele et al. 1998), the BNST, the hippocampus, the nucleus accumbens, some thalamic nuclei, substantia nigra, raphe nuclei, locus coeruleus, periaqueductal gray, and cerebellum (Swanson et al. 1983).

2.2.1 New Members of the Growing CRH Family

With the recent discovery of more endogenous ligands than CRH, the concept is dawning that CRH, its congeners, and their receptors form an intricate network in the brain that potentially provides a variety of targets for drug interventions (Reul and Holsboer 2002). So far, three other neuropeptides of the CRH family have been discovered, urocortin (UCN; Vaughan et al. 1995), urocortin II (UCN II; Reyes et al. 2001), and urocortin III (UCN III; Lewis et al. 2001) the latter of which is also called stresscopin (Hsu and Hsueh 2001). CRHR1 and CRHR2 differ in their ligand affinities for CRH, UCN, UCN II, and UCN III (CRHR1: UCN>CRH; CRHR2: UCN>UCN II>UCN III>>>CRH) (Chalmers et al. 1996; Donaldson et al. 1996; Lovenberg et al. 1995b; Reul and Holsboer 2002). Compared to CRH, UCN has an approximately 40-fold higher affinity for CRHR2 and a roughly sixfold higher affinity for CRHR1 (Vaughan et al. 1995). The CRH-related peptides share about 45% sequence homology with CRH. UCN has many of the effects of CRH, such as a high ACTH secretagogue potency (Asaba et al. 1998). Both UCN II (also termed stresscopin-related peptide) and UCN III (stresscopin) bind selectively to CRHR2, with no appreciable activity on CRHR1 (Reyes et al. 2001; Lewis et al. 2001; Hsu and Hsueh 2001). UCN has been shown to be widely expressed in the brain, with high expression levels in various neocortical areas and the Edinger–Westphal nucleus, and with moderate levels in the hippocampus, basal ganglia, medial septum, medial and cortical amygdaloid nuclei, the PVN and the ventromedial nucleus of the hypothalamus, the superior colliculus, substantia nigra, and cerebellum (Iino et al. 1999; Kozicz et al. 1998; Wong et al. 1998; Yamamoto et al. 1998). There seems to be only limited overlap between the distribution of CRH and UCN (Morin et al. 1999). UCN II mRNA displays a limited subcortical distribution in the rodent brain that is unique, although ostensibly overlapping in part with those of CRH (paraventricular nucleus) and UCN (brainstem and spinal motor nuclei). Of particular interest is the fact that UCN II is expressed in cell groups involved in stress-related physiological and behavioral functions. This includes the locus coeruleus, the hypothalamic paraventricular nucleus, and the arcuate nucleus (Reyes et al. 2001).

2.2.2
CRH Receptors and CRH Binding Protein

The biological actions of CRH, UCN, UCN II, and UCN III/stresscopin are mediated by specific, high-affinity, G protein-coupled membrane receptors. To date, two distinct receptor subtypes have been characterized: CRHR1 and CRHR2 display a markedly different tissue distribution and pharmacological specificity (Chalmers et al. 1995). CRHR1 has been proposed to mediate the effects of CRH on HPA system function and anxiety-related behavior (Liebsch et al. 1995, 1999; Skutella et al. 1998), whereas CRHR2 might be predominantly involved in the regulation of feeding behavior (Spina et al. 1996), cardiovascular function, and the recovery phase of the HPA response (Coste et al. 2000). The CRHR2 receptor family has additional diversity in that two isoforms have been described: CRHR2α and CRHR2β (Chalmers et al. 1995; Lovenberg et al. 1995a). The CRHR2α receptor is expressed primarily in subcortical neuronal populations, whereas the CRHR2β isoform is expressed in non-neuronal cells in the central nervous system (e.g., cerebral arterioles and choroid plexus). Peripherally, CRHR2β mRNA is found in heart, lung, and skeletal muscle (Chalmers et al. 1995; Lovenberg et al. 1995a). In humans only, a third splice variant, CRHR2γ, has been identified which is expressed in selected brain areas, such as the septum and hippocampus and at lower levels in the amygdala, nucleus accumbens, midbrain, and frontal cortex (Kostich et al. 1998).

Another important regulator of HPA system function is the CRH-binding protein (CRH-BP), a secreted glycoprotein which binds CRH and other CRH-related peptides with considerably high affinity (for review see Kemp et al. 1998). Neither UCN II nor UCN III bind to CRH-BP, whereas CRH and UCN do (Lewis et al. 2001). Binding of CRH/UCN by CRH-BP results in CRH/UCN inactivation and decreased ACTH release in vitro (Cortright et al. 1995). Besides pituitary corticotropes, the CRH-BP is predominantly expressed in the cerebral cortex, subcortical limbic structures, the raphe nuclei, and several brainstem nuclei (Potter et al. 1992). CRH-BP is present in both neurons and astrocytes, and its expression overlaps with CRH and CRH receptor expression in several areas. These areas of co-expression suggest that the CRH-BP exerts a modulatory effect on CRH–CRH receptor-interaction, which is corroborated by 40%–60% of CRH in the brain being bound to the CRH-BP (Behan et al. 1995b).

2.3 Anxiogenesis: Activation of CRHR1 and Dual Mode of Action of CRHR2

Numerous investigations in animals have described anxiogenic-like effects after CRH administration (Dunn and Berridge 1990). These effects are likely to be mediated through the CRH1 receptor, as CRHR1 antagonistic approaches have anxiolytic-like properties in most, but not all anxiety paradigms (e.g., Liebsch et al. 1995; Griebel et al. 1998; Keck et al. 2001a). The effectiveness of CRHR1 blockade to reduce anxiety is likely to depend on the animal's stress level, as it has been shown that CRHR1 antagonists acted anxiolytic-like after stress exposure, but not under basal conditions in "normal" rats (e.g., Griebel et al. 1998). However, in selectively bred rats with innate hyperanxiety and a hyperactive HPA system, administration of a CRHR1 antagonist exerted anxiolytic properties under basal conditions (Keck et al. 2001). On the other hand, studies looking at the effects of CRH-BP inhibitors, which will increase CRH activity specifically in cortical areas including the hippocampus, failed to find significant increases in anxiety-related behavior (Behan et al. 1995a,b), suggesting that CRH mediates anxiety at a subcortical level, e.g., within the lateral septum, the central amygdala and periaqueductal gray.

Beyond CRHR1, recent pharmacological experiments point towards a complex involvement of the CRHR2 in anxiety. Central administration of UCN, an endogenous ligand for CRHR2, has been shown to induce a variety of effects, including behavioral consequences such as increased anxiety (Moreau et al. 1997; Slawecki et al. 1999). However, as UCN can bind and activate both CRH receptor subtypes, i.c.v. administered UCN might activate receptors non-selectively in areas where endogenous UCN may not exist. Interestingly, activation of the CRHR2 can result in either anxiolysis or anxiogenesis depending on when the animal is tested and, possibly, where the receptor is localized (Takahashi 2001; Reul and Holsboer 2002). CRHR2 activation in the lateral septum increased anxiety-like behavior after 30 min, which could be prevented by pretreatment with the CRHR2 antagonist anti-sauvagine-30 (Radulovic et al. 1999). In contrast, i.c.v. administration of the selective CRHR2 agonist UCN II had no short-term effects, but after 4 hours resulted in reduced anxiety-related behavior (Valdez et al. 2002). Thus, CRHR2 in the brain is capable of reducing anxiety in a delayed fashion. The anxiogenic and anxiolytic properties of CRHR2 are certainly not paradoxical, because they operate in different time domains after stress.

Combining molecular genetics with behavioral pharmacology, however, studies with antisense probes that selectively reduce CRH receptor subtype levels and transgenic mouse models have indicated that CRHR1 might be the primary target of interest at which selective compounds should be directed to treat pathological anxiety (Keck and Holsboer 2001; Liebsch et al. 1995, 1998; Skutella et al. 1998; Timpl et al. 1998). One compound recently examined is R121919, a high-affinity non-peptide CRHR1 antagonist (Keck et al. 2001, 2003b; Lancel et al. 2002; Heinrichs et al. 2002; Gutman et al. 2003). In an open-label trial in patients suffering from major depression, a dose escalation strategy was employed which led to a 50% reduction in anxiety and depression scores comparable to that obtained with the selective serotonin reuptake inhibitor paroxetine (Keck and Holsboer 2001). Such effects were achieved at dosages that did not hamper the ACTH and cortisol response to CRH stimulation (Zobel et al. 2000; Künzel et al. 2003).

However, the possibility has to be kept in mind that, in addition to CRHR1 hyperfunction, CRHR2 hypofunction might play an important role in anxiety disorders and that an impaired CRHR2-mediated anxiolysis might result in an extended state of anxiety and arousal.

2.4
Vasopressin: Increasingly Recognized Importance

Beyond hyperactivity of central CRH neuropeptidergic circuits, AVP neuronal systems are thought to play a causal role in the etiology and symptomatology of anxiety disorders (Hökfelt et al. 2000; Keck and Holsboer 2001; Keck et al. 2002, 2003b). In support of this, AVP has been shown to exert both behavioral effects, such as increased anxiety following intracerebroventricular administration, and to increase CRH-induced ACTH secretion from pituitary corticotrope cells (Antoni 1993; Landgraf et al. 1998; Bhattacharya et al. 1998; Insel and Young 2000). After prolonged stress, AVP is increasingly expressed and released from hypothalamic neurons in both humans and rodents (e.g., Antoni 1993; Keck et al. 2000). In a clinical study, plasma AVP concentrations were found to be significantly correlated with anxiety-related symptoms in healthy volunteers in response to an anxiogenic drug challenge (Abelson et al. 2001). Moreover, administration of the non-peptide AVP 1b (V_{1b}) receptor antagonist SSR149415 was shown to display anxiolytic and antidepressant-like effects in rodents (Griebel et al. 2002). Similarly, the AVP 1a (V_{1a}) receptor, which is highly expressed in the rat lateral septum, thalamic nuclei, and the amygdalostriatal transition area (Barberis and Tribollet 1996), is well known to play a role in a variety of behaviors such as the modulation of emotionality and stress coping (review: Landgraf et al. 1998). Specifically, septal AVP has been shown to increase anxiety-related behavior in rats (Landgraf et al. 1995; Liebsch et al. 1996; Ebner et al. 1999). Accordingly, in rats displaying an innately increased anxiety-related behavior, V_{1a}-binding sites were higher in the lateral septum when compared to low-anxiety rats (Keck et al. 2003b). In these high-anxiety rats, elevated levels of intra-PVN AVP also were found and chronic administration of paroxetine, a clinically well-established antidepressant, normalized aberrant behavioral and neuroendocrine patterns in this psychopathological animal model (Keck et al. 2003b). Since it was recently demonstrated that a hypothalamic vasopressinergic hyperdrive accounts for the disturbance in HPA system regulation prevalent in these rats (Keck et al. 2002), the paroxetine-induced reduction of vasopressinergic overexpression indicates that this neuropeptidergic system may be critically involved in the action of antidepressant drugs known to be effective in the treatment of anxiety disorders (Keck et al. 2003b).

3 Gene Targeting: Promises and Caveats

In recent years, it has become possible to genetically alter the mouse genome with nucleotide precision (Müller 1999; Müller and Keck 2002). The modification of genetic information opens up the possibility to study the biological organization of an organism following such manipulation: transgenic mice have become an invaluable tool to dissect the functions of individual components of complex biological systems. Modifying genes that encode components of the HPA system allows investigation of how the organism organizes certain aspects of neuroendocrine and behavioral functions in response to this change (Nelson 1997). However, as with all techniques, there are some limitations to this approach. For example, the products of many genes are essential to normal function, and inactivating the gene may prove lethal or induce gross or even subtle morphological or physiological abnormalities that can complicate interpretation of discrete behavioral effects. Unexpected compensatory or redundancy mechanisms might be activated when a gene is missing, clouding interpretation of the normal contribution of the gene to behavior (Moran et al. 1996). Behavioral tests study the effects of the missing gene (and gene product), not the effects of the gene directly—a conceptual problem that is shared with all ablation studies. Moreover, deletion of individual genes is not providing animal models for certain behavioral pathologies, as these are caused by a manifold of minor changes in a series of so-called susceptibility genes.

Another important problem is the question of the genetic background in transgenic mouse lines. The issue of background variability and its impact on the analysis of mouse mutants has gained widespread attention (e.g., Lathe 1996) and recommendations concerning appropriate strain derivation have been proposed (Silva 1997). Both naturally occurring and targeted mutations can have diverse phenotypes when studied on different genetic backgrounds. The mixed genetic background of the vast majority of knockout mice may also affect the outcome of behavioral testing (Crabbe et al. 1999).

3.1
CRH Overexpression: Novel Insights into Old Problems

In 1992, a transgenic mouse line overexpressing CRH was developed (Stenzel-Poore et al. 1992). These animals exhibit prominent endocrine abnormalities involving the HPA system, such as high plasma levels of ACTH and corticosterone. CRH transgenic mice display physical changes similar to those of patients with Cushing's syndrome, such as excess fat accumulation, muscle atrophy, thin skin, and alopecia (Stenzel-Poore et al. 1992). Behavioral analysis revealed increased anxiety-related behavior when transgenic mice were tested in a light/dark box (Heinrichs et al. 1997) or on the elevated plus-maze. In the latter paradigm, increased anxiety-related behavior in transgenic animals could be reversed by administration of the non-selective CRH antagonist α-helical CRH (Stenzel-Poore et al. 1994). Interestingly, adrenalectomy did not attenuate the anxiogenic effect of CRH overproduction, although it normalized plasma corticosterone levels in these animals (Heinrichs et al. 1997), suggesting that the behavioral effects of CRH overexpression are mediated centrally and via the CRH receptor rather than being an effect of enhanced GR activation due to increased corticosterone levels. CRH overexpressing transgenic mice have been reported to be impaired in learning forced-choice alternation and water maze place navigation tasks. Interestingly, however, the place navigation deficit seen in transgenic mice was attenuated by administration of the potent anxiolytic benzodiazepine chlordiazepoxide. This in turn suggests that the navigation deficit seen in transgenic animals was possibly confounded by heightened anxiety or over-arousal (Heinrichs et al. 1996).

3.2
CRH Knockout

To further evaluate the role of CRH in both neuroendocrine and behavioral functions, a mammalian model of CRH deficiency has been generated by targeted mutation in embryonic stem cells (Muglia et al. 1995). CRH-deficient mice reveal a fetal glucocorticoid requirement for lung maturation. Postnatally, they display marked glucocorticoid deficiency and an impaired endocrine response to stress (Jacobson et al. 2000; Muglia et al. 1995).

Surprisingly, no gross behavioral abnormalities have been reported with these animals. Anxiety-related behavior is comparable between mutants and wildtype animals, regardless of whether it is measured under basal conditions or after stress exposure (Dunn and Swiergiel 1999; Weninger et al. 1999), despite the fact that glucocorticoid levels are greatly reduced, which might have been expected to influence anxiety-related and cognitive behavior (Korte 2001).

Interestingly, both the non-selective CRH antagonist α-helical CRH and the CRHR1 antagonist CP-154,526 showed comparable anxiolytic activities in mutant and wildtype mice when tested in fear conditioning (Weninger et al. 1999). This in turn suggests that blockade of CRHR1 is anxiolytic, but that another CRH-related peptide or yet-unidentified CRH receptor ligand could have compensated for the CRH deficiency in these animals.

3.3 Urocortin Knockout

Mice carrying a null mutation of the UCN gene were generated in 2002 (Vetter et al. 2002). UCN-deficient animals showed heightened anxiety-like behaviors in different behavioral testing paradigms while the HPA response to stress was found to be normal. The latter finding supports the view that endogenous UCN is not involved in regulation of the HPA system in response to acute stress, or has a minor or redundant role in such responses. A discrepancy observed between UCN-null mice and studies in which UCN was administered centrally concerns the role of UCN in anxiety. Central administration of UCN in rats can elicit anxiety-like behavior (Moreau et al. 1997; Slawecki et al. 1999) and it has been hypothesized that this could be due to activation of CRHR1. The discrepancy between the pharmacological studies and genetically engineered mice may be explained by considering that centrally administered UCN might non-selectively activate several receptor systems in the brain, resulting in the described anxiety-like behavior. Another explanation could be the fact that UCN knockouts have a reduction in levels of CRHR2 mRNA in the lateral septum (Vetter et al. 2002). Anatomical studies show that abundant UCN-expressing fibers originating from UCN-immunoreactive fibers in the Edinger–Westphal nucleus terminate in the lateral septum (Bittencourt et al. 1999). It has been demonstrated that a significant increase of UCN mRNA in the Edinger–Westphal nucleus and of CRHR2 mRNA in the lateral septum occurs in rats treated chronically with benzodiazepine anxiolytics (Skelton et al. 2000). Therefore, these data derived from UCN-null mutants and from pharmacological studies suggest that UCN-immunoreactive neurons in the Edinger–Westphal nucleus may modulate anxiety in opposition to the actions of CRH itself, and possibly through CRHR2 in the lateral septum.

3.4 CRH-Binding Protein

3.4.1 CRH-Binding Protein-Overexpressing Mice

Two different transgenic mouse mutants overexpressing CRH-BP have been independently generated (Burrows et al. 1998; Lovejoy et al. 1998). The mouse line by Lovejoy and colleagues shows increased CRH-BP level in brain and plasma. In contrast to wildtype mice, these animals also express CRH-BP ectopically in peripheral tissues including liver, kidneys, and spleen (Lovejoy et al. 1998). Basal plasma ACTH and corticosterone levels in these mutants are indistinguishable from those of wildtype littermates, but a significantly lower ACTH secretion was seen in male but not female transgenic mice following HPA system challenge with lipopolysaccharide (LPS). In contrast, no difference in corticosterone levels could be detected following LPS administration (Lovejoy et al. 1998). However, as pointed out by the authors, it should be noted that the expression profile for CRH-BP in these transgenic mice is very different from expression pattern in wildtype mice, which could lead to numerous unforeseen alterations. Unfortunately, behavioral data on this transgenic mouse line have not been reported.

A second CRH-BP transgenic mouse line with overexpression of CRH-BP has been published by Burrows and colleagues (1998). These animals express CRH-BP under the control of the pituitary glycoprotein hormone α-subunit (α-GSU) promoter, which is thought to limit transgene expression to the developing anterior pituitary, although occasional expression was also detected in additional brain regions such as the lateral septum. Given that excess CRH-BP will bind more CRH at the level of the anterior pituitary, these transgenic animals should suffer from attenuation of CRH receptor activation on pituitary corticotropes, what might be expected to lead to decreased activity of the HPA system. However, these transgenic animals have normal plasma ACTH and corticosterone levels under basal conditions and following restraint stress. Hypothalamic CRH and AVP expression are increased in the PVN, most likely reflecting potential compensatory mechanisms to maintain HPA system activity (Burrows et al. 1998).

Behavioral analyses revealed that the mice overexpressing CRH-BP exhibit increased locomotor activity in a novel environment (Burrows et al. 1998). In addition, a tendency towards decreased anxiety-related behavior was observed on the elevated plus maze, which would be in line with limited availability of free CRH due to enhanced binding by CRH-BP in these animals.

3.4.2 *CRHBP* Knockout

CRH-BP is a 37-kDa secreted glycoprotein that binds, as was stated earlier, CRH and other CRH-related peptides with high affinity (Kemp et al. 1998).

Binding of CRH by CRH-BP results in CRH inactivation and decreased ACTH release in vitro (Cortright et al. 1995). To investigate directly the CRH-BP function, a mouse model of CRH-BP deficiency has been created by gene targeting (Karolyi et al. 1999). Under basal conditions as well as following stress exposure, HPA axis function is normal in these animals. However, increased anxiety-like behavior was observed in an open field, on the elevated plus maze, and in the dark–light test (see also chapter by Ohl, this volume), consistent with the possibility that lack of CRH-BP would increase free CRH and UCN, but this was not directly demonstrated. However, the data suggest that the increased anxiety-like behavior in CRH-overexpressing transgenic mice (Heinrichs et al. 1997; Stenzel-Poore et al. 1994) is likely due to central effects, as CRH-BP knockout mice showed altered anxiety-related behavior despite unaltered HPA axis activity.

3.5 *CRHR1* Knockout

3.5.1 Conventional *CRHR1* Knockout: Decreased Anxiety-Related Behavior and Alcohol Problems

To investigate the physiological role of CRHR1 in both anxiety-related behavior and HPA system regulation, two mouse lines deficient for *CRHR1* have been independently generated (Smith et al. 1998; Timpl et al. 1998). Their phenotype confirms the obligatory role of CRHR1 in both the stress-associated response of the HPA system and anxiety: in particular, homozygous *CRHR1* mutants display a severe impairment of stress-induced HPA system activation and marked glucocorticoid deficiency. In addition, homozygous mutants exhibit increased exploratory activity and significantly reduced anxiety-related behavior under both basal conditions and following alcohol withdrawal (Timpl et al. 1998).

There is a relation between stress, anxiety disorders, and alcohol drinking. Stressful life events and maladaptive responses to stress influence alcohol drinking and relapse behavior (e.g., Kreek and Koob 1998) and there is a substantial comorbidity between anxiety disorders and alcohol abuse (Kessler et al. 1996). Mice lacking a functional CRHR1 represent a useful animal model to address the question of whether or not a dysfunctional CRH/CRHR1 system influences the individual vulnerability for alcohol drinking. *CRHR1* knockout mice did not differ from wildtype animals in alcohol intake and preference under stress-free housing conditions. After repeated stress, however, the mutant mice markedly increased their alcohol consumption, which persisted at an elevated level throughout their life. This behavior in knockout mice was found to be associated with enhanced protein levels of the *N*-methyl-D-aspartate receptor subunit NR2B, which is an ethanol-sensitive site and is also influenced by stress (Sillaber et al. 2002). Alterations in the CRH receptor 1 gene-, therefore, may constitute a genetic risk factor for stress-induced alcohol drinking and alcoholism.

Despite the lack of functional CRHR1 on pituitary corticotropes, basal plasma ACTH concentrations in homozygous CRHR1 mutants are similar to those found in wildtype controls (Timpl et al. 1998), suggesting that basal ACTH secretion is stimulated via signaling pathways other than CRH/CRHR1. Since the discovery of CRH by Vale et al. (1981), it was rapidly established that AVP potently synergizes with CRH to stimulate pituitary ACTH release; when CRH and AVP are given together, hormone output is well above the added effects of the two peptides alone, both in rodents and in humans (Gillies et al. 1982; von Bardeleben et al. 1985). This CRH/AVP synergism is known to be functionally relevant under both physiological (Keck et al. 2000; Rivier and Vale 1983a) and pathophysiological conditions such as stress (De Goeij et al. 1992; Rivier and Vale 1983b), glucocorticoid deficiency (Kiss et al. 1984; Kovács et al. 2000), or altered innate emotionality (Keck et al. 2002). Indeed, evidence was provided that the hypothalamic vasopressinergic system is significantly activated to maintain pituitary ACTH secretion in homozygous CRHR1 mutants (Müller et al. 2000). Following continuous treatment with corticosterone, plasma AVP levels in homozygous CRHR1-knockout mice were indistinguishable from those of wildtype littermates, thus providing evidence that glucocorticoid deficiency is the major driving force behind compensatory activation of the vasopressinergic system in CRHR1 mutants.

Selective serotonin reuptake inhibitors such as paroxetine and citalopram are a first-line treatment of anxiety disorders. Studies on the interaction between CRH and serotonergic neurotransmission, therefore, may be relevant for the understanding of the etiology of stress-related psychiatric conditions such as anxiety disorders. By use of in vivo microdialysis studies it could be shown that CRHR1 deficiency resulted in an enhanced synthesis of serotonin during basal conditions and in an augmented release of hippocampal serotonin in response to stress (Penalva et al. 2002). These findings underline the intricate relationship between CRH and serotonin and the important role of the CRHR1 herein and suggest that CRHR1 inactivation might represent a new avenue to modulate serotonergic neurotransmission in anxiety disorders.

3.5.2
Conditional CRHR1 Knockout

To further dissect CRH/CRHR1 central nervous system pathways modulating behavior from those regulating neuroendocrine function, a region-specific, conditional knockout mouse line (*CRHR1$^{loxP/loxP}$ CaMKII Cre*) was generated at the Max Planck Institute of Psychiatry in Munich. In this mouse line, CRHR1 function is inactivated postnatally in forebrain and limbic brain structures while sparing hypothalamic and pituitary expression sites to leave HPA system regulation intact. Selective disruption of CRH/CRHR1 signaling pathways in behaviorally relevant limbic neuronal circuitries significantly reduced anxiety-related behavior. The anxiety-reduced phenotype of *CRHR1$^{loxP/loxP}$ CaMKIIαCre* conditional mutants was confirmed in two different behavioral paradigms based on the natural avoidance behavior of mice, the light/dark box paradigm, and the elevated plus-maze test. Basal activity of the HPA system was normal in the conditional mutant mice (Müller et al. 2003). Thus, CRHR1 deficiency outside the CRH system responsible for HPA system regulation is able to reduce anxiety. This finding underlines the fact that central neuropeptidergic circuits other than those driving the peripherally accessible HPA system act independently and could be therapeutic targets of antagonist actions (Holsboer 2003).

3.6
CRHR2 Knockout: Increased Anxiety-Related Behavior?

Compared to CRHR1 mutants, behavioral and endocrine analysis of *CRHR2* knockout mice has provided a less clear picture. Three different knockout lines deficient for CRHR2 have been independently created (Bale et al. 2000; Coste et al. 2000; Kishimoto et al. 2000). Interestingly, significant differences in aspects of both the endocrine and behavioral phenotype were described between the three knockout mouse lines, pointing towards the fact that most likely the genetic background of genetically engineered mice might play a crucial role, especially when dealing with subtle behavioral alterations (e.g., Lathe 1996).

The neuroendocrine analyses of CRHR2-deficient mice suggest that CRHR2 supplies regulatory features to the HPA system's stress response (Coste et al. 2000); although initiation of the stress response appears to be normal, *CRHR2* knockout mice show early termination of ACTH release, suggesting that CRHR2 is involved in maintaining HPA system drive. CRHR2 also appears to modify the recovery phase of the HPA response, as corticosterone levels remain significantly elevated 90 min after the end of the stressor in *CRHR2* mutants. These endocrine findings were replicated in a second, independently generated CRHR2-deficient mouse line (Bale et al. 2000). Kishimoto and colleagues, in contrast, failed to detect any significant phenotype in basal and stress-induced HPA system regulation in their *CRHR2* knockout line. This finding, however, is most likely because their endocrine analysis was limited to one single time point after stress exposure (Kishimoto et al. 2000).

Taken together, these changes in HPA responses to stress suggest that CRHR1 and CRHR2 act in an antagonistic manner: CRHR1 activates and CRHR2 attenuates the stress response. It has been suggested that UCN and the CRH2α receptor may represent an "antiparallel" stress system to the CRH/CRHR1 system (Skelton et al. 2000). The sites of these antagonistic actions are currently unknown, but might include the pituitary gland, the PVN, and brain areas providing afferent input to the PVN, such as the amygdala (Reul and Holsboer 2002).

The physiological role of CRHR2 in mediating anxiety-like behavior has been the subject of a controversial discussion. Indeed, the behavioral performance reveals significant differences between the three independently created CRHR2-deficient mouse lines. Whereas Coste et al. (2000) found no differences in anxiety-related behavior, Bale and co-workers (2000) and Kishimoto et al. (2000) detected a significant increase in anxiety-like behavior in their *CRHR2* mutants. Interestingly, the latter behavioral phenotype could be observed only in male, but not in female CRHR2-deficient mice.

3.7
CRHR1/CRHR2 Double Knockout: Life Without CRH Receptors

CRHR2/CRHR2 double knockout mice were first generated in 2001. Studies on the HPA axis in these animals confirmed the data obtained with the single gene mutants, although the *CRHR1* mutation had a dominating influence (Preil et al. 2001). It could be shown that mice lacking both known CRH receptors are still viable, again pointing towards the importance of compensatory pathways maintaining and activating HPA system activity under both basal and stress-associated conditions. In line with this, AVP mRNA levels were found to be increased in double-mutant mice. Later, these data were confirmed by others (Bale et al. 2002). Results from testing for anxiety-like behaviors showed that the double-mutant mice are sexually dichotomous. Although the female double-mutant mice displayed less anxiety-like behavior, the male double-mutants showed more anxiety-like behavior compared with the females (Bale et al. 2002).

3.8
V_{1b} Receptor Knockout: Reduced Aggression

In the rat, extrahypothalamic AVP-containing neurons have been characterized mainly in the medial amygdala and the BNST, which innervate limbic structures such as the lateral septum and the ventral hippocampus (e.g., Caffé et al. 1987). In these brain regions, AVP acts as a neurotransmitter, exerting its action by binding to specific G protein-coupled receptors, i.e., V_{1a} and V_{1b} (Barberis and Tribollet 1996; Hernando et al. 2001). To further elucidate the role of V_{1b} receptors, null mutant mice were generated. These V_{1b} receptor-null mutant mice show significantly reduced aggression when tested in a resident-intruder paradigm (Wersinger et al. 2002). This finding might also be indicative of a decreased anxiety-related behavior. Anxiety-related behavior on the elevated plus-maze, however, was found to be indistinguishable between knockouts and wildtype mice. This observation is in line with the finding that infusion of an V_{1a} antisense oligodeoxynucleotide into the rat lateral septum has been shown to exert anxiolytic effects (Landgraf et al. 1995) pointing towards the fact that in the context of anxiety-related behavior the V_{1a}

receptor subtype might be more important. With respect to HPA system regulation, basal and stress-induced plasma corticosterone concentrations showed no difference between V_{1b} receptor-null mutant mice and their wildtype littermates. Data on ACTH release, however, have not been published so far. Concerning the important role of AVP in learning and memory, V_{1b} knockouts displayed a slight impairment in the social recognition test (olfactory-cued memory) but not in the Morris water maze task (spatial memory) (Wersinger et al. 2002).

3.9
GR: Myth and Reality

3.9.1
GR Antisense Transgenics: Born to Be Brave

To further elucidate the role of impaired GR signaling in psychiatric disorders, a transgenic mouse expressing antisense to GR mRNA was generated (Pepin et al. 1992). The transgene is driven by a neurofilament promoter and is therefore primarily active in neuronal tissue, resulting in a markedly reduced GR mRNA expression in the brain. Accordingly, HPA axis regulation in the antisense GR transgenic mice is heavily disturbed, as shown by a reduced glucocorticoid negative feedback efficiency, enhanced CRH- and stress-induced increases in plasma ACTH and adrenocortical hyperresponsiveness to ACTH (Barden et al. 1997; Montkowski et al. 1995). Paradoxically, activity of hypothalamic parvocellular CRH neurons is reduced in the PVN, herewith suggesting that a number of phenotypic changes in the physiology and behavior of these mice may be a consequence of hypothalamic CRH hypoactivity rather than of altered GR function per se (Dijkstra et al. 1998). In this context, the decreased anxiety-related behavior of GR antisense transgenic mice (Montkowski et al. 1995; Rochford et al. 1997) may be secondary to neuronal CRH hypoactivity (Dijkstra et al. 1998).

Complex interactions between serotonergic systems and the HPA axis have been described, and alterations of both systems have been evidenced in anxiety disorders. In order to gain insight into the role of GR in the regulation of serotonergic neurotransmission, a number of studies have been conducted in GR transgenics. GR-impaired mice were shown to have a reduced basal serotonin metabolism in the hippocampus and a delayed stress-induced stimulation of serotonin metabolism in the brain stem and hippocampus (Farisse et al. 1999). In contrast, in another study, as measured by in vivo microdialysis, no basal differences in hippocampal serotonin were found, whereas the stress-induced rise in serotonin was increased in GR-impaired mice. This finding suggests that lifelong GR impairment evolves in hyperresponsiveness of the raphe-hippocampal serotonergic system (Linthorst et al. 2000). As GR-impaired mice display reduced activity of hypothalamic CRH neurons (Dijkstra et al. 1998), this hyperresponsiveness of hippocampal serotonin may result from adaptations of the serotonergic system to a long-term hypoactivity of the CRH system. In support of this is the finding that long-term elevated levels of central CRH cause hyporesponsiveness of hippocampal serotonin to an acute stressor (Linthorst et al. 1997).

3.9.2
Conditional GR Knockouts: Reduced Versus Unchanged Anxiety-Related Behavior

Since conventional disruption of GR signaling is lethal shortly after birth due to lung failure (Cole et al. 1995), a conditional *GR*-knockout where GR function is selectively disrupted in the nervous system has been generated (Tronche et al. 1999). Using the Cre/loxP system to achieve tissue-specific gene inactivation, expression of Cre under the control of the rat nestin promoter and enhancer results in a selective disruption of GR in neuronal and glial cell precursors (Tronche et al. 1999). Mutant mice (GR^{NesCre}) display several symptoms characteristic of patients suffering from Cushing's syndrome, a disease characterized by elevated levels of glucocorticoids. In GR^{NesCre} mice, basal morning plasma corticosterone levels were significantly increased, whereas plasma ACTH concentrations were moderately reduced. Stress-induced plasma ACTH and corticosterone levels were unaltered in conditional GR mutants. CRH expression in the paraventricular nucleus was found to be increased, most likely because negative feedback on paraventricular CRH neurons, predominantly mediated via GR, is disrupted.

Further, disruption of the GR in the nervous system results in reduced anxiety-like behavior. In two tests based on the natural avoidance behavior of mice (dark–light-emergency task and elevated zero-maze; see also chapter Ohl, this volume), significantly reduced anxiety-like behavior was recorded while the general locomotor activity of mutant and control mice was similar (Tronche et al. 1999).

Ligand-activated GRs control transcription either by binding as homodimers or heterodimers (together with a MR molecule) to positive or negative GC response elements (pGREs or nGREs) in the promoter region of GC-regulated target genes. Alternatively, the GR can act as a monomer by interacting with a transcription factor through protein–protein interactions. By introducing a point mutation in one of the dimerization domains of the GR, formation of GR–GR dimers is no longer possible, allowing dissection of GR effects that require DNA binding from effects upon gene activity through GR interaction with transcription factors (Reichardt et al. 1998). In these mice, where dimerization of GRs is abandoned, CRH mRNA and CRH were normal, suggesting that GR regulates CRH not through dimers, but either through a nGRE that downregulates target gene activity via GR monomers or via protein–protein interactions, i.e., a GR binding to a transcription factor. These animals display normal anxiety-related behavior, but impaired spatial memory, suggesting that these two behaviorbehavioral features are regulated through different nuclear mechanisms of the GR (Oitzl et al. 2001).

3.9.3 Conditional GR Overexpression: Preliminary Results

So far, data on conditional overexpression of GR under the brain-region specific promoter calcium–calmodulin-dependent kinase IIα (CaMKIIα) have been presented only in abstract form (Wei et al. 2001). Those conditional GRs overexpressing mice were reported to display normal locomotor activity, but increased anxiety-related behavior in the dark–light test. Basal plasma ACTH and corticosterone levels were reported to be indistinguishable from wildtype littermates. The expression of CRH mRNA was increased in the central nucleus of the amygdala, rostral part, whereas CRH levels in the PVN and BNST remained unchanged. It was suggested that conditional GR-overexpressing mice may provide a suitable model for increased anxiety-related behavior not secondary to altered levels of circulating stress hormones.

4 Summary

The cumulative evidence makes a strong case that the neuroendocrine and behavioral phenotypes of anxiety disorders are at least in part mediated via modulation of CRH and AVP neurocircuitry and that normalization of an altered neurotransmission after treatment may lead to restoration of disease-related alterations. Although this concept was originally derived from peripheral HPA assessments in depressed patients, it is now clear that central CRH and AVP neuropeptidergic circuits other than those driving the peripherally accessible HPA system may well be overactive and could be therapeutic targets of antagonist actions (Holsboer 2003). The combination of molecular genetics with behavioral pharmacology has indicated that CRHR1 might be the primary target of interest at which selective compounds should be directed to treat pathological anxiety. In addition to CRHR1 hyperfunction, however, CRHR2 hypofunction may play an important role in anxiety disorders, since an impaired CRHR2-mediated anxiolysis is likely to result in an extended state of anxiety and arousal. Moreover, there is increasing evidence that dysfunction of AVP neuronal circuitries including both receptor V_{1a} and receptor V_{1b} could result in altered anxiety states.

It is important to point out that deletion of individual genes is not providing animal models for certain behavioral pathologies that are caused by a manifold of minor changes in a series of so-called susceptibility genes. To make a clinical phenotype overt, a number of exogenous factors, e.g., stressful life events and a susceptible genetic endowment, need to interact in at least most of the cases

Gesing A, Bilang-Bleuel A, Droste SK, Linthorst ACE, Holsboer F, Reul JMHM (2001) Psychological stress increases hippocampal mineralocorticoid receptor levels: involvement of corticotropin-releasing hormone. J Neurosci 21:4822–4829
Gillies GE, Linton EA, Lowry PJ (1982) Corticotropin releasing activity of the new CRF is potentiated several times by vasopressin. Nature 299:355–357
Griebel G, Perrault G, Sanger DJ (1998) Characterization of the behavioral profile of the non-peptide CRF receptor antagonist CP-154,526 in anxiety models in rodents. Psychopharmacology (Berl) 138:55–66
Griebel G, Simiand J, Serradeil-Le Gal C, Wagnon J, Pascal M, Scatton B, Maffrand JP, Soubrié P (2002) Anxiolytic- and antidepressant-like effects of the non-peptide vasopressin V1b receptor antagonist, SSR 149415, suggest an innovative approach for the treatment of stress-related disorders. Proc Natl Acad Sci USA 99:6370–6375
Gutman DA, Owens MJ, Skelton KH, Thrivikraman KV, Nemeroff CB (2003) The corticotropin-releasing factor 1 receptor antagonist R121919 attenuates the behavioral and endocrine responses to stress. J Pharmacol Exp Ther 304:874–880
Heinrichs SC, Stenzel-Poore MP, Gold LH, Battenberg E, Bloom FE, Koob GF, Vale WW, Pich EM (1996) Learning impairment in transgenic mice with central overexpression of corticotropin-releasing factor. Neuroscience 74:303–311
Heinrichs SC, Min H, Tamraz S, Carmouche M, Boehme SA, Vale WW (1997b) Anti-sexual and anxiogenic behavioral consequences of corticotropin-releasing factor overexpression are centrally mediated. Psychoneuroendocrinology 22:215–224
Heinrichs SC, De Souza EB, Schulteis G, Lapsansky JL, Grigoriadis DE (2002) Brain penetrance, receptor occupancy and antistress in vivo efficacy of a small molecule corticotropin releasing factor type 1 receptor selective antagonist. Neuropsychopharmacology 27:194–202
Hernando F, Schoots O, Lolait SJ, Burbach JP (2001) Immunohistochemical localization of the vasopressin V1b receptor in the rat brain and pituitary gland: anatomical support for its involvement in the central effects of vasopressin. Endocrinology 142:1659–1668
Hettema JM, Neale MC, Kendler KS (2001) A review and meta-analysis of the genetic epidemiology of anxiety disorders. Am J Psychiatry 158:1568–1578
Heuser I, Yassouridis A, Holsboer F (1994) The combined dexamethasone/CRH test—a refined laboratory test for psychiatric disorders. J Psychiatr Res 28:341–356
Hökfelt T, Broberger C, Xu ZQD, Sergeyev V, Ubink R, Diez M (2000) Neuropeptides—an overview. Neuropharmacology 39:1337–1356
Holsboer F (1999) The rationale for corticotropin-releasing hormone receptor (CRH-R) antagonists to treat depression and anxiety. J Psychiatr Res 33:181–214
Holsboer F (2000) The corticosteroid receptor hypothesis of depression. Neuropsychopharmacology 23:477–501
Holsboer F (2003) Corticotropin-releasing hormone modulators and depression. Curr Opin Investig Drugs 4:46–50
Hsu SY, Hsueh AJW (2001) Human stresscopin and stresscopinn-related peptide are selective ligands for the corticotropin-releasing hormone type 2 receptor. Nat Med 7:605–611
Iino K, Sasano H, Oki Y, Andoh N, Shin RW, Kitamoto T, Takahashi K, Suzuki H, Tezuka F, Yoshimi T, Nagura H (1999) Urocortin expression in the human central nervous system. Clin Endocrinol (Oxf) 50:107–114
Insel TR, Young LJ (2000) Neuropeptides and the evolution of social behavior. Curr Opin Neurobiol 10:784–789
Jacobson L, Sapolsky R (1991) The role of the hippocampus in feedback regulation of the hypothalamic-pituitary-adrenocortical axis. Endocr Rev 12:118–134

Jacobson L, Muglia LJ, Weninger SC, Pacak K, Majzoub JA (2000) CRH deficiency impairs but does not block pituitary-adrenal responses to diverse stressors. Neuroendocrinology 71:79–87

Jaenisch R (1988) Transgenic animals. Science 240:1468–1472

Jeong KH, Jacobson L, Widmaier EP, Majzoub JA (1999) Normal suppression of the reproductive axis following stress in corticotropin-releasing hormone-deficient mice. Endocrinology 140:1702–1708

Joels M, de Kloet ER (1992) Control of neuronal excitability by corticosteroid hormones. Trends Neurosci 15:25–30

Karanth S, Linthorst AC, Stalla GK, Barden N, Holsboer F, Reul JM (1997) Hypothalamic-pituitary-adrenocortical axis changes in a transgenic mouse with impaired glucocorticoid receptor function. Endocrinology 138:3476–3485

Karolyi IJ, Burrows HL, Ramesh TM, Nakajima M, Lesh JS, Seong E, Camper SA, Seasholtz AF (1999) Altered anxiety and weight gain in corticotropin-releasing hormone-binding protein-deficient mice. Proc Natl Acad Sci USA 96:11595–11600

Keck ME, Holsboer F (2001) Hyperactivity of CRH neuronal circuits as a target for therapeutic interventions in affective disorders. Peptides 22:835–844

Keck ME, Hatzinger M, Wotjak CT, Holsboer F, Landgraf R, Neumann ID (2000) Ageing alters intrahypothalamic release patterns of vasopressin and oxytocin in rats. Eur J Neurosci 12:1487–1494

Keck ME, Welt T, Wigger A, Renner U, Engelmann M, Holsboer F, Landgraf R (2001) The anxiolytic effect of the CRH1 receptor antagonist R121919 depends on innate emotionality in rats. Eur J Neurosci 13:373–380

Keck ME, Wigger A, Welt T, Müller MB, Gesing A, Reul JMHM, Holsboer F, Landgraf R, Neumann ID (2002) Vasopressin mediates the response of the combined dexamethasone/CRH test in hyper-anxious rats: implications for pathogenesis of affective disorders. Neuropsychopharmacology 26:94–105

Keck ME, Welt T, Müller MB, Landgraf R, Holsboer F (2003a) The high-affinity non-peptide CRH1 receptor antagonist R121919 attenuates stress-induced alterations in plasma oxytocin, prolactin, and testosterone in rats. Pharmacopsychiatry 6:27–31

Keck ME, Welt T, Müller MB, Uhr M, Ohl F, Wiger A, Toschi N, Holsboer F, Landgraf R (2003b) Reduction of hypothalamic vasopressinergic hyperdrive contributes to clinically relevant behavioral and neuroendocrine effects of chronic paroxetine treatment in a psychopathological rat model. Neuropsychopharmacology 28:235–243

Kemp CF, Woods RJ, Lowry PJ (1998) The corticotrophin-releasing factor-binding protein: an act of several parts. Peptides 19:1119–1128

Kessler RC (2000) The epidemiology of pure and comorbid generalized anxiety disorder: a review and evaluation of recent research. Acta Psychiatr Scand 102:7–13

Kessler RC, Nelson CB, McGonagle KA, Edlund MJ, Frank RG, Leaf PJ (1996) The epidemiology of co-occurring addictive and mental disorders: implications for prevention and service utilizations. Am J Orthopsychiatry 66:17–31

Kishimoto T, Radulovic J, Radulovic M, Lin CR, Schrick C, Hooshmand F, Hermanson O, Rosenfeld MG, Spiess J (2000) Deletion of CRHR2 reveals an anxiolytic role for corticotropin-releasing hormone receptor-2. Nat Genet 24:415–419

Kiss JZ, Mezey E, Skirboll L (1984) Corticotropin-releasing factor immunoreactive neurons of the paraventricular nucleus become vasopressin positive after adrenalectomy. Proc Natl Acad Sci USA 84:1854–1858

Korte SM (2001) Corticosteroids in relation to fear, anxiety and psychopathology. Neurosci Biobehav Rev 25:117–142

Kostich WA, Chen A, Sperle K, Largent BL (1998) Molecular identification and analysis of a novel human corticotropin-releasing factor (CRF) receptor: the CRF2gamma receptor. Mol Endocrinol 12:1077–1085

Kovács KJ, Földes A, Sawchenko PE (2000) Glucocorticoid negative feedback selectively targets vasopressin transcription in parvocellular neurosecretory neurons. J Neurosci 20:3843–3852

Kozicz T, Yanaihara H, Arimura A (1998) Distribution of urocortin-like immunoreactivity in the central nervous system of the rat. J Comp Neurol 391:1–10

Kreek MJ, Koob GF (1998) Drug dependence: stress and dysregulation of brain reward pathways. Drug Alcohol Depend 51:23–47

Kühn R, Schwenk F, Aguet M, Rajewsky K (1995) Inducible gene targeting in mice. Science 259:1427–1429

Künzel HE, Zobel AW, Nickel T, Ackl N, Uhr M, Sonntag A, et al (2003) Treatment of depression with the corticotropin-releasing hormone-1-receptor antagonist R121919: endocrine changes and side effects. J Psychiatr Res 37:525–533

Lancel M, Müller-Preuss P, Wigger A, Landgraf R, Holsboer F (2002) The CRH1 receptor antagonist R121919 attenuates stress-elicited sleep disturbances in rats, particularly in those with high innate anxiety. J Psychiatr Res 36:197–208

Landgraf R, Gerstberger R, Montkowski A, Probst JC, Wotjak CT, Holsboer F, Engelmann M (1995) V1 vasopressin receptor antisense oligodeoxynucleotide into septum reduces vasopressin binding, social discrimination abilities, and anxiety-related behavior in rats. J Neurosci 15:4250–4258

Landgraf R, Wotjak CT, Neumann ID, Engelmann M (1998) Release of vasopressin within the brain contributes to neuroendocrine and behavioral regulation. In: Urban IJA, Burbach JPH, De Wied D (eds) Progress in brain research. Elsevier Science, Amsterdam, pp 201–220

Lathe R (1996) Mice, gene targeting and behaviour: more than just genetic background. Trends Neurosci 19:183–186

Lewis K, Li C, Perrin MH, Blount A, Kunitake K, Donaldson C, Vaughan J, Reyes TM, Gulyas J, Fischer W, Bilezikjian L, Rivier J, Sawchenko PE, Vale WW (2001) Identification of urocortin III, an additional member of the corticotropin-releasing factor (CRF) family with high affinity for the CRF2 receptor. Proc Natl Acad Sci USA 98:7570–7575

Liebsch G, Landgraf R, Gerstberger R, Probst JC, Wotjak CT, Engelmann M, Holsboer F, Montkowski A (1995) Chronic infusion of a CRH1 receptor antisense oligodeoxynucleotide into the central nucleus of the amygdala reduced anxiety-related behavior in socially defeated rats. Regul Pept 59:229–239

Liebsch G, Wotjak CT, Landgraf R, Engelmann M (1996) Septal vasopressin modulates anxiety-related behaviour in rats. Neurosci Lett 217:101–104

Liebsch G, Landgraf R, Engelmann M, Lörscher P, Holsboer F (1999) Differential behavioural effects of chronic infusion of CRH 1 and CRH 2 receptor antisense oligonucleotides into the rat brain. J Psychiatr Res 33:153–163

Linthorst ACE, Flachskamm C, Hopkins SH, Hoadley ME, Labeur MS, Holsboer F, Reul JMHM (1997) Long-term intracerebroventricular infusion of corticotropin-releasing hormone alters neuroendocrine, neurochemical, autonomic, behavioral, and cytokine responses to a systemic inflammatory challenge. J Neurosci 17:4448–4460

Linthorst ACE, Karanth S, Barden N, Holsboer F, Reul JMHM (1999) Impaired glucocorticoid receptor function evolves in aberrant physiological responses to bacterial endotoxin. Eur J Neurosci 11:178–186

Linthorst ACE, Flachskamm C, Barden N, Holsboer F, Reul JMHM (2000) Glucocorticoid receptor impairment alters CNS responses to a psychological stressor: an in vivo microdialysis study in transgenic mice. Eur J Neurosci 12:283–291

Lovejoy DA, Aubry JM, Turnbull A, Sutton S, Potter E, Yehling J, Rivier C, Vale WW (1998) Ectopic expression of the CRF-binding protein: minor impact on HPA axis regulation but induction of sexually dimorphic weight gain. J Neuroendocrinol 10:483–491

Lovenberg TW, Chalmers DT, Liu C, De Souza EB (1995a) CRF2 alpha and CRF2 beta receptor mRNAs are differentially distributed between the rat central nervous system and peripheral tissues. Endocrinology 136:4139–4142

Lovenberg TW, Liaw CW, Grigoriadis DE, Clevenger W, Chalmers DT, De Souza EB, Oltersdorf T (1995b) Cloning and characterization of a functionally distinct corticotropin-releasing factor receptor subtype. Proc Natl Acad Sci U S A 92:836–840

Montkowski A, Barden N, Wotjak C, Stec I, Ganster J, Meaney M, Engelmann M, Reul JMHM, Landgraf R, Holsboer F (1995) Long-term antidepressant treatment reduces behavioural deficits in transgenic mice with impaired glucocorticoid receptor function. J Neuroendocrinol 7:841–845

Moran TH, Reeves RH, Rogers D, Fisher E (1996) Ain't misbehavin'—it's genetic. Nat Genet 12:115–116

Moreau JL, Kilpatrick G, Jenck F (1997) Urocortin, a novel neuropeptide with anxiogenic-like properties. Neuroreport 8:1697–1701

Morin SM, Ling N, Liu XJ, Kahl SD, Gehlert DR (1999) Differential distribution of urocortin- and corticotropin-releasing factor-like immunoreactivities in the rat brain. Neuroscience 92:281–291

Muglia L, Jacobson L, Dikkes P, Majzoub JA (1995) Corticotropin-releasing hormone deficiency reveals major fetal but not adult glucocorticoid need. Nature 373:427–432

Muglia L, Jacobson L, Weninger SC, Luedke CE, Bae DS, Jeong KH, Majzoub JA (1997) Impaired diurnal adrenal rhythmicity restored by constant infusion of corticotropin-releasing hormone in corticotropin-releasing hormone-deficient mice. J Clin Invest 99:2923–2929

Müller MB, Keck ME (2002) Genetically engineered mice for studies of stress-related clinical conditions. J Psychiatr Res 36:53–76

Müller MB, Landgraf R, Sillaber I, Kresse AE, Keck ME, Zimmermann S, Holsboer F, Wurst W (2000) Selective activation of the hypothalamic vasopressinergic system in mice deficient for the corticotropin-releasing hormone receptor 1 is dependent on glucocorticoids. Endocrinology 141:4262–4269

Müller MB, Holsboer F, Keck ME (2002) Genetic modification of corticosteroid receptor signaling: novel insights into pathophysiology and treatment strategies of human affective disorders. Neuropeptides 36:117–131

Müller MB, Zimmermann S, Sillaber I, Hagemeyer TP, Deussing JM, Timpl P, et al (2003) Limbic corticotropin-releasing hormone receptor 1 mediates anxiety-related behaviour and is required for hormonal adaptation to stress. Nat Neurosci 6:1100–1107

Müller U (1999) Ten years of gene targeting: targeted mouse mutants, from vector design to phenotype analysis. Mech Dev 82:3–21

Nelson RJ (1997) The use of genetic "knockout" mice in behavioral endocrinology research. Horm Behav 31:188–196

Oitzl MS, De Kloet ER (1992) Selective corticosteroid antagonists modulate specific aspects of spatial orientation learning. Behav Neurosci 106:62–71

Oitzl MS, de Kloet ER, Joels M, Schmid W, Cole TJ (1997) Spatial learning deficits in mice with a targeted glucocorticoid receptor gene disruption. Eur J Neurosci 9:2284–2296

Penalva RG, Flachskamm C, Zimmermann S, Wurst W, Holsboer F, Reul JMHM, Linthorst ACE (2002) Corticotropin-releasing hormone receptor type 1-deficiency enhances hippocampal serotonergic neurotransmission: an in vivo microdialysis study in mutant mice. Neuroscience 109:253–266

Pepin MC, Pothier F, Barden N (1992) Impaired type II glucocorticoid-receptor function in mice bearing antisense RNA transgene. Nature 355:725–728

Potter E, Behan DP, Linton EA, Lowry PJ, Sawchenko PE, Vale WW (1992) The central distribution of a corticotropin-releasing factor (CRF)-binding protein predicts multiple sites and modes of interaction with CRF. Proc Natl Acad Sci U S A 89:4192–4196

Preil J, Müller MB, Gesing A, Reul JMHM, Sillaber I, van Gaalen M, Landgrebe J, Stenzel-Poore M, Holsboer F, Wurst W (2001) Regulation of the hypothalamic-pituitary-adrenocortical system in mice deficient for corticotropin-releasing hormone receptor 1 and 2. Endocrinology 142:4946–4955

Radulovic J, Ruhmann A, Liepold T, Spiess J (1999) Modulation of learning and anxiety by corticotropin-releasing factor (CRF) and stress: differential roles of CRF receptors 1 and 2. J Neurosci 19:5016–5025

Reichardt H M, Kaestner KH, Tuckermann J, Kretz O, Wessely O, Bock R, Gass P, Schmid W, Herrlich P, Angel P, Schütz G (1998) DNA binding of the glucocorticoid receptor is not essential for survival. Cell 93:531–541

Reul JMHM, de Kloet ER (1985) Two receptor systems for corticosterone in the rat brain: microdistribution and differential occupation. Endocrinology 117:2505–2512

Reul JMHM, Holsboer F (2002) Corticotropin-releasing factor receptors 1 and 2 in anxiety and depression. Curr Opin Pharmacol 2:23–33

Reyes TM, Lewis K, Perrin MH, Kunitake KS, Vaughan J, Arias CA, Hogenesch JB, Gulyas J, Rivier J, Vale WW, Sawchenko PE (2001) Urocortin II: a member of the corticotropin-releasing factor (CRF) neuropeptide family that is selectively bound by type 2 CRF receptors. Proc Natl Acad Sci USA 98:2843–2848

Rivier C, Vale W (1983a) Interaction of corticotropin-releasing factor and arginine-vasopressin in adrenocorticotropin secretion in vivo. Endocrinology 113:939–942

Rivier C, Vale W (1983b) Modulation of stress-induced ACTH release by corticotropin-releasing factor, catecholamines and vasopressin. Nature 305:325–327

Rochford J, Beaulieu S, Rousse I, Glowa JR, Barden N (1997) Behavioral reactivity to aversive stimuli in a transgenic mouse model of impaired glucocorticoid (type II) receptor function: effects of diazepam and FG-7142. Psychopharmacology (Berl) 132:145–152

Rousse I, Beaulieu S, Rowe W, Meaney MJ, Barden N, Rochford J (1997) Spatial memory in transgenic mice with impaired glucocorticoid receptor function. Neuroreport 8:841–845

Sauer B (1998) Inducible gene targeting in mice using the Cre/lox system. Methods Enzymol 14:381–392

Schreiber W, Lauer CJ, Krumrey K, Holsboer F, Krieg JC (1996) Dysregulation of the hypothalamic-pituitary-adrenocortical system in panic disorder. Neuropsychopharmacology 15:7–15

Schulkin J, Gold PW, McEwen BS (1998) Induction of corticotropin-releasing hormone gene expression by glucocorticoids: implication for understanding the states of fear and anxiety and allostatic load. Psychoneuroendocrinology 23:219–243

Sillaber I, Rammes G, Zimmermann S, Mahal B, Zieglgänsberger W, Wurst W, Holsboer F, Spanagel R (2002) Enhanced and delayed stress-induced alcohol drinking in mice lacking functional CRH1 receptors. Science 296:931–933

Silva AJ (1997) Mutant mice and genetic background: recommendations concerning genetic background. Banbury conference on genetic background in mice. Neuron 19:755–759

Skelton KH, Nemeroff CB, Knight DL, Owen MJ (2000) Chronic administration of the triazolobenzodiazepine alprazolam produces opposite effects on corticotropin-releasing factor and urocortin neuronal systems. J Neurosci 20:1240–1248

Skutella T, Probst JC, Renner U, Holsboer F, Behl C (1998) Corticotropin-releasing hormone receptor (type I) antisense targeting reduces anxiety. Neuroscience 85:795–805

Slawecki CJ, Somes C, Rivier JE, Ehlers CL (1999) Neurophysiological effects of intracerebroventricular administration of urocortin. Peptides 20:211–218

Smith GW, Aubry J-M, Dellu F, Contarino A, Bilezikijan LM, Gold LH, Chen R, Marchuk Y, Hauser C, Bentley CA, Sawchenko PE, Koob GF, Vale WW, Lee KF (1998) Corticotropin releasing factor receptor 1-deficient mice display decreased anxiety, impaired stress response, and aberrant neuroendocrine development. Neuron 20:1093–1102

Smythe JW, Murphy D, Timothy C, Costall B (1997) Hippocampal mineralocorticoid, but not glucocorticoid, receptors, modulate anxiety-like behavior in rats. Pharmacol Biochem Behav 56:507–513

Spina M, Merlo-Pich E, Chan RKW, Basso AM, Rivier J, Vale WW, Koob GF (1996) Appetite-suppressing effects of urocortin, a CRF-related neuropeptide. Science 273:1561–1564

Stenzel-Poore MP, Cameron VA, Vaughan J, Sawchenko PE, Vale W (1992) Development of Cushing's syndrome in corticotropin-releasing factor transgenic mice. Endocrinology 130:3378–3386

Stenzel-Poore MP, Heinrichs SC, Rivest S, Koob GF, Vale WW (1994) Overproduction of corticotropin-releasing factor in transgenic mice: a genetic model of anxiogenic behavior. J Neurosci 14:2579–2584

Stratakis CA, Karl M, Schulte HM, Chrousos GP (1994) Glucocorticosteroid resistance in humans. Elucidation of the molecular mechanisms and implications for pathophysiology. Ann N Y Acad Sci 746:362–374

Swanson LW, Sawchenko PE, Rivier J, Vale WW (1983) Organization of ovine corticotropin-releasing factor immunoreactive cells and fibers in the rat brain: an immunohistochemical study. Neuroendocrinology 36:165–186

Takahashi LK (2001) Role of CRF1 and CRF2 receptors in fear and anxiety. Neurosci Biobehav Rev 25:627–636

Timpl P, Spanagel R, Sillaber I, Kresse A, Reul JMHM, Stalla GK, Blanquet V, Steckler T, Holsboer F, Wurst W (1998) Impaired stress response and reduced anxiety in mice lacking a functional corticotropin-releasing hormone receptor 1. Nat Genet 19:162–166

Tronche F, Kellendonk C, Kretz O, Gass P, Anlag K, Orban PC, Bock P, Klein R, Schütz G (1999) Disruption of the glucocorticoid receptor gene in the nervous system results in reduced anxiety. Nat Genet 23:99–103

Turnbull AV, Smith GW, Lee S, Vale WW, Lee KF, Rivier C (1999) CRF type 1 receptor deficient mice exhibit a pronounced pituitary-adrenal response to local inflammation. Endocrinology 140:1013–1017

Valdez GR, Inoue K, Koob GF, Rivier J, Vale WW, Zorilla EP (2002) Human urocortin II: mild locomotor suppressive and delayed anxiolytic effects of a novel corticotropin-releasing factor related peptide. Brain Res 943:142–150

Vale W, Spiess J, Rivier C, Rivier J (1981) Characterization of a 41-residue ovine hypothalamic peptide that stimulates secretion of corticotropin and β-endorphin. Science 213:1394–1397

Van Bockstaele EJ, Colago EE, Valentino RJ (1998) Amygdaloid corticotropin-releasing factor targets locus coeruleus dendrites: substrate for the co-ordination of emotional and cognitive limbs of the stress response. J Neuroendocrinol 10:743–757 van Haarst AD, Oitzl M S, Workel JO, de Kloet ER (1996) Chronic brain glucocorticoid receptor blockade enhances the rise in circadian and stress-induced pituitary-adrenal activity. Endocrinology 137:4935–4943

Vaughan J, Donaldson C, Bittencourt J, Perrin MH, Lewis K, Sutton S, Chan R, Turnbull AV, Lovejoy D, Rivier C, Rivier J, Sawchenko PE, Vale W (1995) Urocortin, a mammalian neuropeptide related to fish urotensin I and to corticotropin-releasing factor. Nature 378:287–292

Vetter DE, Li C, Zhao L, Contarino A, Liberman MC, Smith GW, Marchuk Y, Koob GF, Heinemann SF, Vale W, Lee KF (2002) Urocortin-deficient mice show hearing impairment and increased anxiety-like behavior. Nat Genet 31:363–369

Watts AG (1996) The impact of physiological stimuli on the expression of corticotropin-releasing hormone (CRH) and other neuropeptide genes. Front Neuroendocrinol 17: 281–326

Wei Q, Schafer GL, Hebda-Bauer E, Shieh KR, Watson S, Seasholtz A, Akil H (2001) Tissue-specific overexpression of the glucocorticoid receptor in the brain. Soc Neuroci (abstract)

Weiss B, Davidkova G, Zhang SP (1997) Antisense strategies in neurobiology. Neurochem Int 31:321–348

Weninger SC, Dunn AJ, Muglia LJ, Dikkes P, Miczek KA, Swiergiel AH, Berridge C W, Majzoub JA (1999) Stress-induced behaviors require the corticotropin-releasing hormone (CRH) receptor, but not CRH. Proc Natl Acad Sci USA 96:8283–8288

Weninger SC, Peters LL, Majzoub JA (2000) Urocortin expression in the Edinger-Westphal nucleus is upregulated by stress and corticotropin-releasing hormone deficiency. Endocrinology 256–263

Wersinger SR, Ginns Ei, O'Carroll AM, Lolait SJ, Young WS (2002) Vasopressin V1b receptor knockout reduces aggressive behavior in male mice. Mol Psychiatry 7:975–984

Wong ML, Al-Shekhlee A, Bongiorno PB, Esposito A, Khatri P, Sterneberg EM (1998) Localization of urocortin messenger RNA in rat brain and pituitary. Mol Psychiatry 1:307–312

Yamamoto H, Maeda T, Fujimura M, Fujimiya M (1998) Urocortin-like immunoreactivity in the substantia nigra, ventral tegmental area and Edinger-Westphal nucleus of rat. Neurosci Lett 243:21–24

Zobel AW, Nickel T, Künzel HE, Ackl N, Sonntag A, Ising M, Holsboer F (2000) Effects of the high affinity corticotropin-releasing hormone receptor 1 antagonist R121919 in major depression: the first 20 patients treated. J Psychiatr Res 34:171–181

HEP (2005) 169:143–162

Mutagenesis and Knockout Models: NK1 and Substance P

A. Bilkei-Gorzo · A. Zimmer (✉)

Laboratory of Molecular Neurobiology, Department of Psychiatry, University of Bonn, Siegmund-Freund-Strasse 25, 53105 Bonn, Germany
a.zimmer@uni-bonn.de

Abstract Tachykinins play an important role as peptide modulators in the CNS. Based on the concentration and distribution of the peptides and their receptors, substance P (SP) and its cognate receptor neurokinin 1 (NK1R) seem to play a particularly important role in higher brain functions. They are expressed at high levels in the limbic system, which is the neural basis of emotional responses. Three different lines of evidence from physiological studies support such a role of SP in the regulation of emotionality: (1) stress is often associated with elevated level of SP in animals and humans; (2) systematic and local injections of SP influence anxiety levels in a dose-dependent and site-specific manner; (3) NK1 receptor antagonists show anxiolytic effects in different animal models of anxiety. Although these studies point to the NK1 receptor as a promising target for the pharmacotherapy of anxiety disorders, high affinity antagonists for the human receptors could not be studied in rats or mice due to species differences in the antagonist binding sites. However, studies on anxiety and depression-related behaviors have now been performed in mouse mutants deficient in NK1 receptor or SP and NKA. These genetic studies have shown that anxiety and depression-related phenotypes are profoundly affected by the tachykinin system. For example, NK1R-deficient mice seem to be less prone depression-related behaviors in models of depression, and one study also provided evidence for reduced anxiety levels. Mice deficient in SP and NKA behaved similarly as the *NK1R* knockouts. In animal models of anxiety they performed like wildtype mice treated with anxiolytic drugs. In behavioral paradigms related to depression they behaved like wildtype animals treated with antidepressants. In summary, the genetic studies clearly show that the SP/NK1 system plays an important role in the modulation of emotional behaviors.

Keywords Substance P · Neurokinin 1 · Tachykinin · Anxiety · Knockout model

1 Introduction

Tachykinin neuropeptides can be found in many animal species from invertebrates to mammals. They have been implicated in a variety of physiological roles including immune, cardiovascular, gastrointestinal, pulmonary, and urogenital functions, as well as nociception (Severini et al. 2002). Much of the interest of the pharmaceutical industry in the tachykinin system has been stimulated by the proposed role of substance P (SP) in the facilitation of nociceptive signaling, and a considerable effort has been focused on the development of neurokinin (NK)1 receptor antagonists as an analgesic drug. However, although several antagonists have been developed and evaluated for the treatment of various pain conditions, none of these compounds showed much analgesic efficacy in clinical studies (Herbert et al. 2002).

Nevertheless, a potential use for these compounds came from the unexpected observation that tachykinin receptor antagonists seemed to be active in animal models of affective disorders. Indeed, a clinical study using the NK1 antagonist MK-869 confirmed these findings and demonstrated a very good efficacy of a tachykinin receptor antagonist for the treatment of depression (Kramer et al. 1998). The result of this first clinical study was recently confirmed in another clinical trial using a different NK1 antagonist, L-759274, in outpatients with major depressive disorder (Kramer et al. 2004).

Before we review some intriguing findings from the analysis of genetically altered animals, we will give a brief overview of the tachykinin system and summarize some pharmacological studies as they relate to its role in the regulation in emotional responses and, possibly, in the pathophysiology of affective disorders.

2 An Overview of the Tachykinin System

Von Euler and Gaddum described in 1931 a new substance from the alcoholic extract of equine brain and intestine that potently increased the rhythmic, spontaneous contractions of jejunum and produced hypotension relaxing the large arteries. They later called this compound substance P. The amino acid sequence of this peptide was established 40 years later by Leeman and coworkers, and it represented the first sequence of a neuropeptide (Chang et al. 1971). We now know that SP and two structurally related peptides, NKA and NKB, make up the family of tachykinin neuropeptides. They all share a common the C-terminal sequence Phe-X-Gly-Leu-Met-NH_2 (Severini et al. 2002).

Tachykinin peptides are encoded by the two genes *tac1* and *tac2*. The *tac1* gene encodes SP and NKA, while the *tac2* gene encodes NKB. Several different transcripts are produced from the *tac1* gene by alternative splicing, and they are translated into four distinct precursor proteins (Cooper et al. 1996). Preprotachykinin-α or PPT-α is the least prevalent and, like PPT-δ, it can only be processed into SP. PPT-β and PPT-γ generate SP, NKA, and elongated forms of NKA—neuropeptide K and neuropeptide-γ. The second tachykinin gene, *tac2*, encodes the precursor of NKB (Hokfelt et al. 2001).

2.1 Tachykinin Receptors

The existence of multiple tachykinin receptors was proposed in consideration of the remarkable differences in the pharmacological properties of different agonists (Erspamer et al. 1980) and antagonists (Folkers et al. 1984; Rosell et al. 1983). The existence of three distinct receptors was subsequently confirmed through the molecular cloning of the receptor genes (Nakanishi 1991). NK1, NK2, and NK3 belong to the superfamily of the G protein-coupled (rhodopsin-like) receptors. The NK2 and NK3 receptors are specifically activated by NKA and NKB respectively, whereas the NK1 receptor is more promiscuous (Maggi 1995). This receptor has two binding sites or, probably, two conformational states. One conformational state is specific for SP, while the other binds with similar affinity SP, NKA, and NKB (Beaujouan et al. 2000; Beaujouan et al. 1999; Saffroy et al. 2001).

2.2 Expression Studies

The tachykinins are sometimes thought to act as a unit in some areas of the nervous system, because (1) SP and NKA are produced from a common precursor, (2) the processing of the pre-proprotein involves the same proteolytic enzymes, (3) different tachykinins can be stored and co-released from the same neuron, (4) they are degraded by the same enzymes, and (5) different peptides may activate the same receptors (Hokfelt et al. 2001). However, the distribution and expression levels of the various PPTs and NK receptors are quite distinct in many brain regions.

Tachykinins are among the most abundant neuropeptides in the central nervous system. Limbic structures, which are important in the control of emotional behaviors, in particular contain tachykinins and neurokinin receptor sites in high density (Honkaniemi et al. 1992; Hurd et al. 1999; Ribeiro-da-

Silva and Hokfelt 2000). SP is found in high concentration in the nuclei of the amygdala (Roberts et al. 1982), in the bed nucleus of the stria terminalis (Gray and Magnuson 1992), in the periaqueductal gray (PAG) matter (Commons and Valentino 2002; Smith et al. 1994), and in the lateral septal nucleus (Gavioli et al. 2002). The localization of NKA is similar to that of SP, but its concentration is generally lower (Severini et al. 2002; Takeda et al. 1990). The NK1 receptor is also found throughout the limbic system (Rothman et al. 1984), while NK2 expression is restricted to the hippocampal CA1 and CA3 areas, to some thalamic nuclei, and to the septal area (Saffroy et al. 2001). Since NKA is produced in many limbic structures that do not express NK2, but only NK1, NKA may activate NK1 rather than NK2 receptors in the limbic system.

NK3 is also abundantly expressed in many limbic structures (Rothman et al. 1984; Shughrue et al. 1996; Spitznagel et al. 2001), while NKB is only produced in discrete areas, including the bed nucleus of the stria terminalis, the septal nuclei, and the central gray (Merchenthaler et al. 1992). It is conceivable that NKB is not the only ligand at NK3 receptors in the limbic system.

3 Pharmacological Studies

3.1 Involvement of SP in Stress Responses

Several lines of evidence point to an important role of SP in the regulation of stress responses. Mild stressors such as isolation (Brodin et al. 1994), sequential removal from the home cage, or short-lasting (1 min) restraint (Rosen et al. 1992) significantly increased the concentration of SP in the PAG region. Moreover, treatment with the anxiolytic drug diazepam produced a reduction of SP concentration in the PAG area and in the rostral hippocampus (Brodin et al. 1994). In contrast, more severe stressors like longer-lasting immobilization reduced SP levels in the septum, striatum, and hippocampus, and produced a concomitant decrease in the density of SP receptors in the septum, amygdala, piriform cortex, and hypothalamus (Takayama et al. 1986). It is possible that these changes represent homeostatic adaptations of the tachykinin system, involving an overactivation-induced depletion of SP stores. Indeed, Smith et al. observed that immobilization for 1 h caused an approximately 60% increase in NK1 receptor endocytosis in the basolateral amygdala, and this process was inhibited by acute pretreatment with the NK1 antagonist L-760735 (Smith et al. 1999).

Clinical studies also supported the importance of SP in stress responses and regulation of anxiety level. Even clinically healthy persons showed increased plasma SP levels in extreme stress situations such as war (Weiss et al. 1996) or parachute jumping (Schedlowski et al. 1995).

It has been suggested that central release of SP also may contribute to the development of vegetative stress responses, such as hypertension after prolonged emotional stress. This hypothesis was based on the observation that the injection of SP into different brain nuclei, such as the central nucleus of the amygdala, the nucleus paraventricularis (PVN), the nucleus ventromedialis, the lateral hypothalamus–perifornical region, or the PAG, elicited a hypertensive response (Ku et al. 1998). Painful stressors that trigger the release of SP, such as formalin injected into the lower leg, also induce a marked increase in mean arterial pressure and heart rate. Intracerebroventricular pretreatment with the NK1 receptor antagonist RP67580 attenuated both the cardiovascular and behavioral stress responses (Culman et al. 1997).

Local SP injections into the PAG of conscious rats elicited an increased sympathoadrenal activity, an increase of blood pressure, heart rate, mesenteric and renal vasoconstriction, hind-limb vasodilatation, and also an increased locomotion and grooming behavior (Culman et al. 1995). Thus, a single injection of SP produced the whole spectrum of cardiovascular, behavioral, and endocrine response observed in rodents after nociceptive stress. Moreover, SP injected into the lateral ventricle elicited grooming behavior (face washing and hind limb grooming) and resulted in a marked c-Fos expression in the paraventricular, dorsomedial and parabrachial nuclei, and in the medial thalamus. These brain areas are known to be involved in the central regulation of cardiovascular and neuroendocrine reactions to stress, or to be involved in the processing of nociceptive responses (Spitznagel et al. 2001).

There is a good correlation between stress and the activity of the immune system. SP is a known mediator of immune responses, and, as we have shown above, it is an important stress mediator. It is therefore not surprising that a direct relation between anxiety level, SP concentration, and activity of the immune system has been confirmed in clinical (Fehder et al. 1997) and animal (Teixeira et al. 2003) studies. It seems possible that SP serves also as a mediator of stress-modulated immune reactions. When animals were treated with the NK1 receptor antagonist RP67580 and subsequently stressed through restraint, adrenocorticotropic hormone (ACTH) and corticosterone levels remained high throughout the 4-h observation period, while in control animals elevated stress hormone level was observed 60 min following initiation of the stress, but had returned to basal levels after 4 h (Jessop et al. 2000). This observation may suggest that SP is involved in the transition from acute to chronic stress, and plays an important role in the termination of stress response.

3.2
SP and NK1 Receptor Antagonists in Animal Models of Anxiety

To test the potential involvement of SP in the modulation of anxiety, some studies used a systemic administration of this peptide. The results were, however, inconclusive. Both anxiogenic (Baretta et al. 2001) and anxiolytic (Hasenohrl et al. 2000) effects were reported. In contrast, SP microinjection into separate brain areas provided more conclusive results and showed that the effect of SP is dependent on the site of injection. SP applied into the lateral septal nucleus elicited anxiogenic responses using the elevated plus-maze test (Gavioli et al. 1999), while administration into the nucleus basalis of the ventral pallidum produced anxiolytic effects (Nikolaus et al. 1999). In addition to the regional specificity, the effects of SP were also dose-dependent: a low dose of SP elicited an anxiolytic while a higher dose produced an anxiogenic effect when each was injected into the nucleus basalis (Hasenohrl et al. 1998).

Although the availability of receptor-selective antagonists provided important insights into the various roles of tachykinin receptor in normal physiology and pathophysiological processes, the pharmacological analysis of NK1 functions has been complicated by the fact that small differences in the amino acid sequence between the human and the mouse or rat receptors dramatically alter antagonist binding affinity (Fong et al. 1992). Thus, antagonists with a high affinity for human receptor bind poorly to the rat and mouse receptors and cannot be used in these species. Also, the most widely used and characterized rodent models were not applicable for the analysis of compounds active in humans. While a few antagonists with nanomolar affinities for the rat and mouse receptor, such as RP67580, GR205171, and SR140333 have been developed (Emonds-Alt et al. 1993; Fong et al. 1992; Gardner et al. 1996), the usefulness of these compounds in vivo suffers from their short half-life and poor brain penetration. Moreover, at bioactive doses these compounds can exhibit unspecific pharmacological effects, including the blockade of ion channels. It was therefore often difficult to ascertain whether behavioral effects of these drugs were due to NK1 receptor blockade, or to unspecific side effects. For example, high doses of GR205171 (30 mg/kg), which is probably the best available antagonist for the murine NK1 receptor (Bergstrom et al. 2000), increased the attack-latency in the resident-intruder test, reduced stress-induced neonatal vocalization, and increased the duration of struggle in the forced-swim test (Rupniak et al. 2000; Rupniak et al. 2001). These effects were similar to those observed after treatment with antidepressant drugs, such as fluoxetine or desipramine (Lucki et al. 2001; Rupniak et al. 2001; Schramm et al. 2001). However, it is not clear whether these behavioral effects of GR205171 were mediated by the NK1 receptor, because animals responded similarly after treatment with the low-affinity enantiomer GR22620600 (Rupniak et al. 2000). Also, in the rat forced-swim test GR205171 (40 mg/kg) remained ineffective (Rupniak et al. 2001). In models of anxiety the activity of NK1 receptor blockers is test -specific (Rodgers et al. 2004; Loiseau et al. 2003) and influenced by the strain and sex of the test animals (Vendruscolo et al. 2003).

4
Generation of Knockout Mice

Mice with targeted deletions of the *NK1* and *tac1* genes have been generated by several labs. *Tac1*$^{-/-}$ and *NK1R*$^{-/-}$ mice are viable and fertile, showing no gross alterations in maternal behaviors. These strains are therefore novel useful tools for the analysis of the physiological functions of the tachykinin system.

4.1
General Phenotype of Tac1-Null Mutant Mice

Two independent *tac1*-null mutant mouse strains have been described, originally on a mixed 129/Sv×C57BL/6J (Zimmer et al. 1998) or CD1×C57BL/6J (Cao et al. 1998), and later another on a pure C57BL/6J genetic background (Bilkei-Gorzo et al. 2002). The first reports (Cao et al. 1998; Zimmer et al. 1998) showed altered pain reactivity in *tac1*-deficient mice. The pain sensitivity of *tac1*$^{-/-}$ mice on a C57BL/6J×129/Sv genetic background was reduced in the hot-plate test, but not in another thermal-pain model, in the tail-flick test. Thus, while spinal pain reflexes as measured in the tail-flick test were normal, supra-spinal pain responses as measured in the hot-plate test seemed to be reduced in the absence of SP. Basbaum reported that the pain phenotype in the hot-plate test was intensity dependent: hypoalgesia was present using 55.5 °C but not lower (52.5 °C) or higher (58.5 °C) plate temperatures (Basbaum 1999). *Tac1*$^{-/-}$ mice on the mixed C57BL/6J genetic background were less reactive to intensive (tail clip), but not to mild mechanical pain stimuli (von Frey test). Responses to noxious chemical stimuli were measured after capsaicin injections into the hind paw, or after intraperitoneal injection of $MgSO_4$ or acetic acid (Cao et al. 1998). These stimuli also induced reduced responses in *tac1*$^{-/-}$ mice. Interestingly, Zimmer and coworkers found no difference between *tac1*$^{+/+}$ and *tac1*$^{-/-}$ mice using a twofold higher acetic acid dose, thus indicating an intensity-dependent phenotype (Zimmer et al. 1998). Indeed, a similar stimulus dependency was found in the formalin test where Cao and colleagues found diminished pain sensitivity only in the first phase, and only with one out of the three tested formalin concentrations. Zimmer et al. also found a reduced pain sensitivity in *tac1*$^{-/-}$ mice in the formalin test (Zimmer et al. 1998). Thus, it has been suggested that SP is involved in the intensity coding of pain.

NK1R$^{-/-}$ mice also showed stimulus-dependent changes in pain responses. Increasing mechanical stimulation failed to elicit increased responses. Repeated activation of C-fibers increases the responses to subsequent stimuli, and this "wind-up" reaction is thought to contribute to increased pain sensitivity e.g., in inflammation-induced hyperalgesia. This reaction was studied by De Felipe and coworkers in wildtype and *NK1R*$^{-/-}$ mice by recording the electromyographic activity in the hind paw (De Felipe et al. 1998) after repeated electrical or mechanical stimulation. The intensity of the responses increased gradually due to repetition of the same stimulus in wildtype mice, but this reaction was completely absent in *NK1R*$^{-/-}$ animals.

Stress-induced analgesia was also assessed in *NK1R*$^{-/-}$ mice after they were forced to swim in either cold (4–15 °C) water, which induces a non-opiate, *N*-methyl-D-aspartate (NMDA)-dependent analgesia, or in warm (33 °C) water, which induced an opiate-dependent analgesia (De Felipe et al. 1998). Interestingly, the non-opiate dependent analgesia was reduced in NK1 receptor-null mutant mice, while the opiate-dependent analgesia remained intact. Zimmer and colleagues (1998) found no difference between testing *tac1*$^{+/+}$ and *tac1*$^{-/-}$ mice after a cold-water swim stress.

When comparing the *NK1R*$^{-/-}$ and *tac1*$^{-/-}$ pain phenotype, it is important to remember that *tac1*$^{-/-}$ mice lacked not only SP, but also NKA. Since NKA is a ligand at the NK2 receptor, the signaling through this receptor should be normal in *NK1R* knockouts, but impaired in *tac1*$^{-/-}$ mice.

An involvement of tachykinins in the regulation of pain sensitivity was not unexpected. However, De Felipe and coworkers also noted an altered stress reactivity of *NK1R* knockouts. These animals showed significantly reduced aggressive behaviors in the resident-intruder test. In this test, wildtype and knockout males were isolated for 28 days, and subsequently intruder mice were placed into the home-cage of the isolated mouse. Wildtype mice attacked the intruders within 50 s, while the attack-latency in *NK1R*$^{-/-}$ mice was five times longer, with an average of 250 s. Moreover, the number of attacks was also significantly reduced in the knockouts (de Felipe et al. 2001). In the open-field test, where total and central activity of the animals was analyzed, no difference was observed between the strains. Because stress-induced analgesia was also reduced in this strain, the authors concluded that the activity of NK1 receptor is important to the organization of stress responses to major stressors like pain or invasion of territory (de Felipe et al. 2001). One might wonder why one neurotransmitter system regulates such distinct responses like stress-induced analgesia and territorial aggression? The common denominator of these behaviors in the natural environment may be found in situations where the individual is exposed to injury, pain, or aggression. In these situations, animals have to make fight-or-flight decisions i.e., choose between offensive or defensive defense strategies.

4.2 Behavior of *Tac1* and *NK1R* Knockout Mice in Models of Depression

The stress-related behavior of *tac1*$^{-/-}$ mice derived from these homozygous animals on a congenic C57BL/6J genetic background was analyzed in models of depression and anxiety (Bilkei-Gorzo et al. 2002). Based on the clinical association of depressive episodes and stressful life events, many of the animal models for the evaluation of antidepressant drug activity assess stress-precipitated behaviors. The two most widely employed animal models for antidepressant drug screening are the behavioral despair models of depression, such as the forced-swim (Porsolt 1997; Porsolt et al. 1977b) and tail-suspension tests (Steru et al. 1985). These models are based on the observation that rodents, when forced into an aversive situation from which they cannot escape, will rapidly cease attempts to escape and become immobile. Although the relationship between immobility and depression remains controversial (Gardier and Bourin 2001), drugs with antidepressant activity generally reduce the time in which the animals remain immobile (Borsini and Meli 1988; Porsolt et al. 1977a). Both the forced-swimming test and the tail-suspension test are conceptually similar, but they seem to be controlled by different sets of genes. A recent quantitative trait loci (QTL) study has identified several genetic links to the propensity of behavioral despair using the same assays (Yoshikawa et al. 2002). Unexpectedly, only a small number of QTLs was shared, and one common QTL on chromosome *8* displayed opposite effects in the two tests. Thus, although the test paradigms appear to be similar, distinct genetic pathways may underlie the despair-like behaviors in these tests. This idea is supported by the common observation that the efficacies of antidepressant drugs are different in the forced-swimming and the tail-suspension tests.

In the forced-swimming test, active escape periods alternated with periods in which the animals were completely inactive, or made only the movements necessary to keep their head above water. *Tac1*$^{-/-}$ mice were more active in this test than *tac1*$^{+/+}$ animals; they spent less time in immobility. These animals behaved like wildtype animals treated with antidepressant drugs, including the tricyclic uptake inhibitors imipramine and amitriptyline, or the selective serotonin reuptake inhibitor fluoxetine. In the tail suspension test, the immobility time in *tac1*$^{-/-}$ mice was also significantly reduced. A significantly decreased immobility time in the forced-swimming test was also observed in mouse strains with genetic deletion of monoamine oxidase A (MAO A) (Cases et al. 1995) or MAO B (Grimsby et al. 1997). Both MAO A and MAO B are key enzymes of the degradation of catecholamines. Pharmacological blockade of these enzymes is used clinically for the treatment of depression. *Tac1*$^{-/-}$ mice were also tested in another model of depression-related behavior, in the bulbectomy-induced hyperactivity test. The bulbectomy test is fundamentally different from the forced-swim or tail-suspension tests. It does not involve the concept of behavioral despair, but is rather based on the observation that bulbectomy will induce behavioral and neuroendocrinal changes similar to those observed in depressive patients, which can be reversed with antidepressant treatment (Jesberger and Richardson 1988). In rodents bulbectomy induces a hypermotility, which can be reversed by chronic antidepressant treatment (Otmakhova et al. 1992). *Tac1*$^{-/-}$ animals did not show any bulbectomy-induced hyperactivity, and they also behaved like animals treated with antidepressant in this test.

NK1R$^{-/-}$ mice also exhibited a reduced proneness to depression in the forced-swimming and tail suspension tests (Rupniak et al. 2001). Also, the selective NK1 receptor antagonist GR205171, as well as the serotonin reuptake inhibitor fluoxetine, was active in the forced-swimming test, proving that either acute pharmacological or life-long genetic blockade of NK1 receptor function has a similar effect as antidepressant treatment. On the other hand, pharmacological blockage of NK1 receptors with GR205171 failed to influence the activity of animals in the tail-suspension test. The authors concluded that in this model the increased activity of *NK1R*$^{-/-}$ mice is not related directly to the null mutation (Rupniak et al. 2001). Rather, desensitization of the presynaptic 5-HT$_{1A}$ receptors, which was observed in this strain (Santarelli et al. 2001), might be responsible for the altered behavior of *NK1R* knockouts in this model.

The results with *tac1*$^{-/-}$ and *NK1R*$^{-/-}$ mice in these models of depression strongly support the idea that the tachykinin system is involved in the pathophysiology of depression.

4.3 Behavior of *Tac1* and *NK1R* Knockout Mice in Models of Anxiety

The behavior of *tac1*$^{-/-}$ mice was also analyzed in several animal models of anxiety. The open-field test is a widely used tool for behavioral research, but less specific for the evaluation of the anxiety state of the animal, because it is a summation of the spontaneous motor and the exploratory activities, and only the latter is influenced by the anxiety level (Choleris et al. 2001). Under aversive environmental conditions (high level of illumination) the animals' activity is strongly affected by the emotional state, while less aversive situations (familiar, dimly lit environment) are useful to assess the general motor activity of mice. Because rodents avoid open areas, the activity of mice in the central part of the open-field arena is inversely correlated to the anxiety level. *Tac1*$^{+/+}$ mice spent only 6.5% of their total activity in the central part, which represented 11% of the total field, indicating that they avoided this aversive area. In contrast, *tac1*$^{-/-}$ mice spent 13.6% of their activity in the central area (Bilkei-Gorzo et al. 2002). The increased central activity of the *tac1*$^{-/-}$ mice indicates that the test situation was anxiogenic for *tac1*$^{+/+}$ animals, but less so for the knockout mice.

The Thatcher–Britton novelty conflict paradigm (Rochford et al. 1997) evaluates the effect of an aversive environment on feeding behavior. In this paradigm, food-deprived mice are placed at the periphery of a well-lit and unfamiliar open-field apparatus in contact with the enclosing walls. Food pellets are placed in the center. This test is similar to the open-field test because the animals also avoid the central area, but hunger rather than the exploration drive is here the positive reinforcer (Rochford et al. 1997). *Tac1*$^{-/-}$ mice began to eat significantly faster than *tac1*$^{+/+}$ mice, thus indicating again a reduced state of anxiety in the absence of SP (Bilkei-Gorzo et al. 2002).

Tac1$^{-/-}$ mice were also studied in the zero-maze test, which is considered to provide a specific and sensitive readout of the anxiety state of an animal. The zero-maze device consists of an elevated annular platform, divided into two open and two enclosed compartments. High levels of anxiety are thought to be associated with a reduced activity of mice in the open areas, and an increased frequency of risk-assessing behaviors, such as stretch-attend postures (Shepherd et al. 1994). *Tac1*$^{-/-}$ mice were more active in the open areas and showed fewer stretch-attend postures than *tac1*$^{+/+}$ animals. They behaved similar to wildtype animals treated with diazepam (1 mg/kg), which was more potent than buspirone (1 mg/kg), but less potent than ethanol (2 g/kg) in this test.

Finally, *tac1*$^{-/-}$ and *tac1*$^{+/+}$ mice were tested in the social activity paradigm (File 1985). In this paradigm two male mice that were unfamiliar to each other are brought into a novel, brightly lit cage. The social drive of these animals is tempered by the adversity of the situation. More anxious animals explore the new environment rather than investigating the partner. Therefore time spent with social interactions is inversely related to the anxiety state of the animal. The social activity was significantly higher in *tac1*$^{-/-}$ mice, indicating again reduced levels of anxiety in these animals. The social behavior of the animals was friendly and exploratory, antagonistic behaviors were not observed. Altogether these results strongly suggested that *tac1*$^{-/-}$ mice were less anxious than *tac1*$^{+/+}$ animals.

The first report describing the NK1 receptor knockout phenotype (de Felipe et al. 1998) did not show any alteration in the anxiety level of knockouts using the open-field test, where total and central activity of the animals was analyzed. However, the authors also found that *NK1R*$^{-/-}$ mice were less aggressive than *NK1R*$^{+/+}$ controls and that they showed a reduced stress-induced analgesia. These results indicated that the emotional reactivity of mice was affected by the NK1 mutation. When the animals were tested in a later study in a modified version of the open-field test and in the plus-maze test, the authors again found no anxiety-related phenotype in this strain. Wildtype and knockout mice showed similar horizontal motor activities, habituation-induced reduction in the motor activity, and visits of unfamiliar objects placed onto the arena in the open field. Furthermore, the time spent and visits to the open part of the plus-maze device were also unaffected by the NK1 deletion (Murtra et al. 2000). The negative result in these models clearly showed that the NK1 receptor activation plays a more restricted role in the regulation of emotional states.

More recently, the influence of pharmacological blockade and genetic deletion of NK1 receptor on the intensity of ultrasonic vocalization in neonatal mice and guinea pigs was tested (Rupniak et al. 2000). This assay is based on the observation that infant rodents emit ultrasonic calls that induce increased care from the mothers. The intensity of ultrasonic calls is influenced by numerous factors, but separation from the mother and littermates, and a cold environment, together with unfamiliar tactile stimuli, seem to be the most important. The intensity of the ultrasonic vocalization is related to the severity of the stress experienced by the pups, and is thought to correlate to the emotionality of the animal.

The pharmacological blockage of NK1 receptors with the highly CNS-penetrant antagonists CP-99,994 or GR205171 but not their less-active enantiomer L-796,325 nor poorly CNS-penetrant derivative CGP49823 reduced the intensity of ultrasonic vocalization after maternal separation significantly, while the NK1 receptor agonist GR73632 elicited vocalization in pups. These pharmacological data strongly suggest that the activity of central NK1 receptors modulates stress sensitivity in this paradigm. This finding was confirmed through the analysis of $NK1R^{-/-}$ mouse pups, because they also showed fewer ultrasonic vocalizations after separation from their mothers.

It should be noted that antidepressants also reduce the number of calls in this paradigm. Indeed, anxiolytics and antidepressants are similarly effective. Rupniak and colleagues carefully compared the effects of pharmacological blockade and genetic ablation of the NK1 receptor with the administration of anxiolytics or antidepressant compounds in several animal models of depression and anxiety. While their findings generally supported the antidepressant character of NK1 receptor blockade or deletion, they did not find evidence for a role of NK1 receptors in the regulation of anxiety (Rupniak et al. 2001). They found no difference between wildtype and $NK1R^{-/-}$ mice in the plus-maze test, and NK1 receptor antagonists had no effect in this model using rats (treated with GR205171 having high affinity to the rodent-type NK1 receptor) or guinea pigs (treated with L-760735 having high affinity to the human-type NK1 receptor). Altogether, the experiments with $NK1R^{-/-}$ mice on a mixed genetic background suggested that NK1 receptor is involved in stress responses in animals. Modulation of the activity of these receptors has a clear effect in the models of depression, but has a limited, if any, role in the regulation of anxiety states.

In contrast, Santarelli and colleagues found significantly decreased anxiety level in three different models of anxiety using knockout mice on a pure 129/Sv genetic background (Santarelli et al. 2001). Knockout pups showed a reduced level of ultrasonic vocalization after maternal separation. Diazepam or the NK1 receptor antagonist RP67580 elicited similar effects when injected into wild-type mice. Mice with a selective deletion of NK1 receptor also had a reduced level of anxiety in the elevated plus maze and in the novelty induced suppression of feeding paradigms. Moreover, the NK1 receptor antagonist RP67580 was effective in the plus-maze test in the same set of experiments as the anxiolytic reference compound diazepam, which shows that antagonism or deletion of NK1 receptor produced a similar effect as treatment with an anxiolytic drug. Not only the behavioral analysis, but also the analysis of stress hormone levels provided evidence for decreased stress reactivity in *NK1R* knockouts. Plasma corticosterone levels were elevated after the plus-maze test in $NK1R^{+/+}$ mice, but less so in $NK1R^{-/-}$ mice. Thus, these studies support a proposed role for NK1 receptors in the modulation of anxiety and suggest that this activity may be dependent on the genetic background. A general overview of the pharmacological and genetic studies is summarized in Table 1. Altogether, the Santarelli study (Santarelli et al. 2001) provided strong evidence for reduced anxiety levels in NK1 receptor-deficient mice, in contrast to what has been shown by Rupniak or Murtra (Rupniak et al. 2000, 2001; Murta et al. 2000). The reason for the discrepancy could be the difference in the genetic background (mixed 129/Sv×C57BL/6J in all studies with negative results; pure 129/Sv in the positive result); however, methodological differences also might have contributed.

It is interesting to note that pharmacological blockage of NK1 receptors also revealed anxiolytic effects in the plus-maze test when mice from the 129/Sv strain (Santarelli et al. 2001), but not from other strains (Rodgers et al. 2004), were used. Thus, the anxiolytic effect of NK1 receptor antagonists seems to be more sensitive to effects of the genetic background as compared to benzodiazepines. It is also possible that the relative contribution of the SP–NK1 system in the modulation of anxiety is situation dependent. The basal corticosterone levels in the blood plasma of $NK1R^{+/+}$ and $NK1^{-/-}$ mice do not differ in low-stress situations, but the increase after the stressful elevated plus-maze test is blunted in the knockout animals (Santarelli et al. 2001).

When we compare phenotype of mice with deletion of the neurotransmitter ($tac1^{-/-}$ mice; Bilkei-Gorzo et al. 2002) or the receptor ($NK1R^{-/-}$ mice; Rupniak et al. 2000, 2001) in animal models of anxiety, we find consistently lower levels in $tac1^{-/-}$ mice, while only one study (Santarelli et al. 2001) found an anxiety-related phenotype in $NK1R^{-/-}$ mice. The reason for this discrepancy could be due to the methods used, to genetic factors, or both. Probably the most important difference in the testing method is the timing of the experiments: studies with $NK1R^{-/-}$ mice were carried out in the light (rest) phase, while tests with the $tac1^{-/-}$ mice were performed in the dark (active) phase. Although nocturnal rodents generally avoid well-illuminated areas, they seem to be more affected by the light conditions during the dark phase. It is tempting to speculate that the testing conditions were highly aversive for $tac1^{-/-}$ mice, a factor that, in turn, may have contributed to the robust anxiety phenotype in these animals.

On the other hand, an effect of the genetic background seems equally possible, because significant differences in emotional behaviors have been observed between $NKR1^{-/-}$ mice on different genetic backgrounds.

Finally, we have to consider the possibility that the ablation of NKA in the $tac1^{-/-}$ mice has contributed to the anxiety phenotype. Although the preferred receptor of NKA, NK2, is not abundantly expressed in the brain and many researchers do not consider it an important modulator of anxiety, a recent study has shown that SR48968, a selective antagonist of NK2 receptor, had an enantioselective effect in animal models of anxiety (Griebel et al. 2001) and depression (Steinberg et al. 2001).

Stress is thought to be an important factor in the pathophysiology of depression and anxiety disorders. It seems possible that the reduced stress reactivity of NK1 receptor- and tac1-deficient mice has contributed to the behavioral phenotypes observed in the animal models of anxiety and depression.

Table 1 Effect of pharmacological and molecular genetic modifications of the SP/NK1 system on anxiety level

Model	Intervention	Species	Anxiety level	Reference(s)
Knockout	NK1 gene deletion	C57×129 mice	No change	De Felipe et al. 1998; Murtra et al. 2000; Rupniak et al. 2001
	NK1 gene deletion	129 mice	Decreased	Santarelli et al. 2001
	Tac1 gene deletion	C57BL/6J mice	Decreased	Bilkei-Gorzo et al. 2002
Treatment with agonist	GR73632 (i.c.v.)	Guinea pig	Increased	Kramer et al. 1998
	SP local (ventr. pallidum.)	Rat	Decreased	Nikolaus et al. 1999
	SP local (ventr. pallidum.)	Rat	Dose-dependent effect	Hasenohrl et al. 1998
	SP local (PAG)	Rat	Dose-dependent effect	Aguiar et al. 1996
	SP local (septal n.)	Rat	Increased	Gavioli et al. 1999
	SP fragments local (PAG)	Rat	Fragment dependent	De Araujo et al. 1999
	SP (i.c.v.)	Mice	Increased	Teixeira et al. 1996
	SP and C terminal (i.c.v.)	Rat	Increased	Duarte et al. 2004
	SP	Rat	Decreased	Echeverry et al. 2001
	SP	Swiss mice	Increased	Baretta et al. 2001
	SP fragment	Marmoset	Decreased	Barros et al. 2002
Treatment with antagonist	FK 888	Mice	Decreased	Teixeira et al. 1996
	GR205171	Mice	Decreased	Rupniak et al. 2000
	GR205171	Rat	No change	Rupniak et al. 2001
	L-760735	Gerbil	Decreased	Cheeta et al. 2001
	L-760735	Guinea pig	Decreased	Kramer et al. 1998
	L-760735 local (amygdala)	Guinea pig	Decreased	Boyce et al. 2001
	NKP608	Gerbil	Decreased	Gentsch et al. 2002
	NKP608	Mice	No change	Rodgers et al. 2004
	NKP608	Rat	Decreased	Vassout et al. 2000
	RP 67580	Mice	Decreased	Santarelli et al. 2001
	RP 67580	Mice	Increased	Zernig et al. 1993
	RP 67580	Rat	No change	Loiseau et al. 2003

5
Mechanism of Change in Phenotype in Tac1 and NK1R-Null Mutant Mice

One of the most important questions that remains to be answered concerns the mechanism leading to the manifestation of the knockout phenotype. The tac1 and NK1R gene deletions could directly modulate stress-related behaviors through an altered SP/NK1 signaling of limbic neuronal circuits that mediate stress-related behaviors, or act indirectly by affecting other systems. A direct involvement of NK1R signaling is suggested by the widespread expression of SP in neurons of many limbic structures, including various nuclei of the amygdala (Gray and Magnuson 1992; Roberts et al. 1982), but also in the periaqueductal gray area (Commons and Valentino 2002; Smith et al. 1994) and the nucleus tractus solitarii (Batten et al. 2002).

It has recently been shown that modulations of the NK1R activity, through selective antagonists (Blier et al. 2004) or by genetic ablations of NK1 receptors (Santarelli et al. 2001), resulted in an increased firing of serotonergic neurons in the dorsal raphe nucleus (DRN). This finding suggests that the serotonergic system is a downstream target of SP/NK1R signaling. Interestingly, similar changes in the activity of limbic serotonergic neurons were observed in animals after chronic administration of clinically effective antidepressant selective serotonin reuptake inhibitors (Froger et al. 2001). However, because only a few serotonergic neurons express NK1 receptors, the increased activity of serotonergic cells in the DRN after NK1 receptor blockade is likely to result from an indirect effect. A possible link between the two systems involves noradrenergic projections from the locus coeruleus, which contain NK1 receptors in high concentration, to the raphe nuclei. Alternatively, glutamatergic cells in the DRN, which also contain NK1 receptor, may be involved in the regulation of serotonergic neurons by SP (Santarelli et al. 2001).

Together, the anatomical data and the results from genetically modified animals support the idea that SP and other tac1-derived neuropeptides, play a major role in stress responses, anxiety, and depression. Therefore, modulation of the SP–NK1 system may have therapeutic value in the treatment of stress-related neuropsychiatric disorders.

References

Aguiar MS, Brandao ML (1996) Effects of microinjections of the neuropeptide substance P in the dorsal periaqueductal gray on the behaviour of rats in the plus-maze test. Physiol Behav 60:1183–1186

Baretta IP, Assreuy J, De Lima TC (2001) Nitric oxide involvement in the anxiogenic-like effect of substance P. Behav Brain Res 121:199–205

Barros M, De Souza Silva MA, Huston JP, Tomaz C (2002) Anxiolytic-like effects of substance P fragment (SP(1–7)) in non-human primates (Callithrix penicillata). Peptides 23:967–973

Basbaum AI (1999) Distinct neurochemical features of acute and persistent pain. Proc Natl Acad Sci U S A 96:7739–7743

Batten TF, Gamboa-Esteves FO, Saha S (2002) Evidence for peptide co-transmission in retrograde- and anterograde-labelled central nucleus of amygdala neurones projecting to NTS. Auton Neurosci 98:28–32

Beaujouan JC, Saffroy M, Torrens Y, Sagan S, Glowinski J (1999) Pharmacological characterization of tachykinin septide-sensitive binding sites in the rat submaxillary gland. Peptides 20:1347–1352

Beaujouan JC, Saffroy M, Torrens Y, Glowinski J (2000) Different subtypes of tachykinin NK(1) receptor binding sites are present in the rat brain. J Neurochem 75:1015–1026

Bergstrom M, Fasth KJ, Kilpatrick G, Ward P, Cable KM, Wipperman MD, Sutherland DR, Langstrom B (2000) Brain uptake and receptor binding of two [11C]labelled selective high affinity NK1-antagonists, GR203040 and GR205171—PET studies in rhesus monkey. Neuropharmacology 39:664–670

Bilkei-Gorzo A, Racz I, Michel K, Zimmer A (2002) Diminished anxiety- and depression-related behaviors in mice with selective deletion of the Tac1 gene. J Neurosci 22:10046–10052

Blier P, Gobbi G, Haddjeri N, Santarelli L, Mathew G, Hen R (2004) Impact of substance P receptor antagonism on the serotonin and norepinephrine systems: relevance to the antidepressant/anxiolytic response. J Psychiatry Neurosci 29:208–218

Borsini F, Meli A (1988) Is the forced swimming test a suitable model for revealing antidepressant activity? Psychopharmacology (Berl) 94:147–160

Boyce S, Smith D, Carlson E, Hewson L, Rigby M, O'Donnell R, Harrison T, Rupniak NM (2001) Intra-amygdala injection of the substance P [NK(1) receptor] antagonist L-760735 inhibits neonatal vocalisations in guinea-pigs. Neuropharmacology 41:130–137

Brodin E, Rosen A, Schott E, Brodin K (1994) Effects of sequential removal of rats from a group cage, and of individual housing of rats, on substance P, cholecystokinin and somatostatin levels in the periaqueductal grey and limbic regions. Neuropeptides 26:253–260

Cao YQ, Mantyh PW, Carlson EJ, Gillespie AM, Epstein CJ, Basbaum AI (1998) Primary afferent tachykinins are required to experience moderate to intense pain. Nature 392:390–394

Cases O, Seif I, Grimsby J, Gaspar P, Chen K, Pournin S, Muller U, Aguet M, Babinet C, Shih JC, et al (1995) Aggressive behavior and altered amounts of brain serotonin and norepinephrine in mice lacking MAOA. Science 268:1763–1766

Chang MM, Leeman SE, Niall HD (1971) Amino-acid sequence of substance P. Nat New Biol 232:86–87

Cheeta S, Tucci S, Sandhu J, Williams AR, Rupniak NM, File SE (2001) Anxiolytic actions of the substance P (NK1) receptor antagonist L-760735 and the 5-HT1A agonist 8-OH-DPAT in the social interaction test in gerbils. Brain Res 915:170–175

Choleris E, Thomas AW, Kavaliers M, Prato FS (2001) A detailed ethological analysis of the mouse open field test: effects of diazepam, chlordiazepoxide and an extremely low frequency pulsed magnetic field. Neurosci Biobehav Rev 25:235–260

Commons KG, Valentino RJ (2002) Cellular basis for the effects of substance P in the periaqueductal gray and dorsal raphe nucleus. J Comp Neurol 447:82–97

Cooper JR, Bloom FE, Roth RH (1996) The biochemical basis of neuropharmacology. Oxford University Press, New York

Culman J, Wiegand B, Spitznagel H, Klee S, Unger T (1995) Effects of the tachykinin NK1 receptor antagonist, RP 67580, on central cardiovascular and behavioural effects of substance P, neurokinin A and neurokinin B. Br J Pharmacol 114:1310–1316

Culman J, Klee S, Ohlendorf C, Unger T (1997) Effect of tachykinin receptor inhibition in the brain on cardiovascular and behavioral responses to stress. J Pharmacol Exp Ther 280:238–246

De Araujo JE, Silva RC, Huston JP, Brandao ML (1999) Anxiogenic effects of substance P and its 7–11 C terminal, but not the 1–7 N terminal, injected into the dorsal periaqueductal gray. Peptides 20:1437–1443

De Felipe C, Herrero JF, O'Brien JA, Palmer JA, Doyle CA, Smith AJ, Laird JM, Belmonte C, Cervero F, Hunt SP (1998) Altered nociception, analgesia and aggression in mice lacking the receptor for substance P. Nature 392:394–397

Echeverry MB, Hasenohrl RU, Huston JP, Tomaz C (2001) Comparison of neurokinin SP with diazepam in effects on memory and fear parameters in the elevated T-maze free exploration paradigm. Peptides 22:1031–1036

Emonds-Alt X, Doutremepuich JD, Heaulme M, Neliat G, Santucci V, Steinberg R, Vilain P, Bichon D, Ducoux JP, Proietto V, et al (1993) In vitro and in vivo biological activities of SR140333, a novel potent non-peptide tachykinin NK1 receptor antagonist. Eur J Pharmacol 250:403–413

Erspamer GF, Erspamer V, Piccinelli D (1980) Parallel bioassay of physalaemin and kassinin, a tachykinin dodecapeptide from the skin of the African frog Kassina senegalensis. Naunyn Schmiedebergs Arch Pharmacol 311:61–65

Fehder WP, Sachs J, Uvaydova M, Douglas SD (1997) Substance P as an immune modulator of anxiety. Neuroimmunomodulation 4:42–48

File SE (1985) Animal models for predicting clinical efficacy of anxiolytic drugs: social behaviour. Neuropsychobiology 13:55–62

Folkers K, Hakanson R, Horig J, Xu JC, Leander S (1984) Biological evaluation of substance P antagonists. Br J Pharmacol 83:449–456

Fong TM, Yu H, Strader CD (1992) Molecular basis for the species selectivity of the neurokinin-1 receptor antagonists CP-96,345 and RP67580. J Biol Chem 267:25668–25671

Froger N, Gardier AM, Moratalla R, Alberti I, Lena I, Boni C, De Felipe C, Rupniak NM, Hunt SP, Jacquot C, Hamon M, Lanfumey L (2001) 5-hydroxytryptamine (5-HT)1A autoreceptor adaptive changes in substance P (neurokinin 1) receptor knock-out mice mimic antidepressant-induced desensitization. J Neurosci 21:8188–8197

Gardier AM, Bourin M (2001) Appropriate use of "knockout" mice as models of depression or models of testing the efficacy of antidepressants. Psychopharmacology (Berl) 153:393–394

Gardner CJ, Armour DR, Beattie DT, Gale JD, Hawcock AB, Kilpatrick GJ, Twissell DJ, Ward P (1996) GR205171: a novel antagonist with high affinity for the tachykinin NK1 receptor, and potent broad-spectrum anti-emetic activity. Regul Pept 65:45–53

Gavioli EC, Canteras NS, De Lima TC (1999) Anxiogenic-like effect induced by substance P injected into the lateral septal nucleus. Neuroreport 10:3399–3403

Gavioli EC, Canteras NS, De Lima TC (2002) The role of lateral septal NK1 receptors in mediating anxiogenic effects induced by intracerebroventricular injection of substance P. Behav Brain Res 134:411–415

Gentsch C, Cutler M, Vassout A, Veenstra S, Brugger F (2002) Anxiolytic effect of NKP608, a NK1-receptor antagonist, in the social investigation test in gerbils. Behav Brain Res 133:363–368

Gray TS, Magnuson DJ (1992) Peptide immunoreactive neurons in the amygdala and the bed nucleus of the stria terminalis project to the midbrain central gray in the rat. Peptides 13:451–460

Griebel G, Perrault G, Soubrie P (2001) Effects of SR48968, a selective non-peptide NK2 receptor antagonist on emotional processes in rodents. Psychopharmacology (Berl) 158:241–251

Grimsby J, Toth M, Chen K, Kumazawa T, Klaidman L, Adams JD, Karoum F, Gal J, Shih JC (1997) Increased stress response and beta-phenylethylamine in MAOB-deficient mice. Nat Genet 17:206–210
Hasenohrl RU, Jentjens O, De Souza Silva MA, Tomaz C, Huston JP (1998) Anxiolytic-like action of neurokinin substance P administered systemically or into the nucleus basalis magnocellularis region. Eur J Pharmacol 354:123–133
Hasenohrl RU, Souza-Silva MA, Nikolaus S, Tomaz C, Brandao ML, Schwarting RK, Huston JP (2000) Substance P and its role in neural mechanisms governing learning, anxiety and functional recovery. Neuropeptides 34:272–280
Herbert MK, Holzer P (2002) [Why are substance P(NK1)-receptor antagonists ineffective in pain treatment?]. Anaesthesist 51:308–319
Hokfelt T, Pernow B, Wahren J (2001) Substance P: a pioneer amongst neuropeptides. J Intern Med 249:27–40
Honkaniemi J, Pelto-Huikko M, Rechardt L, Isola J, Lammi A, Fuxe K, Gustafsson JA, Wikstrom AC, Hokfelt T (1992) Colocalization of peptide and glucocorticoid receptor immunoreactivities in rat central amygdaloid nucleus. Neuroendocrinology 55:451–459
Hurd YL, Keller E, Sotonyi P, Sedvall G (1999) Preprotachykinin-A mRNA expression in the human and monkey brain: an in situ hybridization study. J Comp Neurol 411:56–72
Jesberger JA, Richardson JS (1988) Brain output dysregulation induced by olfactory bulbectomy: an approximation in the rat of major depressive disorder in humans? Int J Neurosci 38:241–265
Jessop DS, Renshaw D, Larsen PJ, Chowdrey HS, Harbuz MS (2000) Substance P is involved in terminating the hypothalamo-pituitary-adrenal axis response to acute stress through centrally located neurokinin-1 receptors. Stress 3:209–220
Kramer MS, Cutler N, Feighner J, Shrivastava R, Carman J, Sramek JJ, Reines SA, Liu G, Snavely D, Wyatt-Knowles E, Hale JJ, Mills SG, MacCoss M, Swain CJ, Harrison T, Hill RG, Hefti F, Scolnick EM, Cascieri MA, Chicchi GG, Sadowski S, Williams AR, Hewson L, Smith D, Rupniak NM, et al (1998) Distinct mechanism for antidepressant activity by blockade of central substance P receptors. Science 281:1640–1645
Kramer MS, Winokur A, Kelsey J, Preskorn SH, Rothschild AJ, Snavely D, Ghosh K, Ball WA, Reines SA, Munjack D, Apter JT, Cunningham L, Kling M, Bari M, Getson A, Lee Y (2004) Demonstration of the efficacy and safety of a novel substance P (NK1) receptor antagonist in major depression. Neuropsychopharmacology 29:385–392
Ku YH, Tan L, Li LS, Ding X (1998) Role of corticotropin-releasing factor and substance P in pressor responses of nuclei controlling emotion and stress. Peptides 19:677–682
Loiseau F, Le Bihan C, Hamon M, Thiebot MH (2003) Distinct effects of diazepam and NK1 receptor antagonists in two conflict procedures in rats. Behav Pharmacol 14:447–455
Lucki I, Dalvi A, Mayorga AJ (2001) Sensitivity to the effects of pharmacologically selective antidepressants in different strains of mice. Psychopharmacology (Berl) 155:315–322
Maggi CA (1995) The mammalian tachykinin receptors. Gen Pharmacol 26:911–944
Merchenthaler I, Maderdrut JL, O'Harte F, Conlon JM (1992) Localization of neurokinin B in the central nervous system of the rat. Peptides 13:815–829
Murtra P, Sheasby AM, Hunt SP, De Felipe C (2000) Rewarding effects of opiates are absent in mice lacking the receptor for substance P. Nature 405:180–183
Nakanishi S (1991) Mammalian tachykinin receptors. Annu Rev Neurosci 14:123–136
Nikolaus S, Huston JP, Hasenohrl RU (1999) Reinforcing effects of neurokinin substance P in the ventral pallidum: mediation by the tachykinin NK1 receptor. Eur J Pharmacol 370:93–99

Otmakhova NA, Gurevich EV, Katkov YA, Nesterova IV, Bobkova NV (1992) Dissociation of multiple behavioral effects between olfactory bulbectomized C57Bl/6J and DBA/2J mice. Physiol Behav 52:441–448

Porsolt RD (1997) Historical perspective on CMS model. Psychopharmacology (Berl) 134:363–4; discussion 371–7

Porsolt RD, Bertin A, Jalfre M (1977a) Behavioral despair in mice: a primary screening test for antidepressants. Arch Int Pharmacodyn Ther 229:327–336

Porsolt RD, Le Pichon M, Jalfre M (1977b) Depression: a new animal model sensitive to antidepressant treatments. Nature 266:730–732

Ribeiro-da-Silva A, Hokfelt T (2000) Neuroanatomical localisation of Substance P in the CNS and sensory neurons. Neuropeptides 34:256–271

Roberts GW, Woodhams PL, Polak JM, Crow TJ (1982) Distribution of neuropeptides in the limbic system of the rat: the amygdaloid complex. Neuroscience 7:99–131

Rochford J, Beaulieu S, Rousse I, Glowa JR, Barden N (1997) Behavioral reactivity to aversive stimuli in a transgenic mouse model of impaired glucocorticoid (type II) receptor function: effects of diazepam and FG-7142. Psychopharmacology (Berl) 132:145–152

Rodgers RJ, Gentsch C, Hoyer D, Bryant E, Green AJ, Kolokotroni KZ, Martin JL (2004) The NK1 receptor antagonist NKP608 lacks anxiolytic-like activity in Swiss-Webster mice exposed to the elevated plus-maze. Behav Brain Res 154:183–192

Rosell S, Bjorkroth U, Xu JC, Folkers K (1983) The pharmacological profile of a substance P (SP) antagonist. Evidence for the existence of subpopulations of SP receptors. Acta Physiol Scand 117:445–449

Rosen A, Brodin K, Eneroth P, Brodin E (1992) Short-term restraint stress and s.c. saline injection alter the tissue levels of substance P and cholecystokinin in the peri-aqueductal grey and limbic regions of rat brain. Acta Physiol Scand 146:341–348

Rothman RB, Herkenham M, Pert CB, Liang T, Cascieri MA (1984) Visualization of rat brain receptors for the neuropeptide, substance P. Brain Res 309:47–54

Rupniak NM, Carlson EC, Harrison T, Oates B, Seward E, Owen S, de Felipe C, Hunt S, Wheeldon A (2000) Pharmacological blockade or genetic deletion of substance P (NK(1)) receptors attenuates neonatal vocalisation in guinea-pigs and mice. Neuropharmacology 39:1413–1421

Rupniak NM, Carlson EJ, Webb JK, Harrison T, Porsolt RD, Roux S, de Felipe C, Hunt SP, Oates B, Wheeldon A (2001) Comparison of the phenotype of NK1R−/− mice with pharmacological blockade of the substance P (NK1) receptor in assays for antidepressant and anxiolytic drugs. Behav Pharmacol 12:497–508

Saffroy M, Torrens Y, Glowinski J, Beaujouan JC (2001) Presence of NK2 binding sites in the rat brain. J Neurochem 79:985–996

Santarelli L, Gobbi G, Debs PC, Sibille ET, Blier P, Hen R, Heath MJ (2001) Genetic and pharmacological disruption of neurokinin 1 receptor function decreases anxiety-related behaviors and increases serotonergic function. Proc Natl Acad Sci U S A 98:1912–1917

Schedlowski M, Fluge T, Richter S, Tewes U, Schmidt RE, Wagner TO (1995) Beta-endorphin, but not substance-P, is increased by acute stress in humans. Psychoneuroendocrinology 20:103–110

Schramm NL, McDonald MP, Limbird LE (2001) The alpha(2a)-adrenergic receptor plays a protective role in mouse behavioral models of depression and anxiety. J Neurosci 21:4875–4882

Severini C, Improta G, Falconieri-Erspamer G, Salvadori S, Erspamer V (2002) The tachykinin peptide family. Pharmacol Rev 54:285–322

Shepherd JK, Grewal SS, Fletcher A, Bill DJ, Dourish CT (1994) Behavioural and pharmacological characterisation of the elevated "zero- maze" as an animal model of anxiety. Psychopharmacology (Berl) 116:56–64

Shughrue PJ, Lane MV, Merchenthaler I (1996) In situ hybridization analysis of the distribution of neurokinin-3 mRNA in the rat central nervous system. J Comp Neurol 372:395–414

Smith DW, Hewson L, Fuller P, Williams AR, Wheeldon A, Rupniak NM (1999) The substance P antagonist L-760,735 inhibits stress-induced NK(1) receptor internalisation in the basolateral amygdala. Brain Res 848:90–95

Smith GS, Savery D, Marden C, Lopez Costa JJ, Averill S, Priestley JV, Rattray M (1994) Distribution of messenger RNAs encoding enkephalin, substance P, somatostatin, galanin, vasoactive intestinal polypeptide, neuropeptide Y, and calcitonin gene-related peptide in the midbrain periaqueductal grey in the rat. J Comp Neurol 350:23–40

Spitznagel H, Baulmann J, Blume A, Unger T, Culman J (2001) C-FOS expression in the rat brain in response to substance P and neurokinin B. Brain Res 916:11–21

Steinberg R, Alonso R, Griebel G, Bert L, Jung M, Oury-Donat F, Poncelet M, Gueudet C, Desvignes C, Le Fur G, Soubrie P (2001) Selective blockade of neurokinin-2 receptors produces antidepressant- like effects associated with reduced corticotropin-releasing factor function. J Pharmacol Exp Ther 299:449–458

Steru L, Chermat R, Thierry B, Simon P (1985) The tail suspension test: a new method for screening antidepressants in mice. Psychopharmacology (Berl) 85:367–370

Takayama H, Ota Z, Ogawa N (1986) Effect of immobilization stress on neuropeptides and their receptors in rat central nervous system. Regul Pept 15:239–248

Takeda Y, Takeda J, Smart BM, Krause JE (1990) Regional distribution of neuropeptide gamma and other tachykinin peptides derived from the substance P gene in the rat. Regul Pept 28:323–333

Teixeira RM, De lima TC (2003) Involvement of tachykinin NK1 receptor in the behavioral and immunological responses to swimming stress in mice. Neuropeptides 37:307–315

Teixeira RM, Santos AR, Ribeiro SJ, Calixto JB, Rae GA, De Lima TC (1996) Effects of central administration of tachykinin receptor agonists and antagonists on plus-maze behavior in mice. Eur J Pharmacol 311:7–14

Vassout A, Veenstra S, Hauser K, Ofner S, Brugger F, Schilling W, Gentsch C (2000) NKP608: a selective NK-1 receptor antagonist with anxiolytic-like effects in the social interaction and social exploration test in rats. Regul Pept 96:7–16

Vendruscolo LF, Takahashi RN, Bruske GR, Ramos A (2003) Evaluation of the anxiolytic-like effect of NKP608, a NK1-receptor antagonist, in two rat strains that differ in anxiety-related behaviors. Psychopharmacology (Berl) 170:287–293

Weiss DW, Hirt R, Tarcic N, Berzon Y, Ben-Zur H, Breznitz S, Glaser B, Grover NB, Baras M, O'Dorisio TM (1996) Studies in psychoneuroimmunology: psychological, immunological, and neuroendocrinological parameters in Israeli civilians during and after a period of Scud missile attacks. Behav Med 22:5–14

Yoshikawa T, Watanabe A, Ishitsuka Y, Nakaya A, Nakatani N (2002) Identification of multiple genetic loci linked to the propensity for "behavioral despair" in mice. Genome Res 12:357–366

Zernig G, Troger J, Saria A (1993) Different behavioral profiles of the non-peptide substance P (NK1) antagonists CP-96,345 and RP 67580 in Swiss albino mice in the black- and-white box. Neurosci Lett 151:64–66

Zimmer A, Zimmer AM, Baffi J, Usdin T, Reynolds K, Konig M, Palkovits M, Mezey E (1998) Hypoalgesia in mice with a targeted deletion of the tachykinin 1 gene. Proc Natl Acad Sci U S A 95:2630–2635

HEP (2005) 169:163–179

Genetic Epidemiology of Anxiety Disorders

K. R. Merikangas[1] (✉) · N. C. P. Low[2]

[1]Section on Developmental Genetic Epidemiology, Mood and Anxiety Disorders Program, National Institute of Mental Health/NIH, Building 35, Room 1A201, 35 Convent Drive, MSC 3720, 20892-3720, Bethesda MD, , USA
kathleen.merikangas@nih.gov

[2]Section on Developmental Genetic Epidemiology, Mood and Anxiety Disorder Program, National Institute of Mental Health/NIH, Department of Health and Human Services, Building 35, Room 1A112, 35 Convent Drive, MSC 3720, Bethesda MD, 20892-3720, USA

Abstract This chapter reviews the genetic epidemiology of the major subtypes of anxiety disorders including panic disorder, phobic disorders, generalized anxiety disorder, and obsessive–compulsive disorder. Controlled family studies reveal that all of these anxiety subtypes are familial, and twin studies suggest that the familial aggregation is attributable in part to genetic factors. Panic disorder and, its spectrum have the strongest magnitude of familial clustering and genetic underpinnings. Studies of offspring of parents with anxiety disorders an increased risk of mood and anxiety disorders, but there is far less specificity of the manifestations of anxiety in children and young adolescents. Although there has been a plethora of studies designed to identify genes underlying these conditions, to date, no specific genetic loci have been identified and replicated in independent samples.

Keywords Genetics · Epidemiology · Anxiety disorders · Linkage

1 Introduction

During the past decade there has been an increasing focus on anxiety disorders, which have emerged as the most prevalent mental disorders in the general population. According to several international community surveys, such as the National Comorbidity Study (NCS) in the United States, the Zurich Cohort Study, and the WHO World Mental Health 2000 Initiative Nemesis Study (Bijl et al. 1998), anxiety disorders affect nearly one in four adults in the population. The magnitude of anxiety disorders in youth is quite similar to that reported in adults, thereby indicating the importance of a life-course approach to the study of anxiety. These diverse investigations have also advanced our understanding of anxiety disorders by portraying the natural history of these disorders in a descriptive sense, raising important issues about the comparability of clinical and community samples concerning treatment utilization, and the universal nature of psychiatric conditions.

2 Assessment and Definitions of Anxiety Disorder

In this chapter, we consider categorical anxiety disorders as defined by the standardized diagnostic criteria of American Psychiatric Association's Diagnostic and Statistical Manual for Psychiatric Disorders [i.e., DSM-III (1980), DSM-III-R (1987), DSM-IV (1994)]. The subtypes of anxiety states included are: panic disorder, agoraphobia, specific phobia, social phobia, generalized anxiety/overanxious disorder, separation anxiety, and obsessive–compulsive disorder.

3 Familial and Genetic Factors

The familial aggregation of all of the major subtypes of anxiety disorders has been well-established (Merikangas and Herrell 2004). As reviewed in the following section, when taken together, the results of more than a dozen controlled family studies of probands with specific subtypes of anxiety disorders demonstrate a three- to five-fold elevated risk of anxiety disorders among first-degree relatives of affected probands compared to those of controls. The

Table 1 Summary of family and twin studies of anxiety disorders

Type of study	Comparison	Number of studies	Average relative risk	Range
Family	Relatives of probands vs relatives of controls	13 Panic	5.4	(4.2–17.8)
		4 Social phobia	3.1	(2.5–9.7)
		3 Generalized anxiety	4.3	(2.7–5.6)
		3 OCD	3.5	(1.0–5.1)
Twin	Monozygotic vs dizygotic	3 Panic	2.4	(2.2–2.5)
		4 Phobias	2.6	(1.4–9.5)
		1 OCD	4.9	NR

NR, no range reported; OCD, obsessive–compulsive disorder.

importance of the role of genetic factors in the familial clustering of anxiety has been demonstrated by numerous twin studies of anxiety symptoms and disorders (Kendler et al. 1996; Kendler et al. 1994). However, the relatively moderate magnitude of heritability also strongly implicates environmental etiologic factors. Table 1 summarizes the results of family and twin studies of anxiety disorders.

4 Review of Family and Twin Studies of Anxiety Disorders in Adults

4.1 Panic Disorder

Of the anxiety subtypes, panic disorder has been shown to have the strongest degree of familial aggregation. A recent review of family studies of panic disorder by Gorwood et al. (1999) cited 13 studies that included 3,700 relatives of 780 probands with panic disorder compared to 3,400 relatives of 720 controls. The lifetime prevalence of panic was 10.7% among relatives of panic disorder probands compared to 1.4% among relatives of controls, yielding a sevenfold relative risk. In addition, early onset panic disorder, panic associated with childhood separation anxiety, and panic associated with respiratory symptoms have each been shown to have a higher familial loading than other varieties of panic disorder (Goldstein et al. 1997).

Although there has been some inconsistency reported by twin studies of panic disorder (see McGuffin et al. 1994), two studies applying DSM-III-R diagnostic criteria demonstrated considerably higher rates in monozygotic twins compared to dizygotic twins (Kendler et al. 1993; Skre et al. 1993). Furthermore, current estimates derived from the Virginia Twin Registry show panic disorder to have the highest heritability of all anxiety disorders at 0.44 (i.e., the proportion of variance attributable to genes) (Kendler et al. 1995).

4.2 Phobic Disorders

Though there are far fewer controlled family and twin studies of the other anxiety subtypes, all of the phobic states (i.e., specific phobia, agoraphobia) have also been shown to be familial (Fyer et al. 1995; Noyes et al. 1987; for review, see Merikangas and Angst 1995). The average relative risk of phobic disorders in the relatives of phobics is 3.1. Stein et al. (1998a) found that the familial aggregation of social phobia could be attributed to the generalized subtype of social phobia. Data from the Virginia Twin Study report the estimated total heritability for phobias to be 0.35 (Kendler et al. 1992).

4.3 Generalized Anxiety Disorder

There is also evidence for both the familial aggregation and heritability of generalized anxiety disorder (GAD) in a limited number of studies. There is a five-fold average increase in the rister of GAD among relatives of probands with GAD compared to that among relatives of controls 5 (Mendlewicz et al. 1993; Noyes et al. 1987) and the heritability of GAD among female twins is 0.32 (Kendler et al. 1992).

4.4 Obsessive–Compulsive Disorder

There are also very few controlled family studies of obsessive–compulsive disorder. Two of the three studies (Pauls et al. 1995) reported familial relative risks of 3–4, whereas a third study (Black et al. 1992) found no evidence for familial aggregation. Nestadt et al. (2000) found that both an early age of onset and obsessions, but not compulsion were associated with greater familiality. Twin studies have yielded weak evidence for heritability of obsessive–compulsive disorder (Bellodi et al. 1992; Carey and Gottesman 1981; Lenane et al. 1990).

5 Review of Linkage and Association Studies of Anxiety Disorders

There has been a plethora of linkage and association studies attempting to identify genes for anxiety disorders. The neurotransmitter systems that have been implicated in anxiety disorders include adenosine, adrenaline, noradrenaline, dopamine, serotonin, cholecystokinin, and γ-aminobutyric acid (GABA). In addition, the enzymes involved in biogenic amine degradation [catechol-*O*-methyltransferase (COMT), monoamine oxidase A (MAO-A)] and in catecholamine synthesis [tryptophan hydroxylase (TPH), tyrosine hydroxylase (TRH)] have also been investigated.

Of all the anxiety disorders, panic disorder has been given the most attention. Despite a long list of linkage and association studies (Benjamin et al. 1997; Crawford et al. 1995; Crowe et al. 1987a,b, 1990, 1997, 2001; Deckert et al. 1997, 1998, 1999, 2000; Fehr et al. 2000a,b, 2001; Gelernter et al. 2001; Gratacos et al. 2001; Hamilton et al. 1999, 2000a,b, 2001, 2002, 2003; Han et al. 1999; Hattori et al. 2001; Inada et al. 2003; Ise et al. 2003; Ishiguro et al. 1997; Kato et al. 1996; Kennedy et al. 1999; Knowles et al. 1998; Matsushita et al. 1997; Mutchler et al. 1990; Nakamura et al. 1999; Ohara et al. 1996, 1998a,b, 1999, 2000; Philibert et al. 2003; Sand et al. 2000; Schmidt et al. 1993; Steinlein et al. 1997; Tadic et al. 2003; Thorgeirsson et al. 2003; Wang et al. 1992, 1998; Weissman et al. 2000; Yamada et al. 2001), few have investigated exactly the same genetic variants, and no studies with positive findings have been replicated. Studies with positive findings have either not been replicated or purpoted replications have not tested the same loci (Hattori et al. 2001; Kennedy et al. 1999; Nakamura et al. 1999; Ohara et al. 1999; Tadic et al. 2003; Benjamin et al. 1997; Deckert et al. 1998, 1999; Fehr et al. 2000a, 2001; Hamilton et al. 2000a,b; Inada et al. 2003; Lappalainen et al. 1998; Yamada et al. 2001). There have been, however, a multitude of replicated negative studies.

The summary of findings on obsessive-compulsive disorder is similarly inconclusive. The few positive findings among the long list of linkage and association studies (Alsobrook et al. 2002; Bengel et al. 1999; Billett et al. 1997, 1998; Brett et al. 1995; Camarena et al. 1998, 2001a,b; Catalano et al. 1994; Cavallini et al. 1998; Cruz et al. 1997; Di Bella et al. 1996, 2002; Enoch et al. 1998; Erdal et al. 2003; Frisch et al. 2000; Han et al. 1999; Hanna et al. 2002; Hemmings et al. 2003; Karayiorgou et al. 1997; Karayiorgou et al. 1999; Kinnear et al. 2001; Kinnear et al. 2000; McDougle et al. 1998; Millet et al. 2003; Mundo et al. 2000, 2002; Nicolini et al. 1996; Niehaus et al. 2001; Novelli et al. 1994; Ohara et al. 1998a,b, 1999; Schindler et al. 2000; Walitza et al. 2002) have been replicated by the same authors a few years later (Camarena et al. 1998, 2001a; Karayiorgou et al. 1997, 1999; Kinnear et al. 2001; Mundo et al. 2000, 2002; Niehaus et al. 2001). Other positive studies have either not been replicated (Bengel et al. 1999; Billett et al. 1997, 1998; Camarena et al. 2001b; Kinnear et al. 2000; Ohara et al. 1998a) or not tested again (McDougle et al. 1998; Ohara et al. 1999). The single positive finding of an association between the serotonin receptor (2A) promoter polymorphism (−438G/A) and obsessive-compulsive disorder by Enoch et al. (1998) was replicated by an independent group, Walitza et al. (2002).

Fewer studies (Fehr et al. 2000b, 2001; Ohara et al. 1998a, 1999; Tadic et al. 2003) have investigated genetic loci for generalized anxiety disorder; and those with positive results either await replication (Ohara et al. 1999; Tadic et al. 2003) or have shown to have no association in previous studies (Lappalainen et al. 1998). The two linkage studies (Kennedy et al. 2001; Stein et al. 1998b) of social phobia have been examined by the same group and found to be negative. One genome scan for simple phobia has been performed with a LOD score of 3.17 on chromosome 14 (Gelernter et al. 2003). One positive-association study (Comings et al. 1996) of posttraumatic stress disorder with the A1 allele of the dopamine (D2) receptor polymorphism was not replicated in a subsequent study (Gelernter et al. 1999).

6 Family Studies and Phenotypic Definitions

The lack of success in identifying specific genes for anxiety disorders is not surprising given their complexity. Similar to several other psychiatric disorders, the anxiety disorders are complicated by (1) etiologic and phenotypic heterogeneity, (2) a lack of valid diagnostic thresholds, (3) unclear boundaries between "discrete" anxiety subtypes, and (4) comorbidity with other forms of psychopathology. The major impediment to estimating genetic influences in youth is the lack of rediability and validity of the measures themselves. For example, heritability estimates differ dramatically by the source of the information (child vs. parent) (Eaves et al. 1997).

The family study approach, particularly when employed with systematic community-based samples, is one of the most powerful strategies to minimize heterogeneity, since etiologic factors for the development of a particular disorder can be assumed to be relatively homotypic within families. There is a dearth of studies that have employed within-family designs to examine either phenotypic expression or some of the putative biologic factors underlying the major anxiety disorders. For example, both Perna et al. (1996, 1995) and Coryell (1997) have shown that healthy relatives of probands with panic disorder have increased sensitivity to CO_2 challenge, suggesting that CO_2 sensitivity may be a promising trait marker for the development of panic. Smoller and Tsuang (1998) discuss the value of family and twin studies in identifying phenotypes for genetic studies.

Both family and twin studies have been used to examine sources of overlap within the anxiety disorders, and between the anxiety disorders and other syndromes, including depression, eating disorders, and substance abuse. Fyer et al. (1996, 1995) have demonstrated the independence of familial aggregation of panic and phobias. With respect to comorbidity, whereas panic disorder, generalized anxiety, and depression have been shown to share common familial and genetic liability (Kendler et al. 1996; Maier et al. 1995; Merikangas et al. 1998b), there is substantial evidence for the independent etiology of anxiety disorders and substance use disorders (Kushner et al. 2000; Merikangas et al. 1998b; Smoller and Tsuang 1998). Results emerging from studies of symptoms of anxiety and depression in youth in which both anxiety and depression present indirecte the source may be a common genetic diathesis (Eley and Stevenson 1999; Thapar and McGuffin 1997).

In a comprehensive consideration of what may be inherited, Marks (1986) reviewed the components of anxiety that have been investigated in both human and animal studies. Evidence from twin studies has indicated that the somatic manifestations of anxiety may result in part from genetic factors. These studies demonstrate that physiologic responses, such as pulse, respiration rate, and galvanic skin response, are more alike in monozygotic twins than in dizygotic twin pairs. Furthermore, twin studies of personality factors have shown high heritability of anxiety reaction. Finally, the results of animal studies have suggested that anxiety or emotionality is under genetic control. Selective breeding experiments with mammals have demonstrated that emotional activity analogous to anxiety is controlled by multiple genes (Marks 1986). These findings suggest that anxiety and fear states are highly heterogeneous and that future studies need to investigate (1) the extent to which the components of anxiety result from common versus unique genetic factors and (2) the role of environmental, biologic, and social factors in either potentiating or suppressing their expression.

7 High-Risk Studies of Anxiety Disorders

Given the early age of onset for anxiety disorders, studies of children of parents with anxiety have become an increasingly important source of information on the premorbid risk factors and early forms of expression of anxiety. Increased rates of anxiety symptoms and disorders among offspring of parents with anxiety disorders have been demonstrated by Turner (1987), Biederman (1991), Sylvester (1988), Last (1991), Warner (1995b), Beidel et al. (1997), Beidel (1988), Capps et al. (1996), Merikangas et al. (1998a), Unnewehr et al. (1998), and Warner et al. (1995a). Table 2 shows that the risk of anxiety disorders among offspring of parents with anxiety disorders compared to those of controls averages 3.5 (range 1.3–13.3), suggesting specificity of parent–child concordance within broad subtypes of anxiety disorders.

However, similar to studies of adults that show common familial and genetic risk factors for anxiety and depression (Kendler et al. 1996; Merikangas 1990; Stavrakaki and Vargo 1986), studies in children have also revealed a lack of specificity of parental anxiety and/or depression (Beidel and Turner 1997; Sylvester et al. 1988; Turner et al. 1987; Warner et al. 1995b).

The high rates of anxiety disorders among offspring of parents with anxiety suggest that there may be underlying psychologic or biologic vulnerability factors for anxiety disorders in general that may already manifest in children prior to puberty. Previous research has shown that children at risk for anxiety dis-

Table 2 Controlled high-risk studies of anxiety

Sample					Relative risk	Study Author (YR)
	Proband		Offspring			
Anxiety	Comorbid Diagnoses		*n*	Age		
Panic	Simple		87	5–15	9.2	Unnewehr et al. 1998
Panic	MDD	–	91	7–17	13.3	Sylvester et al. 1987
Panic/MDD	Panic	Early onset MDD	145	6–29	1.3	Warner et al. 1995
Agor	Agor/Panic	–	43	8–24	–	Capps et al. 1996
Agor/OCD	Dysthymia	–	43	7–12	4.8	Turner et al. 1987
Panic/Social Phobia	Alcohol or Drugs	Substance +Anxiety	192	7–17	2.0	Merikangas et al. 1998
Anxiety+Dep	Anxiety	MDD	129	7–12	4.0	Beidel et al. 1997

Agor, agoraphobia; Dep, depression; Dx, diagnosis; MDD, major depressive disorder; OCD, obsessive-compulsive disorder; Not eval, not evaluated.

orders throughout life are characterized by behavioral inhibition (Rosenbaum et al. 1988), autonomic reactivity (Beidel 1988; Merikangas et al. 1999), somatic symptoms (Reichler et al. 1988; Turner et al. 1987), social fears (Sylvester et al. 1988; Turner et al. 1987), enhanced startle reflex (Merikangas et al. 1999), and respiratory sensitivity (Pine et al. 2000).

8
Future Directions for Research on Anxiety

There are several directions that will be fruitful for future research. Better comprehension of the phenomenology of the specific anxiety disorders and their overlap among each other and with other forms of psychopathology should guide the development of the next phase of diagnostic categories of anxiety. In addition, as neuroscience and genetics inform our knowledge regarding neural processes underlying anxiety disorders and the role of genetic and environmental factors in their evolution, studies of treatment and prevention strategies will assume increasing importance in reducing the magnitude and burden of this major source of mental disorders.

Some specific areas of future research should address the following issues:

- Establish more accurate and developmentally sensitive methods of assessment of anxiety with a focus on developing objective measures of the components of anxiety

- Apply within-family design to minimize etiologic heterogeneity and to refine diagnostic boundaries and thresholds
- Investigate the specificity of putative markers with respect to other psychiatric disorders and the longitudinal stability of specific subtypes of anxiety disorders
- Examine the mechanisms for the onset of panic attacks associated with substance use
- Develop research on hormonally mediated neurobiologic function in order to understand gender differences predisposing women to experience decreased resiliency to fear-provoking stimuli
- Investigate the mechanisms for comorbidity of specific medical disorders with anxiety symptoms and disorders.

References

Alsobrook JP 2nd, Zohar AH, Leboyer M, Chabane N, Ebstein RP, Pauls DL (2002) Association between the COMT locus and obsessive-compulsive disorder in females but not males. Am J Med Genet 114:116–120

Beidel DC (1988) Psychophysiological assessment of anxious emotional states in children. J Abnorm Psychol 97:80–82

Beidel DC, Turner SM (1997) At risk for anxiety. I. Psychopathology in the offspring of anxious parents. J Am Acad Child Adolesc Psychiatry 36:918–924

Bellodi L, Sciuto G, Diaferia G, Ronchi P, Smeraldi E (1992) Psychiatric disorders in the families of patients with obsessive-compulsive disorder. Psychiatry Res 42:111–120

Bengel D, Greenberg BD, Cora-Locatelli G, Altemus M, Heils A, Li Q, Murphy DL (1999) Association of the serotonin transporter promoter regulatory region polymorphism and obsessive-compulsive disorder. Mol Psychiatry 4:463–466

Benjamin J, Gulman R, Osher Y, Ebstein RP (1997) Dopamine D4 receptor polymorphism associated with panic disorder. Am J Med Genet (Neuropsychiatr Genet) 74:613

Biederman J, Rosenbaum JF, Bolduc EA, Faraone SV, Hirshfeld DR (1991) A high risk study of young children of parents with panic disorder and agoraphobia with and without comorbid major depression. Psychiatry Res 37:333–348

Bijl RV, van Zessen G, Ravelli A, de Rijk C, Langendoen Y (1998) The Netherlands Mental Health Survey and Incidence Study (NEMESIS): objectives and design. Soc Psychiatry Psychiatr Epidemiol 33:581–586

Billett EA, Richter MA, King N, Heils A, Lesch KP, Kennedy JL (1997) Obsessive compulsive disorder, response to serotonin reuptake inhibitors and the serotonin transporter gene. Mol Psychiatry 2:403–406

Billett EA, Richter MA, Sam F, Swinson RP, Dai XY, King N, Badri F, Sasaki T, Buchanan JA, Kennedy JL (1998) Investigation of dopamine system genes in obsessive-compulsive disorder. Psychiatr Genet 8:163–169

Black DW, Noyes R Jr, Goldstein RB, Blum N (1992) A family study of obsessive-compulsive disorder. Arch Gen Psychiatry 49:362–368

Brett PM, Curtis D, Robertson MM, Gurling HM (1995) Exclusion of the 5-HT1A serotonin neuroreceptor and tryptophan oxygenase genes in a large British kindred multiply affected with Tourette's syndrome, chronic motor tics, and obsessive–compulsive behavior. Am J Psychiatry 152:437–440

Camarena B, Cruz C, de la Fuente JR, Nicolini H (1998) A higher frequency of a low activity-related allele of the MAO-A gene in females with obsessive–compulsive disorder. Psychiatr Genet 8:255–257

Camarena B, Rinetti G, Cruz C, Gomez A, de La Fuente JR, Nicolini H (2001a) Additional evidence that genetic variation of MAO-A gene supports a gender subtype in obsessive–compulsive disorder. Am J Med Genet 105:279–282

Camarena B, Rinetti G, Cruz C, Hernandez S, de la Fuente JR, Nicolini H (2001b) Association study of the serotonin transporter gene polymorphism in obsessive–compulsive disorder. Int J Neuropsychopharmacol 4:269–272

Capps L, Sigman M, Sena R, Henker B, Whalen C (1996) Fear, anxiety and perceived control in children of agoraphobic parents. J Child Psychol Psychiatry 37:445–452

Carey G, Gottesman I (1981) Twin and family studies of anxiety, phobic and obsessive disorders. In: Rabkin JG (ed) Anxiety: new research and changing concepts. Raven Press, New York, pp 117–136

Catalano M, Sciuto G, Di Bella D, Novelli E, Nobile M, Bellodi L (1994) Lack of association between obsessive–compulsive disorder and the dopamine D3 receptor gene: some preliminary considerations. Am J Med Genet 54:253–255

Cavallini MC, Di Bella D, Pasquale L, Henin M, Bellodi L (1998) 5HT2C CYS23/SER23 polymorphism is not associated with obsessive–compulsive disorder. Psychiatry Res 77:97–104

Comings DE, Muhleman D, Gysin R (1996) Dopamine D2 receptor (DRD2) gene and susceptibility to posttraumatic stress disorder: a study and replication. Biol Psychiatry 40:368–372

Coryell W (1997) Hypersensitivity to carbon dioxide as a disease-specific trait marker. Biol Psychiatry 41:259–263

Crawford F, Hoyne J, Diaz P, Osborne A, Dorotheo J, Sheehan D, Mullan M (1995) Occurrence of the Cys311 DRD2 variant in a pedigree multiply affected with panic disorder. Am J Med Genet 60:332–334

Crowe RR, Noyes R Jr, Persico AM (1987a) Pro-opiomelanocortin (POMC) gene excluded as a cause of panic disorder in a large family. J Affect Disord 12:23–27

Crowe RR, Noyes R Jr, Wilson AF, Elston RC, Ward LJ (1987b) A linkage study of panic disorder. Arch Gen Psychiatry 44:933–937

Crowe RR, Noyes R Jr, Samuelson S, Wesner R, Wilson R (1990) Close linkage between panic disorder and alpha-haptoglobin excluded in 10 families. Arch Gen Psychiatry 47:377–380

Crowe RR, Wang Z, Noyes R Jr, Albrecht BE, Darlison MG, Bailey ME, Johnson KJ, Zoega T (1997) Candidate gene study of eight GABAA receptor subunits in panic disorder. Am J Psychiatry 154:1096–1100

Crowe RR, Goedken R, Samuelson S, Wilson R, Nelson J, Noyes R Jr (2001) Genomewide survey of panic disorder. Am J Med Genet 105:105–109

Cruz C, Camarena B, King N, Paez F, Sidenberg D, de la Fuente JR, Nicolini H (1997) Increased prevalence of the seven-repeat variant of the dopamine D4 receptor gene in patients with obsessive–compulsive disorder with tics. Neurosci Lett 231:1–4

Deckert J, Catalano M, Heils A, Di Bella D, Friess F, Politi E, Franke P, Nothen MM, Maier W, Bellodi L, Lesch KP (1997) Functional promoter polymorphism of the human serotonin transporter: lack of association with panic disorder. Psychiatr Genet 7:45–47

Merikangas KR, Herrell R (2004) Contributions of epidemiology to the neurobiology of mental illness. In: Charney D, Nestler EJ (eds) Neurobiology of mental illness. Oxford University Press, Oxford, pp 103–111

Merikangas KR, Dierker LC, Szatmari P (1998a) Psychopathology among offspring of parents with substance abuse and/or anxiety disorders: a high risk study. J Child Psychol Psychiatry 39:711–720

Merikangas KR, Stevens DE, Fenton B, Stolar M, O'Malley S, Woods SW, Risch N (1998b) Co-morbidity and familial aggregation of alcoholism and anxiety disorders. Psychol Med 28:773–788

Merikangas KR, Avenevoli S, Dierker L, Grillon C (1999) Vulnerability factors among children at risk for anxiety disorders. Biol Psychiatry 46:1523–1535

Millet B, Chabane N, Delorme R, Leboyer M, Leroy S, Poirier MF, Bourdel MC, Mouren-Simeoni MC, Rouillon F, Loo H, Krebs MO (2003) Association between the dopamine receptor D4 (DRD4) gene and obsessive–compulsive disorder. Am J Med Genet 116B:55–59

Mundo E, Richter MA, Sam F, Macciardi F, Kennedy JL (2000) Is the 5-HT(1Dbeta) receptor gene implicated in the pathogenesis of obsessive–compulsive disorder? Am J Psychiatry 157:1160–1161

Mundo E, Richter MA, Zai G, Sam F, McBride J, Macciardi F, Kennedy JL (2002) 5HT1Dbeta Receptor gene implicated in the pathogenesis of obsessive–compulsive disorder: further evidence from a family-based association study. Mol Psychiatry 7:805–809

Mutchler K, Crowe RR, Noyes R Jr, Wesner RW (1990) Exclusion of the tyrosine hydroxylase gene in 14 panic disorder pedigrees. Am J Psychiatry 147:1367–1369

Nakamura M, Ueno S, Sano A, Tanabe H (1999) Polymorphisms of the human homologue of the Drosophila white gene are associated with mood and panic disorders. Mol Psychiatry 4:155–162

Nestadt G, Samuels J, Riddle M, Bienvenu OJ 3rd, Liang KY, LaBuda M, Walkup J, Grados M, Hoehn-Saric R (2000) A family study of obsessive–compulsive disorder. Arch Gen Psychiatry 57:358–363

Nicolini H, Cruz C, Camarena B, Orozco B, Kennedy JL, King N, Weissbecker K, de la Fuente JR, Sidenberg D (1996) DRD2, DRD3 and 5HT2A receptor genes polymorphisms in obsessive–compulsive disorder. Mol Psychiatry 1:461–465

Niehaus DJ, Kinnear CJ, Corfield VA, du Toit PL, van Kradenburg J, Moolman-Smook JC, Weyers JB, Potgieter A, Seedat S, Emsley RA, Knowles JA, Brink PA, Stein DJ (2001) Association between a catechol-o-methyltransferase polymorphism and obsessive–compulsive disorder in the Afrikaner population. J Affect Disord 65:61–65

Novelli E, Nobile M, Diaferia G, Sciuto G, Catalano M (1994) A molecular investigation suggests no relationship between obsessive–compulsive disorder and the dopamine D2 receptor. Neuropsychobiology 29:61–63

Noyes R Jr, Clarkson C, Crowe RR, Yates WR, McChesney CM (1987) A family study of generalized anxiety disorder. Am J Psychiatry 144:1019–1024

Ohara K, Xie DW, Ishigaki T, Deng ZL, Nakamura Y, Suzuki Y, Miyasato K (1996) The genes encoding the 5HT1D alpha and 5HT1D beta receptors are unchanged in patients with panic disorder. Biol Psychiatry 39:5–10

Ohara K, Nagai M, Suzuki Y, Ochiai M (1998a) Association between anxiety disorders and a functional polymorphism in the serotonin transporter gene. Psychiatry Res 81:277–279

Ohara K, Nagai M, Suzuki Y, Ochiai M (1998b) No association between anxiety disorders and catechol-O-methyltransferase polymorphism. Psychiatry Res 80:145–148

Ohara K, Suzuki Y, Ochiai M, Tsukamoto T, Tani K (1999) A variable-number-tandem-repeat of the serotonin transporter gene and anxiety disorders. Prog Neuropsychopharmacol Biol Psychiatry 23:55–65

Ohara K, Suzuki Y, Ochiai M, Terada H (2000) Polymorphism in the promoter region of the alpha(2A)-adrenergic receptor gene and panic disorders. Psychiatry Res 93:79–82

Pauls DL, Alsobrook JP 2nd, Goodman W, Rasmussen S, Leckman JF (1995) A family study of obsessive-compulsive disorder. Am J Psychiatry 152:76–84

Perna G, Cocchi S, Bertani A, Arancio C, Bellodi L (1995) Sensitivity to 35% CO2 in healthy first-degree relatives of patients with panic disorder. Am J Psychiatry 152:623–625

Perna G, Bertani A, Caldirola D, Bellodi L (1996) Family history of panic disorder and hypersensitivity to CO2 in patients with panic disorder. Am J Psychiatry 153:1060–1064

Philibert RA, Nelson JJ, Bedell B, Goedken R, Sandhu HK, Noyes R Jr, Crowe RR (2003) Role of elastin polymorphisms in panic disorder. Am J Med Genet 117B:7–10

Pine DS, Cohen E, Cohen P, Brook JS (2000) Social phobia and the persistence of conduct problems. J Child Psychol Psychiatry 41:657–665

Reichler RJ, Sylvester CE, Hyde TS (1988) Biological studies on offspring of panic disorder probands. In: Barrett JE (ed) Relatives at risk for mental disorders. Raven Press, New York

Rosenbaum JF, Biederman J, Gersten M, Hirshfeld DR, Meminger SR, Herman JB, Kagan J, Reznick JS, Snidman N (1988) Behavioral inhibition in children of parents with panic disorder and agoraphobia. A controlled study. Arch Gen Psychiatry 45:463–470

Sand PG, Godau C, Riederer P, Peters C, Franke P, Nothen MM, Stober G, Fritze J, Maier W, Propping P, Lesch KP, Riess O, Sander T, Beckmann H, Deckert J (2000) Exonic variants of the GABA(B) receptor gene and panic disorder. Psychiatr Genet 10:191–194

Schindler KM, Richter MA, Kennedy JL, Pato MT, Pato CN (2000) Association between homozygosity at the COMT gene locus and obsessive compulsive disorder. Am J Med Genet 96:721–724

Schmidt SM, Zoega T, Crowe RR (1993) Excluding linkage between panic disorder and the gamma-aminobutyric acid beta 1 receptor locus in five Icelandic pedigrees. Acta Psychiatr Scand 88:225–228

Skre I, Onstad S, Torgersen S, Lygren S, Kringlen E (1993) A twin study of DSM-III-R anxiety disorders. Acta Psychiatr Scand 88:85–92

Smoller JW, Tsuang MT (1998) Panic and phobic anxiety: defining phenotypes for genetic studies. Am J Psychiatry 155:1152–1162

Stavrakaki C, Vargo B (1986) The relationship of anxiety and depression: a review of the literature. Br J Psychiatry 149:7–16

Stein MB, Chartier MJ, Hazen AL, Kozak MV, Tancer ME, Lander S, Furer P, Chubaty D, Walker JR (1998a) A direct-interview family study of generalized social phobia. Am J Psychiatry 155:90–97

Stein MB, Chartier MJ, Kozak MV, King N, Kennedy JL (1998b) Genetic linkage to the serotonin transporter protein and 5HT2A receptor genes excluded in generalized social phobia. Psychiatry Res 81:283–291

Steinlein OK, Deckert J, Nothen MM, Franke P, Maier W, Beckmann H, Propping P (1997) Neuronal nicotinic acetylcholine receptor alpha 4 subunit (CHRNA4) and panic disorder: an association study. Am J Med Genet 74:199–201

Sylvester CE, Hyde TS, Reichler RJ (1988) Clinical psychopathology among children of adults with panic disorder. In: Barrett JE (ed) Relatives at Risk for Mental Disorder, vol 48. Raven Press, New York, pp 928–934

Tadic A, Rujescu D, Szegedi A, Giegling I, Singer P, Moller HJ, Dahmen N (2003) Association of a MAOA gene variant with generalized anxiety disorder, but not with panic disorder or major depression. Am J Med Genet 117B:1–6
Thapar A, McGuffin P (1997) Anxiety and depressive symptoms in childhood—a genetic study of comorbidity. J Child Psychol Psychiatry 38:651–656
Thorgeirsson TE, Oskarsson H, Desnica N, Kostic JP, Stefansson JG, Kolbeinsson H, Lindal E, Gagunashvili N, Frigge ML, Kong A, Stefansson K, Gulcher JR (2003) Anxiety with panic disorder linked to chromosome 9q in Iceland. Am J Hum Genet 72:1221–1230
Turner SM, Beidel DC, Costello A (1987) Psychopathology in the offspring of anxiety disorders patients. J Consult Clin Psychol 55:229–235
Unnewehr S, Schneider S, Florin I, Margraf J (1998) Psychopathology in children of patients with panic disorder or animal phobia. Psychopathology 31:69–84
Walitza S, Wewetzer C, Warnke A, Gerlach M, Geller F, Gerber G, Gorg T, Herpertz-Dahlmann Niehaus, Schulz E, Remschmidt H, Hebebrand J, Hinney A (2002) 5-HT2A promoter polymorphism −1438G/A in children and adolescents with obsessive-compulsive disorders. Mol Psychiatry 7:1054–1057
Wang Z, Valdes J, Noyes R, Zoega T, Crowe RR (1998) Possible association of a cholecystokinin promotor polymorphism (CCK-36CT) with panic disorder. Am J Med Genet 81:228–234
Wang ZW, Crowe RR, Noyes R Jr (1992) Adrenergic receptor genes as candidate genes for panic disorder: a linkage study. Am J Psychiatry 149:470–474
Warner LA, Kessler RC, Hughes M, Anthony JC, Nelson CB (1995a) Prevalence and correlates of drug use and dependence in the United States. Results from the National Comorbidity Survey. Arch Gen Psychiatry 52:219–229
Warner V, Mufson L, Weissman MM (1995b) Offspring at high and low risk for depression and anxiety: mechanisms of psychiatric disorder. J Am Acad Child Adolesc Psychiatry 34:786–797
Weissman MM, Warner V, Wickramaratne P, Moreau D, Olfson M (1997) Offspring of depressed parents. 10 years later. Arch Gen Psychiatry 54:932–940
Weissman MM, Fyer AJ, Haghighi F, Heiman G, Deng Z, Hen R, Hodge SE, Knowles JA (2000) Potential panic disorder syndrome: clinical and genetic linkage evidence. Am J Med Genet 96:24–35

HEP (2005) 169:181–204

Interactions Between Corticotropin-Releasing Hormone and Serotonin: Implications for the Aetiology and Treatment of Anxiety Disorders

A. C. E. Linthorst

Henry Wellcome Laboratories for Integrative Neuroscience and Endocrinology, Department of Clinical Science at South Bristol, University of Bristol, Dorothy Hodgkin Building, Whitson Street, Bristol BS1 3NY, UK
Astrid.Linthorst@bristol.ac.uk

Abstract The amount of evidence for a role of aberrant serotoninergic neurotransmission in the aetiology of anxiety disorders, such as generalised anxiety and panic disorder, has been increasing steadily during the past several years. Although the picture is far from complete yet—partly due to the large number of serotonin (5-HT) receptors and the often-disparate effects of receptor agonists and antagonists in animal models of anxiety—SSRIs and the 5-HT_{1A} agonist buspirone have now earned their place in the treatment of anxiety disorders. However, these drugs show—as they do in depressed patients—a delayed onset of improvement. Therefore, new therapeutical strategies are being explored. Corticotropin-releasing hormone (CRH), which plays a key role in the autonomic, neuroendocrine and behavioural responses to stress, is a strong anxiogenic neuropeptide and a promising candidate for therapeutical intervention in anxiety disorders. The neuroanatomical localisation of CRH, its congeners (the urocortins) and their receptors within the serotoninergic raphé nuclei suggests that interactions between the CRH system and 5-HT may play a role in fear and anxiety. In this chapter, I will discuss studies from my own and other laboratories showing that CRH and the urocortins influence several aspects of serotoninergic neurotransmission, including the firing rate of 5-HT neurones and the release and synthesis of this monoamine. Moreover, the interactions between CRH and 5-HT during psychologically stressful challenges will be discussed. Finally, I will review data showing that long-term alterations in the CRH system lead to aberrant functioning of serotoninergic neurotransmission under basal and/or stressful conditions. From this growing set of data the picture is emerging that the CRH system exerts a vast modulatory influence on 5-HT neurotransmission. An aberrant cross-talk between CRH and 5-HT may be of crucial importance in the neurobiology of anxiety disorders and represents, therefore, a promising goal for therapeutical intervention in these psychiatric diseases.

Keywords Serotonin · Corticotropin-releasing hormone · Hippocampus · Stress · In vivo microdialysis

Abbreviations

5-HIAA	5-hydroxyindoleacetic acid
5-HT	Serotonin (5-hydroxytryptamine)
CRH	Corticotropin-releasing hormone
CRH1	CRH receptor type 1
CRH2	CRH receptor type 2
DRN	Dorsal raphé nucleus
GABA	γ-aminobutyric acid
HPA	Hypothalamic–pituitary–adrenocortical
i.c.v.	Intracerebroventricular
MRN	Median raphé nucleus
SSRI	Selective serotonin reuptake inhibitor
TPH	Tryptophan hydroxylase

1 Introduction

Stress-related anxiety disorders, such as generalised anxiety and panic disorder, are heavily debilitating diseases. The number of people suffering from these and other stress-related psychiatric illnesses, for instance major depression, seems to increase steadily in Western society. Benzodiazepines have been, already for decades, widely used in the treatment of anxiety. The side-effects of these γ-aminobutyric acidergic (GABAergic) neurotransmission-modulating drugs, including drug dependency and sedation, have urged researchers to search for alternative therapeutic strategies. During the past several years, the neurotransmitter serotonin (5-HT) has attracted much attention. This attention stems, on the one hand, from the principal role of 5-HT in the regulation of behavioural responses to stress [e.g. the regulation of fear (Millan 2003)] and, on the other hand, from the disturbances in this system observed in depressed and anxious subjects (Mann 1998; Maes and Meltzer 1995; Ressler and Nemeroff 2000). Moreover, it has been found that the therapeutic efficacy of drugs influencing serotoninergic neurotransmission can be extended from major depression to several anxiety disorders. The antidepressant selective serotonin reuptake inhibitors (SSRIs; for instance fluoxetine and paroxetine) and the 5-HT_{1A} receptor agonist buspirone are now commonly used in the treatment of anxiety (Millan 2003). The delay in improvement of the disease remains, as it is with the treatment of depression, a major disadvantage of these classes of drugs. Unfortunately, the further development of anxiolytic drugs affecting the 5-HT system is complicated by the very heterogeneous psychopathology of anxiety disorders. Hence, it has been speculated that the role of 5-HT and the neuroanatomical pathways involved may be very different for generalised anxiety disorder as compared to panic disorder (Graeff et al. 1996).

The complex role of 5-HT in anxiety and depression, not in the last place also caused by the high number of 5-HT receptors known at present and by the conflicting results of agonist and antagonist studies, have stimulated investigations on other putative mediators in psychiatric diseases. The neuropeptide corticotropin-releasing hormone (CRH) plays a key role in various aspects of the body's responses to stressful physical and psychological challenges. During stress, CRH not only activates the hypothalamic–pituitary–adrenocortical (HPA) axis and the sympathetic nervous system but also initiates behavioural strategies to cope with the stressor. This latter feature places CRH in a prime position to regulate the behavioural responses to fear-inducing stressful stimuli. Indeed, it has now been well established that CRH has anxiogenic characteristics in various animal models (Dunn and Berridge 1990). Moreover, hyperactivity of the CRH system has been implicated in the pathophysiology of depression and anxiety (Nemeroff et al. 1984; Raadsheer et al. 1994; Arborelius et al. 1999). These observations have led to the development of specific CRH receptor antagonists for the treatment of these psychiatric disorders. Recently, the first open study has demonstrated anxiolytic and antidepressant properties of the CRH receptor type 1 (CRH1) antagonist NBI 30775 (formerly known as R121919) in 20 depressed patients (Zobel et al. 2000).

The above-described observations indicate that aberrations in both CRH and 5-HT functioning are involved in the aetiology of depression and anxiety. Based on this assumption, we have hypothesised that changes in the CRH system may eventually evolve downstream in altered serotoninergic neurotransmission. In this chapter, I will review studies of my own and other laboratories showing that CRH affects the synthesis, release and metabolism of 5-HT after central administration and during psychologically stressful challenges. Moreover, the effects of various anxiety tests on 5-HT in different brain structures will be described and evidence will be put forward showing that long-term changes in the CRH system (e.g. chronic infusion of CRH, CRH1-deficiency and long-term administration of a CRH1 antagonist) result in altered serotoninergic neurotransmission under basal and/or stressful conditions in rodents. The description of the effects of CRH on the 5-HT system will be preceded by an overview of studies that have demonstrated interactions between CRH and 5-HT at the neuroanatomical level.

2
Neuroanatomical Basis for the Interactions Between CRH and Serotonin

The raphé nuclei in the brainstem, containing the majority of serotoninergic neurones (Dahlström and Fuxe 1964; Steinbusch 1981), can be neuroanatomically divided in a rostral and a caudal group (Jacobs and Azmitia 1992; Törk 1990). The caudal raphé nuclei (raphé pallidus nucleus, raphé obscurus nucleus, raphé magnus nucleus and serotoninergic neurones in the ventral lateral medulla) largely project to the spinal cord. In contrast, the rostral raphé nuclei, i.e. the dorsal raphé nucleus (DRN), the median raphé nucleus (MRN), the caudal linear nucleus and the supralemniscal region, give rise to a dense innervation of the forebrain. Regarding depression and anxiety, the DRN and the MRN have drawn the most attention, based on their innervation of higher brain structures. However, given the prime role of the caudal raphé nuclei in the regulation of visceral parameters together with the pronounced vegetative complaints in depressed and anxious subjects, future research will also need to implement these 5-HT cell body regions.

Already in the early 1980s, CRH-immunoreactive cell bodies and fibres had been described in the rostral and caudal raphé nuclei of rats (Cummings et al. 1983; Swanson et al. 1983; Sakanaka et al. 1987). A recent study on post-mortem human brain sections demonstrated CRH nerve terminals in close vicinity of serotoninergic somata and primary dendrites (Ruggiero et al. 1999). Also in the rat, CRH-immunoreactive fibres are found in association with neurones containing tryptophan hydroxylase (TPH) (Lowry et al. 2000) or 5-HT (Kirby et al. 2000; Valentino et al. 2001). Fibres immunoreactive for the CRH-like neuropeptide urocortin 1 (Vaughan et al. 1995) have been found in the DRN and to a lesser extent in the MRN and the caudal raphé nuclei (Bittencourt et al. 1999). Moreover, cell bodies containing urocortin 1 were visible in the rat DRN and MRN after colchicine treatment (Kozicz et al. 1998; Bittencourt et al. 1999). Regarding the recently discovered new members of the CRH family, the neuroanatomical localisation of urocortin 2 (Reyes et al. 2001) in the raphé nuclei has not been studied so far, and no—or very few—urocortin 3 (Lewis et al. 2001) immunoreactive fibres have been observed in the DRN (Li et al. 2002).

The serotoninergic neurones in the DRN and MRN do not form a homogeneous population, as was pointed out excellently in a recent article by Christopher Lowry (2002). Based on differences in morphology, electrophysiological properties and stress-responsiveness, several subpopulations of serotoninergic neurones can be recognised (see also Jacobs and Azmitia 1992 and Beck et al. 2004). A concept is now arising that such subpopulations may have distinct functional properties and are differentially regulated by afferent inputs. In this light, the recent studies on the topographical distribution of CRH-immunoreactive fibres in the DRN are highly relevant. CRH-immunoreactive fibres show a clear rostral to caudal innervation pattern of the DRN (Kirby et al. 2000; Lowry et al. 2000; Valentino et al. 2001). CRH innervation is most dense in the interfascicular and ventromedial regions at rostral to medial levels. In the caudal DRN this innervation changes to the more dorsal and dorsolateral parts. Interestingly, the innervation pattern of CRH does not match one-to-one with the distribution of 5-HT neurones in the DRN. For example, ample 5-HT cell bodies are found in the ventromedial/interfascicular region of the caudal DRN, but CRH innervation is relatively sparse in this region. This innervation pattern may underlie the important observation that CRH also innervates non-serotoninergic neurones and that CRH receptors have been found on GABA-immunopositive neurones (Roche et al. 2003).

Modulation of 5-HT neurones by neuropeptides of the CRH family is also supported by the localisation of CRH receptors in the raphé nuclei. At present, two types of CRH receptors have been characterised, CRH1 and CRH receptor type 2 (CRH2). These receptors show distinct binding affinities for CRH and the urocortins (see Reul and Holsboer 2002). Whereas CRH binds relatively selectively to CRH1 (Chen et al. 1993; Lovenberg et al. 1995), urocortin 2 and urocortin 3 are selective ligands for CRH2 (Lewis et al. 2001; Reyes et al. 2001). In contrast, urocortin 1 binds with high affinity to both receptor types (Vaughan et al. 1995). The rostral and caudal groups of raphé nuclei express low to moderate levels of CRH1 mRNA (Chalmers et al. 1995; Bittencourt and Sawchenko 2000; Van Pett et al. 2000). Whereas no CRH2 mRNA has been found in the caudal raphé nuclei, moderate CRH2 mRNA levels were detected in the MRN and higher levels in the DRN (Chalmers et al. 1995; Bittencourt and Sawchenko 2000; Van Pett et al. 2000). Day and colleagues (2004) showed CRH2 mRNA expression especially at the middle and caudal levels of the DRN. In contrast, CRH2-immunoreactive neuronal profiles have been described in both the rostral and caudal raphé nuclei (Lowry et al. 2002). Double-labelling immunocytochemistry studies have shown that CRH1 co-localises with TPH (Lowry et al. 2002). Moreover, in the dorsolateral part of the DRN, CRH receptors are expressed on GABAergic neurones (Roche et al. 2003). This study could not, however, discriminate between CRH1 and CRH2. Interestingly, at caudal DRN levels CRH2 seems to be expressed on both 5-HT and GABAergic neurones (Day et al. 2004).

The neuroanatomical data collected so far indicate that the CRH system is clearly in the position to modulate serotoninergic neurotransmission at the level of the raphé nuclei, not only directly but also indirectly via effects targeted to GABAergic interneurones. More detailed studies, however, will be needed to clarify the modulatory role of CRH and its related neuropeptides on the functionally different subpopulations of 5-HT neurones in the raphé nuclei. Their results will also contribute to our understanding of the role of CRH-5-HT interactions in anxiety and depression.

3 Effects of Central Administration of CRH and Related Peptides on Serotoninergic Neurotransmission

The anxiogenic properties of CRH after central administration in rodents have been extensively described (see Dunn and Berridge 1990; Arborelius et al. 1999). Moreau and colleagues have shown that central administration of urocortin 1 also induces anxiety-like behaviour in various classical tests for anxiety in rats and mice (Moreau et al. 1997). In contrast, two recent reports have found no effect or increased anxiety in urocortin 1-deficient mice (Wang et al. 2002; Vetter et al. 2002). The role of the CRH system in anxiety has become more complex with the recent observations that urocortin 2 and urocortin 3 may exert anxiolytic properties. Urocortin 3 displays an acute (10-min pre-treatment interval) anxiolytic effect in rats tested on the elevated plus maze (Valdez et al. 2003). Interestingly, the anxiolytic effects of urocortin 2 were delayed, i.e. this neuropeptide increases open arm exploration in the elevated plus maze not earlier than 4 h after administration (Valdez et al. 2002). Therefore, the question now arises whether the neuropeptides of the CRH family also exert distinct effects and/or effects with differential time courses within the 5-HT system. Unfortunately, this question cannot be fully answered at present, as only very limited data are available on the effects of the urocortins on neurotransmitters. The effects of central administration of CRH and the urocortins (as far as available) on different aspects of serotoninergic neurotransmission will be discussed in the following section.

3.1 Activation of Serotonin Neurones as Indicated by Expression of *c*-Fos

Expression of the immediate early gene product *c*-fos is often used as an indicator of neuronal activation. Various anxiety-inducing challenges, such as the elevated plus maze (Silveira et al. 1993), social defeat (Martinez et al. 1998), forced swimming (Cullinan et al. 1995) and inescapable tail shock (Grahn et al. 1999) all induce the expression of *c*-fos in the DRN and/or MRN. Recent evidence shows that anxiogenic drugs of different chemical classes activate a specific subset of neurones in the DRN (Abrams et al. 2002). Intracerebroventricular (i.c.v.) administration of CRH and urocortin 1 causes a profound increase in the expression of *c*-fos in the DRN, whereas only moderate effects in the MRN and in some caudal nuclei were observed (Bittencourt and Sawchenko 2000). Urocortin 2 did not influence *c*-fos expression in the raphé nuclei (Reyes et al. 2001), whereas the effects of urocortin 3 on this parameter have not been studied so far. Interestingly, swim stress-induced expression of *c*-fos in the DRN (dorsolateral part) is reduced after pre-treatment with the CRH1 antagonist antalarmin (Roche et al. 2003). The majority of neurones expressing *c*-fos in this study, however, were doubly labelled for GABA.

3.2 Firing Rate of Serotonin Neurones

The effects of CRH on the firing rate of 5-HT neurones in the DRN have been studied during two conditions: the in vivo firing rate under halothane anaesthesia and the in vitro firing rate in brain slices. The effects of i.c.v. administration of CRH on the in vivo firing rate depend on the dose used. Low doses of CRH result in a decrease in the firing rate of neurones in the rostral and medial aspects of the DRN, whereas higher doses are without effect or cause an increase (Price et al. 1998; Kirby et al. 2000). Similar results were obtained after intra-raphé application of CRH (Price et al. 1998). In contrast, bath application of CRH causes a clear increase in the in vitro firing rate of neurones located in the ventral and interfascicular region of the caudal DRN, but is without effect in the dorsomedial region at the same rostral-caudal level (Lowry et al. 2000). It cannot be excluded that the anaesthesia in the in vivo and the absence of innervation in the in vitro preparation contribute to the differential effects of CRH as observed in these studies. However, alternatively, these results may point to the existence of distinct subpopulations of 5-HT neurones. Unfortunately, little information is yet available on the topography of 5-HT neurones activated during anxiogenic challenges. Of interest in this respect is the observation that forced swimming in water of 25 °C especially activates GABAergic neurones in the dorsolateral DRN (Roche et al. 2003). Because these GABAergic neurones are enveloped by CRH fibres, it may be speculated that the effects of CRH on 5-HT neurones may depend on the final balance between direct stimulatory effects and inhibitory effects caused by the release of GABA. The dose-dependency of the effects of CRH on in vivo firing rate may also point to a differential involvement of CRH1 and CRH2. The first study on the effects of CRH2 ligands by Rita Valentino and colleagues has shown that injection of urocortin 2 into the DRN inhibits 5-HT neurones at a dose of 0.1–10 ng, but results in an activation of such neurones at a dose of 30 ng, possibly as a result of inhibition of non-serotoninergic (GABA?) neurones (Pernar et al. 2004). Based on the dose-dependent effects of CRH and urocortin 2, these authors have postulated that CHR1 and CRH2 may exert opposing effects on the firing rate of 5-HT neurones in the DRN via activation and inhibition of GABAergic neurones respectively (Pernar et al. 2004). A detailed picture of the localisation of CRH1 and CRH2 within the different types of neurones in the raphé nuclei and of their distribution in the different subregions of these brain structures will be of utmost importance for a better understanding of the effects of CRH and its congeners on 5-HT neuronal activity.

3.3
Synthesis of Serotonin

The essential amino acid L-tryptophan is the precursor for the synthesis of 5-HT. The rate-limiting step in the synthesis of 5-HT is the hydroxylation step from L-tryptophan into 5-hydroxytryptophan by the enzyme TPH [existing in two isoforms, of which TPH2 is the form acting in the brain (Walther et al. 2003)]. The effects of stress on TPH activity have been reviewed before (Boadle-Biber 1993). Loud sound stress (a fear-inducing procedure) causes an increase in TPH activity in the cortex and midbrain of rats (Boadle-Biber et al. 1989). The effects of sound stress on TPH activity in the raphé nuclei, but also of forced swimming and tail shock, seem to be confined to the MRN (Dilts and Boadle-Biber 1995; Daugherty et al. 2001; Corley et al. 2002). Data on the effects of CRH on TPH activity are scarce. Administration (i.c.v.) of CRH mimics the effect of sound stress, i.e. activation of TPH activity in the cortex and midbrain (Singh et al. 1992). However, CRH seems to be without effect on enzyme activity in the mediobasal hypothalamus (Van Loon et al. 1982).

3.4
Extracellular Levels of Serotonin and 5-Hydroxyindoleacetic Acid

In a recent in vivo microdialysis study we have assessed the effects of CRH and urocortin 1 on the extracellular levels of 5-HT and its metabolite 5-hydroxyindoleacetic acid (5-HIAA) in the hippocampus of conscious, freely moving, male Wistar rats. We focused on the hippocampus based on its intricate involvement in the regulation of neuroendocrine and behavioural responses to stress. Moreover, the hippocampus is thought to play a critical role in the pathophysiology of anxiety (and affective) disorders. Intracerebroventricular injections of low and high doses of CRH and urocortin 1 (dose range 0.03–10 μg) increase extracellular levels of both 5-HT and 5-HIAA in the hippocampus (Linthorst et al. 2002). These results have recently been confirmed in Sprague–Dawley rats (1.0 μg CRH) by another research group (Kagamiishi et al. 2003). The CRH2-selective neuropeptides urocortin 2 and urocortin 3 also enhance hippocampal extracellular levels of 5-HT and its metabolite after central administration, but these responses were much shorter in duration compared to the responses induced by CRH and urocortin 1 (De Groote et al. 2005). In agreement with these findings are the observations that i.c.v. injection of D-Phe-CRH_{12-41} (a non-specific CRH receptor antagonist) and the CRH1 antagonist CP-154,526 (moderately) decrease extracellular levels of 5-HT in the rat hippocampus (Isogawa et al. 2000; Linthorst et al. 2002). As will be discussed in more detail in Sect. 5 of this chapter, basal hippocampal extracellular levels of 5-HT and 5-HIAA are not affected by chronic central administration of CRH.

The research group of Irwin Lucki has extensively investigated the effects of i.c.v. administration of CRH on serotoninergic neurotransmission in the striatum and lateral septum. Interestingly, in contrast to our findings in the hippocampus, CRH has biphasic effects in these brain structures. Low doses of CRH (0.1 and 0.3 μg) were found to decrease extracellular levels of 5-HT. Higher doses of CRH (1.0 and 3.0 μg) have, however, no effect or increase 5-HT levels in the striatum and lateral septum (Price et al. 1998; Price and Lucki 2001). The biphasic effects on levels of 5-HT in these brain regions may be related to the dose-dependent effects of CRH on the firing rate of DRN 5-HT neurones as described above. Interestingly, local injection of CRH in the DRN results in a decrease in extracellular 5-HT in the striatum and septum (Price and Lucki 2001).

The effects of central administration of CRH on 5-HT and 5-HIAA levels have hardly been studied in other brain regions. An increase in the extracellular concentrations of 5-HIAA has been observed in the medial prefrontal cortex [although CP-154,526 was without effect in this brain structure (Isogawa et al. 2000)] and in the medial hypothalamus after i.c.v. injection of CRH (Lavicky and Dunn 1993). Of utmost relevance would be to clarify the effects of CRH and CRH-like neuropeptides on serotoninergic neurotransmission in the (different subnuclei of the) amygdala and the periaqueductal grey, two brain areas of central importance in the regulation of fear and anxiety.

4
Interactions Between CRH and Serotonin Under Anxiogenic and Psychologically Stressful Conditions: What Can We Learn from In Vivo Microdialysis Studies?

Ample studies applying various 5-HT receptor agonists and antagonists have shown that 5-HT exerts anxiogenic and anxiolytic effects depending on the brain structure(s) and specific 5-HT receptor(s) involved as well as on the route of administration (for an excellent, comprehensive review see Millan 2003). Whereas from a clinical point of view, it may be sufficient to know whether a (new) compound shows anxiolytic properties after oral administration, it is known that the anxiolytic properties of SSRIs and of the 5-HT$_{1A}$ antagonist buspirone show a delayed onset and are only clinically efficacious after prolonged treatment. The development of new and more rapid-onset treatments may, therefore, benefit highly from a more in-depth characterisation of serotoninergic neurotransmission, i.e. release, metabolism and synthesis, in different brain structures during different forms of anxiety (including panic) and psychological stress. As described above, peptides of the CRH family (and their two known receptor types) have a clear neuroanatomical localisation within the 5-HT system and influence serotoninergic neurotransmission after central administration in a dose- and region-dependent way. Based on the anxiogenic properties of CRH and urocortin 1, but also on the putative (time-dependent) role of CRH2 in anxiolysis, the question arises whether interactions between CRH and 5-HT also play a crucial role during fear and anxiety. Although studies on the role of CRH in anxiety- and stress-related responses of the 5-HT system have started to emerge recently, a complete and satisfactory answer cannot be given yet. In this section I will review the changes in serotoninergic neurotransmission during anxiety tests and psychologically stressful challenges. I will focus on the effects of these manipulations on extracellular levels of 5-HT and 5-HIAA in different brain structures as assessed by in vivo microdialysis. As far as information is available, the role of CRH in anxiety- and stress-induced 5-HT changes will also be discussed.

4.1 Anxiety Tests Involving Unconditioned Responses

Many tests employed to assess putative anxiolytic characteristics of new drug compounds or anxiety profiles of mutant animals make use of innate fear and/or the perceiving of conflict in rodents. Although, on the one hand, rats and mice want to explore their environment, they are on the other hand afraid of open spaces [open field paradigm, elevated plus (or X) maze], of heights (elevated plus maze) and of brightly lit areas (light–dark box). Clearly, these animals will experience fear when encountering a possible predator [predator exposure, fear/defence test battery (rat), anxiety/defence test battery (rat) and mouse defence test battery (see for comprehensive review Blanchard et al. 2003)].

Microdialysis studies on serotoninergic neurotransmission during most unconditioned anxiety tests are scarce or absent. However, a picture starts to emerge from studies performing microdialysis during the elevated plus maze test and during exposure to a predator. The group of Charles Marsden was the first to show that exposure of Lister hooded rats to an elevated X-maze design causes an increase in hippocampal 5-HT levels, without effects on the levels of 5-HIAA (Wright et al. 1992). Exposure of Sprague–Dawley rats to an elevated plus maze also results in increased extracellular levels of 5-HT (but not of noradrenaline) in the hippocampus, an effect that is significantly augmented in a transgenic rat line with indices of increased anxiety in this test (Voigt et al. 1999). A recent study demonstrated that both the stimulating effect of CRH on hippocampal 5-HT levels and the anxiogenic effects of this neuropeptide in the plus-maze paradigm can be blocked by the 5-HT_{1A} receptor agonist 8-OH-DPAT (Kagamiishi et al. 2003). Based on these observations, the authors argue that CRH-induced increases in hippocampal 5-HT may mediate anxious behaviour in the plus maze test. Unfortunately, these experiments were performed on separate groups of animals, making firm conclusions at this stage impossible. The elevated plus maze has also been found to induce increases in extracellular levels of 5-HT in the frontal cortex of guinea-pigs (Rex et al. 1993) and rats (Kanno et al. 2003).

Exposure to a predator represents a test paradigm for anxiety (and panic, depending on the exact experimental design) that is receiving increased attention (Blanchard et al. 2003). Data from in vivo microdialysis studies performed during exposure to a predator (Rueter and Jacobs 1996; Linthorst et al. 2000; M. Beekman et al. 2005) clearly show increased serotoninergic neurotransmission in higher limbic brain structures such as the frontal cortex, hippocampus and amygdala in rats (exposed to a cat) and in mice (exposed to a rat). Studies assessing the turnover of 5-HT (post-mortem ratio between the tissue levels of 5-HIAA and 5-HT) also indicate increases in serotoninergic neurotransmission in the hippocampus and frontal cortex of mice *after* predator stress (Hayley et al. 2001; Belzung et al. 2001). There are conflicting results regarding the effects of predator stress on 5-HT in the striatum. Exposure of rats to a cat during the dark phase of the light–dark cycle has been found to moderately increase extracellular levels of 5-HT in the striatum (Rueter and Jacobs 1996), whereas the exposure of C57Bl6/N mice to a rat (during the light phase) has no effect on the levels of 5-HT and 5-HIAA in this brain structure (M. Beekman et al. 2005). Recently, we observed that mutant mice with an impaired glucocorticoid receptor functioning show an enhanced response in hippocampal levels of 5-HT during exposure to a rat (Linthorst et al. 2000). This is especially interesting given the alterations in behavioural coping strategies (more investigation along the separation wall) and the absence of an activation of the HPA axis in these mice. At present it still has to be resolved whether the changes in behavioural and neurochemical responses to predator stress are related to the impairment of the glucocorticoid receptor or to the changes in the CRH system also observed in this mutant mouse line (Dijkstra et al. 1998).

4.2 Anxiety Tests Involving Conditioned Responses

Anxiety tests involving conditioned (trained) responses are used extensively not only to screen for drugs with anxiolytic properties but also to elucidate the neurobiological mechanisms underlying fear and anxiety. Although the microdialysis technique may be of relevance especially for the latter purpose, until now it has not been widely applied in this field of research. Wilkinson et al. (1996) found that conditioned fear stress causes a rise in the extracellular levels of 5-HT in the hippocampus of rats, which seems to be related to the contextual aversive cues (and not to the conditioned discrete stimulus). Increased levels of 5-HT (Yoshioka et al. 1995; Hashimoto et al. 1999) and an increased turnover of this neurotransmitter (Inoue et al. 1994) were also observed in the rat prefrontal cortex during fear conditioning. Moreover, during the Vogel conflict test (punishment of drinking behaviour by an electric shock) elevated levels of 5-HT were observed in the (dorsal) hippocampus of rats (Matsuo et al. 1996); interestingly, the benzodiazepine midazolam was found to block both the conflict behaviour and the rise in hippocampal 5-HT in this anxiety test.

To increase the successful use of microdialysis in paradigms with extensive training, adaptations in surgical procedures and experimental equipment may be indispensable (e.g. sterile surgery, swivel systems with low torque, specially designed home cages and test chambers). Moreover, to fully appreciate the changes in neurotransmission during all phases of training and testing, emphasis should be put on the development of highly sensitive (HPLC) methods to measure neurotransmitter levels in dialysates sampled both in (very) short intervals (i.e. $\leq$3 min) and over extensive periods of time. Beyond doubt, more sophisticated microdialysis and analysis paradigms will be of enormous value to further our understanding of the neurobiology of fear and anxiety.

4.3 Psychologically Stressful Challenges

In the following two sections I will discuss the effects of two psychologically stressful challenges on serotoninergic neurotransmission. Although these challenges are not used as tests for anxiety per se (and are even often used in research on symptoms of depression), they certainly involve aspects of fear and anxiety and will activate brain circuits involved in these responses. Importantly, attempts have been made by various research groups to elucidate the role of the CRH system in the changes in serotoninergic neurotransmission induced by psychological stress.

4.3.1 Forced Swimming

The forced swim test, as developed by Porsolt, is often used to screen new compounds for putative antidepressant characteristics. However, forced swimming has also been found to be a very useful paradigm in assessing the effects of stress on various aspects of brain functioning (neurotransmission, *c*-fos and P-CREB expression, HPA axis regulation; see Bilang-Bleuel et al. 2002) in rats and (mutant) mice. Forced swimming is a stressor with physical (activity, body temperature changes) and psychological (cognition, coping strategies) components. Because it represents a putative life-threatening situation for the animal it will also induce fear. Hence, forced swim stress will lead to a coordinated response of lower brain structures involved in the regulation of homeostasis and higher limbic brain structures organising the cognitive and emotional responses to this form of stress. Indeed *c*-fos expression studies showed a widespread activation of the brain after forced swim stress (Cullinan et al. 1995). The complex nature of forced swim stress is underscored by its multifarious effects on serotoninergic neurotransmission.

To assess the effects of forced swim stress on hippocampal serotoninergic neurotransmission and the role of CRH herein we performed a microdialysis study in rats. Rats were connected to a swivel system via a plastic collar around their neck to allow for free movement during forced swimming (and in the home cage). During a 15-min period of forced swimming in water of 25 °C, hippocampal levels of 5-HT rose to about 900% of baseline (Linthorst et al. 2002). Careful inspection of the data identified two groups of animals, i.e. high and low responders. Comparison of the rises in 5-HT and the behaviour of the animals during the swim session revealed that the exaggerated increase in hippocampal 5-HT was only found in rats that dive during the test (with a maximum of 1,500% of baseline on average; Fig. 1). Most interestingly, this dramatic rise in hippocampal 5-HT levels in diving animals could be prevented by i.c.v. pre-treatment of animals with the CRH receptor antagonist D-Phe-CRH_{12-41} (Linthorst et al. 2002). These observations may be of high relevance for anxiety disorders. We hypothesised that the exaggerated 5-HT response in diving animals is related to a different appraisal of the situation because of the collar around their neck, possibly causing a panic-like response. This hypothesis is supported by the finding that such dramatic responses in hippocampal 5-HT are not observed in diving rats connected to the swivel

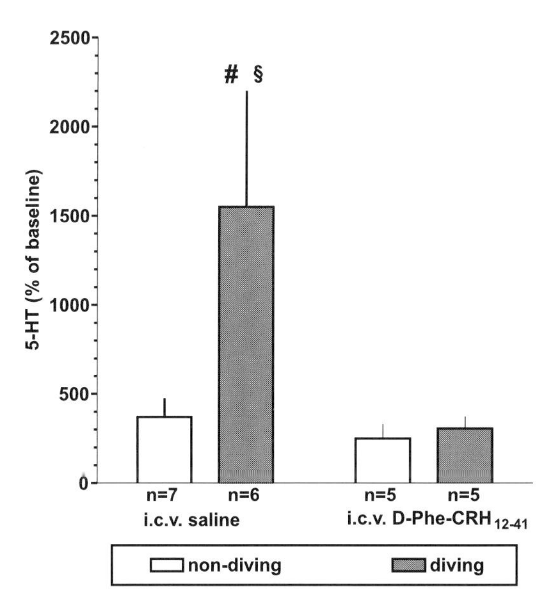

Fig. 1 Effects of forced swimming (15 min, water temperature 25 °C) on hippocampal extracellular levels of 5-HT (expressed as percentage of baseline) in male Wistar rats as assessed by in vivo microdialysis. Extracellular levels of 5-HT showed a dramatic increase in rats that dived during the forced swimming procedure. This effect could be prevented by i.c.v. pretreatment of the rats with the CRH receptor antagonist D-Phe-CRH_{12-41} (5 µg). #, $p<0.05$ as compared to non-diving saline-pretreated rats; §, $p<0.05$ as compared to diving D-Phe-CRH_{12-41}-treated rats (Scheffé post-hoc comparisons). For experimental details, please see text and original paper. (From Linthorst et al. 2002, with permission of *European Journal of Neuroscience*)

system via a peg on their head (A.C.E. Linthorst, unpublished observations). Moreover, forced swimming in mice, which never dive during the test, induces rises in hippocampal 5-HT levels of maximally 140%–240% of baseline (Peñalva et al. 2002; Fujino et al. 2002; Oshima et al. 2003). Further supporting our postulate, the panic-inducing substance *m*-chloro-phenylpiperazine (mCPP) has been found to increase extracellular levels of 5-HT in the hippocampus up to 1,400% of baseline (Eriksson et al. 1999).

A comprehensive series of microdialysis studies by Lucki and colleagues in rats shows that forced swimming in water of 21–22 °C for 30 min results in brain region-specific changes in 5-HT levels. Whereas swim stress has no effect on 5-HT levels in the hippocampus and frontal cortex, a rise and a decrease of the levels of this neurotransmitter is found in the striatum and lateral septum (and amygdala), respectively (Kirby et al. 1995, 1997). In contrast, a decrease in the levels of 5-HIAA was observed in all brain regions studied. These authors showed that the swim stress-induced decrease in 5-HT in the lateral septum could be blocked by D-Phe-CRH_{12-41}, pointing to a prominent role of the CRH family in the swim stress-induced 5-HT changes also in this brain structure (Price et al. 2002). This is an important finding given the role of the lateral septum and septal CRH receptors in fear and anxiety. Of interest in this respect is our recent observation that exposure to a predator results in an immediate rise in extracellular levels of 5-HT in the lateral septum as assessed by in vivo microdialysis in mice (M. Beekman et al. 2005).

As mentioned above, forced swimming does not solely involve cognitive and coping processes but will also affect homeostasis. Often neglected in the interpretation of the results of forced swim stress is the decrease in body temperature during the procedure. This decrease in body temperature depends on the temperature of the water and the duration of the test session. Using a biotelemetry method, we found that the body temperature of rats drops to about 29 °C at the end of a 15-min swim session in water of 25 °C. Swimming at 35 °C has little effect on body temperature during a 15-min swim session (drop of 1–2 °C), but swimming at 19 °C results in body temperatures of about 24 °C. In a follow-up study, we were able to demonstrate that the hippocampal 5-HT and 5-HIAA responses to swim stress are dependent on the water temperature used (Linthorst et al. 2001). This may also offer an explanation for the observations that forced swimming for 30 min in water of 21–22 °C results in a decrease in hippocampal levels of 5-HIAA without affecting 5-HT levels (Kirby et al. 1995, 1997), whereas a similar procedure in water of 30–35 °C increases extracellular levels of both 5-HT and 5-HIAA in different forebrain regions, among them the hippocampus (Rueter and Jacobs 1996).

Taken together, the data reviewed here show that forced swim stress induces a highly differentiated (putatively CRH-dependent) response in serotoninergic neurotransmission in higher brain structures, with the final outcome of the manipulation depending on the exact experimental design. Moreover, the dramatic CRH receptor-dependent increase in hippocampal 5-HT as observed in our study may represent a new phenomenon in the interactions between CRH and 5-HT and be of high relevance for elucidating the pathophysiology of panic disorder.

4.3.2
Electric Shock

Electric foot or tail shock is often used as a combined intense physical/psychological stressor and as part of models to study the neurobiology of anxiety and depression, such as fear conditioning (see above) and learned helplessness. There are only few data available on the effects of electric shock on serotoninergic neurotransmission in rodents. From the literature (Amat et al. 1998a, 1998b; Dunn 1988; Inoue et al. 1993, 1994; Hajos-Korcsok et al. 2003), the picture emerges that electric shock stimulates serotoninergic neurotransmission in the forebrain (hippocampus, amygdala, prefrontal cortex and hypothalamus) of rats and mice. Interestingly, the possibility to cope with this stressor seems to influence the outcome of the 5-HT response, albeit in a brain structure-dependent manner. Hence, it has been demonstrated that inescapable, but not escapable, foot and tail shocks cause a rise in extracellular levels of 5-HT in the rat ventral hippocampus (Amat et al. 1998a), the amygdala (Amat et al. 1998b), the frontal cortex (Heinsbroek et al. 1991; Petty et al. 1994) and DRN (Maswood et al. 1998). In contrast, only escapable shocks result in an increase in dialysate levels of 5-HT in the periaqueductal grey (Amat et al. 1998a). (It should be noted, however, that Amat and colleagues compared the effects of escapable and inescapable tail shock on 5-HT levels against those found in restraint-stressed animals and not unstressed controls.) The different effects of escapable and inescapable shock are of special interest given the differential involvement of the periaqueductal grey and the hippocampus/frontal cortex in the coordination of the behavioural aspects of fear. Steven Maier and colleagues performed a series of elegant studies to elucidate the role of CRH–5-HT interactions in the effects of inescapable tail shocks on behaviour (the behavioural consequences of inescapable shock are termed behavioural depression or learned helplessness). They showed, using intra-raphé administration of CRH1- and CRH2-specific ligands and antagonists, that CRH2 in the (caudal) DRN is the key mediator of the behavioural responses to inescapable stress (Hammack et al. 2002, 2003).

In summary, the available data show that electric shock induces an activation of serotoninergic neurotransmission in various brain structures, particularly when the shock is uncontrollable. Interactions between CRH and 5-HT may play an important role in the behavioural consequences of inescapable shocks via the activation of CRH2 in the DRN. The exact interactions between CRH1 and CRH2 and the detailed neuroanatomical localisation of these interactions within the raphé nuclei, however, need further investigation.

5 Consequences of Long-Term Changes in the CRH System for Hippocampal Serotoninergic Neurotransmission

Until now I have focussed on the consequences of acute manipulations of the CRH system for serotoninergic neurotransmission. We have collected, however, during the past several years, data showing that long-term changes in the CRH system evolve in changes/adaptations in serotoninergic neurotransmission. In the first study, rats were infused i.c.v. with CRH via a miniosmotic pump (Linthorst et al. 1997). Long-term elevation of central levels of CRH has no effects on basal levels of hippocampal 5-HT and 5-HIAA. However, the response of hippocampal 5-HT to a stressful challenge (intraperitoneal administration of bacterial endotoxin) is significantly diminished in long-term CRH-treated animals, which may involve desensitisation of CRH receptors (Linthorst et al. 1997). Serotoninergic neurotransmission seems also to be affected in CRH-overexpressing mice. In female CRH transgenic mice, a reduced stimulation of the HPA axis but a normal hypothermia response was found after subcutaneous administration of the 5-HT_{1A} receptor agonist 8-OH-DPAT (Van Gaalen et al. 2002). Hence, with presynaptic 5-HT_{1A} receptor functioning intact, postsynaptic 5-HT_{1A} receptors seem to be desensitised as a consequence of life-long elevated CRH levels.

Recently we performed in vivo microdialysis experiments in mice with a life-long deficiency of CRH1 (Peñalva et al. 2002). Homozygous CRH1-deficient mice (Timpl et al. 1998) show elevated levels of 5-HIAA, but not of 5-HT, over the diurnal rhythm. Importantly, forced swim stress (10 min, 25 °C) induces an augmented hippocampal 5-HT response in CRH1-deficient mice (homo- and heterozygous) as compared to wild-type littermates (Peñalva et al. 2002). Given the putative anxiogenic properties of 5-HT at the level of the hippocampus, the enhanced response of hippocampal 5-HT in less anxious (Timpl et al. 1998; Smith et al. 1998) CRH1-deficient mice may at first sight seem contradictory. However, at present it cannot be excluded that, given the life-long deficiency in CRH1 in this mutant mouse model, compensatory mechanisms may have developed. This possibility is underscored by our observations in C57Bl6/N mice that were orally treated with the CRH1 antagonist NBI 30775 [this compound had anxiolytic properties in an open study in 20 depressed patients (Zobel et al. 2000)] for 15–16 days. Basal levels of 5-HT and 5-HIAA are normal over the complete diurnal rhythm in NBI 30775-treated mice. However, NBI 30775-treated mice show a significantly diminished rise in hippocampal extracellular 5-HT to forced swim stress (Oshima et al. 2003), possibly contributing to the anxiolytic properties of the compound (Fig. 2A). Interestingly, the forced swimming-induced increase in hippocampal extracellular levels of 5-HIAA is prolonged in these animals (Fig. 2B).

Taken together the above-described studies clearly indicate that chronic changes within the CRH system evolve in altered serotoninergic neurotrans-

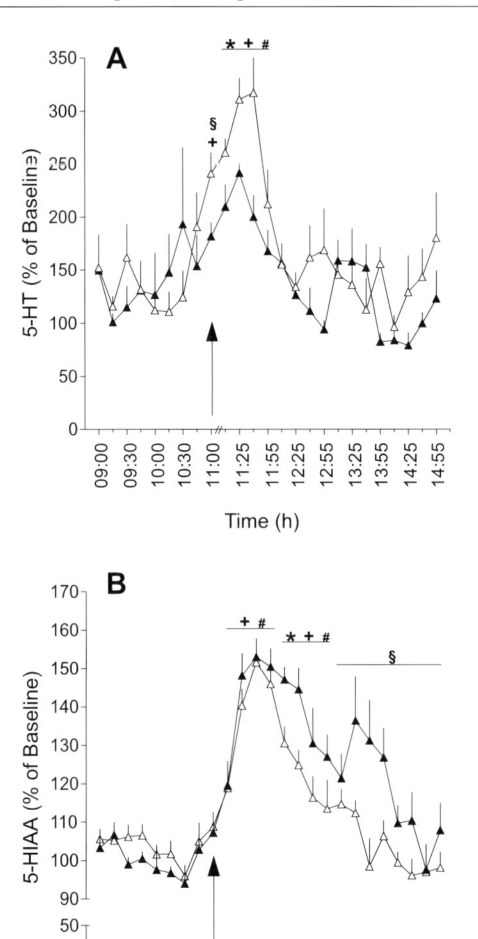

Fig. 2A, B Effects of long-term treatment (16 days, orally, about 19 mg/kg body weight per day) with the CRH1 antagonist NBI 30775 on forced swim stress-induced changes in extracellular levels of 5-HT (percentage of baseline; **A**) and 5-HIAA (percentage of baseline; **B**) in the hippocampus of C57Bl6/N mice as assessed by in vivo microdialysis. Mice were forced to swim for 10 min in water of 25 °C. Chronic treatment with NBI 30775 (*closed triangles*) resulted in a diminished and prolonged response of hippocampal 5-HT and 5-HIAA to forced swim stress, respectively, as compared to control-treated mice (*open triangles*). The *arrow* indicates the start of the 10-min forced swim period (11:00–11:10 A.M.). Other symbols: *, significant difference between control- and NBI 30775-treated mice; §, $p=0.04$ for the difference between the two treatment groups; +, significantly different from baseline for control mice; #, significantly different from baseline for NBI 30775-treated animals (post-hoc tests with contrasts and Bonferroni correction; for results of ANOVA analyses see original paper). (From Oshima et al. 2003, with permission of *Neuropsychopharmacology*)

mission under basal and/or stressful conditions. However, at this moment it is still difficult to draw, based on the data available, a complete picture of (1) how long-term changes in CRH affect the 5-HT system and (2) the putative consequences for anxiety/anxiolysis. Future studies should not only implement different brain structures and anxiety tests but should also look in detail at the differential contributions of CRH1 and CRH2. It is easily conceivable that differences in the balance between these two receptors may determine the outcome of manipulations of CRH function for serotoninergic neurotransmission in forebrain regions.

6 Conclusion

As may be taken from the studies summarised in this chapter, aberrant interactions between the CRH system and serotoninergic neurotransmission may play an important role in the aetiology of (stress-related) anxiety disorders. However, the picture is far from complete at the moment. Further studies on the consequences of aberrations in the CRH system and on the balance between CRH1 and CRH2 effects will increase our understanding of the neurobiology of anxiety and may boost the development of new therapeutical strategies. In vivo microdialysis in rats and (mutant) mice will be an important tool to pursue these goals. By applying rapid sampling techniques over extended periods, such studies should aim at obtaining a more detailed picture of changes in neurotransmitters (or the interactions between neurotransmitters) during anxiety (tests) and of the mechanisms underlying anxiolysis and stress-coping strategies.

Acknowledgements I thank the present and former members of my group for their unsurpassed contribution to our studies on stress and serotonin. I am especially indebted to Ms. Cornelia Flachskamm for her continuous and excellent technical support. The studies of the author described in this chapter have been performed at the Max Planck Institute of Psychiatry, Munich, Germany and have been supported by the Max Planck Society and the Volkswagen Foundation.

References

Abrams JK, Johnson PL, Shekhar A, Lowry CA (2002) Different anxiogenic drugs activate a common, topographically distinct subpopulation of serotonergic neurones in the rat dorsal raphe nucleus. Program No 75.1. 2002 abstract viewer/itinerary planner. Society for Neuroscience, Washington. http://sfn.scholarone.com/itin2002/index.html. Cited 20 December 2004

Amat J, Matus-Amat P, Watkins LR, Maier SF (1998a) Escapable and inescapable stress differentially and selectively alter extracellular levels of 5-HT in the ventral hippocampus and dorsal periaqueductal gray of the rat. Brain Res 797:12–22

Amat J, Matus-Amat P, Watkins LR, Maier SF (1998b) Escapable and inescapable stress differentially alter extracellular levels of 5-HT in the basolateral amygdala of the rat. Brain Res 812:113–120

Arborelius L, Owens MJ, Plotsky PM, Nemeroff CB (1999) The role of corticotropin-releasing factor in depression and anxiety disorders. J Endocrinol 160:1–12

Beck SG, Pan YZ, Akanwa AC, Kirby LG (2004) Median and dorsal raphe neurons are not electrophysiologically identical. J Neurophysiol 91:994–1005

Beckman M,Flachskamm C, Linthorst ACE (2005) Effects of exposure to a predator on behaviour and serotonergic neurotransmission in different brain regions of C57BL/6N mice. Eur J Neurosci in press

Belzung C, El Hage W, Moindrot N, Griebel G (2001) Behavioral and neurochemical changes following predatory stress in mice. Neuropharmacology 41:400–408

Bilang-Bleuel A, Rech J, De Carli S, Holsboer F, Reul JMHM (2002) Forced swimming evokes a biphasic response in CREB phosphorylation in extrahypothalamic limbic and neocortical brain structures in the rat. Eur J Neurosci 15:1048–1060

Bittencourt JC, Sawchenko PE (2000) Do centrally administered neuropeptides access cognate receptors? An analysis in the central corticotropin-releasing factor system. J Neurosci 20:1142–1156

Bittencourt JC, Vaughan J, Arias C, Rissman RA, Vale WW, Sawchenko PE (1999) Urocortin expression in rat brain: evidence against a pervasive relationship of urocortin-containing projections with targets bearing type 2 CRF receptors. J Comp Neurol 415:285–312

Blanchard DC, Griebel G, Blanchard RJ (2003) The Mouse Defense Test Battery: pharmacological and behavioral assays for anxiety and panic. Eur J Pharmacol 463:97–116

Boadle-Biber MC (1993) Regulation of serotonin synthesis. Prog Biophys Mol Biol 60:1–15

Boadle-Biber MC, Corley KC, Graves L, Phan TH, Rosecrans J (1989) Increase in the activity of tryptophan hydroxylase from cortex and midbrain of male Fischer 344 rats in response to acute or repeated sound stress. Brain Res 482:306–316

Chalmers DT, Lovenberg TW, De Souza EB (1995) Localization of novel corticotropin-releasing factor receptor (CRF_2) mRNA expression to specific subcortical nuclei in rat brain: comparison with CRF_1 receptor mRNA expression. J Neurosci 15:6340–6350

Chen R, Lewis KA, Perrin MH, Vale WW (1993) Expression cloning of a human corticotropin-releasing-factor receptor. Proc Natl Acad Sci U S A 90:8967–8971

Corley KC, Phan TH, Daugherty WP, Boadle-Biber MC (2002) Stress-induced activation of median raphe serotonergic neurons in rats is potentiated by the neurotensin antagonist, SR 48692. Neurosci Lett 319:1–4

Cullinan WE, Herman JP, Battaglia DF, Akil H, Watson SJ (1995) Pattern and time course of immediate early gene expression in rat brain following acute stress. Neuroscience 64:477–505

Cummings S, Elde R, Ells J, Lindall A (1983) Corticotropin-releasing factor immunoreactivity is widely distributed within the central nervous system of the rat: an immunohistochemical study. J Neurosci 3:1355–1368

Dahlström A, Fuxe K (1964) Evidence for the existence of monoamine-containing neurons in the central nervous system. Acta Physiol Scand 62:2–55

Daugherty WP, Corley KC, Phan TH, Boadle-Biber MC (2001) Further studies on the activation of rat median raphe serotonergic neurons by inescapable sound stress. Brain Res 923:103–111

Day HE, Greenwood BN, Hammack SE, Watkins LR, Fleshner M, Maier SF, Campeau S (2004) Differential expression of 5HT-1A, alpha 1b adrenergic, CRF-R1, and CRF-R2 receptor mRNA in serotonergic, gamma-aminobutyric acidergic, and catecholaminergic cells of the rat dorsal raphe nucleus. J Comp Neurol 474:364–378

De Groote L, Penalva RG, Flachkamm C, Linthorst, ACE (2005) Differential monoaminergic, neuroendocrine and behavioural responses after central administration of corticotropin-releasing factor receptor type 1 and 2 agonist. J Neurochem in press

Dijkstra I, Tilders FJH, Aguilera G, Kiss A, Rabadandiehl C, Barden N, Karanth S, Holsboer F, Reul JMHM (1998) Reduced activity of hypothalamic corticotropin-releasing hormone neurons in transgenic mice with impaired glucocorticoid receptor function. J Neurosci 18:3909–3918

Dilts RP, Boadle-Biber MC (1995) Differential activation of the 5-hydroxytryptamine-containing neurons of the midbrain raphe of the rat in response to randomly presented inescapable sound. Neurosci Lett 199:78–80

Dunn AJ (1988) Stress-related changes in cerebral catecholamine and indoleamine metabolism: lack of effect of adrenalectomy and corticosterone. J Neurochem 51:406–412

Dunn AJ, Berridge CW (1990) Physiological and behavioral responses to corticotropin-releasing factor administration—Is CRF a mediator of anxiety or stress responses. Brain Res Brain Res Rev 15:71–100

Eriksson E, Engberg G, Bing O, Nissbrandt H (1999) Effects of mCPP on the extracellular concentrations of serotonin and dopamine in rat brain. Neuropsychopharmacology 20:287–296

Fujino K, Yoshitake T, Inoue O, Ibii N, Kehr J, Ishida J, Nohta H, Yamaguchi M (2002) Increased serotonin release in mice frontal cortex and hippocampus induced by acute physiological stressors. Neurosci Lett 320:91–95

Graeff FG, Guimaraes FS, Deandrade TGCS, Deakin JFW (1996) Role of 5-HT in stress, anxiety, and depression. Pharmacol Biochem Behav 54:129–141

Grahn RE, Will MJ, Hammack SE, Maswood S, McQueen MB, Watkins LR, Maier SF (1999) Activation of serotonin-immunoreactive cells in the dorsal raphe nucleus in rats exposed to an uncontrollable stressor. Brain Res 826:35–43

Hajos-Korcsok E, Robinson DD, Yu JH, Fitch CS, Walker E, Merchant KM (2003) Rapid habituation of hippocampal serotonin and norepinephrine release and anxiety-related behaviors, but not plasma corticosterone levels, to repeated footshock stress in rats. Pharmacol Biochem Behav 74:609–616

Hammack SE, Richey KJ, Schmid MJ, LoPresti ML, Watkins LR, Maier SF (2002) The role of corticotropin-releasing hormone in the dorsal raphe nucleus in mediating the behavioral consequences of uncontrollable stress. J Neurosci 22:1020–1026

Hammack SE, Schmid MJ, LoPresti ML, Der-Avakian A, Pellymounter MA, Foster AC, Watkins LR, Maier SF (2003) Corticotropin releasing hormone type 2 receptors in the dorsal raphe nucleus mediate the behavioral consequences of uncontrollable stress. J Neurosci 23:1019–1025

Hashimoto S, Inoue T, Koyama T (1999) Effects of conditioned fear stress on serotonin neurotransmission and freezing behavior in rats. Eur J Pharmacol 378:23–30

Hayley S, Borowski T, Merali Z, Anisman H (2001) Central monoamine activity in genetically distinct strains of mice following a psychogenic stressor: effects of predator exposure. Brain Res 892:293–300

Heinsbroek RP, van Haaren F, Feenstra MG, Boon P, van de Poll NE (1991) Controllable and uncontrollable footshock and monoaminergic activity in the frontal cortex of male and female rats. Brain Res 551:247–255

Inoue T, Koyama T, Yamashita I (1993) Effect of conditioned fear stress on serotonin metabolism in the rat brain. Pharmacol Biochem Behav 44:371–374

Inoue T, Tsuchiya K, Koyama T (1994) Regional changes in dopamine and serotonin activation with various intensity of physical and psychological stress in the rat brain. Pharmacol Biochem Behav 49:911–920

Isogawa K, Akiyoshi J, Hikichi T, Yamamoto Y, Tsutsumi T, Nagayama H (2000) Effect of corticotropin releasing factor receptor 1 antagonist on extracellular norepinephrine, dopamine and serotonin in hippocampus and prefrontal cortex of rats in vivo. Neuropeptides 34:234–239

Jacobs BL, Azmitia EC (1992) Structure and function of the brain serotonin system. Physiol Rev 72:165–229

Kagamiishi Y, Yamamoto T, Watanabe S (2003) Hippocampal serotonergic system is involved in anxiety-like behavior induced by corticotropin-releasing factor. Brain Res 991:212–221

Kanno M, Matsumoto M, Togashi H, Yoshioka M, Mano Y (2003) Effects of repetitive transcranial magnetic stimulation on behavioral and neurochemical changes in rats during an elevated plus-maze test. J Neurol Sci 211:5–14

Kirby LG, Allen AR, Lucki I (1995) Regional differences in the effects of forced swimming on extracellular levels of 5-hydroxytryptamine and 5-hydroxyindoleacetic acid. Brain Res 682:189–196

Kirby LG, Chou-Green JM, Davis K, Lucki I (1997) The effects of different stressors on extracellular 5-hydroxytryptamine and 5-hydroxyindoleacetic acid. Brain Res 760:218–230

Kirby LG, Rice KC, Valentino RJ (2000) Effects of corticotropin-releasing factor on neuronal activity in the serotonergic dorsal raphe nucleus. Neuropsychopharmacology 22:148–162

Kozicz T, Yanaihara H, Arimura A (1998) Distribution of urocortin-like immunoreactivity in the central nervous system of the rat. J Comp Neurol 391:1–10

Lavicky J, Dunn AJ (1993) Corticotropin-releasing factor stimulates catecholamine release in hypothalamus and prefrontal cortex in freely moving rats as assessed by microdialysis. J Neurochem 60:602–612

Lewis K, Li C, Perrin MH, Blount A, Kunitake K, Donaldson C, Vaughan J, Reyes TM, Gulyas J, Fischer W, Bilezikjian L, Rivier J, Sawchenko PE, Vale WW (2001) Identification of urocortin III, an additional member of the corticotropin-releasing factor (CRF) family with high affinity for the CRF2 receptor. Proc Natl Acad Sci U S A 98:7570–7575

Li C, Vaughan J, Sawchenko PE, Vale WW (2002) Urocortin III-immunoreactive projections in rat brain: partial overlap with sites of type 2 corticotrophin-releasing factor receptor expression. J Neurosci 22:991–1001

Linthorst ACE, Flachskamm C, Hopkins SJ, Hoadley ME, Labeur MS, Holsboer F, Reul JMHM (1997) Long-term intracerebroventricular infusion of corticotropin-releasing hormone alters neuroendocrine, neurochemical, autonomic, behavioral, and cytokine responses to a systemic inflammatory challenge. J Neurosci 17:4448–4460

Linthorst ACE, Flachskamm C, Barden N, Holsboer F, Reul JMHM (2000) Glucocorticoid receptor impairment alters CNS responses to a psychological stressor: an in vivo microdialysis study in transgenic mice. Eur J Neurosci 12:283–291

Linthorst ACE, Flachskamm C, Reul JMHM (2001) Hippocampal serotonin responses to forced swim stress in rats: influence of water temperature. Proceedings of the 31st Annual Meeting of the Society for Neuroscience, San Diego

Linthorst ACE, Peñalva RG, Flachskamm C, Holsboer F, Reul JMHM (2002) Forced swim stress activates rat hippocampal serotonergic neurotransmission involving a corticotropin-releasing hormone receptor-dependent mechanism. Eur J Neurosci 16:2441–2452

Lovenberg TW, Liaw CW, Grigoriadis DE, Clevenger W, Chalmers DT, De Souza EB, Oltersdorf T (1995) Cloning and characterization of a functionally distinct corticotropin-releasing factor receptor subtype from rat brain. Proc Natl Acad Sci U S A 92:836–840

Lowry CA (2002) Functional subsets of serotonergic neurones: implications for control of the hypothalamic-pituitary-adrenal axis. J Neuroendocrinol 14:911–923

Lowry CA, Rodda JE, Lightman SL, Ingram CD (2000) Corticotropin-releasing factor increases in vitro firing rates of serotonergic neurons in the rat dorsal raphe nucleus: evidence for activation of a topographically organized mesolimbocortical serotonergic system. J Neurosci 20:7728–7736

Lowry CA, Johnson PL, Hathway NJA, Lightman SL (2002) Distribution of corticotropin-releasing factor receptor 1-(CRFR1) and CRFR2-immunoreactivity in limbic and caudal brainstem raphe nuclei. Program No 867.6. 2002 abstract viewer/itinerary planner. Society for Neuroscience, Washington http://sfn.scholarone.com/itin2002/index.html. Cited 20 December 2004

Maes M, Meltzer HY (1995) The serotonin hypothesis of major depression. In: Bloom FE, Kupfer DJ (eds) Psychopharmacology: the fourth generation of progress. pp 933–944

Mann JJ (1998) The neurobiology of suicide. Nat Med 4:25–30

Martinez M, Phillips PJ, Herbert J (1998) Adaptation in patterns of c-fos expression in the brain associated with exposure to either single or repeated social stress in male rats. Eur J Neurosci 10:20–33

Maswood S, Barter JE, Watkins LR, Maier SF (1998) Exposure to inescapable but not escapable shock increases extracellular levels of 5-HT in the dorsal raphe nucleus of the rat. Brain Res 783:115–120

Matsuo M, Kataoka Y, Mataki S, Kato Y, Oi K (1996) Conflict situation increases serotonin release in rat dorsal hippocampus: In vivo study with microdialysis and Vogel test. Neurosci Lett 215:197–200

Millan MJ (2003) The neurobiology and control of anxious states. Prog Neurobiol 70:83–244

Moreau JL, Kilpatrick G, Jenck F (1997) Urocortin, a novel neuropeptide with anxiogenic-like properties. Neuroreport 8:1697–1701

Nemeroff CB, Widerlov E, Bissette G, Walleus H, Karlsson I, Eklund K, Kilts CD, Loosen PT, Vale W (1984) Elevated concentrations of CSF corticotropin-releasing factor-like immunoreactivity in depressed patients. Science 226:1342–1344

Oshima A, Flachskamm C, Reul JMHM, Holsboer F, Linthorst ACE (2003) Altered serotonergic neurotransmission but normal hypothalamic-pituitary-adrenocortical axis activity in mice chronically treated with the corticotropin-releasing hormone receptor type 1 antagonist NBI 30775. Neuropsychopharmacology 28:2148–2159

Peñalva RG, Flachskamm C, Zimmermann S, Wurst W, Holsboer F, Reul JMHM, Linthorst ACE (2002) Corticotropin-releasing hormone receptor type 1-deficiency enhances hippocampal serotonergic neurotransmission: an in vivo microdialysis study in mutant mice. Neuroscience 109:253–266

Pernar L, Curtis AL, Vale WW, Rivier JE, Valentino RJ (2004) Selective activation of corticotropin-releasing factor-2 receptors on neurochemically identified neurons in the rat dorsal raphe nucleus reveals dual actions. J Neurosci 24:1305–1311

Petty F, Kramer G, Wilson L, Jordan S (1994) In vivo serotonin release and learned helplessness. Psychiatry Res 52:285–293

Price ML, Lucki I (2001) Regulation of serotonin release in the lateral septum and striatum by corticotropin-releasing factor. J Neurosci 21:2833–2841

Price ML, Curtis AL, Kirby LG, Valentino RJ, Lucki I (1998) Effects of corticotropin-releasing factor on brain serotonergic activity. Neuropsychopharmacology 18:492–502

Price ML, Kirby LG, Valentino RJ, Lucki I (2002) Evidence for corticotropin-releasing factor regulation of serotonin in the lateral septum during acute swim stress: adaptation produced by repeated swimming. Psychopharmacology (Berl) 162:406–414

Raadsheer FC, Hoogendijk WJG, Stam FC, Tilders FJH, Swaab DF (1994) Increased numbers of corticotropin-releasing hormone expressing neurons in the hypothalamic paraventricular nucleus of depressed patients. Neuroendocrinology 60:436–444

Ressler KJ, Nemeroff CB (2000) Role of serotonergic and noradrenergic systems in the pathophysiology of depression and anxiety disorders. Depress Anxiety 12:2–19

Reul JMHM, Holsboer F (2002) Corticotropin-releasing hormone receptors 1 and 2 in anxiety and depression. Curr Opin Pharmacol 2:23–33

Rex A, Marsden CA, Fink H (1993) 5 HT1A receptors and changes in extracellular 5-HT in the guinea-pig prefrontal cortex—involvement in aversive behaviour. J Psychopharmacol 7:338–345

Reyes TM, Lewis K, Perrin MH, Kunitake KS, Vaughan J, Arias CA, Hogenesch JB, Gulyas J, Rivier J, Vale WW, Sawchenko PE (2001) Urocortin II: a member of the corticotropin-releasing factor (CRF) neuropeptide family that is selectively bound by type 2 CRF receptors. Proc Natl Acad Sci U S A 98:2843–2848

Roche M, Commons KG, Peoples A, Valentino RJ (2003) Circuitry underlying regulation of the serotonergic system by swim stress. J Neurosci 23:970–977

Rueter LE, Jacobs BL (1996) A microdialysis examination of serotonin release in the rat forebrain induced by behavioral environmental manipulations. Brain Res 739:57–69

Ruggiero DA, Underwood MD, Rice PM, Mann JJ, Arango V (1999) Corticotropic-releasing hormone and serotonin interact in the human brainstem: behavioral implications. Neuroscience 91:1343–1354

Sakanaka M, Shibasaki T, Lederis K (1987) Corticotropin releasing factor-like immunoreactivity in the rat brain as revealed by a modified cobalt-glucose oxidase-diaminobenzidine method. J Comp Neurol 260:256–298

Silveira MC, Sandner G, Graeff FG (1993) Induction of Fos immunoreactivity in the brain by exposure to the elevated plus-maze. Behav Brain Res 56:115–118

Singh VB, Hao-Phan T, Corley KC, Boadle-Biber MC (1992) Increase in cortical and midbrain tryptophan hydroxylase activity by intracerebroventricular administration of corticotropin releasing factor: block by adrenalectomy, by RU 38486 and by bilateral lesions to the central nucleus of the amygdala. Neurochem Int 20:81–92

Smith GW, Aubry JM, Dellu F, Contarino A, Bilezikjian LM, Gold LH, Chen RP, Marchuk Y, Hauser C, Bentley CA, Sawchenko PE, Koob GF, Vale W, Lee KF (1998) Corticotropin releasing factor receptor 1-deficient mice display decreased anxiety, impaired stress response, and aberrant neuroendocrine development. Neuron 20:1093–1102

Steinbusch HWM (1981) Distribution of serotonin-immunoreactivity in the central nervous system of the rat-cell bodies and terminals. Neuroscience 6:557–618

Swanson LW, Sawchenko PE, Rivier J, Vale WW (1983) Organization of ovine corticotropin-releasing factor immunoreactive cells and fibers in the rat brain: an immunohistochemical study. Neuroendocrinology 36:165–186

Timpl P, Spanagel R, Sillaber I, Kresse A, Reul JMHM, Stalla GK, Blanquet V, Steckler T, Holsboer F, Wurst W (1998) Impaired stress response and reduced anxiety in mice lacking a functional corticotropin-releasing hormone receptor 1. Nat Genet 19:162–166

Törk I (1990) Anatomy of the serotonergic system. Ann N Y Acad Sci 600:9–35

Valdez GR, Inoue K, Koob GF, Rivier J, Vale W, Zorrilla EP (2002) Human urocortin II: mild locomotor suppressive and delayed anxiolytic-like effects of a novel corticotropin-releasing factor related peptide. Brain Res 943:142–150

Valdez GR, Zorrilla EP, Rivier J, Vale WW, Koob GF (2003) Locomotor suppressive and anxiolytic-like effects of urocortin 3, a highly selective type 2 corticotropin-releasing factor agonist. Brain Res 980:206–212

Valentino RJ, Liouterman L, Van Bockstaele EJ (2001) Evidence for regional heterogeneity in corticotropin-releasing factor interactions in the dorsal raphe nucleus. J Comp Neurol 435:450–463

Van Gaalen MM, Reul JMHM, Gesing A, Stenzel-Poore MP, Holsboer F, Steckler T (2002) Mice overexpressing CRH show reduced responsiveness in plasma corticosterone after a 5-HT1A receptor challenge. Genes Brain Behav 1:174–177

Van Loon GR, Shum A, Ho D (1982) Lack of effect of corticotropin releasing factor on hypothalamic dopamine and serotonin synthesis turnover rates in rats. Peptides 3:799–803

Van Pett K, Viau V, Bittencourt JC, Chan RKW, Li HY, Arias C, Prins GS, Perrin M, Vale W, Sawchenko PE (2000) Distribution of mRNAs encoding CRF receptors in brain and pituitary of rat and mouse. J Comp Neurol 428:191–212

Vaughan J, Donaldson C, Bittencourt J, Perrin MH, Lewis K, Sutton S, Chan R, Turnbull AV, Lovejoy D, Rivier C, Rivier J, Sawchenko PE, Vale W (1995) Urocortin, a mammalian neuropeptide related to fish urotensin I and to corticotropin-releasing factor. Nature 378:287–292

Vetter DE, Li C, Zhao L, Contarino A, Liberman MC, Smith GW, Marchuk Y, Koob GF, Heinemann SF, Vale W, Lee KF (2002) Urocortin-deficient mice show hearing impairment and increased anxiety-like behavior. Nat Genet 31:363–369

Voigt JP, Rex A, Sohr R, Fink H (1999) Hippocampal 5-HT and NE release in the transgenic rat TGR(mREN2)27 related to behavior on the elevated plus maze. Eur Neuropsychopharmacol 9:279–285

Walther DJ, Peter JU, Bashammakh S, Hörtnagl H, Voits M, Fink H, Bader M (2003) Synthesis of serotonin by a second tryptophan hydroxylase isoform. Science 299:76

Wang X, Su H, Copenhagen LD, Vaishnav S, Pieri F, Shope CD, Brownell WE, De Biasi M, Paylor R, Bradley A (2002) Urocortin-deficient mice display normal stress-induced anxiety behavior and autonomic control but an impaired acoustic startle response. Mol Cell Biol 22:6605–6610

Wilkinson LS, Humby T, Killcross S, Robbins TW, Everitt BJ (1996) Dissociations in hippocampal 5-hydroxytryptamine release in the rat following pavlovian aversive conditioning to discrete and contextual stimuli. Eur J Neurosci 8:1479–1487

Wright IK, Upton N, Marsden CA (1992) Effect of established and putative anxiolytics on extracellular 5-HT and 5-HIAA in the ventral hippocampus of rats during behaviour on the elevated X-maze. Psychopharmacology (Berl) 109:338–346

Yoshioka M, Matsumoto M, Togashi H, Saito H (1995) Effects of conditioned fear stress on 5-HT release in the rat prefrontal cortex. Pharmacol Biochem Behav 51:515–519

Zobel AW, Nickel T, Künzel HE, Ackl N, Sonntag A, Ising M, Holsboer F (2000) Effects of the high-affinity corticotropin-releasing hormone receptor 1 antagonist R121919 in major depression: the first 20 patients treated. J Psychiatr Res 34:171–181

HEP (2005) 169:205–223

Anxiety Disorders: Noradrenergic Neurotransmission

A. Neumeister[3] (✉) · R. J. Daher[1] · D. S. Charney[2]

[1]Mood and Anxiety Disorders Research Program, National Institute of Mental Health/NIH, 15K North Drive, MSC 2670, Bethesda MD, 20892-2670, USA

[2]Mount Sinai School of Medicine, One Gustave L. Levy Place, Box 1102, New York NY , 10029, USA

[3]Yale University School of Medicine, Molecular Imaging Program of the Clinical Neuroscience Division, 950 Campbell Avenue, West Haven NY , 10029, USA
Alexander.Neumeister@yale.edu

Abstract The past decade has seen a rapid progression in our knowledge of the neurobiological basis of fear and anxiety. Specific neurochemical and neuropeptide systems have been demonstrated to play important roles in the behaviors associated with fear and anxiety-producing stimuli. Long-term dysregulation of these systems appears to contribute to the development of anxiety disorders, including panic disorder, posttraumatic stress disorder (PTSD), and social anxiety disorder. These neurochemical and neuropeptide systems have been shown to have effects on distinct cortical and subcortical brain areas that are relevant to the mediation of the symptoms associated with anxiety disorders. Moreover, advances in molecular genetics portend the identification of the genes that underlie the neurobiological disturbances that increase the vulnerability to anxiety disorders. This chapter reviews clinical research pertinent to the neurobiological basis of anxiety disorders. The implications of this synthesis for the discovery of anxiety disorder vulnerability genes and novel psychopharmacological approaches will also be discussed.

Keywords Fear · Anxiety · Pathophysiology · Circuitry · Neurochemistry · Treatments

1 Neural Mechanisms of Anxiety and Fear

Classical fear conditioning is a form of associative learning in which subjects come to express fear responses to neutral conditioned stimuli (CS) that are paired with an aversive unconditioned stimulus (US). The CS, as a consequence of this pairing, acquire the ability to elicit a spectrum of behavioral, autonomic, and endocrine responses that normally would only occur in the context of danger (Blair et al. 2001). Fear conditioning can be adaptive and enable efficient behavior in dangerous situations. The individual who can accurately predict threat can engage in the appropriate behaviors in the face of danger. In the clinical situation, specific environmental features (CS) may be linked to a traumatic event, spontaneous panic attack, or embarrassing social situation (US), such that re-exposure to a similar environment produces a recurrence of symptoms of anxiety and fear. Patients often generalize these cues and experience a continuous perception of threat to the point that they become conditioned to context (Table 1 outlines the five neural mechanisms of anxiety and fear).

Cue-specific CS are transmitted to the thalamus by external and visceral pathways. Afferents then reach the lateral amygdala (LA) via two parallel circuits: a rapid subcortical path directly from the dorsal (sensory) thalamus and a slower regulatory cortical pathway encompassing primary somatosensory cortices, the insula, and the anterior cingulate/prefrontal cortex. Contextual CS are projected to the LA from the hippocampus and perhaps the bed nucleus of the stria terminalis. The long loop pathway indicates that sensory information relayed to the amygdala undergoes substantial higher level processing, thereby enabling assignment of significance, based upon prior experience, to complex stimuli. Cortical involvement in fear conditioning is clinically relevant because it provides a mechanism by which cognitive factors will influence whether symptoms are experienced or not, following stress exposure (LeDoux 2000).

During the expression of fear-related behaviors, the LA engages the central nucleus of the amygdala (CEA), which, as the principal output nucleus, projects to areas of the hypothalamus and brain stem that mediate the autonomic, endocrine, and behavioral responses associated with fear and anxiety (Schafe et al. 2001). The molecular and cellular mechanisms that underlie synaptic plasticity in amygdala-dependent learned fear is an area of very active investigation (Shumyatsky et al. 2002). Long-term potentiation (LTP) in the LA appears to be a critical mechanism for storing memories of the CS–US associa-

Table 1 Neural mechanisms related to pathophysiology and treatment of anxiety disorders

Mechanism	Neurochemical systems	Brain regions	Pathophysiology	Treatment development
Pavlovian (cue specific) fear conditioning	Glutamate, NMDA receptors, VGCCs	Medial prefrontal cortex, sensory cortex, anterior cingulate, dorsal thalamus, lateral amygdala, central nucleus of amygdala	May account for common clinical observation in panic disorder, PTSD that sensory and cognitive stimuli associated with or resembling the frightening experience elicit panic attacks, flashbacks, and autonomic symptoms	Treatment with NMDA receptor antagonist and VGCC antagonist may attenuate acquisition of fear
Inhibitory avoidance (contextual fear)	NE/β-adrenergic receptor, cortisol/ glucocorticoid receptor, CRH, GABA, opioids, acetylcholine	Medial prefrontal cortex, basolateral amygdala, hippocampus, BNST entorhinal cortex	Excessive stress-mediated release of CRH, cortisol, and NE will facilitate development of indelible fear memories. Chronic anxiety and phobic symptoms may result from excessive contextual fear conditioning	CRH antagonists and β-adrenergic receptor agonists may have preventative effects
Reconsolidation	Glutamate, NMDA receptors, NE, β-adrenergic receptors, CREB	Amygdala, hippocampus	Repeated reactivation and reconsolidation may further strengthen the memory trace and lead to persistence of trauma and phobia-related symptoms	Treatment with NMDA receptor and β-adrenergic receptor antagonists after memory reactivation may reduce the strength of the original anxiety provoking memory
Extinction	Glutamate, NMDA receptors, VGCCs, NE, dopamine, GABA	Medial prefrontal sensory cortex, amygdala	Failure in neural mechanisms of extinction may relate to persistent traumatic memories, re-experiencing symptoms, autonomic hyperarousal, and phobic behaviors	Psychotherapies need to be developed that facilitate extinction through the use of conditioned inhibitors and the learning of "new memories" The combination of extinction based psychotherapy and D-cycloserine may be a particularly effective treatment
Sensitization	Dopaminergic, noradrenergic NMDA receptors	Nucleus accumbens, amygdala, striatum, hypothalamus	May explain the adverse effects of early life trauma on subsequent responses to stressful like events. May play a role in the chronic course of many anxiety disorders and, in some cases, the worsening of the illness over time	Suggests the efficacy of treatment may vary according to the state of evolution of the disease process. Emphasizes the importance of early treatment intervention

BNST, bed nucleus of the stria terminalis; CREB, cyclic AMP response element-binding protein; CRH, corticotrophin-releasing hormone; CS, conditioned stimuli; GABA, γ-aminobutyric acid; GCC, voltage-gated calcium channels; NE, norepinephrine; NMDA, *N*-methyl-D-aspartate; PAG, periaqueductal gray.

tion (Blair et al. 2001). A variety of behavioral and electrophysiological data has led LeDoux and colleagues to propose a model to explain how neural responses to the CS and US in the LA could influence LTP-like changes that store memories during fear conditioning. This model proposes that calcium entry through *N*-methyl-D-aspartate (NMDA) receptors and voltage gated calcium channels (VGCCs) initiates the molecular processes to consolidate synaptic changes into long-term memory (Blair et al. 2001). Short-term memory requires calcium entry only through NMDA receptors and not VGCCs.

This hypothesis leads to several predictions that may have relevance to the discovery of novel therapeutics for anxiety disorders. It suggests that blocking NMDA receptors in the amygdala during learning should impair short- and long-term fear memory. This has been demonstrated in rodents (Walker et al. 2000; Rodrigues et al. 2001). Valid human models of fear conditioning and the availability of the NMDA receptor antagonist memantine should permit this hypothesis to be tested clinically (Grillon 2002). If memantine impairs the acquisition of fear in humans, it may have utility in the prevention and treatment of anxiety disorders such as posttraumatic stress disorder (PTSD), and panic disorder. Blockade of VGCCs appears to block long-term but not short-term memory (Bauer et al. 2002). Therefore, clinically available calcium channel blockers such as verapamil and nimodipine may be helpful for in diminishing the intensity and impact of recently acquired fear memory and perhaps in preventing PTSD as well.

The discussion above has focused primarily upon the neural mechanisms related to the coincident learning of the US–CS association (i.e., Pavlovian fear conditioning) in the LA. However, there is significant evidence that a broader neural circuitry underlies fear memory that is modulated by amygdala activity. The inhibitory avoidance paradigm is used to examine memory consolidation for aversively motivated tasks and involves intentional instrumental choice behavior. Studies using inhibitory avoidance learning procedures have been used to support the view that the amygdala is not the sole site for fear learning; this view posits that the amygdala can modulate the strength of memory storage in other brain structures (McGaugh 2002).

Specific drugs and neurotransmitters infused into the basolateral amygdala (BLA) influence consolidation of memory for inhibitory avoidance training. Post-training peripheral or intra-amygdala infusions of drugs affecting γ-aminobutyric acid (GABA), opioid, glucocorticoid, and muscarinic acetylcholine receptors have dose- and time-dependent effects on memory consolidation (McGaugh 2002). Norepinephrine (NE) infused directly into the BLA after inhibitory avoidance training enhances memory consolidation, indicating that the degree of activation of the noradrenergic system within the amygdala by an aversive experience may predict the extent of the long-term memory for the experience (McIntyre et al. 2002).

Interactions among corticotropin-releasing hormone (CRH), cortisol, and NE have very important effects on memory consolidation, which is likely to be relevant to the effects of traumatic stress on memory. Extensive evidence indicates that glucocorticoids influence long-term memory consolidation via stimulation of glucocorticoid receptors (GR). The glucocorticoid effects on memory consolidation require activation of the BLA, and lesions of the BLA block retention enhancement of intrahippocampal infusions of a GR agonist. Additionally, the BLA is a critical locus of interaction between glucocorticoids and NE in modulating memory consolidation (McGaugh et al. 2002).

There is extensive evidence consistent with a role for CRH in mediating stress effects on memory consolidation. Activation of CRH receptors in the BLA by CRH released from the CEA facilitates stress effects on memory consolidation. Memory enhancement produced by CRH infusions in the hippocampus are blocked by propranolol, suggesting CRH, through a presynaptic mechanism, stimulates NE release in the hippocampus (Roozendaal et al. 2002).

These results support the concept that CRH via an interaction with glucocorticoids interacts with the noradrenergic system to consolidate traumatic memories. Individuals with excessive stress-induced release of CRH, cortisol, and NE are likely to be prone to the development of indelible traumatic memories and associated re-experiencing symptoms. Administration of CRH antagonists, glucocorticoid receptor antagonists, and β-adrenergic receptor antagonists may prevent these effects in vulnerable subjects.

2 Reconsolidation

Reconsolidation is a process in which old, reactivated memories undergo another round of consolidation (Debiec et al. 2002; Milekic et al. 2002; Myers et al. 2002). The process of reconsolidation is extremely relevant to both vulnerability and resiliency to the effects of extreme stress. It is the rule rather than the exception that memories are reactivated by cues associated with the original trauma. Repeated reactivation of these memories may serve to strengthen the memories and facilitate long-term consolidation (Przbyslawski et al. 1999; Sara 2000). Each time a traumatic memory is retrieved, it is integrated into an ongoing perceptual and emotional experience and becomes part of a new memory. Moreover, recent preclinical studies indicate that consolidated memories for auditory fear conditioning, which are stored in the amygdala (Nader et al. 2000a), hippocampal-dependent contextual fear memory (Debiec et al. 2002), and hippocampal-dependent memory associated with inhibitory avoidance (Milekic et al. 2002) are sensitive to disruption upon reactivation by administration with a protein synthesis inhibitor directly into the amygdala and hippocampus, respectively. The reconsolidation process, which has enormous clinical implications, results in reactivated memory trace that returns to a state of lability and must undergo consolidation once more if it is to remain in long-term storage. Some controversies persist regarding the temporal persistence of systems reconsolidation. Debiec and colleagues found that intrahippocampal infusions of anisomycin caused amnesia for a consolidated hippocampal-dependent memory if the memory was reactivated even up to 45 days after training (Debiec et al. 2002). Milekic and Alberini (2002), however, found that the ability of intrahippocampal infusion of anisomycin to produce amnesia for an inhibitory avoidance task was evident only when the memory was recent (up to 7 days). Further work is needed to resolve this very important question (Myers et al. 2002).

The reconsolidation process involves NMDA receptors, β-adrenergic receptors, and requires cyclic AMP response element binding protein (CREB) induction. The CREB requirement suggests that nuclear protein synthesis is necessary (Kida et al. 2002). NMDA receptor antagonists and β-receptor antagonists impair reconsolidation (Przbyslawski et al. 1997, 1999). The effect of the β-receptor antagonist propranolol was greater after memory reactivation than when administered immediately after initial training. These results suggest that reactivation of memory initiates a cascade of intracellular events that involve both NMDA receptor and β-receptor activation in a fashion similar to post-acquisition consolidation.

This remarkable lability of a memory trace, which permits a reorganization of an existing memory in a retrieval environment, provides a theoretical basis for both psychotherapeutic and pharmacotherapeutic intervention for traumatic stress exposure as well as other anxiety disorders. Administration of β-receptor and NMDA receptor antagonists shortly after trauma exposure or spontaneous panic attacks as well as after reactivation of memory associated with the anxiety-inducing event may reduce the strength of the original memory.

3 Extinction

When the CS is presented repeatedly in the absence of the US, a reduction in the condition fear response occurs. This process is called extinction. It forms the basis for exposure-based psychotherapies for the treatment of anxiety disorders characterized by exaggerated fear responses. Individuals who show an ability to quickly attenuate learned fear through a powerful and efficient extinction processes are likely to function more effectively under dangerous conditions.

Extinction is characterized by many of the same neural mechanisms as in fear acquisition. Activation of amygdala NMDA receptors by glutamate is essential (Myers and Davis 2004) and L-type VGCCs also contribute to extinction plasticity (Cain et al. 2002). Long-term extinction memory is altered by a number of different neurotransmitters systems including GABA, NE, and dopamine (DA) in a manner similar to fear acquisition (McGaugh et al. 1990; Willick et al. 1995).

Destruction of the medial prefrontal cortex (mPFC) blocks recall of fear extinction (Quirk et al. 2000; Morgan et al. 1993), indicating that the mPFC might store long-term extinction memory. Infralimbic neurons, which are part of the mPFC, fire only when rats are recalling extinction—greater firing correlates with reduced fear behaviors (Milad et al. 2002). It has been suggested that the consolidation of extinction involves potentiation of inputs into the mPFC by means of NMDA-dependent plasticity. The BLA sends direct excitatory inputs to the mPFC, and NMDA antagonists infused into BLA blocks extinction. The ability of the mPFC to modulate fear behaviors is probably related to projections from the mPFC via GABA interneurons to the BLA (Royer et al. 2000).

Failure to achieve an adequate level of activation of the mPFC after extinction might lead to persistent fear responses (Herry et al. 2002). Individuals with the capacity to function well following states of high fear may have potent mPFC inhibition of amygdala responsiveness. In contrast, patients with PTSD exhibit depressed ventral mPFC activity which correlated with increased autonomic arousal after exposure to traumatic reminders (Bremner et al. 1999). Consistent with this hypothesis, we recently showed that PTSD patients had increased left amygdala activation during fear acquisition and decreased mPFC/anterior cingulate activity during extinction (Bremner et al. 2003). It has been proposed that potentiating NMDA receptors using the glycine agonist, D-cycloserine, may facilitate the extinction process when given in combination with behavioral therapy in patients with anxiety disorders (Davis 2002).

These preclinical investigations suggest that clinical research paradigms capable of evaluating the mechanisms of fear conditioning in clinical populations would be of great value. Psychophysiological studies in PTSD patients have been reviewed recently and have consistently demonstrated increased electrophysiological and autonomic responses to trauma related stimuli (Orr et al. 2002). The startle reflex has been used to study fear conditioning in humans. Startle is a useful method for examining fear responding in experimental studies involving both animals and humans that is mediated by the amygdala and connected structures. There is evidence of elevation of baseline startle in almost all anxiety disorders (Grillon 2002), suggestive of increased contextual fear. This is consistent with hyperexcitability of neural structures underlying contextual fear such as the BNST. Vulnerability to anxiety disorders may relate to startle responses. Girls at high risk for developing anxiety disorders are overly sensitive to contextual threat, but exhibit normal fear-potentiated startle. High-risk boys, on the other hand, exhibit elevated potentiated startle and normal contextual responses. Cue fear learning is an adaptive process by which undifferentiated fear becomes cue specific. Deficits in cue fear learning may lead non-adaptive aversive expectancies and a state of chronic anxiety.

4 The Neurochemical Basis of Fear and Anxiety

Specific neurotransmitters and neuropeptides act on brain areas noted above in the mediation of fear and anxiety responses. These neurochemicals are released during stress, and chronic stress results in long-term alterations in function of these systems. Stress axis neurochemical systems prepare the organism for threat in multiple ways, through increased attention and vigilance, modulation of memory (in order to maximize the utilization of prior experience), planning, and preparation for action. In addition, these systems have peripheral effects, which include increased heart rate and blood pressure (catecholamines) and rapid modulation of the body's use of energy (cortisol). The neurobiological responses to threat and severe stress are clearly adaptive and have survival value, but they also can have maladaptive consequences when they become chronically activated. Examination of the preclinical data concerning neurochemical substrates of the stress response, the long-term impact of early life exposure to stress, and possible stress-induced neurotoxicity provide a context to consider clinical investigations of the pathophysiology of the anxiety disorders.

5 Noradrenergic System

Stressful stimuli of many types produce marked increases in brain noradrenergic function. Stress produces regional selective increases in NE turnover in the locus coeruleus (LC), limbic regions (hypothalamus, hippocampus, and amygdala), and cerebral cortex. These changes can be elicited with immobilization stress, foot-shock stress, tail-pinch stress, and conditioned fear. Exposure to stressors from which the animal cannot escape results in behavioral deficits termed learned helplessness. The learned helplessness state is associated with depletion of NE, probably reflecting the point where synthesis cannot keep up with demand. These studies have been reviewed elsewhere in detail (Bremner et al. 1996a,b).

The LC is a compact nucleus containing noradrenergic neurons as well as peptide neurotransmitters (e.g., hypocretin and CRH) that influence its activity. LC neurons can fire in either a tonic or phasic pattern, and electrotonic coupling between neurons can be influenced by neurotransmitters. Release of NE can be accompanied by co-release of the peptide neurotransmitter galanin, which is inhibitory and may alter the firing rate of DA neurons, thus altering its hedonic tone. Shifts in the pattern of firing of LC neurons are thought to be of great importance in understanding attentional processes, often disrupted in depression. The LC neurons have long dendritic processes for synaptic contact to influence its activity, and LC neurons may be strongly influenced by anterior cingulate cortex. It has recently been suggested that the A2 group of the medulla may innervate important structures such as the amygdala and nucleus accumbens and thus may be important in affect regulation. Receptors for NE are grouped into $\alpha 1$, $\alpha 2$, $\beta 1$, and $\beta 2$ subtypes.

Chronic therapy with antidepressants results in adaptive receptor alterations in the noradrenergic system. Three genes code for the expression of $\alpha 1$ subtypes, and these three receptor subtypes (A–C) have distinctive pharmacological properties. Recent data suggest that chronic antidepressants and electroconvulsive stimulation (ECS) may increase frontal cortex expression of mRNA specifically for the $\alpha 1$A-adrenoreceptor subtype and, as such, this receptor may be involved in the action of noradrenergic antidepressants (Nalepa et al. 2002). Repeated administration of antidepressants has been observed to increase behavioral responsivity to $\alpha 1$-adrenergic agonists (such as aggressiveness and hyperexploration) as well as increasing agonist-binding affinity for $\alpha 1$-adrenoreceptors.

Electrophysiological studies in the hippocampus also support enhanced $\alpha 1$ responses after chronic antidepressant treatments. Recent data indicate that the novel antidepressant tianeptine, which may increase serotonin reuptake when given chronically, also increases responsiveness of the $\alpha 1$-adrenergic system (Rogoz et al. 2001).

Single unit recording from the LC indicates that chronic administration of multiple classes of antidepressants and electroconvulsive stimulation reduce LC baseline and sensory-stimulated firing rates (Grant et al. 2001). It has been hypothesized that reducing the firing rate of noradrenergic neurons may be therapeutic, especially in anxiety disorders and also subgroups of patients with depression with certain clinical features, such as psychomotor retardation, by reducing the release of inhibitory co-released galanin neuropeptide onto DA neurons in the ventral tegmental area (VTA) (Weiss et al. 1998). However, reduced firing in the noradrenergic neurons of the LC could simply be a function of increased levels of synaptic or extracellular NE resulting in feedback inhibition of LC firing (Grant et al. 2001).

The $\alpha 2$ antagonist, yohimbine, has been observed to augment the speed of response to fluoxetine (Sanacora et al 2004), and in a very small study (n=14) of bipolar depression the $\alpha 2$ antagonist idazoxan appeared to have antidepressant effects equal to bupropion (Grossman et al 1999). Upregulation of immunolabeled $\alpha 2$A receptors and associated G proteins (Gi) are observed postmortem in suicide victims (Garcia-Sevilla et al. 1999). This is of interest, given the critical importance of these receptors in stress and in regulating levels of monoamines via autoreceptors and heteroreceptors. As well, the efficacy of antidepressants such as mirtazapine and mianserin may in part depend on these receptors. Recent transgenic experiments suggest that the $\alpha 2$ receptor may act as a "suppressor of depression," Knockout of the gene for the $\alpha 2$ receptor increases immobility in the forced swim test and eliminates the augmentation of forced swim test activity by imipramine (Schramm et al. 2001).

In contrast, other recent experiments suggest that mice lacking α2C receptors perform on the forced swim test in the *same* fashion as mice treated with antidepressants (Sallinen et al. 1999). Thus, the α2A and the α2C receptors may have complementary and opposing roles in the regulation of mood and anxiety and have a complementary role in NE responses in the heart (Schramm et al. 2001). If reducing α2C activity is to be used as an antidepressant strategy it may require some method of targeting only those receptors in the CNS, since an α2CDel322–325 polymorphism that reduces feedback inhibition of sympathetic NE released in the heart is associated a markedly a increased risk of heart disease (Small et al. 2002). Since some individuals with depression also have memory disturbance, recent evidence that mutation of the α2A receptor impairs working memory could also help us understand the cognitive symptoms observed in depression (Franowicz et al. 2002).

Crosstalk between the catecholamine system and steroids may be another novel mechanism through which NE and epinephrine—by increasing the sensitivity of glucocorticoid receptors to ligand activation—could alter mood and anxiety symptoms. A recent study found that amitriptyline prevented the appearance of impairment in spatial memory in aged rats and reduced glucocorticoid levels, and this effect is most likely secondary to NE-mediated alteration in glucocorticoid signaling (Yau et al. 2002). Augmentation effects of catecholamines on GR signaling may thus be important in cognitive and emotional processing. The PI3-K signaling pathway activation through β-receptors appears to be responsible for this putative enhancement of glucocorticoid activity, and it is tempting to conjecture that antidepressants that are known to downregulate β-receptors and influence PI3-K signaling could act by glucocorticoid receptor sensitization (Schmidt et al. 2001).

As can be seen in Table 2, chronic symptoms experienced by anxiety disorder patients, such as panic attacks, insomnia, startle, and autonomic hyperarousal, are characteristic of increased noradrenergic function (Charney et al. 1984, 1987a). Potential drugs of abuse, such as alcohol, opiates, and benzodiazepines (but not cocaine), decrease firing of noradrenergic neurons. Increases in abuse of these substances parallels increased anxiety symptoms, providing evidence for self-medication of these symptoms that is explainable based on animal studies of noradrenergic function. In addition, patients with anxiety disorders frequently report significant improvement of symptoms of hyperarousal and intrusive memories with alcohol, benzodiazepines, and opiates, which decrease LC firing, but worsening of these symptoms with cocaine, which increases LC firing.

There is strong evidence that function of the brain noradrenergic system is involved in mediating fear conditioning (Rasmussen et al. 1986; Charney and Deutch 1996). Neutral stimuli paired with shock (CS) produce increases in brain NE metabolism and behavioral deficits similar to those elicited by the shock alone (Cassens et al. 1981) as well as increased firing rate of cells in the LC (Rasmussen et al. 1986). An intact noradrenergic system appears

Table 2 Evidence for altered catecholaminergic function in anxiety disorders[a]

	PTSD	Panic disorder
Increased resting heart rate and blood pressure	+/–	+/–
Increased heart rate and blood pressure response to traumatic reminders/panic attacks	+ + +	+ +
Increased resting urinary NE and E	+	+/–
Increased resting plasma NE or MHPG	–	–
Increased plasma NE with traumatic reminders/panic attacks	+	+/–
Increased orthostatic heart rate response to exercise	+	+
Decreased binding to platelet α_2 receptors	+	+/–
Decrease in basal and stimulated activity of cAMP	+/–	+
Decrease in platelet MAO activity	+	NS
Increased symptoms, heart rate and plasma MHPG with yohimbine noradrenergic challenge	+ +	+ + +
Differential brain metabolic response to yohimbine	+	+

[a] One or more studies do not support this finding (with no positive studies), or the majority of studies does not support this finding; +/–, an equal number of studies support and do not support this finding; +, at least one study supports and no studies do not support the finding, or the majority of studies supports the finding; + +, two or more studies support and no studies do not support the finding; + + +, three or more studies support and no studies do not support the finding; cAMP, cyclic adenosine 3',5';-monophosphate; E, epinephrine; MAO, monoamine oxidase; MHPG, 3-methosy-4-hydroxyphenylglycol; NE, norepinephrine; NS, not studied; PTSD, posttraumatic stress disorder.

to be necessary for the acquisition of fear-conditioned responses (Cose and Robbins 1987).

Many patients with anxiety disorders experience an increased susceptibility to psychosocial stress. Behavioral sensitization may account for these clinical phenomena. In the laboratory model of sensitization, single or repeated exposure to physical stimuli or pharmacological agents sensitizes an animal to subsequent stressors (reviewed in Charney et al. 1993). For example, in animals with a history of prior stress, there is a potentiated release of NE in the hippocampus with subsequent exposure to stressors (Nisenbaum et al. 1991). Similar findings were observed in medial prefrontal cortex (Finlay and Abercrombie 1991). The hypothesis that sensitization is underlying neural mechanism contributing to the course of anxiety disorders is supported by clinical studies demonstrating that repeated exposure to traumatic stress is an important risk factor for the development of anxiety disorders, particularly PTSD (Table 1).

6
Posttraumatic Stress Disorder

There is extensive clinical evidence that NE plays a role in human anxiety. Well-designed psychophysiological studies have been conducted that have documented heightened autonomic or sympathetic nervous system arousal in combat veterans with chronic PTSD. Because central noradrenergic and peripheral sympathetic systems function in concert (Aston-Jones et al. 1991), the data from these psychophysiology investigations are consistent with the hypothesis that noradrenergic hyperreactivity in patients with PTSD may be associated with the conditioned or sensitized responses to specific traumatic stimuli.

There is some evidence that baseline levels of NE are consistently altered in combat-related PTDS. Women with PTSD secondary to childhood sexual abuse had significantly elevated levels of catecholamines (NE, epinephrine, DA) and cortisol in 24-h urine samples (Lemieux and Coe 1995). Sexually abused girls excreted significantly greater amounts of catecholamine metabolites, metanephrine, vanilmandelic acid, and homovanillic acid (HVA) than girls who were not sexually abused (DeBellis et al. 1994). Plasma levels of NE were elevated throughout a 24-h period collection period (Yehuda et al. 1995b) as were CSF levels of NE in PTSD patients (Baker et al. 1997). In the latter case, exposure to traumatic reminders in the form of combat films resulted in increased epinephrine (McFall et al. 1992) and NE (Blanchard et al. 1991) release.

Studies of peripheral NE receptor function have also shown alterations in $\alpha 2$ receptor and cyclic adenosine 39,59-monophosphate (cAMP) function in patients with PTSD. Decreases in platelet adrenergic $\alpha 2$-receptor number (Perry et al. 1987), platelet basal adenosine, isoproterenol, forskolin-stimulated cAMP signal transduction (Lerer et al. 1987), and basal platelet monoamine oxidase (MAO) activity (Davidson et al. 1985) have been found in PTSD. These findings may reflect chronic high levels of NE release which lead to compensatory receptor down-regulation and decreased responsiveness.

Patients with combat-related PTSD compared to healthy controls had enhanced behavioral, biochemical, and cardiovascular responses to the $\alpha 2$ antagonist yohimbine, which stimulates central NE release (Southwick et al. 1993, 1997). Moreover, a positron emission tomography study demonstrated that PTSD patients have a cerebral metabolic response to yohimbine consistent with increased NE release (Bremner et al. 1997b).

7
Panic Disorder

There is considerable evidence that abnormal regulation of brain noradrenergic systems is also involved in the pathophysiology of panic disorder. Panic disorder patients are very sensitive to the anxiogenic effects of yohimbine in addition to having exaggerated plasma 3-methoxy-4-hydroxyphenylethyleneglycol (MHPG), cortisol, and cardiovascular responses (Charney et al. 1984, 1987a, 1992; Gurguis and Uhde 1990; Albus et al. 1992; Yeragani et al. 1992). Children with a variety of anxiety disorders exhibit greater anxiogenic responses to yohimbine than normal comparison children (Sallee et al. 2000). The responses to the α2-adrenergic receptor agonist clonidine are also abnormal in panic disorder patients. Clonidine administration caused greater hypotension, greater decreases in plasma MHPG, and less sedation in panic patients than in controls (Uhde et al. 1988; Nutt 1989; Coplan et al. 1995a,b; Marshall et al. 2002).

8 Phobic Disorders

Few studies have examined noradrenergic function in patients with phobic disorders. In patients with specific phobias, increases in subjective anxiety and increased heart rate, blood pressure, plasma NE, and epinephrine have been associated with exposure to the phobic stimulus (Nesse et al. 1985). This finding may be of interest from the standpoint of the model of conditioned fear, reviewed above, in which a potentiated release of NE occurs in response to a reexposure to the original stressful stimulus. Patients with social phobia have been found to have greater increases in plasma NE in comparison to healthy controls and patients with panic disorder (Stein et al. 1992). In contrast to panic disorder patients, the density of lymphocyte α-adrenoceptors is normal in social phobic patients (Stein et al. 1993). The growth hormone response to intravenous clonidine (a marker of central α2-receptor function) is blunted in social phobia patients (Tancer et al. 1990).

9 Conclusion

There is emerging evidence that links the role of genetic factors to the vulnerability to stress-related psychopathology, such as PTSD. An investigation of twin pairs from the Vietnam Twin Registry reported that inherited factors accounted for up to 32% of the variance of PTSD symptoms beyond the contribution of trauma severity (True et al. 1993). The molecular neurobiological abnormalities that underlie these findings have not been elucidated. Two relatively small association studies which evaluated *D_2 dopamine receptor* polymorphisms in PTSD yielded contradictory results (Comings et al. 1996; Gelernter et al. 1999). A preliminary study found an association between the dopamine transporter (*DAT*) polymorphism and PTSD (Gelernter et al. 1999). Volumetric magnetic resonance imaging investigations demonstrated a smaller hippocampal volume in PTSD patients (Bremner et al. 1995; Bremner et al. 1997; Gurvits et al.

1996). A study of monozygotic twins discordant for trauma exposure found evidence that smaller hippocampal volume may constitute a risk factor for the development of stress-related psychopathology (Gilbertson et al. 2002). The recent identification of functional polymorphisms for the glucocorticoid receptor (DeRijk et al. 2002), the α2C adrenergic receptor subtype (Small et al. 2002), and for NPY synthesis (Kallio et al. 2001) provide opportunities to investigate the genetic basis of the neurochemical response patterns to stress.

Work is commencing to examine the genetic basis of the neural mechanisms of fear conditioning. There have been several recent advances in understanding the genetic contribution and molecular machinery related to amygdala-dependent learned fear. A gene encoding gastrin-releasing peptide (*Grp*) has been identified in the LA. The Grp receptor (GRPR) is expressed in GABAergic interneurons and mediates their inhibition of principal neurons. In GRPR knockout mice, this inhibition is reduced and LTP enhanced. These mice have enhanced and prolonged fear memory for auditory and contextual cues, indicating that the GRP signaling pathway may serve as an inhibitory feedback constraint on learned fear (Walker et al. 2000). The work further supports the role of GABA in fear and anxiety states (Goddard et al. 2001) and suggests the genetic basis of vulnerability to anxiety may relate to GRP, GRPR, and GABA (Ishikawa-Brush et al. 1997). Other preclinical studies indicate that there may be a genetically determined mesocortical and mesoaccumbens dopamine response to stress that relates to learned helplessness (Ventura et al. 2002). Recently, it was demonstrated that healthy subjects with the serotonin transporter polymorphism that has been associated with reduced 5-HT expression and function and increased fear and anxiety behaviors, exhibit increased amygdala neuronal activity in response to fear-inducing stimuli (Hariri et al. 2002; Garpenstrand et al. 2001; Holmes et al. 2002). These preclinical and clinical data suggest that multidisciplinary studies that use neurochemical, neuroimaging, and genetic approaches have the potential to clarify the complex relationships among genotype, phenotype, and psychobiological responses to stress.

References

Albus M, Zahn TP, Brier A (1992) Anxiogenic properties of yohimbine: behavioral, physiological and biochemical measures. Eur Arch Psychiatry Clin Neurosci 241:337–344

Aston-Jones G, Shipley MT, Chouvet G, Ennis M, VanBockstaele EJ, Pieribone V, Shiekhattar R (1991) Afferent regulation of locus coeruleus neurons: anatomy, physiology and pharmacology. Prog Brain Res 88:47–75

Baker DG, West SA, Orth DN, Hill KK, Nicholson WE, Ekhator NN, Bruce AB, Wortman MD, Keck PE, Geracioti JD (1997) Cerebrospinal fluid and plasma beta endorphin in combat veterans with post traumatic stress disorder. Psychoneuroendocrinology 22:517–529

Bauer EP, Schafe GE, LeDoux JE (2002) NMDA receptors and L-type voltage-gated calcium channels contribute to long-term potentiation and different components of fear memory formation in the lateral amygdala. J Neurosci 22:5239–5249

Schmidt P, Holsboer F, Spengler D (2001) Beta(2)-adrenergic receptors potentiate glucocorticoid receptor transactivation via G protein beta gamma-subunits and the phosphoinositide 3-kinase pathway. Mol Endocrinol 15:553–564

Schramm NL, McDonald MP, Limbird LE (2001) The alpha(2a)-adrenergic receptor plays a protective role in mouse behavioral models of depression and anxiety. J Neurosci 21:4875–4882

Shumyatsky G, Tsvetkov E, Malleret G, Vronskaya S, Horton M, Hampton L, Battey JF, Dulac C, Kandel ER, Bolshakov VY (2002) Identification of a signaling network in lateral nucleus of amygdala important for inhibiting memory specifically related to learned fear. Cell 111:905–918

Small KM, Wagoner LE, Levin AM, Kardia SL, Liggett SB (2002) Synergistic polymorphisms of β1and α2c adrenergic receptors the risk of congestive heart failure. N Engl J Med 347:1135–1142

Southwick SM, Krystal JH, Morgan CA, Johnson D, Nagy LM, Nicolaou A, Heninger GR, Charney DS (1993) Abnormal noradrenergic function in posttraumatic stress disorder. Arch Gen Psychiatry 50:266–274

Southwick SM, Krystal JH, Bremner JD, Morgan CA, Nicolaou A, Nagy LM, Johnson DR, Heninger GR, Charney DS (1997) Noradrenergic and serotonergic function in posttraumatic stress disorder. Arch Gen Psychiatry 54:749–758

Stein MB, Tancer ME, Uhde TW (1992) Heart rate and plasma norepinephrine responsivity to orthostatic challenge in anxiety disorders. Comparison of patients with panic disorder and social phobia and normal control subjects. Arch Gen Psychiatry 49:311–317

Stein MB, Huzel LL, Delaney SM (1993) Lymphocyte b-adrenoceptors in social phobia. Biol Psychiatry 34:45–50

Tancer ME, Stein MB, Uhde TW (1990) Effects of thyrotropin-releasing hormone on blood pressure and heart rate in phobic and panic patients: a pilot study. Biol Psychiatry 27:781–783

True WR, Rice J, Eisen SA, Heath AC, Goldberg J, Lyons MJ, Nowak J, (1993) A twin study of genetic and environmental contributions to liability for posttraumatic stress symptoms. Arch Gen Psychiatry 50:257–264

Uhde T, Joffe RT, Jimerson DC, Post RM (1988) Normal urinary free cortisol and plasma MHPG in panic disorder: clinical and theoretical implications. Biol Psychiatry 23:575–585

Ventura R, Cabib S, Puglisi-Allegra S (2002) Genetic susceptibility of mesocortical dopamine to stress determines liability to inhibition of mesoaccumbens dopamine and to behavioral despair in a mouse model of depression. Neuroscience 115:99–1007

Walker DL, Davis M (2000) Involvement of NMDA receptors within the amygdala in short-versus long-term memory for fear conditioning as assessed with fear-potentiated startle. Behav Neurosci 114:1019–1033

Weiss JM, Bonsall RW, Demetrikopoulos MK, Emery MS, West CH Galanin (1998) A significant role in depression? Ann N Y Acad Sci 863:364–382

Willick ML, Kokkinidis L (1995) Cocaine enhances the expression of fear-potentiated startle: evaluation of state-dependent extinction and the shock-sensitization of acoustic startle. Behav Neurosci 109:929–938

Yau JL, Noble J, Hibberd C, Rowe WB, Meaney MJ, Morris RG, Seckl JR (2002) Chronic treatment with the antidepressant amitriptyline prevents impairments in water maze learning in aging rats. J Neurosci 22:1436–1442

Yehuda R, Boisoneau D, Lowy MT, Giller EL Jr (1995b) Dose response changes in plasma cortisol and lymphocyte glucocoriticooid receptors following dexamethasone administration in combat veterans with and without posttraumatic stress disorder. Arch Gen Psychiatry 52:583–593

HEP (2005) 169:225–247

Pathophysiology and Pharmacology of $GABA_A$ Receptors

H. Möhler[1,2,3] (✉) · J.-M. Fritschy[1] · K. Vogt[1] · F. Crestani[1] · U. Rudolph[1]

[1]Institute of Pharmacology and Toxicology, University of Zurich, Winterthurerstr. 190, 8057 Zurich, Switzerland
mohler@pharma.unizh.ch

[2]Department of Chemistr and Applied Biosciences, Swiss Federal Institute of Technology (ETH), Winterthurerstr. 190, 8057 Zurich, Switzerland

[3]Collegium Helveticum, Schmelzenbergstr. 25, 8092 Zurich, Switzerland

Abstract By controlling spike timing and sculpting neuronal rhythms, inhibitory interneurons play a key role in brain function. GABAergic interneurons are highly diverse. The respective $GABA_A$ receptor subtypes, therefore, provide new opportunities not only for understanding GABA-dependent pathophysiologies but also for targeting of selective neuronal circuits by drugs. The pharmacological relevance of $GABA_A$ receptor subtypes is increasingly being recognized. A new central nervous system pharmacology is on the horizon. The development of anxiolytic drugs devoid of sedation and of agents that enhance hippocampus-dependent learning and memory has become a novel and highly selective therapeutic opportunity.

Keywords Anxiolytics · Hypnotics · Memory · Schizophrenia · Epilepsy

1 Inhibitory Interneurons

The dynamics of neural networks are largely shaped by the activity pattern of interneurons, most of which are GABAergic (Buzsaki and Chrobak 1995; Paulsen and Moser 1998; Freund and Buzsaki 1996; Miles 2000; Klausberger et al. 2002; Klausberger et al. 2003). The activity of these interneurons is thought to set the spatio-temporal conditions required for different patterns of network oscillations that may be critical for information processing (O'Keefe and Recce 1993; Skaggs et al. 1996; Paulsen and Moser 1998; Engel et al. 2001; Harris et al. 2002; Mehta et al. 2002; Traub et al. 2002; Klausberger et al. 2003).

For instance, cortical interneuron networks may generate both slow and fast cortical oscillatory activity (Whittington et al. 1995, 1997; Buhl et al. 1998; Fisahn et al. 1998; Penttonen 1998; Zang et al. 1998; Traub et al. 2002). Similarly, inhibitory neurons of the thalamic reticular and perigeniculate nuclei generate the synchronized activity of thalamocortical networks (McCormick and Bal 1997). Furthermore gamma oscillations (30–100 Hz) occur in various brain structures and can occur over large distances. They could, therefore, provide a substrate for "binding" together spatially separated areas of cortex, a hypothetical process whereby disparate aspects of a complex object, for example, are combined to form a unitary perception of it (Traub et al. 1996; Laurent 1996; Engel et al. 2001; Singer and Gray 1995). In addition the activity of interneurons sets the spatio-temporal conditions required for synaptic plasticity as shown most clearly for hippocampus-dependent learning and memory (O'Keefe and Nadel 1978; O'Keefe and Reece 1993; Paulsen and Moser 1998; Csicsvari et al. 1999; Miles 2000; Maccaferri et al. 2000; Burgess et al. 2002; Brun et al. 2002).

1.1 Diversity of Interneurons

To achieve a strict time control of principal cells, GABAergic interneurons display several remarkable features. (1) Their action potential is traditionally faster than that of pyramidal cells and the kinetics of synaptic events that excite inhibitory cells are faster than those that excite pyramidal cells (Martina et al. 1998; Geiger et al. 1997). (2) The GABAergic interneurons are morphologically highly diverse, which reflects their multiple functions in neuronal networks

(Gupta et al. 2000). (3) The interneurons show a domain-specific innervation of principal cells. Thus, depending on the type of interneuron, particular input domains of pyramidal cells can be selectively regulated. Similarly, the output of pyramidal cells can be specifically regulated by axo-axonic GABAergic interneurons. (4) The response properties of interneuron signalling are shaped by the type of $GABA_A$ receptor expressed synaptically or extrasynaptically. For instance, the soma of hippocampal pyramidal cells is innervated by two types of basket cells. The fast-spiking parvalbumin-containing basket cells form synapses containing $\alpha_1 GABA_A$ receptors, which display fast kinetics of deactivation (Nyiri et al. 2001; Klausberger et al. 2002; Freund and Buzsaki 1996; Pawelzik et al. 2002). In contrast, the synapses of the regular-spiking cholecystokinin (CCK)-positive basket cells contain $\alpha_2 GABA_A$ receptors, which display slower kinetics than α_1 receptors (Nyiri et al. 2001; Brussaard and Herbison 1997; Hutcheon et al. 2000; Jüttner et al. 2001; Vicini et al. 2001). Axon initial segments of principal cells also contain α_2 receptors, which appear to be kinetically sufficient for simple on/off signalling. Furthermore, distinct $GABA_A$ receptors are segregated to synaptic and extrasynaptic membranes (Nusser et al. 1998; Fritschy and Brünig 2003). Thus, functionally specialized interneurons operate with the kinetically appropriate $GABA_A$ receptor subtypes to regulate network behaviour (Figs. 1 and 2). Since GABAergic interneurons are operative throughout the brain, a highly diverse repertoire of $GABA_A$ receptors is required.

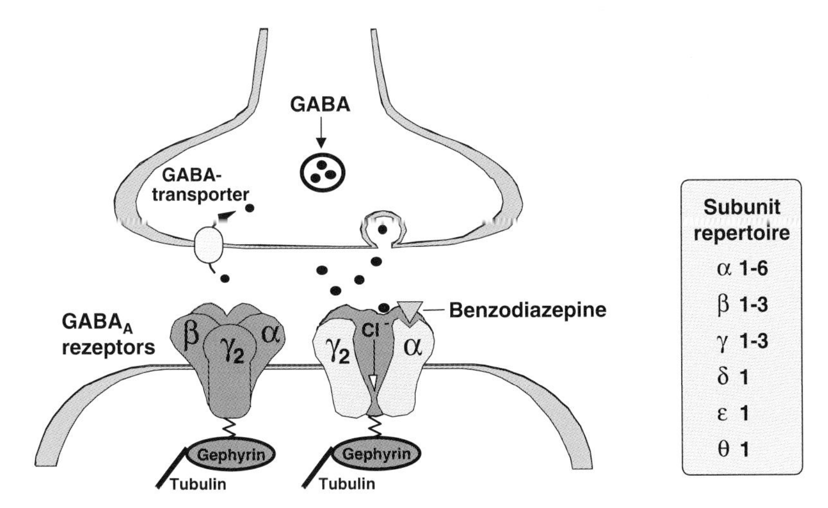

Fig. 1 Scheme of GABAergic synapse, depicting major elements of signal transduction. The $GABA_A$ receptors are heteromeric membrane proteins that are linked, by a yet-unknown mechanism, to the synaptic anchoring protein gephyrin and the cytoskeleton (Sassoe-Pognetto et al. 2000; Fritschy and Brünig 2003)

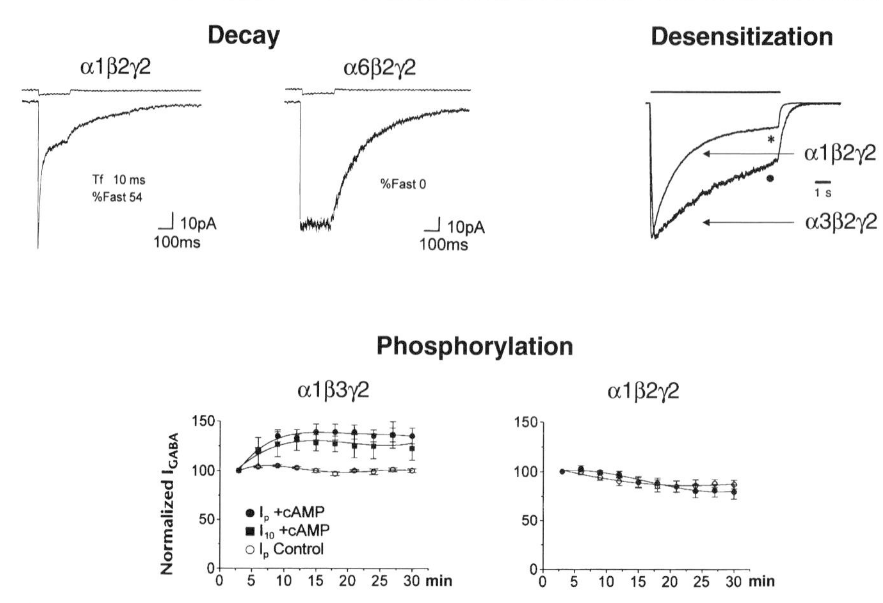

Fig. 2 Variation in kinetic properties of recombinant $GABA_A$ receptor subtypes. Decay kinetics (Gingrich et al. 1995), desensitization kinetics (Tia et al. 1996) and sensitivity to cAMP-induced phosphorylation (McDonald et al. 1998) vary with the subunit composition

1.2
Retrograde Regulation of GABAergic Interneurons

Retrograde signalling adds another level of complexity to the regulation of interneuron activity. The terminals of CCK-positive GABAergic basket cells in hippocampus and amygdala contain CB1-cannabinoid receptors (Katona et al. 1999, 2001). These receptors mediate depolarization-induced suppression of inhibition (DSI) (Pitler and Alger 1994; Alger and Pitler 1995). This phenomenon is due to endocannabinoids that emanate from the postsynaptic cell and act as a retrograde signal (Wilson and Nicoll 2001; Maejima et al. 2001). The depolarization of hippocampal pyramidal cells (Pitler and Alger 1992) and of cerebellar Purkinje cells (Llano et al. 1991) results in a transient decrease in the release of GABA from inhibitory terminals that contain CB1-receptors and synapse onto the depolarized cells (Vincent and Marty 1993; Pitler and Alger 1994; Alger and Pitler 1995).

Both the DSI in the hippocampus (Wilson and Nicoll 2001; Maejima et al. 2001) and the cerebellum (Kreitzer and Regehr 2001a,b) are the result of activity-dependent de novo synthesis and release of endocannabinoids from the postsynaptic neuron. By interacting with CB1 receptors, the calcium influx into the presynaptic terminal is reduced (Kreitzer and Regehr 2001a; Caulfield and Brown 1992) resulting in a decrease of GABA release. Pharmacologically, the inhibition of the degradation of anandamide resulted in a CB1 receptor-mediated anxiolytic response (Kathuria et al. 2003). Recently, endocannabinoids were shown to mediate not only transient, but also long-term changes in inhibitory synaptic transmission. Endocannabinoid production, stimulated through metabotropic glutamate receptor activation in hippocampal pyramidal cells, caused a long-lasting reduction in GABAergic signalling onto the pyramidal cells (Chevaleyre and Castillo 2003). A similar change in long-term synaptic plasticity of the GABAergic system was observed in the amygdala. The release of endocannabinoids in the basolateral amygdala contributed to the extinction of aversive memory based on a long-lasting decrease of GABAergic signalling (Marsicano et al. 2002). Thus, CB1 receptor activation in the hippocampus, amygdala and possibly other parts of the brain results in reduced levels of anxiety. This is due to either a transient depression of GABA release or a modulation of long-term plasticity at the respective synapses. In addition, endocannabinoids act as retrograde signals at excitatory glutamatergic synapses where they mediate a depolarization-induced suppression of excitation (Kreitzer and Regehr 2001a) that may also contribute to their behavioural effects.

2 Diversity of $GABA_A$ Receptors

The physiological significance of the structural diversity of $GABA_A$ receptors lies in the provision of receptors which differ in their channel kinetics, affinity for GABA, rate of desensitization and ability for transient chemical modification such as phosphorylation. In addition, $GABA_A$ receptor subtypes can show a cell type-specific expression and—in the case of multiple receptor subtypes being present in a neuron—a domain-specific location. Based on the presence of 7 subunit families comprising at least 18 subunits in the CNS (α_{1-6}, β_{1-3}, γ_{1-3}, δ, ε, θ, ϱ_{1-3}) the $GABA_A$ receptors display an extraordinary structural heterogeneity. Most $GABA_A$ receptors subtypes in vivo are considered heteropentamers composed of isoforms of α, β and γ subunits (Fig. 1; for review Barnard et al. 1998; Whiting et al. 2000; Sieghart and Sperk 2002; Möhler et al. 2000, 2002; Möhler 2001, 2002; Fritschy et al. 2004).

2.1 Diazepam-Sensitive $GABA_A$ Receptors

Receptors containing the α_1, α_2, α_3 or α_5 subunit in combination with any of the β-subunits and the γ_2 subunit are most prevalent in the brain (Fig. 3). These receptors are sensitive to benzodiazepine modulation. The major receptor subtype is assembled from the subunits $\alpha_1\beta_2\gamma_2$, with only a few brain regions lacking this receptor (granule cell layer of the olfactory bulb, reticular nucleus of the thalamus, spinal cord motoneurons) (Fritschy and Mohler 1995; Pirker et al. 2000; Fritschy and Brunig 2003) (Table 1).

Table 1 $GABA_A$—receptor subtypes (from Möhler et al. 2002 and Fritschy and Brünig 2003)

Composition	Pharmacological characteristics	Regional and neuronal localization	Subcellular localization
$\alpha_1\beta_2\gamma_2$	Major subtype (60% of all $GABA_A$ receptors). Mediates the sedative, amnestic and—to a large extent—the anticonvulsant action of benzodiazepine site agonists. High affinity for classical benzodiazepines, zolpidem and the antagonist flumazenil	Cerebral cortex (layer I–VI, selected interneurons and principal cells); hippocampus (selected interneurons and principal cells); pallidum striatum (interneurons); thalamic relay nuclei; olfactory bulb (mitral cells and interneurons); cerebellum (Purkinje cells and granule cells); deep cerebellar nuclei; amygdala; basal forebrain; substantia nigra pars reticulata; inferior colliculus; brainstem	Synaptic (soma and dendrites) and extrasynaptic in all neurons with high expression
$\alpha_2\beta_3\gamma_2$	Minor subtype (15%–20%). Mediates anxiolytic action of benzodiazepine site agonists. High affinity for classical benzodiazepine agonists and the antagonist flumazenil. Intermediate affinity for zolpidem	Cerebral cortex (layers I–IV). Hippocampal formation, (principal cells mainly on the axon initial segment); olfactory bulb (granule cells); striatum (spiny stellate cells); inferior olivary neurons (mainly on dendrites); hypothalamus; amygdala (principal cells); superior colliculus; motor neurons	Mainly synaptic, enriched in axon initial segment of cortical and hippocampal pyramidal cells
$\alpha_3\beta_n\gamma_2$	Minor subtype (10%–15%). High affinity for classical benzodiazepine agonists and the antagonist flumazenil. Intermediate affinity for zolpidem	Cerebral cortex (principal cells in particular in layers V and VI; some axon initial segments); hippocampus (some hilar cells); olfactory bulb (tufted cells); thalamic reticular neurons; cerebellum (Golgi type II cells); medullary reticular formation; inferior olivary neurons; amygdala; superior colliculus; brainstem; spinal cord; medial septum; basal forebrain cholinergic neurons; raphé and locus coeruleus (serotoninergic and catecholaminergic neurons)	Mainly synaptic, including some axon initial segments; extrasynaptic in inferior olivary neurons
$\alpha_4\beta_n\delta$	Less than 5% of all receptors. Insensitive to classical benzodiazepine agonists and zolpidem. Lacks benzodiazepine site	Dentate gyrus (granule cells); thalamus	Extrasynaptic (no direct morphological evidence)
$\alpha_5\beta_3\gamma_2$	Less than 5% of all receptors; High affinity for classical benzodiazepine agonists and the antagonist flumazenil. Very low affinity for zolpidem	Hippocampus (pyramidal cells); olfactory bulb (granule cells, periglomerular cells); cerebral cortex; amygdala; hypothalamus; superior colliculus; superior olivary neurons; spinal trigeminal neurons; spinal cord	Extrasynaptic in hippocampus, cerebral cortex, and olfactory bulb; synaptic and extrasynaptic in spinal trigeminal nucleus and superior olivary nucleus
$\alpha_6\beta_{2,3}\gamma_2$ $\alpha_6\beta_{2,3}\delta$	Less than 5% of all receptors. Insensitive to classical benzodiazepine agonists and zolpidem. Minor population. Lacks benzodiazepine site	Cerebellum (granule cells); dorsal cochlear nucleus	Synaptic (cerebellar glomeruli) and extrasynaptic on granule cell dendrites and soma

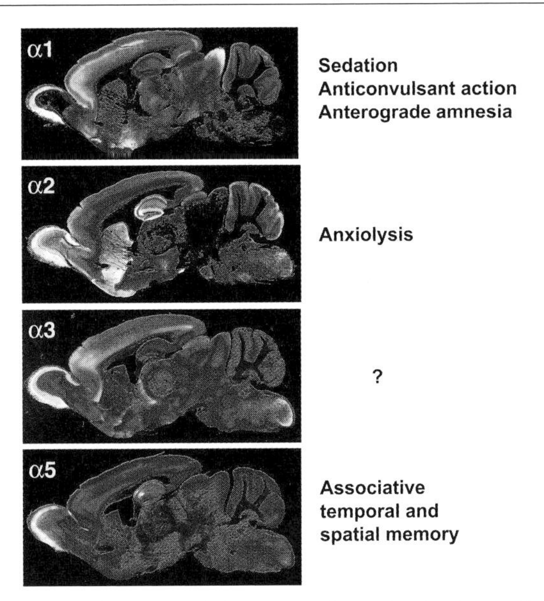

Fig. 3 $GABA_A$ receptor subtypes and the regulation of behaviour. The four classes of diazepam-sensitive $GABA_A$ receptors, visualized immunohistochemically, show distinct expression patterns. Distinct pharmacological functions (α_1, α_2, α_3 receptor subtypes) and physiological functions (α_5 receptor subtype) are indicated

Receptors containing the α_2 or α_3 subunit are considerably less abundant and are highly expressed in brain areas where the α_1 subunit is absent or present at low levels (Table 1). The α_2 and α_3 subunits are frequently coexpressed with the β_3 and γ_2 subunits; this is particularly evident in hippocampal pyramidal neurons ($\alpha_2\beta_3\gamma_2$) and in cholinergic neurons of the basal forebrain ($\alpha_3\beta_3\gamma_2$). The α_3 $GABA_A$ receptors are the main subtypes expressed in monoaminergic and basal forebrain cholinergic cells (Gao et al. 1993) and are, in addition, strategically located in the thalamic reticular nucleus for modulating the thalamo cortical circuit (Huntsmann et al. 1999). Marked differences in desensitization kinetics have been reported between synaptic α_2 and extrasynaptic α_3 receptors whereby the latter desensitize very slowly (Devor et al. 2001). The factors regulating $GABA_A$ receptor kinetics at synaptic and extrasynaptic sites are yet unknown (Moss and Smart 2001). The ligand-binding profile of the α_2 and α_3 receptors differs from that of $\alpha_1\beta_2\gamma_2$ by having a considerably lower displacing potency for ligands such as βCCM, CL 218,872 and zolpidem (Table 1).

Receptors containing the α_5 subunit are of minor abundance in the brain (Table 1) but are expressed to a significant extent in the hippocampus, where they comprise 15%–20% of the diazepam-sensitive $GABA_A$ receptor population, predominately co-assembled with the β_3 and γ_2 subunits. Pharmacologically, the α_5 receptors are differentiated from $\alpha_1\beta_2\gamma_2$, $\alpha_2\beta_3\gamma_2$ and $\alpha_3\beta_3\gamma_2$ receptors by a lower affinity to CL 218, 872 and near-insensitivity to zolpidem (Table 1).

The subunits γ_1 and γ_3 characterize a small population of receptors that contain various types of α- and β-subunits. Due to their reduced affinity for the classical benzodiazepines, these receptors do not appear to contribute to any great extent to benzodiazepine pharmacology in vivo.

2.2 Diazepam-Insensitive $GABA_A$ Receptors

$GABA_A$ receptors that do not respond to clinically used ligands of the benzodiazepine site such as diazepam, flunitrazepam, clonazepam, and zolpidem are of low abundance in the brain and are largely characterized by the α_4 and α_6 subunits (Table 1). Receptors containing the α_4 subunit are generally expressed at very low abundance but more prominently in thalamus and dentate gyrus (Pirker et al. 2000); those containing the α_6 subunit are restricted to the granule cell layer of the cerebellum (about 30% of all $GABA_A$ receptors in the cerebellum; Nusser et al. 1996b). Both receptor populations are structurally heterogeneous, and the majority of the α_6-containing receptors are of the $\alpha_6\beta_2\gamma_2$ combination (Table 1). The benzodiazepine-site profile of α_4 and α_6 receptors is characterized by a low affinity for flumazenil and bretazenil, and a switch in the efficacy of Ro 15-4513 from an inverse agonist to an agonist. The δ-subunit is frequently co-assembled with the α_4 or the α_6 subunit in benzodiazepine insensitive receptors (Möhler et al. 2000; Whiting et al. 2000; Möhler 2001). Receptors containing the δ-subunit are located exclusively at extrasynaptic sites as shown in dentate gyrus and cerebellum. They are tailor made for tonic inhibition, due to their high affinity for GABA and slow desensitization kinetics (Brickley et al. 1996; Mody and Nusser 2000; Brickley et al. 2001).

In the retina, homomeric receptors consisting of the ϱ-subunit represent a particular class of GABA-gated chloride channels. Their GABA site is insensitive to bicuculline and baclofen and they are not modulated by barbiturates or benzodiazepines. Due to these distinctive features, the receptors are sometimes termed $GABA_C$ receptors (Bormann 2000), although they are be considered a homomeric class of $GABA_A$ receptors (Barnard et al. 1998).

3 Pathophysiology of the GABA System

If the balance between excitatory and inhibitory activity is shifted pharmacologically in favour of GABAergic transmission, then anxiolysis, sedation, amnesia and ataxia arise. On the other hand, an attenuation of the GABAergic system results in arousal, anxiety, restlessness, insomnia, exaggerated reactivity and even seizures. These pharmacological manifestations point to the contribution of inhibitory neurotransmission to the pathophysiology of brain disorders. A GABAergic deficit is particularly apparent in anxiety disorders, epilepsy and schizophrenia (Olsen and DeLorey 1999; Mohler 2002).

3.1 Anxiety Disorders

Anxiety disorders have a high prevalence and are the most common cause of medical intervention in primary care (Weiller et al. 1998). The pharmacology of the GABA system supports the view that GABAergic dysfunctions are causally related to symptoms of anxiety. For instance, pentylenetetrazole acts by blocking $GABA_A$ receptor function and produces extreme anxiety, traumatic memories and extreme avoidance behaviour when used clinically (Kalueff and Nutt 1997). Conversely, enhancing GABAergic transmission, e.g. by benzodiazepines, is a powerful mechanism to inhibit the experience of anxiety and its aversive reinforcement.

Neuroimaging has given fresh insight into the role of GABAergic inhibition in anxiety disorders. In a recent positron emission tomography (PET) study using ^{11}C-flumazenil, a significant global reduction in flumazenil binding to $GABA_A$ receptors was apparent throughout the brain in patients with panic disorder (Malizia et al. 1998). The greatest decrease observed occurred in areas thought to be involved in the experience of anxiety such as the orbitofrontal and temporal cortex. Single photon emission computed tomography (SPECT) studies, using the related radioligand ^{123}I-iomazenil, have shown similar decreases in binding (Malizia 1999). A localized reduction in benzodiazepine binding in the temporal lobe has also been reported in generalized anxiety disorders (Tiihonen et al. 1997). Furthermore, using magnetic resonance spectroscopy, decreased cortical levels of GABA were observed in patients with panic disorders (Goddard et al. 2001). These findings are consistent with the view that at least some anxiety disorders are linked to a defective GABAergic neuroinhibitory process (Nutt and Malizia 2001).

Anxiety in man frequently arises at the interface between a genetic predisposition and experience. Recently, the hypothesis was tested whether a partial $GABA_A$ receptor deficit would be sufficient to generate an anxiety state. Using molecular biological techniques the $GABA_A$ receptor deficit seen in patients with anxiety disorders was reproduced in an animal model (Crestani et al. 1999). The γ_2 subunit of the $GABA_A$ receptor is known to anchor the receptors in the subsynaptic membrane. By reducing the gene dosage for the γ_2-subunit in mice—heterozygosity for the γ_2-subunit gene—the synaptic clustering of $GABA_A$ receptors was reduced. A partial receptor deficit was apparent throughout most of the brain including the areas that are known to be involved in the processing of anxiety responses, such as the cerebral cortex, amygdala and hippocampus. The animals behaved normally in a wide range of behavioural tests except when exposed to aversive situations caused by either natural or conditioned fear stimuli. Under such conditions enhanced anxiety responses and a bias for threat cues were observed (Crestani et al. 1999). The bias of the animals for threat cues was especially significant, since this behaviour corresponds to the cognitive deficit contributing to the inability of anxious individuals to distinguish an ambiguous from a threatening situation (Eysenck 1992). Thus, a $GABA_A$ receptor deficit is considered a predisposition for anxiety disorders in humans. Anxiety symptoms are a sensitive manifestation of an impaired GABAergic neurotransmission (Nutt and Malizia 2001; Crestani et al. 1999; Möhler 2002).

3.2 Epilepsy

Genetic evidence has provided the most direct link of epilepsy to $GABA_A$ receptor dysfunction. A K289M mutation located in the extracellular loop of the γ_2-subunit was associated with generalized epilepsy with febrile seizures (Baulac et al. 2001). Another mutation in the γ_2 subunit of the $GABA_A$ receptor was linked to childhood absence epilepsy and febrile seizures with a conserved arginine residue being mutated to glutamine (R43Q) (Wallace et al. 2001). However, since childhood absence epilepsy is not inherited in a simple Mendelian manner, the latter point mutation is not considered sufficient (by itself) to cause this phenotype. Furthermore, a single nucleotide exchange at the splice donor site of intron 6 of the γ_2 subunit (Kananura et al. 2002), resulting most likely in a non-functional allele, was associated with childhood absence epilepsy and febrile seizures. Finally, a loss-of-function mutation of the $\alpha_1 GABA_A$ receptor (A322D) was found in a family with an autosomal dominant form of juvenile myoclonic epilepsy (Cossette et al. 2002). The functional consequence of the γ_2 mutations *K289M* and *R43Q* are controversial (Bianchi et al. 2002).

In temporal lobe epilepsy profound changes in $GABA_A$ receptor expression have been observed in patients and animal models (Olsen and DeLorey 1999; Coulter 2001; Treimann 2001; Snead et al. 1999). While the extensive cell loss in CA1 is accompanied with a loss of $GABA_A$ receptors, receptor staining is increased in the surviving neurons with subtype-specific changes in their subcellular distribution (Loup et al. 2000). In a recent mouse model, the cellular pathophysiology of human temporal lobe epilepsy is largely reproduced (Bouilleret et al. 1999; Riban et al. 2002; Bouilleret et al. 2000). A marked increase in α_1, α_2, α_5 and γ_2 subunit immunoreactivity was found pointing to a potential sprouting of GABAergic axons in the epileptic dentate gyrus.

Finally, interictal activity in human temporal lobe epilepsy was associated with depolarizing GABAergic synaptic events at pyramidal cells (Cohen et al. 2002). Indeed, under certain circumstances, GABA responses can be depolarizing and facilitate action potential generation (Gulledge and Stuart 2003). The depolarizing GABA response appears to be due to the very negative resting potential of the particular cells and not due to a change in chloride gradient (Stein and Nicoll 2003). These new developments are of major pharmacological interest since approximately half the antiepileptic drugs in clinical use are thought to owe their efficacy either totally or partially to potentiating GABAergic inhibitory effects (Meldrum and Whiting 2001).

3.3 Schizophrenia

Alterations in cortical GABAergic systems have been reported in post-mortem brain of schizophrenic patients, including reduced uptake and release of GABA and a reduced activity of glutamic acid decarboxylase. Most conspicuously, the axon terminals of GABAergic chandelier neurons were altered in the prefrontal cortex, as shown by a 40% decrease of GABA-transporter 1 (GAT-1) staining (Woo et al. 1998). Chandelier neurons innervate the axon initial segments of pyramidal cells. They are therefore strategically positioned to powerfully regulate the excitatory output of these cells and consequently affect the pattern of neuronal activity in the prefrontal cortex and its projection areas (Woo et al. 1998). A compensatory upregulation of α_2GABA_A receptor in the axon initial segment of pyramidal cells of patients was observed, pointing to a synapse-specific deficit of GABAergic transmission in schizophrenia (Volk et al. 2002). Altered ratios of subunit splice variants of $GABA_A$ receptors were also found in prefrontal cortex of schizophrenics (Huntsman et al. 1998). In addition, benzodiazepine receptor inverse agonists are associated with psychotogenic effects (Sarter et al. 2001). Furthermore, in primate brain, D_4 dopamine receptors (a member of the D_2 receptor family with a high affinity for clozapine) modulate GABAergic interneurons in critical brain areas (cerebral cortex, hippocampus, thalamic reticular nucleus, globus pallidus). Thus, the beneficial effects of clozapine in schizophrenia may be achieved, in part, through D_4-mediated GABA modulation (Mrzljak et al. 1996). Finally, GABAergic neurons have been found to be especially vulnerable to glucocorticoid hormones and to glutamatergic excitotoxicity, which may explain the increased number of certain glutamatergic neurons in, for example, the cingulate gyrus of schizophrenic brains. This, in conjunction with a postulated role of stress in the pathogenesis of schizophrenia, would strengthen the assumption of an important role of a GABAergic deficit in schizophrenia (Carlsson et al. 2001). A GABAergic dysfunction that might arise in the course of the disorder may result in long-lasting and perhaps lifelong sensitivity changes.

4 Pharmacology of $GABA_A$ Receptor Subtypes

The pharmacological relevance of $GABA_A$ receptor subtypes for the spectrum of benzodiazepine effects was recently identified based on a genetic approach (Rudolph et al. 1999, 2001; Löw et al. 2000; McKernan et al. 2000;

Whiting et al. 2000; Möhler 2002; Möhler et al. 2002; Whiting 2003). Experimentally, the $GABA_A$ receptor subtypes were rendered diazepam-insensitive by replacing a conserved histidine residue with an arginine residue in the respective α-subunit gene [α_1(H101R), α_2(H101R), α_3(H126R) and α_5(H105R)] (Rudolph et al. 1999; Löw et al. 2000). This strategy permitted the allocation of the benzodiazepine drug actions to the α_1, α_2, α_3 and α_5 $GABA_A$ receptor subtypes (Rudolph et al. 2001; Crestani et al. 2002). In addition, it implicated the neuronal networks expressing the particular receptor in mediating the corresponding drug actions.

4.1 Receptor for Sedation

Among α_1-, α_2- and α_3-point-mutated mice, only the α_1(H101R) mutants were resistant to the depression of motor activity by diazepam and zolpidem (Rudolph et al. 1999; Löw et al. 2000; Crestani et al. 2000). This effect was specific for ligands of the benzodiazepine site, since pentobarbital or a neurosteroid remained as effective in α_1(H101R) mice as in wild-type mice in inducing sedation. An α_1(H101R) mouse line was also generated by McKernan et al. (2000), confirming that sedation is linked to α_1 $GABA_A$ receptors and differs mechanistically from the anxiolytic action of benzodiazepines.

4.2 Receptor for Amnesia

Anterograde amnesia is a classical side-effect of benzodiazepine drugs. The memory impairing effect of diazepam, analysed in a step-through passive avoidance paradigm, was strongly reduced in the α_1(H101R) mice compared to wild-type mice (Rudolph et al. 1999). This effect was not due to a potential nonspecific impairment, since the ability of a muscarinic antagonist to induce amnesia was retained in the α_1(H101R) mice. These results demonstrate that the diazepam-induced anterograde amnesia is mediated by α_1 receptors.

4.3 Receptor for Protection Against Seizures

The anticonvulsant activity of diazepam, assessed by its protection against pentylenetetrazole-induced tonic convulsions, was strongly reduced in α_1-(H101R) mice compared to wild-type animals (Rudolph et al. 1999). Sodium phenobarbital remained fully effective as anticonvulsant in α_1(H101R) mice. Thus, the anticonvulsant activity of benzodiazepines is partially but not fully mediated by α_1 receptors. The anticonvulsant action of zolpidem is exclusively mediated by α_1 receptors, since its anticonvulsant action is completely absent in α_1(H101R) mice (Crestani et al. 2000).

4.4 Receptor for Anxiolysis

New strategies for the development of daytime anxiolytics that are devoid of drowsiness and sedation are of high priority. Experimentally, the anxiolytic-like action of diazepam is due to the modulation of α_2 $GABA_A$ receptors as shown by the lack of tranquillizing action of diazepam in α_2(H102R) mice (elevated plus maze; light/dark choice test). The α_2 $GABA_A$ receptors, which comprise only about 15% of all diazepam-sensitive $GABA_A$ receptors, are mainly expressed in the amygdala and in principal cells of the cerebral cortex and the hippocampus, with particularly high densities on their axon initial segments (Nusser et al. 1996a; Fritschy et al. 1998a,b). Thus, the inhibition of the output of these principal neurons appears to be a major mechanism of anxiolysis.

It had previously been postulated that the anxiolytic action of diazepam is based on the dampening of the reticular activating system, which is mainly represented by noradrenergic and serotonergic neurons of the brain stem. These neurons express exclusively α_3 $GABA_A$ receptors. The analysis of the α_3 point-mutated mice [α_3(H126R)] indicated that the anxiolytic effect of benzodiazepine drugs was unaffected (Löw et al. 2000). The reticular activating system therefore does not appear to be a major contributor to anxiolysis. The role of α_3 $GABA_A$ receptors remains to be identified.

4.5 Receptor for Myorelaxation

The muscle relaxant effect of diazepam is largely mediated by α_2 $GABA_A$ receptors, as shown by the failure of diazepam to induce changes in muscle tone in the α_2 point-mutated mouse line (Crestani et al. 2001). α_2 $GABA_A$ receptors in the spinal cord, notably in the superficial layer of the dorsal horn and in motor neurons (Bolhalter et al. 1996), are most likely implicated in this effect. The muscle-relaxant effect requires considerably higher doses of diazepam than its anxiolytic-like activity, which is mediated by α_2 $GABA_A$ receptors located in the limbic system (see above). It was only at very high doses of diazepam that α_3 and α_5 $GABA_A$ receptors were also implicated in mediating myorelaxation (Crestani et al. 2001; Crestani et al. 2002).

4.6 Receptor for Associative Learning and Memory

The acquisition of spatial and temporal memory is associated with excitatory synaptic plasticity involving hippocampal *N*-methyl-D-aspartate (NMDA) receptors (Morris et al. 1986, 1989; Davis et al. 1992; McHugh et al. 1996; Tsien et al. 1996; Huerta et al. 2000; Nakazawa et al. 2002; Tang et al. 1999). Recently, an additional GABAergic control component became apparent involving α_5 $GABA_A$ receptors.

The α_5 $GABA_A$ receptor subtype has a privileged site of expression on hippocampal pyramidal cells, being located extrasynaptically at the base of the spines that receive the excitatory input and on the adjacent shaft of the dendrite (Fritschy et al. 1998b, Crestani et al. 2002). The α_5 $GABA_A$ receptors were therefore considered able to modulate the transduction of the signal arising at excitatory synapses and, by doing so, would operate as control element of learning and memory in their own right.

In α_5(H105R) mice, the point mutation was without major effect on the pharmacology of diazepam. Surprisingly, however, the content of α_5 $GABA_A$ receptors was reduced by 30%–40% exclusively in the hippocampus (Crestani et al. 2002). The remaining hippocampal α_5 receptors showed a normal wild-type distribution. There was no indication for adaptive changes of other $GABA_A$ receptors expressed in the same pyramidal cells (Crestani et al. 2002). Behaviourally, the partial deficit of hippocampal α_5 $GABA_A$ receptors resulted in an improved performance in trace fear conditioning, a hippocampus-dependent task in which a conditioned stimulus has to be memorized for a period of time sufficiently long to be associated with an unconditioned stimulus given after a silent time lag (trace of 1 s). These results pointed to a role of α_5 $GABA_A$ receptors in the function of temporal memory. When the α_5 $GABA_A$ receptors were deleted in the entire brain by targeting the α_5 subunit gene (Collinson et al. 2002; Whiting 2003) a significantly improved performance in a water maze model of spatial learning was observed. In addition, the amplitude of hippocampal inhibitory postsynaptic potential (IPSC) was decreased and the paired-pulse facilitation of field excitatory postsynaptic potential (EPSP) amplitudes was enhanced. These data strongly suggest that α_5 $GABA_A$ receptors play a crucial role in cognitive processes of hippocampal learning and memory.

It is striking that the behavioural consequences of an impairment of α_5 $GABA_A$ receptors are opposite to those of a NMDA receptor deficit. While mice with a deficit in hippocampal NMDA receptors (NRI-CA1 knockout) show a deficit in the formation of spatial and temporal memory (Tsien et al. 1996; Tang et al. 1999), the mice with a deficit in α_5 $GABA_A$ receptors display an improvement in hippocampal spatial and temporal memory performance. Thus, it appears that these two receptor systems play a complementary role in controlling neuronal processing in the hippocampus.

5
Novel Subtype Selective Benzodiazepine Site Ligands

Among the clinically used ligands of the benzodiazepine site, only the hypnotic zolpidem displays a pronounced preferential subtype selectivity (Langer et al.

1992) (Table 1). Additional benzodiazepine site ligands with subtype selectivity are under experimental or clinical investigation (Hood et al. 2000).

5.1 Hypnotics

Zaleplon This (CL284,846) is a pyrazolopyrimidine developed for the treatment of insomnia (Sanger et al. 1996). At recombinant receptors, zaleplon binds preferentially to α_1 receptors ($\alpha_1\beta_2\gamma_2$) and to receptors containing the γ_3 subunit, but binds 8- to 20-fold less to α_2, α_3 and $\alpha_5$receptors (Dämgen and Lüddens 1999). Thus, zaleplon is largely a ligand with preference for α_1 receptors, which is in keeping with its preponderant hypnotic activity. The contribution of its interaction with γ_3 receptors is unclear, since these receptors are of low abundance in the brain.

5.2 Anxiolytics

L-838,417 This benzodiazepine site ligand displays a dramatic subtype selective efficacy. L-838,417 failed to modulate the GABA response at α_1 receptors but enhanced the GABA response at α_2, α_3 and α_5 receptors (McKernan et al. 2000). L-838,417 showed a high potency in anxiolytic tests (elevated plus maze and fear-potentiated startle) and in anticonvulsant tests [pentylenetetrazole (PTZ), audiogenic seizures]. However, L-838,417 failed to impair the motor performance (rotarod test, chain pulling test) (McKernan et al. 2000). Thus, ligands with subtype-selective efficacy as factor that distinguishes α_2, α_3 and α_5 receptors from α_1 receptors, provide a new way to develop selective anxiolytics without sedative component. A further improvement of anxiolytic efficacy may be achieved by focusing the ligand affinity or efficacy more specifically on α_2 receptors.

SL65.1498 The pyrido-indole-4-carboxamide derivative SL65.1498 shows higher affinity for α_1, α_2 and α_3 $GABA_A$ receptors compared to α_5 receptors. In addition, it acts as a full agonist at α_2 and α_3 receptors but as partial agonist at α_1 $GABA_A$ receptors. In line with its selectivity for the activation of α_2 and α_3 receptors, the compound showed potent anxiolytic action in animal models (punished lever pressing, punished drinking, elevated plus maze, light/dark test) but did not impair motor coordination (e.g. rotarod) or working memory (Morris water maze) (Scatton et al. 2000).

5.3 Memory Enhancers

A deficit in α_5 $GABA_A$ receptor activation is associated with an improved hippocampal performance in temporal and spatial memory tasks (see above) and can conceivably be mimicked pharmacologically. Partial inverse agonists binding with preferential affinity to the α_5 $GABA_A$ receptors would be expected to enhance hippocampus-dependent learning and memory functions. Indeed, a α_5 partial inverse agonist from a 6,7-dihydro-2-benzothiophen-4-ones series enhanced the cognitive performance in a water maze test without proconvulsant or convulsant activity (Chambers et al. 2003; Whiting 2003). Such ligands open new avenues for the treatment of memory disorders.

References

Alger BE, Pitler TA (1995) Retrograde signaling at $GABA_A$-receptor synapses in the mammalian CNS. Trends Neurosci 18:333–340

Barnard EA, Skolnick P, Olsen RW, Möhler H, Sieghart W, Biggio G, Braestrup C, Bateson AN, Langer SZ (1998) Subtypes of γ-aminobutyric acid A receptors: classification on the bases of subunit structure and receptor function. Pharmacol Rev 50:291–313

Baulac S, Huberfeld G, Gourfinkel-An I, Mitropoulou G, Beranger A, Prud'homme JF, Baulac M, Brice A, Bruzzone R, LeGuern E (2001) First genetic evidence of $GABA_A$ receptor dysfunction in epilepsy: a mutation in the γ2-subunit gene. Nat Genet 28:46–48

Bianchi MT, Song L, Zhang H, Macdonald RL (2002) Two different mechanisms of disinhibition produced by $GABA_A$ receptor mutations linked to epilepsy in humans. J Neurosci 22:5321–5327

Bohlhalter S, Weinmann O, Möhler H, Fritschy JM (1996) Laminar compartmentalization of $GABA_A$-receptor subtypes in the spinal cord: an immunohistochemical study. J Neurosci 16:283–297

Bormann J (2000) The "ABC" of GABA receptors. Trends Pharmacol Sci 21:16–19

Bouilleret V, Ridoux V, Depaulis A, Marescaux C, Nehlig A, Le Gal La Salle G (1999) Recurrent seizures and hippocampal sclerosis following intrahippocampal kainite injection in adult mice: EEG, histopathology and synaptic reorganization similar to mesial temporal lobe epilepsy. Neuroscience 89:7171–7729

Bouilleret V, Loup F, Kiener T, Marescaux C, Fritschy JM (2000) Early loss of interneurons and delayed subunit-specific changes in $GABA_A$-receptor expression in a mouse model of mesial temporal lobe epilepsy. Hippocampus 10:305–324

Brickley SG, Cull-Candy SG, Farrant M (1996) Development of a tonic form of synaptic inhibition in rat cerebellar granule cells resulting from persistent activation of $GABA_A$ receptors. J Physiol 497:753–759

Brickley SG, Revilla V, Cull-Candy SG, Wisden W, Farrant M (2001) Adaptive regulation of neuronal excitability by a voltage independent potassium conductance. Nature 409:88–92

Brun VH, Otnaess MK, Molden S, Steffenach HA, Witter MP, Moser MB, Moser EI (2002) Place cells and place recognition maintained by direct entorhinal-hippocampal circuitry. Science 296:2243–2246

Brussaard AB, Herbison AE (2000) Long-term plasticity of postsynaptic $GABA_A$-receptor function in the adult brain: insights from the oxytocin neurone. Trends Neurosci 23:190–195

Buhl EH, Tamas G, Fisahn A (1998) Cholinergic activation and tonic excitation induce persistent gamma oscillations in mouse somatosensory cortex in vitro. J Physiol (Lond) 513:117–126

Burgess N, Maguire EA, O'Keefe J (2002) The human hippocampus and spatial and episodic memory. Neuron 35:625–641

Buzsaki G, Chrobak JJ (1995) Temporal structure in spatially organized neuronal ensembles: a role for interneuronal networks. Curr Opin Neurobiol 5:504–510

Carlsson A, Waters N, Holm-Waters S, Tedroff J, Nilsson M, Carlsson ML (2001) Interactions between monoamines, glutamate, and GABA in schizophrenia: new evidence. Annu Rev Pharmacol Toxicol 41:237–260

Caulfield MP, Brown DA (1992) Cannabinoid receptor agonists inhibit Ca current in NG108-15 neuroblastoma cells via a pertussis toxin-sensitive mechanism. Br J Pharmacol 106:231–232

Chambers MS, attack JR, Broughton HB, Collinson N, Cook S, Dawson GR, Hobbs SC, Marshall G, Maubach KA, Pillai GV, Reeve AJ, MacLeod AM (2003) Identification of a novel, selective $GABA_A$ $\alpha5$ receptor inverse agonist which enhances cognition. J Med Chem 46:2227–2240

Chevaleyre V, Castillo PE (2003) Heterosynaptic LTD of hippocampal GABAergic synapses: a novel role of endocannabinoids in regulating excitability [comment]. Neuron 38:461–472

Cohen I, Navarro V, Clemenceau S, Baulac M, Miles R (2002) On the origin of interictal activity in human temporal lobe epilepsy in vitro. Science 298:1418–1421

Collinson N (2002) Enhanced learning and memory and altered GABAergic synaptic transmission in mice lacking the $\alpha5$ subunit of the $GABA_A$ receptor. J Neurosci 22:5572–5580

Cossette P, Liu L, Brisebois K, Dong H, Lortie A, Vanasse M, Saint-Hilaire JM, Carmant L, Verner A, Lu WY, Wang YT, Rouleau GA (2002) Mutation of GABRA1 in an autosomal dominant form of juvenile myoclonic epilepsy. Nat Genet 31:184–189

Coulter DA (2001) Epilepsy-associated plasticity in γ-aminobutyric acid receptor expression, function, and inhibitory synaptic properties. Int Rev Neurobiol 45:237–252

Crestani F, Lorez M, Baer K, Essrich C, Benke D, Laurent JP, Belzung C, Fritschy JM, Luscher B, Möhler H (1999) Decreased $GABA_A$-receptor clustering results in enhanced anxiety and a bias for threat cues. Nat Neurosci 2:833–839

Crestani F, Martin JR, Möhler H, Rudolph U (2000) Mechanism of action of the hypnotic zolpidem in vivo. Br J Pharmacol 131:1251–1254

Crestani F, Löw K, Keist R, Mandelli M, Möhler H, Rudolph U (2001) Molecular targets for the myorelaxant action of diazepam. Mol Pharmacol 59:442–445

Crestani F, Keist R, Fritschy JM, Benke D, Vogt K, Prut L, Bluethmann H, Möhler H, Rudolph U (2002) Trace fear conditioning involves hippocampal $\alpha5$ $GABA_A$ receptors. Proc Natl Acad Sci U S A 99:8980–8985

Csicsvari J, Hirase H, Czurko A, Mamiya A, Buzsaki G (1999) Oscillatory coupling of hippocampal pyramidal cells and interneurons in the behaving rat. J Neurosci 19:274–287

Damgen K, Luddens H (1999) Zaleplon displays a selectivity to recombinant $GABA_A$ receptors different from zolpidem, zopiclone and benzodiazepines. Neurosci Res Commun 25:139–148

Davis S, Butcher SP, Morris RGM (1992) The NMDA receptor antagonist D-2-amino-5-phosphonopentanoate (D-AP5) impairs spatial learning and LTP in vivo at intracerebral concentrations comparable to those that block LTP in vitro. J Neurosci 12:21–34

Devor A, Fritschy JM, Yarom Y (2001) Synaptic and extrasynaptic $GABA_A$ receptors in the inferior olivary nucleus differ in their spatial distribution, desensitization kinetics and subunit composition. J Neurophysiol 85:1686–1696

Engel AK, Fries P, Singer W (2001) Dynamic predictions: oscillations and synchrony in top-down processing. Nat Rev Neurosci 2:704–716

Eysenck MW (1992) The nature of anxiety. In: Gale A, Eysenck MW (eds) Handbook of individual differences: biological perspectives. Wiley and Sons, New York, pp 157–178

Fisahn A, Pike FG, Buhl EH, Paulsen O (1998) Cholinergic induction of network oscillations at 40 Hz in the hippocampus in vitro. Nature 394:186–189

Freund TF, Buzsaki G (1996) Interneurons of the hippocampus. Hippocampus 6:345–470

Fritschy JM, Brünig I (2003) Formation and plasticity of GABAergic synapses: physiological mechanisms and pathophysiological implications. Pharmacol Ther 98:299–323

Fritschy JM, Möhler H (1995) $GABA_A$ receptor heterogeneity in the adult rat brain: differential regional and cellular distribution of seven major subunits. J Comp Neurol 359:154–194

Fritschy JM, Weinmann O, Wenzel A, Benke D (1998a) Synapse-specific localization of NMDA and $GABA_A$ receptor subunits revealed by antigen-retrieval immunohistochemistry. J Comp Neurol 390:194–210

Fritschy JM, Johnson DK, Möhler H, Rudolph U (1998b) Independent assembly and subcellular targeting of $GABA_A$ receptor subtypes demonstrated in mouse hippocampal and olfactory neurons in vivo. Neurosci Lett 249:99–102

Fritschy JM, Crestani F, Rudolph U, Möhler H (2004) $GABA_A$ receptor subtypes with special reference to memory function and neurological disorders. In: Hensch T (ed) Excitatory inhibitory balance: synapses, circuits and systems plasticity. Kluver Academic Press, pp 215–228

Gao B, Fritschy JM, Benke D, Möhler H (1993) Neuron-specific expression of $GABA_A$ receptor subtypes: differential associations of the α1- and α3-subunits with serotonergic and GABAergic neurons. Neuroscience 54:881–892

Geiger JR, Lubke J, Roth A, Frotscher M, Jonas P (1997) Submillisecond AMPA receptor-mediated signalling at a principal neuron-interneuron synapse. Neuron 18:1009–1023

Gingrich KJ, Roberts WA, Kass RS (1995) Dependence of the $GABA_A$ receptor gating kinetics on the α-subunit isoform: implications for structure-function relations and synaptic transmission. J Physiol 489:529–543

Goddard AW, Mason GF, Almai A, Rothman DL, Behar KL, Petroff OA, Chamey DS, Krystal JH (2001) Reductions in occipital cortex GABA levels in panic disorder detected 1H MRS. Arch Gen Psychiatry 58:556–561

Gulledge AT, Stuart GJ (2003) Excitatory actions of GABA in the cortex. Neuron 37:299–309

Gupta A, Wang Y, Markam H (2000) Organizing principles for a diversity of GABAergic interneurons and synapses in the neocortex. Science 287:273–278

Harris KD, Henze DA, Hirase H, Leinekugel X, Dragoi G, Czurko A, Buzsaki G (2002) Spike train dynamics predicts theta-related phase precession in hippocampal pyramidal cells. Nature 417:738–741

Hood SD, Argyropoulos SV, Nutt DJ (2000) Agents in development for anxiety disorders. Current status and future potential. CNS Drugs 13:421–431

Huerta PT, Sun LD, Wilson MA, Tonegawa S (2000) Formation of temporal memory requires NMDA receptors within CA1 pyramidal neurons. Neuron 25:473–480

Huntsman MM, Tran BV, Potkin SG, Bunney Wejr, Jones EG (1998) Altered ratios of alternatively spliced γ2 subunit of mRNAs of $GABA_A$ receptors in prefrontal cortex of schizophrenics. Proc Natl Acad Sci U S A 95:15066–15071

Huntsmann MM, Porcello DM, Homanics GE, DeLorey TM, Huguenard JR (1999) Reciprocal inhibitory connections and network synchrony in the mammalian thalamus. Science 283:541–543

Hutcheon B, Morley P, Poulter MO (2000) Developmental change in $GABA_A$ receptor desensitization kinetics and its role in synapse function in rat cortical neurons. J Physiol 522:3–17

Jüttner R, Meier J, Grantyn R (2001) Slow IPSC kinetics, low levels of α1 subunit expression and paired-pulse depression are distinct properties of neonatal inhibitory GABAergic synaptic connections in the mouse superior colliculus. Eur J Neurosci 13:2088–2098

Kalueff A, Nutt DJ (1997) Role of GABA in memory and anxiety. Depress Anxiety 4:100–110

Kananura C, Haug K, Sander T, Runge U, Gu W, Hallmann K, Rebstock J, Heils A, Steinlein OK (2002) A splice-site mutation in GABRG2 associated with childhood absence epilepsy and febrile convulsions. Arch Neurol 59:1137–1141

Kathuria S, Gaetani S, Fegley D, Valino F, Duranti A, Tontini A, Mor M, Tarzia G, La Rana G, Calignano A, Giustino A, Tattoli M, Palmery M, Cuomo V, Piomelli D (2003) Modulation of anxiety through blockade of anandamide hydrolysis. Nat Med 9:76–81

Katona I, Sperlagh B, Sik A, Käfalvi A, Vizi ES, Mackie K, Freund TF (1999) Presynaptically located CB1 cannabinoid receptors regulate GABA release from axon terminals of specific hippocampal interneurons. J Neurosci 19:4544–4558

Katona I, Rancz EA, Acsady L, Ledent C, Mackie K, Hajos N, Freund TF (2001) Distribution of CB1 cannabinoid receptors in the amygdala and their role in the control of GABAergic transmission. J Neurosci 21:9506–9518

Klausberger T, Roberts JD, Somogyi P (2002) Cell type- and input-specific differences in the number and subtypes of synaptic $GABA_A$ receptors in the hippocampus. J Neurosci 22:2513–2521

Klausberger T, Magill PJ, Marton LF, Roberts JDB, Cobden PM, Buzsaki G, Somogyi P (2003) Brain state-and cell type-specific firing of hippocampal interneurons in vivo. Nature 421:844–848

Kreitzer AC, Regehr WG (2001a) Retrograde inhibition of presynaptic calcium influx by endogenous cannabinoids at excitatory synapses onto Purkinje cells [comment]. Neuron 29:717–727

Kreitzer AC, Regehr WG (2001b) Cerebellar depolarization-induced suppression of inhibition is mediated by endogenous cannabinoids. J Neurosci 21:RC174

Langer SZ, Faure-Halley C, Seeburg P, Graham D, Arbilla S (1992) The selectivity of zolpidem and alpidem for the α1-subunit of the $GABA_A$ receptor. Eur Neuropsychopharmacol 2:232–234

Laurent G (1996) Dynamical representation of odors by oscillating and evolving neural assemblies. Trends Neurosci 19:489–496

Llano I, Leresche N, Marty A (1991) Calcium entry increases the sensitivity of cerebellar Purkinje cells to applied GABA and decreases inhibitory synaptic currents. Neuron 6:565–574

Loup F, Wieser HG, Yonekawa Y, Aguzzi A, Fritschy JM (2000) Selective alterations in $GABA_A$ receptor subtypes in human temporal lobe epilepsy. J Neurosci 20:5401–5419

Löw K, Crestani F, Keist R, Benke D, Brunig I, Benson JA, Fritschy JM, Rulicke T, Bluethmann H, Möhler H, Rudolph U (2000) Molecular and neuronal substrate for the selective attenuation of anxiety. Science 290:131–134

Maccaferri G, Roberts JDB, Szucs P, Cottingham CA, Somogyi P (2000) Cell surface domain specific postsynaptic currents evoked by identified GABAergic neurons in rat hippocampus in vitro. J Physiol 524:91–116

Maejima T, Ohno-Shosaku T, Kano M (2001) Endogenous cannabinoid mediate retrograde signals from depolarized postsynaptic neurons to presynaptic terminals. Neuron 29:729–738

Malizia AL (1999) What do brain imaging studies tell us about anxiety disorders? J Psychopharmacol 13:372–378

Malizia AL, Cunningham VJ, Bell CJ, Liddle PF, Jones T, Nutt DJ (1998) Decreased brain $GABA_A$-benzodiazepine receptor binding in panic disorders: preliminary results from a quantitative PET study. Arch Gen Psychiatry 55:715–720

Marsicano G, Wotjak CT, Azad SC, Bisogno T, Rammes G, Cascio MG, Hermann H, Tang J, Hofmann C, Zieglgansberger W, Di Marzo V, Lutz B (2002) The endogenous cannabinoid system controls extinction of aversive memories. Nature 418:530–534

Martina M, Schultz JH, Ehmke H, Monyer H, Jonas P (1998) Functional and molecular differences between voltage-gated K+ channels of fast-spiking interneurons and pyramidal neurons of rat hippocampus. J Neurosci 18:1811–1825

McCormick DA, Bal T (1997) Sleep and arousal: thalamocortical mechanisms. Annu Rev Neurosci 20:185–215

McDonald BJ, Amato A, Connolly CN, Benke D, Moss SJ, Smart TG (1998) Adjacent phosphorylation sites on $GABA_A$ receptor β subunits determine regulation by cAMP-dependent protein kinase. Nat Neurosci 1:23–28

McHugh TJ, Blum KI, Tsien JZ, Tonegawa S, Wilson MA (1996) Impaired hippocampal representation of space in CA1-specific NMDAR1 knockout mice. Cell 87:1339–1349

McKernan RM, Rosahl TW, Reynolds DS, Sur C, Wafford KA, Atack JR, Farrar S, Myers J, Cook G, Ferris P, Garrett L, Bristow L, Marshall G, Macaulay A, Brown N, Howell O, Moore KW, Carling RW, Street LJ, Castro JL, Ragan CI, Dawson GR, Whiting PJ (2000) Sedative but not anxiolytic properties of benzodiazepines are mediated by the $GABA_A$ receptor α1 subtype. Nat Neurosci 3:587–592

Meldrum BS, Whiting P (2001) Anticonvulsants acting on the GABA system. In: Möhler H (ed) Pharmacology of GABA and glycine neurotransmission. Springer-Verlag, Berlin, Heidelberg, New York, pp 173–194

Metha MR, Lee AK, Wilson MA (2002) Role of experience and oscillations in transforming a rate code into a temporal code. Nature 417:741–746

Miles R (2000) Perspectives: neurobiology. Diversity in inhibition. Science 287:244–246

Minassian BA, DeLorey TM, Olsen RW, Philippart M, Bronstein Y, Zhang Q, Guerrini R, Van Ness P, Livet MO, Delgado-Escueta AV (1998) Angelman syndrome: correlations between epilepsy phenotypes and genotypes. Ann Neurol 43:485–493

Mody I, Nusser Z (2000) Differential activation of synaptic and extrasynaptic $GABA_A$ receptors. Eur J Neurosci 12 Suppl 11:398

Möhler H (2001) Functions of GABA receptors: pharmacology and pathophysiology. In: Möhler H (ed) Pharmacology of GABA and glycine neurotransmission. Springer-Verlag, Berlin, Heidelberg, New York, pp 101–116

Möhler H (2002) Pathophysiological aspects of diversity in neuronal inhibition: a new benzodiazepine pharmacology. Dialogues Clin Neurosci 4:261–269 (this reference served as major source for the present article)

Möhler H, Benke D, Fritschy JM, Benson J (2000) The benzodiazepine site of $GABA_A$ receptors. In: Martin DL, Olsen RW (eds) GABA in the nervous system: the view at fifty years. Lippincott, Philadelphia, pp 97–112

Möhler H, Fritschy JM, Rudolph U (2002) A new benzodiazepine pharmacology. J Pharmacol Exp Ther 300:2–8 (This reference served as major source for the present article)

Morris RGM (1989) Synaptic plasticity and learning: selective impairment of learning in rats and blockade of long-term potentiation in vivo by the N-methyl-D-aspartate receptor antagonist AP5. J Neurosci 9:3040–3057

Morris RGM, Anderson A, Lynch GS, Baudry M (1986) Selective impairment of learning and blockade of long-term potentiation by an N-methyl-D-aspartate receptor antagonist AP5. Nature 319:774–776

Moss SJ, Smart TG (2001) Constructing inhibitory synapses. Nat Rev Neurosci 2:240–250
Mrzljak L, Bergson C, Pappy M, Huff R, Levenson R, Goldman-Rakic PS (1996) Localization of D4 receptors in GABAergic neurons in primate brain. Nature 381:245–248
Nakazawa K, Quirk MC, Chitwood RA, Watanabe M, Yeckel MF, Sun LD, Kato A, Carr CA, Johnston D, Wilson MA, Tonegawa S (2002) Requirement for hippocampal CA3 NMDA receptors in associative memory recall. Science 297:211–218
Nusser Z, Sieghart W, Benke D, Fritschy JM, Somogyi P (1996a) Differential synaptic localization of two major γ-aminobutyric acid type: a receptor α subunits on hippocampal pyramidal cells. Proc Natl Acad Sci U S A 93:11939–11944
Nusser Z, Sieghart W, Stephenson FA, Somogyi P (1996b) The α6 subunit of the $GABA_A$ receptor is concentrated in both inhibitory and excitatory synapses on cerebellar granule cells. J Neurosci 16:103–114
Nusser Z, Sieghart W, Somogyi P (1998) Segregation of different $GABA_A$ receptors to synaptic and extrasynaptic membranes of cerebellar granule cells. J Neurosci 18:1693–1703
Nutt DJ, Malizia AL (2001) New insights into the role of the $GABA_A$ receptor in psychiatric disorders. Brit J Psychiatry 179:390–396
Nyíri G, Freund TF, Somogyi P (2001) Input-dependent synaptic targeting of α2 subunit containing $GABA_A$ receptors in hippocampal pyramidal cells of the rat. Eur J Neurosci 13:428–442
O'Keefe J, Recce ML (1993) Phase relationship between hippocampal place units and the EEG theta rhythm. Hippocampus 3:317–330
Olsen RW, DeLorey TM, Gordey M, Kang MH (1999) GABA receptor function and epilepsy. Adv Neurol 79:499–510
Paulsen O, Moser EI (1998) A model of hippocampal memory encoding and retrieval: GABAergic control of synaptic plasticity. Trends Neurosci 21:273–278
Pawelzik H, Hughes DI, Thomson AM (2002) Physiological and morphological diversity of immunocytochemically defined parvalbumin- and cholecystokinin-positive interneurons in CA1 of the adult rat hippocampus. J Comp Neurol 443:346–367
Penttonen M (1998) Gamma frequency oscillation in the hippocampus of the rat: intracellular analysis in vivo. Eur J Neurosci 10:718–728
Pirker S, Schwarzer C, Wieselthaler A, Sieghart W, Sperk G (2000) $GABA_A$ receptors: immunocytochemical distribution of 13 subunits in the adult rat brain. Neuroscience 101:815–850
Pitler TA, Alger BE (1992) Postsynaptic spike firing reduces synaptic $GABA_A$ responses in hippocampal pyramidal cells. J Neurosci 12:4122–4132
Pitler TA, Alger BE (1994) Depolarization-induced suppression of GABAergic inhibition in rat hippocampal pyramidal cells: G protein involvement in a presynaptic mechanism. Neuron 13:1447–1455
Riban V, Bouilleret V, Pham-Lé BT, Fritschy JM, Marescaux C, Depaulis A (2002) Evolution of hippocampal epileptic activity during the development of hippocampal sclerosis in a mouse model of temporal lobe epilepsy. Neuroscience 112:101–111
Rudolph U, Crestani F, Benke D, Brünig I, Benson J, Fritschy JM, Martin JR, Bluethmann H, Mohler H (1999) Benzodiazepine actions mediated by specific γ-aminobutyric acid A receptor subtypes. Nature 401:796–800
Rudolph U, Crestani F, Möhler H (2001) $GABA_A$ receptor subtypes: dissecting their pharmacological functions. Trends Pharmacol Sci 22:188–194
Sanger DJ, Morel E, Perrault G (1996) Comparison of the pharmacological profiles of the hypnotic drugs, zaleplon and zolpidem. Eur J Pharmacol 313:35–42

Sarter M, Bruno JP, Berntson GG (2001) Psychotogenic properties of benzodiazepine receptor inverse agonists. Psychopharmacology (Berl) 156:1–13

Sassoè-Pognetto M, Panzanelli P, Sieghart W, Fritschy JM (2000) Co-localization of multiple $GABA_A$ receptor subtypes with gephyrin at postsynaptic sites. J Comp Neurol 420:481–498

Scatton B, Depoortere H, George P, Sevrin M, Benavides J, Schoemaker H, Perrault (2000) Selectivity for $GABA_A$ receptor α subunits as a strategy for developing hypnoselective and anxioselective drugs. Int J Neuropsychopharmacol 3:S41.3

Sieghart W, Sperk G (2002) Subunit composition, distribution and function of $GABA_A$ receptor subtypes. Curr Top Med Chem 2:795–816

Singer W, Gray CM (1995) Visual feature integration and the temporal correlation hypothesis. Annu Rev Neurosci 18:555–586

Skaggs WE, McNaughton BL, Wilson MA, Barnes CA (1996) Theta phase precession in hippocampal neuronal populations and the compression of temporal sequences. Hippocampus 6:149–172

Snead OC, Depaulis A, Vergues M, Marescaux C (1999) Absence epilepsy: advances in experimental animal models. Adv Neurol 79:253–278

Stein V, Nicoll RA (2003) GABA generates excitement. Neuron 37:375–378

Tang YP, Shimizu E, Dube GR, Rampon C, Kerchner GA, Zhuo M, Liu G, Tsien JZ (1999) Genetic enhancement of learning and memory in mice. Nature 401:63–69

Tia S, Wang JF, Kotchabhakdi N, Vicini S (1996) Distinct deactivation and desensitization kinetics of recombinant $GABA_A$ receptors. Neuropharmacology 35:1375–1382

Tiihonen J, Kuikka J, Rasanen P, Lepola U, Koponen H, Liuska A, Lehmusvaara A, Vainio P, Kononen M, Bergstrom K, Yu M, Kinnunen I, Akerman K, Karhu J (1997) Cerebral benzodiazepine receptor binding and distribution in generalized anxiety disorders: a fractal analysis. Mol Psychiatry 6:463–471

Traub RD, Whittington MA, Stanford IM, Jefferys JG (1996) A mechanism for generation of long-range synchronous fast oscillations in the cortex. Nature 383:621–624

Traub RD, Draguhn A, Whittington MA, Baldeweg T, Bibbig A, Buhl EH, Schmitz D (2002) Axonal gap junctions between principal neurons: a novel source of network oscillations, and perhaps epileptogenesis. Rev Neurosci 13:1–30

Treimann DM (2001) GABAergic mechanisms in epilepsy. Epilepsia 42:8–12

Tsien JZ, Huerta PT, Tonegawa S (1996) The essential role of hippocampal CA1 NMDA receptor-dependent synaptic plasticity in spatial memory. Cell 87:1327–1338

Vicini S, Ferguson C, Prybylowski K, Kralic J, Morrow AL, Homanics GE (2001) $GABA_A$ receptor α1 subunit deletion prevents developmental changes of inhibitory synaptic currents in cerebellar neurons. J Neurosci 21:3009–3016

Vincent P, Marty A (1993) Neighboring cerebellar Purkinje cell communicate via retrograde inhibition of common presynaptic interneurons. Neuron 11:885–893

Volk DW, Fritschy JM, Pierri JN, Auh S, Sampson AR, Lewis DA (2002) Reciprocal alterations in pre- and postsynaptic inhibitory markers at chandelier cell inputs to pyramidal neurons in schizophrenia. Cereb Cortex 12:1063–1070

Wallace RH, Marini C, Petrou S, Harkin LA, Bowser DN, Panchal RG, Williams DA, Sutherland GR, Mulley JC, Scheffer IE, Berkovic SF (2001) Mutant $GABA_A$ receptor γ2-subunit in childhood absence epilepsy and febrile seizures. Nat Genet 28:49–52

Weiller E, Bisserbe JC, Maier W, Lecrubier Y (1998) Prevalence and recognition of anxiety syndromes in five European primary care settings. A report from the WHO study on Psychological Problems in General Health Care. Br J Psychiatry Suppl 173:18–23

Whiting P, Wafford KA, McKernan RM (2000) Pharmacologic subtypes of $GABA_A$ receptors based on subunit composition. In: Martin DL, Olsen RW (eds) GABA in the nervous system: the view at fifty years. Lippincott, Philadelphia, pp 113–126

Whiting PJ (2003) $GABA_A$ receptor subtypes in the brain: a paradigm for CNS drug discovery? Drug Discov Today 8:445–450

Whittington MA, Traub RD, Jeffreys JG (1995) Synchronized oscillations in interneuron network driven by metabotropic glutamate receptor activation. Nature 373:612–615

Whittington MA, Traub RD, Faulkner HJ, Stauford IM, Jeffreys JG (1997) Recurrent excitatory postsynaptic potentials induced by synchronized fast cortical oscillations. Proc Natl Acad Sci U S A 94:12'198–12'203

Wilson RI, Nicoll RA (2001) Endogenous cannabinoids mediate retrograde signalling at hippocampal synapses. [Erratum appears in Nature 2001 Jun 21;411(6840):974]. Nature 410:588–592

Woo TU, Whitehead RE, Melchitzky DS, Jewis DA (1998) A subclass of prefrontal GABA axon terminals are selectively altered in schizophrenia. Proc Natl Acad Sci U S A 95:5341–5346

Zhang Y, Perez Velazquez JL, Tian GF, Wu CP, Skinner FK, Caslen PL, Zhang L (1998) Slow oscillations ($\leq$1 Hz) mediated by GABAergic interneuronal networks in rat hippocampus. J Neurosci 18:9256–9268

HEP (2005) 169:249–303

Excitatory Amino Acid Neurotransmission

C.G. Parsons[2] · W. Danysz[2] · W. Zieglgänsberger[1] (✉)

[1]Max Planck Institute of Psychiatry, Kraepelinstrasse 2-10, 80804 Munchen, Germany
wzg@mpipsykl.mpg.de

[2]Merz Pharmaceuticals GmbH, Eckenheimer Landstrasse 100, 60318 Frankfurt/Main, Germany

Abstract In recent years great progress has been made in understanding the function of ionotropic and metabotropic glutamate receptors; their pharmacology and potential therapeutic applications. It should be stressed that there are already *N*-methyl-D-aspartate (NMDA) antagonists in clinical use, such as memantine, which proves the feasibility of their therapeutic potential. It seems unlikely that competitive NMDA receptor antagonists and high-affinity channel blockers will find therapeutic use due to limiting side-effects, whereas agents acting at the glycine$_B$ site, NMDA receptor subtype-selective agents and moderate-affinity channel blockers are far more promising. This is supported by the fact that there are several glycine$_B$ antagonists, NMDA moderate-affinity channel blockers and NR2B-selective agents under development. Positive and negative modulators of AMPA receptors such as the AMPAkines and 2,3-benzodiazepines also show more promise than e.g. competitive antagonists. Great progress has also been made in the field of metabotropic glutamate receptors since the discovery of novel, allosteric modulatory sites for these receptors. Selective agents acting at these transmembrane sites have been developed that are more drug-like and have a much better access to the central nervous system than their competitive counterparts. The chapter will critically review preclinical and scarce clinical experience in the development of new ionotropic and metabotropic glutamate receptor modulators according to the following scheme: rational, preclinical findings in animal models and finally clinical experience, where available.

Keywords Glutamate receptors · Ionotropic · Metabotropic · AMPA · NMDA · Kainate · Stroke · Traumatic brain injury · Alzheimer's disease · Parkinson's disease · Huntington's disease · Amyotrophic lateral sclerosis · Anxiety · Depression · Pain · Schizophrenia · Drug tolerance · Drug abuse

1 Introduction

Glutamate is the principal excitatory neurotransmitter in the mammalian central nervous system (CNS) and is involved in virtually all functions of the CNS (Mayer and Armstrong 2004). This provides the basis, on the one hand, for therapeutic intervention in many brain dysfunctions, but on the other hand, for potential side-effects. After release from presynaptic terminals, glutamate binds to both ionotropic and metabotropic receptors to mediate fast, slow, and persistent effects on synaptic transmission and integrity (McFeeters and Oswald 2004). Many studies have expanded the functional repertoire of glutamate by showing that glutamate receptors are also present in a variety of non-excitable cells such as astrocytes (Nedergaard et al. 2002).

Synaptic strength at glutamatergic synapses shows a remarkable degree of use-dependent plasticity and such modifications may represent a physiological correlate to learning and memory. Two prominent examples are long-term potentiation (LTP) and long-term depression (LTD), whose mechanisms have been the subject of considerable scrutiny over the past few decades. Dynamic regulation of synaptic efficacy is thought to play a crucial role also in formation of neuronal connections and experience-dependent modification of neural circuitry (Lamprecht and LeDoux 2004). The rodent whisker-to-barrel system is used as a model of activity-dependent cortical plasticity. Specialized anatomical configurations called 'barrels' are structurally and functionally linked to individual whiskers. Much like the reorganization of the human cortex after amputation, peripheral injury resulting from ablation of a single whisker follicle produces atrophy of the cortical barrel connected to it, and enhanced growth of surrounding barrels. The rearrangement of synaptic connections during normal and deprived development is though to be controlled by correlations in glutamatergic afferent impulse activity (Schierloh et al. 2004).

The molecular and cellular mechanisms by which synaptic changes are triggered and expressed are the focus of intense interest (Song and Huganir 2002). Many of the proteins involved in the physiology of glutamatergic synapses have been cloned and their functional role is currently defined. Rapid changes in cytoskeletal and adhesion molecules after learning contribute to short-term plasticity and memory, whereas later changes, which depend on de novo protein synthesis as well as the early modifications, seem to be required for the persistence of long-term memory (Sheng and Kim 2002). The mechanisms of this structural plasticity are still poorly understood, but recent findings are beginning to provide clues (Xu-Friedman and Regehr 2004).

Various aspects of synaptic ultra structure have also been implicated in the mechanisms of short-term plasticity (Dodt et al. 1999, 2002; Eder et al. 2003) and information gating (Govindaiah and Cox 2004) or rhythmic activity (Hughes et al. 2004) mediated by ionotropic and metabotropic glutamate receptors (mGluRs). There is increasing evidence that dendritic spines undergo an activity-dependent structural remodelling and memories are created by alterations in glutamate-dependent excitatory synaptic transmission on dendritic spines (Kasai et al. 2003). The changes in synaptic transmission are initiated by elevations in intracellular calcium and consequent activation of second messenger signalling pathways in the postsynaptic neuron. A large family of interacting proteins regulates glutamate receptor turnover at synapses and thereby influences synaptic strength (Carroll and Zukin 2002). Neuronal activity controls this highly dynamic process of synaptic receptor targeting and trafficking. The targeting, trafficking and internalization mechanisms of glutamate receptors are organized at synapses by cytoskeletal proteins containing multiple protein-interacting domains. The physical transport of glutamate receptors in and out of the synaptic membrane contributes to several forms of long-lasting synaptic plasticity. These modifications are then actively stabilized, over hours or days, by structural changes downstream from glutamate receptors (McGee and Bredt 2003). Recent studies demonstrate that these 'scaffolding' proteins within the postsynaptic specialization do not only play a role in the synaptic delivery and maintenance of the receptor assembly but have also the capacity to promote synaptic maturation, influence synapse size, and modulate glutamate receptor function. Given that glutamate receptors are widely expressed throughout the CNS, regulation of their activity-dependent redistribution provides a potentially important way to modulate efficacy of synaptic transmission (Malinow and Malenka 2002; Bredt and Nicoll 2003). Furthermore, recent results indicate that these distinct protein–protein interactions may be subtly regulated by phosphorylation/dephosphorylation of the intracellular domains of receptor subunits (Flajolet et al. 2003).

Excitatory amino acid transporters (EAATs) are the primary regulators of extracellular glutamate concentrations in the CNS. Glutamate clearance (and consequently glutamate concentration and diffusion in the extracellular space) is associated with the degree of astrocytic coverage of its neurons (Oliet et al. 2001). The genes encoding glutamate transporter proteins have been cloned both from rats and humans (Arriza et al. 1994; Malandro and Kilberg 1996). The human transporters EAAT1 and EAAT2 (rat equivalents GLAST and GLT1) are found in astroglia and microglia and are widely distributed in the CNS. Human EAAT3 (rat EAAC1) is restricted to neurons but is also found outside of the CNS. Human EAAT4 is expressed by cerebellar neurons.

The vesicular glutamate transporter (VGLUT) is responsible for the active transport of L-glutamate in synaptic vesicles and thus is a potential marker for the glutamatergic phenotype. VGLUT comprises three isoforms, VGLUT1, 2 and 3. Recent studies indicated that VGLUT is also expressed in non-neuronal cells, and localized with various organelles such as synaptic-like microvesicles in the pineal gland, and hormone containing secretory granules in endocrine cells. L-Glutamate is stored in these organelles, secreted upon various forms of stimulation, and then acts as a paracrine-like modulator. Thus, VGLUTs highlight a novel framework of glutamatergic signalling and reveal its diverse modes of action (Amara and Fontana 2002).

Under various conditions neurons can become so sensitive to glutamate that it actually kills them ('excitotoxicity'; Rothman and Olney 1987) through receptor-mediated depolarization and calcium influx (Parsons et al. 1998). It has been implied that excitotoxicity is involved in many types of acute and chronic insults to the CNS (Choi 1995). Recent studies suggest that synaptic *N*-methyl-D-aspartate (NMDA) receptors may also be involved in neuroprotective mechanisms (Lu et al. 2003). Disturbance of glutamate homeostasis probably plays a pivotal role in the execution of pathological changes in many disease states and may be triggered by a wide variety of factors that facilitate the neurotoxic potential of endogenous glutamate. These factors include: increase in glutamate release, malfunctioning of neuronal and glial uptake, energy deficits, neuronal depolarization, changes in glutamate receptor properties or expression patterns, and free radical formation (Danysz et al. 1995; Beal 1995; Parsons et al. 1998). Such excitotoxic effects can be pronounced during acute events, such as ischaemic stroke and trauma, or milder but prolonged in chronic neurodegenerative diseases such as Alzheimer's disease, Parkinson's disease, Huntington's disease and amyotrophtropic lateral sclerosis (ALS) (Starr 1995; Beal 1995; Plaitakis et al. 1996; Parsons and Danysz 2002). Glutamatergic dysfunction is also involved in the symptomatology of disorders such as schizophrenia, anxiety, and depression (Danysz et al. 1995; Parsons et al. 1998), as well as in the development of disorders associated with long-term plastic changes in the CNS such as chronic pain, drug tolerance, dependence, addiction, partial complex seizures and tardive dyskinesia (Danysz et al. 1995; Trujillo and Akil 1995; Dickenson 1997; Parsons et al. 1998). Both environmental and genetic glutamate receptor manipulations enhance learning and memory (Tang et al. 2000a,b). Genetic manipulation conferring enhanced cognitive abilities may also provide unintended traits, such as increased susceptibility to persistent pain (Tang et al. 2001).

2 Glutamate Receptors

Glutamate receptors are divided into ionotropic receptors (directly coupled to an ion channel) and metabotropic receptors (coupled to intracellular signalling cascades).

2.1 Ionotropic Glutamate Receptors

There are three types of ionotropic glutamate receptors: NMDA, α-amino-3-hydroxy-5-methyl-4-isoxazoleproprionic acid (AMPA), and kainate receptors (Fig. 1). Each is principally activated by the agonist bearing its name and is permeable to cationic flux; hence, their activation results in membrane depolarization. Ionotropic glutamate receptors were originally classified based on three selective, synthetic agonists: quisqualate, kainate and NMDA. After the discovery of metabotropic receptors, it became clear that quisqualate also interacts with them. Since that time, quisqualate-sensitive ionotropic receptors have been classified by the more selective agonist AMPA.

All ionotropic glutamate receptors can form heteromeric subunit assemblies that have different physiological and pharmacological properties and are differentially distributed throughout the CNS (Mcbain and Mayer 1994; Danysz et al. 1995; Parsons et al. 1998; Danysz and Parsons 1998). Both AMPA (Rosenmund et al. 1998) and NMDA receptors (Laube et al. 1998) are probably largely formed from tetrameric, heteromeric assemblies of different subunits (Mansour et al. 2001). The four distinct subunits are believed to be topologically arranged with three transmembrane-spanning and one pore-lining (hairpin loop) domain (Madden 2002).

High-resolution studies of ionotropic glutamate receptor (iGluR) extracellular domains are beginning to bridge the gap between structure and function. Crystal structures have defined the ligand-binding pocket well beyond what was suggested by mutational analysis and homology models alone, providing initial suggestions about the mechanisms of channel gating and desensitization. Nuclear magnetic resonance (NMR)-derived backbone dynamics and

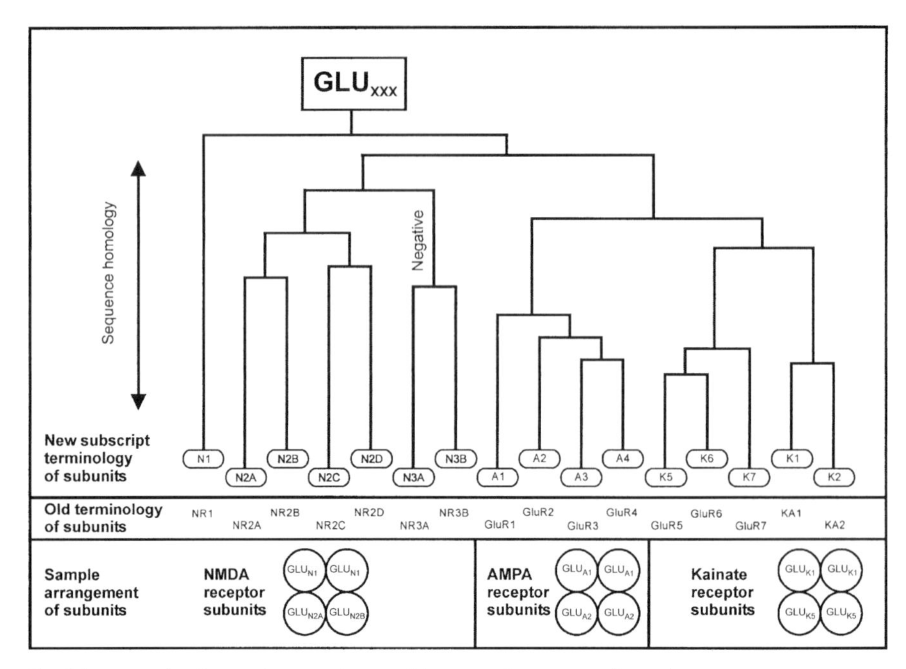

Fig. 1 Ionotropic glutamate receptors: subtypes and sequence homology

molecular dynamics simulations have added further insights into the role of protein dynamics in receptor function.

2.1.1
AMPA Receptors

AMPA receptors are involved in mediating most forms of fast glutamatergic neurotransmission. There are four known subunits, GluR1 to GluR4—sometimes termed GluRA to GluRD—which are widely, but differentially, distributed throughout the CNS (Fig. 2; Parsons et al. 1998). The types of subunits forming these receptors determine their biophysical properties and pharmacological sensitivity. AMPA is selective for GluR1- to GluR4-containing receptors and induces strong desensitization. Two alternative splice variants of GluR1 to GluR4 subunits designated as 'flip' and 'flop' have been shown to differ in their expression throughout the brain and during development and to impart different pharmacological properties (Sommer et al. 1990; Monyer et al. 1991).

The GluR2 subunit imparts particular properties to heteromeric AMPA receptors. Receptors containing this subunit show low Ca^{2+} permeability, linear current-voltage relationships and low sensitivity to block by polyamines and spider toxins. Receptors lacking this subunit show relatively high Ca^{2+} permeability (Burnashev 1996), strong rectification, i.e. non-linear current-voltage relationships (Verdoorn et al. 1991) mediated by channel blockade via intracel-

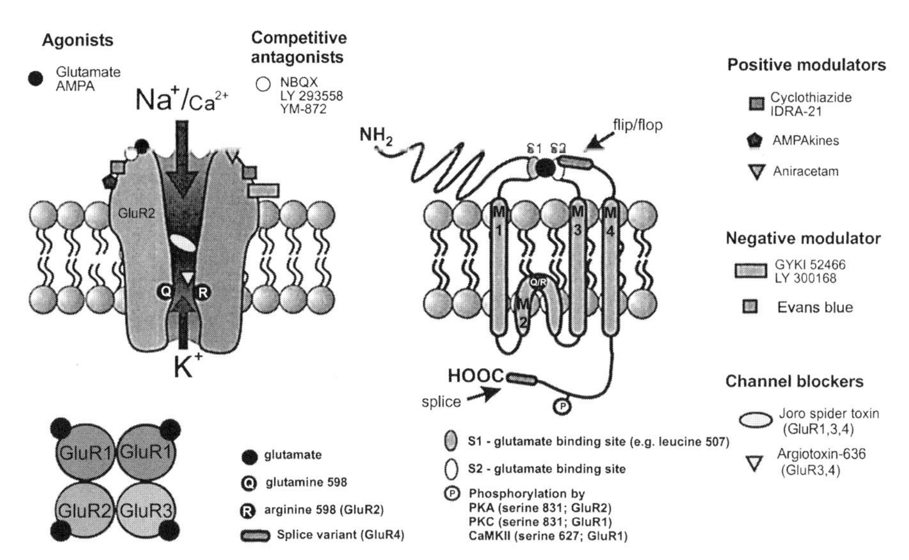

Fig. 2 Schematic of AMPA receptor pharmacology, subtypes and topology

lular polyamines such as spermine (Bowie and Mayer 1995), and are sensitive to block by toxins such as Joro spider toxin, philanthotoxin-343 and argiotoxin-636 (see Parsons et al. 1998). The GluR2 subunit shows developmentally distinct edited and unedited—posttranslational modified protein—forms (Burnashev 1996), and it is the presence of a positively charged arginine (R) residue in the second membrane-inserted segment (MIS, position 586) of edited receptors that renders them Ca^{2+} impermeable. Unedited homomeric GluR2 receptors are also much more sensitive to the positive modulatory effects of cyclothiazide. Cyclothiazide is a selective positive modulator of AMPA receptors, whereas concanavalin-A is much more effective on kainate-preferring receptors. 2,3-Benzodiazepines, such as GYKI 52466, are non-competitive AMPA receptor antagonists and are much less active at kainate receptors (Bleakman et al. 1996). Although the 2,3-benzodiazepines and cyclothiazide show strong allosteric interactions, it is now clear that these effects are mediated at different recognition sites (Rammes et al. 1996, 1998). In general, AMPA receptor flip isoforms show somewhat slower desensitization kinetics and are more sensitive to the positive modulatory effects of cyclothiazide. Glutamate receptors form 'hot spots' on the apical dendrite of neocortical neurons of the rat (Frick et al. 2001). Along this structure AMPA and NMDA receptors are differentially distributed.

Ca^{2+}-permeable receptors are most prominent at early stages of development and show a much more limited distribution in the adult brain (Pellegrini-Giampietro et al. 1992). There are some indications that Ca^{2+}-permeable AMPA receptors are expressed at higher levels under certain pathological conditions such as global ischaemia (Goldberg et al. 1996). However, they also play an important physiological role on inhibitory γ-aminobutyric acid (GABA) interneurons, and selective blockade could lead to excitotoxicity via disinhibition (Racca et al. 1996). Moreover, Ca^{2+}-permeable AMPA receptors seem to have an important role for correct structural and functional relations between Bergman glia and glutamatergic synapses in the cerebellum, such as the removal of synaptically released glutamate (Iino et al. 2001). Selective antagonists proved to be useful in the prevention and treatment of a variety of neurological and non-neurological diseases (Gitto et al. 2004).

2.1.2
AMPA Receptor-Positive Modulators

Cyclothiazide is a positive modulator of AMPA receptors that potentiates agonist-induced currents by reducing or essentially eliminating desensitization (Service 1994; Fricker 1997; Yamada 1998). Such findings underlie the hypothesis that prolongation of AMPA–EPSC decay by inhibition of AMPA receptor desensitization might increase the ability of synaptically released glutamate to depolarize target neurons sufficiently to remove the Mg^{2+} blockade of NMDA receptors and thus facilitate the induction of LTP and learning. More recent data indicate that cyclothiazide also prolongs AMPA receptor deactivation kinetics, i.e. decreases current decay after agonist removal (Rammes et al. 1996, 1998) and this mechanism has been suggested to be more important for AMPA receptor-positive modulators (Yamada 1998; see Rammes et al. 1999).

2.1.3
Kainate Receptors

Physiological studies have identified both post- and presynaptic roles for ionotropic kainate receptors. Kainate receptors contribute to excitatory postsynaptic currents in many regions of the CNS including hippocampus, cortex, spinal cord and retina. In some cases, postsynaptic kainate receptors are co-distributed with AMPA and NMDA receptors, but there are also synapses where transmission is mediated exclusively by postsynaptic kainate receptors: for example, in the retina at connections made by cones onto off bipolar cells. Extrasynaptically located postsynaptic kainate receptors are most likely activated by 'spill-over' glutamate (Eder et al. 2003). Modulation of transmitter release by presynaptic kainate receptors can occur at both excitatory and inhibitory synapses. The depolarization of nerve terminals by current flow through ionotropic kainate receptors appears sufficient to account for most examples of presynaptic regulation; however, a number of studies have provided evidence for metabotropic effects on transmitter release that can be initiated by activation of kainate receptors. The hyperexcitability evoked by locally applied kainate, which is quite effectively reduced by endocannabinoids, is probably mediated preferentially via an activation of postsynaptic kainate receptors (Marsicano et al. 2003).

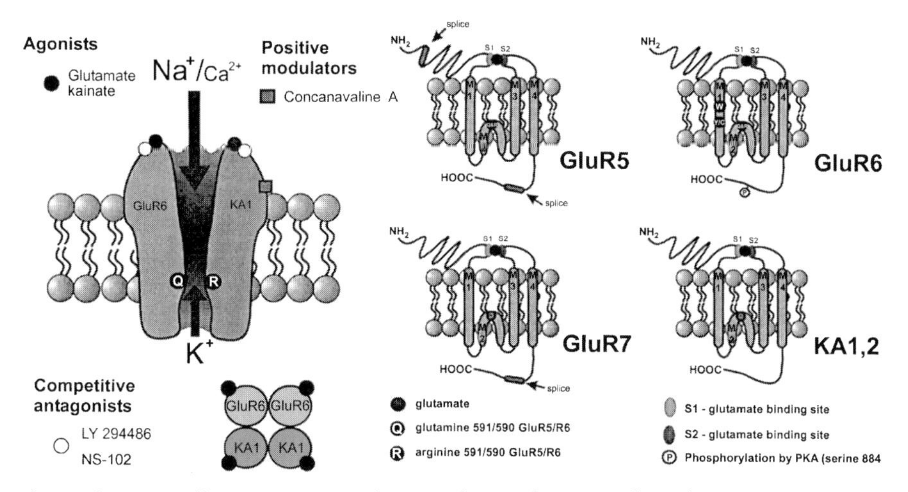

Fig. 3 Schematic of kainate receptor pharmacology, subtypes and topology

Recent analysis of knockout mice lacking one or more of the subunits that contribute to kainate receptors, as well as studies with subunit-selective agonists and antagonists, have revealed the important roles that kainate receptors play in short- and long-term synaptic plasticity. The subunits GluR5, GluR6, KA1 and KA2 form receptor assemblies previously designated as high-affinity kainate receptors (Fig. 3). Kainate receptors were previously believed to be largely presynaptic; for example, they are expressed in the dorsal root ganglia, and activation of these kainate receptors has been shown to facilitate transmitter release (Schmitz et al. 2001). LTP and short-term synaptic facilitation is reduced in knockout mice lacking the GluR6, but not the GluR5, kainate receptor subunit, suggesting that kainate receptors act as presynaptic autoreceptors on mossy fibre terminals to facilitate synaptic transmission (Contractor et al. 2001). Postsynaptic kainate receptors are involved in neurotransmission in some pathways (Wilding and Huettner 1997; Lerma et al. 1997). Kainate receptor activation shows rapid and profound desensitization of GluR5-, GluR6-, GluR7-, KA1- and KA2-containing receptors. SYM 2081, previously assumed to be a kainate receptor antagonist, is actually an agonist (Jones et al. 1997) which produces profound and rapid kainate receptor desensitization and thereby acts as a functional antagonist when continuously present (Wilding and Huettner 1997).

2.1.4 NMDA Receptors

NMDA receptors are highly permeant for Ca^{2+}, show slower gating kinetics than AMPA receptors and the channel is blocked in a voltage- and use-dependent manner by physiological concentrations of Mg^{2+} ions (Mcbain and

Mayer 1994). These properties make them ideally suited for their role as a coincidence detector underlying Hebbian processes in synaptic plasticity such as learning, chronic pain, drug tolerance and dependence (Collingridge and Singer 1990; Trujillo and Akil 1995; Danysz and Parsons 1995; Collingridge and Bliss 1995; Dickenson 1997). Novel techniques revealed a differential distribution of NMDA receptors along apical dendrites of neocortical neurons (Frick et al. 2001) and suggest a very localized generation of glutamate-induced synaptic plasticity (Dodt et al. 1999; Frick et al. 2004). Two major subunit families designated NR1 and NR2, as well as a modulatory subunit designated NR3, have been cloned. Most functional receptors in the mammalian CNS are formed by combination of NR1 and NR2 subunits that express the glycine and glutamate recognition sites, respectively (Hirai et al. 1996; Laube et al. 1997).

2.1.4.1
NR1 Subunits

Alternative splicing generates eight isoforms for the NR1 subfamily (Fig. 4; Zukin and Bennett 1995). The variants arise from splicing at three exons. One encodes a 21-amino acid insert in the N-terminal domain (N1, exon 5), and two encode adjacent sequences of 37 and 38 amino acids in the C-terminal domain (C1, exon 21 and C2, exon 22). NR1 variants are sometimes denoted by

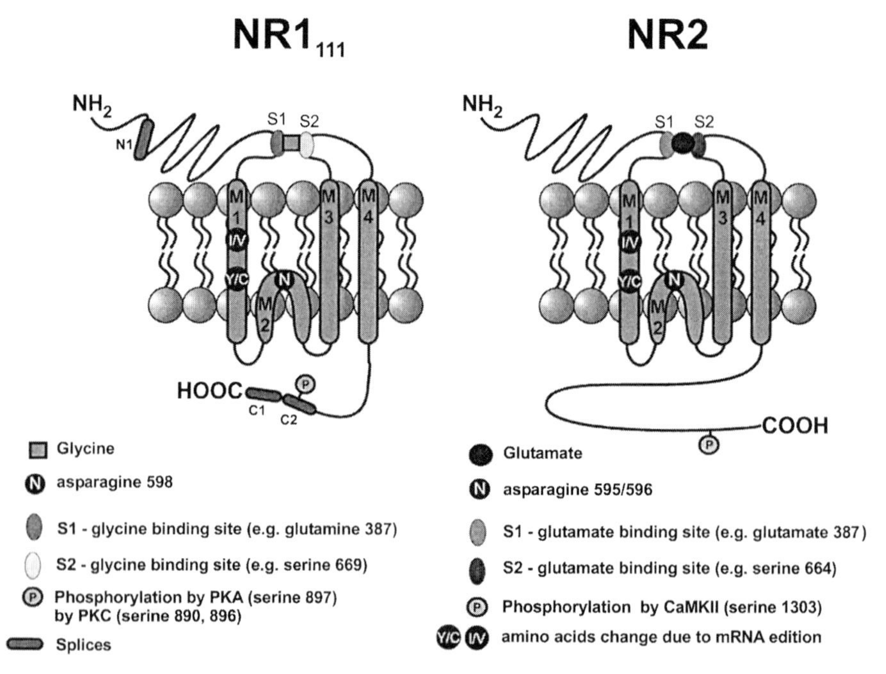

Fig. 4 Schematic representation of the NMDA receptor

the presence or absence of these three alternatively spliced exons (from N to C1 to C2). $NR1_{111}$ has all three exons, $NR1_{000}$ has none, and $NR1_{100}$ has only the N-terminal exon. The variants from $NR1_{000}$ to $NR1_{111}$ are alternatively denoted as NMDA receptor 1-4a, -2a, -3a, -1a, -4b, -2b, -3b and -1b respectively. The mRNA for double splice variants in the C1/C2 regions, such as $NR1_{011}$ (NR1a), show an almost complementary pattern to those lacking both of these inserts, such as $NR1_{100}$ (NR1b); the former are more concentrated in rostral structures such as cortex, caudate, and hippocampus, while the latter are principally found in more caudal regions such as thalamus, colliculi, locus coeruleus and cerebellum (Laurie et al. 1995). NMDA receptors cloned from murine CNS have a different terminology from those in the rat: ζ1 remains the terminology for the mouse equivalent of NR1, and ε1 to ε4 represent NR2A to 2D subunits respectively.

2.1.4.2 NR2 Subunits

The NR2 subfamily consists of four individual subunits, NR2A to NR2D (Figs. 4 and 5). Various heteromeric NMDA receptor channels formed by combinations of NR1 and NR2 subunits are known to differ in gating properties, Mg^{2+} sensitivity and pharmacological profile (Parsons et al. 1998). The heteromeric assembly of NR1 and NR2C subunits, for instance, has a lower sensitivity to Mg^{2+} but increased sensitivity to glycine (see Sect. 2.2) and a very restricted distribution in the brain. In situ hybridization has revealed overlapping but

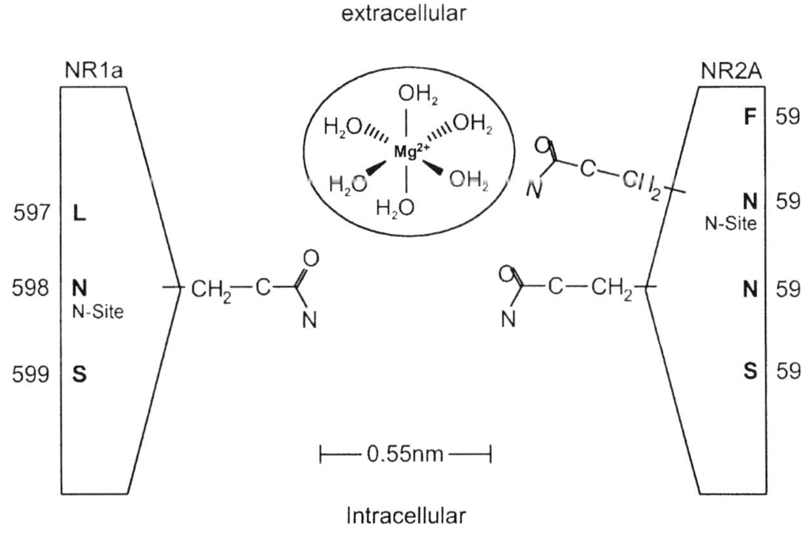

Fig. 5 Schematic of the NMDA receptor channel binding domain for Mg^{2+}

different expression for NR2 mRNA, e.g. NR2A mRNA is distributed ubiquitously like NR1 with highest densities occurring in hippocampal regions and NR2B is expressed predominantly in forebrain but not in cerebellum where NR2C predominates. The spinal cord expresses high levels of NR2C and NR2D and these may form heteroligomeric receptors with NR1 plus NR2A, which would provide a basis for the development of drugs selectively aimed at spinal cord disorders.

The overexpression of NR2B in the forebrains of transgenic mice leads to enhanced activation of NMDA receptors, facilitating synaptic potentiation in response to stimulation at 10–100 Hz (Tang et al. 1999). These mice exhibit superior ability in learning and memory in various behavioural tasks, showing that NR2B is critical for plasticity and memory formation. Environmental enrichment significantly increases protein levels of GluR1, NR2B and NR2A (Rampon et al. 2000).

2.1.4.3
NR3 Subunits

NR3 (NRL or Chi-1) is expressed predominantly in the developing CNS and does not seem to form functional homomeric glutamate-activated channels, but co-expression of NR3 with NR1 plus NR2 subunits decreases response magnitude (Sucher et al. 1995; Matsuda et al. 2002). However, NR3A or NR3B does co-assemble with NR1 alone in *Xenopus* oocytes to form excitatory glycine receptors that are unaffected by glutamate or NMDA, Ca^{2+}-impermeable and resistant to blockade by Mg^{2+} but inhibited by the glycine co-agonist D-serine (Chatterton et al. 2002).

2.2
Glycine as a Co-agonist

Glycine is a co-agonist at NMDA receptors at a strychnine-insensitive recognition site (glycine$_B$), and its presence at moderate nanomolar concentrations is a prerequisite for channel activation by glutamate or NMDA (Danysz and Parsons 1998) and the prevention of NMDA receptor desensitization. Recently it has been suggested that D-serine may be more important than glycine as an endogenous co-agonist at NMDA receptors in the telencephalon and developing cerebellum. There is still some debate as to whether the glycine$_B$ site is saturated in vivo (Danysz and Parsons 1998), but it seems likely that the degree of NMDA receptor activation varies depending on regional differences in receptor subtype expression and local glycine or D-serine concentrations. Moreover, glycine concentrations at synaptic NMDA receptors could be finely modulated by local expression of specific glycine transporters such as GLYT1 (Danysz and Parsons 1998).

2.2.1
Polyamines

The polyamines spermine and spermidine have multiple effects on the activity of NMDA receptors (Johnson 1996; Williams 1997). These include an increase in the magnitude of NMDA-induced whole-cell currents seen in the presence of saturating concentrations of glycine, an increase in glycine affinity, a decrease in glutamate affinity, and voltage-dependent inhibition at higher concentrations. Endogenous polyamines could act as a bi-directional gain control of NMDA receptors by dampening toxic chronic activation by low concentrations of glutamate—through changes in glutamate affinity and voltage-dependent blockade—but enhancing transient synaptic responses to millimolar concentrations of glutamate (Williams 1997; Zhang and Shi 2001).

2.2.2
Competitive NMDA Receptor Antagonists

Antagonists such as D-(−)-2-amino-5-phosphono-valeric acid (D-APV), which competitively block NMDA receptors, cause numerous side-effects such as memory impairment, psychotomimetic effects, ataxia and motor dis-coordination, since they also impair normal synaptic transmission. The challenge has therefore been to develop NMDA receptor antagonists that prevent the pathological activation of NMDA receptors but allow their physiological activation.

2.2.3
Uncompetitive NMDA Receptor Antagonists

It has been suggested that uncompetitive NMDA receptor antagonists with rapid unblocking kinetics but somewhat less pronounced voltage-dependency than Mg^{2+} should be able to antagonize the pathological effects of the sustained, but relatively small increases in extracellular glutamate concentration but, like Mg^{2+}, leave the channel as a result of strong depolarization following physiological activation by transient release of millimolar concentrations of synaptic glutamate (Parsons et al. 1993a; Parsons et al. 1999; Jones et al. 2001). As such, uncompetitive NMDA receptor antagonists with moderate, rather than high affinity may be desirable. Memantine and dextromethorphan are clinically used agents that belong to this category. Several promising agents have unfortunately been abandoned at late stages of development, possibly due to the choice of the wrong, too ambitious, clinical indications such as stroke and trauma.

2.3
Glycine Site Antagonists

Most full glycine$_B$ antagonists (i.e. those without intrinsic partial agonist activity) show very poor penetration to the CNS, although some agents with improved, but by no means optimal, pharmacokinetic properties have now been developed. Glycine$_B$ antagonists have been reported to lack many of the side-effects classically associated with NMDA receptor blockade such as no neurodegenerative changes in the cingulate/retrosplenial cortex even after high doses and no psychotomimetic-like or learning impairing effects at anticonvulsive doses (see Danysz and Parsons 1998). The Merck compound L-701,324 has even been proposed to have atypical antipsychotic effects (Bristow et al. 1996). The improved neuroprotective therapeutic profile of glycine$_B$ full antagonists could be due to their ability to reveal glycine-sensitive desensitization (Parsons et al. 1993b).

Kynurenic acid is an endogenous glycine$_B$ antagonist, but it seems unlikely that concentrations are sufficient to interact with NMDA receptors under normal conditions (Danysz and Parsons 1998; Stone 2001). However, concentrations are raised under certain pathological conditions (Danysz and Parsons 1998; Stone 2001) and interactions with other receptors such as α7 neuronal nicotinic have been reported at lower concentrations (Hilmas et al. 2001). Strategies aimed at increasing kynurenic acid concentrations by, for example, giving its precursor 4-Cl-kynurenine, inhibiting brain efflux with probenecid or inhibiting its metabolism have been proposed to be of therapeutic potential (Danysz and Parsons 1998; Stone 2001).

D-cycloserine and (+R)-HA-966 are partial agonists at the glycine$_B$ site with different levels of intrinsic activity (Karcz-Kubicha et al. 1997). Although these systemically active partial agonists do not induce receptor desensitization they have favourable therapeutic profiles in some in vivo models (Lanthorn 1994; see Danysz and Parsons 1998). This may, in part, be due to their own intrinsic activity as agonists at the glycine$_B$ site, which would serve to preserve a certain level of NMDA receptor function even at very high concentrations (Danysz and Parsons 1998).

D-cycloserine shows agonist-like features at low in vivo doses, while with increasing dosing antagonistic effects predominate (Lanthorn 1994). The consistent biphasic effects of D-cycloserine seen in vivo may be related to different affinities and intrinsic activities at NMDA receptor subtypes. D-cycloserine is a partial agonist for the murine equivalents of NR1/2A and NR1/2B heteromers but is more effective than glycine at NR1/2C (O'Connor et al. 1996). This effect is accompanied by higher affinity at NR1/2C receptors—NR1/2C>NR1/2D>> NR1/2B>NR1/2A (O'Connor et al. 1996). As such, it is likely that the biphasic effects seen in vivo are due to agonistic actions at NR1/2C receptors at lower doses and inhibition of NR1/2A and NR1/2B containing receptors at higher doses. This receptor subtype selectivity and differential intrinsic activity could well underlie its promising preclinical profile in some animals models.

There is a very large body of literature confirming that D-cycloserine enhances learning in a wide variety of models (see Danysz and Parsons 1998 for review plus Harper 2000; Lelong et al. 2001). Taken together, these data indicate that the acute positive effects of D-cycloserine on learning in animal models are only seen at a very narrow range of doses and that even these effects are lost upon repetitive or chronic administration.

Although 1-amino-cyclopropane-carboxylic acid (ACPC) has been reported to be a partial agonist with very high intrinsic activity, it is probably really a full agonist at the glycine$_B$ site and actually behaves as an antagonist in some in vivo models (neuroprotection, anticonvulsive effects) that are likely to be mediated via competitive antagonistic properties at higher concentrations (Skolnick et al. 1989). The consistent observation that chronic treatment with ACPC is neuroprotective could be because it desensitizes or uncouples NMDA receptors (Skolnick et al. 1992; Papp and Moryl 1996), or it may be related to an increase in the relative levels of NR2C expression (Danysz and Parsons 1998).

N-Acetyl-aspartyl-glutamate (NAAG), which is abundant in the mammalian CNS, acts as a weak partial agonist at NMDA receptors with low intrinsic activity and an agonist at mGluR3 receptors (Neale et al. 2000). Inhibition of NAALADase (*N*-acetyl-a-linked-acidic dipeptidase, or glutamate carboxypeptidase II, E.C. 3.4.17.21) has been proposed to be useful in numerous CNS disorders associated with disturbances in glutamatergic transmission by decreasing the concentration of glutamate and increasing the concentration of NAAG.

2.3.1 NR2B-Selective Antagonists

Ifenprodil and its analogue eliprodil block NMDA receptors in a spermine-sensitive manner and were originally proposed to be polyamine antagonists. It is now clear that both agents are selective for NR2B subunits (Legendre and Westbrook 1991) and bind to a site that is distinct from the polyamine recognition site, but interact allosterically with this site and the glycine$_B$ site. NR2B-selective agents may also offer a promising approach to minimize side-effects, as agents would not produce maximal inhibition of responses of neurons expressing heterogeneous receptors. Thus, cortical and hippocampal neurons express both NR2A and NR2B receptors in approximately similar proportions, but very little NR2C or NR2D. NR2B-selective agents therefore block NMDA receptor-mediated responses of such neurons to a maximal level of around 30%–50% of control. Several studies have shown that ifenprodil and eliprodil reduce seizures and are effective neuroprotectants against focal and global ischaemia and trauma at doses that do not cause ataxia or impair learning (Parsons et al. 1998). An unfortunate new side-effect has recently been reported, i.e. that some of these agents may produce a prolongation of the Q-T interval in the cardiac action potential due to blockade of human ether-a-go-go-related gene (hERG) potassium channels (Gill et al. 1999). Several substances selective for NR2B NMDA receptor subtypes such as traxoprodil, Ro-25-6981 and EMD-95885 have been claimed to have a good neuroprotective profile, and traxoprodil is in clinical development. Other promising agents have been abandoned at various stages of clinical development, but often due to side-effects unrelated to NMDA receptor antagonism.

2.3.2
δ-Glutamate Receptors

While most subtypes of glutamate receptors have been studied extensively, less is known about the δ-glutamate receptors, δ1 and δ2, which are abundant only in parallel fibre synapses on Purkinje cells. Although neither forms functional channels when expressed in heterologous cells, genetic analyses have demonstrated the physiological significance of δ2. A δ-glutamate receptor-binding protein was identified in the rat, which is widely expressed in both brain and peripheral tissues, including high expression in brainstem and enrichment in the postsynaptic density. Morphological changes in this binding protein may regulate the δ-glutamate receptor clustering on the dendritic spines, and may affect synaptic efficacy and plasticity (Hirai 2000; Ly et al. 2002).

2.4
Metabotropic Glutamate Receptors

Eight genes encoding mGluRs have been identified. They are G protein (guanine nucleotide-binding protein)-coupled receptors linked to second-messenger systems. Each metabotropic glutamate receptor is composed of seven transmembrane-spanning domains, with only minor homology to GABA(B)-type receptors, and appears evolutionarily distinct from the other members of the superfamily of metabotropic G protein-coupled receptors (GCPRs), which includes noradrenergic, muscarinic, acetylcholinergic, dopaminergic and serotonergic (except other type III GPCRs such as GABAB and Ca^{2+} sensing receptors) receptors. mGluRs are divided into three major groups, I–III, which are involved in the generation of slow excitatory and inhibitory synaptic potentials, and modulation of synaptic transmission, as well as synaptic and neuronal plasticity and development (Nakanishi et al. 1998). In addition to glutamate, mGluRs are activated by ibotenate and quisqualate. They undergo constitutive internalization after activation by glutamate. This glutamate-induced internalization involves an arrestin- and clathrin-dependent pathway and is inhibited by receptor-inverse agonists (Pula et al. 2004). Functional mGluRs are probably dimers linked via extracellular disulphide bridges. The extreme intracellular C-terminus of the α-subunit of the G protein is important for the Gi/Go-coupled group II and III mGluRs, and the second loop is probably coupled to phospholipase C (PLC). The first and the third loops are highly conserved (Pin et al. 1998). MGluRs interact with Homer proteins via a proline-rich motif within the intracellular C-terminus. The Homer 1a isoform, which is upregulated during seizures, stimulates group I mGluRs in the absence of an agonist (Ango et al. 2001). mGluRs couple to mitogen-activated protein kinase (MAPK), and extracellular signal regulated kinase (ERK)1/2 can activate some immediate early genes (IEGs) encoding transcription factors (Peavy and Conn 1998; Ferraguti et al. 1999).

Group I of mGluRs consists of two receptor subtypes, mGluR1, which has four splice variants, and mGluR5, which has two splice variants. The group I mGluRs are mainly localized to the somatodendritic membrane, and their postsynaptic activation most commonly evokes excitation in neurons. Activation of presynaptic group I mGluRs reduces glutamate release and increases paired-pulse facilitation, e.g. in hippocampal neurons (Manzoni and Bockaert 1995). Postsynaptic group I of mGluRs are positively coupled to PLC. PLC promotes the conversion of phosphatidylinositol 4,5-bisphosphate (PIP_2) to diacylglycerol (DAG) and IP_3. DAG activates membrane-bound protein kinase (PK)C that in turn can phosphorylate ionotropic glutamate receptors. IP_3 has numerous intracellular effects including stimulation of Ca^{2+} release from intracellular stores. Purkinje cells in the cerebellum carry exclusively mGluR1 including the two splice variants targeted to the perisynaptic regions (Mateos et al. 2000) but no mGluR5. In the cerebellum, AMPA receptors mediate the fast synaptic responses to glutamate released from parallel fibres. Only tetanic parallel fibre stimulation evokes a slow mGluR1-mediated excitation. There is evidence that this slow time course is not due to a slow diffusion of synaptically released glutamate but to an indirect signal transduction mechanism different from that active in hippocampal pyramidal neurons (Reichelt and Knöpfel 2002). The depolarizing current following mGluR1 activation in Purkinje cells might be carried largely by Na^+-ions (Tempia et al. 2001). In ventral midbrain dopamine neurons, the activation of mGluR1 triggers a slow inhibitory synaptic potential. This synaptic event is caused by a potassium conductance that is activated by an increase in intracellular Ca^{2+} concentration due to mobilization from intracellular stores (Fiorillo and Williams 1998). The first neurophysiological recordings revealed that group I mGluR activation blocks a potassium conductance in the hippocampus in vitro that was triggered by the increase in intracellular Ca^{2+} following the opening of voltage-gated Ca^{2+} channels during action potential generation. In the presence of the antagonist, action potential was generated throughout the depolarizing input. Voltage-clamp experiments demonstrated that the calcium-activated potassium current (I_{AHP}) decreased while the Ca^{2+} dynamics remained unchanged (Charpak et al. 1990). The two different splice variants of the mGluR1 seem to be involved in different signalling pathways (Mannaioni et al. 2001).

Metabotropic glutamate receptor (mGluR) antagonists co-applied with dopamine block LTD in rat prefrontal cortex (PFC). This suggests that the dopamine-facilitated LTD requires synaptic activation of groups I and II mGluRs during tetanus (Otani et al. 1999). Activation of mGluRs causes membrane hyperpolarization in midbrain dopamine neurons. This hyperpolarization results from the opening of Ca^{2+}-sensitive K^+ channels, which is mediated by the release of Ca^{2+} from intracellular stores. The mGluR-mediated hyperpolarization was shown to induce a transient pause in the spontaneous firing of dopamine neurons. The mGluR-mediated Ca^{2+} mobilization in dopamine neurons is caused by multiple intracellular pathways to exert an inhibitory control on the excitability of dopamine neurons (Morikawa et al. 2003). Dopamine neurons of the substantia nigra pars compacta receive a prominent serotonin (5-HT) projection from the dorsal raphé nucleus. MgluR-evoked postsynaptic currents are inhibited by an activation of 5-HT_{2A} and 5-HT_4 receptors (Paolucci et al. 2003).

Group II (mGluR2/3) and group III (mGluR4/6/7/8; mGluR4/7/8 have two splice variants) receptors differ in their sequence homology but are both coupled to a different effector system, i.e. they decrease the activity of adenylate cyclase. Both group II and III mGluRs are located largely on presynaptic neurons and glia and modulate the release of glutamate as well as other, e.g. inhibitory, transmitters such as GABA (Salt et al. 1999). The activation of group II and III mGluRs evokes predominantly inhibitory effects on neuronal excitability. However, 4-aminopyrrolidine-2,4-dicarboxylic acid (APDC), a selective and potent group II mGluR agonist, reversibly increased NMDA receptor currents in acutely dissociated PFC pyramidal neurons. Selective group II mGluR antagonists, but not group I mGluR antagonists, blocked APDC-induced enhancement of NMDA receptor currents, suggesting the mediation by mGluR2/3 receptors. Inhibiting PKC or dialysis with Ca^{2+} chelators largely blocked the mGluR2/3 modulation of NMDA receptor currents. Moreover, treatment of PFC slices with APDC significantly increased the PKC activity and PKC phosphorylation of NMDA receptors. These findings suggest that activation of mGluR2/3 receptors potentiates NMDA receptor channel functions in PFC through a PKC-dependent mechanism (Tyszkiewicz et al. 2004).

A very interesting finding is that AMPA receptor activation inhibits ADP-ribosylation and forskolin-stimulated activity of adenylate cyclase in rat cortical neurons (Wang et al. 1997). These effects were independent of Ca^{2+} and Na^+ influx, suggesting that the ionotropic AMPA receptor is also directly coupled to metabotropic processes. This is supported by the finding that AMPA receptors activate a G protein (Kawai and Sterling 1999). Brief kainate exposure caused long-lasting inhibition of a post-spike potassium current (I sAHP) in CA1 pyramidal cells and this inhibition did not require ionotropic action or network activity, but was blocked by an inhibitor of pertussis toxin-sensitive G proteins (ethylmaleimide), or the PKC inhibitor calphostin C (Melyan et al. 2002). Agonist-dependent downregulation of recombinant NR1/2A receptors by tyrosine dephosphorylation independent of ion flux has also recently been reported (Vissel et al. 2001).

Recent data suggest that a progressive increase in tonic mGluR activity during postnatal development contributes to a reduction of release probability of glutamate in excitatory cortical synapses (Chen and Roper 2004). Group I metabotropic glutamate receptor activation produces a direct excitation of

Table 1 Metabotropic glutamate receptors

	Group I	Group II	Group III	mGIuRs connected with PLD
Subtypes	**mGIuR1**	**MgIuR2**	**mGIuR4**	Non-identified
	$mGIuR1_a$	**MgIuR3**	$mGIuR4_a$	
	$mGIuR1_b$		$mGIuR4_b$	
	$mGIuR1_c$		**mGIuR6**	
	$mGIuR1_d$		**mGIuR7**	
	mGIuR5		$mGIuR7_a$	
	$mGIuR5_a$		$mGIuR7_b$	
	$mGIuR5_b$		**mGIuR8**	
Second messenger system	↑ **PLC**	↓ **AC**	↓ **AC**	↑ **PLD**

↑, Increase; ↓, decrease; AC, adenylate cyclase; PLC, phospholipase C; PLD, phospholipase D.

identified septohippocampal cholinergic neurons (Hajszan et al. 2004). Septohippocampal cholinergic neurons innervate the hippocampus and provide it with almost its entire acetylcholine. These findings may be of significance in treatment of cognitive deficits associated with neurodegenerative disorders, as mGluR-mediated activation of septohippocampal cholinergic neurons would enhance the release of acetylcholine both in the hippocampus and in the septum. NMDA receptor-independent LTP has been shown to be mediated by activation of group II mGluRs (Wu et al. 2004).

The prominent involvement of mGluRs in genomic responses to synaptic stimulation is considered to play a pivotal role in a variety of neurological disorders. Available data indicate that the eight subtypes of mGluRs have distinct effects on gene expression. The group I subtypes facilitate, whereas group II and III subtypes inhibit, gene expression. Due to their significance in regulating drug action, mGluRs have been considered as promising targets for the development of novel therapeutic drugs for the treatment of drug addiction. It seems likely that allosteric modulation of mGluRs may provide a valid strategy for the development of new pharmaceuticals in the near future (Gasparini et al. 2002). Numerous highly selective agonists and antagonists are now available (Knöpfel et al. 1995; Schoepp et al. 1999; Table 1).

3 Synaptic Plasticity

NMDA receptor channels are only activated in the presence of a local strong depolarization induced by strong AMPA receptor activation and concurrent

GABAergic disinhibition via feedback effects of GABA on $GABA_B$ autoreceptors. As a result, the Mg^{2+} blockade of NMDA receptors is transiently fully relieved allowing Ca^{2+} to flow into the postsynaptic neuron. This Ca^{2+} influx triggers a cascade of secondary messengers that ultimately activate a number of enzymes such as PKC, phospholipase A_2 (PLA_2), PLC and Ca^{2+}/calmodulin-dependent protein kinase II (CaM kinase II) (Abraham and Tate 1997; Grant and Silva 1994; Lisman 1994; Benowitz and Routtenberg 1997; Lan et al. 2001; Bayer et al. 2001). Consequently, these processes lead to fixation of changes in postsynaptic AMPA receptors such as an increase in their affinity and/or number (Benke et al. 1998) and, possibly through retrograde signals (arachidonic acid, nitric oxide), modulate presynaptic glutamatergic terminals influencing transmitter release (Collingridge and Bliss 1995).

There is accumulating evidence that LTP and LTD share some common mechanisms, although LTD occurs with increases in postsynaptic Ca^{2+} that are insufficient to induce LTP (Artola and Singer 1993; Christie et al. 1994; Cummings et al. 1996; Derrick and Martinez 1996; Hansel et al. 1996; Kirkwood et al. 1996; Tsumoto and Yasuda 1996). Although hippocampal LTP and spatial learning are impaired by NMDA receptor blockade, learning deficits can be almost completely prevented if rats are pretrained in a different water maze (Bannerman et al. 1995; Saucier and Cain 1995). NMDA receptors may therefore not be required for encoding the spatial representation of a specific environment, but rather other forms of memory that are important for learning the water maze task (Morris 1996). Recent evidence indicates that LTP is not only important for synaptic plasticity in the mature CNS but also in the formation of conducting glutamatergic synapses in the developing mammalian brain (Durand et al. 1996).

There is still considerable debate as to the site at which the increase in synaptic strength is expressed (Collingridge and Bliss 1995). Presynaptic mechanisms should be reflected in a change in release probability. This can be measured at excitatory synapses on cultured hippocampal neurons by analysis of the progressive block of NMDA receptor-mediated synaptic currents by the essentially irreversible open channel blocker dizocilpine [(+)MK-801] (Rosenmund et al. 1998). This technique was used to demonstrate that release probability was not affected after the induction of LTP, making a presynaptic mechanism unlikely (Manabe and Nicoll 1994). Moreover, recent reports indicate that a high proportion of synapses in hippocampal area CA1 transmit with NMDA receptors but not AMPA receptors, making these synapses effectively non-functional at normal resting potentials due to Mg^{2+} blockade (Liao et al. 1995; Nicoll and Malenka 1995; Montgomery et al. 2001; Montgomery and Madison 2002). These silent synapses acquire AMPA-type responses following LTP induction. Furthermore, this form of LTP is accompanied by an increase in the conductance of postsynaptic AMPA receptors. Taken together, these findings challenge the view that LTP in CA1 involves a presynaptic modification,

hindlimb paresis (Rothstein et al. 1996). In fact, after glutamate loading there is a significantly higher increase in glutamate and aspartate levels in plasma in ALS patients than in matched controls (Gredal and Moller 1995).

Over a decade ago, it was shown that the pattern of neuronal loss in the spinal cord in patients suffering from ALS resembles that obtained after excitotoxic lesions induced by kainate in animals (Hugon et al. 1989). Injection of kainate to the spinal cord produces damage to motoneurons, while NMDA lesions affect mainly dorsal horn neurons (Ikonomidou et al. 1996). Similarly, short exposure to kainate in vitro results in selective Ca^{2+}-dependent death of motoneurons expressing Ca^{2+}-permeant AMPA receptors, while dorsal horn neurons are unaffected (Van Den Bosch et al. 2000). α-Motoneurons in organotypic cultures of rat spinal cord are considerably more sensitive to kainate and quisqualate than to NMDA toxicity (Saroff et al. 2000). This selective motoneuron death is completely inhibited by the AMPA receptor antagonists LY300164 and Joro spider toxin (selective for Ca^{2+} permeant receptors) (Van Den Bosch et al. 2000). This suggests non-NMDA (AMPA or kainate) receptor involvement.

One of the crucial questions is why certain motoneuron populations are particularly prone to death in ALS. Although there is no clear-cut answer, one of the possibilities is weak Ca^{2+} buffering capacity connected with lowered levels of cytoplasmic proteins responsible for such buffering, such as parvalbumin and calbindin (Krieger et al. 1994; Roy et al. 1998).

The competitive AMPA receptor antagonist RPR 119990 was found to be active in a transgenic mouse model of familial ALS (SOD1-G93A) where it was able to improve grip muscle strength and glutamate uptake from spinal synaptosomal preparations, and prolong survival (Canton et al. 2001). Most of the evidence presented above points to a primary involvement of AMPA receptors in neurodegeneration in ALS. In fact, data on NMDA receptors role are scarce and clinical evidence rather discouraging.

Riluzole—the only drug registered as a disease-modifying agent for ALS—delayed the development of motor impairment and prolonged life span in superoxide dismutase (SOD) transgenic mice (Gurney et al. 1998) (Kennel et al. 2000). In spite of extensive studies, the precise mechanism of action of riluzole (RP 54274) remains elusive. Riluzole clearly decreases the synaptic release of glutamate and other neurotransmitters (Bryson et al. 1996) and this effect is probably secondary to inhibition of voltage-activated Na^{+} channels (VASCs) (Doble 1996) and voltage-activated Ca^{2+} channels (VASCs, P/Q>N>>>L-type channels) (Stefani et al. 1997; Huang et al. 1997; Yokoo et al. 1998).

The evidence given above is probably sufficient to consider glutamate antagonists as plausible neuroprotective treatment of ALS. Unfortunately, most clinical trials with glutamate antagonists completed to date have not been encouraging. Dextromethorphan (NMDA channel blocker) showed no benefit (Blin et al. 1996; Gredal et al. 1997).

4.5 Parkinson's Disease

There is some evidence that neurodegeneration of dopaminergic pathways of the substantia nigra pars compacta (SNc) in Parkinson's disease involves excitotoxicity (Schmidt et al. 1990; Greenamyre and O'Brien 1991; Blandini and Greenamyre 1998). In rats, NMDA receptor antagonists protect against damage of dopaminergic neurons induced by the dopaminomimetic methamphetamine (Sonsalla et al. 1991). In vitro, MPTP (1-methyl-4-phenyl-1,2,3,6-tetrahydropyridine) application inhibits the astroglial glutamate transporter (Hazell et al. 1997), probably through free radicals. MPTP induces toxicity and Parkinsonian symptoms in rats and monkeys, and this is prevented either by NMDA receptor antagonists or by lesion of the descending cortico-striatal glutamatergic pathway (see Blandini and Greenamyre 1998; Parsons et al. 1998).

It is now widely accepted that NMDA receptor antagonists might manifest their symptomatic anti-Parkinsonian effects by attenuating an imbalance between dopaminergic and glutamatergic pathways within the basal ganglia network (Schmidt and Kretschmer 1997; Danysz et al. 1997). Based on preclinical data, one could speculate that NMDA receptor antagonists (and maybe AMPA receptor antagonists) should provide some degree of neuroprotection in Parkinson's patients (Danysz et al. 1997).

Amantadine has a long history in the symptomatic treatment of Parkinson's disease. Several recent double-blind, placebo-controlled studies have confirmed the impressive acute antidyskinetic effects of amantadine (Rajput et al. 1998; Verhagen Metman et al. 1998; Luginger et al. 2000; Del Dotto et al. 2001; Shoghi-Jadid et al. 2002). One study also indicates that amantadine's antidyskinetic benefit is maintained for at least 1 year (Metman et al. 1999). The related compound memantine was found, as in the MPTP monkey, to have no effect on dyskinesia in a double-blind study; but it did improve parkinsonian symptoms (Merello et al. 1999). This indicates that the antidyskinetic effects of amantadine may be unrelated to NMDA receptor antagonism (Danysz et al. 1997).

4.6 Huntington's Disease

Evidence suggests that overactivation of NMDA receptors contributes to selective degeneration of medium-sized spiny striatal neurons in Huntington's disease. The pattern of neuronal loss in the striatum in Huntington's disease is similar to that obtained after excitotoxic lesions in animals (Schwarcz and Köhler 1983). Striatal neurodegeneration produced by mitochondrial toxins 3-NP and malonate (inhibitors of complex II–III), producing a similar type of damage, also is attenuated by lesions of the glutamatergic inputs, the glutamate release inhibitor lamotrigine and/or NMDA receptor antagonists such as dizocilpine and memantine (Greene and Greenamyre 1995; Schulz et al. 1996; Lee et al. 2000). Hence, it is likely that mitochondrial dysfunction evoked by these toxins triggers a chain of reactions including excitotoxicity. In fact, there are data indicating a deficit of mitochondrial complex II–III activity in the brains of Huntington's patients (Browne et al. 1997). HEK293 cells co-expressing mutant *huntingtin* with polyglutamine expansion (htt-138Q) and either NR1A/NR2A- or NR1A/NR2B showed a significant increase in apoptotic cell death in the presence of NMDA, and this difference was larger for NR1A/NR2B (Zeron et al. 2001).

4.7 Tardive Dyskinesia

According to current concepts, tardive dyskinesia seen after long-term treatment with some neuroleptics involves progressive neuronal damage resulting from excitotoxicity and free radical production (DeKeyser 1991; Cadet and Kahler 1994). At the level of the striatum, chronic blockade of D_2 inhibitory dopaminergic receptors localized on glutamatergic terminals from the cortex may lead to a persistent, enhanced release of glutamate that eventually damages output neurons (DeKeyser 1991; Gunne and Andren 1993).

4.8 Glaucoma

Excitotoxicity has also been implicated in glaucoma. Mild chronic intravitreal elevation in glutamate concentration by serial intravitreal glutamate injections resulted in death of retinal ganglion cells. Concurrent daily injections of memantine completely prevented this cell death (Vorwerk et al. 1996). As such, the memantine follow-up compound neramexane (MRZ 2/579) could also be useful for the treatment of glaucoma and this possibility is presently being tested as a proof concept for the neuroprotective effects of moderate affinity uncompetitive NMDA receptor antagonists in chronic neurodegenerative diseases. Retinal ischaemia induced in rats by elevating intraocular pressure also caused an elevation in the mean vitreous concentration of glutamate and glycine and caused pronounced loss of retinal ganglion cells (Lagreze et al. 1998).

4.9 HIV Dementia

There are a number of indications that glutamate might be involved in some aspects of acquired immunodeficiency syndrome (AIDS)-related neurological deficits (Lipton 1992b). Gp120 (HIV coat protein) produces toxicity in vitro that is attenuated by NMDA receptor antagonists such as dizocilpine (MK-801) and memantine (Lipton 1992a). Neurodegeneration in transgenic mice over expressing gp120 was attenuated by memantine (Toggas et al. 1996). This toxicity is probably secondary to glutamate release from glial cells rather than a direct agonistic effect of Gp120.

4.10
Multiple Sclerosis

In a mouse model of autoimmune encephalomyelitis (EAE) there is a deficit of astroglial enzymes (glutamate dehydrogenase and glutamine synthase) responsible for degradation of glutamate taken up from the extracellular space (HardinPouzet et al. 1997). This may lead to an increase in extracellular glutamate and neurotoxicity seen in this disease. Memantine dose-dependently ameliorated neurological deficits in rodents with EAE. This was not via interactions with the immune system per se and implies that effector mechanisms responsible for reversible neurological deficits in EAE may involve NMDA receptors (Wallstrom et al. 1996). Oligodendrocyte excitotoxicity via overactivation of AMPA and kainate receptors could also be involved in the pathogenesis of such demyelinating disorders (Matute et al. 2001). Brain damage in multiple sclerosis includes both glial activation and pathological changes in axons. The expression patterns of both group I and II mGluRs in multiple sclerosis tissue differed significantly from the patterns in control tissue. Changes in mGluR immunoreactivity were also observed in glia. A diffuse increase in the expression of mGluR5 and mGluR2/3 was detected in reactive astrocytes in multiple sclerosis lesions (Geurts et al. 2003).

4.11
Astroglioma

Implanted astroglioma cells secrete glutamate in vivo, and those showing high glutamate release have a distinct growth advantage in host brain that is not present in vitro. Treatment with dizocilpine or memantine slowed the growth of glutamate-secreting tumours in situ, suggesting that activation of NMDA receptors facilitates tumour expansion (Takano et al. 2001). Glutamate antagonists have also been shown to inhibit proliferation of various human tumour cells in vitro, but the concentrations required were very high and probably not of therapeutic relevance (Rzeski et al. 2001). This suggests that astroglioma cells secrete glutamate to destroy healthy surrounding tissue and thereby make space for tumour expansion. These findings support a new approach for therapy of brain tumours, based upon antagonizing glutamate secretion or its target receptors.

4.12
Epilepsy

Although epilepsy was one of the first suggested therapeutic applications of NMDA receptor antagonists (Czuczwar and Meldrum 1982; Meldrum 1985) only few such agents reached clinical testing for this indication and these failed to show sufficient benefits and produced serious side-effects (Troupin et al. 1986; Leppik et al. 1988; Sveinbjornsdottir et al. 1993).

Preclinical data indicate that NMDA receptor antagonists will probably not be useful as a monotherapy in epilepsy. However, the observation that NMDA receptor antagonists greatly enhance the anticonvulsive potency of AMPA receptor antagonists (Löscher et al. 1993) and conventional anti-epileptic drugs (Czuczwar et al. 1996), thereby allowing significant dose reductions, opens new avenues in this regard. Drugs that interact with more than one anticonvulsant target may show synergistic anticonvulsant actions but may not have increased toxicity. Indeed, highly effective, broad-spectrum antiepileptic drugs such as felbamate and topiramate, may act through such multiple mechanisms.

Transient stimulation of group I mGluRs induces persistent prolonged epileptiform discharges via an activation of ERK1/2 in the hippocampus (Zhao et al. 2004). Group I mGluR antagonists show anticonvulsant efficacy against pentylenetetrazole-induced seizures, kindling, and kindling-related learning deficits. It was concluded that mGluR1 and mGluR5 play a specific role in the convulsive component of kindling and that the beneficial action of the antagonists on kindling-induced impairments in shuttle-box learning may be associated with their effect on glutamatergic synaptic activity (Nagaraja et al. 2004). Mice lacking mGluR7 exhibit an increased susceptibility to seizures, suggesting a preponderance of excitation (Sansing et al. 2001).

4.13 Chronic Pain

Despite intensive research on the neurobiological mechanisms of chronic pain, this therapeutic area remains one of the least satisfactorily covered by current drugs. There is considerable preclinical evidence that hyperalgesia and allodynia following peripheral tissue or nerve injury is not only due to an increase in the sensitivity of primary afferent nociceptors at the site of injury but also depends on NMDA receptor-mediated central changes in synaptic excitability (Zieglgänsberger and Tolle 1993; Sandkühler and Liu 1998; Eide 2000; Parsons 2001; Fundytus 2001).

The uncompetitive NMDA receptor antagonist ketamine has been available for clinical use as an anaesthetic for 40 years (Domino et al. 1965). Ketamine is effective in various animal models of hyperalgesia and allodynia and has been reported to have antinociceptive effects in some of these models at doses devoid of obvious side-effects. Others, however, have reported that the effects of ketamine are only seen at doses producing ataxia (see Parsons 2001 for review). Ketamine reportedly inhibits the area of secondary hyperalgesia induced by chemical (Park et al. 1995) or thermal stimuli (Ilkjaer et al. 1996; Warncke et al. 1997) and inhibits temporal summation of repeated mechanical (Warncke et al. 1997) and electrical stimuli (Arendtnielsen et al. 1995; Andersen et al.

1996). There is evidence for tonic NMDA receptor activation in inflammatory hyperalgesia (Boxall et al. 1998) and in the mechanisms underlying the reduced effectiveness of opioids in chronic neuropathic pain states (Mao et al. 1995; Cai et al. 1997; Fan et al. 1998; Yashpal et al. 2001). Several controlled trials in patients with peripheral neuropathic pain have shown positive effects of acute injections of ketamine on spontaneous ongoing pain (Eide et al. 1994; Max et al. 1995; Nikolasjen et al. 1996; Felsby et al. 1996) central neuropathic pain (Eide et al. 1995b), fibromyalgia (Pud et al. 1998), and chronic ischaemic pain (Persson et al. 1998). After pre- and peri-operative treatment with low-dose ketamine, postoperative pain relief has been evidenced by reduced morphine consumption, but the intensity of ongoing post-operative pain was less affected (Roytblat et al. 1993; Ngan Kee et al. 1997; Fu et al. 1997). Oral treatments with NMDA receptor antagonists have thus far been disappointing and often associated with side-effects (e.g. Haines and Gaines 1999). Nevertheless, there is great potential in the combination of therapeutically safe NMDA receptor antagonists with opioids in the treatment of chronic pain (Elliott et al. 1995; Eide et al. 1995a; Schmid et al. 1999).

Memantine also blocks and reverses thermal hyperalgesia, mechanical allodynia in rat models of painful mononeuropathy without obvious effects on motor reflexes following systemic (Carlton and Hargett 1995; Eisenberg et al. 1993, 1995; Suzuki et al. 2001) and local spinal administration (Chaplan et al. 1997). Although the hypothesis underlying the ability of this moderate affinity open channel blocker to differentiate between phasic physiological and tonic excitotoxic pathological activation of NMDA receptors has gained relatively wide acceptance (Mealing et al. 1997; Kornhuber and Weller 1997; Parsons et al. 1999) it is still unclear how such compounds could differentiate between normal and abnormal synaptic activation of NMDA receptors.

Another promising target for NMDA receptor antagonism is the glycine$_B$ modulatory site (Danysz and Parsons 1998). Indeed, systemically active glycine$_B$ antagonists have good therapeutic indices following systemic administration in models of hyperalgesia and allodynia (see Parsons 2001 for references; plus Quartaroli et al. 2001).

The NR2B-selective agent traxoprodil has also been reported to be effective in suppressing hyperalgesia in animal models of chronic pain (carrageenan, capsaicin and allodynia in neuropathic rats) at doses devoid of negative side-effects in motor co-ordination or behaviour (Taniguchi et al. 1997; Boyce et al. 1999). A good separation was also reported for (+/−)-Ro 25–6981 (Boyce et al. 1999), indicating that NR2B-selective antagonists may also have clinical utility for the treatment of neuropathic and other pain conditions in man with a reduced side-effect profile (Chizh et al. 2001).

Recent data also indicate that peripheral NMDA receptors are involved in inflammatory somatic and visceral pain (Leem et al. 2001). Peripheral glutamate receptors are associated with unmyelinated axons (Carlton et al. 1995), and the number of somatic sensory axons containing ionotropic glutamate receptors increases during peripheral sensitization due to inflammation (Carlton and Coggeshall 1999; Coggeshall and Carlton 1999). Immunohistochemical studies indicate that NR1 subunits are expressed on the cell bodies and peripheral terminals of primary afferent nerves innervating the colon and may provide a novel mechanism for development of peripheral sensitization and visceral hyperalgesia (McRoberts et al. 2001).

Evidence from experimental pain research has revealed that mGluRs play a pivotal role in nociceptive processing, inflammatory pain and hyperalgesia. mGluRs have been implicated in dorsal horn neuronal nociceptive responses and pain associated with short-term inflammation (Neugebauer 2002) as well as its emotional component involving limbic structures such as the amygdala (Han et al. 2004).

Peripheral group II mGluRs reduce inflammation-induced mechanical allodynia and may mediate endogenous anti-allodynia effects, which speed recovery from inflammation-induced hypersensitivity (Yang and Gereau 2003). mGluR subtypes are differentially expressed in spinal cord dorsal horn in response to persistent inflammation (Dolan et al. 2003). Recent findings suggest that the induction of LTP in the spinal dorsal horn by high-frequency, high-intensity stimulation of afferent C fibres requires the activation of mGluR1/5 (Azeku et al. 2003). Antisense oligonucleotide knockdown of spinal mGluR(1) attenuates thermal hyperalgesia and mechanical allodynia in rats injected with CFA in one hindpaw, suggesting a role for mGluR(1) in persistent inflammatory nociception (Fundytus et al. 2002). Group I mGluR antagonism, and Group II or III mGluR agonism, can effectively decrease the development of mechanical and cold hypersensitivity associated with chronic constriction injury (CCI) in rats. The results can be interpreted to suggest that activation of spinal group I mGluRs contributes to spinal plasticity leading to the development of neuropathic pain, and that this effect is offset by activation of groups II and III mGluRs (Fisher et al. 2002).

4.14 Addiction

It is believed that phenomena such as sensitization, tolerance and drug-dependence might also involve synaptic plasticity. In fact, numerous studies indicate that NMDA receptor antagonists block sensitization to amphetamine and cocaine as well as tolerance and dependence to ethanol and opioids in animal models (Trujillo and Akil 1991; Pasternak and Inturrisi 1995; Trujillo and Akil 1995; Mao 1999). Recent studies indicate that the uncompetitive NMDA receptor antagonists dextromethorphan, memantine and neramexane not only prevent the development of morphine tolerance, but also reverse established tolerance in the continuing presence of this opioid, prevent the expression of withdrawal symptoms in rats (Popik and Skolnick 1996; Popik and Danysz 1997; Popik and Kozela 1999; Houghton et al. 2001) and attenuate the expression of opioid physical dependence in morphine-dependent humans elicited by administration of naloxone (Bisaga et al. 2001). Likewise, systemically active glycine$_B$ antagonists attenuate both physical dependence to morphine and the development of tolerance to the antinociceptive effects of opioids following repeated administration (Pasternak and Inturrisi 1995; Popik et al. 1998; Belozertseva et al. 2000a; Belozertseva et al. 2000b; for review see Danysz and Parsons 1998).

Under certain conditions NMDA receptor antagonists (e.g. dizocilpine) inhibit sensitization (locomotor response) to repetitive administration of cocaine (Carey et al. 1995; see Karler and Calder 1992; Sripada et al. 2001). The sensitization to cocaine in rats seems to be related to an increase in AMPA receptor sensitivity in the nucleus accumbens and an increased glutamate release in response to cocaine challenge (Pierce et al. 1996). NMDA receptor antagonists block the acquisition of cocaine-induced place preference, but not its expression (Cervo and Samanin 1995). Also, established self-administration of cocaine, and its development in rats is decreased by dextromethorphan (Pulvirenti et al. 1992; Ranaldi et al. 1997), probably through blockade of NMDA receptors in the nucleus accumbens (Pulvirenti et al. 1992). A recent study identified group II mGluRs as a pharmacotherapeutic target for craving and relapse prevention associated with cocaine cue exposure (Baptista et al. 2004).

In the nucleus accumbens, a key structure for the effects of all addictive drugs, presynaptic cannabinoid CB_1 receptors and postsynaptic mGluRs play a pivotal role in LTD (Marsicano et al. 2002; Azad et al. 2004). Also, in the CA1 region of the hippocampus activation of postsynaptic group I metabotropic glutamate receptors evoked LTD and the release of endocannabinoids from pyramidal cells. Since the effect of the mGluR agonist was still present in the *CB_1* knockout mouse, it is suggested that endocannabinoids, acting on a non-CB_1 cannabinoid receptor, contribute to the depression of mGluR induced LTD (Rouach and Nicoll 2003).

A single in vivo cocaine administration abolishes endocannabinoid-dependent LTD. This effect of cocaine was not present in mice lacking D_1 dopamine receptors and was blocked by a selective D_1 receptor and NMDA receptor antagonist, suggesting the involvement of D_1 and NMDA receptors (Fourgeaud et al. 2004).

Ethanol is an NMDA receptor antagonists (e.g. Mirshahi and Woodward 1995) and the affinity is within the range seen in the brains of alcohol abusers. It has been suggested that the effects of ethanol may be related to selective actions at NR2B receptors (Yang et al. 1996; Lovinger and Zieglgansberger 1996; Follesa and Ticku 1996; Smothers et al. 2001).

A post-mortem study on the brains of alcoholics showed a modest increased binding for [3H]glutamate and [3H]CGP-39653—a competitive NMDA receptor antagonist (Freund and Anderson 1996). In humans with a history of alcohol abuse, an increase in immunoreactivity toward AMPA GluR2 and GluR3 subunits was also found (Breese et al. 1995). In rodents lacking functional corticotropin-releasing hormone (CRH)1 receptors, the delayed and enhanced stress-induced alcohol drinking was associated with an upregulation of NR2B subunits (Sillaber et al. 2002).

Blockade of NMDA receptors inhibits some aspects of alcohol dependence. Memantine given before ethanol administration prevented the development of ethanol dependence induced by intragastric administration of ethanol, supporting the notion that NMDA receptors are involved in the development of ethanol dependence (Kotlinska 2001). In an alcohol 'craving' model, memantine and neramexane infused s.c. failed to change alcohol intake under normal conditions, but completely inhibited alcohol consumption during the relapse phase (Holter et al. 1996; Holter et al. 2000). There are several possibilities how NMDA receptor antagonists exert this effect: (1) produce alcohol-like effects—memantine shows partial generalization to the ethanol cue in rats trained to discriminate ethanol (Bienkowski et al. 1997; Hundt et al. 1998); (2) block recognition of the alcohol cue; (3) inhibit association of environmental cues with alcohol use; or, (4) block the reinforcing action of ethanol. Neramexane is currently in phase II clinical trials for alcohol dependence.

Acamprosate has been approved as an anti-craving drug in most European countries for several years and most recently also in the USA. Acamprosate and the opioid receptor antagonist naltrexone represent the first agents to show positive results in properly conducted clinical trials (Littleton and Zieglgänsberger 2003). The pharmacology of acamprosate is still poorly understood. It has been reported that acamprosate modulates NMDA receptor function after binding to a spermidine-sensitive site (Naassila et al. 1998). Unexpectedly, in a recent patch clamp study, acamprosate reversed polyamine potentiation on NMDA- or glutamate-induced currents only in a subset of cultured neurons (Popp and Lovinger 2000). Studies performed in cultured hippocampal neurons and in *Xenopus* oocytes or HEK-293 cells expressing NR1a/2A and NR1a/2B receptor assemblies were also unable to show any interaction of acamprosate with the polyamine site or influence on agonist affinity (Rammes et al. 2001). However, in this same study, acamprosate produced similar increases in NR1 and NR2B receptor expression in vivo to those seen following acute treatment with dizocilpine or memantine, indicating that acamprosate may produce changes in the CNS that are similar to those seen following NMDA receptor antagonists, and these changed may, in turn, underlie the effects of both kinds of drugs in the treatment of alcohol abuse.

4.15 Anxiety

Various compounds that decrease glutamatergic transmission via blockade of NMDA or group I mGlu receptors produce anxiolytic- and antidepressant-like actions in animal tests and models. Anxiolytic activity resulting from NMDA receptor antagonism was reported as early as 1986 (Bennett and Amrick

1986; Stephens et al. 1986). Later, anxiolytic activity of uncompetitive and competitive NMDA receptor antagonists have been shown in the conflict test, social interaction test, elevated plus maze, separation-induced vocalization in rat pups and by blockade of fear potentiated startle (see Danysz and parsons 1998). Similarly, partial agonists of the glycine$_B$ site have been shown to posses anxiolytic potential. However, this effect is not seen in all animal models and is not related to intrinsic activity of these agents (Karcz-Kubicha et al. 1997). Moreover, full antagonists of the glycine$_B$ site showing good penetration to the brain (Licostinel; MRZ 2/576, L-701,324) failed to show a consistent anxiolytic profile (Wiley et al. 1995; Karcz-Kubicha et al. 1997). Moreover, there is no good reason to believe that NMDA receptor antagonists will be better than classical anxiolytic agents such as benzodiazepines.

More promising preclinical data have been generated with the competitive metabotropic mGluR2 agonist LY-354740 (Chojnacka-Wojcik et al. 2001). This compound seems to cause only mild sedation in mice, does not disturb motor coordination and has no potential to cause dependence. Therefore, similar ligands or positive modulators of mGluR2 receptors may become the anxiolytics of the future, free from the side-effects characteristic of benzodiazepine. However, this enthusiasm might be dampened by problems with ADME (absorption, distribution, metabolism, excretion). Recently, structure–activity relationship studies led to the discovery of a new, orally active mGlu5 receptor antagonist with anxiolytic activity (Roppe et al. 2004). In addition, highly potent orally active group II and group III mGluR agonists have proved to effective in animal models for anxiety and psychosis (Palucha et al. 2004; Collado et al. 2004). Metabotropic glutamate receptors (mGluRs) have also been implicated in regulating anxiety, stress responses and the neurobehavioural effects of psychostimulants, suggesting group II mGluRs as a pharmacotherapeutic target for craving and relapse prevention associated with cocaine cue exposure (Baptista et al. 2004).

4.16 Schizophrenia

Several lines of evidence have implicated NMDA receptor hypofunction in the pathophysiology of schizophrenia. The administration of certain, but not all, uncompetitive NMDA receptor antagonists exacerbates psychotic symptoms in schizophrenics and mimics schizophrenia in non-psychotic subjects (Coyle et al. 2003; Konradi and Heckers 2003).

Recent studies have identified abnormalities associated with schizophrenia that interfere with the activation of the glycine modulatory site of the NMDA receptor (Coyle and Tsai 2004). Further, the use of NMDA receptor glycine site agonists such as glycine, D-serine or D-cycloserine in clinical trials has demonstrated some efficacy in ameliorating the negative symptoms and cognitive disabilities in schizophrenics (Coyle and Tsai 2004).

The mGluRs are highly enriched in PFC—a brain region critically involved in the regulation of cognition and emotion. A disturbance of glutamatergic transmission has been suggested to contribute to the development of schizophrenic pathophysiology based primarily on the ability of ionotropic glutamate receptor antagonists to induce schizophrenic-like symptoms. Emerging evidence suggests that mGluRs are viable drug targets for neuropsychiatric disorders associated with reduced glutamatergic function in the PFC (Marek et al. 2000). Group II mGluR agonists have been reported to reduce the behavioural and neurochemical effects of phencyclidine (PCP) administration (Olszewski et al. 2004). PCP administration elicits positive and negative symptoms that resemble those of schizophrenia and is widely accepted as a model for the study of this human disorder.

4.17 Depression

In humans, the antidepressant activity of NMDA receptor antagonists has not been evaluated extensively (Skolnick 1999). In animal models of depression, NMDA receptor antagonists have been reported to exert positive effects in most studies (Trullas 1997). This concerns mainly the forced swim test (Maj 1992; Moryl et al. 1993; Przegalinski et al. 1997) and stress-induced anhedonia (Papp and Moryl 1994). Amantadine but not memantine was effective against reserpine-induced hypothermia (Moryl et al. 1993). In the forced swim test, both amino-adamantanes produced specific antidepressive-like activity (Moryl et al. 1993).

Possible synergistic interactions between classical antidepressants and uncompetitive NMDA receptor antagonists in the forced swim test were recently investigated. Fluoxetine, which was inactive when given alone, showed a positive effect when combined with amantadine, memantine or neramexane, suggesting that the combination of traditional antidepressant drugs and NMDA receptor antagonists may produce enhanced antidepressive effects (Rogoz et al. 2002). This was proposed to be of particular relevance for antidepressant-resistant patients. In clinical trials the glutamate release inhibitor lamotrigine has also been shown to have antidepressant activity (Nikolasjen et al. 1996; Sporn and Sachs 1997; Eide 2000). In summary, clinical evidence supporting the antidepressant efficacy of NMDA antagonists is scarce. However, if effective, NMDA antagonists may show unique rapidity of clinical efficacy. Thus, further evaluation of this drug class in the treatment of depression would seem to be reasonable.

Compounds that decrease glutamatergic transmission via blockade of group I mGluRs produce anxiolytic- and antidepressant-like action in animal tests and models. In recent studies, group II and III mGluR-agonists that reduce glutamate release have been suggested to play a role in the therapy of both anxiety and depression (Palucha et al. 2004; Chaki et al. 2004). CB_1 receptor activation reduces both the release and uptake of glutamate in vitro; and both mechanisms are dependent on mGluR activation (Brown et al. 2003b). The relevance of these findings for the anxiolytic properties of cannabis remains to be shown.

5 Summary

At present there seems to be a consensus that competitive AMPA and NMDA receptor antagonists a low chance of finding therapeutic applications. Antagonists showing moderate affinity and satisfactory selectivity for certain NMDA receptor subtypes seem to have a more favourable profile. From the therapeutic point of view, the real challenge is not only to improve the symptoms of diseases, but also to interfere with their pathomechanism, i.e. prevent progression. The most promising symptomatic indications for NMDA receptor antagonists seem to be various forms of dementia, alcohol abuse, and possibly some forms of chronic pain such as phantom pain and postoperative pain, in particularly in combination with opioids. Clinical evidence for neuroprotective activity of glutamate antagonists in chronic neurodegenerative diseases is scarce. Results with NMDA receptor antagonists in clinical trials of stroke and trauma have been very disappointing. Some degree of hope remains for NR2B-selective NMDA receptor antagonists, AMPA and/or kainate receptor antagonists and combinations thereof with, for example, additional blockade of voltage-activated channels.

In terms of clinical proof of neuroprotective effects in chronic neurodegenerative diseases, so far there are promising clinical results in ALS only with riluzole, and even then, the increase of survival obtained was only modest. The failure of remacemide in a recent study in Huntington's disease was clearly a big setback. On the other hand, the moderate-affinity NMDA receptor antagonist memantine provides clear symptomatic improvement in dementia in both clinical and preclinical situations, and the preclinical data predict neuroprotective effects, substantiated by numerous animal models.

AMPA receptor antagonists acting at the 2,3-benzodiazepine modulatory site seem to have a better safety profile than competitive agents such as NBQX, probably due to their better solubility and associated reduced side-effects such as renal toxicity. It is still not clear for which indications they might be useful, although their effects in animal models of acute and chronic neurodegenerative diseases look quite promising.

Some of the modulators of glutamate function indeed still have a promising future as therapeutics for numerous CNS diseases. The importance of glutamate receptors in both physiological and pathological processes emphasizes the need to develop agents that selectively modulate the latter (Parsons et al. 1999; Hardingham and Bading 2003).

References

Abraham WC, Tate WP (1997) Metaplasticity: a new vista across the field of synaptic plasticity. Prog Neurobiol 52:303–323

Aiba A, Chen C, Herrup K, et al (1994) Reduced hippocampal long-term potentiation and context-specific deficit in associative learning in mGluR1 mutant mice. Cell 79:365–375

Amara SG, Fontana ACK (2002) Excitatory amino acid transporters: keeping up with glutamate. Neurochem Int 41:313–318

Andersen OK, Felsby S, Nicolaisen L, et al (1996) The effect of ketamine on stimulation of primary and secondary hyperalgesic areas induced by capsaicin—a double-blind, placebo-controlled, human experimental study. Pain 66:51–62

Ango F, Pin JP, Tu JC, et al (2000) Dendritic and axonal targeting of type 5 metabotropic glutamate receptor is regulated by Homer1 proteins and neuronal excitation. J Neurosci 20:8710–8716

Arendtnielsen L, Petersenfelix S, Fischer M, et al (1995) The effect of N-methyl-D-aspartate antagonist (ketamine) on single and repeated nociceptive stimuli: a placebo-controlled experimental human study. Anesth Analg 81:63–68

Areosa SA, Sherriff F (2003) Memantine for dementia (Cochrane Review). Cochrane Database Syst Rev 1:CD003154

Arriza JL, Fairman WA, Wadiche JI, et al (1994) Functional comparisons of three glutamate transporter subtypes cloned from human motor cortex. J Neurosci 14:5559–5569

Artola A, Singer W (1993) Long-term depression of excitatory synaptic transmission and its relationship to long-term potentiation. Trends Neurosci 16:480–487

Azad SC, Monory K, Marsicano G, Cravatt BF, Lutz B, Zieglgänsberger W, Rammes G (2004) Circuitry for associative plasticity in the amygdala involves endocannabinoid signaling. J Neurosci (in press)

Azkue JJ, Liu XG, Zimmermann M, Sandkühler J (2003) Induction of long-term potentiation of C fibre-evoked spinal field potentials requires recruitment of group I, but not group II/III metabotropic glutamate receptors. Pain 106:373–379

Bannerman DM, Good MA, Butcher SP, et al (1995) Distinct components of spatial learning revealed by prior training and NMDA receptor blockade. Nature 37:182–186

Bao WL, Williams AJ, Faden AI, Tortella FC (2001) Selective mGluR5 receptor antagonist or agonist provides neuroprotection in a rat model of focal cerebral ischemia. Brain Res 922:173–179

Baptista MAS, Martin-Fardon R, Weiss F (2004) Preferential effects of the metabotropic glutamate 2/3 receptor agonist LY379268 on conditioned reinstatement versus primary reinforcement: comparison between cocaine and a potent conventional reinforcer. J Neurosci 24:4723–4727 2004

Bayer KU, De Koninck P, Leonard AS, et al (2001) Interaction with the NMDA receptor locks CaMKII in an active conformation. Nature 411:801–805

Beal MF (1995) Aging, energy, and oxidative stress in neurodegenerative diseases. Ann Neurol 38:357–366

Belozertseva IV, Danysz W, Bespalov AY (2000a) Effects of short-acting NMDA receptor antagonist MRZ 2/576 on morphine tolerance development in mice. Naunyn Schmiedebergs Arch Pharmacol 361:573–577

Belozertseva IV, Danysz W, Bespalov AY (2000b) Short-acting NMDA receptor antagonist MRZ 2/576 produces prolonged suppression of morphine withdrawal in mice. Naunyn Schmiedebergs Arch Pharmacol 361:279–282

Benke TA, Luthi A, Isaac JTR, Collingridge GL (1998) Modulation of AMPA receptor unitary conductance by synaptic activity. Nature 393:793–797

Bennett DA, Amrick CL (1986) 2-Amino-7-phosphoheptanoic acid (AP7) produces discriminative stimuli and anticonflict effects similar to diazepam. Life Sci 39:2455–2461

Benowitz LI, Routtenberg A (1997) GAP-43: an intrinsic determinant of neuronal development and plasticity. Trends Neurosci 20:84–91

Benveniste H, Drejer J, Schusboe A, Diemer NH (1984) Elevation of the extracellular concentrations of glutamate and aspartate in rat hippocampus during transient cerebral ischemia monitored by intracerebral microdialysis. J Neurochem 43:1369–1374

Bienkowski P, Stefanski R, Kostowski W (1997) Discriminative stimulus effects of ethanol: lack of antagonism with N-methyl-D-aspartate and D-cycloserine. Alcohol 14:345–350

Bisaga A, Comer SD, Ward AS, et al (2001) The NMDA antagonist memantine attenuates the expression of opioid physical dependence in humans. Psychopharmacology (Berl) 157:1–10

Blandini F, Greenamyre JT (1998) Prospects of glutamate antagonists in the therapy of Parkinson's disease. Fundam Clin Pharmacol 12:4–12

Bleakman D, Ballyk BA, Schoepp DD, et al (1996) Activity of 2,3-benzodiazepines at native rat and recombinant human glutamate receptors in vitro: Stereospecificity and selectivity profiles. Neuropharmacology 35:1689–1702

Blin O, Azulay JP, Desnuelle C, et al (1996) A controlled one-year trial of dextromethorphan in amyotrophic lateral sclerosis. Clin Neuropharmacol 19:189–192

Bowie D, Mayer ML (1995) Inward rectification of both AMPA and kainate subtype glutamate receptors generated by polyamine-mediated ion channel block. Neuron 15:453–462

Boxall SJ, Berthele A, Tölle TR, et al (1998) mGluR activation reveals a tonic NMDA component in inflammatory hyperalgesia. Neuroreport 9:1201–1203

Boyce S, Wyatt A, Webb JK, et al (1999) Selective NMDA NR2B antagonists induce antinociception without motor dysfunction: correlation with restricted localisation of NR2B subunit in dorsal horn. Neuropharmacology 38:611–623

Bredt DS, Nicoll RA (2003) AMPA receptor trafficking at excitatory synapses. Neuron 40:361–379

Breese CR, Freedman R, Leonard SS (1995) Glutamate receptor subtype expression in human postmortem brain tissue from schizophrenics and alcohol abusers. Brain Res 674:82–90

Bristow LJ, Flatman KL, Hutson PH, et al (1996) The atypical neuroleptic profile of the glycine/N-methyl-D-aspartate receptor antagonist, l-701,324, in rodents. J Pharmacol Exp Ther 277:578–585

Brown QB, Baude AS, Gilling K, et al (2003a) Memantine and neramexane protect against semi-chronic 3-NP toxicity in organotypic hippocampal cultures. Soc Neurosci Abstr 29

Brown TM, Brotchie JM, Fitzjohn SM (2003b) Cannabinoids decrease corticostriatal synaptic transmission via an effect on glutamate uptake. J Neurosci 23:11073–11077

Browne SE, Bowling AC, MacGarvey U, et al (1997) Oxidative damage and metabolic dysfunction in Huntington's disease: selective vulnerability of the basal ganglia. Ann Neurol 41:646–653

Bruno V, Battaglia G, Copani A, et al (2001) Metabotropic glutamate receptor subtypes as targets for neuroprotective drugs. J Cereb Blood Flow Metab 21:1013–1033

Bryson HM, Fulton B, Benfield P (1996) Riluzole—a review of its pharmacodynamic and pharmacokinetic properties and therapeutic potential in amyotrophic lateral sclerosis. Drugs 52:549–563

Bullock MR, Lyeth BG, Muizelaar JP (1999) Current status of neuroprotection trials for traumatic brain injury: lessons from animal models and clinical studies. Neurosurgery 45:207–217

Bullock R, Zauner A, Woodward J, Young HF (1995) Massive persistent release of excitatory amino acids following human occlusive stroke. Stroke 26:2187–2189

Burnashev N (1996) Calcium permeability of glutamate-gated channels in the central nervous system. Curr Opin Neurobiol 6:311–317

Cadet JL, Kahler LA (1994) Free radical mechanisms in schizophrenia and tardive dyskinesia. Neurosci Biobehav Rev 18:457–467

Cai YC, Ma L, Fan GH, et al (1997) Activation of N-methyl-D-aspartate receptor attenuates acute responsiveness of delta-opioid receptors. Mol Pharmacol 51:583–587

Cameron HA, McEwen BS, Gould E (1995) Regulation of adult neurogenesis by excitatory input and NMDA receptor activation in the dentate gyrus. J Neurosci 15:4687–4692

Cameron HA, Tanapat P, Gould E (1998) Adrenal steroids and N-methyl-D-aspartate receptor activation regulate neurogenesis in the dentate gyrus of adult rats through a common pathway. Neuroscience 82:349–354

Canton T, Bohme GA, Boireau A, et al (2001) RPR 119990, a novel alpha-amino-3-hydroxy-5-methyl-4-isoxazolepropionic acid antagonist: synthesis, pharmacological properties, and activity in an animal model of amyotrophic lateral sclerosis. J Pharmacol Exp Ther 299:314–322

Carey RJ, Dai HL, Krost M, Huston JP (1995) The NMDA receptor and cocaine: evidence that MK-801 can induce behavioral sensitization effects. Pharmacol Biochem Behav 51:901–908

Carlton SM, Coggeshall RE (1999) Inflammation-induced changes in peripheral glutamate receptor populations. Brain Res 820:63–70

Carlton SM, Hargett GL (1995) Treatment with the NMDA antagonist memantine attenuates nociceptive responses to mechanical stimulation in neuropathic rats. Neurosci Lett 198:115–118

Carlton SM, Hargett GL, Coggeshall RE (1995) Localization and activation of glutamate receptors in unmyelinated axons of rat glabrous skin. Neurosci Lett 197:25–28

Carroll RC, Zukin RS (2002) NMDA-receptor trafficking and targeting: implications for synaptic transmission and plasticity. Trends Neurosci 25:571–577

Cervo L, Samanin R (1995) Effects of dopaminergic and glutamatergic receptor antagonists on the acquisition and expression of cocaine conditioning place preference. Brain Res 673:242–250

Chaki S, Yoshikawa R, Hirota S, et al (2004) MGS0039: a potent and selective group II metabotropic glutamate receptor antagonist with antidepressant-like activity. Neuropharmacology 46:457–467

Chaplan SR, Malmberg AB, Yaksh TL (1997) Efficacy of spinal NMDA receptor antagonism in formalin hyperalgesia and nerve injury evoked allodynia in the rat. J Pharmacol Exp Ther 280:829–838

Charpak S, Gähwiler BH, Do KQ, Knöpfel T (1990) Potassium conductances in hippocampal neurons blocked by excitatory aminoacid transmitters. Nature 347:765–767

Chatterton JE, Awobuluyi M, Premkumar LS, et al (2002) Excitatory glycine receptors containing the NR3 family of NMDA receptor subunits. Nature 415:793–798

Chen HX, Roper SN (2004) Tonic activity of metabotropic glutamate receptors is involved in developmental modification of short-term plasticity in the neocortex. J Neurophysiol 92:838–844

Chizh BA, Headley PM, Tzschentke TM (2001) NMDA receptor antagonists as analgesics: focus on the NR2B subtype. Trends Pharmacol Sci 22:636–642

Choi DW (1995) Calcium: still center-stage in hypoxic-ischemic neuronal death. Trends Neurosci 18:58–60

Chojnacka-Wojcik E, Klodzinska A, Pilc A (2001) Glutamate receptor ligands as anxiolytics. Curr Opin Investig Drugs 2:1112–1119
Christie BR, Kerr DS, Abraham WC (1994) Flip side of synaptic plasticity: long-term depression mechanisms in the hippocampus. Hippocampus 4:127–135
Coggeshall RE, Carlton SM (1999) Evidence for an inflammation-induced change in the local glutamatergic regulation of postganglionic sympathetic efferents. Pain 83:163–168
Collado I, Pedregal C, Bueno AB, et al (2004) (2S,1′S,2′R,3′R)-2-(2′-carboxy-3′-hydroxymethylcyclopropyl) glycine is a highly potent group II and III metabotropic glutamate receptor agonist with oral activity. J Med Chem 47:456–466
Collingridge GL, Bliss TVP (1995) Memories of NMDA receptors and LTP. Trends Neurosci 18:54–56
Contractor A, Swanson G, Heinemann SF (2001) Kainate receptors are involved in short- and long-term plasticity at mossy fiber synapses in the hippocampus. Neuron 29:209–216
Coyle JT, Tsai G (2004) NMDA receptor function, neuroplasticity, and the pathophysiology of schizophrenia. Int Rev Neurobiol 59:491–515
Coyle JT, Tsai G, Goff D (2003) Converging evidence of NMDA receptor hypofunction in the pathophysiology of schizophrenia. Ann NY Acad Sci 1003:318–327
Cummings JA, Mulkey RM, Nicoll RA, Malenka RC (1996) Ca(2+) signaling requirements for long-term depression in the hippocampus. Neuron 16:825–833
Czuczwar SJ, Meldrum BS (1982) Protection against chemically induced seizures by 2-amino-7-phosphonoheptanoic acd. Eur J Pharmacol 83:335–338
Czuczwar SJ, Turski WA, Kleinrok Z (1996) Interactions of excitatory amino acid antagonists with conventional antiepileptic drugs. Metab Brain Dis 11:143–152
Danysz W, Parsons CG (1995) NMDA receptor antagonists—multiple modes of action on learning processes. Behav Pharmacol 6:619
Danysz W, Parsons CG (1998) Glycine and N-methyl-D-aspartate receptors: Physiological significance and possible therapeutic applications. Pharmacol Rev 50:597–664
Danysz W, Parsons CG, Bresink I, Quack G (1995) Glutamate in CNS disorders—a revived target for drug development. Drug News Perspect 8:261–277
Danysz W, Parsons CG, Kornhuber J, et al (1997) Aminoadamantanes as NMDA receptor antagonists and antiparkinsonian agents—preclinical studies. Neurosci Biobehav Rev 21:455–468
Danysz W, Parsons CG, Möbius HJ, et al (2000) Neuroprotective and symptomatological action of memantine relevant for Alzheimer's disease—an unified glutamatergic hypothesis on the mechanism of action. Neurotox Res 2:85–97
Davis SM, Lees KR, Albers GW, et al (2000) Selfotel in acute ischemic stroke: possible neurotoxic effects of an NMDA antagonist. Stroke 31:347–354
DeKeyser J (1991) Excitotoxic mechanisms may be involved in the pathophysiology of tardive dyskinesia. Clin Neuropharmacol 14:562–565
Del Dotto P, Pavese N, Gambaccini G, et al (2001) Intravenous amantadine improves levodopa-induced dyskinesias: an acute double-blind placebo-controlled study. Mov Disord 16:515–520
Dempsey RJ, Baskaya MK, Dogan A (2000) Attenuation of brain edema, blood-brain barrier breakdown, and injury volume by ifenprodil, a polyamine-site N-methyl-D-aspartate receptor antagonist, after experimental traumatic brain injury in rats. Neurosurgery 47:399–404
Deng WB, Wang H, Rosenberg PA, et al (2004) Role of metabotropic glutamate receptors in oligodendrocyte excitotoxicity and oxidative stress. Proc Natl Acad Sci USA 101:7751–7756

Derrick BE, Martinez Jr JL (1996) Associative, bidirectional modifications at the hippocampal mossy fibre-CA3 synapse. Nature 381:429–434
Dickenson A (1997) Mechanisms of central hypersensitivity: excitatory amino acid mechanisms and their control. In: Dickenson A, Besson JM (eds) Pharmacology of pain. Springer-Verlag, Berlin, pp 167–210
Doble A (1996) The pharmacology and mechanism of action of riluzole. Neurology 47:S233–S241
Dodt H-U, Eder M, Frick A, Zieglgänsberger W (1999) Precisely localized LTD in the neocortex revealed by infrared-guided laser stimulation. Science 286:110–113
Dodt H-U, Eder M, Schierloh A, Zieglgänsberger W (2002) Infrared-guided laser stimulation of neurons in brain slices. Science STKE 120:PL2
Dolan S, Kelly JG, Monteiro AM, Nolan AM (2003) Up-regulation of metabotropic glutamate receptor subtypes 3 and 5 in spinal cord in a clinical model of persistent inflammation and hyperalgesia. Pain 106:501–512
Domino EF, Chodoff P, Corssen G (1965) Pharmacologic effects of CI-581, a new dissociative anesthetic, in man. Clin Pharmacol Ther 6:279–291
Doraiswamy PM (2002) Non-cholinergic strategies for treating and preventing Alzheimer's disease. CNS Drugs 16:811–824
Durand GM, Kovalchuk Y, Konnerth A (1996) Long-term potentiation and functional synapse induction in developing hippocampus. Nature 381:71–75
Eder M, Becker K, Rammes G, et al (2003) Distribution and properties of functional postsynaptic kainate receptors on neocortical layer V pyramidal neurons. J Neurosci 23:6660–6670
Eide PK (2000) Wind-up and the NMDA receptor complex from a clinical perspective. Eur J Pain 4:5–15
Eide PK, Jorum E, Stubhaug A, et al (1994) Relief of post-herpetic neuralgia with the N-methyl-D-aspartic acid receptor antagonist ketamine: a double-blind, cross-over comparison with morphine and placebo. Pain 58:347–354
Eide PK, Stubhaug A, Oye I (1995a) The NMDA-antagonist ketamine for prevention and treatment of acute and chronic post-operative pain. In: Breivik H (ed) Baillieres clinical anaesthesiology international practice and research. Tindall, London, pp 539–554
Eide PK, Stubhaug A, Oye I, Breivik H (1995b) Continuous subcutaneous administration of the N-methyl-D-aspartic acid (NMDA) receptor antagonist ketamine in the treatment of post-herpetic neuralgia. Pain 61:221–228
Eisenberg E, Vos BP, Strassman AM (1993) The NMDA antagonist memantine blocks pain behavior in a rat model of formalin-induced facial pain. Pain 54:301–307
Eisenberg E, Lacross S, Strassman AM (1995) The clinically tested N-methyl-D-aspartate receptor antagonist memantine blocks and reverses thermal hyperalgesia in a rat model of painful mononeuropathy. Neurosci Lett 187:17–20
Elliott K, Kest B, Man A, Kao B, Inturrisi C E (1995) N-methyl-D-aspartate (NMDA) receptors, mu and kappa opioid tolerance, and perspectives on new analgesic drug development. Neuropsychopharmacology 13:347–356
Fan GH, Zhao J, Wu YL, Lou LG, Zhang Z, Jing Q, Ma L, Pei G (1998) N-Methyl-D-aspartate attenuates opioid receptor-mediated G protein activation and this process involves protein kinase C. Mol Pharmacol 53:684–690
Felsby S, Nielsen J, Arendtnielsen L, Jensen TS (1996) NMDA receptor blockade in chronic neuropathic pain: a comparison of ketamine and magnesium chloride. Pain 64:283–291
Ferraguti F, Baldani-Guerra B, Corsi M, Nakanishi S, Corti C (1999) Activation of the extracellular signal-regulated kinase 2 by metabotropic glutamate receptors. Eur J Neurosci 11:2073–2082

Fiorillo CD, Williams JT (1998) Glutamate mediates an inhibitory postsynaptic potential in dopamine neurons. Nature 394:78–82

Fisher K, Lefebvre C, Coderre TJ (2002) Antinociceptive effects following intrathecal pretreatment with selective metabotropic glutamate receptor compounds in a rat model of neuropathic pain. Pharmacol Biochem Behav 73:411–418

Flajolet M, Rakhilin S, Wang H, et al (2003) Protein phosphatase 2C binds selectively to and dephosphorylates metabotropic glutamate receptor 3. Proc Natl Acad Sci USA 100:16006–16011

Follesa P, Ticku MK (1996) NMDA receptor upregulation: molecular studies in cultured mouse cortical neurons after chronic antagonist exposure. J Neurosci 16:2172–2178

Fourgeaud L, Mato S, Bouchet D, et al (2004) A single in vivo exposure to cocaine abolishes endocannabinoid-mediated long-term depression in the nucleus accumbens. J Neurosci 24:6939–6945

Fray AE, Ince PG, Banner SJ, et al (1998) The expression of the glial glutamate transporter protein EAAT2 in motor neuron disease: an immunohistochemical study. Eur J Neurosci 10:2481–2489

Freund G, Anderson KJ (1996) Glutamate receptors in the frontal cortex of alcoholics. Alcohol Clin Exp Res 20:1165–1172

Frick A, Zieglgänsberger W, Dodt H-U (2001) Glutamate receptors form hot spots on apical dendrites of neocortical pyramidal neurons. J Neurophysiol 86:1412–1421

Frick A, Magee J, Johnston D (2004) LTP is accompanied by an enhanced local excitability of pyramidal neuron dendrites. Nat Neurosci 7:126–135

Fricker J (1997) From mechanisms to drugs in Alzheimer's disease. Lancet 349:480

Fu ES, Miguel R, Scharf JE (1997) Preemptive ketamine decreases postoperative narcotic requirements in patients undergoing abdominal surgery. Anesth Analg 84:1086–1090

Fundytus ME (2001) Glutamate receptors and nociception: implications for the drug treatment of pain. CNS Drugs 15:29–58

Fundytus ME, Osborne MG, Henry JL, et al (2002) Antisense oligonucleotide knockdown of mGluR(1) alleviates hyperalgesia and allodynia associated with chronic inflammation. Pharmacol Biochem Behav 73:401–410

Gasparini F, Kuhn R, Pin JP (2002) Allosteric modulators of group I metabotropic glutamate receptors: novel subtype-selective ligands and therapeutic perspectives. Curr Opin Pharmacol 2:43–49

Geurts JJG, Wolswijk G, Bo L, van der Valk P, et al (2003) Altered expression patterns of group I and II metabotropic glutamate receptors in multiple sclerosis. Brain 126:1755–1766

Gill R, Kemp JA, Richards JG, Kew JNC (1999) NMDA receptor antagonists: past disappointments and future prospects as neuroprotective agents. Curr Opin Cardiovasc Pulm Ren Invest Drugs 1:576–591

Ginsberg MD (1995) Neuroprotection in brain ischemia: an update (part I). Neuroscientist 1:95–103

Gitto R, Barreca ML, De Luca L, et al (2004) New trends in the development of AMPA receptor antagonists. Expert Opin Ther Pat 14:1199–1213

Globus MYT, Busto R, Dietrich WD, et al (1988) Effect of ischemia on the in vivo release of striatal dopamine, glutamate, and y-aminobutyric acid studied in intracerebral microdialysis. J Neurochem 51:1455–1464

Goldberg YP, Kalchman MA, Metzler M, et al (1996) Absence of disease phenotype and intergenerational stability of the CAG repeat in transgenic mice expressing the human Huntington disease transcript. Hum Mol Genet 5:177–185

Gould E, McEwen BS, Tanapat P, et al (1997) Neurogenesis in the dentate gyrus of the adult tree shrew is regulated by psychosocial stress and NMDA receptor activation. J Neurosci 17:2492–2498

Govindaiah, Cox CL (2004) Synaptic activation of metabotropic glutamate receptors regulates dendritic outputs of thalamic interneurons. Neuron 41:611–623

Grant SGN, Silva AJ (1994) Targeting learning. Trends Neurosci 17:71–75

Gredal O, Moller SE (1995) Effect of branched-chain amino acids on glutamate metabolism in amyotrophic lateral sclerosis. J Neurol Sci 129:40–43

Gredal O, Werdelin L, Bak S, et al (1997) A clinical trial of dextromethorphan in amyotrophic lateral sclerosis. Acta Neurol Scand 96:8–13

Greenamyre JT, O'Brien CF (1991) N-methyl-D-aspartate antagonists in the treatment of Parkinson's disease. Arch Neurol 48:977–981

Greene JG, Greenamyre JT (1995) Characterization of the excitotoxic potential of the reversible succinate dehydrogenase inhibitor malonate. J Neurochem 64:430–436

Gunne LM, Andren PE (1993) An animal model for coexisting tardive dyskinesia and tardive parkinsonism—a glutamate hypothesis for tardive dyskinesia. Clin Neuropharmacol 16:90–95

Gurney ME, Fleck TJ, Himes CS, Hall ED (1998) Riluzole preserves motor function in a transgenic model of familiar amyotrophic lateral sclerosis. Neurology 50:62–67

Haines DR, Gaines SP (1999) N of 1 randomised controlled trials of oral ketamine in patients with chronic pain. Pain 83:283–287

Hajszan T, Xu CQ, Leranth C, Alreja M (2004) Group I metabotropic glutamate receptor activation produces a direct excitation of identified septohippocampal cholinergic neurons. J Neurophysiol 92:1216–1225

Han JS, Bird GC, Neugebauer V (2004) Enhanced group III mGluR-mediated inhibition of pain-related synaptic plasticity in the amygdala. Neuropharmacology 46:918–926

Hansel C, Artola A, Singer W (1996) Different threshold levels of postsynaptic [Ca2+](i) have to be reached to induce LTP and LTD in neocortical pyramidal cells. J Physiol Paris 90:317–319

Hardingham GE, Bading H (2003) The Yin and Yang of NMDA receptor signalling. Trends Neurosci 26:81–89

HardinPouzet H, Krakowski M, Bourbonniere L, et al (1997) Glutamate metabolism is downregulated in astrocytes during experimental allergic encephalomyelitis. Glia 20:79–85

Harper DN (2000) An assessment and comparison of the effects of oxotremorine, D-cycloserine, and bicuculline on delayed matching-to-sample performance in rats. Exp Clin Psychopharmacol 8:207–215

Hazell AS, Itzhak Y, Liu HB, Norenberg MD (1997) 1-Methyl-4-phenyl-1,2,3,6-tetrahydropyridine (MPTP) decreases glutamate uptake in cultured astrocytes. J Neurochem 68:2216–2219

Hilmas C, Pereira EF, Alkondon M (2001) The brain metabolite kynurenic acid inhibits alpha7 nicotinic receptor activity and increases non-alpha7 nicotinic receptor expression: physiopathological implications. J Neurosci 21:7463–7473

Hirai H (2000) Clustering of delta glutamate receptors is regulated by the actin cytoskeleton in the dendritic spines of cultured rat Purkinje cells. Eur J Neurosci 12:563–570

Hirai H, Kirsch J, Laube B, et al (1996) The glycine binding site of the N-methyl-D-aspartate receptor subunit NR1: identification of novel determinants of co-agonist potentiation in the extracellular m3-m4 loop region. Proc Natl Acad Sci USA 93:6031–6036

Holter SM, Danysz W, Spanagel R (1996) Evidence for alcohol anti-craving properties of memantine. Eur J Pharmacol 314:R1–R2

Holter SM, Danysz W, Spanagel R (2000) Novel uncompetitive N-methyl-D-aspartate (NMDA)-receptor antagonist MRZ 2/579 suppresses ethanol intake in long-term ethanol-experienced rats and generalizes to ethanol cue in drug discrimination procedure. J Pharmacol Exp Ther 292:545–552

Houghton AK, Parsons CG, Headley PM (2001) MRZ 2/579, a fast kinetic NMDA receptor antagonist, delays the development of morphine tolerance in awake rats. Pain 91:201–207

Huang CS, Song JH, Nagata K, et al (1997) Effects of the neuroprotective agent riluzole on the high voltage-activated calcium channels of rat dorsal root ganglion neurons. J Pharmacol Exp Ther 282:1280–1290

Hughes SW, Lorincz M, Cope DW, et al (2004) Synchronized oscillations at alpha and theta frequencies in the lateral geniculate nucleus. Neuron 42:253–268

Hugon J, Vallat JM, Spencer PS, et al (1989) Kainic acid induces early and delayed degenerative neuronal changes in rat spinal cord. Neurosci Lett 104:258–262

Hundt W, Danysz W, Holter SM, Spanagel R (1998) Ethanol and N-methyl-D-aspartate receptor complex interactions: A detailed drug discrimination study in the rat. Psychopharmacology (Berl) 135:44–51

Ichise T, Kano M, Hashimoto K, et al (2000) mGIuR1 in cerebellar Purkinje cells essential for long-term depression, synapse elimination, and motor coordination. Science 288:1832–1835

Iino M, Goto K, Kakegawa W, et al (2001) Glia-synapse interaction through Ca2+-permeable AMPA receptors in Bergmann glia. Science 292:926–929

Ikonomidou C, Qin Qin Y, Labruyere J, Olney J W (1996) Motor neuron degeneration induced by excitotoxin agonists has features in common with those seen in the SOD-1 transgenic mouse model of amyotrophic lateral sclerosis. J Neuropathol Exp Neurol 55:211–224

Ilkjaer S, Petersen KL, Brennum J, et al (1996) Effect of systemic N-methyl-D-aspartate receptor antagonist (ketamine) on primary and secondary hyperalgesia in humans. Br J Anaesth 76:829–834

Johnson TD (1996) Modulation of channel function by polyamines. Trends Pharmacol Sci 17:22–27

Jones KA, Wilding TJ, Huettner JE, Costa AM (1997) Desensitization of kainate receptors by kainate, glutamate and diastereomers of 4-methylglutamate. Neuropharmacology 36:853–863

Jones MW, McClean M, Parsons CG, Headley PM (2001) The in vivo significance of the varied channel blocking properties of uncompetitive NMDA receptor antagonists. Neuropharmacology 41:50–61

Kanthan R, Shuaib A (1995) Clinical evaluation of extracellular amino acids in severe head trauma by intracerebral in vivo microdialysis. J Neurol Neurosurg Psychiatry 59:326–327

Karcz-Kubicha M, Jessa M, Nazar M, et al (1997) Anxiolytic activity of glycine-B antagonists and partial agonists—no relation to intrinsic activity in the patch clamp. Neuropharmacology 36:1355–1367

Karler R, Calder LD (1992) Excitatory amino acids and the actions of cocaine. Brain Res 582:143–146

Kasai H, Matsuzaki M, Noguchi J, et al (2003) Structure-stability-function relationships of dendritic spines. Trends Neurosci 26:360–368

Kawai F, Sterling P (1999) AMPA receptor activates a G-protein that suppresses a cGMP-gated current. J Neurosci 19:2954–2959

KawasakiYatsugi S, Ichiki C, Yatsugi S, et al (2000) Neuroprotective effects of an AMPA receptor antagonist YM872 in a rat transient middle cerebral artery occlusion model. Neuropharmacology 39:211–217

Kennel P, Revah F, Bohme GA, et al (2000) Riluzole prolongs survival and delays muscle strength deterioration in mice with progressive motor neuropathy (pmn). J Neurol Sci 180:55–61
Kirkwood A, Rioult MG, Bear MF (1996) Experience-dependent modification of synaptic plasticity in visual cortex. Nature 381:526–528
Knöpfel T, Kuhn R, Allgeier H (1995) Metabotropic glutamate receptors: novel targets for drug development. J Med Chem 38:1417–1426
Konradi C, Heckers S (2003) Molecular aspects of glutamate dysregulation: implications for schizophrenia and its treatment. Pharmacol Ther 97:153–179
Kornhuber J, Weller M (1997) Psychotogenicity and n-methyl-D-aspartate receptor antagonism: implications for neuroprotective pharmacotherapy. Biol Psychiatry 41:135–144
Kotlinska J (2001) NMDA antagonists inhibit the development of ethanol dependence in rats. Pol J Pharmacol 53:47–50
Krieger C, Jones K, Kim SU, Eisen AA (1994) The role of intracellular free calcium in motor neuron disease. J Neurol Sci 124:27–32
LaBella V, Goodman JC, Appel SH (1997) Increased CSF glutamate following injection of ALS immunoglobulins. Neurology 48:1270–1272
Lagreze WA, Knorle R, Bach M, Feuerstein TJ (1998) Memantine is neuroprotective in a rat model of pressure-induced retinal ischemia. Invest Ophthalmol Vis Sci 39:1063–1066
Lamprecht R, LeDoux J (2004) Structural plasticity and memory. Nat Rev Neurosci 5:45–54
Lan JY, Skeberdis VA, Jover T, et al (2001) Protein kinase C modulates NMDA receptor trafficking and gating. Nat Neurosci 4:382–390
Lanthorn TH (1994) D-Cycloserine: agonist turned antagonist. Amino Acids 6:247–260
Laube B, Hirai H, Sturgess M, et al (1997) Molecular determinants of agonist discrimination by NMDA receptor subunits: analysis of the glutamate binding site on the NR2B subunit. Neuron 18:493–503
Laube B, Kuhse J, Betz H (1998) Evidence for a tetrameric structure of recombinant NMDA receptors. J Neurosci 18:2954–2961
Laurie DJ, Putzke J, Zieglgänsberger W, et al (1995) The distribution of splice variants of the NMDA receptor1 subunit mRNA in adult rat brain. Mol Brain Res 32:94–108
Lee WT, Shen YZ, Chang C (2000) Neuroprotective effect of lamotrigine and MK-801 on rat brain lesions induced by 3-nitropropionic acid: evaluation by magnetic resonance imaging and in vivo proton magnetic resonance spectroscopy. Neuroscience 95:89–95
Leem JW, Hwang JH, Hwang SJ, et al (2001) The role of peripheral N-methyl-D-aspartate receptors in Freund's complete adjuvant induced mechanical hyperalgesia in rats. Neurosci Lett 297:155–158
Lees GJ (1993) Contributory mechanisms in the causation of neurodegenerative disorders. Neuroscience 54:287–322
Lees GJ (2000) Pharmacology of AMPA/kainate receptor ligands and their therapeutic potential in neurological and psychiatric disorders. Drugs 59:33–78
Legendre P, Westbrook GL (1991) Ifenprodil blocks N-Methyl-D-aspartate receptors by a two-component mechanism. Mol Pharmacol 40:289–298
Lelong V, Dauphin F, Boulouard M (2001) RS 67333 and D-cycloserine accelerate learning acquisition in the rat. Neuropharmacology 41:517–522
Leppik IE, Marienau K, Graves NM, Rask CA (1988) MK-801 for epilepsy: a pilot study. Neurology 38 (Suppl 1):405
Lerma J, Morales M, Vicente MA, Herreras O (1997) Glutamate receptors of the kainate type and synaptic transmission. Trends Neurosci 20:9–12
Liao DZ, Hessler NA, Malinow R (1995) Activation of postsynaptically silent synapses during pairing-induced LTP in CA1 region of hippocampal slice. Nature 375:400–404

Lipton SA (1992a) Memantine prevents HIV coat protein induced neuronal injury in vitro. Neurology 42:1403–1405
Lipton SA (1992b) Models of neuronal injury in AIDS—another role for the NMDA receptor. Trends Neurosci 15:75–79
Lisman J (1994) The cam kinase II hypothesis for the storage of synaptic memory. Trends Neurosci 17:406–412
Littleton J, Zieglgänsberger W (2003) Pharmacological mechanisms of naltrexone and acamprosate in the prevention of relapse in alcohol dependence. Am J Addict 12 (Suppl 1):S3–S11
Lodder J (2000) Neuroprotection in stroke—analysis of failure, and alternative strategies. Neurosci Res Commun 26:173–179
Löscher W, Rundfeldt C, Honack D (1993) Low doses of NMDA receptor antagonists synergistically increase the anticonvulsant effect of the AMPA receptor antagonist NBQX in the kindling model of epilepsy. Eur J Neurosci 5:1545–1550
Lovinger DM, Zieglgansberger W (1996) Interactions between ethanol and agents that act on the NMDA-type glutamate receptor. Alcoholism 20:A187–A191
Lu J, Goula D, Sousa N, Almeida OFX (2003) Ionotropic and metabotropic glutamate receptor mediation of glucocorticoid-induced apoptosis in hippocampal cells and the neuroprotective role of synaptic N-methyl-D-aspartate receptors. Neuroscience 121:123–131
Lu YM, Jia Z, Janus C, et al (1997) Mice lacking metabotropic glutamate receptor 5 show impaired learning and reduced CA1 long-term potentiation (LTP) but normal CA3 LTP. J Neurosci 17:5196–5205
Luginger E, Wenning GK, Bösch S, Poewe W (2000) Beneficial effects of amantadine on L-dopa-induced dyskinesias in Parkinson's disease. Mov Disord 15:873–878
Ly CD, Roche KW, Lee HK, Wenthold RJ (2002) Identification of rat EMAP, a delta-glutamate receptor binding protein. Biochem Biophys Res Commun 291:85–90
Madden DR (2002) The structure and function of glutamate receptor ion channels. Nat Rev Neurosci 3:91–101
Maj M, Bruno V, Dragic Z, et al (2003) (−)-PHCCC, a positive allosteric modulator of mGluR4: characterization, mechanism of action, and neuroprotection. Neuropharmacology 45:895–906
Malandro MS, Kilberg MS (1996) Molecular biology of mammalian amino acid transporters. Annu Rev Biochem 65:305–336
Malinow R, Malenka RC (2002) AMPA receptor trafficking and synaptic plasticity. Annu Rev Neurosci 25:103–126
Manabe T, Nicoll RA (1994) Long-term potentiation: evidence against an increase in transmitter release probability in the CA1 region of the hippocampus. Science 265:1888–1892
Mannaioni G, Marino MJ, Valenti O, et al (2001) Metabotropic glutamate receptors 1 and 5 differentially regulate CA1 pyramidal cell function. J Neurosci 21:5925–5934
Mansour M, Nagarajan N, Nehring RB, et al (2001) Heteromeric AMPA receptors assemble with a preferred subunit stoichiometry and spatial arrangement. Neuron 32:841–853
Manzoni O, Bockaert J (1995) Metabotropic glutamate receptors inhibiting excitatory synapses in the CA1 area of rat hippocampus. Eur J Neurosci 7:2518–2523
Mao JR (1999) NMDA and opioid receptors: their interactions in antinociception, tolerance and neuroplasticity. Brain Res Brain Res Rev 30:289–304
Mao JR, Price DD, Mayer DJ (1995) Experimental mononeuropathy reduces the antinociceptive effects of morphine: implications for common intracellular mechanisms involved in morphine tolerance and neuropathic pain. Pain 61:353–364

Marek GJ, Wright RA, Schoepp DD, et al (2000) Physiological antagonism between 5-hydroxytryptamine(2A) and group II metabotropic glutamate receptors in prefrontal cortex. J Pharmacol Exp Ther 292:76–87
Marsicano G, Wotjak CT, Azad SC, et al (2002) The endogenous cannabinoid system controls extinction of aversive memories. Nature 418:530–534
Marsicano G, Goodenough S, Monory K, et al (2003) CB1 cannabinoid receptors and on-demand defense against excitotoxicity. Science 302:84–88
Mateos JM, Benitez R, Elezgarai I, et al (2000) Immunolocalization of the mGluR 1b splice variant of the metabotropic glutamate receptor 1 at parallel fiber-Purkinje cell synapses in the rat cerebellar cortex. J Neurochem 74:1301–1309
Matsuda K, Kamiya Y, Matsuda S, Yuzaki M (2002) Cloning and characterization of a novel NMDA receptor subunit NR3B: a dominant subunit that reduces calcium permeability. Mol Brain Res 100:43–52
Matute C, Alberdi E, Domercq M, et al (2001) The link between excitotoxic oligodendroglial death and demyelinating diseases. Trends Neurosci 24:224–230
Max MB, Byassmith MG, Gracely RH, Bennett GJ (1995) Intravenous infusion of the NMDA antagonist, ketamine, in chronic posttraumatic pain with allodynia: a double-blind comparison to alfentanil and placebo. Clin Neuropharmacol 18:360–368
Mayer ML, Armstrong N (2004) Structure and function of glutamate receptor ion channels. Annu Rev Physiol 66:161–181
McFeeters RL, Oswald RE (2004) Emerging structural explanations of ionotropic glutamate receptor function. FASEB J 18:428–438
McGee AW, Bredt DS (2003) Assembly and plasticity of the glutamatergic postsynaptic specialization. Curr Opin Neurobiol 13:111–118
McRoberts JA, Coutinho SV, Marvizon JC, et al (2001) Role of peripheral N-methyl-D-aspartate (NMDA) receptors in visceral nociception in rats. Gastroenterology 120:1737–1748
Mealing GAR, Lanthorn TH, Small DL, et al (1997) Antagonism of N-methyl-D-aspartate-evoked currents in rat cortical cultures by ARL 15896AR. J Pharmacol Exp Ther 281:376–383
Meldrum B (1985) Possible therapeutic applications of antagonists of excitatory amino acid neurotransmitters. Clin Sci 68:113–122
Merello M, Nouzeilles MI, Cammarota A, Leiguarda R (1999) Effect of memantine (NMDA antagonist) on Parkinson's disease: a double-blind crossover randomized study. Clin Neuropharmacol 22:273–276
Metman LV, DelDotto P, LePoole K, et al (1999) Amantadine for levodopa-induced dyskinesias—a 1-year follow-up study. Arch Neurol 56:1383–1386
Mirshahi T, Woodward JJ (1995) Ethanol sensitivity of heteromeric NMDA receptors: effects of subunit assembly, glycine and NMDA receptor1 mg2+-insensitive mutants. Neuropharmacology 34:347–355
Möbius HJ, Stöffler A (2002) New approaches to clinical trials in vascular dementia: memantine in small vessel disease. Cerebrovasc Dis 13 (Suppl 2):61–66
Montgomery JM, Madison DV (2002) State-dependent heterogeneity in synaptic depression between pyramidal cell pairs. Neuron 33:765–777
Montgomery JM, Pavlidis P, Madison DV (2001) Pair recordings reveal all-silent synaptic connections and the postsynaptic expression of long-term potentiation. Neuron 29:691–701
Monyer H, Seeburg PH, Wisden W (1991) Glutamate-operated channels—developmentally early and mature forms arise by alternative splicing. Neuron 6:799–810

Morikawa H, Khodakhah K, Williams JT (2003) Two intracellular pathways mediate metabotropic glutamate receptor-induced Ca2+ mobilization in dopamine neurons. J Neurosci 23:149–157

Morris RG (1996) Further studies of the role of hippocampal synaptic plasticity in spatial learning: is hippocampal LTP a mechanism for automatically recording attended experience? J Physiol Paris 90:333–334

Moryl E, Danysz W, Quack G (1993) Potential antidepressive properties of amantadine, memantine and bifemelane. Pharmacol Toxicol 72:394–397

Naassila M, Hammoumi S, Legrand E, et al (1998) Mechanism of action of acamprosate. Part I. Characterization of spermidine-sensitive acamprosate binding site in rat brain. Alcoholism 22:802–809

Nacher J, Alonso-Llosa G, Rosell DR, McEwen BS (2003) NMDA receptor antagonist treatment increases the production of new neurons in the aged rat hippocampus. Neurobiol Aging 24:273–284

Nagaraja RY, Grecksch G, Reymann KG, et al (2004) Metabotropic glutamate receptors interfere in different ways with pentylenetetrazole seizures, kindling, and kindling-related learning deficits. Naunyn Schmiedebergs Arch Pharmacol 370:26–34

Nakanishi S, Nakajima Y, Masu M, et al (1998) Glutamate receptors: brain function and signal transduction. Brain Res Brain Res Rev 26:230–235

Neale JH, Bzdega T, Wroblewska B (2000) N-acetylaspartylglutamate: The most abundant peptide neurotransmitter in the mammalian central nervous system. J Neurochem 75:443–452

Nedergaard M, Takano T, Hansen AJ (2002) Beyond the role of glutamate as a neurotransmitter. Nat Rev Neurosci 3:748–755

Neugebauer V (2002) Metabotropic glutamate receptors—important modulators of nociception and pain behavior. Pain 98:1–8

Ngan Kee WD, Khaw KS, Ma ML, et al (1997) Postoperative analgesic requirement after cesarean section: a comparison of anesthetic induction with ketamine or thiopental. Anesth Analg 85:1294–1298

Nicoll RA, Malenka RC (1995) Contrasting properties of two forms of long-term potentiation in the hippocampus. Nature 377:115–118

Nikam SS, Meltzer LT (2002) NR2B selective NMDA receptor antagonists. Curr Pharm Des 8:845–855

Nikolasjen L, Hansen CL, Nielsen J, et al (1996) The effect of ketamine on phantom pain: a central neuropathic disorder maintained by peripheral input. Pain 67:69–77

Nilsson M, Perfilieva E, Johansson U, et al (1999) Enriched environment increases neurogenesis in the adult rat dentate gyrus and improves spatial memory. J Neurobiol 39:569–578

O'Connor AJ, Evalchogiannnis G, Moskal J (1996) Subunit specific effects of D-cycloserine on NMDA receptor receptors expressed in Xenopus oocytes. Soc Neurosci Abstr 22:604.15

Okiyama K, Smith DH, Gennarelli TA, et al (1995) The sodium channel blocker and glutamate release inhibitor BW1003c87 and magnesium attenuate regional cerebral edema following experimental brain injury in the rat. J Neurochem 64:802–809

Okiyama K, Smith DH, White WF, et al (1997) Effects of the novel NMDA antagonists CP-98,113, CP-101,581 and CP-101,606 on cognitive function and regional cerebral edema following experimental brain injury in the rat. J Neurotrauma 14:211–222

Oliet SH, Piet R, Poulain DA (2001) Control of glutamate clearance and synaptic efficacy by glial coverage of neurons. Science 292:923–926

Olszewski RT, Bukhari N, Zhou J, et al (2004) NAAG peptidase inhibition reduces locomotor activity and some stereotypes in the PCP model of schizophrenia via group II mGluR. J Neurochem 89:876–885

Otani S, Auclair N, Desce JM, et al (1999) Dopamine receptors and groups I and II mGluRs cooperate for long-term depression induction in rat prefrontal cortex through converging postsynaptic activation of MAP kinases. J Neurosci 19:9788–9802

Palmer AM, Gershon S (1990) Is the neuronal basis of Alzheimer's disease cholinergic or glutamatergic? FASEB J 4:2745–2752

Palmer AM, Marion DW, Botscheller ML, et al (1994) Increased transmitter amino acid concentration in human ventricular CSF after brain trauma. Neuroreport 6:153–156

Palucha A, Tatarczynska E, Branski P, et al (2004) Group III mGlu receptor agonists produce anxiolytic- and antidepressant-like effects after central administration in rats. Neuropharmacology 46:151–159

Paolucci E, Berretta N, Tozzi A, et al (2003) Depression of mGluR-mediated IPSCs by 5-HT in dopamine neurons of the rat substantia nigra pars compacta. Eur J Neurosci 18:2743–2750

Papp M, Moryl E (1994) Antidepressant activity of non-competitive and competitive NMDA receptor antagonists in a chronic mild stress model of depression. Eur J Pharmacol 263:1–7

Papp M, Moryl E (1996) Antidepressant-like effects of 1-aminocyclopropanecarboxylic acid and D-cycloserine in an animal model of depression. Eur J Pharmacol 316:145–151

Park KM, Max MB, Robinovitz E, et al (1995) Effects of intravenous ketamine, alfentanil, or placebo on pain, pinprick hyperalgesia, and allodynia produced by intradermal capsaicin in human subjects. Pain 63:163–172

Parsons CG (2001) NMDA receptors as targets for drug action in neuropathic pain. Eur J Pharmacol 429:71–78

Parsons CG, Gruner R, Rozental J, et al (1993a) Patch clamp studies on the kinetics and selectivity of N-methyl-D-aspartate receptor antagonism by memantine (1-amino-3,5-dimethyladamantan). Neuropharmacology 32:1337–1350

Parsons CG, Zong XG, Lux HD (1993b) Whole cell and single channel analysis of the kinetics of glycine-sensitive N-methyl-D-aspartate receptor desensitization. Br J Pharmacol 109:213–221

Parsons CG, Danysz W, Quack G (1998) Glutamate in CNS disorders as a target for drug development. An update. Drug News Perspect 11:523–569

Parsons CG, Danysz W, Quack G (1999) Memantine is a clinically well tolerated N-methyl-D-aspartate (NMDA) receptor antagonist—a review of preclinical data. Neuropharmacology 38:735–767

Parsons CG, Danysz, W (2002) Amyotrophic lateral sclerosis (ALS). In: Lodge D, Danysz W, Parsons CG (eds) Therapeutic potential of ionotropic glutamate receptor antagonists and modulators. FP Graham Publishing, New York, pp 540–562

Pasternak GW, Inturrisi CE (1995) Pharmacological modulation of opioid tolerance. Expert Opin Investig Drugs 4:271–281

Peavy RD, Conn PJ (1998) Phosphorylation of mitogen-activated protein kinase in cultured rat cortical glia by stimulation of metabotropic glutamate receptors. J Neurochem 71:603–612

Pellegrini-Giampietro DE, Bennett MVL, Zukin RS (1992) Are Ca-2+-permeable kainate/AMPA receptors more abundant in immature brain. Neurosci Lett 144:65–69

Persson J, Hasselstrom J, Wiklund B, et al (1998) The analgesic effect of racemic ketamine in patients with chronic ischemic pain due to lower extremity arteriosclerosis obliterans. Acta Anaesthesiol Scand 42:750–758

Pierce RC, Bell K, Duffy P, Kalivas PW (1996) Repeated cocaine augments excitatory amino acid transmission in the nucleus accumbens only in rats having developed behavioral sensitization. J Neurosci 16:1550–1560

Pin JP, Parmentier ML, Joly C, et al (1998) Coupling of metabotropic glutamate receptors to G-proteins: differences from and similarities with other G-protein-coupled receptors. In: Moroni F, Nicoletti F, Pellegrini-Giampietro DE (eds) Metabotropic glutamate receptors and brain function. Portland Press, Colchester, pp 9–18

Pitsikas N, Brambilla A, Besozzi C, et al (2001) Effects of cerestat and NBQX on functional and morphological outcomes in rat focal cerebral ischemia. Pharmacol Biochem Behav 68:443–447

Plaitakis A, Fesdjian CO, Shashidharan P (1996) Glutamate antagonists in amyotrophic lateral sclerosis: a review of their therapeutic potential. CNS Drugs 5:437–456

Popik P, Danysz W (1997) Inhibition of reinforcing effects of morphine and motivational aspects of naloxone-precipitated opioid withdrawal by N-methyl-D-aspartate receptor antagonist, memantine. J Pharmacol Exp Ther 280:854–865

Popik P, Kozela E (1999) Clinically available NMDA antagonist, memantine, attenuates tolerance to analgesic effects of morphine in a mouse tail flick test. Pol J Pharmacol 51:223–231

Popik P, Skolnick P (1996) The NMDA antagonist memantine blocks the expression and maintenance of morphine dependence. Pharmacol Biochem Behav 53:791–797

Popik P, Mamczarz J, Fraczek M, et al (1998) Inhibition of reinforcing effects of morphine and naloxone-precipitated opioid withdrawal by novel glycine site and uncompetitive NMDA receptor antagonists. Neuropharmacology 37:1033–1042

Popp RL, Lovinger DM (2000) Interaction of acamprosate with ethanol and spermine on NMDA receptors in primary cultured neurons. Eur J Pharmacol 394:221–231

Przegalinski E, Tatarczynska E, Derenwesolek A, Chojnackawojcik E (1997) Antidepressant-like effects of a partial agonist at strychnine-insensitive glycine receptors and a competitive NMDA receptor antagonist. Neuropharmacology 36:31–37

Pud D, Eisenberg E, Spitzer A, et al (1998) The NMDA receptor antagonist amantadine reduces surgical neuropathic pain in cancer patients: a double blind, randomized, placebo controlled trial. Pain 75:349–354

Pula G, Mundell SJ, Roberts PJ, Kelly E (2004) Agonist-independent internalization of metabotropic glutamate receptor 1a is arrestin- and clathrin-dependent and is suppressed by receptor inverse agonists. J Neurochem 89:1009–1020

Pulvirenti L, Maldonadolopez R, Koob GF (1992) NMDA receptors in the nucleus accumbens modulate intravenous cocaine but not heroin self-administration in the rat. Brain Res 594:327–330

Quartaroli M, Fasdelli N, Bettelini L, et al (2001) GV196771A, an NMDA receptor/glycine site antagonist, attenuates mechanical allodynia in neuropathic rats and reduces tolerance induced by morphine in mice. Eur J Pharmacol 430:219–227

Racca C, Catania MV, Monyer H, Sakmann B (1996) Expression of AMPA-glutamate receptor b subunit in rat hippocampal GABAergic neurons. Eur J Neurosci 8:1580–1590

Rajput AH, Rajput A, Lang AE, et al (1998) New use for an old drug: amantadine benefits levodopa-induced dyskinesia. Mov Disord 13:851

Rammes G, Swandulla D, Collingridge GL, et al (1996) Interactions of 2,3-benzodiazepines and cyclothiazide at AMPA receptors: patch clamp recordings in cultured neurones and area CA1 in hippocampal slices. Br J Pharmacol 117:1209–1221

Rammes G, Swandulla D, Spielmanns P, Parsons CG (1998) Interactions of GYKI 52466 and NBQX with cyclothiazide at AMPA receptors: experiments with outside-out patches and EPSCs in hippocampal neurones. Neuropharmacology 37:1299–1320

Rammes G, Zeilhofer HU, Collingridge GL, et al (1999) Expression of early hippocampal CA1 LTP does not lead to changes in AMPA-EPSC kinetics or sensitivity to cyclothiazide. Pflugers Arch 437:191–196

Rammes G, Mahal B, Putzke J, et al (2001) The anti-craving compound acamprosate acts as a weak NMDA-receptor antagonist, but modulates NMDA-receptor subunit expression similar to memantine and MK-801. Neuropharmacology 40:749–760
Rampon C, Jiang CH, Dong H, et al (2000) Effects of environmental enrichment on gene expression in the brain. Proc Natl Acad Sci USA 97:12880–12884
Ranaldi R, Bauco P, Wise RA (1997) Synergistic effects of cocaine and dizocilpine (MK-801) on brain stimulation reward. Brain Res 760:231–237
Rao VLR, Dogan A, Todd KG, et al (2001) Neuroprotection by memantine, a non-competitive NMDA receptor antagonist after traumatic brain injury in rats. Brain Res 911:96–100
Reichelt W, Knöpfel T (2002) Glutamate uptake controls expression of a slow postsynaptic current mediated by mGluRs in cerebellar Purkinje cells. J Neurophysiol 87:1974–1980
Rogoz Z, Skuza G, Maj J, Danysz W (2002) Synergistic effect of uncompetitive NMDA receptor antagonists and antidepressant drugs in the forced swimming test in rats. Neuropharmacology 42:1024–1030
Roppe JR, Wang BW, Huang DH, et al (2004) 5-[(2-methyl-1,3-thiazol-4-yl)ethynyl]-2,3′-bipyridine: a highly potent, orally active metabotropic glutamate subtype 5 (mGlu5) receptor antagonist. Bioorg Med Chem Lett 14:3993–3996
Rosenmund C, Stern Bach Y, Stevens CF (1998) The tetrameric structure of a glutamate receptor channel. Nature 280:1596–1599
Rothman SM, Olney JW (1987) Excitotoxicity and the NMDA receptor. Trends Neurosci 10:299–302
Rothstein JD, Martin LJ, Kuncl RW (1992) Decreased glutamate transport by the brain and spinal cord in amyotrophic lateral sclerosis. N Engl J Med 326:1464–1468
Rothstein JD, Vankammen M, Levey AI, et al (1995) Selective loss of glial glutamate transporter GLT-1 in amyotrophic lateral sclerosis. Ann Neurol 38:73–84
Rothstein JD, Dykes-Hoberg M, Pardo CA, et al (1996) Knockout of glutamate transporters reveals a major role for astroglial transport in excitotoxicity and clearance of glutamate. Neuron 16:675–686
Rouach N, Nicoll RA (2003) Endocannabinoids contribute to short-term but not long-term mGluR-induced depression in the hippocampus. Eur J Neurosci 18:1017–1020
Roy J, Minotti S, Dong LC, et al (1998) Glutamate potentiates the toxicity of mutant Cu/Zn-superoxide dismutase in motor neurons by postsynaptic calcium-dependent mechanisms. J Neurosci 18:9673–9684
Roytblat L, Korotkoruchko A, Katz J, et al (1993) Postoperative pain: the effect of low-dose ketamine in addition to general anesthesia. Anesth Analg 77:1161–1165
Rzeski W, Turski L, Ikonomidou C (2001) Glutamate antagonists limit tumor growth. Proc Natl Acad Sci USA 98:6372–6377
Salt TE, Binns KE, Turner JP, et al (1999) Antagonism of the mGlu5 agonist 2-chloro-5-hydroxyphenylglycine by the novel selective mGlu5 antagonist 6-methyl-2-(phenylethynyl)-pyridine (MPEP) in the thalamus. Br J Pharmacol 127:1057–1059
Sandkühler J, Liu XG (1998) Induction of long-term potentiation at spinal synapses by noxious stimulation or nerve injury. Eur J Neurosci 10:2476–2480
Saroff D, Delfs J, Kuznetsov D, Geula C (2000) Selective vulnerability of spinal cord motor neurons to non-NMDA toxicity. Neuroreport 11:1117–1121
Sasaki S, Komori T, Iwata M (2000) Excitatory amino acid transporter 1 and 2 immunoreactivity in the spinal cord in amyotrophic lateral sclerosis. Acta Neuropathol (Berl) 100:138–144
Saucier D, Cain DP (1995) Spatial learning without NMDA receptor-dependent long-term potentiation. Nature 378:186–189

Schielke GP, Kupina NC, Boxer PA, et al (1999) The neuroprotective effect of the novel AMPA receptor antagonist PD152247 (PNQX) in temporary focal ischemia in the rat. Stroke 30:1472–1477
Schierloh A, Eder M, Zieglgänsberger W, Dodt H-U (2004) Effects of sensory deprivation on columnar organization of neuronal circuits in the rat barrel cortex. Eur J Neurosci 20:1118–1124
Schmid RL, Sandler AN, Katz J (1999) Use and efficacy of low-dose ketamine in the management of acute postoperative pain: a review of current techniques and outcomes. Pain 82:111–125
Schmidt WJ, Kretschmer BD (1997) Behavioural pharmacology of glutamate receptors in the basal ganglia. Neurosci Biobehav Rev 21:381–392
Schmidt WJ, Bubser M, Hauber W (1990) Excitatory amino acids and Parkinson's disease. Trends Neurosci 13:46
Schmitz D, Mellor J, Nicoll RA (2001) Presynaptic kainate receptor mediation of frequency facilitation at hippocampal mossy fiber synapses. Science 291:1972–1976
Schoepp DD, Jane DE, Monn JA (1999) Pharmacological agents acting at subtypes of metabotropic glutamate receptors. Neuropharmacology 38:1431–1476
Schulz JB, Matthews RT, Henshaw DR, Beal MF (1996) Neuroprotective strategies for the treatment of lesions produced by mitochondrial toxins: implications for neurodegenerative diseases. Neuroscience 71:1043–1048
Schwarcz R, Köhler C (1983) Differential vulnerability of central neurons of the rat to quinolinic acid. Neurosci Lett 38:85–90
Service RF (1994) Neuroscience—will a new type of drug make memory-making easier? Science 266:218–219
Sheng M, Kim MJ (2002) Postsynaptic signaling and plasticity mechanisms. Science 298: 776–780
Shoghi-Jadid K, Small GW, Agdeppa ED, et al (2002) Localization of neurofibrillary tangles and beta-amyloid plaques in the brains of living patients with Alzheimer disease. Am J Geriatr Psychiatry 10:24–35
Sillaber I, Rammes G, Zimmermann S, et al (2002) Enhanced and delayed stress-induced alcohol drinking in mice lacking functional CRH1 receptors. Science 296:931–933
Skolnick P (1999) Antidepressants for the new millennium. Eur J Pharmacol 375:31–40
Skolnick P, Marvizon JCG, Jackson BW, et al (1989) Blockade of N-methyl-D-aspartate induced convulsions by 1-aminocyclopropanecarboxylates. Life Sci 45:1647–1656
Skolnick P, Miller R, Young A, et al (1992) Chronic treatment with 1-aminocyclopropanecarboxylic acid desensitizes behavioral responses to compounds acting at the N-methyl-D-aspartate receptor complex. Psychopharmacology (Berl) 107:489–496
Smothers CT, Clayton R, Blevins T, Woodward JJ (2001) Ethanol sensitivity of recombinant human N-methyl-D-aspartate receptors. Neurochem Int 38:333–340
Sommer B, Keinanen K, Verdoorn TA, et al (1990) Flip and Flop: a cell-specific functional switch in glutamate-operated channels of the CNS. Science 249:1580–1584
Song I, Huganir RL (2002) Regulation of AMPA receptors during synaptic plasticity. Trends Neurosci 25:578–588
Sonsalla PK, Riordan DE, Heikkila RE (1991) Competitive and noncompetitive antagonists at N-methyl-D-aspartate receptors protect against methamphetamine-induced dopaminergic damage in mice. J Pharmacol Exp Ther 256:506–512
Sporn J, Sachs G (1997) The anticonvulsant lamotrigine in treatment-resistant manic-depressive illness. J Clin Psychopharmacol 17:185–189
Sripada S, Gaytan O, Swann A, Dafny N (2001) The role of MK-801 in sensitization to stimulants. Brain Res Brain Res Rev 35:97–114

Starr MS (1995) Antiparkinsonian actions of glutamate antagonists—alone and with L-dopa: a review of evidence and suggestions for possible mechanisms. J Neural Transm Park Dis Dement Sect 10:141–185
Stefani A, Spadoni F, Bernardi G (1997) Differential inhibition by riluzole, lamotrigine, and phenytoin of sodium and calcium currents in cortical neurons: Implications for neuroprotective strategies. Exp Neurol 147:115–122
Stephens D N, Meldrum B S, Weidmann R, et al (1986) Does the excitatory amino acid receptor antagonist 2-APH exhibit anxiolytic activity? Psychopharmacology (Berl) 90:166–169
Stone TW (2001) Kynurenines in the CNS: from endogenous obscurity to therapeutic importance. Prog Neurobiol 64:185–218
Sucher NJ, Akbarian S, Chi CL, et al (1995) Developmental and regional expression pattern of a novel NMDA receptor-like subunit (NMDA receptor-l) in the rodent brain. J Neurosci 15:6509–6520
Suzuki R, Matthews EA, Dickenson AH (2001) Comparison of the effects of MK-801, ketamine and memantine on responses of spinal dorsal horn neurones in a rat model of mononeuropathy. Pain 91:101–109
Sveinbjornsdottir S, Sander JWAS, Upton D, et al (1993) The excitatory amino acid antagonist d-CPP-ene (sdz eaa-494) in patients with epilepsy. Epilepsy Res 16:165–174
Takano T, Lin JH, Arcuino G, et al (2001) Glutamate release promotes growth of malignant gliomas. Nat Med 7:1010–1015
Tang Y-P, Shimizu E, Dube G, et al (1999) Genetic enhancement of learning and memory in mice. Nature 401:63–69
Tang Y-P, Rampon C, Goodhouse J, et al (2000a) Enrichment induces structural changes and recovery from nonspatial memory deficits in CA1 NMDA receptor1-knockout mice. Nat Neurosci 3:238–244
Tang Y-P, Shimizu E, Rampon C, Tsien JZ (2000b) NMDA receptor-dependent synaptic reinforcement as a crucial process for memory consolidation. Science 290:1170–1174
Tang Y-P, Shimizu E, Tsien JZ (2001) Do 'smart' mice feel more pain, or are they just better learners? Nat Neurosci 4:453–454
Taniguchi K, Shinjo K, Mizutani M, et al (1997) Antinociceptive activity of CP-101,606, an NMDA receptor NR2B subunit antagonist. Br J Pharmacol 122:809–812
Tempia F, Alojado ME, Strata P, Knöpfel T (2001) Characterization of the mGIuR(1)-mediated electrical and calcium signaling in Purkinje cells of mouse cerebellar slices. J Neurophysiol 86:1389–1397
Toggas SM, Masliah E, Mucke L (1996) Prevention of HIV-1 gp120-induced neuronal damage in the central nervous system of transgenic mice by the NMDA receptor antagonist memantine. Brain Res 706:303–307
Troupin AS, Nendius JR, Cheng F, Risinger MW (1986) MK-801. In: Meldrum BS, Porter RJ (eds) New anticonvulsant drugs. John Libbey, London, pp 191–201
Trujillo KA, Akil H (1991) Inhibition of morphine tolerance and dependence by the NMDA receptor antagonist MK-801. Science 251:85–87
Trujillo KA, Akil H (1995) Excitatory amino acids and drugs of abuse: a role for N-methyl-D-aspartate receptors in drug tolerance, sensitization and physical dependence. Drug Alcohol Depend 38:139–154
Tsuchida E, Rice M, Bullock R (1997) The neuroprotective effect of the forebrain-selective NMDA antagonist CP101,606 upon focal ischemic brain damage caused by acute subdural hematoma in the rat. J Neurotrauma 14:409–417
Tsumoto T, Yasuda H (1996) A switching role of postsynaptic calcium in the induction of long-term potentiation or long-term depression in visual cortex. Semin Neurosci 8:311–319

Tyszkiewicz JP, Gu ZL, Wang X, et al (2004) Group II metabotropic glutamate receptors enhance NMDA receptor currents via a protein kinase C-dependent mechanism in pyramidal neurones of rat prefrontal cortex. J Physiol (Lond) 554:765–777

Van Den Bosch L, Vandenberghe W, Klaassen H, et al (2000) Ca(2+)-permeable AMPA receptors and selective vulnerability of motor neurons. J Neurol Sci 180:29–34

Verdoorn TA, Burnashev N, Monyer H, et al (1991) Structural determinants of ion flow through recombinant glutamate receptor channels. Science 252:1715–1718

Verhagen Metman L, Dotto PD, Munckhof PVD, et al (1998) Amantadine as treatment for dyskinesias and motor fluctuations in Parkinson's disease. Neurology 50:1323–1329

Vissel B, Krupp JJ, Heinemann SF, Westbrook GL (2001) A use-dependent tyrosine dephosphorylation of NMDA receptors is independent of ion flux. Nat Neurosci 4:587–596

Vorwerk CK, Lipton SA, Zurakowski D, et al (1996) Chronic low-dose glutamate is toxic to retinal ganglion cells—toxicity blocked by memantine. Invest Ophthalmol Vis Sci 37:1618–1624

Wahl F, Renou E, Mary V, Stutzmann JM (1997) Riluzole reduces brain lesions and improves neurological function in rats after a traumatic brain injury. Brain Res 756:247–255

Wallstrom E, Diener P, Ljungdahl A, et al (1996) Memantine abrogates neurological deficits, but not CNS inflammation, in lewis rat experimental autoimmune encephalomyelitis. J Neurol Sci 137:89–96

Wang Y, Small DL, Stanimirovic DB, et al (1997) AMPA receptor-mediated regulation of a Gi-protein in cortical neurons. Nature 389:502–504

Warncke T, Jorum E, Stubhaug A (1997) Local treatment with the N-methyl-D-aspartate receptor antagonist ketamine, inhibit development of secondary hyperalgesia in man by a peripheral action. Neurosci Lett 227:1–4

Wilding TJ, Huettner JE (1997) Activation and desensitization of hippocampal kainate receptors. J Neurosci 17:2713–2721

Wiley JL, Cristello AF, Balster RL (1995) Effects of site-selective NMDA receptor antagonists in an elevated plus-maze model of anxiety in mice. Eur J Pharmacol 294:101–107

Williams K (1997) Modulation and block of ion channels: a new biology of polyamines. Cell Signal 9:1–13

Wu JQ, Rowan MJ, Anwyl R (2004) An NMDA receptor-independent LTP mediated by group II metabotropic glutamate receptors and p42/44 MAP kinase in the dentate gyrus in vitro. Neuropharmacology 46:311–317

Xu-Friedman MA, Regehr WG (2004) Structural contributions to short-term synaptic plasticity. Physiol Rev 84:69–85

Yamada KA (1998) Modulating excitatory synaptic neurotransmission: Potential treatment for neurological disease? Neurobiol Dis 5:67–80

Yang DN, Gereau RW (2003) Peripheral group II metabotropic glutamate receptors mediate endogenous anti-allodynia in inflammation. Pain 106:411–417

Yang H, Criswell HE, Simson P, et al (1996) Evidence for a selective effect of ethanol on N-methyl-D-aspartate responses: ethanol affects a subtype of the ifenprodil-sensitive N-methyl-D-aspartate receptors. J Pharmacol Exp Ther 278:114–124

Yashpal K, Fisher K, Chabot JG, Coderre TJ (2001) Differential effects of NMDA and group I mGluR antagonists on both nociception and spinal cord protein kinase C translocation in the formalin test and a model of neuropathic pain in rats. Pain 94:17–29

Yokoo H, Shiraishi S, Kobayashi H, et al (1998) Selective inhibition by riluzole of voltage-dependent sodium channels and catecholamine secretion in adrenal chromaffin cells. Naunyn Schmiedebergs Arch Pharmacol 357:526–531

Zauner A, Bullock R (1995) The role of excitatory amino acids in severe brain trauma: opportunities for therapy: a review. J Neurotrauma 12:547–554
Zeron MM, Chen N, Moshaver A, et al (2001) Mutant huntingtin enhances excitotoxic cell death. Mol Cell Neurosci 17:41–53
Zhang XX, Shi WX (2001) Dynamic modulation of NMDA-induced responses by ifenprodil in rat prefrontal cortex. Synapse 39:313–318
Zhao WF, Bianchi R, Wang M, Wong RKS (2004) Extracellular signal-regulated kinase 1/2 is required for the induction of group I metabotropic glutamate receptor-mediated epileptiform discharges. J Neurosci 24:76–84
Zieglgänsberger W, Tölle TR (1993) The pharmacology of pain signalling. Curr Opin Neurobiol 3:611–618

HEP (2005) 169:305–334

Neurobiology and Treatment of Anxiety: Signal Transduction and Neural Plasticity

C. H. Duman · R. S. Duman (✉)

Laboratory of Molecular Psychiatry, Departments of Psychiatry and Pharmacology, Yale University School of Medicine, 34 Park Street, New Haven CT, 06508, USA
ronald.duman@yale.edu

Abstract The stress-dependence and chronic nature of anxiety disorders along with the anxiolytic effectiveness of antidepressant drugs suggests that neuronal plasticity may play a role in the pathophysiology of anxiety. Intracellular signaling pathways are known in many systems to be critical links in the cascades from surface signals to the molecular alterations that result in functional plasticity. Chronic antidepressant treatments can regulate intracellular signaling pathways and can induce molecular, cellular, and structural changes over time. These changes may be important to the anxiolytic effectiveness of these drugs. In addition, the signaling proteins implicated in the actions of chronic antidepressant action, such as cAMP response element binding protein (CREB), have also been implicated in conditioned fear and in anxiety. The cellular mechanisms underlying conditioned fear indicate roles for additional signaling pathways; however, less is known about such mechanisms in anxiety. The challenge to identify intracellular signaling pathways and related molecular and structural changes that are critical to the etiology and treatment of anxiety will further establish the importance of mechanisms of neuronal plasticity in functional outcome and improve treatment strategies.

Keywords Antidepressant · Calcium signaling · cAMP · cGMP–NOS signaling · CREB · Fear conditioning

1 Introduction

Anxiety disorders can develop as a result of exposure to stress and can be chronic in nature, suggesting the relevance of experience-dependent processes and persistent functional alterations in the etiology of anxiety. Time-dependent adaptive processes might also be important in the treatment of anxiety. The regulation of intracellular signaling mechanisms and associated regulation of gene expression has been suggested to underlie the time-dependent effects of several classes of drugs including antidepressants and addictive drugs (Duman et al. 2000; Nestler et al. 2002). Antidepressant drugs are effective upon chronic, but not after acute administration for the treatment of both depression and anxiety, suggesting that drug-induced plasticity may contribute to the effectiveness of these drugs in both types of affective disorders (Duman et al. 2000; Manji et al. 2001). Chronic antidepressant treatment is known to regulate intracellular signaling pathways and can induce cellular, molecular, and structural changes over time (Duman et al. 2000; Nestler et al. 2002). This chapter will discuss intracellular signaling mechanisms that may be important to the effectiveness of these drugs with the goal of understanding how regulation of signaling could alter molecular processes important to functional outcome. In addition, this chapter will discuss components of intracellular signaling pathways that are critical to conditioned fear and provide examples of regulation of intracellular signaling that might be important to the actions of other classes of drugs effective in the treatment of anxiety, including γ-aminobutyric acid (GABA)-A/benzodiazepine receptor ligands and corticotrophin releasing factor (CRF) receptor antagonists. Understanding the critical signaling mechanisms that contribute to neuronal plasticity and functional outcome will allow for the targeting of novel anxiety treatments to cellular processes that might be more directly responsible for the development and persistence of anxiety than are the more nonspecific targets of existing anxiety medications.

2 Overview of Anxiety-Induced Signal Transduction and Neural Plasticity

Exposure to situations that cause anxiety and/or stress results in activation of many neurotransmitter and neuropeptide systems. The actions of these extracellular signaling systems are subsequently mediated by activation of metabotropic (i.e., G protein-coupled receptors of many classes) and ionotropic receptors for amino acid neurotransmitters (i.e., γ-amino butyric acid and glutamate). The metabotropic receptors in turn result in activation of intracellular signal transduction cascades, including the cyclic adenosine monophosphate (cAMP), cyclic guanosine monophosphate (cGMP), inositol triphosphate (IP_3) and diacylglycerol pathways. Ionotropic receptors gate ions (e.g., Cl^-, Na^+, and Ca^{2+}) that influence the charge of neurons but that also lead to regulation of signaling cascades. In most cases the activity of these signaling systems occurs via regulation of protein phosphorylation. This occurs through the addition (via protein kinases) or the removal (via phosphatases) of phosphate groups from target proteins. In addition, neurotrophic factors and cytokines act on transmembrane receptors that contain extracellular binding sites and intracellular kinase domains that can directly phosphorylate target proteins. These intracellular signal systems ultimately influence all aspects of neuronal function including rapid effects via regulation of ion channels, and short- and long-term modulatory effects, including regulation of gene expression. These signal cascades and the way that neuronal systems are altered when they are regulated by external stimuli represent a form of neural plasticity that underlies the ability of an organism to respond to the same or related stimuli in the future. This section will provide a brief overview of the common signal transduction cascades that are influenced by anxiety and stress.

2.1 G Protein-Coupled Pathways

There are several intracellular second messengers that are activated by metabotropic receptors, also referred to as G protein-coupled receptors (Duman and Nestler 1999). These receptors couple with G proteins that are heterotrimers made up of α-, β- and γ-subunits. Interaction of the G protein heterotrimer with activated receptor increases the exchange of guanosine triphosphate (GTP) for bound guanosine diphosphate (GDP), resulting in dissociation of the heterotrimer into free α and βγ subunits that in turn can regulate second messenger effectors (Fig. 1). After activation of effector proteins, the intrinsic GTPase activity in the Gα subunits hydrolyzes GTP to GDP and promotes the association of the Gα with the Gβγ subunits. In many cases the Gα subunits interact with effector proteins to regulate the synthesis of second messengers; however, the free βγ subunits can also influence second messenger production. In addition to regulation of second messengers, G proteins can also directly regulate certain types of ion channels, notably activation of inward rectifying K^+ channels (GIRKs) and inhibition of voltage-gated Ca^{2+} channels. It is important to remember that the type of second messengers and/or ion channels regulated depends on the cell type and the complement of signaling machinery expressed in that particular cell. This section will discuss a few of the best-characterized second messenger pathways and ones most relevant to anxiety and stress.

2.1.1 cAMP Second Messenger Cascade

One of the best-characterized effectors and second messenger systems is the cAMP cascade that can be either activated or inhibited by neurotransmitter/neuropeptide receptors, including those implicated in anxiety/stress such as CRF. Receptors that activate cAMP synthesis couple with the stimulatory G protein, Gsα, and those that inhibit this second messenger couple with the inhibitory G protein, Giα, and these either stimulate or inhibit adenylyl cyclase, the effector enzyme responsible for synthesis of cAMP (Duman and Nestler 1999). There are at least nine different forms of adenylyl cyclase that have been identified by molecular cloning, each with a unique distribution in the brain. The different types of adenylyl cyclase are activated by Gsα as well as the diterpene forskolin, but are differentially regulated by Giα, the βγ subunits, Ca^{2+}, and by phosphorylation. This provides for fine control of adenylyl cyclase enzyme activity and regulation by other effector pathways.

The actions of cAMP occur almost exclusively via regulation of cAMP-dependent protein kinase (PKA), which consists of catalytic and regulatory subunits (Nestler and Duman 1999). There are three different isoforms of the catalytic subunit and four isoforms of the regulatory subunit. In the inactive state, in the absence of cAMP, PKA exists as a dimer of two catalytic and two regulatory subunits. Upon binding of cAMP to the regulatory subunits, the catalytic subunits are released and can phosphorylate substrate proteins. The types of cellular proteins that serve as substrates for PKA include metabolic enzymes, receptors, ion channels, effector proteins, and gene transcription factors.

The actions of cAMP are terminated by phosphodiesterases (PDEs) that catalyze the breakdown of cAMP to 5′-AMP (Duman and Nestler 1999). There are at least 11 different forms of PDEs that are characterized based on their affinity and selectivity for cAMP, as well as cGMP. In addition, the PDEs are differentially regulated by the cyclic nucleotides themselves, by phosphorylation, and by increased expression of certain splice variants (e.g., some isoforms are

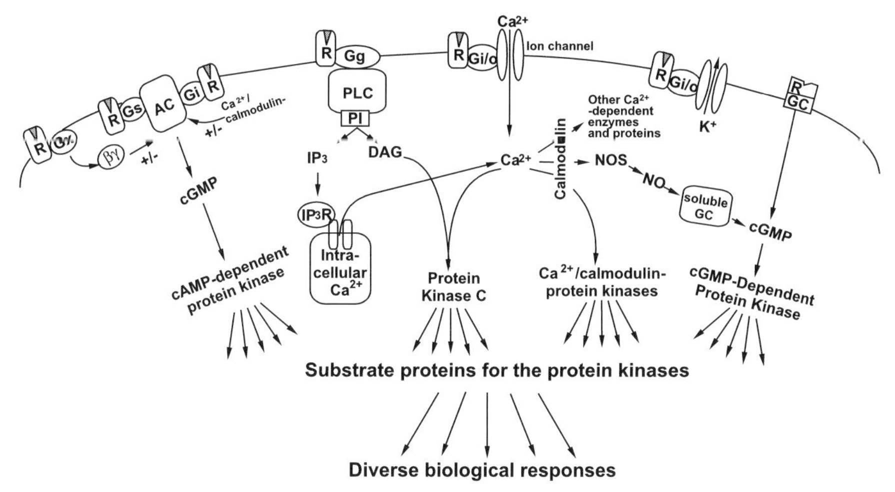

Fig. 1 Schematic diagram of the major second messenger pathways in the brain. Gs and Gi/o couple with neurotransmitter receptors (*R*) and mediate activation or inhibition of adenylyl cyclase, the enzyme that catalyzes the synthesis of cAMP. Also shown is the ability of G protein βγ subunits that are released from any type of G protein, or Ca^{2+}/calmodulin, to stimulate or inhibit different forms of adenylyl cyclase. Gq and possibly Gi/o underlie the ability of neurotransmitter receptors to regulate phospholipase C (*PLC*), which metabolizes phosphatidylinositol (*PI*) into the second messengers inositol triphosphate (IP_3) and diacylglycerol (*DAG*). IP_3 then acts on specific IP_3 receptors (IP_3R) to increase intracellular levels of free Ca^{2+} by releasing Ca^{2+} from internal stores. Increased levels of intracellular Ca^{2+} also result from the flux of Ca^{2+} across the plasma membrane through Ca^{2+} and other ion channels, which can be stimulated by nerve impulses and certain neurotransmitters. Increased Ca^{2+} levels activate nitric oxide synthase (*NOS*) and increase levels of nitric oxide, which in turn leads to the activation of cytoplasmic guanylyl cyclase (the enzyme that catalyzes the synthesis of cGMP). Membrane-bound receptors for certain peptides also have intrinsic guanylyl cyclase activity. Second messenger cascades produce cellular effects via activation of specific types of protein kinases. Brain contains one major type of cAMP-dependent protein kinase and cGMP-dependent protein kinase. These enzymes phosphorylate a specific array of substrate proteins, which can be considered third messengers. Brain contains two major classes of Ca^{2+}-dependent protein kinase. One is activated by Ca^{2+} and calmodulin and is referred to as Ca^{2+}/calmodulin-dependent protein kinase. The other major class is activated by Ca^{2+} in conjunction with DAG and various phospholipids and is referred to as Ca^{2+}/DAG-dependent protein kinase or protein kinase C. Phosphorylation alters the physiological activity of substrate proteins and thereby represents the biological responses of the extracellular messengers

rapidly induced at the level of gene expression). In addition, certain types of PDEs are regulated by G protein subunits or by Ca^{2+}/calmodulin. As discussed for the multiple types of cAMP synthesis regulation, the wide range and types of PDEs provide additional mechanisms for fine-tuning the cellular levels of these important signaling molecules. The PDEs also represent an important class of target for the development of a wide range of therapeutic agents. For example, rolipram, an inhibitor of cAMP-specific PDE type IV (PDE4) is reported to have efficacy in behavioral models of depression and in clinical trials (Duman et al. 2000).

2.1.2
cGMP and Nitric Oxide Second Messenger Cascades

The regulation of cGMP formation is very different from cAMP, and there are two primary mechanisms that have been identified (Duman and Nestler 1999; Nestler and Duman 1999). One mechanism involves activation of a membrane receptor that contains intrinsic guanylyl cyclase enzyme activity. There are a small number of peptides, such as atrial natriuretic factor, that have receptors with intrinsic guanylyl cyclase activity. The more common mechanism for formation of cGMP is via activation of cytosolic guanylyl cyclase. Cytosolic or soluble guanylyl cyclase is activated by nitric oxide (NO) that is formed by NO synthase (NOS). There are three major forms of NOS, neuronal, endothelial and inducible NOS, which are named based on tissue enrichment or mechanism of activation. However, all three forms of NOS are differentially expressed throughout the brain. NOS is activated by Ca^{2+} and the Ca^{2+}-binding protein calmodulin. and activation of neurotransmitter systems that increase the influx of Ca^{2+} or release this divalent cation from intracellular stores can activate NOS and cGMP formation. Like cAMP, the actions of cGMP are mediated by activation of cGMP-dependent protein kinase. NOS and cGMP signaling is one pathway that has been implicated in the anxiogenic actions of benzodiazepines and nitrous oxide and is discussed in more detail in the following sections.

2.1.3
Ca^{2+} and Phosphatidylinositol Signaling

Ca^{2+} is a key signaling molecule in the nervous system and has been implicated in many regulatory functions and neuronal plasticity, including learning and memory as well as conditioned fear, as discussed in Sect. 4.1 below. There are two general mechanisms for regulation of Ca^{2+} levels that involve either influx of extracellular Ca^{2+} or release of Ca^{2+} from intracellular stores (Fig. 1) (Duman and Nestler 1999). Influx of Ca^{2+} occurs through a number of mechanisms, including ionotropic receptors (e.g., activated nicotinic and *N*-methyl-D-aspartate or NMDA receptors), G protein-coupled receptor regulation of GIRKs, and neuronal depolarization which results in activation of voltage-gated Ca^{2+} channels. The control of intracellular Ca^{2+} release occurs primarily via activation of phospholipase C (PLC). Many types of G protein-coupled receptors can activate the β-form of PLC via interactions with specific G protein subtypes, most commonly $Gq\alpha$ but also $Gi\alpha$ and $Go\alpha$. Once activated, PLC catalyzes the breakdown of phosphatidylinositol into IP_3 and diacylglycerol (DAG), both of which act as second messengers. This second messenger system can influence the release of Ca^{2+} via activation of IP_3 receptors localized on intracellular organelles such as the endoplasmic reticulum. DAG also acts as a second messenger, in conjunction with Ca^{2+}, by activation of protein kinase C (PKC) (Nestler and Duman 1999). In addition, Ca^{2+}/calmodulin can also activate Ca^{2+}/calmodulin-dependent protein kinase (CaMK). There are multiple forms of both PKC and CaMK with different expression patterns and mechanisms of regulation (for additional information see Duman and Nestler 1999; Nestler and Duman 1999). As discussed for PKA, PKC and CaMK have a multitude of substrates spanning all types of cellular proteins and functions.

2.2
Neurotrophic Factor Signaling Pathways

The signaling mechanisms activated by neurotrophic factors, which include nerve growth factor (NGF), brain derived neurotrophic factor (BDNF) and neurotrophin-3 (NT-3) are fundamentally different from those discussed for G protein-coupled receptors and Ca^{2+} (Russell and Duman 2002). The neurotrophic factors bind to specific receptors, TrkA, TrkB, and TrkC (the name Trk is derived from their identification as troponin/receptor kinases from colon carcinoma) (Fig. 2). The Trk receptors contain an extracellular binding domain, a transmembrane domain, and an intracellular tyrosine kinase domain. Two neurotrophic factor molecules are required for activation of a Trk receptor dimer, resulting in activation of the tyrosine kinase domains and phosphorylation of substrate proteins as well as autophosphorylation of the Trk receptor itself.

There are at least three major effector pathways that are activated by neurotrophic factor–Trk receptors. The best-characterized pathway is the extracellular-regulated kinase (ERK) cascade, which is regulated by activation of Ras, a small membrane-bound G protein. Activation of Ras occurs when activated Trk receptor associates with adaptor proteins and a GTP exchange factor (see Russell and Duman 2002 for details). Ras in turn recruits and activates a serine threonine kinase, Raf, to the membrane resulting in the activation of ERK kinase (also referred to as MEK) and ERK (also known as mitogen activated protein kinase or MAPK). Activation of the Ras-Raf-MEK-ERK cascade can lead to regulation of many cellular proteins, including ribosomal S6-kinase (RSK).

Other pathways that are activated by Trk include phospholipase C-γ (PLC-γ) and phosphatidylinositol-3′-OH-kinase (PI-3-K). As discussed for PLC, PLC-γ also cleaves phosphatidylinositol into IP_3 and DAG, resulting in release of intracellular Ca^{2+} and activation of PKC. The mechanisms underlying the regulation of PI-3-K are not as well characterized. PI-3-K is able to bind phosphorylated Trk receptors, but it is more likely that it interacts with another family of proteins, insulin receptor substrate (IRS) proteins that bind to autophosphorylated Trk receptors. IRS proteins, including IRS1, IRS2, and IRS4

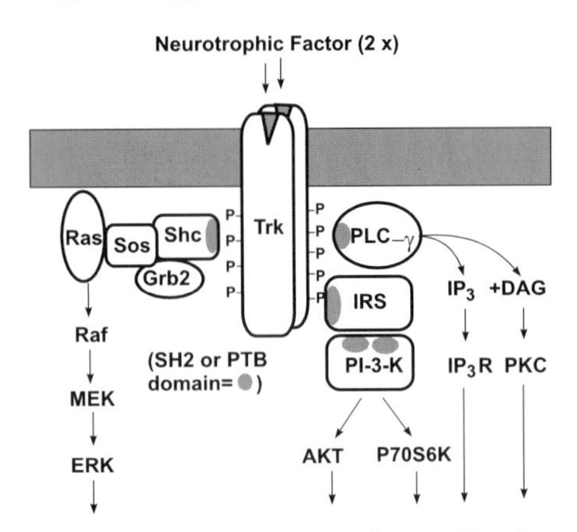

Fig. 2 Schematic illustration of neurotrophic factor signal transduction pathways in the brain. This scheme represents the three major signaling pathways that are activated by Trk-like tyrosine kinase receptors. The ligand, here a neurotrophic factor dimer, binds to its receptor and activates the tyrosine kinase activity. This results in autophosphorylation of the receptor as well as other substrate proteins (indicated by *P*). Phosphorylation also leads to formation of a protein complex in which docking proteins bind to the autophosphorylated receptor and are activated. These docking proteins bind to the receptor phosphorylation sites via src homology domain-2 (*SH2*) domains (i.e., for PLC-γ) or phosphotyrosine binding protein (*PTB*) domains (i.e., for IRS and Shc). The three pathways shown here are: (1) Ras becomes activated via the stimulation of the GDP/GTP exchange activity of son of sevenless (*Sos*), which is in a complex with Shc and Grb2. Activated Ras then turns the extracellular regulated kinase (*ERK*) cascade, including Raf, ERK kinase (*MEK*), and ERK. ERKs stimulate many known effectors, including ribosomal S6 kinase (RSK). (2) The phospholipase C-γ (*PLC*-γ) pathway leads to the production of inositol triphosphate (IP_3) and diacylglycerol (*DAG*) and regulation of the IP_3 receptor, intracellular Ca^{2+}, and protein kinase C (*PKC*). (3) Phosphatidylinositol-3-kinase (*PI-3-K*) binds to an insulin receptor substrate (*IRS*)-like adaptor protein and becomes activated. The IRS protein then interacts with the Trk receptor and then activates other signaling proteins. The PI-3-K then activates AKT and p70 S6-kinase (*P70S6K*). AKT has an anti-apoptotic effect. Each of these pathways then exerts a number of nuclear and non-nuclear actions with acute and long-term consequences for the cells

that are found in brain, bind to phosphorylated Trk and then the IRS proteins are phosphorylated themselves on tyrosine residues. PI-3-K then binds to the phosphorylated IRS tyrosine residues and become activated. The lipid product of PI-3-K, phosphatidylinositol-3′-phosphate, can then activate other protein kinases (see Russell and Duman 2002 for details).

2.3
Regulation of Transcription Factors and Gene Expression

Regulation of gene transcription factors by signal transduction pathways represents one of the primary mechanisms for effecting long-term changes in neuronal function. Transcription factors bind to specific sequences of DNA in the promoter regions of genes and can either increase or inhibit gene expression. The unique pattern of promoter elements governs the expression pattern of a particular gene in the adult and during development. There are two major mechanisms for regulation of transcription factors: activation of existing transcription factors (usually by phosphorylation) and induction of transcription factor expression. The best example of the first case is the cAMP response element binding protein (CREB). CREB is normally expressed and located in the nucleus, but transcriptional activity is significantly increased upon phosphorylation. CREB can be phosphorylated and activated by several different protein kinases discussed in this section, including PKA, PKC, CaMK, and RSK, underscoring the important role this transcription factor plays in neuronal function. Other transcription factors, including many members of the Fos family (i.e., c-Fos), are expressed at very low levels but are rapidly induced by a variety of extracellular stimuli, hence the term immediate early gene transcription factor. The newly transcribed c-Fos protein is then available to regulate expression of genes that contain activator protein-1 (AP-1) elements.

3
Cellular Mechanisms of Stress-Induced Structural Plasticity

Chronic or severe stress is a well-known precipitant of some forms of anxiety and depressive disorders, and it is likely that stress-induced alterations in neuronal plasticity may underlie functional changes. Stress exposure in experimental animals can result in dendritic remodeling and atrophy in hippocampal CA3 pyramidal neurons (Margarinos et al. 1996; Sousa et al. 2000; Watanabe et al. 1992) as well as decreased hippocampal volume and neurogenesis (Czeh et al. 2001; Gould et al. 1997). These findings, along with observations of decreased hippocampal volume seen clinically in association with PTSD and depression (Bremner et al. 1995; Sheline et al. 1996), and the observation that clinically effective antidepressant treatments can reverse stress-induced changes in neuronal structure (Magarinos et al. 1999), have suggested the potential relevance of stress-induced structural alterations to affective disorders (Duman et al. 2000; Manji et al. 2001; Nestler et al. 2002).

As in the hippocampus, dendritic remodeling occurs in the amygdala in response to stress exposure (Vyas et al. 2002). However, unlike the hippocampus where dendritic atrophy occurs, the amygdala shows dendritic hypertrophy in response to chronic immobilization stress. Exposure to immobilization stress also results in heightened emotionality in animals. Little is known about the signaling pathways that underlie the actions of stress in the hippocampus or amygdala. McEwen and colleagues have reported that decreased hippocampal neurogenesis by stress is dependent on NMDA receptor activation (i.e., the decrease is blocked by pretreatment with a NMDA receptor antagonist), which suggests that it is Ca^{2+} dependent (McEwen 1999). However, it is also possible that this effect of NMDA occurs outside the hippocampus. Activation of the cAMP-CREB cascade has been shown to increase neurogenesis and dendritic arborization of hippocampal neurons (Nakagawa et al. 2002). A role for extracellular proteolysis in mediating such structural changes has been suggested, and the serine protease tissue-plasminogen activator (tPA) has been studied in learning and activity-dependent plasticity in the hippocampus (Huang et al. 1996; Madini et al. 1999). Extracellular proteolysis mediated by tPA may also be important in the stress-induced regulation of plasticity including dendritic remodeling in the amygdala (Pawlak et al. 2003). Increased tPA in the central and medial nuclei of the amygdala has been shown after acute restraint stress. Mice in which *tPA* has been knocked out ($tPA^{-/-}$ mice) lack the stress-induced phosphorylation of ERK1/2 that is seen in wild-type mice, and stress-induced increases in GAP-43, which is a presynaptic protein used as a marker of axonal plasticity, are also absent in $tPA^{-/-}$ mice (Pawlak et al. 2003). Additionally, stress-induced increases in anxiety behavior do not occur in $tPA^{-/-}$ mice even though the hypothalamic-pituitary-adrenal (HPA) stress response of these mice is normal as assessed by corticosterone levels. These results suggest that tPA acts in the amygdala to facilitate stress-induced anxiety behavior and to promote cellular mechanisms of plasticity. The amygdala is thought to exert an excitatory drive on HPA axis function while the hippocampus inhibits the HPA axis (Allen and Allen 1974; Herman et al. 1989). It is suggested that the contrasting patterns of dendritic remodeling in response to stress in the amygdala and hippocampus could contribute to dysregulation of HPA axis function and comprise a candidate cellular substrate for behavioral consequences of chronic stress exposure.

4
Intracellular Signaling Pathways Involved in Fear Memory

Anxiety represents a state of heightened vigilance and fear, but pathological anxiety can be distinguished from fear in that it is inappropriately evoked and may persist in the absence of real threat or danger. The study of conditioned fear has provided detailed information on the neural circuitry and intracellular mechanisms that are important to fear responses and their long-term retention. The description of neural circuitry and the mechanisms underlying disorders of fear memory such as posttraumatic stress disorder (PTSD) may also be relevant to other anxiety states that share common neural substrates.

In classical Pavlovian fear conditioning, an initially neutral cue (conditioned stimulus, CS), through temporal pairing with an aversive unconditioned stimulus (US), acquires the ability to elicit a fear response in the absence of the US. This acquired ability represents a type of associative learning and implies that plasticity mechanisms underlying fear learning may be similar to those underlying other types of learning. Long-term potentiation (LTP) of synaptic transmission is favored as a cellular model for the plasticity underlying associative memory (Bliss and Collingridge 1993; Brown et al. 1988; Martin et al. 2000), and the cellular mechanisms of LTP are hypothesized to underlie the plasticity that is responsible for associative fear memory formation in the amygdala (Blair et al. 2001; Chapman et al. 1990; Huang and Kandel 1998; Rogan and LeDoux 1995; Schafe et al. 2001). Here we summarize the cellular mechanisms that are thought to be relevant to LTP and fear memory, including intracellular signaling involving cAMP–CREB, Ca^{2+}/calmodulin kinase, and MAPK pathways.

The lateral nucleus of the amygdala (LA) is the primary termination site for afferent pathways to the amygdala that carry sensory information used during fear conditioning, and the LA has been implicated as a critical site for the convergence of sensory information and for plasticity mechanisms underlying fear learning (LeDoux 2000; Schafe et al. 2001). Commonality between LTP and fear conditioning is suggested by the similarity in the enhancement of synaptic transmission in the sensory input pathways after fear conditioning and after artificial induction of LTP and by their similar sensitivities to stimulus contingencies (Bauer et al. 2001; McKernan and Shinnick-Gallagher 1997; Rogan et al. 1997).

4.1 Fear Conditioning and Ca^{2+}-Signaling

Calcium is an important mediator of neuronal plasticity and increased intracellular Ca^{2+} is a key intermediate in activity-induced activation of protein kinases in many systems. Regulation of Ca^{2+} signaling in the amygdala appears to be important for fear memory, and dependence on NMDA receptors and voltage-gated calcium channels (VGCC) has been demonstrated for fear conditioning and for LTP in the amygdala. The acquisition of fear conditioned responses is inhibited by NMDA receptor blockade (Campeau et al. 1992; Fanselow and Kim 1994; Miserendino et al. 1990) and this can be shown with a NR2B subunit-specific antagonist that does not appear to disrupt normal synaptic transmission (Rodrigues et al. 2001). Mice with regulated expression of a CaMKII transgene or that are deficient in a CaMKII or CaMKIV signaling have impaired long-term memory of fear conditioning (Kang et al. 2001; Mayford et al. 1996; Silva et al. 1996). Bauer et al. (2002) have shown that blocking NMDA receptors in the LA impairs both short- and long-term fear memory, while blockade of VGCCs selectively impairs long-term fear memory. LTP in the LA also depends on Ca^{2+} entry, and both VGCC-dependent and NMDA receptor-dependent components have been demonstrated (Bauer et al. 2002; Huang and Kandel 1998; Weisskopf et al. 1999). Bauer et al. (2002) demonstrate distinct NMDA receptor and VGCC-dependent forms of LTP in the LA in vitro and suggest that a combination of both contribute to the formation of fear memories in vivo. LeDoux and colleagues (Bauer et al. 2002; Blair et al. 2001; Schafe et al. 2001) have described cellular events during fear conditioning as including calcium entry through both NMDARs and L-type VGCCs in LA principal cells as a result of the associative pairing of CS and US. They suggest involvement of Ca^{2+} entry through NMDA receptors in short-term memory (STM), and Ca^{2+} entry through L-type VGCCs as critical to processes involved in long-term fear memory formation.

4.2
Fear Conditioning and cAMP-Signaling

cAMP signaling is important to many forms of neural plasticity including learning and memory (Silva et al. 1998), and evidence suggests that this signaling pathway is an regulator of the RNA and protein synthesis that regulate long-term fear memory and late-phase LTP in the amygdala. As with other forms of memory, fear memory can be divided into two temporal phases; short-term (STM), which is protein synthesis independent and a long-lasting form, which requires RNA and protein synthesis (Davis and Squire 1984). LTP similarly occurs in distinct temporal phases also with only the late phase (L-LTP) requiring macromolecular synthesis (Huang et al. 1994; Kandel 1997). The consolidation of fear memory but not STM, is blocked by inhibitors of protein synthesis or PKA (Bourtchuladze et al. 1998; Schafe et al. 1999), and this has been shown specifically in the lateral/basolateral amygdala (Schafe and LeDoux 2000). Long-term fear conditioning is enhanced when concentrations of cAMP are experimentally increased by the PDE inhibitor rolipram (Barad et al. 1998). PKA is also implicated in LTP. PKA is important in hippocampal L-LTP (Frey et al. 1993; Nguyen and Kandel 1996; Roberson and Sweatt 1996), and expression of L-LTP in the amygdala has likewise been shown to involve PKA (Huang et al. 2000). Selective reductions in the long-term forms of conditioned fear memory and LTP were shown in transgenic mice with reduced hippocampal PKA activity (Abel et al. 1997). The reduced PKA activity had no effects on STM or early-phase LTP, supporting the involvement of PKA-dependent processes specifically in protein synthesis-dependent phases. Another fear memory task, inhibitory avoidance, is also sensitive to modulation of cAMP signaling. Long-term but not short-term memory of an inhibitory avoidance task is modulated by D1/D5 receptor cAMP signaling in the hippocampus, also supporting a role for cAMP/PKA signaling in long-term fear memory (Bernabeu et al. 1997).

PKC signaling is also involved in the plasticity underlying fear learning and memory. Inhibition of PKC in rats impairs acquisition of fear conditioning (Goosens et al. 2000; Li et al. 2002) and mice with a knockout of the isoform of protein kinase C have deficient cued and contextual fear conditioning (Weeber et al. 2000). Inhibitors of PKC also block LTP (Malinow et al. 1989; Reyman et al. 1988).

CREB mediates experience-dependent plasticity in a variety of systems including a role in learning and memory (Silva et al. 1998), and this transcription factor is implicated in the translation of cAMP signaling into changes in the expression of genes that participate in fear memory consolidation. The phosphorylation of CREB and CRE-mediated transcription are associated with long-term memory and occur during fear conditioning (Bernabeu et al. 1997; Impey et al. 1998). An early phase of CREB phosphorylation occurs after fear conditioning, which may be a result of stress exposure, while a later phase is thought to relate to memory consolidation (Stanciu et al. 2001). CREB is also activated by LTP-inducing stimulation (Impey et al. 1996). Studies using viral vector-mediated overexpression of CREB in the amygdala showed CREB-dependent facilitation of long- but not short-term memory of fear-potentiated startle (Josselyn et al. 2001). Studies of mutant mice with deletions of CREB isoforms indicate that CREB function is required for intact long-term fear memory, while acquisition and STM are not altered (Bourtchuladze et al. 1994; Kogan et al. 1996). Transgenic mice that express an inducible CREB repressor also show a role for CREB in determining fear memory consolidation (Kida et al. 2002). These studies suggest that CREB-induced initiation of transcriptional changes as a result of fear experience are critical to the regulation of gene expression underlying fear memory. This function of CREB may be similar to its role in other forms of learning and memory in that it translates intracellular signaling changes resulting from neuronal activity into regulation of transcription within the nucleus. Other signaling pathways that regulate fear memory such as those involving CaMK and MAP kinase can also activate CREB, and their effects on fear memory may be partially due to regulation of CREB-dependent transcription.

4.3 Fear Conditioning and MAP Kinase Signaling

ERK/MAPK are activated by fear conditioning and following stimulation that induces L-LTP (Atkins et al. 1998; English and Sweatt 1996). Blockade of ERK/MAPK signaling in the amygdala impairs fear conditioning (Schafe et al. 2000) and also impairs L-LTP (English and Sweatt 1997; Huang et al. 2000). Evidence from mutant mice also suggests that Ras-GRF signaling via the Ras/MAP kinase pathway may be involved in the formation of long-term fear memory in the amygdala. Mice that lack the neuronal-specific exchange factor Ras-GRF are impaired in avoidance and fear conditioning tasks that require the amygdala but perform hippocampal-dependent behavioral tasks normally (Brambilla et al. 1997). These mice also show abnormal LTP in the basolateral amygdala (Brambilla et al. 1997). The involvement of MAPK signaling in fear memory further supports a role for MAPK signaling in mechanisms of plasticity that underlie a variety of types of learning and memory (Berman et al. 1998; Blum et al. 1999; Selcher et al. 1999; Sweatt 2001). PI-3-K is another intracellular transducer of growth factor signaling and it plays a role in the regulation of cell survival (Miller et al. 1997; Yao and Cooper 1995). PI-3-K is selectively activated in the amygdala following fear conditioning and after LTP-inducing tetanic stimulation (Lin et al. 2001). Inhibition of PI-3-K impairs fear memory and LTP, and blocks the activation of MAPK and CREB phosphorylation induced by tetanic stimulation, forskolin or fear conditioning (Lin et al. 2001). This suggests that a mechanism for PI-3-K involvement in fear conditioning might be through the activation of CREB and MAPK.

A model for the cellular processes underlying fear memory formation in the LA has been suggested by LeDoux and colleagues and is depicted in Fig. 3

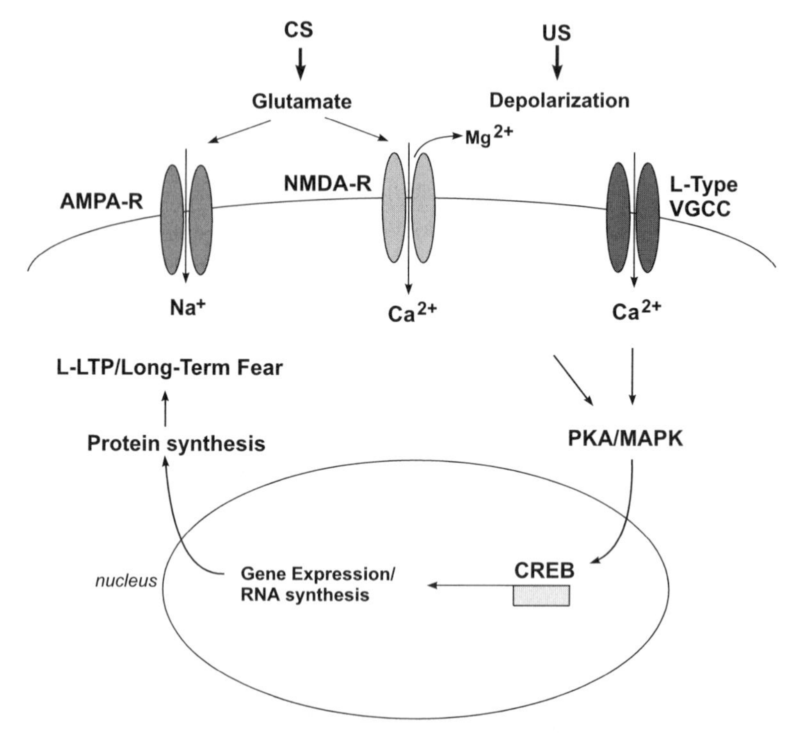

Fig. 3 A model for the cellular processes underlying fear memory formation in the lateral amygdala. Calcium entry through NMDA receptors (*NMDA-R*) and L-type voltage gated Ca channels (*VGCC*) is thought to occur in the lateral amygdala principal cells as a result of the associative pairing of a conditioned stimulus (*CS*) and an unconditioned stimulus (*US*). Ca^{2+} entry through NMDA receptors may contribute to STM. Ca^{2+} entry through L-type VGCCs and the consequent activation of protein kinases including PKA, CaM kinases and ERK/MAP kinases contribute to CREB phosphorylation and CREB-dependent gene expression. Intermediate steps going from Ca^{2+} to PKA/MAPK have not been fully elucidated, although several different cascades have been demonstrated in cell lines. The resulting changes in RNA and protein synthesis are thought to be responsible for the synaptic and behavioral plasticity that is experimentally observed as LTP and conditioned fear memory

(Bauer et al. 2002; Blair et al. 2001; Schafe et al. 2001). Calcium entry through NMDARs and L-type VGCCs is thought to occur in LA principal cells as a result of the associative pairing of a CS and US. Ca^{2+} entry through NMDA receptors may contribute to STM. Ca^{2+} entry through L-type VGCCs and consequent activation of protein kinases including PKA, CaMK, and ERK/MAP kinase are thought to contribute to CREB phosphorylation and CREB-dependent gene expression (Deisseroth et al. 1998; Mermelstein et al. 2000; Dolmetsch et al. 2001; Hardingham et al. 2001). The resulting changes in RNA and protein synthesis are thought to be responsible for the synaptic and behavioral plasticity that is experimentally observed as LTP and conditioned fear memory.

4.4 Extinction of Fear Memory

The extinction of fear memory occurs when a CS is presented without pairing with a US and results in a gradual decrement in conditioned responding. Extinction is thought to be an active process utilizing mechanisms similar to those involved in the acquisition of conditioned fear. Extinction is facilitated by enhanced signaling via NMDA receptors, PKA, MAPK, and CaMKII (Falls et al. 1992; Lu et al. 2001; Miserendino et al. 1990; Szapiro et al. 2003; Vianna et al. 2001), suggesting mechanistic similarities with long-term fear memory formation. Phosphatase activity may also be important in memory consolidation and extinction (Mansuy et al. 1998). Fear training induces phosphorylation of specific protein substrates and extinction training is accompanied by a reduction in this phosphorylation and increased phosphatase activity (Lin et al. 2003). These authors have suggested a role of calcineurin and have shown that inhibition of calcineurin impairs the extinction of fear memory and prevents the dephosphorylation of specific protein substrates. They suggest that Ca^{2+} influx through NMDA receptors in the amygdala during extinction training results in increased intracellular Ca^{2+} and consequent activation of protein kinases such as PI-3-K and MAPK. Subsequent activation of CREB and/or other transcription factors could promote calcineurin synthesis resulting in decreased phosphorylation of specific substrates, thereby weakening fear memory.

4.5 Modulation of Fear Learning

The work of McGaugh and colleagues on inhibitory avoidance (IA) memory has demonstrated that this memory can be modulated by mediators of the stress response including substances that are normally released by arousal, such as norepinephrine (NE), epinephrine, and glucocorticoids (reviewed in McGaugh et al. 2000). The effects of peripheral epinephrine and glucocorticoids on storage of IA memory are mediated by the release of NE in the amygdala consequent to the actions of these substances in the brain stem.

Other substances that exert modulatory influences on IA memory such as opioids or GABAergics also regulate consolidation of IA memory by regulating NE within the amygdala. Within the amygdala, the basolateral complex (BLA) has been implicated in mediating the neuromodulatory influences on memory storage. These studies have indicated that the amygdala is necessary for modulatory influences but does not appear to be a critical site for the consolidation of IA memory (Cahill and McGaugh 1998).

Signaling mechanisms involved in the modulation or consolidation of IA memory include activation of adenylyl cyclase (AC) through direct coupling with *B* adrenergic receptors, and $\alpha 1$ adrenergic receptor activity in the BLA may facilitate this effect (McGaugh et al. 2000). The activation of CaMKII in the amygdala and hippocampus appears to participate in the consolidation of IA memory, and may interact with PKA-dependent mechanisms in the hippocampus (Barros et al. 1999; Wolfman et al. 1994).

The ways in which learning and memory mechanisms are recruited by fear-inducing stimuli as well as knowledge about how a conditioned cue can come to *lose* its ability to elicit a fear response are of great significance to strategies for the clinical management of anxiety states, and may be particularly relevant to PTSD. Continued investigation of the intracellular/molecular mechanisms that allow fear responses to come under the control of cues in the environment and the elements that are critical to the persistence of such responses will help define new, more selective targets for pharmacological intervention.

5
Intracellular Signaling Pathways Involved in Anxiety

The cAMP/PKA pathway is the most studied signaling system with respect to long-term functional changes induced by drugs of a variety of classes and is also implicated in functional alterations induced by some anxiolytic drugs. cAMP-dependent pathways may also regulate experience-dependent plasticity mechanisms that promote anxiety. An example in support of this comes from knockout mice that are deficient in the AC8 isoform of adenylyl cyclase. AC8 is a calcium-stimulated AC that is normally present in brain areas involved in neuroendocrine and behavioral responses to stress such as thalamus, habenula, and paraventricular nucleus (PVN). AC8 knockout mice do not show the typical stress-induced increase in anxiety and do not have the stress-induced increases in phosphorylation of CREB seen in wild-type mice after restraint stress, suggesting that AC8 is required for these stress responses (Schaefer et al. 2000). Hippocampal LTD is also reduced in these mice. AC8 activation and consequent regulation of CREB-dependent genes could transduce changes in intracellular Ca^{2+} that occur after stress exposure which alter the induction of LTD and modify behavioral function (Schaefer et al. 2000).

Regulation of the cAMP-PKA pathway might also be important in the modulation of anxiety by CRF receptors. Clinical observations and the known role for CRF in the stress response have suggested that CRF might be important to stress-induced anxiety disorders and their treatment (Koob and Heinrichs 1999; Nemeroff et al. 1984; Vale et al. 1981). CRF activates Gs and cAMP signaling in hippocampus (Chen et al. 1986; Huag and Storm 2000). CRF-induced activation of the cAMP-PKA pathway can occur via interaction with Gs, Gq/11, and Gi; however, CRF-induced activation of the PLC-PKC path (via interaction with Gq/11) can also occur depending on the mouse strain (Blank et al. 2003). CRF-induced increases in hippocampal neuronal activity can occur via PKC or PKA also in a strain-dependent manner (Blank et al. 2003). The extent to which different subtypes of CRF receptors vs differential coupling of a given receptor are responsible for the differential activation of AC and PKC is unknown. Activation of these signaling pathways by CRF receptors could provide a basis for cellular and molecular plasticity that regulates behavioral responses to stress. Dysfunction or loss of plasticity in these pathways could lead to inappropriate responses to stress exposure.

6
Intracellular Signal Transduction Pathways in the Treatment of Anxiety

The most commonly prescribed treatments for anxiety are benzodiazepines and antidepressants. These therapeutic agents have very different mechanisms of action, time course of effect and side effect profiles. Benzodiazepines enhance the responsiveness of GABA, the major inhibitor neurotransmitter in the brain, by binding to the GABA/benzodiazepine (GABA/BZ) receptor complex. This results in rapid anxiolytic activity in rodent models and in humans, but can be accompanied by sedation. Antidepressants most often used for anxiety include the serotonin selective reuptake inhibitors (SSRIs), which block the serotonin (5-HT) transporter and increase synaptic levels of this monoamine. However, the therapeutic action of antidepressants for the treatment of anxiety, as well as depression, requires several weeks and sometimes months. This has led to the widely held hypothesis that adaptations, or neural plasticity changes, to the acute elevation of 5-HT are necessary for a therapeutic effect. Because of this hypothesis, there has been a great deal of research directed toward identification of the molecular and cellular adaptations to repeated antidepressant treatment. In contrast, there is much less known about the intracellular signaling that is relevant to the actions of benzodiazepines. This section will discuss the signaling adaptations resulting from antidepressants, as well as benzodiazepines. These effects will be discussed with regard to the signaling pathways regulated by fear and stress.

7
Intracellular Pathways Regulated by Antidepressant Treatment

Early studies to identify the long-term adaptations that underlie the actions of antidepressant treatment were focused on alterations in levels of monoamine receptors and transporters. This work has provided useful information demonstrating that multiple 5-HT and NE receptor subtypes are regulated by antidepressant treatment, presumably as a result of elevated synaptic levels of these monoamines. More recent studies have examined adaptations of intracellular signal transduction cascades that could represent common targets for antidepressants. This section will review the work on the cAMP and Ca^{2+} signaling pathways and how these pathways could contribute to the anxiolytic actions of antidepressants. This is not meant to be a comprehensive review of the intracellular pathways regulated by antidepressants, but is a focused discussion of these signaling cascades, which to date have received the most attention. It is very likely that there are many other pathways that are regulated by and that are important mediators of the actions of antidepressant treatment, and these other pathways will be the focus of future investigations.

7.1
Antidepressant Treatment Upregulates the cAMP-CREB Cascade

Antidepressant drugs block the reuptake or metabolism of NE and 5-HT and thereby increase the function of receptors coupled to these monoamines. One of the signal transduction cascades regulated by NE and 5-HT and linked to the actions of antidepressant drugs is the cAMP-CREB cascade (Fig. 4). There are several classes of NE and 5-HT receptors that directly couple to and activate the cAMP second messenger cascade, including β_1- and β_2-adrenergic receptors and 5-HT_4, 5-HT_6, and 5-HT_7 receptor subtypes. These receptors couple with the stimulatory G protein, Gs, and thereby activate adenylyl cyclase, the enzyme that catalyzes the formation of the second messenger cAMP. The cellular actions of cAMP are in turn mediated by activation of PKA, which can phosphorylate and regulate the function of many cellular proteins, including neurotransmitter receptors, ion channels, synthetic enzymes, other kinases, and transcription factors. Antidepressant treatment increases the levels of PKA in limbic brain regions, including the cerebral cortex (Nestler et al. 1989; Perez et al. 1989). This effect is dependent on chronic treatment, consistent with the time course for the therapeutic action of antidepressants. Different classes of antidepressants, including NE and SSRIs, increase PKA, indicating that this kinase is a common target of antidepressants.

There are many cellular targets of PKA that could contribute to the actions of antidepressants. Due to the requirement for long-term treatment, there has been interest in identifying the influence of antidepressants on gene expression. Early studies have demonstrated that antidepressant treatment increases the

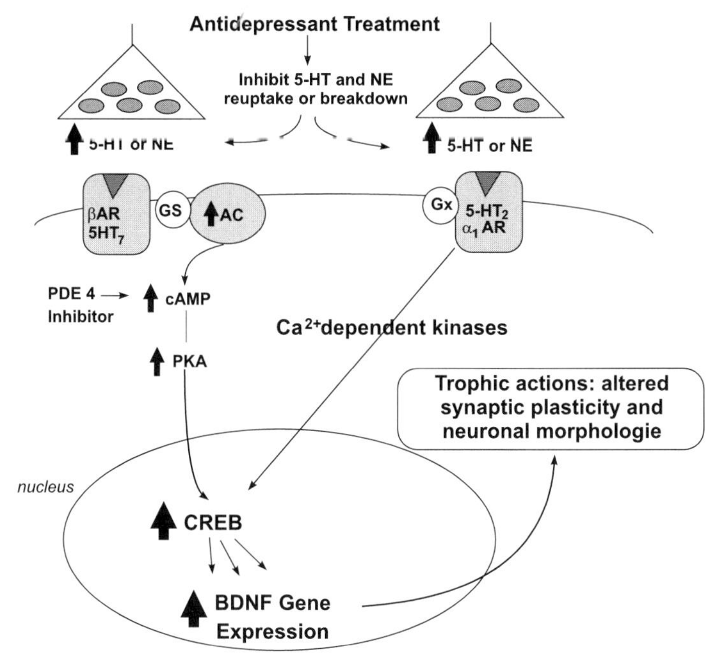

Fig. 4 Influence of antidepressant treatment on the cAMP–CREB cascade. Antidepressant treatment increases synaptic levels of NE and 5-HT via blocking the reuptake or breakdown of these monoamines. This results in activation of monoamine receptors, including β_1AR and 5-HT_7 receptors that are positively coupled to the cAMP-protein kinase A (*PKA*) cascade and α_1AR and 5-HT_2 receptors that can stimulate Ca^{2+}-dependent kinases. Chronic antidepressant treatment increases Gs coupling to adenylyl cyclase, particulate levels of cAMP-dependent protein kinase (PKA), and the function and expression of cAMP response element binding protein (*CREB*). Also shown is the action of a phosphodiesterases type IV (*PDE4*) inhibitor, which increases the function of the cAMP-PKA-CREB cascade by blocking the breakdown of cAMP. PDE4 inhibitors such as rolipram have been shown to have antidepressant efficacy in humans and animal models. CREB can also be phosphorylated by Ca^{2+}-dependent protein kinases, which can be activated by the phosphatidylinositol pathway as shown in Figs. 1 and 2, or by glutamate ionotropic receptors. Glutamate receptors and Ca^{2+}-dependent protein kinases are also involved in neural plasticity. One gene target of antidepressant treatment and the cAMP-CREB cascade is *BDNF*, which contributes to the cellular processes underlying neuronal plasticity and cell survival

translocation of PKA from a cytosolic fraction to a nuclear fraction, indicating that there may be phosphorylation and regulation of transcription factors. This possibility has been supported by studies demonstrating that the expression and function of CREB, as well as other gene transcription factors, is increased by antidepressant treatment. As with PKA, upregulation of CREB is dependent on chronic antidepressant treatment and is observed with different classes of antidepressants. The regulation of CREB has been examined using several different approaches. First, chronic antidepressant treatment increases levels of

CREB mRNA and protein in cerebral cortex and hippocampus (Nibuya et al. 1996). Second, antidepressant treatment increases levels of CREB phosphorylation (Thome et al. 2000). Third, antidepressant treatment increases levels of CRE-mediated gene expression, determined by analysis of CRE-reporter transgenic mice (Thome et al. 2000). These studies support the hypothesis that CREB is a common downstream target of antidepressant treatment.

7.2 Antidepressant Treatment Increases CaMKII

Antidepressant treatment is also reported to regulate CaMKII in limbic brain structures. Antidepressant treatment increases levels of CaMKII activity in presynaptic elements of hippocampus but not cerebral cortex (Popoli et al. 1995). This effect is dependent on chronic antidepressant treatment and is seen with selective 5-HT reuptake blockers, as well as nonselective agents. The effect was found to be selective for synaptic terminals as there was no change in levels of CaMKII activity in preparations that comprise primarily postsynaptic elements. Subsequent studies have also demonstrated that upregulation of CaMKII activity is accompanied by increased phosphorylation of markers of synaptic vesicles, synapsin I and synaptotagmin (Popoli et al. 1997a,b; Verona et al. 1998). These studies report a regional difference between 5-HT and NE selective reuptake inhibitors: SSRIs increase CaMKII activity and substrate phosphorylation in hippocampus only, while NE selective agents have effects in hippocampus and cerebral cortex. These studies suggest that antidepressant treatment may influence synaptic transmission via effects on presynaptic terminals. Moreover, because of the magnitude of the change and the widespread expression of CaMKII, it is likely that different types of transmitter function are influenced.

7.3 Influence of CREB in Models of Depression

Upregulation of CREB could play an important role in the actions of antidepressant treatment. To directly test this hypothesis, the influence of *CREB* expression in models of depression has been examined. For these studies, the expression of CREB or a dominant-negative mutant of *CREB* (mCREB) is increased by viral mediated gene transfer or by inducible transgenic expression. In addition, the influence of null mutation of one isoform, α-CREB, on behavior has been studied. Viral expression of wild-type CREB in the hippocampus, where antidepressant treatment increases CREB, results in an antidepressant-like effect in the forced swim (FST) and learned helplessness (LH) paradigms (Chen et al. 2001). The results demonstrate that increased expression of CREB is sufficient to produce an antidepressant effect and support the hypothesis that this transcription factor is an important target of antidepressant treatment.

The influence of CREB expression in other brain regions, most notably the nucleus accumbens, on behavioral models of depression has also been examined. In contrast to the hippocampus, viral mediated gene expression of CREB in the nucleus accumbens results in a pro-depressive effect in the FST and LH models of depression (Newton et al. 2002; Pliakas et al. 2001). Expression of a dominant-negative CREB in this region produces an antidepressant-like response. Studies of inducible transgenic mice that express a dominant negative mutant of *CREB* in the nucleus accumbens, as well as null mutation of *CREB*, support these findings (Conti et al. 2002; Newton et al. 2002). The difference between the nucleus accumbens and hippocampus could be related to different target genes in these two regions. In the hippocampus one of the target genes of CREB is *BDNF*, and infusions of BDNF into hippocampus also produce an antidepressant response in the FST and LH models (Shirayama et al. 2002). In the nucleus accumbens CREB regulates the expression of dynorphin, and increased expression of this neuropeptide could contribute to the aversive aspects of exposure to these models of depression (i.e., stress associated with exposure to water or uncontrollable footshock).

These studies demonstrate that CREB can influence behavior in these stress-related models used for identifying antidepressant agents, and that the effect depends on the brain region examined. It is also possible that CREB expression in these brain regions, as well as other regions such as the amygdala, will also influence behavior in anxiety models. A role for CREB in fear conditioning has been discussed, and it is possible that CREB and changes in gene expression and neural plasticity resulting from activation of CREB also influence anxiety. Additional studies will be necessary to further characterize the role of CREB in the etiology and treatment of anxiety behavior.

7.4 Therapeutic Actions of PDE4 Inhibitors

A role for the cAMP cascade in the actions of antidepressant treatment is supported by earlier studies demonstrating that inhibitors of phosphodiesterases type IV (PDE4), the high-affinity cAMP specific PDE that catalyzes the breakdown of cAMP, also produces an antidepressant response (see Duman et al. 2000). The PDE4 selective inhibitor, rolipram, has been shown to have antidepressant efficacy in animal models and in clinical trials (Duman et al. 1997; Duman et al. 2000). Rolipram was not further developed as a therapeutic agent because of its side effect profile, most noticeably nausea. However, there are several different isoforms of PDE4, PDE4A, PDE4B, and PDE4D, in the brain, and each of these has multiple splice variants that are differentially expressed. It is possible that one of these isoforms underlies the therapeutic actions of rolipram, and that another produces the side effects. Studies are currently underway to identify if this is the case and to produce isoforms selective PDE4 inhibitors. It is also interesting to speculate that differential expression of these isoforms could explain why rolipram produces an antidepressant response even when expression of CREB has different effects, depending on the brain region in which it is expressed.

There are also a few studies that have examined the influence of rolipram in rodent models of anxiety. This work demonstrates that rolipram administration is capable of producing anxiolytic effects in the light–dark test and the elevated plus maze (Griebel et al. 1991; Silvestre et al. 1999). This is not surprising given the therapeutic actions of antidepressant drugs, which also upregulate the cAMP-CREB cascade. Additional work will be necessary to determine the brain region and target genes that underlie the anxiolytic actions of rolipram.

8
Intracellular Pathways Regulated by Benzodiazepines

Mediation of the anxiolytic-like action of benzodiazepines through the enhancement of GABAergic transmission at the GABA-A/benzodiazepine receptor complex is well known. Benzodiazepines act at the γ-subunit of the GABA receptor complex to enhance chloride influx and thereby cause hyperpolarization of neurons. However, there is much less known about the role of intracellular signal transduction in the actions of benzodiazepines.

Cellular signaling involving nitric oxide (NO) has been implicated in the anxiolytic effects resulting from GABA/BZ receptor occupation and has also been implicated in the behavioral effects of the anesthetic gas nitrous oxide. NOS-like activity is found in brain regions associated with anxiety (Dinerman et al. 1994). Inhibition of NO production by pharmacologic inhibition of NOS can decrease the anxiolytic effects of a benzodiazepine, GABA-A agonists, or nitrous oxide (Caton et al. 1994; Quock and Nguyen 1992) and inhibition of NO function has a similar effect (Li et al. 2003a). These effects can be shown in different paradigms of experimental anxiety including the mouse elevated plus maze, staircase, or light–dark tests. The neuronal isoform of NOS, referred to as nNOS, is implicated in these effects (Li et al. 2001, 2003b). A role for NO production in mediating anxiolysis is supported by studies showing that pharmacologic inhibition of NOS in the absence of other drugs is anxiogenic (Lino de Oliveira et al. 1997; Vale et al. 1998), although opposite results were found in other studies (Dunn et al. 1998; Volke et al. 1997). NO is thought to have a neurotransmitter-like function and is implicated in the control of blood flow and in the regulation of neuronal excitability, as well as other functions including activity-dependent plasticity and learning and memory (Bogdan 2001; Nelson et al. 1997b; Prast and Philippu 2001; Schuman and Madison 1991). It is possible that the effects of NO signaling in modulating behavioral anxiety are related to plasticity functions of NO within neurons. Alternatively, the altered anxiety that results from knockout of another form of NOS expressed in endothelial cells, (eNOS) suggests that non-neuronal processes such as alterations in vascular function could also contribute to behavioral effects (Frisch et al. 2000).

9 Summary and Conclusions

Significant progress has been made in understanding the neurobiology of fear memory, including the signal transduction cascades that are activated and required for the formation of long-term fear memory. This has resulted in hypothesis-based novel targets and strategies for fear-related disorders, including PTSD. This progress has resulted in part from the well-defined models and neural circuitry for fear conditioning, which provides a system for testing the role of specific signaling molecules. Progress on anxiety has been limited to information resulting from the use of effective anxiolytic drugs, primarily the benzodiazepines and behavioral models that were designed to test for drugs of this class. However, this has not led to fundamental information regarding the neurobiology of anxiety. Development of better models of anxiety will provide information on the neural circuitry underlying this disorder. This information is critical for further characterization of the neurotransmitters and signaling cascades that control these neural pathways, and ultimately better therapeutic agents.

Acknowledgements This work is supported by USPHS grants MH45481 and 2 PO1 MH25642, a Veterans Administration National Center Grant for PTSD, and by the Connecticut Mental Health Center.

References

Abel T, Nguyen PV, Marad M, Deuel TAS, Kandel ER, Bourtchouladze R (1997) Genetic demonstration of a role for PKA in the late phase of LTP and in hippocampus-based long-term memory. Cell 88:615–626

Allen J, Allen CF (1974) Role of the amygdaloid complexes in the stress-induced release of ACTH in the rat. Neuroendocrinology 15:220–230

Atkins C, Selcher JC, Petraitis JJ, Trzaskos JM, Sweatt JD (1998) The MAPK cascade is required for mammalian associative learning. Nat Neurosci 1:602–609

Barad M, Bourtchouladze R, Winder DG, Golan H, Kandel E (1998) Rolipram, a type IV-specific phosphodiesterase inhibitor, facilitates the establishment of long-lasting long-term potentiation and improves memory. Proc Natl Acad Sci U S A 95:15020–15025

Barros D, Izquierdo LA, Sant'Anna MK, Quevedo J, Medina JH, McGaugh JL, Izquierdo I (1999) Stimulators of the cAMP cascade reverse amnesia induced by intra-amygdala but intrahippocampal KN-62 administration. Neurobiol Learn Mem 71:94–103

Bauer E, LeDoux JE, Nader K (2001) Fear conditioning and LTP in the lateral amygdala are sensitive to the same stimulus contingencies. Nat Neurosci 4:687–688

Bauer E, Schafe GE, LeDoux J (2002) NMDA receptors and L-type voltage-gated calcium channels contribute to long-term potentiation and different components of fear memory formation in the lateral amygdala. J Neurosci 22:5239–5249
Berman D, Hazvi S, Rosenblum K, Seger R, Dudai Y (1998) Specific and differential activation of mitogen-activated protein kinase cascades by unfamiliar taste in the insular cortex of the behaving rat. J Neurosci 18:10037–10044
Bernabeu R, Bevilaqua L, Ardenghi P, Bromberg E, Schmitz P, Bianchin M, Izquierdo I, Medina JH (1997) Involvement of hippocampal cAMP/cAMP-dependent protein kinase signaling pathways in a late memory consolidation phase of aversively motivated learning in rats. Proc Natl Acad Sci U S A 94:7041–7046
Blair H, Schafe GE, Bauer EP, Rodrigues SM, LeDoux JE (2001) Synaptic plasticity in the lateral amygdala: a cellular hypothesis of fear conditioning. Learn Mem 8:229–242
Blank T, Nijholt I, Grammatopoulos DK, Randeva HS, Hillhuse EW, Spiess J (2003) Corticotropin-releasing factor receptors couple to multiple G-proteins to activate diverse intracellular signaling pathways in mouse hippocampus: role in neuronal excitability and associative learning. J Neurosci 23:700–707
Bliss T, Collingridge G (1993) A synaptic model of memory: long-term potentiation in the hippocampus. Nature 361:31–39
Blum S, Adams AN, Dash PK (1999) A mitogen-activated protein kinase cascade in the CA1/CA2 subfield of the dorsal hippocampus is essential for long-term spatial memory. J Neurosci 19:3535–3544
Bogdan C (2001) Nitric oxide and the regulation of gene expression. Trends Cell Biol 11:66–75
Bourtchuladze R, Frenguelli B, Blendy J, Cioffi D, Schutz G, Silva AJ (1994) Deficient long-term memory in mice with a targeted mutation of the cAMP-responsive element-binding protein. Cell 79:59–68
Bourtchuladze R, Abel T, Berman N, Gordon R, Lapidus K, Kandel ER (1998) Different training procedures recruit either one or two critical periods for contextual memory consolidation, each of which requires protein synthesis and PKA. Learn Mem 5:365–374
Brambilla R, Gnesutta N, Minichiello L, White G, Roylance AJ, Herron CE, Ramsey M, Wolfer DP, Cestari V, Rossi-Arnaud C, Grant SG, Chapman PF, Lipp HP, Sturani E, Klein R (1997) A role for the Ras signalling pathway in synaptic transmission and long-term memory. Nature 390:281–286
Bremner JD, Randall P, Scott TM, Bronen RA, Seibyl JP, Southwick SM, Delaney RC, McCarthy G, Charney DS, Innis RB (1995) MRI-based measurement of hippocampal volume in patients with combat-related posttraumatic stress disorder. Am J Psychiatry 152:973–981
Brown T, Chapman PF, Kairiss EW, Keenan CL (1988) Long-term synaptic potentiation. Science 242:724–728
Cahill L, McGaugh JL (1998) Mechanisms of emotional arousal and lasting declarative memory. Trends Neurosci 21:294–299
Campeau S, Miserendino MJD, Davis M (1992) Intra-amygdala infusion of the N-methyl-D-asparate receptor antagonist AP5 blocks acquisition but not expression of fear-potentiated startle to an auditory conditioned stimulus. Behav Neurosci 106:569–574
Caton P, Tousman SA, Quock RM (1994) Involvement of nitric oxide in nitrous oxide anxiolysis in the elevated plus-maze. Pharmacol Biochem Behav 48:689–692
Chapman P, Kairiss EW, Keenan CL, Brown TH (1990) Long-term synaptic potentiation in the amygdala. Synapse 6:271–278

Popoli M, Zanotti S, Radaelli R, Gaggianesi C, Verona M, Brunello N, Racagni G (1997b) The neurotransmitter release machinery as a site of action for psychotropic drugs: effect of typical and atypical antidepressants. Soc Neurosci Abstr 2325

Prast H, Philippu A (2001) Nitric oxide as modulator of neuronal function. Prog Neurobiol 64:51 68

Quock R, Nguyen E (1992) Possible involvement of nitric oxide in chlordiazepoxide-induced anxiolysis in rats. Life Sci 51:255–260

Reyman K, Brodemann R, Kase H, Matthies H (1988) Inhibitors of calmodulin and protein kinase C block different phases of hippocampal long-term potentiation. Brain Res 461:388–392

Roberson E, Sweatt JD (1996) Transient activation of cyclic AMP-dependent protein kinase during hippocampal long-term potentiation. J Biol Chem 271:30436–30441

Rodrigues S, Schafe GE, LeDoux JE (2001) Intra-amygdala blockade of the NR2B subunit of the NMDA receptor disrupts the acquisition but not the expression of fear conditioning. J Neurosci 21:6889–6896

Rogan M, LeDoux JE (1995) LTP is accompanied by commensurate enhancement of auditory-evoked responses in a fear conditioning circuit. Neuron 15:127–136

Rogan M, Staubli U, LeDoux J (1997) Fear conditioning induces associative long-term potentiation in the amygdala. Nature 390:604–607

Russell DS, Duman RS (2002) Neurotrophic factors and intracellular signal transduction pathways. In: Davis KL, Charney D, Coyle JT, Nemeroff C (eds) Neuropsychopharmacology: the fifth generation of progress. Lippincott Williams and Wilkins, Philadelphia, pp 207–215

Schaefer M, Wong ST, Wozniak DF, Muglia LM, Liauw JA, Zhuo M, Nardi A, Hartman RE, Vogt SK, Luedke CE, Storm DR, Muglia LJ (2000) Altered stress-induce anxiety in adenylyl cyclase type VIII-deficient mice. J Neurosci 20:4809–4820

Schafe G, LeDoux JE (2000) Memory consolidation of auditory Pavlovian fear conditioning requires protein synthesis and protein kinase A in the amygdala. J Neurosci 20:1–5

Schafe G, Nadel NV, Sullivan GM, Harris A, LeDoux JE (1999) Memory consolidation for contextual and auditory fear conditioning is dependent on protein synthesis, PKA and MAP Kinase. Learn Mem 6:97–110

Schafe G, Atkins CM, Swank MW, Bauer EP, Sweatt JD, LeDoux JE (2000) Activation of ERK/MAP Kinase in the amygdala is required for memory consolidation of Pavlovian fear conditioning. J Neurosci 20:8177–8187

Schafe G, Nader K, Blair HT, LeDoux JE (2001) Memory consolidation of Pavlovian fear conditioning: a cellular and molecular perspective. Trends Neurosci 24:540–546

Schuman E, Madison DV (1991) A requirement for the intercellular messenger nitric oxide in long-term potentiation. Science 254:1503–1506

Selcher J, Atkins CM, Trzaskos JM, Paylor R, Sweatt JD (1999) A necessity for MAP kinase activation in mammalian spatial learning. Learn Mem 6:478–490

Sheline Y, Wany P, Gado MH, Csernansky JG, Vannier MW (1996) Hippocampal atrophy in recurrent major depression. Proc Natl Acad Sci U S A 93:3908–3913

Shirayama Y, Nakagawa S, Chen AC-H, Russel DS, and Duman RS (2002) Brain derived neurotrophic factor produces antidepressant effects in behavioral models of depression. J Neurosci 22:3251–3261

Silva A, Rosahl TW, Chapman PF, Marowitz Z, Friedman E, Frankland PW, Cestari V, Cioffi D, Sadhof TC, Bourtchuladze R (1996) Impaired learning in mice with abnormal short-lived plasticity. Curr Biol 6:1509–1518

Silva A, Kogan JH, Frankland PW, Kida S (1998) CREB and memory. Annu Rev Neurosci 21:127–148

Silvestre J, Fernandez AG, Palacios JN (1999) Effects of rolipram on the elevated plus-maze test in rats: a preliminary study. J Psychopharmacol 13:274–277

Sousa N, Lukoyanov NV, Madeira MD, Almeida OF, Paula-Barbosa MM (2000) Reorganization of the morphology of hippocampal neurites and synapsed after stress-induced damage correlates with behavioral improvements. Neuroscience 97:253–266

Stanciu M, Radulovic J, Spiess J (2001) Phosphorylated cAMP response element binding protein in the mouse brain after fear conditioning: relationship to Fos production. Mol Brain Res 94:15–24

Sweatt J (2001) The neuronal MAP kinase cascade: a biochemical signal integration system subserving synaptic plasticity and memory. J Neurochem 76:1–10

Szapiro G, Vianna MRM, McGaugh JL, Medina JH, Izquierdo I (2003) The role of NMDA glutamate receptors, PKA, MAPK, and CAMKII in the hippocampus in extinction of conditioned fear. Hippocampus 13:53–58

Thome J, Sakai N, Shin KH, Steffen C, Zhang Y-J, Impey S, Storm DR, Duman RS (2000) cAMP response element-mediated gene transcription is upregulated by chronic antidepressant treatment. J Neurosci 20:4030–4036

Vale A, Green S, Montgomery AM, Shafi S (1998) The nitric oxide synthesis inhibitor L-NAME produces anxiogenic-like effects in the rat elevated plus-maze test, but not in the social interaction test. J Psychopharmacol 12:268–272

Vale W, Spiess J, Rivier C, Rivier J (1981) Characterization of a 41-residue ovine hypothalamic peptide that stimulates secretion of corticotropin and β-endorphin. Science 213:1394–1397

Verona M, Zanotti S, Gggianesi C, Schafer T, Racagni G, Popoli M (1998) Modulation of protein-protein interaction in presynaptic terminals: a possible sit of action for psychotropic drugs. Int J Neuorpsychopharm Suppl 1

Vianna M, Szapiro G, McGaugh JL, Medina JH, Izquierdo I (2001) Retrieval of memory for fear-motivated training initiates extinction requiring protein synthesis in the rat hippocampus. Proc Natl Acad Sci U S A 98:12251–12254

Volke V, Soosaar A, Koks S, Bourin M, Mannisto PT, Vasar E (1997) 7-Nitroindazole, a nitric oxide synthase inhibitor, has anxiolytic-like properties in exploratory models of anxiety. Psychopharmacology (Berl) 131:399–405

Vyas A, Mitra R, Shankaranarayana Rao BS, Chattarji S (2002) Chronic stress induces contrasting patterns of dendritic remodeling in hippocampal and amygdaloid neurons. J Neurosci 22:6810–6818

Watanabe Y, Gould E, McEwen BS (1992a) Stress induces atrophy of apical dendrites of hippocampal CA3 pyramidal neurons. Brain Res 588:341–345

Weeber E, Atkins CM, Selcher JC, Varga AW, Mirnikjoo B, Paylor R, Leitges M, Sweatt JD (2000) A role for the β isoform of protein kinase C in fear conditioning. J Neurosci 20:5906–5914

Weisskopf M, Bauer EP, LeDoux JE (1999) L-Type voltage-gated calcium channels mediate NMDA-independent associative long-term potentiation at thalamic input synapses to the amygdala. J Neurosci 19:10512–10519

Wolfman C, Fin C, Dias M, Bianchin M, Da Silva RC, Schmitz PK, Medina JH, Izquierdo I (1994) Intrahippocampal or intraamygdala infusion of KN62, a specific inhibitor of calcium/calmodulin-dependent protein kinase II, causes retrograde amnesia in the rat. Behav Neural Biol 61:203–205

Yao R, Cooper GM (1995) Requirement for phosphatidyl-inositol 3-kinase in the prevention of apoptosis by nerve growth factor. Science 267:2003–2006

HEP (2005) 169:335–369

Neuropeptides in Anxiety Modulation

R. Landgraf

Max Planck Institute of Psychiatry, Kraepelinstr. 2-10, 80804 Munich, Germany
landgraf@mpipsykl.mpg.de

Abstract This review is focused on the involvement of neuropeptides in the modulation of physiological and pathological anxiety. Neuropeptides play a major role as endogenous modulators of complex behaviours, including anxiety-related behaviour and psychopathology, particularly due to their high number and diversity, the dynamics of release patterns in distinct brain areas and the multiple and variable modes of interneuronal communication they are involved in. Manipulations of central neuropeptidergic systems to reveal their role in anxiety (and often comorbid depression-like behaviour) include a broad spectrum of loss-of-function and gain-of-function approaches. This article concentrates on those neuropeptides for which an involvement as endogenous anxiolytic or anxiogenic modulators is well established by such complementary approaches. Particular attention is paid to corticotropin-releasing hormone (CRH) and vasopressin (AVP) which, closely linked to stress, neuroendocrine regulation, social behaviour and learning/memory, play critical roles in the regulation of anxiety-related behaviour of rodents. Provided that their neurobiology, neuroendocrinology and molecular–genetic background are well characterized, these and other neuropeptidergic systems may be promising targets for future anxiolytic strategies.

Keywords Fear · Vasopressin · CRH · Oxytocin · NPY · CCK

1 Neuropeptides

Neuropeptides are biologically active sequences of amino acids that are produced in and released from distinct populations of neurons and are capable of influencing functional parameters of target neurons via G protein-coupled receptors. Their role in interneuronal communication (including feed-back actions on their own neurons) is based on actions as neurotransmitters, neuromodulators or both; additionally, secreted into the systemic circulation, neuropeptides may act as hormones (for review see Landgraf and Neumann 2004). As transmitters, neuropeptides contribute to the synaptic mode of information transfer, which refers to fast point-to-point signalling including transient actions, which are limited to postsynaptic sites. As modulators, they are non-synaptically released from multiple sites of the neuronal membrane, particularly from dendrites, and act on relatively distant targets. The distribution of a neuropeptide, in addition to its widespread release, is achieved because it persists in the extracellular fluid for long periods and is thus able to diffuse considerable distances (range of hundreds of micrometers). Accordingly, this signalling is not primarily defined by its topology, but by the chemistry of the modulator and the distribution of its receptor(s) on the target neurons. Through this type of information transfer (also called volume transmission, Fuxe and Agnati 1991), the brain is liberated from the constraints of wiring, since neuropeptides can reach any point of target neurons, thus enormously increasing the information handling capacity of neurons. This explains why a few "classical" neurotransmitters are sufficient for the myriads of wired pathways, whereas a large number of different neuropeptides are essential to define volume transmission by their chemistry.

Genomic mechanisms regulating neuropeptides and their receptors are paramount. As primary products of protein biosynthesis, both neuropeptides and their receptors are prone to direct structural changes by mutations. In addition to genetic polymorphisms in the coding region, resulting in structurally changed gene products, even subtle variations in the promoter structure or other components of the transcriptional machinery of genes can alter the pattern of neuropeptide release and/or receptor distribution in the brain with consequent changes in neuroendocrine and behavioural characteristics. Due to their resulting remarkable number and diversity, the dynamics of their intracerebral release patterns, and the multiplicity of receptors to which they bind, neuropeptides are considered ideal neuromodulator candidates.

While the distinction between transmitters, modulators and hormones has its heuristic value explaining, for instance, high speed, spatial precision, and a theoretically unlimited variability in signalling, probably even simple information transfers use a combination of these modes of communication. Responses to a neuropeptide, in other words, are likely to reflect a combination, from synaptic through non-synaptic to hormonal actions, often in a synergistic manner (Landgraf 1995; Landgraf and Neumann 2004). While it is debatable, in this context, whether neuropeptides first evolved as transmitters, modulators or hormones, it seems to be clear that they pre-date the "classical" transmitters as neuronal signalling molecules.

Whatever the underlying modes of action may be, endogenous neuropeptides are known to be involved in a broad range of behavioural regulation, including anxiety. Dependent on the multiple mechanisms of action of neuropeptides, their structural diversity and the complexity of the behavioural feature of interest, the methodological approaches to shed more light on this involvement differ and need consideration, particularly to justify comparability among different findings. A major approach in this context is to manipulate the neuropeptide of interest and its receptor(s), including neuropeptide–receptor interactions, and subsequently to examine behavioural consequences. This includes administration of synthetic neuropeptides, their receptor agonists or antagonists, antisense targeting, virally mediated gene transfer, RNA interference, knockout and transgenic strategies, all of them having their own advantages and limitations. In this context it is of note that it is generally difficult to mimic the dynamic temporal and spatial pattern of central neuropeptide release simply by administration of the synthetic neuropeptide. The nonspecific effects triggered in this way are often indistinguishable from specific ones. But even the more physiological approach of receptor antagonist administration may bear the risk of changing receptor characteristics (e.g. upregulation or supersensitivity, which may cancel the desired effect), cross-reacting with related receptors and having mixed antagonist/agonist-like properties. Another major approach alternative to external manipulation aims at monitoring neuropeptide release patterns within the brain and their physiological significance (Landgraf 1995). An appropriate experiment includes, for example, microdialysis of a distinct brain area of a freely behaving animal to monitor local neuropeptide release patterns prior to, during and after exposure to an anxiogenic stimulus. Inverse microdialysis of the corresponding receptor antagonist might then reveal the physiological significance of the centrally released neuropeptide for behavioural regulation, including stress coping strategies.

Measurements of neuropeptide contents in distinct brain areas, often used to describe neuropeptide responses to stressor exposure, should be interpreted with caution. As a matter of fact, contents are determined by a variety of variables including synthesis, transport, storage, release and degradation and are, thus, hardly to be interpreted in terms of dynamics of central synthesis and release that are much better reflected, for instance, by a combination of in situ hybridization and microdialysis in vivo. Although, in clinical trials, there is often no alternative other than determining neuropeptide levels in the cerebrospinal fluid, such trials essentially reflect the more or less "global" activity of the corresponding neuropeptide in the brain. Furthermore, once in the cerebrospinal fluid following transport via bulk flow and diffusion from the extracellular fluid, the neuropeptide is probably no longer of biological significance (Landgraf and Neumann 2004).

Although sometimes difficult, it is important to distinguish between physiological and pharmacological approaches. While not likely to contain biologically active endogenous neuropeptides, the cerebrospinal fluid may be used as a vehicle to transport exogenous neuropeptides against the naturally occurring concentration gradient across the ependyma into the brain parenchyma following intracerebroventricular administration (Bittencourt and Sawchenko 2000). Likewise, the role of the blood–brain barrier varies depending on physiological versus pharmacological conditions. The former provide a barrier that the endogenous neuropeptides cannot penetrate in physiologically relevant amounts, the more so as neuropeptide concentrations in the extracellular fluid of the brain may be orders of magnitude higher than in plasma (Landgraf 1995), thus excluding the need of a blood-to-brain transport. This is supported by the capacity of the brain to release neuropeptides centrally (i.e. within the brain) and peripherally (i.e. into the systemic circulation) in an independently (but often co-ordinated) regulated manner (Neumann et al. 1993; Landgraf 1995; Wotjak et al. 1998; Ludwig et al. 2002); this regulatory capacity probably coevolved with the blood–brain barrier. If, however, pharmacological doses are peripherally administered, the exogenous neuropeptide may reach the brain parenchyma in functionally significant amounts, particularly via circumventricular organs such as the subfornical organ (Ermisch et al. 1985, 1992). Simultaneously, the transport across the blood–brain barrier, unlikely to be relevant for endogenous neuropeptides, might become significant if the plasma concentration of a given exogenous neuropeptide reaches a certain threshold.

Compelling evidence for a critical role of a neuropeptide in complex behaviours such as anxiety is only provided by a combined approach comprising, for instance, changes in anxiety-related behaviour after selective manipulation of the respective neuropeptidergic system and, vice versa, responses of the endogenous system to an adequate behavioural challenge. This short review will predominantly focus on those neuropeptides that fulfil the requirements of having been tested in multiple approaches, thus providing the basis for comparability.

2 Anxiety-Related Behaviour

Negative emotions such as anxiety are founded on circuits in the brain that evolved to facilitate survival and reproduction in a dangerous and challenging environment. Trait anxiety reflects a genetic predisposition, i.e. hard-wired basal anxiety including reactions to danger. While some animals rely mainly on these reactions, mammals, including rodents and humans, are able to make the transition from reaction to action, including anticipation. Based particularly on forebrain expansion and the development of the emotional brain, action comprises both emotional and cognitive factors that have trait (i.e. genetically predisposed) roots. The wide range of anxiety-related behaviour is orchestrated by a system of many genes, each probably with small effects (Glatt and Freimer 2002; Tabor et al. 2002). Accordingly, all signalling circuits in probably all brain areas are involved directly or indirectly in regulatory patterns underlying anxiety. There is no doubt that neuropeptides fulfil the requirements to play a major role in this context. Their characteristics, including the high number and diversity of neuropeptides and receptors and their role in multiple and variable modes of interneuronal communication (Landgraf and Neumann 2004), make a co-evolution with the ever-growing number of emotional facets likely.

Anxiety may be interpreted as an emotional anticipation of an aversive scenario, difficult to predict and control, that is likely to occur. Fear is not seen as a basal state, but a complex response (including freezing, startle, increased vigilance) elicited during danger to facilitate appropriate defensive behaviours that can reduce danger or injury (e.g. avoidance, escape). Emotionality, often used as a synonym for anxiety as well as fearfulness, may be seen in a broader sense, comprising both inborn anxiety and stimulus-related fear. Despite anxiety being seen by some authors as independent of fear, the distinction between these two constructs is often difficult.

How representative are animal models to study the interaction between anxiety and neuropeptides? Most importantly, the same circuits underlie both physiological and pathological anxiety; in other words, pathological anxiety evolves from normal anxiety and fear (Rosen and Schulkin 1998). Furthermore, there are many neuroanatomical parallels in rodent and human anxiety, including neuropeptidergic circuits in hypothalamic and limbic brain areas.

The choice of the proper genotype and the dependent endophenotype is becoming one of the major problems in behavioural neuroscience. Although it is far from being trivial to definitely determine anxiety in subjects that, unlike humans, cannot self-report their emotional status or comply with questionnaires, animal models provide the advantage that the substrate for neuropeptide-anxiety interactions, namely the brain, is accessible. Additionally, the multidimensionality of emotionality (Ramos and Mormede 1999), a sometimes ignored behavioural phenomenon, may be determined in a variety of appropriate behaviour tests, including the elevated plus maze, the open field, and the light–dark test, all of which represent naturally occurring paradigms where rodents are challenged by the conflict between anxiety/fear on the one hand, and exploratory curiosity on the other. In particular, the amount of time spent in the risky environment relative to the safe one is used as an index of the animal's level of anxiety/fear. Caution in interpreting behavioural data is suggested because of methodological differences across studies, different animal models used, etc.

3 Neuropeptides Involved in the Regulation of Anxiety-Related Behaviour

3.1 Corticotropin-Releasing Hormone

Among the various neuropeptide systems that have been implicated in the regulation of anxiety-related behaviour, the corticotropin-releasing hormone (CRH) system plays a major role. CRH-containing circuits mainly originate in the parvocellular subdivision of the hypothalamic paraventricular nucleus (PVN); CRH neurons with a neuroendocrine role in the regulation of the hypothalamo–pituitary–adrenocortical axis (HPA) co-produce vasopressin (AVP) and terminate at the median eminence to secrete the corticotropin (ACTH) secretagogues into the portal blood; ACTH in turn stimulates the secretion of glucocorticoids from the adrenal cortex. CRH neurons involved in behavioural regulation are supposed to produce CRH alone and project to other brain areas, including the noradrenergic locus coeruleus and the central nucleus of the amygdala, areas of recognized importance in anxiety (Charney et al. 1998). Most extra-hypothalamic CRH neurons are located within the central nucleus of the amygdala, from where they project to, among other targets, the PVN (Gray 1993), the bed nucleus of the stria terminalis (Sakanaka et al. 1986) and the locus coeruleus (Koegler-Muly et al. 1993). In contrast to the PVN, there are glucocorticoid-mediated positive effects on the amygdaloid CRH system (Makino et al. 2002).

Similar to CRH, its receptor subtypes are widely distributed throughout the brain, including areas that have been implicated in the mediation of anxiety-related behaviour. Two types of CRH receptors, CRHR-1 and CRHR-2, the latter with different splice variants, have been identified in the brain (Perrin and Vale 1999). While the CRHR-1 is the main receptor subtype at the anterior pituitary level, CRHR-2α predominates at the level of the PVN (van Pett et al. 2000). CRH binds with high affinity to the CRHR-1, while binding with 15-fold lower affinity to the CRHR-2 (Vaughan et al. 1995). Urocortin, a second CRH-related neuropeptide, hypothesized to be an endogenous ligand for the CRHR-2, binds with equally high affinity to both receptor subtypes (Vaughan et al. 1995). Recently, urocortin II/stresscopin-related peptide, a 38 amino acid member of the CRH neuropeptide family, was identified in the mouse brain and has been shown to be equipotent in its binding affinity at the CRHR-2, but 1,000-fold more selective in binding to the latter compared to urocortin (Reyes et al. 2001; Skelton et al. 2000b). Furthermore, urocortin III/stresscopin with high affinity for the CRHR-2 has been identified (Hsu and Hsueh 2001; Lewis et al. 2001). In addition to its receptor subtypes, CRH has been shown to bind to the CRH-binding protein (Potter et al. 1991), postulated to function as an endogenous buffer for the actions of the CRH family of ligands at their receptors (Behan et al. 1996).

The presence of multiple CRH-related neuropeptides and CRH receptors gives this system enormous versatility and plasticity. A plethora of preclinical and clinical data indicates that the CRH system is involved in mediating behavioural (Heinrichs and Koob 2004; Koob et al. 1993), neuroendocrine (Vale et al. 1981), and autonomic (Dunn and Berridge 1990) responses to stressors. Central administration of CRH, remarkably shown to increase both CRH mRNA and CRHR-1 mRNA in the PVN (Imaki et al. 1996; Mansi et al. 1996; but see Makino et al. 1997, 2002), mimics the behavioural responses to stress, leading to increased anxiety-related behaviour (Heinrichs and Joppa 2001; Smagin et al. 2001) and locomotor activation (Sutton et al. 1982). Importantly, these behavioural effects appear to occur independently of the HPA axis (Britton et al. 1986; Smith et al. 1998), suggesting a central site of action.

While centrally administered CRH produced changes analogous to those seen in both anxiety and depression, central administration of (1) a CRH receptor antagonist (Dunn and Berridge 1990; Koob et al. 1993), (2) a CRH mRNA antisense oligodeoxynucleotide (Skutella et al. 1994) or (3) CRH receptor mRNA antisense oligodeoxynucleotides (Liebsch et al. 1995, 1999; Heinrichs et al. 1997; Skutella et al. 1998) produced anxiolytic effects in the rat (for review see Takahashi 2001). Recent studies by Heinrichs et al. (1997) and Liebsch et al. (1999) comparing CRHR-1 and CRHR-2 antisense oligodeoxynucleotides provided evidence that the anxiogenic actions of CRH are predominantly mediated by CRHR-1 rather than CRHR-2, the latter probably being involved in stress coping (Liebsch et al. 1999). The hypothesis suggesting different roles of the CRH receptor subtypes in mediating stress-induced behavioural effects has recently been confirmed by Skelton et al. (2000a), who showed that alprazolam, a benzodiazepine agonist, decreases CRHR-1 levels but upregulates both CRHR-2 levels and its putative ligand urocortin, indicating that an aspect of the efficacy of benzodiazepines in treating anxiety disorders is their ability to increase the activity of a neurobiological system involved in coping effectively in stressful and anxiety-provoking situations. Steckler and Holsboer (1999) suggested that CRHR-1 may be more concerned with cognitive aspects of behaviour, including learning and memory, emotionality, and attention, whereas CRHR-2 primarily influences processes necessary for survival, including feeding, reproduction and defence.

The time-dependent induction of anxiogenic-like effects of CRH and related neuropeptides has recently been addressed by Spina et al. (2002). Unlike CRH, urocortin did not produce anxiogenic-like effects when rats were tested 5 min after i.c.v. administration. However, 30 min after administration, urocortin was more potent than CRH in inducing anxiogenic-like effects, a finding supported also by Sajdyk et al. (1999). Spina et al. (2002) hypothesized that CRH and urocortin possibly maintain an independent physiological role in non-stressful conditions, and their roles overlap during anxiogenic, stressful events. Urocortin II was shown not to share the motor activating and anxiogenic-like properties of CRH, but induced a delayed anxiolytic-like action as measured on the elevated plus maze (Valdez et al. 2002). While this finding is consistent with a functional antagonism between CRHR-1 and CRHR-2 in regulating anxiety-related behaviour, there are lines of evidence conflicting with this hypothesis, suggesting that activation of both CRH receptor subtypes leads to increased anxiety. A recent study, for instance, showed that urocortin II produces a dose-dependent increase in anxiety-related behaviour in the plus maze test (Pelleymounter et al. 2002). Likewise, i.c.v. injection of antisauvagine-30, a CRHR-2 antagonist, produced anxiolytic-like effects in the mouse (Pelleymounter et al. 2002) and rat (Takahashi et al. 2001). Along the same lines, lesions of the lateral septum, an area high in CRHR-2 expression (Lovenberg et al. 1995), produced anxiolytic-like effects (Menard and Treit 1996), and antisense targeting of CRHR-2 in this area also attenuated fear conditioning (Ho et al. 2001). Astressin, a non-selective CRH receptor antagonist, but not antisauvagine-30, impaired CRH-enhanced fear conditioning when injected into the hippocampus; both of these antagonists, however, attenuated fear conditioning when injected into the lateral septum (Radulovic et al. 1999). Mice carrying a null mutation of the urocortin gene showed heightened anxiety-related behaviours in several tests (Vetter et al. 2002). In this particular case, the behavioural effect was not due to over-expression of CRH or CRHR-1. The authors hypothesized that urocortin-immunoreactive neurons in the Edinger–Westphal nucleus may modulate anxiety in opposition to the actions of CRH itself, and possibly through CRHR-2 in the lateral septum. Bakshi et al. (2002) recently described that acute and selective antagonism of CRHR-2 in the lateral septum reduced stress-induced defensive behaviour. CRHR-2 might thus represent a potential target for the development of novel CRH system anxiolytics (Bakshi et al. 2002).

The anxiogenic effects of endogenous and exogenous CRH have been hypothesized to be mediated through actions on both the locus coeruleus noradrenergic and the caudal dorsal raphé nucleus serotonergic systems. In both the locus coeruleus and the dorsal raphé nucleus, CRH is presumably acting at the CRHR-1 subtype (van Pett et al. 2000). The former involves synaptic contacts between CRH terminals originating in the amygdala (Koegler-Muly et al. 1993) and dendrites of noradrenergic cells; accordingly stressor exposure modulated increases in CRH immunoreactivity and efficacy in the locus coeruleus (van Bockstaele et al. 1996; Curtis et al. 1999). As in the amygdala, the locus coeruleus has long been implicated as an essential component of the neural substrates underlying anxiety and fear (Charney et al. 1995). Administration of anxiolytic benzodiazepines did not only lead to a reduction of HPA axis activity, but also to a decreased CRH content in the locus coeruleus (Owens et al. 1993). Similar to the noradrenergic system, CRH may play a role in regulating serotonergic neurotransmission originating in the dorsal raphé nucleus known to be involved in anxiety-related behaviour (Lowry et al. 2000; Hammack et al. 2002; Umriukhin et al. 2002). Lowry et al. (2000) have recently identified a population of serotonin neurons in the caudal dorsal raphé nucleus that are involved in the modulation of anxiety and are potently excited by CRH. Interestingly, chronic treatment with antidepressant drugs, including dual serotonin/noradrenaline reuptake inhibitors, resulted in a diminished sensitivity of CRH neurons to stressor exposure (Stout et al. 2002).

A major approach for studying the involvement of neuropeptides in anxiety-related behaviour has focused on characterizing the changes in anxiety following either over-expression or under-expression of a particular gene product. In line with pharmacologically oriented findings, transgenic mice over-expressing CRH exhibited behavioural effects associated with acute CRH administration, including increased anxiogenic behaviour (Stenzel-Poore et al. 1994; van Gaalen et al. 2002). These effects were potently blocked by administration of the CRH receptor antagonist α-helical CRH. CRH transgenic mice also showed a profound decrease in sexual behaviours and deficits in learning; higher order functions such as those are typically abolished when a situation is found threatening and anxiety-related behaviours are recruited (Heinrichs et al. 1997). Along the same lines is the finding that CRH-binding protein knockout mice showed increased stress-like and anxiety-related behaviour (Karolyi et al. 1999). Thus, as with the CRH-over-expressing mice, CRH-binding protein knockouts may represent a genetically engineered model of an anxiety-related endophenotype.

CRH knockout mice appear to be nearly indistinguishable from their genetically unaltered wild-type control mice, since stress-induced responses in freezing behaviour and paradigms such as the elevated plus maze and open field were not different in *CRH* knockout mice relative to the wild-type (Weninger et al. 1999). Thus, while increased levels of urocortin gene expression in *CRH* knockout mice are not always seen, it has been suggested that an alternative CRH-like ligand could be subserving effects on anxiety-related behaviour (Weninger et al. 1999).

An anxiolytic action has recently been reported in transgenic mice lacking CRHR-1 (Smith et al. 1998; Timpl et al. 1998). Interestingly, similar to *CRH* knockout mice (Muglia et al. 2000), animals deficient for the CRHR-1 showed signs of AVP mRNA over-expression in the PVN (Fig. 1; Müller et al. 2000b). An over-expression of AVP in the PVN has recently been reported to induce anxiogenic-like effects in rats (Murgatroyd et al. 2004; Wigger et al. 2004) and could, thus, mitigate the anxiolysis observed in *CRHR-1* knockout mice. This indicates that development with a missing gene may lead to an induction of compensatory systems, making the interpretation of results with transgenic and knockout mice difficult.

While the reduced anxiety profile in *CRHR-1* knockout mice has been consistently observed by separate laboratories and across a number of different behavioural paradigms, suggesting that it is a fairly robust and reliable phenomenon, *CRHR-2* knockout mice seem to display a different and also less consistent phenotype. Behavioural effects following *CRHR-2* null mutation were described to include decreased open arm entries in the elevated plus maze in

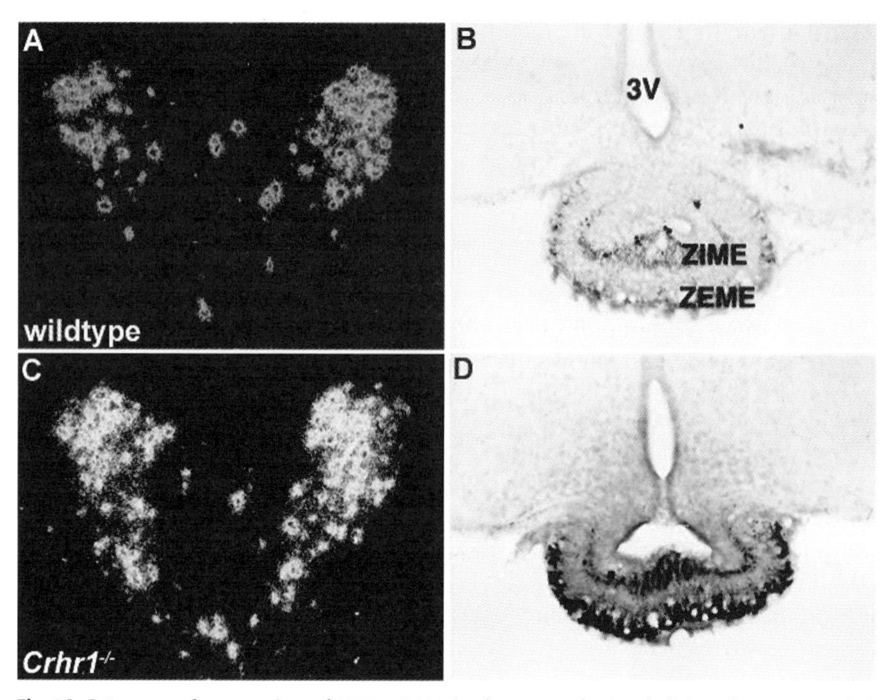

Fig. 1A–D Increased expression of AVP mRNA in the PVN of mice deficient for CRHR-1 (**C**) relative to wild-type mice (**A**). Accordingly, AVP immunoreactivity is markedly enhanced in the external zone of the median eminence (*ZEME*) of mutants (**D**) compared to the wild-type (**B**). (Adapted from Müller et al. 2000b)

both male and female mice (Bale et al. 2000) and in males only (Kishimoto et al. 2000), while others failed to exhibit any change in this measure of anxiety, irrespective of gender (Coste et al. 2000). Moreover, entries into the centre of an open field were found decreased in *CRHR-2* knockouts by Bale et al. (2000), but increased by Kishimoto et al. (2000), suggesting that anxiety-related facets seem to be highly variable. In these animals, basal levels of CRH and urocortin gene expression were elevated, possibly underlying the observed increases in anxiety-related behaviour, again indicating a compensatory response of the system to *CRHR-2* deletion (Bale et al. 2000; Coste et al. 2000). An important aspect in this context appears to be the timing of the gene deletion; novel inducible-knockout technologies will help to clarify the developmental versus acute changes of CRHR-2 in relation to anxiety.

Another important though often ignored aspect is the variability in trait and state anxiety among the experimental animals used. This variability, likely to influence behavioural and neuroendocrine responses to stressor exposure and drugs, is determined genetically (e.g. Wigger et al. 2001; Landgraf and Wigger 2003) and epigenetically by non-genomic mechanisms (e.g. Caldji et al.

1998). Accordingly, Roman high and low avoidance rats, selected and bred for rapid learning versus non-acquisition of active avoidance behaviour, responded differentially to central CRH, only the former showing long-lasting behavioural consequences (Wiersma et al. 1998). Two studies have shown that α-helical CRH 9-41 produced anxiolytic-like effects in the elevated plus maze only after animals had been stressed by exposure to conspecific aggression (Heinrichs et al. 1992; Menzaghi et al. 1994). In line with these findings is the paper by Conti et al. (1994), who showed that α-helical CRH 9-41 was more efficacious in BALB/c mice described to be "emotional" than in three "non-emotional" strains. Studies with novel CRH receptor antagonists further support the hypothesis that state/trait anxiety is critical when studying the behavioural effects of such compounds. The nonpeptide antagonist CP-154,526, for example, elicited anxiolytic-like effects in the rat only if mean baseline levels of exploration of the aversive parts of the maze were low (Lundkvist et al. 1996; Griebel 1999). Antalarmin, a structurally related analogue of CP-154,526, blocked the anxiogenic-like effect of CRH in the elevated plus maze without affecting anxiety-like behaviour in vehicle-treated animals (Zorrilla et al. 2002). Likewise, rats bred for extremely high trait anxiety (HAB) were shown to respond to the CRHR-1 antagonist R121919 with reduced anxiety-related behaviour, while non-anxious rats (LAB) failed to alter that level of anxiety (Keck et al. 2001), thus confirming that CRH antagonists may require a background of hyper-anxiety to show anxiolytic-like effects. Interestingly, the anxiolytic effects of R121919 in hyper-anxious rats were independent of the dose-dependent attenuation of the HPA axis activity in the same animals, further confirming that behavioural alterations and HPA axis activity are not necessarily linked to each other. Inversely, in the same psychopathological rat model, synthetic CRH given i.c.v. induced marked anxiogenic effects in HAB, but not LAB animals (A. Wigger and K. Michael, unpublished data).

Finally, CRH mRNA levels have been found to increase upon exposure to anxiogenic stimuli (Lightmann and Young III 1988; Hsu et al. 1998; Makino et al. 2002), particularly in the PVN, and to decrease upon treatment with clinically effective anxiolytics and antidepressants (Brady et al. 1992; Imaki et al. 1995). In addition to CRH mRNA, acute exposure to various stressors or intracerebroventricular CRH has been shown to upregulate CRHR-1, but not CRHR-2, mRNA in the rat PVN (Makino et al. 1997; Jezova et al. 1999; Arima and Aguilera 2000; van Pett et al. 2000). In rats bred for either high (HAB) or low (LAB) trait anxiety, more inter-line differences were reported in CRHR-2 than CRHR-1 expression, with more CRHR-2 mRNA found, for example, in the PVN and central amygdala of HAB animals. In contrast to AVP mRNA, however, CRH mRNA levels failed to differ between HAB and LAB animals (Wigger et al. 2004).

Disturbance of the prenatal environment by stressing the mother can lead to increased CRH gene expression in the fetal PVN (Fujioka et al. 1999) and amygdala (Ward et al. 2000). These findings support the notion that mother–infant interactions may be a critical factor in determining the future disposition of the offspring to anxiety. Despite the HPA axis stimulation, repeated maternal deprivation in the postpartum period seemed to cause a decrease in the level of CRH gene expression in the PVN (Hatalski et al. 1998; Dent et al. 2000). Interestingly, the nature of the separation determined the direction of the long-term changes in CRH system gene expression and behaviour. Handling, i.e. short periods of separation from the mother, decreased hypothalamic CRH gene expression and stress vulnerability, whereas longer periods of maternal separation seemed to have the opposite effects, including increased CRH gene expression, exaggerated HPA axis responses to stress and increased anxiety-related behaviour (Plotsky and Meaney 1993; Rots et al. 1996; Caldji et al. 1998; Wigger and Neumann 1999; Meaney 2001). Likewise, offspring from low licking/grooming mothers showed increased CRH gene expression in the central nucleus of the amygdala and increased CRH receptor levels in the locus coeruleus (Caldji et al. 2000).

Many studies implicated alterations of the central CRH system in the aetiology of human stress disorders, particularly anxiety and depression (Arborelius et al. 1999; Holsboer 1999; Bakshi and Kalin 2000; Dautzenberg and Hauger 2002; Heinrichs and Koob 2004). In this context it is of note that Zobel et al. (2000) recently succeeded in alleviating symptoms in depressed patients after treatment with the newly developed CRHR-1 antagonist R121919. Clearly, further investigation focusing on site-specific injections, receptor subtype-specific antagonists, time-independent effects and proper genetic manipulations will assist in resolving the controversy of physiological roles of CRH-related neuropeptide and their receptor subtypes in anxiety-related behaviour and psychopathology.

3.2 Vasopressin

AVP and oxytocin are neuropeptides closely related to the CRH neuropeptide family. They are mainly synthesized in parvo- and magnocellular neurons of the hypothalamic PVN and supraoptic nucleus. Upon appropriate stimulation, AVP and oxytocin are secreted from axon terminals into the systemic circulation. This secretion occurs at the level of the posterior pituitary, where axons from magnocellular neurons terminate, or of the eminentia mediana, where neuropeptides of parvo- and magnocellular (Wotjak et al. 2002; Engelmann et al. 2004) origin are secreted into the portal blood system. There are lines of evidence suggesting that AVP, oxytocin and neuropeptides of the CRH family are closely interrelated in these peripherally projecting neuronal systems. A direct action of CRH and urocortin on magnocellular neurons (Bruhn et al. 1986; Kakiya et al. 1998) is confirmed by the presence of CRHR-1 and CRHR-2 on these neurons (Arima and Aguilera 2000). In neuroendocrine parvocellular neurons of the PVN, AVP and CRH are co-localized and act synergistically to release ACTH; their gene transcription may be differentially regulated (Itoi et al. 1999; Helmreich et al. 2001).

Dependent on the quality and intensity of the stressor, the CRH/AVP ratio released into the portal blood may vary (Aguilera et al. 2002). CRH appears to be the dominant trigger for HPA axis activation during acute stress, while AVP may be more important in mediating chronic and repeated stress (Jessop 1999; Aguilera and Rabadan-Diehl 2000; Makino et al. 2002). In this context it is of note that the sensitivity of CRH and AVP transcription to glucocorticoid negative feedback is different, with PVN (1) AVP mRNA levels being more sensitive than are the levels of CRH mRNA (Makino et al. 2002) and (2) CRHR-1 mRNA levels being reduced, and V1b receptor mRNA levels and coupling to phospholipase C being stimulated by elevated glucocorticoids (Ma et al. 1999; Aguilera and Rabadan-Diehl 2000). These effects may contribute to the refractoriness of AVP-stimulated ACTH secretion to glucocorticoid feedback.

The suggestion that AVP is critical for sustaining corticotrope responsiveness in the presence of elevated glucocorticoid levels is supported by Müller et al. (2000b). These authors succeeded in showing that in mice deficient in the CRHR-1, a selective compensatory activation of the AVP system occurs that maintains basal ACTH secretion and HPA axis activity (Fig. 1). In addition, AVP released within the PVN (Wotjak et al. 1996) and the supraoptic nucleus (Wotjak et al. 2002) may contribute to HPA axis regulation, suggesting an involvement of magnocellular AVP at multiple levels in the fine-tuned regulation of ACTH secretion. Oxytocin has been described to attenuate the HPA axis activity in rats (Neumann et al. 2000; Legros 2001). Importantly, although being essential components of the HPA axis, the most salient behavioural effects of CRH, AVP, and oxytocin are mediated outside the axis.

In addition to peripheral secretion, both AVP and oxytocin are centrally released in a differentiated manner. Depending upon the intensity and quality of the stimulus, neurons within both the PVN and SON are apparently capable of regulating their peripheral (from axon terminals) and central (predominantly from dendrites) release of AVP/oxytocin in either a co-ordinated or independent manner (Neumann et al. 1993; Landgraf 1995; Ludwig 1998; Wotjak et al. 1998; Ludwig et al. 2002). The central release is known to contribute to behavioural regulation, including learning and memory processes, emotionality, stress coping, and affiliation (Engelmann et al. 1994, 1996; Ebner et al. 1999; Young 2001; Wigger et al. 2003), and to the control of HPA axis activity (Wotjak et al. 1998, 2002; Keck et al. 2002).

In the brain, the effects of AVP are mediated through G protein-coupled receptors, which have been classified as V1a and V1b subtypes, the former being expressed—*inter alia*—at the level of the amygdala, septum and hypothalamus (Ostrowski et al. 1992; Tribollet et al. 1999). While the V1b receptor is primarily localized in the anterior pituitary, it has also been detected in various brain areas, including the amygdala, the hypothalamus and the hippocampus (Lolait et al. 1995; Hernando et al. 2001), and has recently been shown—in addition to the V1a receptor (Landgraf et al. 1995)—to be involved in the regulation of anxiety- and depression-related behaviour (Griebel et al. 2002). *V1b receptor* knockout mice, however, did not differ in elevated plus maze-related parameters from their wild-type littermates (Wersinger et al. 2002) raising concerns about possible developmental confounds. Interestingly, both V1a and V1b receptors are essentially expressed in magnocellular vasopressinergic neurons themselves (Hurbin et al. 2002), suggesting an involvement in both positive feed-back action (Wotjak et al. 1994; Hurbin et al. 2002) and HPA axis regulation (Wotjak et al. 1996, 2002).

While not studied in as much depth as CRH and while the central AVP system is thought to be primarily involved in cognition (Engelmann et al. 1996; McEwen 2004), AVP is likely to play a role in emotionality. Using antisense targeting, we provided compelling evidence for septal V1a receptors being involved in the regulation of anxiety-related behaviour of rats as scored on the elevated plus maze. Transient and selective downregulation in septal V1a receptor density resulted not only in marked cognitive deficits, but also in reduced anxiety compared to controls (vehicle, mixed bases, sense) (Landgraf et al. 1995). In a follow-up experiment (Liebsch et al. 1996), synthetic AVP was administered by inverse microdialysis to mimic intraseptal release patterns as closely as possible. While a dose of AVP as low as 0.25 ng (delivered over a 30-min retrodialysis period) failed to alter plus-maze behaviour, rats treated with a V1a/b receptor antagonist (5 ng over a 30-min retrodialysis period) made significantly more entries into and spent more time on the open arms of the maze, indicating reduced anxiety-related behaviour. Neither AVP nor its antagonist influenced locomotor activity of the rats. These data give rise to the hypothesis that AVP acts at the level of the septum to co-ordinate different central functions such as learning, memory and emotionality, which, in concert, determine adequate behavioural responses of an animal to environmental demands.

As mentioned before, both V1a (Landgraf et al. 1995) and V1b (Griebel et al. 2002) receptors are critically involved in mediating anxiogenic-like activity of the centrally released neuropeptide in rats and mice. Accordingly, male prairie voles, over-expressing the V1a receptor in their ventral pallidum, exhibited a decrease in time spent in the open arm of the elevated plus maze and higher levels of affiliation as measured by increased time investigating and huddling with a juvenile (Pitkow et al. 2001). The authors claimed that the increases in anxiety and affiliation are probably regulated by different mechanisms. Whereas Bhattacharya et al. (1998) and Ronan et al. (2001) confirmed an anxiogenic effect of central AVP in the rat, Appenrodt et al. (1998) described anxiolytic effects of both centrally and peripherally administered AVP.

Again, the physiological impact of this neuropeptide has to be confirmed in a proper animal model. Indeed, under both basal and stressful conditions, more AVP mRNA was detectable in the PVN of hyper-anxious HAB than LAB animals without any difference in oxytocin mRNA (Murgatroyd et al. 2004; Wigger et al. 2004). mRNA and peptide levels, however, are not necessarily regulated in the same manner by a particular manipulation or event. Therefore, in addition to its increased expression, AVP release patterns under basal conditions and upon stimulation have recently been measured by microdialysis in freely behaving rats. It turned out that more AVP is released within the PVN of HAB versus LAB rats (Fig. 2; Wigger et al. 2004), suggesting centrally released AVP plays a major role in the hyper-reactive HPA axis of the former (Neumann et al. 1998; Landgraf et al. 1999). This is strongly supported by Keck et al. (2002), who succeeded in showing the pathophysiological relevance of an over-production of AVP in HAB rats. In more detail, the pathological outcome of the dexamethasone suppression/CRH challenge test in HABs (i.e. both elevated plasma levels of ACTH and response to synthetic CRH despite prior dexamethasone administration) could be abolished by co-administration of a V1a/b receptor antagonist. In addition to HPA axis regulation, intra-PVN over-expression of AVP was suspected to be critically involved in the regulation of anxiety-related behaviour. Indeed, bilateral PVN administration of a V1a/b antagonist by inverse microdialysis resulted in an attenuation of hyper-anxiety/depression, thus making the behaviour of HAB more similar to that of LAB rats (Wigger et al. 2004). This finding was subsequently supported by long-term treatment of both HAB and LAB animals with the antidepressant paroxetine, which induced a shift towards more active stress-coping in the former only. Interestingly, this antidepressive effect was associated with a normalization of AVP over-expression at the level of the PVN of HAB animals (Keck et al. 2003). Likewise, the development of a depressive illness in a man with an olfactory neuroblastoma, associated with elevated AVP levels, resolved following surgical resection of the tumour and subsequent normalisation of AVP levels (Müller et al. 2000a). These findings, substantiating previous results (Altemus et al. 1992; de Bellis et al. 1996), make the vasopressinergic system a target for future anxiolytic and antidepressant drugs (Landgraf and Wigger 2003). In this context, both V1a (Liebsch et al. 1996; Wigger et al. 2004) and V1b (Griebel et al. 2002) receptor antagonists may be of therapeutic benefit.

Based on comprehensive neuroendocrine and behavioural phenotyping of HAB/LAB rats, the AVP gene was considered a candidate gene of trait anxiety/depression (Landgraf and Wigger 2002, 2003). Indeed, a variety of molecular–genetic approaches succeeded in identifying a single nucleotide polymorphism (SNP; a single nucleotide base within a DNA sequence is replaced with another) in the promoter sequence of the AVP but not CRH gene of HABs, likely to underlie AVP over-expression and, thus, the behavioural and neuroendocrine phenomena of hyper-anxiety (Murgatroyd et al. 2004). This, together with polymorphisms in the AVP receptor promoter sequence (Young 2001), may prove useful for determining the molecular-genetic mechanisms underlying psychiatric disorders.

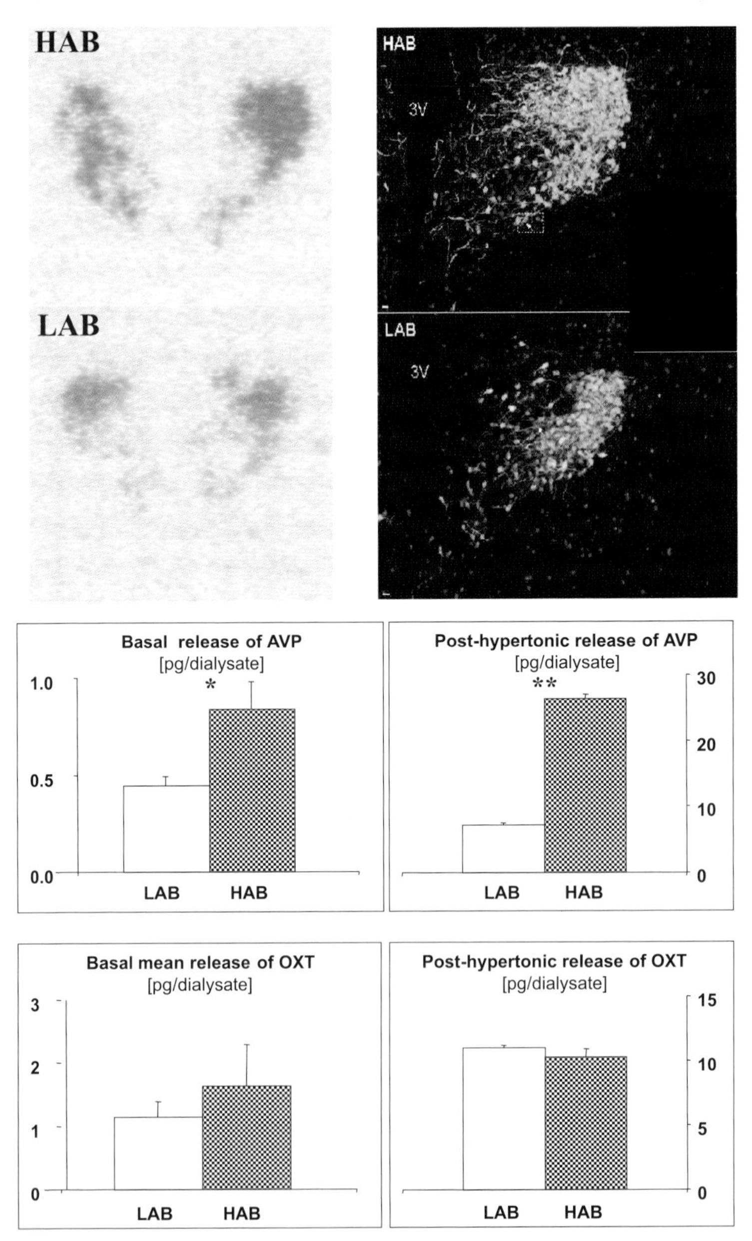

Fig. 2 Increased synthesis, content and release of vasopressin (*AVP*) in the PVN of high-anxiety (*HAB*) vs low-anxiety (*LAB*) rats under basal circumstances. *Above left*: in situ hybridization. *Above right*: immunocytochemistry (courtesy of Dr. N. Singewald, University of Innsbruck). *Middle and below*: intra-PVN release of AVP and oxytocin (*OXT*) measured by in vivo microdialysis under basal conditions and in response to hypertonic stimulation to reveal the releasable neuropeptide pool. *$p<0.05$, **$p<0.01$ vs LAB. (Adapted from Wigger et al. 2004)

Further studies are needed to characterize the mechanisms underlying AVP over-expression and over-release in more detail. The same holds true for the hyper- and hypo-emotional Roman low- and high-avoidance rats, originally selected and bred for poor versus rapid acquisition of two-way active avoidance response (Steimer and Driscoll 2003). In the former, more ACTH and corticosterone were secreted upon a mild stressor, again indicating a hyper-reactive HPA axis. This was shown to be associated with higher AVP mRNA levels in the PVN, whereas CRH mRNA did not differ between the lines (Aubry et al. 1995).

In view of the wide range of CRH and AVP effects on anxiety-related behaviour, one can hypothesize that—similar to but independent of HPA axis regulation—both neuropeptides in varying ratios may shape emotionality, dependent on the intensity and quality of the anxiogenic stimulus. Although the role of a given neuropeptide has to be strictly considered in context to its co-players, the pivotal role of AVP is becoming increasingly appreciated (Landgraf and Wigger 2002, 2003; Scott and Dinan 2002).

3.3 Oxytocin

In addition to its well-known reproductive functions, oxytocin released within the mammalian brain is known to be involved in a variety of regulatory pathways including those underlying social, sexual, maternal (Argiolas and Gessa 1991; Richard et al. 1991; Landgraf 1995; Bielsky and Young 2004) and depression-like behaviour (Arletti and Bertolini 1987). Following peripheral (Uvnäs-Moberg et al. 1994) or central (Windle et al. 1997) administration, an anxiolytic-like effect has been described in rats. Oxytocin infused into the central nucleus of the amygdala, but not the ventromedial nucleus of the hypothalamus, was anxiolytic, indicating brain region-specific effects (Bale et al. 2001). Recently, Neumann et al. (2000) succeeded in showing that a specific oxytocin antagonist given centrally significantly enhanced the anxiety-related behaviour in both pregnant (Fig. 3A) and lactating rats, without exerting similar effects in virgin female or male animals. Thus, the anxiogenic effects of the oxytocin antagonist appear to depend on the cycle-stage, revealing an anxiolytic action of central oxytocin only at a time when the brain oxytocin system is highly activated. Along the same lines, reduced anxiety-related behaviour has been shown previously in lactating rats (Hard and Hansen 1984). These reproduction-dependent behavioural alterations might be related to the complex pattern of maternal behaviour, which includes an increased aggressive behaviour toward conspecifics in order to protect the offspring (Erskine et al. 1978). Thus, activation of the central oxytocin system during the peripartum period (Landgraf et al. 1992) may be related not only to the onset of maternal behaviour to provide nutritional and social support for the young, but also to the reduction in anxiety-related behaviour necessary for their protection (Neumann et al. 2000).

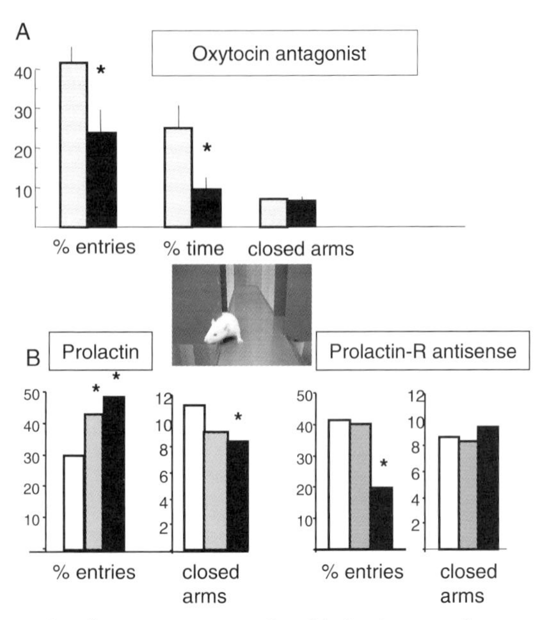

Fig. 3A, B Neuropeptide effects on anxiety-related behaviour. **A** The oxytocin receptor antagonist (*black bars*) administered intracerebroventricularly (i.c.v.) increased indices of anxiety-related behaviour in pregnant rats as measured on the elevated plus maze. Entries into the closed arms indicate unchanged locomotor activity. **B** Prolactin is an anxiolytic neuropeptide in female rats as revealed by i.c.v. administration of synthetic prolactin (*grey* and *black bars* represent two different doses) and by antisense targeting of the prolactin receptor (*R*). Vehicle (*white bars*) vs mixed bases (*grey bars*) and antisense oligodeoxynucleotide (*black bars*). *$p<0.05$ vs vehicle (*white bars*). (Adapted from Neumann et al. 2000) and Torner et al. 2001)

Unlike AVP, oxytocin does not seem to be involved in trait anxiety, as its expression and release in the PVN did not differ between HAB and LAB rats (Wigger et al. 2004).

3.4 Cholecystokinin

Cholecystokinin (CCK) is one of the brain–gut peptides. Its most abundant form in the brain is the C-terminal sulphated octapeptide fragment CCK8, which interacts with the same affinity with both CCK receptor subtypes, CCK-A and CCK-B. Extensive pharmacological studies have been carried out over the last few years suggesting that CCK may participate in the neuroendocrine responses to stress (Harro et al 1993; Daugé and Léna 1998). Interestingly, CCK8 and CRH are co-localized in neurons of the hypothalamic PVN (Mezey et al. 1985).

Together with CRH, CCK belongs to the most extensively studied neuropeptides in anxiety models. Generally, CCK is thought to induce anxiogenic-like effects, although the results of those studies have been highly variable and sometimes contradictory (Griebel 1999). This is presumably because the behavioural profile observed after CCK administration depends on baseline anxiety levels (Harro et al. 1993; Daugé and Léna 1998) as has already been shown for CRH and CRH antagonists (see above). Furthermore, anxiogenic-like effects have been revealed, in large part, by behavioural tests based on exploratory activity, suggesting that these paradigms are more suitable for the investigation of CCK than those based on punished responses. The heterogeneity of behavioural responses produced by CCK can further be explained by the fact that the neuropeptide has been infused in different brain areas in order to delineate the anatomical substrate of CCK-induced anxiogenic-like effects. Local administration of CCK8 directly into the amygdala, for example, produced hyper-anxiety (Frankland et al. 1997), but not in the anterior nucleus accumbens (Daugé et al. 1990). Regional differences in CCK receptors are probably responsible for this discrepancy. Interestingly, CCK-B antagonists have been shown to induce antidepressant-like effects in rodents (Izumi et al. 1996). In this context, limbic brain structures, particularly the nucleus accumbens and the central amygdala, seem to play a role in the interaction between CCK-B receptors and enkephalins to produce antidepressant-like effects. Decreased amygdala CCK-B receptor binding has been demonstrated in rats with high anxiety-like responses on the plus maze, possibly as a compensation for increased CCK activity (Wunderlich et al. 2002). As has recently been shown in CCK receptor gene knockout mice, however, the role of the receptor subtypes in anxiety-related behaviour is still controversial (Miyasaka et al. 2002).

CCK8 concentrations were found to be lower in panic patients than in normal control subjects (Brambilla et al. 1993) and the CCK-B receptors were hyper-sensitive in panic disorders (Akiyoshi et al. 1996). Accordingly, CCK-B receptor agonists such as pentagastrin or CCK-4 have panic-like anxiogenic effects in humans (Radu et al. 2002). Clinical trials, however, have provided inconclusive data about the anxiolytic potential of CCK-B antagonists (Shlik et al. 1997).

3.5 Neuropeptide Y

Neuropeptide Y (NPY) is a highly conserved 36 amino acid peptide of the pancreatic polypeptide family that is widely distributed throughout the mammalian brain. Y1 and Y2 receptors represent the major subtypes expressed in brain areas known to be activated upon anxiogenic stimulation, thus providing the rationale for studying the involvement of NPY and its receptor subtypes in anxiety-related behaviour (Kask et al. 2002). Other receptor subtypes in the rat (Y4, Y5) have not been linked to anxiety regulation with certainty (Sajdyk et al. 2002a).

Central administration in rats has shown that NPY produces anxiolytic-like and antidepressant-like actions in a variety of behavioural tests (Broqua et al. 1995; Sajdyk et al. 1999; Stogner and Holmes 2000; Redrobe et al. 2002a,b), and *NPY* knockout mice have been reported to display an anxiogenic-like phenotype (Bannon et al. 2000). Accordingly, the elevated expression of NPY mRNA in limbic regions of rat brains was found to be associated with less anxiety-like behaviours (Miller et al. 2002), and NPY-transgenic rats show signs of behavioural insensitivity to stress and fear (Thorsell et al. 2000; Thorsell and Heilig 2002). Consistent with the co-localization of NPY with norepinephrine in many brain areas is the notion that norepinephrine, but not the γ-aminobutyric acid (GABA)/benzodiazepine receptor complex, may be involved in the anxiolytic effects of NPY (Griebel 1999). Similar to NPY, high-affinity Y1 agonists induced anxiolysis, whereas a highly selective nonpeptide Y1 receptor antagonist was found to produce anxiogenic-like effects (Broqua et al. 1995; Kask et al. 1998), thereby confirming the critical involvement of the Y1 receptor subtype in the regulation of anxiety-related behaviour. This is further supported by the finding that downregulation of the Y1 receptor expression by antisense targeting resulted in anxiogenic-like effects (Wahlestedt et al. 1993) and blocked the anxiolytic action of bilateral NPY administration in the amygdala (Heilig 1995). Unfortunately, in the latter study the behavioural specificity of intra-amygdala NPY effects in the plus maze remained unclear, as there was no report regarding treatment effects on closed arm entries indicative of locomotor activity.

Unlike Y1 receptor agonists, NPY analogues that bind selectively to the Y2 receptor subtype failed to influence anxiety-related behaviour (Britton et al. 1997) or appeared to produce anxiogenic responses (Sajdyk et al. 2002b). Similarly, mice lacking Y2 receptors displayed an anxiolytic-like phenotype, suggesting that this receptor subtype may play an inhibitory role in the anxiolytic-like effects of NPY (Redrobe et al. 2003; Tschenett et al. 2003).

Heilig et al. (1994) as well as Britton et al. (2000) and Kask et al. (2001) suggested NPY to be an endogenous neuropeptide that "buffers" against the stressor-induced release of CRH. Interestingly, the latter has been shown to stimulate the release of NPY from the rat PVN (Morris and Pavia 1998), further supporting an interaction between central CRH and NPY in vivo.

Recently, Kask et al. (2002) provided evidence indicating that alterations in NPY synthesis and release in a variety of brain areas may be relevant for the pathogenesis of psychiatric diseases.

3.6 Substance P

Substance P, the most abundant neurokinin in the brain, is widely distributed; its NK1 receptor is highly expressed in areas traditionally implicated in the regulation of emotionality such as the amygdala. Central administration of substance P has been found to induce anxiogenic effects in the elevated plus maze (Teixeira et al. 1996; de Araujo 1999), whereas administration of substance P antagonists produced anxiolysis in different tests of anxiety and animal models (File 2000; Varty et al. 2002). Again, it seems that these effects depend on both the neuropeptide dose and the specific brain region (Hasenöhrl et al. 2000). Recently, diminished anxiety- and depression-related behaviours were described in mice with selective deletion of the *Tac1* gene, which encodes substance P and neurokinin A (Bilkei-Gorzo et al. 2002). Furthermore, both genetic disruption and pharmacological blockade of the NK1 receptor in mice resulted in a reduction of anxiety and stress-related responses as well as in a selective desensitization of 5-HT_{1A} inhibitory autoreceptors, which resembles the effect of sustained antidepressant treatment (Santarelli et al. 2001). Also NK2 receptor antagonists were suggested to have a potential in the treatment of some forms of anxiety disorders (Griebel et al. 2001).

In a 6-week double-blind, placebo-controlled trial in patients diagnosed with comorbid anxiety and depression, Kramer et al. (1998) described the anxiolytic and antidepressant effects of the NK1 antagonist, MK-869.

3.7 Prolactin

Convincing evidence of an anxiolytic action of both exogenous and endogenous prolactin has recently been provided by Torner et al. (2001). While intracerebral infusion of this neuropeptide exerted anxiolytic effects on the elevated plus maze in a dose-dependent manner in both male and female rats (Fig. 3B), downregulation of the long form of the prolactin receptor by highly efficient antisense targeting resulted in increased anxiety-related behaviour. In addition to its behavioural action, prolactin was shown to attenuate the responsiveness of the HPA axis to an emotional stressor (Torner et al. 2001). These actions seem to be of particular relevance in females during lactation (Torner and Neumann 2002).

3.8 Other Neuropeptides

Another neuropeptide that is involved in anxiogenic and fearful responses is calcitonin gene-related peptide (Poore and Helmstetter 1996). Likewise, central administration of angiotensin II in the rat increases anxiety-related behaviour (Wright and Harding 1992). Accordingly, transgenic rats accumulating angiotensin II in the brain exhibit signs of hyper-anxiety (Wilson et al. 1996). On the other hand, mice lacking the angiotensin II type-2 receptor displayed increased anxiety-related behaviour (Okuyama et al. 1999), but mice lacking angiotensinogen did not differ in their anxiety-related behaviour from controls

(Walther et al. 1999). Increased levels of anxiety after central administration of the glucagon-like peptide-1 (7-36) amide have been reported by Kinzig et al. (2003).

Melanin-concentrating hormone, an orexigenic hypothalamic neuropeptide, has recently been shown to be likely to induce anxiogenic effects (Borowski et al. 2002; but see Monzón and deBarioglio 1999).

Other neuropeptides thought to induce anxiolytic effects include nociceptin (orphanin FQ; Jenck et al. 2000), galanin (Bing et al. 1993), atrial natriuretic peptide (Bhattacharya et al. 1996; Ströhle et al. 1997; Wiedemann et al. 2000), neuropeptide S (Xu et al. 2004) and enkephalin (Bilkei-Gorzo et al. 2001). While κ-receptors do not seem to be involved in emotionality, μ- and δ-opioid receptors act oppositely in behavioural regulation, the latter contributing to hypo-anxiety (Filliol et al. 2000). More work is needed to reveal and to confirm the physiological significance of these neuropeptides in the regulation of anxiety-related behaviour.

4 Summary

The ability to quickly and unambiguously identify and interpret threatening stimuli is of an obvious selectional advantage for an individual and the species, as is the termination for the anxiety-inducing signal in order to prevent overshooting and disrupting homeostasis. Particularly due to their high number and diversity, the dynamics of their central release and the multiple and variable modes of interneuronal communication they are involved in, neuropeptides play a major role in the regulation of anxiety-related behaviour and psychopathology.

Methodological differences may account for discrepant results found with the same neuropeptide. A major weakness of many of the studies mentioned in this review is that they examined only one element of a complex neuropeptidergic system in a single brain region at a certain time point. Future studies in which multiple components of neuropeptidergic systems are all studied at once and in various brain areas in their dynamics will aid in understanding how the individual elements of this system co-ordinate their activity in regulating anxiety-related behaviour.

Neuropeptides central to behavioural regulation and psychopathology may interact with other neuromodulators/neurotransmitters, including neuropeptides (Holmes et al. 2003). If so, they may act in the same direction; AVP, for example, can synergize with and potentiate the anxiogenic effects of CRH. On the other hand, neuropeptides such as AVP versus oxytocin or CRH versus NPY might be released simultaneously, indicating opposing processes in the fine-tuned regulation of anxiety-related behaviour. Therefore, the frequently asked question of whether a particular neuropeptide is anxiolytic or anxiogenic in its physiological effects can often only be answered in the context of its co-players.

Manipulations of a given neuropeptidergic system have to consider the dynamics of local release patterns and subsequent multiple and variable modes of neuropeptide communication (Landgraf and Neumann 2004). Despite the immense progress in the field of neuropeptides and anxiety, we are far away from mimicking these processes by simply administering synthetic agonists or receptor antagonists, which reflects the challenge of treating and curing anxiety disorders and comorbid depression.

References

Aguilera G, Rabadan-Diehl C (2000) Vasopressinergic regulation of the hypothalamic-pituitary-adrenal axis: implications for stress adaptation. Regul Pept 96:23–29

Aguilera G, Lightman SL, Ma XM (2002) Transcriptional and post-transcriptional regulation of corticotropin releasing hormone and vasopressin expression by stress and glucocorticoids. In: McCarty R, Aguilera G, Sabban G, Kvetnansky R (eds) Stress. Neural, endocrine and molecular studies. Taylor and Francis, New York, pp 91–97

Akiyoshi J, Moriyama T, Isogawa K, Miyamoto M, Sasaki I, Kuga K, Yamamoto H, Yamada K, Fugii I (1996) CCK-4-induced calcium mobilization in T cells is enhanced in panic disorder. J Neurochem 66:1610–1615

Altemus M, Cizza G, Gold P (1992) Chronic fluoxetine treatment reduces hypothalamic vasopressin secretion in vitro. Brain Res 593:311–313

Appenrodt E, Schnabel R, Schwarzberg H (1998) Vasopressin administration modulates anxiety-related behavior in rats. Physiol Behav 64:543–547

Arborelius L, Owens MJ, Plotsky PM, Nemeroff CB (1999) The role of corticotropin-releasing factor in depression and anxiety disorders. J Endocrinol 160:1–12

Argiolas A, Gessa GL (1991) Central functions of oxytocin. Neurosci Biobehav Rev 15:217–231

Arima H, Aguilera G (2000) Vasopressin and oxytocin neurones of hypothalamic surpaoptic and paraventricular nuclei co-express mRNA for type-1 and type-2 corticotropin-releasing hormones receptors. J Neuroendocrinol 12:833–842

Arletti R, Bertolini A (1987) Oxytocin acts as an antidepressant in two animal models of depression. Life Sci 41:1725–1730

Aubry J-M, Bartanusz V, Driscoll P, Schulz P, Steimer T, Kiss JZ (1995) Corticotropin-releasing factor and vasopressin mRNA levels in Roman high- and low-avoidance rats: response to open-field exposure. Neuroendocrinology 61:89–97

Bakshi VP, Kalin NH (2000) Corticotropin-releasing hormone and animal models of anxiety: gene-environment interactions. Biol Psychiatry 48:1175–1198

Bakshi VP, Smith-Roe S, Newman SM, Grigoriadis DE, Kalin NH (2002) Reduction of stress-induced behavior by antagonism of corticotropin-releasing hormone 2 (CRH2) receptors in lateral septum or CRH1 receptors in amygdala. J Neurosci 22:2926–2935

Bale TL, Contarino A, Smith GW, Chan R, Gold LH, Sawchenko PE, Koob G, Vale WW, Lee KL (2000) Mice deficient for corticotropin-releasing hormone receptor-2 display anxiety-like behaviour and are hypersensitive to stress. Nat Genet 24:410–414

Bale TL, Davis AM, Auger AP, Dorsa DM, McCarthy MM (2001) CNS region-specific oxytocin receptor expression: importance in regulation of anxiety and sex behavior. J Neurosci 21:2546–2552

Bannon AW, Seda J, Carmouche M, Francis JM, Norman MH, Karbon B, McCaleb ML (2000) Behavioral characterization of neuropeptide Y knockout mice. Brain Res 868:79–87

Behan DP, Grigoriadis DE, Lovenberg T, Chalmers D, Heinrichs S, Liaw C, De Souza EB (1996) Neurobiology of corticotropin-releasing factor (CRF) receptors and CRF-binding protein: implications for the treatment of CNS disorders. Mol Psychiatry 1:265–277

Bhattacharya SK, Chakrabarti A, Sandler M, Glover V (1996) Anxiolytic activity of intraventricularly administered atrial natriuretic peptide in the rat. Neuropsychopharmacology 15:199–206

Bhattacharya SK, Bhattacharya AR, Chakrabarti AMIT (1998) Anxiogenic activity of intraventricularly administered arginine–vasopressin in the rat. Biogenic Amines 14:367–385

Bielsky IF, Young LJ (2004) Oxytocin, vasopressin, and social recognition in mammals. Peptides 25:1565–1574

Bilkei-Gorzo A, Schuetz B, Zimmer A (2001) Influence of the genetic background on the reactivity of pre-proenkephaline knockout mice in models of anxiety. Soc Neurosci Abstr 955.9

Bilkei-Gorzo A, Racz I, Michel K, Zimmer A (2002) Diminished anxiety- and depression-related behaviors in mice with selective deletion of the Tac1 Gene. J Neurosci 22:10046–10052

Bing O, Möller C, Engel JA, Söderpalm B, Heilig M (1993) Anxiolytic-like action of centrally administered galanin. Neurosci Lett 164:17–20

Bittencourt JC, Sawchenko PE (2000) Do centrally administered neuropeptides access cognate receptors? an analysis in the central corticotropin-releasing factor system. J Neurosci 20:1142–1156

Borowsky B, Durkin MM, Ogozalek K, Marzabadi MR, DeLeon J, Heurich R, Lichtblau R, Shaposhnik Z, Daniewska I, Blackburn TP, Branchek TA, Gerald C, Vaysse PJ, Forray C (2002) Antidepressant, anxiolytic and anorectic effects of a melanin-concentrating hormone-1 receptor antagonist. Nat Med 8:825–830

Brady LS, Gold PW, Herkenham M, Lynn AB, Whitfield HJ Jr (1992) The antidepressants fluoxetine, idazoxan and phenelzine alter corticotropin-releasing hormone and tyrosine hydroxylase mRNA levels in rat brain: therapeutic implications. Brain Res 572:117–125

Brambilla F, Bellodi L, Perna G, Garberi A, Panerai A, Sacerdote P (1993) T cell cholecystokinin concentrations in panic disorder. Am J Psychiatry 150:1111–1113

Britton KT, Lee G, Dana R, Risch SC, Koob GF (1986) Activating and 'anxiogenic' effects of corticotropin releasing factor are not inhibited by blockade of the pituitary-adrenal system with dexamethasone. Life Sci 39:1281–1286

Britton KT, Southerland S, VanUden E, Kirby D, Rivier J, Koob G (1997) Anxiolytic activity of NPY receptor agonists in the conflict test. Psychopharmacology (Berl) 132:6–13

Britton KT, Akwa Y, Spina MG, Koob GF (2000) Neuropeptide Y blocks anxiogenic-like behavioral action of corticotropin-releasing factor in an operant conflict test and elevated plus maze. Peptides 21:37–44

Broqua P, Wettstein JG, Rocher MN, Gauthiermartin B, Junien JL (1995) Behavioral effects of neuropeptide Y receptor agonists in the elevated plus-maze and fear-potentiated startle procedures. Behav Pharmacol 6:215–222

Bruhn TO, Sutton SW, Plotsky PM, Vale WW (1986) Central administration of corticotropin-releasing factor modulates oxytocin secretion in the rat. Endocrinology 119:1558–1563

Caldji C, Tannenbaum B, Sharma S, Francis D, Plotsky PM, Meaney MJ (1998) Maternal care during infancy regulates the development of neural systems mediating the expression of fearfulness in the rat. Proc Natl Acad Sci USA 95:5335–5340

Caldji C, Diorio J, Meaney MJ (2000) Variations in maternal care in infancy regulate the development of stress reactivity. Biol Psychiatry 48:1164–1174

Charney DS, Bremner JD, Redmond DE (1995) Noradrenergic neural substrates for anxiety and fear: clinical associations based on preclinical research. In: Bloom FE, Kupfer DJ (eds) Psychopharmacology: the fourth generation of progress. Raven, New York, pp 387–395

Charney DS, Grillon C, Bremner JD (1998) The neurobiological basis of anxiety and fear: circuits, mechanisms, and neurochemical interactions. Neuroscientist 4:35–44

Conti LH, Costello DG, Martin LA, White MF, Abreu ME (1994) Mouse strain differences in the behavioral effects of corticotropin-releasing factor (CRF) and the CRF antagonist alpha-helical CRF 9-41. Pharmacol Biochem Behav 48:497–503

Coste SC, Kesterson RA, Heldwein KA, Stevens SL, Heard AD, Hollis JH, Murray SE, Hill JK, Pantely GA, Hohimer AR, Hatton DC, Phillips TJ, Finn DA, Low MJ, Rittenberg MB, Stenzel P, Stenzel-Poore MP (2000) Abnormal adaptations to stress and impaired cardiovascular function in mice lacking corticotropin-releasing hormone receptor-2. Nat Genet 24:403–409

Curtis AL, Pavcovich LA, Valentino RJ (1999) Long-term regulation of locus ceruleus sensitivity to corticotropin-releasing factor by swim stress. J Pharmacol Exp Ther 289:1211–1219

Daugé V, Léna I (1998) CCK in anxiety and cognitive processes. Neurosci Biobehav Rev 22:815–825

Daugé V, Bohme GA, Crawley JN, Durieux C, Stutzmann JM, Feger J, Blanchard JC, Roques BP (1990) Investigation of behavioral and electrophysiological responses induced by selective stimulation of CCKB receptors by using a new highly potent CCK analog, BC 264. Synapse 6:73–80

Dautzenberg FM, Hauger RL (2002) The CRF peptide family and their receptors: yet more partners discovered. Trends Pharmacol Sci 23:71–77

De Araujo JE, Silva RCB, Huston JP, Brandao ML (1999) Anxiogenic effects of substance P and its 7–11 C terminal, but not the 1–7 N terminal, injected into the dorsal periaqueductal gray. Peptides 20:1437–1443

De Bellis M, Gold PW, Geriacoti T, Listwak S, Kling M (1996) An association of fluoxetine treatment with reductions in CSF corticotropin-releasing hormone and arginine vasopressin in patients with depression. Am J Psychiatry 150:656–657

Dent GW, Okimoto DK, Smith MA, Levine S (2000) Stress-induced alterations in corticotropin-releasing hormone and vasopressin gene expression in the paraventricular nucleus during ontogeny. Neuroendocrinology 71:333–342

Dunn AJ, Berridge CW (1990) Physiological and behavioral responses to corticotropin-releasing factor administration: is CRF a mediator of anxiety or stress responses? Brain Res Brain Res Rev 15:71–100

Ebner K, Wotjak CT, Holsboer F, Landgraf R, Engelmann M (1999) Vasopressin released within the septal brain area during swim stress modulates the behavioral stress response in rats. Eur J Neurosci 11:997–1002

Engelmann M, Ludwig M, Landgraf R (1994) Simultaneous monitoring of intracerebral release and behavior: vasopressin improves social recognition. J Neuroendocrinol 6:391–395

Engelmann M, Wotjak CT, Ludwig M, Neumann ID, Landgraf R (1996) Behavioral consequences of intracerebral vasopressin and oxytocin: focus on learning and memory. Neurosci Biobehav Rev 20:341–358

Engelmann M, Landgraf R, Wotjak CT (2004) Interaction between the hypothalamic-neurohypophysial system (HNS) and the hypothalamic-pituitary-adrenal (HPA) axis under stress—an old concept revisited. Front Neuroendocrinol 25:132–149
Ermisch A, Rühle HJ, Landgraf R, Heß J (1985) Blood-brain barrier and peptides. J Cereb Blood Flow Metab 5:350–357
Ermisch A, Landgraf R, Rühle HJ (eds) (1992) Circumventricular organs and brain fluid environment. Prog Brain Res 91. Elsevier, Amsterdam
Erskine MS, Barfield RJ, Goldman BD (1978) Intraspecific fighting during late pregnancy and lactation in rats and effects of litter removal. Behav Biol 23:206–213
File SE (2000) NKP608, an NK1 receptor antagonist, has an anxiolytic action in the social interaction test in rats. Psychopharmacology (Berl) 152:105–109
Filliol D, Ghozland S, Chluba J, Martin M, Matthes HWD, Simonin F, Befort K, Gavériaux-Ruff C, Dierich A, LeMeur M, Valverde O, Maldonado R, Kieffer BL (2000) Mice deficient for δ- and μ-opioid receptors exhibit opposing alterations of emotional responses. Nat Genet 25:195–200
Frankland PW, Josselyn SA, Bradwejn J, Vaccarino FJ, Yeomans JS (1997) Activation of amygdala cholecystokinin B receptors potentiates the acoustic startle response in the rat. J Neurosci 17:1838–1847
Fujioka T, Sakata Y, Yamaguchi K, Shibasaki T, Kato H, Nakamura S (1999) The effects of prenatal stress on the development of hypothalamic paraventricular neurons in fetal rats. Neuroscience 92:1079–1988
Fuxe K, Agnati LF (eds) (1991) Volume transmission in the brain: novel mechanisms for neural transmission. Raven Press, New York
Glatt CE, Freimer NB (2002) Association analysis of candidate genes for neuropsychiatric disease: the perpetual campaign. Trends Genet 18:307–312
Gray TS (1993) Amygdaloid CRF pathways. Role in autonomic, neuroendocrine, and behavioral responses to stress. Ann NY Acad Sci 697:53–60
Griebel G (1999) Is there a future for neuropeptide receptor ligands in the treatment of anxiety disorders? Pharmacol Ther 82:1–61
Griebel G, Perrault G, Soubrié P (2001) Effects of SR48968, a selective non-peptide NK2 receptor antagonist on emotional processes in rodents. Psychopharmacology (Berl) 158:241–251
Griebel G, Simiand J, Serradeil-LeGal C, Wagnon J, Pascal M, Scatton B, Maffrand JP, Soubrie P (2002) Anxiolytic- and antidepressant-like effects of the non-peptide vasopressin V1b receptor antagonist, SSR 149415, suggest an innovative approach for the treatment of stress-related disorders. Proc Natl Acad Sci USA 99:6370–6375
Hammack SE, Richey KJ, Schmid MJ, LoPresti ML, Watkins LR, Maier SF (2002) The role of corticotropin-releasing hormone in the dorsal raphe nucleus in mediating the behavioral consequences of uncontrollable stress. J Neurosci 22:1020–1026
Hard E, Hansen S (1984) Reduced fearfulness in the lactating rat. Physiol Behav 35:641–643
Harro J, Vasar E, Bradwejn J (1993) Cholecystokinin in animal and human research of anxiety. Trends Pharmacol Sci 14:244–249
Hasenöhrl RU, De Souza-Silva MA, Nikolaus S, Tomaz C, Brandao ML, Schwarting RKW, Huston JP (2000) Substance P and its role in neural mechanisms governing learning, anxiety and functional recovery. Neuropeptides 34:272–280
Hatalski CG, Guirguis C, Baram TZ (1998) Corticotropin-releasing factor mRNA expression in the hypothalamic paraventricular nucleus and the central nucleus of the amygdala is modulated by repeated acute stress in the immature rat. J Neuroendocrinol 10:663–669

Heilig M (1995) Antisense inhibition of neuropeptide Y (NPY)-Y1 receptor expression blocks the anxiolytic-like action of NPY in amygdala and paradoxically increases feeding. Regul Pept 59:201–205
Heilig M, Koob GF, Ekman R, Britton KT (1994) Corticotropin-releasing factor and neuropeptide Y: role in emotional integration. Trends Neurosci 17:80–85
Heinrichs SC, Joppa M (2001) Dissociation of arousal-like from anxiogenic-like actions of brain corticotrophin-releasing factor receptor ligands in rats. Behav Brain Res 122:43–50
Heinrichs SC, Koob GF (2004) Corticotropin-releasing factor in brain: a role in activation, arousal, and affect regulation. J Pharmacol Exp Ther 311:427–440
Heinrichs SC, Pich EM, Miczek KA, Britton KT, Koob GF (1992) Corticotropin-releasing factor antagonist reduces emotionality in socially defeated rats via direct neurotropic action. Brain Res 581:190–197
Heinrichs SC, Lapsansky J, Lovenberg TW, De Souza EB, Chalmers DT (1997) Corticotropin-releasing factor CRF1, but not CRF2, receptors mediate anxiogenic-like behavior. Regul Pept 71:15–21
Helmreich DL, Itoi K, Lopez-Figueroa MO, Akil H, Watson SJ (2001) Norepinephrine-induced CRH and AVP gene transcription within the hypothalamus: differential regulation by corticosterone. Mol Brain Res 88:62–73
Hernando F, Schoots O, Lolait SJ, Burbach JPH (2001) Immunohistochemical localization of the vasopressin V1b receptor in the rat brain and pituitary gland: anatomical support for its involvement in the central effects of vasopressin. Endocrinology 142:1659–1668
Ho SP, Takahashi LK, Livanov V, Spencer K, Lesher T, Maciag C, Smith MA, Rohrbach KW, Hartig PR, Arneric SP (2001) Attenuation of fear conditioning by antisense inhibition of brain corticotropin releasing factor-2 receptor. Mol Brain Res 89:29–40
Holmes A, Heilig M, Rupniak NMJ, Steckler T, Griebel G (2003) Neuropeptide systems as novel therapeutic targets for depression and anxiety disorders. Trends Pharmacol Sci 24:580–588
Holsboer F (1999) The rationale for corticotropin-releasing hormone receptor (CRH-R) antagonists to treat depression and anxiety. J Psychiatr Res 33:181–214
Hsu DT, Chen F-L, Takahashi LK, Kalin NH (1998) Rapid stress-induced elevations in corticotropin-releasing hormone mRNA in rat central amygdala nucleus and hypothalamic paraventricular nucleus: an in situ hybridization analysis. Brain Res 788:305–310
Hsu SY, Hsueh AJ (2001) Human stresscopin and stresscopin-related peptide are selective ligands for the type 2 corticotropin-releasing hormone receptor. Nat Med 7:605–611
Hurbin A, Orcel H, Alonso G, Moos F, Rabié A (2002) The vasopressin receptors colocalize with vasopressin in the magnocellular neurons of the rat supraoptic nucleus and are modulated by water balance. Endocrinology 143:456–466
Imaki T, Xiao-Quan W, Shibasaki T, Harada S, Chikada N, Takahashi C, Naruse M, Demura H (1995) Chlordiazepoxide attenuates stress-induced activation of neurons, corticotropin-releasing factor (CRF) gene transcription and CRF biosynthesis in the paraventricular nucleus (PVN). Mol Brain Res 32:261–270
Imaki T, Naruse M, Harada S, Chikada N, Imaki J, Onodera H, Demura H, Vale W (1996) Corticotropin-releasing factor up-regulates its own receptor mRNA in the paraventricular nucleus of the hypothalamus. Mol Brain Res 38:166–170
Itoi K, Helmreich DL, Lopez-Figueroa MO, Watson SJ (1999) Differential regulation of corticotropin-releasing hormone and vasopressin gene transcription in the hypothalamus by norepinephrine. J Neurosci 19:5464–5472

Izumi T, Inoue T, Tsuchiya K, Hashimoto S, Ohmori T, Koyama T (1996) Effect of the selective CCKB receptor antagonist LY288513 on conditioned fear stress in rats. Eur J Pharmacol 300:25–31

Jenck F, Ouagazzal AM, Pauly-Evers M, Moreau JL (2000) Orphanin FQ: role in behavioral fear responses and vulnerability to stress? Mol Psychiatry 5:572–574

Jessop DS (1999) Stimulatory and inhibitory regulators of the hypothalamo-pituitary-adrenocortical axis. Baillieres Clin Endocrinol Metab 13:491–501

Jezova D, Ochedalski T, Glickman M, Kiss A, Aguilera G (1999) Central corticotropin-releasing hormone receptors modulate hypothalamic-pituitary-adrenocortical and sympathoadrenal activity during stress. Neuroscience 94:797–802

Kakiya S, Yokoi H, Arima H, Iwasaki Y, Oki Y, Oiso Y (1998) Central administration of urocortin inhibits vasopressin release in conscious rats. Neurosci Lett 248:144–146

Karolyi IJ, Burrows HL, Ramesh TM, Nakajima M, Lesh JS, Seong E, Camper SA, Seasholtz AF (1999) Altered anxiety and weight gain in corticotropin-releasing hormone-binding protein-deficient mice. Proc Natl Acad Sci USA 96:11595–11600

Kask A, Rägo L, Harro J (1998) NPY Y1 receptors in the dorsal periaqueductal gray matter regulate anxiety in the social interaction test. Neuroreport 9:2713–2716

Kask A, Nguyen HP, Pabst R, von Hörsten S (2001) Neuropeptide Y Y1 receptor-mediated anxiolysis in the dorsocaudal lateral septum: functional antagonism of corticotropin-releasing hormone-induced anxiety. Neuroscience 104:799–806

Kask A, Harro J, von Hörsten S, Redrobe JP, Dumont Y, Quirion R (2002) The neurocircuitry and receptor subtypes mediating anxiolytic-like effects of neuropeptide Y. Neurosci Biobehav Rev 26:259–283

Keck ME, Welt T, Wigger A, Renner U, Engelmann M, Holsboer F, Landgraf R (2001) The anxiolytic effect of the CRH1 receptor antagonist R121919 depends on innate emotionality in rats. Eur J Neurosci 13:373–380

Keck ME, Wigger A, Welt T, Müller MB, Gesing A, Reul JHMH, Holsboer F, Landgraf R, Neumann ID (2002) Vasopressin mediates the response of the combined dexamethasone/CRH test in hyper-anxious rats: implications for pathogenesis of affective disorders. Neuropsychopharmacology 26:94–105

Keck ME, Welt T, Müller MB, Uhr M, Ohl F, Wigger A, Toschi N, Holsboer F, Landgraf R (2003) Reduction in hypothalamic vasopressinergic hyperdrive contributes to clinically relevant behavioral and neuroendocrine effects of chronic paroxetine. Neuropsychopharmacology 28:235–243

Kinzig KP, D'Alessio DA, Herman JP, Sakai RR, Vahl TP, Figueiredo HF, Murphy EK, Seeley RJ (2003) CNS glucagon-like peptide-1 receptors mediate endocrine and anxiety responses to interoceptive and psychogenic stressors. J Neurosci 23:616–6170

Kishimoto T, Radulovic J, Radulovic M, Lin CR, Schrick C, Hooshmand F (2000) Deletion of CRHR2 reveals an anxiolytic role for corticotropin-releasing hormone receptor-2. Nat Genet 24:415–419

Koegler-Muly SM, Owens MJ, Ervin GN, Kilts CD, Nemeroff CB (1993) Potential corticotropin-releasing factor pathways in the rat brain as determined by bilateral electrolytic lesions of the central amygdaloid nucleus and paraventricular nucleus of the hypothalamus. J Neuroendocrinol 5:95–98

Koob GF, Heinrichs SC, Pich EM, Menzaghi F, Baldwin H, Miczek K, Britton KT (1993) The role of corticotropin-releasing factor in behavioral responses to stress. In: De Souza EB, Nemeroff CB (eds) Corticotropin-releasing factor: basic and clinical studies of a neuropeptide. CIBA Foundation Symposium 172. John Wiley and Sons, Chichester, pp 277–295

Smagin GN, Heinrichs SC, Dunn AJ (2001) The role of CRH in behavioral responses to stress. Peptides 22:713–724
Smith GW, Aubry J-M, Dellu F, Contarino A, Bilezikijan LM, Gold LH, Chen R, Marchuk Y, Hauser C, Bentley CA, Sawchenko PE, Koob GF, Vale W, Lee K-F (1998) Corticotropin releasing factor receptor 1-deficient mice display decreased anxiety, impaired stress response, and aberrant neuroendocrine development. Neuron 20:1093–1102
Spina MG, Merlo-Pich E, Akwa Y, Baducci C, Basso AM, Zorrilla EP, Britton KT, Rivier J, Vale WW, Koob GF (2002) Time-dependent induction of anxiogenic-like effects after central infusion of urocortin or corticotropin-releasing factor in the rat. Psychopharmacology (Berl) 160:113–121
Steckler T, Holsboer F (1999) Corticotropin-releasing hormone receptor subtypes and emotion. Biol Psychiatry 46:1480–1508
Steimer T, Driscoll P (2003) Divergent stress responses and coping styles in psychogenetically selected Roman high- (RHA) and low- (RLA) avoidance rats: behavioural, neuroendocrine and developmental aspects. Stress 6:87–100
Stenzel-Poore MP, Heinrichs SC, Rivest S, Koob G, Vale WW (1994) Overproduction of corticotropin-releasing factor in transgenic mice: a genetic model of anxiogenic behavior. J Neurosci 14:2579–2584
Stogner KA, Holmes PV (2000) Neuropeptide-Y exerts antidepressant-like effects in the forced swim test in rats. Eur J Pharmacol 387:R9–R10
Stout SC, Owens MJ, Nemeroff CB (2002) Regulation of corticotropin-releasing factor neuronal systems and hypothalamic-pituitary-adrenal axis activity by stress and chronic antidepressant treatment. J Pharmacol Exp Ther 300:1085–1092
Ströhle A, Jahn H, Montkowski A, Liebsch G, Boll E, Landgraf R, Holsboer F, Wiedemann K (1997) Central and peripheral administration of atriopeptin is anxiolytic in rats. Neuroendocrinology 65:210–215
Sutton RE, Koob GF, Le Moal M, Rivier J, Vale W (1982) Corticotropin releasing factor produces behavioural activation in rats. Nature 297:331–333
Tabor HK, Risch NJ, Myers RM (2002) Candidate-gene approaches for studying complex genetic traits: practical considerations. Nat Rev Genet 3:1–7
Takahashi LK (2001) Role of CRF1 and CRF2 receptors in fear and anxiety. Neurosci Biobehav Rev 25:627–636
Takahashi LK, Ho SP, Livanov V, Graciani N, Arneric SP (2001) Antagonism of CRF(2) receptors produces anxiolytic behavior in animal models of anxiety. Brain Res 902:135–142
Teixeira RM, Santos ARS, Ribeiro SJ, Calixto JB, Rae GA, DeLima TCM (1996) Effects of central administration of tachykinin receptor agonists and antagonists on plus-maze behavior in mice. Eur J Pharmacol 311:7–14
Thorsell A, Heilig M (2002) Diverse functions of neuropeptide Y revealed using genetically modified animals. Neuropeptides 36:182–193
Thorsell A, Michalkiewicz M, Dumont Y, Quirion R, Caberlotto L, Rimondini R, Mathé AA, Heilig M (2000) Behavioral insensitivity to restraint stress, absent fear suppression of behavior and impaired spatial learning in transgenic rats with hippocampal neuropeptide Y overexpression. Proc Natl Acad Sci USA 97:12852–12857
Timpl P, Spanagel R, Sillaber I, Kresse A, Reul JMHM, Stalla GK, Blanquet V, Steckler T, Holsboer F, Wurst W (1998) Impaired stress response and reduced anxiety in mice lacking a functional corticotropin-releasing hormone receptor 1. Nat Genet 19:162–166
Torner L, Neumann ID (2002) Increased hypothalamic expression of prolactin in lactation: involvement in behavioural and neuroendocrine stress responses. Eur J Neurosci 15:1381–1389

Torner L, Toschi N, Pohlinger A, Landgraf R, Neumann ID (2001) Anxiolytic and anti-stress effects of brain prolactin: improved efficacy of antisense targeting of the prolactin receptor by molecular modeling. J Neurosci 21:3207–3214

Tribollet E, Raufatse D, Maffrand JP, Serradeil-le Gal C (1999) Binding of the non-peptide vasopressin V1a receptor antagonist SR-49059 in the rat brain: an in vitro and in vivo autoradiographic study. Neuroendocrinology 69:113–120

Tschenett A, Singewald N, Carli M, Balducci C, Salchner P, Vezzani A, Herzog H, Sperk G (2003) Reduced anxiety and improved stress coping ability in mice lacking NPY-Y2 receptors. Eur J Neurosci 18:143–148

Umriukhin A, Wigger A, Singewald N, Landgraf R (2002) Hypothalamic and hippocampal release of serotonin in rats bred for hyper- or hypo-anxiety. Stress 5:299–305

Uvnäs-Moberg K, Ahlenius S, Hillgaart V, Alster P (1994) High doses of oxytocin cause sedation and low doses cause an anxiolytic-like effect in male rats. Pharmacol Biochem Behav 49:101–106

Valdez GR, Inoue K, Koob GF, Rivier J, Vale W, Zorrilla EP (2002) Human urocortin II: mild locomotor suppressive and delayed anxiolytic-like effects of a novel corticotropin-releasing factor related peptide. Brain Res 943:142–150

Vale W, Spiess J, Rivier C, Rivier J (1981) Characterization of a 41-residue ovine hypothalamic peptide that stimulates secretion of corticotropin and β-endorphin. Science 213:1394–1397

Van Bockstaele EJ, Colago EE, Valentino RJ (1996) Corticotropin-releasing factor-containing axon terminals synapse onto catecholamine dendrites and may presynaptically modulate other afferents in the rostral pole of the nucleus locus coeruleus in the rat brain. J Comp Neurol 364:523–534

Van Gaalen MM, Stenzel Poore MP, Holsboer F, Steckler T (2002) Effects of transgenic overproduction of CRH on anxiety-like behaviour. Eur J Neurosci 15:2007–2015

Van Pett K, Viau V, Bittencourt JC, Chan RKW, Li HY, Arias C, Prins GS, Perrin M, Vale W, Sawchenko PE (2000) Distribution of mRNAs encoding CRF receptors in brain and pituitary of rat and mouse. J Comp Neurol 428:191–212

Varty GB, Cohen-Williams ME, Morgan CA, Pylak U, Duffy RA, Lachowicz JE, Carey GJ, Coffin VL (2002) The gerbil elevated plus-maze II: anxiolytic-like effects of selective neurokinin NK1 receptor antagonists. Neuropsychopharmacology 27:371–379

Vaughan J, Donaldson C, Bittencourt J, Perrin MH, Lewis K, Sutton S, Chan R, Turnbull AV, Lovejoy D, Rivier C, et al (1995) Urocortin, a mammalian neuropeptide related to fish urotensin I and to corticotropin-releasing factor. Nature 378:287–292

Vetter DE, Li C, Zhao L, Contarino A, Liberman MC, Smith GW, Marchuk Y, Koob GF, Heinemann SF, Vale W, Lee KF (2002) Urocortin-deficient mice show hearing impairment and increased anxiety-like behavior. Nat Genet 31:363–369

Wahlestedt C, Pich EM, Koob GF, Yee F, Heilig M (1993) Modulation of anxiety and neuropeptide Y-Y1 receptors by antisense oligodeoxynucleotides. Science 259:528–531

Walther T, Voigt JP, Fukamizu A, Fink H, Bader M (1999) Learning and anxiety in angiotensin-deficient mice. Behav Brain Res 100:1–4

Ward HE, Johnson EA, Salm AK, Birkle DL (2000) Effects of prenatal stress on defensive withdrawal behavior and corticotropin releasing factor systems in rat brain. Physiol Behav 70:359–366

Weninger SC, Dunn AJ, Muglia LJ, Dikkes P, Miczek KA, Swiergiel AH, Berridge CW, Majzoub JA (1999) Stress-induced behaviors require the corticotropin-releasing hormone (CRH) receptor, but not CRH. Proc Natl Acad Sci USA 96:8283–8288

Wersinger SR, Ginns EI, O'Carroll AM, Lolait SJ, Young III WS (2002) Vasopressin V1b receptor knockout reduces aggressive behavior in male mice. Mol Psychiatry 7:975–984

Wiedemann K, Jahn H, Kellner M (2000) Effects of natriuretic peptides upon hypothalamo-pituitary-adrenocortical system activity and anxiety behaviour. Exp Clin Endocrinol Diabetes 108:5–13

Wiersma A, Konsman JP, Knollema S, Bohus B, Koolhaas JM (1998) Differential effects of CRH infusion into the central nucleus of the amygdala in the Roman high-avoidance and low-avoidance rats. Psychoneuroendocrinology 23:261–274

Wigger A, Neumann ID (1999) Periodic maternal deprivation induces gender-dependent alterations in behavioural and neuroendocrine responses to emotional stress in adult rats. Physiol Behav 66:293–302

Wigger A, Loerscher P, Weissenbacher P, Holsboer F, Landgraf R (2001) Cross-fostering and cross-breeding of HAB and LAB rats: a genetic rat model of anxiety. Behav Genet 31:371–382

Wigger A, Sánchez MM, Mathys KC, Ebner K, Liu D, Kresse A, Neumann ID, Holsboer F, Plotsky PM, Landgraf R (2004) Alterations in central neuropeptide expression, release, and receptor binding in rats bred for high anxiety: critical role of vasopressin. Neuropsychopharmacology 29:1–14

Wilson W, Voigt P, Bader M, Marsden CA, Fink H (1996) Behaviour of the transgenic (mREN2)27 rat. Brain Res 729:1–9

Windle RJ, Shanks N, Lightman SL, Ingram CD (1997) Central oxytocin administration reduces stress-induced corticosterone release and anxiety behavior in rats. Endocrinology 138:2829–2834

Wotjak CT, Ludwig M, Landgraf R (1994) Vasopressin facilitates its own release within the rat supraoptic nucleus in vivo. Neuroreport 5:1181–1184

Wotjak CT, Kubota M, Liebsch G, Montkowski A, Holsboer F, Neumann I, Landgraf R (1996) Release of vasopressin within the rat paraventricular nucleus in response to emotional stress: a novel mechanism of regulating adrenocorticotropic hormone secretion? J Neurosci 16:7725–7732

Wotjak CT, Ganster J, Kohl G, Holsboer F, Landgraf R, Engelmann M (1998) Dissociated central and peripheral release of vasopressin, but not oxytocin, in response to repeated swim stress: new insights into the secretory capacities of peptidergic neurons. Neuroscience 85:1209–1222

Wotjak CT, Ludwig M, Ebner K, Russell JA, Singewald N, Landgraf R, Engelmann M (2002) Vasopressin from hypothalamic magnocellular neurons has opposite actions at the adenohypophysis and in the supraoptic nucleus on ACTH secretion. Eur J Neurosci 16:477–485

Wright JW, Harding JW (1992) Regulatory role of brain angiotensins in the control of physiological and behavioral responses. Brain Res Brain Res Rev 17:227–262

Wunderlich GR, Raymond R, DeSousa NJ, Nobrega JN, Vaccarino FJ (2002) Decreased CCKB receptor binding in rat amygdala in animals demonstrating greater anxiety-like behavior. Psychopharmacology (Berl) 164:193–199

Young LJ (2001) Oxytocin and vasopressin as candidate genes for psychiatric disorders: lessons from animal models. Am J Med Genet 105:53–54

Xu Y-L, Reinscheid RK, Huitron-Resendiz S, Clark SD, Wang Z, Lin SH, Brucher FA, Zeng J, Ly NK, Henriksen SJ, de Lecea L, Civelli O (2004) Neuropeptide S: a neuropeptide promoting arousal and anxiolytic-like effects. Neuron 43:487–497

Zobel A, Nickel T, Künzel HE, Ackl N, Sonntag A, Ising M, Holsboer F (2000) Effects of the high-affinity corticotropin-releasing hormone receptor 1 antagonist R121919 in major depression: the first 20 patients. J Psychiatr Res 34:171–181

HEP (2005) 169:371–403

Neuroendocrine Aspects of PTSD

R. Yehuda

Psychiatry Department and Division of Traumatic Stress Studies, Mount Sinai School of Medicine and Bronx Veterans Affairs, 130 West Kingsbridge Road, Bronx NY, 10468, USA
Rachel.yehuda@med.va.gov

Abstract This chapter discussed how neuroendocrine findings in posttraumatic stress disorder (PTSD) potentially inform hypothalamic-pituitary-adrenal (HPA) alterations in PTSD and highlight alterations relevant to the identification of targets for drug development. Most studies demonstrate alterations consistent with an enhanced negative feedback inhibition of cortisol on the pituitary, an overall hyperreactivity of other target tissues (adrenal gland, hypothalamus), or both in PTSD. However, findings of low cortisol and increased reactivity of the pituitary in PTSD are also consistent with reduced adrenal output. The observations in PTSD are part of a growing body of neuroendocrine data providing evidence of insufficient glucocorticoid signaling in stress-related neuropsychiatric disorders.

Keywords Posttraumatic stress disorder · Cortisol · Neuroendocrine alterations · Negative feedback inhibition · Glucocorticoid receptors · CRF

1 Introduction

The development of drugs that might be effective in treating anxiety disorders in part depends on the ability of clinical neuroscience to identify biologic alterations that might serve as targets for drug development. Unfortunately, an observable biologic change—even one that is directly correlated with severity of symptoms or the absence or presence of a disorder—does not always constitute a core pathophysiologic process requiring biologic "repair." Biologic alterations may be present in specific anxiety disorders because they are correlates or proxies for other pathophysiologic processes, or even because they represent compensatory mechanisms of adaptation.

The study of the neuroendocrinology of posttraumatic stress disorder (PTSD) has been illuminating in highlighting alterations that have not historically been associated with pathologic processes. The most infamous of these findings—low cortisol levels—has been subjected to much discussion and scrutiny, likely because it has been a counterintuitive result, given modern interpretations of the damaging effects of stress hormones. Indeed, the initial observation of low cortisol in a disorder precipitated by extreme stress directly contradicted the popular formulation of hormonal responses to stress, the "glucocorticoid cascade hypothesis" (Sapolsky 1986), which was emerging as a cogent rationale for antiglucocorticoid treatments in depression, and other psychiatric disorders thought to be driven by hypercortisolism.

This chapter discusses how cortisol findings in PTSD potentially inform hypothalamic–pituitary–adrenal (HPA) alterations in PTSD and highlights what might be true targets of drug development. The observations in PTSD are part of a growing body of neuroendocrine data providing evidence of insufficient glucocorticoid signaling in stress-related neuropsychiatric disorders (Raison and Miller 2003). The majority of studies demonstrates alterations consistent with an enhanced negative feedback inhibition of cortisol on the pituitary, an overall hyperreactivity of other target tissues (adrenal gland, hypothalamus), or both in PTSD. This model explains most of the reported observations in PTSD. Theoretically, however, findings of low cortisol and increased reactivity of the pituitary in PTSD are also consistent with reduced adrenal output (Maes et al. 1998; Heim et al. 2000), but this latter model is only supported by the minority of HPA alterations observed in PTSD.

It may be that models of enhanced negative feedback, increased HPA reactivity, and reduced adrenal capacity explain different facets of the neuroendocrinology of PTSD, or that the tendency for reduced adrenal output may represent a pre-existing risk factor related to certain types of early experiences, at least in certain persons who develop PTSD. On the other hand, alterations associated with enhanced negative feedback inhibition may develop over time in response to the complex biologic demands of extreme trauma and its aftermath. Moreover, the findings of increased HPA reactivity may reflect a more nonspecific response to ongoing environmental challenges associated with having chronic PTSD. Furthermore, the absence of cortisol alterations in some studies imply that alterations associated with low cortisol and enhanced negative feedback are only present in a biologic subtype of PTSD. The observations in the aggregate, and the alternative models of pathology or adaptation suggested by them, must be clearly understood in using neuroendocrine data in PTSD to identify targets for drug development.

2 Basal HPA Hormone Levels in PTSD

The first report on cortisol levels in PTSD was that of Mason et al. who found that the mean 24-h urinary excretion of cortisol was significantly lower in combat Vietnam veterans with PTSD compared to psychiatric patients in four other diagnostic groups (Mason et al. 1986). The authors noted surprise at the fact that cortisol levels were low, since "certain clinical features such as depression and anxiety [in PTSD] might have been expected to be associated with increased activity of the pituitary-adrenal cortical system." Since this initial observation, the majority of the evidence supports the conclusion that cortisol alterations in PTSD are different from those observed in acute and chronic stress, and major depression, but more importantly, that the HPA axis appears to be regulated differently.

2.1 Urinary Cortisol Levels in PTSD

The initial report of sustained, lower urinary cortisol levels in PTSD highlighted the disassociation between cortisol and catecholamine levels in PTSD. Norepinephrine and epinephrine levels assayed from the same urine specimens revealed elevations in both of these catecholamines, while cortisol levels in PTSD fell within the "normal range" of 20–90 µg/day, indicating that the alteration was not in the "hypoadrenal" or endocrinopathologic range (Mason et al. 1986). This finding established the expectation that alterations in basal levels of cortisol might be subtle, and not easily differentiated from normal values (Mason et al. 1986).

Table 1 shows that this is in fact the case. Whereas the majority of studies has found evidence of low cortisol in PTSD, it is clear that group differences are not always present between subjects with and without PTSD. The inconsistency in published reports examining urinary 24-h cortisol levels has been widely noted. There are numerous sources of potential variability in such studies related to the selection of subjects and comparison groups, adequate sample size, and sample inclusion/exclusion criteria, as well as considerations that are specific to the methods of collecting and assaying cortisol levels, that can explain the discrepant finding. However, the simplest explanation for disparate observations is that cortisol levels can show day to day fluctuations, making it difficult to consistently observe group differences.

Table 1 Summary of data from studies of 24-h urinary cortisol excretion in adults with PTSD

Author(s), year	Trauma survivors with PTSD Cortisol μg/day (*n*)		Trauma w/out PTSD Cortisol μg/day (*n*)		Normal comparison Cortisol μg/day (*n*)		Psychiatric comparison Cortisol μg/day (*n*)	
Mason et al. 1986*	33.3	(9)					48.5	(35)
Kosten et al. 1990*	50.0	(11)			55.0	(28)	70.0	(18)
Pitman and Orr 1990**	107.3	(20)	80.5	(15)				
Yehuda et al. 1990*	40.9	(16)			62.8	(16)		
Yehuda et al. 1993a*	38.6	(8)					69.4	(32)
Yehuda et al. 1995b*	32.6	(22)	62.7	(25)	51.9	(15)		
Lemieux and Coe 1995**	111.8	(11)	83.1	(8)	87.8	(9)		
Maes et al. 1998**	840.0	(10)			118	(17)	591.0	(10)
Thaller et al. 1999*	130.9	(34)			213.9	(17)		
Baker et al. 1999	84.4	(11)			76.2	(12)		
DeBellis et al. 1999**	57.3	(18)			43.6	(24)	56.0	(10)
Yehuda et al. 2000*	48.3	(22)			65.1	(15)		
Rasmusson et al. 2001	42.8	(12)			34.6	(8)		
Glover and Poland 2001*[a]	9.8	(14)	16.5	(7)	12.8	(8)		

*Denotes findings in which cortisol levels were significantly lower than comparison subjects, or, in the case of Kosten et al., from depression only. **Denotes findings in which cortisol levels were significantly higher than comparison subjects. [a]Results are from a 12-h rather than 24-h urine collection and are expressed as μg/12 h.

2.2 Cortisol Levels Over the Diurnal Cycle in PTSD

Among the many potential methodologic problems associated with 24-h urine collections is the possibility that persons who are asked to collect 24-h samples at home may not provide complete collections. To the extent that there may be a systematic bias in protocol nonadherence between subjects with and without PTSD, in that the former might be more likely to miss collections than the latter, this could contribute to observed low cortisol levels. One of the initial rationales for performing a comprehensive circadian rhythm analysis was to corroborate and extend findings from the 24-h urine excretion studies and those using single-point estimates (Yehuda et al. 1990). An initial study of circadian parameters in PTSD was conducted by obtaining 49 consecutive blood samples from three groups of subjects—Vietnam combat veterans with PTSD, subjects (largely veterans) with major depression, and non-psychiatric comparison subjects—every 30 min over a 24-h period under carefully controlled laboratory conditions.

Mean basal cortisol release was found to be significantly lower in the PTSD, and cortisol levels were also reduced, at several points during the circadian period, primarily in the late evening and early morning hours compared to the other groups. The major difference between PTSD and non-PTSD groups was that cortisol levels were lower in the late night and very early A.M., and remained lower for a longer period of time in PTSD during hours when subjects are normally sleeping. By the time of awakening, the peak cortisol release, was comparable in PTSD subjects and age-matched subjects. In a second study, these findings were replicated and extended in a sample of 52 women with and without a history of early childhood sexual abuse and PTSD. Cortisol levels obtained every 15 min over a 24-h period demonstrated significantly low cortisol levels, this time in the afternoon and evening hours in the PTSD group.

Thaller et al. also reported that PTSD subjects seemed to show a greater dynamic range as evidenced by a greater disparity between 8:00 A.M. and 5:00 P.M. cortisol levels compared to those of normal controls (Thaller et al. 1999). In PTSD, mean cortisol levels were 21.6 μg/dl in the A.M. and 8.8 μg/dl in the P.M. compared to 21.4 μg/dl in the A.M. and 14.6 μg/dl at 5:00 P.M. for comparison subjects. These findings are consistent with those obtained from the more comprehensive circadian rhythm analysis, indicating that cortisol levels are comparable at their peak, but lower at the nadir in PTSD. In contrast, Hoffman et al. also reported a greater A.M. to P.M. decline in PTSD, but in this case subjects with PTSD went from 18.2 μg/dl to 10.1 μg/dl, compared to control subjects who diminished from 14.1 μg/dl to 9.9 μg/dl (Hoffman et al. 1989).

In Yehuda et al., the raw cortisol data were then subjected to single and multi-oscillator cosinor analyses to determine circadian rhythm parameters (Yehuda et al. 1996b). An increased amplitude-to-mesor (midline estimating statistic of rhythm) ratio reflected the fact that PTSD subjects displayed a greater dynamic range of cortisol compared to controls. That is, although the cortisol peak among individuals without PTSD was not statistically different from the peak among individuals with PTSD, the lower trough among those with PTSD, and the longer period spent at the nadir, resulted in a decreased mesor. Considering differences in the peak of cortisol relative to the mesor also provides an estimate of the "signal-to-noise" ratio of the system. In contrast, depressed patients showed a less dynamic circadian release of cortisol, reflected in an increased mesor of cortisol release over the 24-h cycle, a decreased amplitude-to-mesor ratio, and an elevated trough (Yehuda et al. 1996b). These findings suggest that the main feature of basal cortisol release in PTSD is potential for a greater reactivity of the system.

2.3 Cortisol Levels in Response to Stress

The potential significance of the findings of an increased range of cortisol is that the HPA axis may be maximally responsive to stress-related cues in PTSD, whereas major depressive disorder may reflect a condition of minimal responsiveness to the environment. That is, an enhanced amplitude-to-mesor ratio describes a system with particularly low background activity and, accordingly, a potentially increased capacity to respond to environmental cues. In support of this, Liberzon et al. observed an increased cortisol [but not increased corticotrophin (ACTH)] response in combat veterans with PTSD compared to controls who were exposed to white noise and combat sounds (Liberzon et al. 1999). Elizinga et al. also observed that women with PTSD related to childhood abuse had substantially higher salivary cortisol levels in response to hearing scripts related to their childhood experiences compared to controls, who had relatively lower cortisol levels in response to hearing scripts of other people's traumatic stories (Elzinga et al. 2003). Similarly, Bremner et al. also observed an increased salivary cortisol response in anticipation of a cognitive challenge test relative to controls in women with PTSD related to childhood abuse (these were a subset of the same women in whom plasma cortisol levels had been low at baseline) (Bremner et al. 2003a). The authors suggest that although cortisol levels were found low at baseline, there did not appear to be an impairment in the cortisol response to stressors in PTSD. These studies demonstrate transient increases in cortisol levels that are consistent with the notion of a more generalized HPA axis reactivity in PTSD.

2.4 Observations About Baseline Cortisol Based on Single Estimates of Plasma or Saliva

Investigations of single plasma and salivary cortisol levels have become increasingly popular in the last decade given the relative ease in acquiring samples.

However, the use of a single sampling of cortisol, particularly at a set time of the day, may not represent an appropriate method for estimating cortisol levels because of moment-to-moment fluctuations in cortisol levels due to transient stressors in the environment (including the actual stress of venipuncture or anticipatory anxiety). Variability in single sampling estimates of cortisol may also reflect individual variation in sleep cycles. Because cortisol levels steadily decline from their peak, which is usually observed at 30 min post-awakening (Hucklebridge et al. 1999), differences in wake-time of several minutes to an hour may increase the variability substantially.

Table 2 provides a summary of cortisol levels in studies that specifically obtained 8:00 A.M. cortisol concentrations, and highlights the lack of uniform findings in relation to cortisol levels, possibly reflecting the above-mentioned methodologic considerations. Of particular note, however, is Boscarino's report of low cortisol in a large epidemiologic sample of over 2,000 Vietnam veterans with PTSD compared to those without PTSD, which implies that to consistently observe low morning cortisol would require an extremely large sample size (Boscarino 1996). The magnitude of difference between PTSD and non-PTSD subjects at 8:00 A.M. was very modest—there was only a 4% difference between veterans with and without current or lifetime PTSD. Cortisol levels were significantly lower in combat veterans with very high exposure (17.9 μg/day) compared to those with no or low exposure (19.1 μg/day). The finding of an inverse relationship between combat exposure severity and 8:00 A.M. cortisol levels had been reported earlier in a much smaller sample of Vietnam veterans (Yehuda et al. 1995a).

The use of salivary assessments has helped supply data in studies of children and adolescents, for whom even a blood draw may be too invasive, and also helped in our evaluation of longitudinal outcomes. King et al. (2001) observed significantly low cortisol levels in children aged 5–7 years who had been sexually abused compared to control subjects. Goenjian et al. (1996) demonstrated a relationship between low salivary cortisol levels and PTSD symptoms in adolescents exposed to the Armenian earthquake. However, both Lipschitz et al. (2003) and Carrion et al. (2002) failed to note differences in salivary cortisol levels at baseline in multiply traumatized adolescents.

Using repeated salivary cortisol assessments in a single individual, Kellner et al. (1997) demonstrated that salivary cortisol decreased dramatically 3 months after a traumatic event, and in the course of further research showed an inverse relation to fluctuating, but gradually improving PTSD symptoms. Post-dexamethasone (post-DEX) cortisol was suppressed below the detection limit early after trauma, and rose again more than 1 year post-trauma. In a similar case report, Heber et al. demonstrated an increase in basal salivary cortisol and an increasingly attenuated cortisol response to dexamethasone (DEX) in PTSD patients who were successfully treated using eye-movement desensitization reprocessing therapy (EMDR) (Heber et al. 2002), suggesting some relationship between low cortisol and PTSD symptoms.

Table 2 Plasma A.M. cortisol levels in PTSD and comparison subjects

Reference	Trauma survivors with PTSD Cortisol µg/day (*n*)	Trauma w/out PTSD Cortisol µg/day (*n*)	Normal comparison Cortisol µg/day (*n*)	Psychiatric comparison Cortisol µg/day (*n*)
Hoffman et al. 1989**	18.2 (21)		14.1 (20)	
Halbreich et al. 1989	7.7 (13)		7.3 (21)	12.3 (23) (MDD)
Yehuda et al. 1991b	14.3 (15)		14.9 (11)	
Yehuda et al. 1993b	14.3 (21)		15.1 (12)	
Yehuda et al. 1995a*	12.7 (14)	16.4 (12)	15.0 (14)	
Yehuda et al. 1996a#	11.6 (15)		14.2 (15)	12.2 (14) (MDD)
Yehuda et al. 1996a#	11.8 (11)		9.8 (8)	
Boscarino 1996*	17.7 (293)	18.4 (2197)		
Jensen et al. 1997*	4.6 (7)		8.9 (7)	9.9 (7) (Panic)
Liberzon et al. 1999**	12.1 (17)	7.9 (11)	9.3 (14)	
Thaller et al 1999	21.6 (34)		21.4 (17)	
Kellner et al. 2000*	7.8 (8)		13.3 (8)	
Kanter et al. 2001*#	7.6 (13)		10.6 (16)	
Atmaca et al. 2002**	12.9 (14)		10.7 (14)	
Gotovac et al. 2003*	14.4 (28)		17.2 (19)	
Seedat et al. 2003*	10.3 (10)	10.6 (12)	13.4 (16)	
Oquendo et al. 2003#*	11.8 (13)		14.8 (24)	16 (45) (MDD)
Lueckeh et al. 2004#*	8.7 (13)		14.4 (47)	(Cancer)
Yehuda et al. 2004a,b				

#No means reported in the text; data estimated from the figures provided. *Significantly lower in PTSD than normal comparison. **Significantly higher in PTSD than normal comparison. MDD, major depressive disorder.

2.5
Correlates of Cortisol in PTSD

Even in cases where there is failure to find group differences, there are often correlations within the PTSD group with indices of PTSD symptom severity. Baker et al. (1999) failed to find group differences between Vietnam veterans with PTSD compared to non-exposed controls, but did report a negative correlation between 24-h urinary cortisol and PTSD symptoms in combat veterans. A negative correlation between baseline plasma cortisol levels and PTSD symptoms, particularly avoidance and hyperarousal symptoms, were observed in adolescents with PTSD (Goenjian et al. 2003). Rasmusson et al. (2003) failed to observe a significant difference in urinary cortisol between premenopausal women with PTSD and healthy women, but noted an inverse correlation between duration since the trauma and cortisol levels, implying that low cortisol is associated with early traumatization. This finding is consisted with Yehuda and colleagues' observation of an inverse relationship between childhood emotional abuse and cortisol levels in adult children of Holocaust survivors (Yehuda et al. 2002a).

Cortisol levels have also been correlated with findings from brain imaging studies in PTSD. In one report, there was a positive relationship between cortisol levels and hippocampal acetylaspartate (NAA)—a marker of cell atrophy presumed to reflect changes in neuronal density or metabolism—in subjects with PTSD, suggesting that rather than having neurotoxic effects, cortisol levels in PTSD may have a trophic effect on the hippocampus (Neylan et al. 2003a). Similarly, cortisol levels in PTSD were negatively correlated with medial temporal lob perfusion, while anterior cingulate perfusion and cortisol levels were positively correlated in PTSD, but negatively correlated in trauma survivors without PTSD (Bonne et al. 2003b). The authors suggest that the negative correlation may result from an augmented negative hippocampal effect secondary to increased sensitivity of brain glucocorticoid receptors (GRs), which would account for the inverse correlation in PTSD despite equal cortisol levels in both the PTSD and non-PTSD groups. On the other hand, the positive correlation between regional cerebral blood flow in the fronto-cingulate transitional cortex and cortisol levels in PTSD may reflect unsuccessful attempts of the fronto-cingulate transitional cortex to terminate the stress response, which has also been linked to low cortisol.

Cortisol may be related to specific, or state-dependent features of the disorder, such as comorbid depression or the time course of the disorder. Mason et al. (2001) have underscored the importance of examining intrapsychic correlates of individual differences in cortisol levels in PTSD, and have hypothesized that cortisol levels in PTSD may be related to different levels of emotional arousal, and opposing antiarousal disengagement defense mechanisms or other coping styles. Further, Wang et al. have posited that adrenal activity may change over time in a predicted manner reflecting stages of decompensation in PTSD (Wang et al. 1996).

2.6 CRF Levels in PTSD

There have been three published reports examining the concentration of corticotrophin-releasing factor (CRF) in cerebrospinal fluid (CSF) in PTSD. The assessment of CSF CRF does not necessarily provide a good estimate of hypothalamic CRF release, but rather, an estimate of both hypothalamic and extrahypothalamic release of this neuropeptide (Yehuda and Nemeroff 1994). An initial report using a single lumbar puncture indicated that CRF levels were elevated in combat veterans with PTSD (Bremner et al. 1997). A second study, examining serial CSF sampling over a 6-h period by means of an indwelling catheter, also reported significantly higher CSF CRF concentrations, but did not observe a relationship between CRF and 24-h urinary cortisol release (Baker et al. 1999). A third report demonstrated that PTSD subjects with psychotic symptoms had significantly higher mean levels of CRF than either subjects with PTSD without psychotic symptoms or controls subjects (Sautter et al. 2003).

2.7 ACTH Levels in PTSD

Among the challenges in assessing pituitary activity under basal conditions is the fact that the normal positive and negative feedback influences on the pituitary can mask the true activity of this gland. Because the pituitary mediates between CRF stimulation from the hypothalamus and the inhibition of ACTH release resulting from the negative feedback of adrenal corticosteroids, baseline ACTH levels may appear to be "normal" even though the pituitary gland may be receiving excessive stimulation from CRF. In most studies ACTH levels in PTSD patients were reported to be comparable to non-exposed subjects.

The majority of studies has reported no detectible differences in ACTH levels between PTSD and comparison subjects even when cortisol levels obtained from the same sample were found to be significantly lower. This pattern was observed in Kellner et al. who reported that cortisol levels were 41% lower, but that ACTH levels were only 7.4% lower in PTSD compared to normals (Kellner et al. 2000), and Hockings et al. who showed that cortisol levels were 12% lower in PTSD but ACTH levels identical to controls (Hockings et al. 1993). Kanter et al. also reported that cortisol levels were substantially lower in PTSD, while ACTH levels were comparable to controls (Kanter et al. 2001). In Yehuda et al. cortisol levels were lower at baseline on the placebo day in PTSD, but not at the baseline time point on the metyrapone day (i.e., prior to metyrapone administration) compared to comparison subjects, but ACTH levels were comparable in both groups on both days (Yehuda et al. 1996a). Similar data were reported by Neylan et al. (2003a).

Lower cortisol levels in the face of normal ACTH levels can reflect a relatively decreased adrenal output. Yet under circumstances of classic adrenal insufficiency, there is usually increased ACTH release compared to normal levels. Thus, in PTSD there may be an additional component of feedback on the pituitary that is acting to depress ACTH levels, making them appear normal. Indeed, elevations in ACTH would be expected not only from a reduced adrenal output but also from increased CRF stimulation (Baker et al. 1999; Bremner et al. 1997). On the other hand, the adrenal output in PTSD may be relatively decreased, but not substantially enough to affect ACTH levels. In any event, the "normal" ACTH levels in PTSD in the context of the other findings suggest a more complex model of the regulatory influences on the pituitary in this disorder than reduced adrenal insufficiency.

In contrast to the above-mentioned findings, Hoffman et al. reported that cortisol levels were 22.5% higher in PTSD, but ACTH was only 4% lower compared to controls (Hoffman et al. 1989). In this report, mean plasma β-endorphin (co-localized and released with ACTH) was reported as lower in PTSD. Liberzon et al. also reported mean cortisol levels to be 33% higher, but ACTH 31% lower in PTSD compared to controls (Liberzon et al. 1999). Smith et al. also reported cortisol levels were 48% higher and ACTH 32% lower in PTSD than controls, but this was in the afternoon (Smith et al. 1989). Although ACTH levels were not significantly different in PTSD compared to controls, the increase in cortisol relative to ACTH is reminiscent of classic models of HPA dysregulation in depression where there is hypercortisolism but a reduced ACTH negative feedback inhibition. Rasmusson et al. (2001) demonstrated a 13% increase in cortisol with no differences in ACTH in PTSD at 8:00 P.M., which is consistent with the idea of an overall, but somewhat mild, HPA hyperactivity.

2.8 Corticosteroid Binding Globulin

Kanter et al. reported an increase concentration of the corticosteroid binding globulin (CBG) (Kanter et al. 2001). Most cortisol is bound to CBG, and is biologically inactive. A greater concentration of CBG is consistent with low levels of measurable free cortisol, and provides a putative explanation for how cortisol levels could be measurably low even though other aspects of HPA axis functioning do not seem hypoactive. However, the extent to which CBG levels are a contributing cause of low cortisol requires further examination.

3
Glucocorticoid Receptors in PTSD

Type II GRs are expressed in ACTH- and CRF-producing neurons of the pituitary, hypothalamus, and hippocampus, and mediate most systemic glucocorticoid effects, particularly those related to stress responsiveness (deKloet et al. 1991). Low circulating levels of a hormone or neurotransmitter can result in increased numbers of available receptors (Sapolsky et al. 1984) that improve response capacity and facilitate homeostasis. However, alterations in the number and sensitivity of both type I (mineralocorticoid) and type II GRs can also significantly influence HPA axis activity, and in particular, can regulate hormone levels by mediating the strength of negative feedback (Svec 1985; Holsboer et al. 2000).

Lymphocyte and brain GRs have been found to share similar regulatory and binding characteristics (Lowy 1989). A greater number of 8:00 A.M., but not 4:00 P.M., mononuclear leukocytes (presumably lymphocyte) type II GRs was reported in Vietnam veterans with PTSD compared to a normal comparison group (Yehuda et al. 1991b). Subsequently, Yehuda et al. reported an inverse relationship between 24-h urinary cortisol excretion and lymphocyte GR number in PTSD and depression (i.e., low cortisol and increased receptor levels were observed in PTSD, whereas in major depressive disorder, elevated cortisol and reduced receptor number were observed) (Yehuda et al. 1993a). Although it is not clear whether alterations in GR number reflect an adaptation to low cortisol levels or some other alteration, the observation of an increased number of lymphocyte GRs provided the basis for the hypothesis of an increased negative feedback inhibition of cortisol secondary to increased receptor sensitivity (Yehuda et al. 1995a).

Following the administration of a 0.25-mg dose of DEX, it was possible to observe that the cortisol response was accompanied by a concurrent decline in the number of cytosolic lymphocyte receptors (Yehuda et al. 1995a). This finding contrasted with the observation of a reduced decline in the number of cytosolic lymphocyte receptors in major depression, implying that the reduced cortisol levels following DEX administration may reflect an enhanced negative feedback inhibition in PTSD (Gormley et al. 1985).

Observations regarding the cellular immune response in PTSD are also consistent with enhanced GR responsiveness in the periphery. In one study, beclomethasone-induced vasoconstriction was increased in women PTSD subjects compared to healthy, non-trauma-exposed comparison subjects (Coupland et al. 2003). Similarly, an enhanced delayed-type hypersensitivity of skin test responses was observed in women who survived childhood sexual abuse vs those who did not (Altemus et al. 2003). Because immune responses, like endocrine ones, can be multiply regulated, these studies provide only indirect evidence of GR responsiveness. However, when considered in the context of the observation that PTSD patients showed increased expression of the receptors in all lymphocyte subpopulations, despite a relatively low quantity of intracellular GR as determined by flow cytometry, and in the face of lower ambient cortisol levels (Gotovac et al. 2003), the findings convincingly support an enhanced sensitivity of the GR to glucocorticoids. Furthermore, Kellner et al. reported an absence of alterations of the mineralocorticoid receptor in PTSD as investigated by examining the cortisol and ACTH response to spironolactone following CRF stimulation (Kellner et al. 2002a).

Finally, a recent study provided the first demonstration of an alteration in target tissue sensitivity in glucocorticoids using an in vitro paradigm. Mononuclear leukocytes isolated from the blood of 26 men with PTSD and 18 men without PTSD were incubated with a series of concentrations of DEX to determine the rate of inhibition of lysozyme activity; a portion of cells was frozen for the determination of GRs. Subjects with PTSD showed evidence of a greater sensitivity to glucocorticoids as reflected by a significantly lower mean lysozyme $IC_{50\text{-}DEX}$ (nM). The lysozyme $IC_{50\text{-}DEX}$ was significantly correlated with age at exposure to the first traumatic event in subjects with PTSD. The number of cytosolic GRs was correlated with age at exposure to the focal traumatic event (Yehuda et al., in press).

4 Cortisol and ACTH Responses to Neuroendocrine Challenge

4.1 The Dexamethasone Suppression Test in PTSD

In contrast to observations regarding ambient cortisol and ACTH levels, results using the DEX suppression test (DST) have presented a more consistent view of reduced cortisol suppression in response to DEX administration. The DST provides a direct test of the effects of GR activation in the pituitary on ACTH secretion, and cortisol levels following DEX administration are thus interpreted an estimate of the strength of negative feedback inhibition, provided that the adrenal response to ACTH is not altered. There are several hundred published studies reporting on the use of the DST in depression, all reporting that approximately 40%–60% of patients with major depression demonstrate a failure to suppress cortisol levels below 5.0 μg/100 dl in response to 1.0 mg of DEX (Ribeiro et al. 1993). Nonsuppression of cortisol results from a reduced ability of DEX to exert negative feedback inhibition on the release of CRF and ACTH (Holsboer 2000).

The initial DST studies in PTSD using the 1.0-mg dose of DEX did not consider the possibility of a hypersuppression to DEX and tested the hypothesis that patients with PTSD might show a nonsuppression of cortisol similar to patients with major depressive disorder. A large proportion of the PTSD subjects studied also met criteria for major depression. Four (Dinan et al.

1990; Halbreich et al. 1989; Kosten et al. 1990; Reist et al. 1995) out of five (Kudler et al. 1987) of the earlier studies noted that PTSD did not appear to be associated with cortisol nonsuppression, using the established criterion of 5 µg/100 ml at 4:00 P.M. A more recent study did not use the established criterion to determine nonsuppression, but nonetheless reported a greater mean cortisol in PTSD compared to normal subjects at 8:00 A.M. (Thaller et al. 1999). In this study, Thaller et al. reported that DEX resulted in 67% suppression in PTSD (n=34) compared to 85% suppression in comparison (n=17) subjects. Similarly, Atmaca et al. showed a significantly higher DST nonsuppression in the PTSD group (63.12%) compared to healthy controls (79.6%) using the 1.0 mg DST (Atmaca et al. 2002).

Although the 1.0-mg DST studies primarily focused on evaluating failure of normal negative feedback inhibition, Halbreich et al. noted that post-DEX cortisol levels in the PTSD group were particularly lower than subjects with depression and even comparison subjects (Halbreich et al. 1987). The mean post-DEX cortisol levels were 0.96±0.63 µg/dl in PTSD compared to 3.72±3.97 µg/dl in depression and 1.37± µg/dl in comparison subjects, raising the possibility that the 1-mg dose produced a "floor effect" in the PTSD group. Based on this observation, and on findings of low cortisol and increased GR number, Yehuda et al. hypothesized that PTSD patients would show an enhanced, rather than reduced, cortisol suppression to DEX and administered lower doses of DEX—0.50 mg and 0.25 mg—to examine this possibility (Yehuda et al. 1993a, 1995a). A hyperresponsiveness to low doses of DEX, as reflected by significantly lower post-DEX cortisol levels, was observed in PTSD patients compared to non-exposed subjects. The enhanced suppression of cortisol was present in combat veterans with PTSD who met the diagnostic criteria for major depressive disorder (Yehuda et al. 1993a) and was not present in combat veterans without PTSD (Yehuda et al. 1995a).

The finding of an exaggerated suppression of cortisol in response to DEX was also observed by Stein et al. who studied adult survivors of childhood sexual abuse (Stein et al. 1997), and by Kellner et al. who evaluated Gulf War soldiers who were still in active duty about a year an a half after their deployment to the Persian Gulf (Kellner et al. 1997). More recently, an exaggerated suppression following 0.50 mg DEX was also observed in older subjects with PTSD (i.e., Holocaust survivors and combat veterans) compared to appropriate comparison subjects (Yehuda et al. in 2002b) in a sample of depressed women with PTSD resulting from early childhood abuse (Newport et al. 2004), and a mixed group of trauma survivors with PTSD (Yehuda et al. 2004b; Table 3).

Results from these studies are expressed as the extent of cortisol suppression, evaluated by the quotient of 8:00 A.M. post-DEX cortisol to 8:00 A.M. baseline cortisol. Expressing the data in this manner accounts for individual differences in baseline cortisol levels and allows for a more precise characterization of the strength of negative feedback inhibition as a continuous rather than as a dichotomous variable. Whereas studies of major depression emphasize the 4:00 P.M. post-DEX value as relevant to the question of nonsuppression (Stokes et al. 1984), studies of PTSD have been concerned with the degree to which DEX suppresses negative feedback at the level of the pituitary, rather than the question of "early escape" from the effects of DEX. Goeinjian et al. observed an enhanced suppression of salivary cortisol at 4:00 P.M. following 0.50 mg of DEX in adolescents who had been closer to the epicenter of an earthquake 5 years earlier (and had more substantial PTSD symptoms) compared to those who had been further from the epicenter (Goeinjian et al. 1996). However, the percentage suppression of cortisol in these two groups was comparable at 8:00 A.M. The authors concluded that the suppression of cortisol to DEX may last longer in PTSD. Unfortunately, the authors were not able to study a non-exposed comparison group. Similarly, Lipschitz et al. failed to observe cortisol hypersuppression at 8:00 A.M. in adolescents with PTSD exposed to multiple traumatic events (Lipschitz et al. 2004). Unfortunately, the authors were not able to obtain data at the 4:00 P.M. time point.

There is some debate about whether DST hypersuppression reflects trauma exposure in psychiatric patients or PTSD per se. Using the combined DEX/CRF challenge in women with borderline personality disorder with and without PTSD relating to sustained childhood abuse, Rinne et al. (2002) demonstrated that chronically abused patients with borderline personality disorder had a significantly enhanced ACTH and cortisol response to the DEX/CRF challenge

Table 3 Summary of data from studies of using the dexamethasone suppression test

Reference	Dex dose/day	PTSD: % supp (*n*)		Comparison: % supp (*n*)	
Yehuda et al. 1993b*	0.5	87.5	(21)	68.3	(12)
Stein et al. 1997*	0.5	89.1	(13)	80.0	(21)
Yehuda et al. 1995a*	0.5	90.0	(14)	73.4	(14)
Yehuda et al. 1995*	0.25	54.4	(14)	36.7	(14)
Kellner et al. 1997***	0.50	90.1	(7)		
Yehuda et al. 2002b*	0.50	89.9	(17)[#]	77.9	(23)
Grossman et al. 2003*[a]	0.50	83.6	(16)	63.0	(36)
Newport et al. 2004*[b]	0.50	92.3	(16*)	77.78	(19)
Yehuda et al. 2004b*	0.50	82.5	(19)	68.9	(10)

[#]Includes subjects without depression; subjects with both PTSD and MDD (*n*=17) showed a percentage suppression of 78.8, which differs from our previous report (Yehuda et al. 1993b) in younger combat veterans. *Significantly more suppressed than controls. **Significantly less suppressed than controls. ***No control group was studied. [a]Comparison subjects were those with personality disorders but without PTSD. [b]It is impossible from this paper to get the correct mean for the actual 15 subjects with PTSD. These 16 subjects had MDD, but 15/16 also had PTSD, so this group also contains 1 subject who had been exposed to early abuse with past, but not current, PTSD.

compared with nonabused subjects, suggested a hyperresponsiveness of the HPA axis. The authors attribute the finding to trauma exposure. On the other hand, Grossman et al. (2003) examined the cortisol response to 0.50 mg DEX in a sample of personality disordered subjects and found that cortisol hypersuppression was related to the comorbid presence of PTSD, but not trauma exposure.

In the study by Newport and colleagues (2004), the authors attempted to determine whether cortisol hypersuppression was related to early abuse in PTSD and major depression. However, insofar as all the exposed subjects with current depression had PTSD (all except one), it was difficult to attribute the observed hypersuppression to PTSD or depression. Recently, however, Yehuda et al. observed cortisol hypersuppression following 0.50 mg DST in PTSD, and subjects with both PTSD and depression, but noted that hypersuppression was particularly prominent in persons with depression comorbidity if there had been a prior traumatic experience. Thus, cortisol hypersuppression in response to DEX appears to be associated with PTSD, but in subjects with depression, hypersuppression may be present as a result of early trauma, and possibly past PTSD (Yehuda et al. 2004b).

4.2 The Cholecystokinin Tetrapeptide Challenge Test in PTSD

Cholecystokinin tetrapeptide (CCK)-4 is a potent stimulator of ACTH. Kellner et al. administered a 50-µg bolus of CCK-4 to subjects with PTSD and found substantially attenuated elevations of ACTH in PTSD, which occurred despite comparable ACTH levels at baseline (Kellner et al. 2000). Cortisol levels were lower in PTSD at baseline, but rose to a comparable level in PTSD and control subjects. However, the rate of decline from the peak was faster, leading to an overall lower total cortisol surge. The attenuated ACTH response to CCK-4 is compatible with the idea of CRF overdrive in PTSD, and is a similar to the administration of CRF. That less ACTH can produce a similar activation of the adrenal gland, but a more rapid decline of cortisol is also consistent with a more sensitive negative feedback inhibition secondary to increased glucocorticoid receptor activity at the pituitary. Although the comparatively greater effects on cortisol relative to ACTH is also compatible with an increased sensitivity of the adrenal gland to ACTH, rather than an enhanced negative feedback sensitivity on the pituitary, this explanation only accounts for the greater rise cortisol, but not the more rapid rate of decline of cortisol, following CCK-4.

4.3 The Metyrapone Stimulation Test

Whereas both the results of the DST and CCK challenge tests are consistent with the idea of an enhanced negative feedback inhibition in PTSD, these alterations do not directly imply that an enhanced negative feedback inhibition is a primary disturbance in PTSD. Yehuda et al. used the metyrapone stimulation test as a way of providing further support for the enhanced negative feedback hypothesis (Yehuda et al. 1996a). Metyrapone prevents adrenal steroidogenesis by blocking the conversion of 11-deoxycortisol to cortisol, thereby unmasking the pituitary gland from the influences of negative feedback inhibition. If a sufficiently high dose of metyrapone is used such that an almost complete suppression of cortisol is achieved, this allows a direct examination of pituitary release of ACTH without the potentially confounding effects of differing ambient cortisol levels. When metyrapone is administered in the morning—when HPA axis activity is relatively high—maximal pituitary activity can be achieved, facilitating an evaluation of group differences in pituitary capability. The administration of 2.5 mg metyrapone in the morning resulted in a similar and almost complete reduction in cortisol levels in both PTSD and normal subjects (i.e., and removal of negative feedback inhibition), but a higher increase in ACTH and 11-deoxycortisol in combat Vietnam veterans with PTSD compared to non-exposed subjects (Yehuda et al. 1996a). In the context of low cortisol levels and increased CSF CRF levels, the findings supported the hypothesis of a stronger negative feedback inhibition in PTSD. Both pituitary and adrenal insufficiency would not likely result in an increased ACTH response to removal of negative feedback inhibition, since the former would be associated with an attenuated ACTH response and reduced adrenal output would not necessarily affect the ACTH response. To the extent that ambient cortisol levels are lower than normal, an increased ACTH response following removal of negative feedback inhibition implies that when negative feedback is intact, it is strong enough to inhibit ACTH and cortisol. The increased ACTH response is most easily explained by increased suprapituitary activation; however, a sufficiently strong negative feedback inhibition would account for the augmented ACTH response even in the absence of hypothalamic CRF hypersecretion.

Kanter et al. failed to find evidence for an exaggerated negative feedback inhibition using a different type of metyrapone stimulation paradigm (Kanter et al. 2002). In this study, a lower dose of metyrapone was used, administered over a 3-h period (750 mg at 7:00 A.M. and 10:00 A.M.), and rather than simply examining the ACTH response to this manipulation, the cortisol levels were introduced by means of an infusion, allowing the effects of negative feedback inhibition to be evaluated more systematically. Under conditions of enhanced negative feedback inhibition, the introduction of cortisol following metyrapone administration should result in a greater suppression of ACTH in PTSD. However, no significant differences in the ACTH response to cortisol infusion between PTSD and comparison subjects (but a non-significant trend, p=.10, for such a reduction) were observed. There was, however, a reduced response of 11-*b*-deoxycortisol. The authors concluded that their findings provided evidence of sub-clinical adrenocortical insufficiency.

In evaluating this finding, it must be noted, as the authors do, that at the dose used, metyrapone did not accomplish a complete suppression of cortisol in this study. Furthermore, the manipulation produced a more robust suppression of cortisol in comparison subjects, suggesting that the control group was significantly more perturbed by the same does of metyrapone prior to the cortisol infusion than the PTSD group. The authors suggest that the lack of decline in ACTH following cortisol infusion in the PTSD group argues against an enhanced negative feedback inhibition. However, insofar as the drug produced a significantly greater decrease in cortisol in the comparison subjects, while not producing a significant difference in ACTH concentrations, it might be that the lack of an ACTH reduction in PTSD following cortisol infusion may have been caused by a floor effect, rather than a demonstration of lack of reactivity of the system. Indeed, because metyrapone at the dose used did not fully suppress cortisol, the endogenous cortisol present may have already been high enough to suppress ACTH secretion in the PTSD group. Interestingly, although metyrapone did not result in as great a decline of cortisol in PTSD, it did result in the same level of cortisol inhibition, implying differences in the activity of the enzyme 11-β-hydroxylase, which merits further investigation.

To the extent that there was a significant reduction of the 11-deoxycortisol response in PTSD in the absence of an attenuated ACTH response, this would indeed support the idea of a reduced adrenal output. However, the trend for an ACTH response suggests that part of the failure to achieve statistical significance may have also occurred because of limited power, particularly given the lack of evidence for increased ambient ACTH levels in PTSD relative to normal controls. Dose–response studies using the higher vs lower dose of metyrapone should certainly be conducted to further address this critical issue.

A third study used metyrapone to evaluate CRF effects on sleep, but in the process also provided information relevant to negative feedback inhibition. Metyrapone (750 mg) was administered at 8:00 A.M. every 4 h for 16 h, and cortisol, 11-deoxycortisol and ACTH levels were measured at 8:00 A.M. the following morning. Cortisol, 11-deoxycortisol, and ACTH levels were increased in the PTSD group relative to the controls, suggesting that the same dose of metyrapone did not produce the same degree of adrenal suppression of cortisol synthesis. Under these conditions, it is difficult to evaluate the true effect on ACTH and 11-deoxycortisol, which depends on achieving complete cortisol suppression, or at least the same degree of cortisol suppression in the two groups. The endocrine response to metyrapone in this study does not support the model of reduced adrenal capacity, since this would have been expected to yield a large ratio of ACTH to cortisol release; yet the mean ACTH/cortisol ratio prior to metyrapone was no different in PTSD vs controls. On the other hand, the mean ACTH/cortisol ratio post-metyrapone was lower, though non-significantly, suggesting, if anything, an exaggerated negative feedback rather than reduced adrenal capacity (Neylan et al. 2003a).

The idea of reduced adrenal capacity as a possible model for PTSD has also been recently raised by Heim et al., who concluded that low cortisol may not be a unique feature of PTSD, but may represent a more universal phenomenon related to bodily disorders, having an etiology related to chronic stress (Heim et al. 2000). There are numerous stress-related disorders such as chronic fatigue syndrome, fibromyalgia, rheumatoid arthritis, chronic pain syndromes, and other disorders that are characterized by hypocortisolism. In one study, Heim et al. showed decreased cortisol responses to low-dose DEX, but failed to observe blunted ACTH responses to CRF in women with chronic pelvic pain, some of whom had PTSD, compared to women with infertility (Heim et al. 1998). Since the data were not analyzed on the basis of the subgroup with and without trauma and/or PTSD, it is not possible to directly compare results of that study to other reports examining PTSD directly.

4.4 The CRF Challenge Test and ACTH Stimulation Test in PTSD

Infusion of exogenous CRF increases ACTH levels and provides a test of pituitary sensitivity. In several studies of major depression, the ACTH response to CRF was shown to be "blunted," reflecting a reduced sensitivity of the pituitary to CRF (e.g., Krishnan 1993). This finding has been widely interpreted as reflecting a downregulation of pituitary CRF receptors secondary to CRF hypersecretion, but may also reflect increased cortisol inhibition of ACTH secondary to hypercortisolism (Krishnan 1993; Yehuda and Nemeroff 1994).

A study of eight PTSD subjects demonstrated that the ACTH response to CRF is also blunted (Smith et al. 1989). However, although the authors noted a uniform blunting of the ACTH response, this did not always occur in the context of hypercortisolism. Furthermore, although the ACTH response was significantly blunted, the cortisol response was not (however, though not statistically significant, it should be noted that the area under the curve for cortisol was 38% less than controls). Bremner et al. also observed a blunted ACTH response to CRF in women with PTSD as a result of early childhood sexual abuse (Bremner et al. 2003b). Yehuda et al. previously suggested that the blunted ACTH response in PTSD might reflect an increased negative feedback inhibition of the pituitary secondary to increased GR number or sensitivity (Yehuda et al. 1995a). This explanation supports the idea of CRF hypersecretion in PTSD, and explains the pituitary desensitization and resultant lack of hypercortisolism as arising from a stronger negative feedback inhibition.

A blunted ACTH response to CRF in the context of a normal cortisol response was also observed in sexually abused girls, but the diagnosis of PTSD was not systematically made in this study (deBellis et al. 1994). When living in the context of ongoing abuse, abused children with depression showed an enhanced ACTH response to CRF in comparison with abused children without depression and normals (Kauffman et al. 1997). Again, although such subjects were considered at greater risk for the development of PTSD, it is difficult to draw direct conclusions from these studies about the neuroendocrinology of PTSD because this variable was not directly measured.

In contrast, Rasmusson et al. (2001) recently reported an augmented ACTH response to CRF in 12 women with PTSD compared to 11 healthy controls. In the same subjects, the authors also performed a neuroendocrine challenge with 250 μg of cosyntropin (ACTHα1-24) to determine the response of the pituitary gland to this maximally stimulating dose. Women with PTSD demonstrated an exaggerated cortisol response to ACTH compared to healthy subjects. Basal assessments did not reveal group differences in either 24-h urinary cortisol levels, or basal plasma cortisol or ACTH levels. The authors concluded that their findings suggested an increased reactivity of both the pituitary and adrenal in PTSD.

What is particularly interesting about the finding of the increased ACTH in response to CRF is that the magnitude of the ACTH response appeared to be much higher than the cortisol response. The ACTH response was 87% greater in the subjects with PTSD, but the cortisol response was only 35% higher. Thus, although ACTH levels were more increased in PTSD than controls, this increased ACTH level did not result in a comparable stimulation of cortisol, suggesting a reduced adrenal capacity or an enhanced inhibition of cortisol. On the other hand, Rasmusson and colleagues' demonstration of an increased cortisol response to cosyntropin in the same patients suggests the opposite. The authors do not discuss the possibility that the results of the CRF test suggest reduced adrenal capacity, nor do they suggest a model that accounts for the co-existence of these two apparently disparate observations.

Attempting to resolve the two discrepant observations in the Rasmusson et al. finding (2001) will by necessity require viewing HPA axis alterations as reflecting a more complex set of processes than are currently described in classical clinical endocrinology, and will be aided, no doubt, by the appearance of currently unavailable information. However, the discrepancy may also result from a methodologic artifact owing to the administration of the cosyntropin at variable times during the day (ranging from 8:15 A.M. to 4:15 P.M.). It may be that if the cortisol data were corrected for time of day of administration of cosyntropin that the findings might no longer be significant, and it would be important to rule this out. Indeed, to conclude that a greater reactivity of the pituitary gland occurs in the context of a more reduced cortisol response would be a simpler observation to contend with.

In fact, the observation of an increased ACTH response to CRF would be compatible with a recent study by Heim et al. who examined such responses in abused women with and without major depressive disorder compared with nonabused depressed women and comparison subjects (Heim et al. 2000). Abused women without depression showed an augmented ACTH response to CRF, but a reduced cortisol response to ACTH compared to other groups. Only a small proportion (4/20) met criteria for PTSD. Abused women with depression (14/15 with PTSD) showed a blunted ACTH response to CRF compared to controls, as did nonabused women with depression. These findings are compatible with those of Smith et al. (1989). Although the study by Heim et al. (2000) did not focus directly on the issue of HPA alterations in PTSD, the model presented by the authors is extremely informative in suggesting the possibility that early abuse may be associated in and of itself with a profile of pituitary-adrenocortical alterations (particularly, low ambient cortisol as a function of a diminished adrenal responsiveness) that are opposite to those seen in depression. However, when depression is present, these alterations may be "overridden" by the results of depression-related CRF hypersecretion. Early trauma exposure is a risk factor not only for depression, but also for PTSD in the absence or presence of depression. It is possible that low cortisol levels resulting from this risk factor may also be influenced by PTSD-related alterations (i.e., increased GR responsiveness and increased responsiveness of negative feedback inhibition).

4.5 The Naloxone Stimulation Test in PTSD

Another strategy for examining CRF activity involves the assessment of ACTH and cortisol after administration of agents that normally block the inhibition of CRF. Naloxone increases CRF release by blocking the inhibition normally exerted by opioids in the hypothalamus. Naloxone was administered to 13 PTSD patients and 7 normal comparison subjects (Hockings et al. 1993). Of the PTSD subjects, 6/7 showed an increased ACTH and cortisol response to naloxone. These findings appear to contradict those of Smith et al. (1989) who showed a blunted ACTH response to CRF; however, here too the absence of information about ambient CRF complicates the interpretation of these findings. This finding is noteworthy for illustrating that only a proportion of subjects in a particular group may exhibit evidence of pituitary adrenocortical alterations.

5 Drawing Conclusions from Challenge Studies: Do They Provide a Window into the Brain?

Although the neuroendocrine challenges described above directly asscss ACTH and cortisol, hypothalamic CRF release may be inferred from some of the results. For example, because metyrapone administration results in the elimination of negative feedback inhibition, its administration allows an exploration of suprapituitary release of ACTH, without the potentially confounding effects of differing ambient cortisol levels. To the extent that metyrapone administration results in a substantially higher increase in ACTH and 11-deoxycortisol in PTSD compared to controls, it is possible to infer that the increase in ACTH results occurs as a direct result of hypothalamic CRF stimulation.

Similarly, the CRF challenge test has also been used to estimate hypothalamic CRF activity, since a blunted ACTH response is suggestive of a downregulation of pituitary receptors secondary to CRF hypersecretion. Using this logic, an augmented ACTH response to CRF would reflect a decreased hypothalamic CRF release, or at least an upregulation of pituitary CRF receptors. Rasmusson et al. (2001) assert that the finding of an increased ACTH response to CRF is analogous to the increased ACTH response to metyrapone obtained by Yehuda et al. (1996a). Although this might not be the most likely explanation for the finding, insofar as the subjects in Rasmusson et al. did not show increases in either basal ACTH or cortisol levels, it is possible that the finding of an augmented ACTH response to CRF does indeed reflect an enhanced negative feedback on the pituitary, particularly in view of the relatively weaker effect of CRF on cortisol relative to ACTH. However, the model of enhanced negative feedback inhibition would not explain the increased cortisol response to ACTH observed in the same patients.

6
Putative Models of HPA Axis Alterations in PTSD

Cortisol levels are most often found to be lower than normal in PTSD, but can also be similar to or greater than those in comparison subjects. Findings of changes in circadian rhythm suggest that there may be regulatory influences that result in a greater dynamic range of cortisol release over the diurnal cycle in PTSD. Together, these findings imply that although cortisol levels may be generally lower, the adrenal gland is certainly capable of producing adequate amounts of cortisol in response to challenge.

The model of enhanced negative feedback inhibition is compatible with the idea that there may be transient elevations in cortisol, but would suggest that when present, these increases would be shorter-lived due to a more efficient containment of ACTH release as a result of enhanced GR activation. This model posits that chronic or transient elevations in CRF release stimulate the pituitary release of ACTH, which in turn stimulates the adrenal release of cortisol. However, an increased negative feedback inhibition would result in reduced cortisol levels under ambient conditions. In contrast to other models of endocrinopathy, which identify specific and usually singular primary alterations in endocrine organs and/or regulation, the model of enhanced negative feedback inhibition in PTSD is in large part descriptive. The model currently offers little explanation for why some individuals show such alterations of the HPA axis following exposure to traumatic experiences while others do not, but it represents an important development in the field of neuroendocrinology of PTSD by accounting for a substantial proportion of the findings observed.

On the other hand, the model of reduced adrenal output accounts for why ambient cortisol levels would be lower than normal, and even for the relatively smaller magnitude of differences in ACTH relative to cortisol, but does not account for why basal ACTH levels are not significantly higher in PTSD than in comparison subjects, particularly in light of evidence of CRF hypersecretion. One of the challenge in elucidating a neuroendocrinology of PTSD is in being able to resolve the apparent paradox that cortisol levels are low when CRF levels appear to be elevated, as well as to accommodate a dynamic process in that accounts for observed diurnal fluctuations and potential responsivity to environmental cues. Heim et al. (2001) have again argued that in response to early trauma, CRF hypersecretion may result in a downregulation of pituitary CRF receptors leading to a decreased ACTH response. However, it is not quite clear according to this why in such cases CRF hypersecretion would lead to pituitary desensitization and low cortisol as opposed to the more classic model of HPA dysfunction articulated for major depressive disorder in which the effect of hypothalamic CRF release on the pituitary would ultimately result in hypercortisolism.

Findings of increased CRF levels in PTSD are important to the theory of enhanced negative feedback inhibition in PTSD, but are not necessarily relevant to theories of adrenal insufficiency. That is, to the extent that there are increases in CRF, these would not necessarily occur as a direct response to reduced adrenal output, but might have a different origin. Under conditions of reduced adrenal output, it is possible, as implied by Heim et al. (2000), that compensatory changes in hypothalamic CRF might occur to the extent that there is a weaker negative feedback inhibition because of decreased cortisol output. But if this were occurring, it would be difficult to find an explanation for why the ACTH response to CRF (Heim et al. 2001) and psychologic stressors (Heim et al. 2000) were augmented in relation to early traumatization.

Findings of the cortisol response to DEX are compatible with both the enhanced negative feedback inhibition model and adrenal insufficiency. However, in the latter case, one would not expect that a reduced cortisol level to result from, or even be accompanied by, changes in the GR, but rather, would reflect reduced adrenal output rather than an enhanced containment of ACTH.

Findings of a blunted ACTH response to CRF are compatible with the enhanced negative feedback model, but not the adrenal insufficiency hypothesis. Adrenal insufficiency would not be expected to result in a blunted ACTH response to CRF. On the contrary, primary adrenal insufficiency is characterized by increased ACTH at baseline and in response to CRF. Findings demonstrating an augmented ACTH to metyrapone are also consistent with enhanced negative feedback inhibition, but not adrenal insufficiency. Adrenal insufficiency is also incompatible with findings showing a greater activation of cortisol in the context of reduced ACTH responses to pituitary challenges.

Table 4 summarizes these HPA findings in PTSD and the explanations compatible with these findings. This table demonstrates that the model of enhanced negative feedback is compatible with 15/21 observations of HPA alterations in PTSD, whereas reduced adrenal capacity is consistent with 9/21 observations.

6.1 Findings of Cortisol in the Acute Aftermath of Trauma

Recent data have provided some support for the idea that low cortisol levels may be an early predictor of PTSD rather than a consequence of this condition. Low cortisol levels in the immediate aftermath of a motor vehicle accident predicted the development of PTSD in a group of 35 accident victims consecutively presenting to an emergency room (Yehuda et al. 1998). Delahanty et al. (2000) also reported that low cortisol levels in the immediate aftermath of a trauma contributed to the prediction of PTSD symptoms at 1 month. In a sample of 115 people who survived a natural disaster, cortisol levels were similarly found to be lowest in those with highest PTSD scores at 1 month post-trauma, however cortisol levels were not predictive of symptoms at 1 year (Anisman et al. 2001). Similarly, lower morning, but higher evening cortisol levels were observed in 15 subjects with high levels of PTSD symptoms 5 days following a mine accident in Lebanon compared to 16 subjects with lower levels of PTSD symptoms (Aardal-Eriksson et al. 2001).

In a study examining the cortisol response in the acute aftermath of rape, low cortisol levels were associated with prior rape or assault, themselves risk factors for PTSD (Resnick et al. 1995), but not with the development of PTSD per se. A post hoc analysis of the data reported in (Yehuda et al. 1998) confirmed the observation that low cortisol levels were also associated with prior trauma exposure in this group as well (A.C. McFarlane et al., personal communication).

These findings imply that cortisol levels might have been lower in trauma survivors who subsequently develop PTSD even before their exposure to trauma, and might therefore represent a pre-existing risk factor. Consistent with this, low 24-h urinary cortisol levels in adult children of Holocaust survivors were specifically associated with the risk factor of parental PTSD. These studies raise the possibility that low cortisol levels represent an index of risk, and may actually contribute to the secondary biologic alterations that ultimately lead to the development of PTSD. Interestingly, the risk factor of parental PTSD in offspring of Holocaust survivors was also associated with an increased incidence of traumatic childhood antecedents (Yehuda et al. 2001). In this study, both the presence of subject-rated parental PTSD and scores reflecting childhood emotional abuse were associated with low cortisol levels in offspring. Thus, it may be that low cortisol levels occur in those who have experienced an adverse event early in life, and then remain different from those not exposed to early adversity. Although there might reasonably be HPA axis fluctuations in the aftermath of stress, and even differences in

Table 4 Summary of data from studies supporting enhanced negative feedback or reduced adrenal output in PTSD

Finding in PTSD	Enhanced negativc feedback	Reduced adrenal output
Lower ambient cortisol levels	Yes	Yes
Normal or variable cortisol levels	Yes	No
Higher cortisol levels	Yes[a]	No
Increased circadian rhythm of cortisol	Yes	No
Decreased circadian rhythm of cortisol	No	Yes
Normal ACTH levels	Yes	No
Low β-endorphin levels	Yes[b]	No
Increased CRF levels in CSF	Yes	Yes
Increased glucocorticoid receptor sensitivity/number	Yes	No[c]
Normal cortisol levels to 1 mg DEX	Yes	Yes
Decreased cortisol levels following 0.5 DEX	Yes	Yes
Increased cortisol levels following 1 mg DEX	No	No
Decreased number of cytosolic glucocorticoid receptors following DEX compared to baseline receptors	Yes	No[d]
Increased ACTH levels to high dose metyrapone	Yes	No
Decreased ACTH levels to low dose metyrapone	No[e]	Yes
Decreased ACTH levels following CRF	Yes	No
Increased ACTH levels following CRF	No	Yes
Increased cortisol responses to ACTH	No	No
Decreased ACTH levels following CCK-4	Yes	No
Increased ACTH levels following naloxone	No	Yes
Increased ACTH levels following stress	Yes	Yes

*Also observed in samples of subjects with early abuse, depression, or somatic illnesses with or without comorbid PTSD. [a]Higher cortisol levels are only consistent with enhanced negative feedback to the extent that they represent transient elevations. [b]To the extent that β-endorphin is co-released with ACTH and reflects ACTH, this finding is compatible. What is problematic is the lack of relationship in this paper between ACTH and β-endorphin, which raises methodologic questions. [c]This conclusion is based on empirical findings from studies of endocrinologic disorders that have generally failed to observe accommodation in glucocorticoid receptors in response to either very high or very low cortisol levels (reviewed in Yehuda 2002). It is theoretically possible, however, that low levels of ambient cortisol would result in an "upregulation" of glucocorticoid receptors. [d]Based on c. [e]See extensive discussion on this paper in text.

the magnitude of such responses compared to those not exposed to trauma early in life, HPA parameters would subsequently recover to their pre-stress baseline.

Low cortisol levels may impede the process of biologic recovery from stress, resulting in a cascade of alterations that lead to intrusive recollections of the event, avoidance of reminders of the event, and symptoms of hyperarousal. This failure may represent an alternative trajectory to the normal process of adaptation and recovery after a traumatic event.

Additionally, it is possible that, within the time frame between several hours or days following a trauma and the development of PTSD at 1 month, there is an active process of adaptation and an attempt at achieving homeostasis, and that PTSD symptoms themselves are determined by biologic responses, rather than the opposite. For example, Hawk et al. (2000) found that at 1 month post-trauma, urinary cortisol levels were elevated among men with PTSD symptoms (but not women). By 6 months, there were no group differences in cortisol, but emotional numbing at 1 month predicted lower cortisol levels 6 months after the accident. Similarly, in a prospective study in which plasma cortisol and continuous measures of PTSD symptoms were obtained from 21 survivors at 1 week and 6 months post-trauma, cortisol levels at 1 week did not predict subsequent PTSD, but cortisol levels at 6 months negatively correlated with self-reported PTSD symptoms within PTSD subjects (Bonne et al. 2003a).

PTSD may arise from any number of circumstances, one of which may be the hormonal milieu at the time of trauma, which may reflect an interaction of pre- and peri-traumatic influences. These responses may be further modified in the days and weeks preceding it by a variety of other influences. For example, under normal circumstances, CRF and ACTH are activated in response to stress, and ultimately culminate in cortisol release, which negatively feeds back to keep the stress response in check. A reduced adrenal capacity might initially lead to a stronger activation of the pituitary due to increased CRF stimulation in synergy with other neuropeptides, such as arginine vasopressin, resulting in a high magnitude ACTH response. This might lead to a greater internal necessity by the pituitary for negative feedback inhibition. Achieving regulation under these conditions might necessitate a progressive decline in the ACTH/cortisol ratio, possibly facilitated by accommodations in the sensitivity of GRs and other central neuromodulators, ultimately leading to an exaggerated negative feedback inhibition. Affecting these hormonal responses might also be the demands made by posttraumatic factors. Although such a model is hypothetical, it is consistent with the adaptational process of allostatic load described by McEwen (1999): that is, that physiologic systems accommodate to achieve homeostasis based on already existing predispositions to stress responses. Thus, the neuroendocrinologic response to trauma of a person with lower cortisol levels at the outset might be fundamentally different from that of someone with a greater adrenal capacity and higher ambient cortisol levels.

One of most compelling lines of evidence supporting the hypothesis that lower cortisol levels may be an important pathway to the development of PTSD symptoms involves results of studies by Schelling et al. (2001) who administered stress doses of hydrocortisone during septic shock and evaluated the effects of this treatment on the development of PTSD and traumatic memories. Indeed, the results of a randomized, double-blind study demonstrated that administration of hydrocortisone in high, but physiologic-stress, doses was associated with reduced PTSD symptoms compared to the group that received saline. These findings support the idea that low cortisol levels may facilitate the development of PTSD in response to an overwhelming biologic demand—at least in some circumstances.

7 Conclusions

The HPA axis alterations in PTSD support the idea that HPA axis alterations are complex and might be associated with different aspects of PTSD, including risk for the development of this disorder. For the findings to coalesce into an integrative neuroendocrine hypothesis of PTSD, it would be necessary to assert that (1) some features of the HPA axis may be altered prior to the exposure to a focal trauma; (2) that components of the HPA axis are not uniformly regulated (e.g., circadian rhythm patterns, tonic cortisol secretion, negative feedback inhibition, and the cortisol response to stress are differentially mediated; (3) that the system is dynamic, and may therefore show transient increases or hyperresponsivity under certain environmental conditions; that (4) other regulatory influences might affect HPA axis regulation in PTSD; and probably (though not necessarily), that (5) there might be different biologic variants of PTSD with relatively similar phenotypic expressions, as is the case with major depressive disorder.

The wide range of observations observed in the neuroendocrinology of PTSD underscores the important observation of Mason et al. (1986) that HPA response patterns in PTSD are fundamentally in the normal range and do not reflect endocrinopathy. In endocrinologic disorders, where there is usually a lesion in one or more target tissues or biosynthetic pathways, endocrine methods can usually isolate the problem with the appropriate test(s), and then obtain rather consistent results. In psychiatric disorders, neuroendocrine alterations may be subtle, and therefore, when using standard endocrine tools to examine these alterations, there is a high probability of failing to observe all the alterations consistent with a neuroendocrine explanation of the pathology in tandem, or of obtaining disparate results within the same patient group owing to a stronger compensation or re-regulation of the HPA axis following challenge.

The next generation of studies should aim to apply more rigorous tests of neuroendocrinology of PTSD based on the appropriate developmental issues and in consideration of the longitudinal course of the disorder, and the individual differences that affect these processes. No doubt such studies will require a closer examination of a wide range of biologic responses, including the cellular and molecular mechanisms involved in adaptation to stress, and an understanding of the relationship between the endocrine findings and other identified biologic alterations in PTSD.

Acknowledgements This work was supported by MH 49555, MH 55-7531, and MERIT review funding to R.Y.

References

Aardal-Eriksson E, Eriksson TE, Thorell L (2001) Salivary cortisol, posttraumatic stress symptoms and general health in the acute phase and during 9-month follow-up. Biol Psychiatry 50:986–993

Altemus M, Cloitre M, Dhabhar FR (2003) Enhanced cellular immune responses in women with PTSD related to childhood abuse. Am J Psychiatry 160:1705–1707

Anisman H, Griffiths J, Matheson K, Ravindran AV, Merali Z (2001) Posttraumatic stress symptoms and salivary cortisol levels. Am J Psychiatry 158:1509–1511

Atmaca M, Kuloglu M, Tezcan E, Onal S, Ustundag B (2002) Neopterin levels and dexamethasone suppression test in posttraumatic stress disorder. Eur Arch Psychiatry Clin Neurosci 252:161–165

Baker DG, West SA, Nicholson WE, Ekhator NN, Kasckow JW, Hill KK, Bruce AB, Orth DN, Geracioti TD Jr (1999) Serial CSF corticotropin-releasing hormone levels and adrenocortical activity in combat veterans with posttraumatic stress disorder [published erratum appears in Am J Psychiatry 1999 Jun;156(6):986]. Am J Psychiatry 156:585–588

Bonne O, Brandes D, Segman R, Pitman RK, Yehuda R, Shalev AY (2003a) Prospective evaluation of plasma cortisol in recent trauma survivors with posttraumatic stress disorder. Psychiatry Res 119:171–175

Bonne O, Gilboa A, Louzoun Y, Brandes D, Yona I, Lester H, Barkai G, Freedman N, Chisin R, Shalev AY (2003b) Resting regional cerebral perfusion in recent posttraumatic stress disorder. Biol Psychiatry 54:1077–1086

Boscarino JA (1996) Posttraumatic stress disorder, exposure to combat, and lower plasma cortisol among Vietnam veterans: findings and clinical implications. J Consult Clin Psychol 64:191–201

Bourne PG, Rose RM, Mason JW (1968) 17 OHCS levels in combat. Special forces "A" team under threat of attack. Arch Gen Psychiatry 17:104–110

Bremner JD, Licinio J, Darnell A, Krystal JH, Owens MJ, Southwick SM, Nemeroff CB, Charney DS (1997) Elevated CSF corticotropin-releasing factor concentrations in posttraumatic stress disorder. Am J Psychiatry 154:624–629

Bremner JD, Vythilingma M, Vermetten E, Adil J, Khan S, Nazeer A, Afzal N, McGlashan T, Elzinga B, Anderson GM, Heninger G, Southwick SM, Charney DS (2003a) Cortisol response to a cognitive stress challenge in posttraumatic stress disorder related to childhood abuse. Psychoneuroendocrinology 28:733–750

Bremner JD, Vythilingam M, Anderson G, Vermetten E, McGlashan T, Heninger G, Rasmusson A, Southwick SM, Charney DS (2003b) Assessment of the hypothalamic-pituitary-adrenal axis over a 24-hour period and in response to neuroendocrine challenges in women with and without childhood sexual abuse and posttraumatic stress disorder. Biol Psychiatry 54:710–718

Carrion VG, Weems CF, Ray RD, Glaser B, Hessl D, Reiss AL (2002) Diurnal salivary cortisol in pediatric posttraumatic stress disorder. Biol Psychiatry 51:575–582

Coupland NJ, Hegadoren KM, Myrholm J (2003) Increased beclomethasone-induced vasoconstriction in women with posttraumatic stress disorder. J Psychiatr Res 37:221–228

De Bellis MD, Chrousos GP, Dorn LD, Burke L, Helmers K, Kling MA, Trickett PK, Putnam FW (1994) Hypothalamic-pituitary-adrenal axis dysregulation in sexually abused girls. J Clin Endocrinol Metab 78:249–255

De Bellis MD, Baum AS, Birmaher B, Keshavan MS, Eccard CH, Boring AM, Jenkins FJ, Ryan ND (1999) A.E. Bennett Research Award. Developmental traumatology. Part I: Biological stress systems [see comments]. Biol Psychiatry 45:1259–1270

de Kloet ER, Joels M, Oitzl M, Sutanto W (1991) Implication of brain corticosteroid receptor diversity for the adaptation syndrome concept. Methods Achiev Exp Pathol 14:104–132

Delahanty DL, Raimonde AJ, Spoonster E (2000) Initial posttraumatic urinary cortisol levels predict subsequent PTSD symptoms in motor vehicle accident victims. Biol Psychiatry 48:940–947

Dinan TG, Barry S, Yatham LN, Mobayed M, Brown I (1990) A pilot study of a neuroendocrine test battery in posttraumatic stress disorder. Biol Psychiatry 28:665–672

Elzinga BM, Schmahl CG, Vermetten E, van Dyck R, Bremner JD (2003) Higher cortisol levels following exposure to traumatic reminders in abuse-related PTSD. Neuropsychopharmacology 28:1656–1665

Glover D, Poland R (2002) Urinary cortisol and catecholamines in mothers of child cancer survivors with and without PTSD. Psychoneuroendocrinology 27:805–819

Goeinjian AK, Pynoos RS, Steinberg Am, Endres D, Abraham K, Geffner ME, Fairbanks LA (2003) Hypothalamic-pituitary-adrenal activity among Armenian adolescents with PTSD symptoms. J Trauma Stress 16:319–323

Goenjian AK, Yehuda R, Pynoos RS, Steinberg AM, Tashjian M, Yang RK, Najarian LM, Fairbanks LA (1996) Basal cortisol, dexamethasone suppression of cortisol and MHPG in adolescents after the 1988 earthquake in Armenia. Am J Psychiatry 153:929–934

Gormley GJ, Lowy MT, Reder AT, Hospelhorn VD, Antel JP, Meltzer HY (1985) Glucocorticoid receptors in depression: relationship to the dexamethasone suppression test. Am J Psychiatry 142:1278–1284

Gotovak K, Sabioncello A, Rabatic S, Berki T, Dekaris D (2003) Flow cytometric determination of glucocorticoid receptor expression in lymphocyte subpopulations: lower quantity of glucocorticoid receptors in patients with posttraumatic stress disorder. Clin Exp Immunol 131:335–339

Grossman R, Yehuda R, New A, Schmeidler J, Silverman J, Mitropoulous V, Sta Maria N, Bgolier J, Siever L (2003) Dexamethasone suppression test findings in subjects with personality disorders: associations with posttraumatic stress disorder and major depression. Am J Psychiatry 160:1291–1298

Halbreich U, Olympia J, Carson S, Glogowski J, Yeh CM, Axelrod S, Desu MM (1989) Hypothalamo-pituitary-adrenal activity in endogenously depressed posttraumatic stress disorder patients. Psychoneuroendocrinology 14; :365–370

Hawk LW, Dougall AL, Ursano RJ, Baum A (2000) Urinary catecholamines and cortisol in recent-onset posttraumatic stress disorder after motor vehicle accidents. Psychosom Med 62:423–434

Heim C, Newport JJ, Heit S, Graham YP, Wilcox M, Bonsall R, Miller AH, Nemeroff CG (2000) Pituitary-adrenal and autonomic responses to stress in women after sexual and physical abuse in childhood. JAMA 284:592–597

Heim C, Ehlert U, Rexhausen J, Hanker JP, Hellhammer DH (1998) Abuse-related posttraumatic stress disorder and alterations of the hypothalamic-pituitary-adrenal axis in women with chronic pelvic pain. Psychosom Med 60:309–318

Heim C, Ehlert U, Hellhammer DH (2000) The potential role of hypocortisolism in the pathophysiology of stress-related bodily disorders. Psychoneuroendocrinology 25:1–35

Heim C, Newport DJ, Bonsall R, Miller AH, Nemeroff CB (2001) Altered pituitary-adrenal axis responses to provocative challenge tests in adult survivors of childhood abuse. Am J Psychiatry 158:575–581

Hockings GI, Grice JE, Ward WK, Walters MM, Jensen GR, Jackson RV (1993) Hypersensitivity of the hypothalamic-pituitary-adrenal axis to naloxone in posttraumatic stress disorder. Biol Psychiatry 33:585–593

Hoffmann L, Burges Watson P, Wilson G, Montgomery J (1989) Low plasma b-endorphin in posttraumatic stress disorder. Aust NZ J Psychiatry 23:269–273

Holsboer F (2000) The corticosteroid receptor hypothesis of depression. Neuropsychopharmacology 23:477–501

Holsboer F, Lauer CJ, Schreiber W, Krieg JC (1995) Altered hypothalamic-pituitary-adrenocortical regulation in healthy subjects at high familial risk for affective disorders. Neuroendocrinology 62:340–347

Hucklebridge FH, Clow A, Abeyguneratne T, Huezo-Diaz P, Evans P (1999) The awakening cortisol response and blood glucose levels. Life Sci 64:931–937

Jensen CF, Keller TW, Peskind ER, McFall ME, Veith RC, Martin D,Wilkinson CW, Raskind M (1997) Behavioral and neuroendocrine responses to sodium lactate infusion in subjects with posttraumatic stress disorder. Am J Psychiatry 154:266–268

Kanter ED, Wilkinson CW, Radant AD, Petrie EC, Dobie DJ, McFall ME, Peskind ER, Raskind MA (2001) Glucocorticoid feedback sensitivity and adrenocortical responsiveness in posttraumatic stress disorder. Biol Psychiatry 50:238–245

Kauffman J, Birmaher B, Perel J, Dahl RE, Moreci P, Nelson B, Wells W, Ryan ND (1997) The corticotrophin-releasing hormone challenge in depressed abused, depressed nonabused, and normal control children. Biol Psychiatry 42:669–679

Kellner M, Baker DG, Yehuda R (1997) Salivary cortisol in Operation Desert Storm returnees. Biol Psychiatry 42:849–850

Kellner M, Wiedemann K, Yassouridis A, Levengood R, Guo LS, Holsboer F, Yehuda R (2000) Behavioral and endocrine response to cholecystokinin tetrapeptide in patients with posttraumatic stress disorder. Biol Psychiatry 47:107–111

Kellner M, Baker DG, Yassouridis A, Bettinger S, Otte C, Naber D, Wiedemann K (2002a) Mineralocorticoid receptor function in patients with posttraumatic stress disorder. Am J Psychiatry 159:1938–1940

Kellner M, Yehuda R, Arlt J, Wiedemann K (2002b) Longitudinal course of salivary cortisol in posttraumatic stress disorder. Acta Psychiatr Scand 105:153–155

King JA, Mandansky D, King S, Fletcher KE, Brewer J (2001) Early sexual abuse and low cortisol. Psychiatry Clin Neurosci 55:71–74

Kosten TR, Wahby V, Giller E, Mason J (1990) The dexamethasone suppression test and thyrotropin-releasing hormone stimulation test in posttraumatic stress disorder. Biol Psychiatry 28:657–664

Krishnan KRR, Rayasam K, Reed D, Smith M, Chapell P, Saunders WB, Ritchie JC, Carroll BJ, Nemeroff CB (1993) The corticotropin releasing factor stimulation test in patients with major depression: relationship to dexamethasone suppression test results. Depression 1:133–136

Kudler H, Davidson J, Meador K, Lipper S, Ely T (1987) The DST and posttraumatic stress disorder. Am J Psychiatry 14:1058–1071

Lemieux AM, Coe CL (1995) Abuse-related posttraumatic stress disorder: evidence for chronic neuroendocrine activation in women. Psychosom Med 57:105–115

Liberzon I, Abelson JL, Flagel SB, Raz J, Young EA (1999) Neuroendocrine and psychophysiologic responses in PTSD: a symptom provocation study. Neuropsychopharmacology 21:40–50

Lipschitz DS, Rasmusson AM, Yehuda R, Wang S, Anyan W, Gueoguieva R, Grilo CM, Fehon DC, Southwick SM (2003) Salivary cortisol response to dexamethasone in adolescents with posttraumatic stress disorder. J Am Acad Child Adolesc Psychiatry 42:1310–1317

Lowy MT (1989) Quantification of type I and II adrenal steroid receptors in neuronal, lymphoid and pituitary tissues. Brain Res 503:191–197

Luecken LJ, Dausche B, Gulla V, Hong R, Compas BE (2004) Alterations in morning cortisol associated with PTSD in women with breast cancer. J Psychosom Res 56:13–15

Maes M, Lin A, Bonaccorso S, van Hunsel F, Van Gastel A, Delmeire L, Biondi M, Bosmans E, Kenis G, Scharpe S (1998) Increased 24-hour urinary cortisol excretion in patients with post-traumatic stress disorder and patients with major depression, but not in patients with fibromyalgia. Acta Psychiatr Scand 98:328–335

Mason JW, Giller EL, Kosten TR, Ostroff RB, Podd L (1986) Urinary free-cortisol levels in posttraumatic stress disorder patients. J Nerv Ment Dis 174:145–159

Mason JW, Wang S, Yehuda R, Riney S, Charney DS, Southwick SM (2001) Psychogenic lowering of urinary cortisol levels linked to increased emotional numbing and a shame-depressive syndrome in combat-related posttraumatic stress disorder. Psychosom Med 63:387–401

McEwen BS, Seeman T (1999) Protective and damaging effects of mediators of stress: elaborating and testing the concepts of allostasis and allostatic load. Ann NY Acad Sci 896:30–47

Newport DJ, Heim C, Bonsall R, Miller AH, Nemeroff CB (2004) Pituitary-adrenal responses to standard and low-dose dexamethasone suppression tests in adult survivors of child abuse. Biol Psychiatry 55:10–20

Neylan TC, Lenoci M, Maglione ML, Rosenlicht NZ, Metlzer TJ, Otte C, Schoenfeld FB, Yehuda R, Marmar CR (2003a) Delta sleep response to metyrapone in posttraumatic stress disorder. Neuropsychopharmacology 28:1666–1676

Neylan TC, Schuff N, Lenoci M, Yehuda R, Weiner MW, Marmar CR (2003b) Cortisol levels are positively correlated with hippocampal N-acetylaspartate. Biol Psychiatry 54:1118–1121

Oquendo MA, Echavarria G, Galfalvy HC, Grunebaum MF, Burke A, Barrera A, Cooper TB, Malone KM, Mann JJ (2003) Lower cortisol levels in depressed patients with comorbid posttraumatic stress disorder. Neuropsychopharmacology 28:591–598

Pitman RK, Orr SP (1990) Twenty-four hour urinary cortisol and catecholamine excretion in combat-related posttraumatic stress disorder. Biol Psychiatry 27:245–247

Raison CL, Miller AH (2003) When not enough is too much: the role of insufficient glucocorticoid signaling in the pathophysiology of stress related disorders. Am J Psychiatry 160:1554–1565

Rasmusson AM, Lipschitz DS, Wang S, Hu S, Vojvoda D, Bremner JD, Southwick SM, Charney D (2001) Increased pituitary and adrenal reactivity in premenopausal women with posttraumatic stress disorder. Biol Psychiatry 12:965–977

Reist C, Kauffmann ED, Chicz-Demet A, Chen CC, Demet EM (1995) REM latency, dexamethasone suppression test, and thyroid releasing hormone stimulation test in posttraumatic stress disorder. Prog Neuropsychopharmacol Biol Psychiatry 19:433–443
Resnick HS, Yehuda R, Pitman RK, Foy DW (1995) Effect of previous trauma on acute plasma cortisol level following rape. Am J Psychiatry 152:1675–1677
Ribeiro SC, Tandon R, Grunhaus L, Greden JF (1993) The DST as a predictor of outcome in depression: a meta-analysis [see comments]. Am J Psychiatry 150:1618–1629
Rinne T, deKloet ER, Wouters L, Goekoop JG, DeRijk RH, van de Brink W (2002) Hyperresponsiveness of hypothalamic-pituitary-adrenal axis to combined dexamethasone/corticotropin-releasing hormone challenge in female borderline personality disorder subjects with a history of sustained childhood abuse. Biol Psychiatry 52:1102–1112
Sapolsky RM, Krey LC, McEwen BS (1984) Stress down-regulates corticosterone receptors in a site-specific manner in the brain. Endocrinology 114:287–292
Sapolsky RM, Krey LC, McEwen BS (1986) The neuroendocrinology of stress and aging: the glucocorticoid cascade hypothesis. Endocr Rev 7:284–301
Sautter FJ, Bisette G, Wiley J, Manguno-Mire G, Schoenbachler B, Myers L, Johnson JE, Cerbone A, Malaspina D (2003) Corticotropin-releasing factor in posttraumatic stress disorder with secondary psychotic symptoms, nonpsychotic PTSD, and health control subjects. Biol Psychiatry 54:1382–1388
Schelling G, Briegel J, Roozendaal B, Stoll C, Rothenhausler HB, Kapfhammer HP (2001) The effect of stress doses of hydrocortisone during septic shock on posttraumatic stress disorder in survivors. Biol Psychiatry 50:978–985
Schulte HM, Chrousos GP, Avgerinos P, Oldfield EH, Gold PW, Cutler GB, Loriaux DL (1984) The corticotrophin-releasing hormone stimulation test: a possible aid in the evaluation of patients with adrenal insufficiency. J Clin Endocrinol metab 58:1064–1067
Seedat S, Stein MB, Kennedy CM, Hauger RL (2003) Plasma cortisol and neuropeptide Y in female victims of intimate partner violence. Psychoneuroendocrinology 28:796–808
Smith MA, Davidson J, Ritchie JC, Kudler H, Lipper S, Chappell P, Nemeroff CB (1989) The corticotropin-releasing hormone test in patients with posttraumatic stress disorder. Biol Psychiatry 26:349–355
Stein MB, Yehuda R, Koverola C, Hanna C (1997) Enhanced dexamethasone suppression of plasma cortisol in adult women traumatized by childhood sexual abuse. Biol Psychiatry 42:680–686
Stokes PE, Stoll PM, Koslow SH, Maas JW, Davis JM, Swann AC, Robins E (1984) Pretreatment DST and hypothalamic-pituitary-adrenocortical function in depressed patients and comparison groups. A multicenter study. Arch Gen Psychiatry 41:257–267
Svec F (1985) Glucocorticoid receptor regulation. Life Sci 36:2359–2366
Thaller V, Vrkljan M, Hotujac L, Thakore J (1999) The potential role of hypocortisolism in the pathophysiology of PTSD and psoriasis. Coll Antropol 23:611–619
Wang S, Wilson JP, Mason JWL (1996) Stages of decompensation in combat-related posttraumatic stress disorder: a new conceptual model. Integr Physiol Behav Sci 31:237–253
Yehud R, Hallig SL, Grossman R (2001) Childhood trauma and risk for PTSD: relationship to intergenerational effects of trauma, parental PTSD, and cortisol excretion. Dev Psychopathol 13:733–753
Yehuda R (1999a) Risk factors for posttraumatic stress disorder. American Psychiatric Press, Washington
Yehuda R (1999b) Biological factors associated with susceptibility to posttraumatic stress disorder. Can J Psychiatry 44:34–39

Yehuda R (2002) Current status of cortisol findings in post-traumatic stress disorder. Psychiatr Clin North Am 25:341–368
Yehuda R, Nemeroff CB (1994) Neuropeptide alterations in affective and anxiety disorders. In: DenBoer JA, Sisten A (eds) Handbook on depression and anxiety: a biological approach. Marcel Dekker, New York, pp 543–571
Yehuda R, Southwick SM, Nussbaum G, Wahby V, Giller EL, Mason JW (1990) Low urinary cortisol excretion in patients with posttraumatic stress disorder. J Nerv Ment Dis 178:366–369
Yehuda R, Giller EL, Southwick SM, Lowy MT, Mason JW (1991a) Hypothalamic-pituitary-adrenal dysfunction in posttraumatic stress disorder. Biol Psychiatry 30:1031–1048
Yehuda R, Lowy MT, Southwick S, Shaffer D, Giller EL (1991b) Lymphocyte glucocorticoid receptor number in posttraumatic stress disorder. Am J Psychiatry 148:499–504
Yehuda R, Boisoneau D, Mason JW, Giller EL (1993a) Glucocorticoid receptor number and cortisol excretion in mood, anxiety, and psychotic disorders. Biol Psychiatry 34:18–25
Yehuda R, Southwick SM, Krystal JH, Bremner D, Charney DS, Mason JW (1993b) Enhanced suppression of cortisol following dexamethasone administration in posttraumatic stress disorder. Am J Psychiatry 150:83–86
Yehuda R, Boisoneau D, Lowy MT, Giller EL (1995a) Dose-response changes in plasma cortisol and lymphocyte glucocorticoid receptors following dexamethasone administration in combat veterans with and without posttraumatic stress disorder. Arch Gen Psychiatry 52:583–593
Yehuda R, Kahana B, Binder-Brynes K, Southwick S, Mason JW, Giller EL (1995b) Low urinary cortisol excretion in Holocaust survivors with posttraumatic stress disorder. Am J Psychiatry 152:982–986
Yehuda R, Levengood RA, Schmeidler J, Wilson S, Guo LS, Gerber D (1996a) Increased pituitary activation following metyrapone administration in post-traumatic stress disorder. Psychoneuroendocrinology 21:1–16
Yehuda R, Teicher MH, Trestman RL, Levengood RA, Siever LJ (1996b) Cortisol regulation in posttraumatic stress disorder and major depression: a chronobiological analysis. Biol Psychiatry 40:79–88
Yehuda R, Bierer LM, Schmeidler J, Aferiat DH, Breslau I, Dolan S (2000) Low cortisol and risk for PTSD in adult offspring of holocaust survivors. Am J Psychiatry 157:1252–1259
Yehuda R, Halligan S, Grossman R (2001) Childhood trauma and risk for PTSD: relationship to intergenerational effects of trauma, parental PTSD and cortisol excretion. Stress Dev 13:731–751
Yehuda R, Halligan SL, Bierer LM (2002a) Cortisol levels in adult offspring of Holocaust survivors: relation to PTSD symptom severity in the parent and child. Psychoneuroendocrinology 27:171–180
Yehuda R, Halligan SL, Grossman R, Golier JA, Wong C (2002b) The cortisol and glucocorticoid receptor response to low dose dexamethasone administration in aging combat veterans and holocaust survivors with and without posttraumatic stress disorder. Biol Psychiatry 52:393–403
Yehuda R, Golier J, Yang RK, Tischler L (2004a) Enhanced sensitivity to glucocorticoids in peripheral mononuclear leukocytes in posttraumatic stress disorder. Biol Psychiatry 55:1110–1116
Yehuda R, Halligan SL, Golier J, Grossman R, Bierer LM (2004b) Effects of trauma exposure on the cortisol response to dexamethasone administration in TPSD and major depressive disorder. Psychoneuroendocrinology 29:389–404

HEP (2005) 169:405–432

Anxiety Disorders: Clinical Presentation and Epidemiology

R. Lieb

Clinical Psychology and Epidemiology, Max-Planck-Institute of Psychiatry,
Kraepelinstr. 2-10, 80804 München, Germany
lieb@mpipsykl.mpg.de

Abstract This chapter gives an overview of the clinical presentation of anxiety disorders and reviews basic epidemiological knowledge on them. The presented knowledge is largely related to the classification of anxiety disorders as presented by the Diagnostic and Statistical Manual of Mental Disorders since its third revision (DSM-III). Without going into detail into the history of the classification of anxiety disorders and into the history and development of the several editions of the Diagnostic Manual of Mental Disorders (DSM) of the American Psychiatric Association (APA) it should just briefly be mentioned that the DSM of the APA has undergone until today four revisions. Within these revisions, the third edition (DSM-III) changed most radically from the forerunning ones. The major change in DSM-III was that the category "anxiety neurosis" was deleted because this term was too general and could not be defined reliably. On the basis of evidence that imipramine can block panic attacks, panic disorder was created as a new diagnosis for the first time in DSM-III. Anxiety states without spontaneous panic attacks were separated from panic disorder and defined as a residual category, generalized anxiety disorder. The revised version of DSM-III, DSM-III-R, was published in 1987, and the fourth and most recent edition, DSM-IV, was published in 1994. More recently, a text revision of DSM-IV has been published that does not entail changes to the diagnostic criteria of disorders, but provides updated empirical reviews for each diagnostic category regarding associated features, cultural, age, and gender features, prevalence, course, familial patterns, and differential diagnosis (DSM-IV-R). Without going into further details of the development and changes across the different editions and revisions of DSM—these have been reviewed comprehensively in other reviews—this chapter gives an overview about the clinical presentations of anxiety disorders by referring mainly to the forth edition of the DSM (DSM-IV 1994). In the second part, the chapter reviews and summarizes selected aspects (prevalence, correlates, risk factors and comorbidity) of epidemiological knowledge on anxiety disorders.

Keywords Phenomenology · Epidemiology · Prevalence · Age of onset · Comorbidity · Correlates

This chapter gives an overview of the clinical presentation of anxiety disorders and reviews basic epidemiological knowledge on them. The presented knowledge is largely related to the classification of anxiety disorders as presented by the Diagnostic and Statistical Manual of Mental Disorders since its third revision (DSM-III 1980). Without going into detail into the history of the classification of anxiety disorders and into the history and development of the several editions of the Diagnostic Manual of Mental Disorders (DSM) of the American Psychiatric Association (APA) it should just briefly be mentioned that the DSM of the APA has undergone until today four revisions. Within these revisions, the third edition (DSM-III 1980) changed most radically from the forerunning ones. The major change in DSM-III was that the category "anxiety neurosis" was deleted because this term was too general and could not be defined reliably. On the basis of evidence that imipramine can block panic attacks (Klein 1964), panic disorder was created as a new diagnosis for the first time in DSM-III. Anxiety states without spontaneous panic attacks were separated from panic disorder and defined as a residual category, generalized anxiety disorder. The revised version of DSM-III, DSM-III-R, was published in 1987, and the fourth and most recent edition, DSM-IV, was published in 1994. More recently, a text revision of DSM-IV has been published that does not entail changes to the diagnostic criteria of disorders, but provides updated empirical reviews for each diagnostic category regarding associated features, cultural, age, and gender features, prevalence, course, familial patterns, and differential diagnosis (DSM-IV-R 2000). Without discussing further details of the development and changes across the different editions and revisions of DSM—these have been reviewed comprehensively in other reviews (see Brown and Barlow 2002; Marshall and Klein 2003)—this chapter gives in the first part an overview about the clinical presentations of anxiety disorders by referring mainly to the forth edition of the DSM (DSM-IV 1994). In the second part, the chapter reviews and summarizes selected aspects (prevalence, correlates, risk factors and comorbidity) of epidemiological knowledge on anxiety disorders.

1 Part I: Clinical Presentation

Overall, the core element of all anxiety disorders is the occurrence of an anxiety reaction that may vary widely in terms of intensity, frequency, persistence, trigger situations, severity and consequences and other qualifying features. Anxiety disorders as defined in the current DSM-IV can be described in terms of the situations, objects or thoughts which provoke anxiety, the specific expression of anxiety in terms of autonomic, and cognitive or motoric features, as well as the specific behaviours used to cope with the provoked anxiety. In some disorders anxiety is expressed mainly in physiological reactions such as heart palpitations (panic disorder), in others primarily by avoidance (specific phobias), and still others by cognitive symptoms such as obsessions or worries (obsessive-compulsive disorder, generalized anxiety disorder). Table 1 gives an overview of the key features of major anxiety disorders included in DSM-IV.

DSM-IV specifies a total of 12 anxiety disorders, but starts by defining panic attacks. Panic attacks are defined separately but are not considered as a separate diagnostic category because they may occur in many of the other anxiety disorders. Likewise, agoraphobia and panic disorder are not considered as specific anxiety diagnoses but rather their combination. In the following, a description of the clinical presentation will be given:

1.1 Panic Attack

Panic attacks are brief, recurrent, unexpected and discrete periods of feelings of intense fear or discomfort. For a diagnosis according to DSM-IV, at least four of the following typical panic symptoms must be present: pounding heart or accelerated heart rate, sweating, trembling or shaking, sensations of shortness of breath or smothering, feeling of choking, chest pain or discomfort, nausea or abdominal stress, feeling dizzy, light-headed or faint, unsteady, derealization (feelings of unreality) or depersonalization, fear of losing control or going crazy, fear of dying, paresthesias, and chills or hot flushes. Spontaneous panic attacks occur "out of the blue" without any obvious environmental or situational triggers. The DSM-IV also identifies (1) situationally bound (cued) panic attacks, in which the panic attack almost invariably occurs immediately on exposure to the situational trigger, and (2) situationally predisposed panic attacks, which are more likely to occur on exposure to the situational cue but

Table 1 Overview of key features major anxiety disorders according to DSM-IV

Code	Diagnosis	Key feature(s)
	Agoraphobia	Anxiety about being in places/situations in which escape might be difficult/embarrassing, or help may not be available in the event of having panic attacks
	Panic disorder	Presence of recurrent, unexpected panic attacks
		Persistent concern about having other attacks
		Worry about possible implications/consequences of the attacks
300.01	Panic disorder without agoraphobia	Criteria for panic disorder are met
		Criteria for agoraphobia are not met
300.21	Panic disorder with agoraphobia	Criteria for panic disorder are met
		Criteria for agoraphobia are met
300.22	Agoraphobia without history of panic disorder	Criteria for agoraphobia are met
		Focus of fear is on the occurrence of embarrassing panic-like symptoms rather than full panic attacks
300.29	Specific phobia	Marked and persistent fear of clearly circumscribed objects or situations
		Exposure to phobic stimulus provokes immediate anxiety response
300.23	Social phobia	Marked and persistent fear of social or performance situations in which embarrassment may occur
		Exposure to situation provokes immediate anxiety response
300.3	Obsessive-compulsive disorder	Recurrent obsessions or compulsions that are severe enough to be time consuming
		Cause-marked distress or impairment
		Person recognizes that obsessions or compulsions are excessive or unreasonable
308.3	Acute stress disorder	Symptoms similar to those of posttraumatic stress disorder that occur immediately in the aftermath of an extremely traumatic event
309.81	Posttraumatic stress disorder	Development of characteristic symptoms following an extreme traumatic event
300.02	Generalized anxiety disorder	Chronic excessive anxiety and worry about a number of events or activities
293.89	Anxiety disorder due to a general medical condition	Symptoms of anxiety that are judged to be a direct physiological consequence of a general medical conditions
	Substance-induced anxiety disorder	Prominent symptoms of anxiety that are judged to be a direct physiological consequence of a drug of abuse, a medication or toxin exposure
300.00	Anxiety disorder not otherwise specified	Anxiety or phobic avoidance that do not meet criteria for any of the specific anxiety disorders

are not invariably associated with this cue and do not necessarily occur immediately after exposure. Panic attacks can occur in a wide range of mental disorders, including other anxiety disorders and mood disorders.

1.2 Panic Disorder and Agoraphobia

In the DSM-IV description of panic disorder, recurrent and at least initially unexpected panic attacks are the key clinical feature, along with persistent concerns about having another attack, worry about the implications or the consequences of the attack, or a remarkable behavioural change related to the attacks. Here again, the panic attacks do not reflect exposure to a situation that always causes anxiety (as in specific phobia) and are not triggered, for example, by social attention. Panic disorder should be classified as either with or without agoraphobia. Agoraphobia is defined as fear of situations from which escape may be difficult or embarrassing or in which help may not be available when panic attacks occur. The person avoids these situations, endures them with anxiety about having another panic attack, or can tolerate them only if another person is present.

According to the DSM-IV criteria, agoraphobia without panic disorder can be described as being in places where help might be difficult or embarrassing, or in which help may not be available in the event of panic-like symptoms, rather than the presence of full panic attacks. Typical situations involve being outside the home, being in a crowd or a line, being on a bridge, or in a bus, subway or car (DSM-IV 1994). For a diagnosis of panic disorder with or without agoraphobia, differential diagnosis must assure that neither the panic attacks nor the avoidance behaviour is part of a physiological condition, medical condition, or other mental disorder.

1.3 Specific Phobia

The key feature of specific phobia is an intense and persistent fear of circumscribed situations or specific stimuli (e.g. exposure to animals, blood). Confrontation with the situation or stimulus provokes almost invariably an immediate anxiety response. Often, the situation or stimulus is therefore avoided or endured with considerable dread. Adolescents and adults with this disorder recognize that this anxiety reaction is excessive or unreasonable, but this may not be the case in children. For a diagnosis according to DSM-IV, the avoidance, fear or anxious anticipation of the phobic stimulus must interfere with the persons daily life or the person must be markedly distressed about having the phobia. Further, the phobic reactions are not better explained by another mental disorder, such as, for example, social phobia.

1.4
Social Phobia

Social phobia can be characterized as overwhelming anxiety and excessive self-consciousness in social situations. The fundamental clinical feature of social phobia is a marked and persistent fear of social or performance situations in the presence of unfamiliar people or when scrutiny by others is possible, even in the context of small groups. Examples would be concern about being unable to speak in public or choking on food when eating in a restaurant. Exposure to such social and performance situations provokes an immediate anxiety response or results in maladaptive avoidance behaviour. Associated features of social phobia frequently include poor social skills, hypersensitivity to criticism and negative evaluation and difficulty of being assertive, as well as low self-esteem and feelings of inferiority. This fear of social situations can be associated with physical symptoms such as blushing, sweating, trembling or heart palpitations. Many people with social phobia recognize that their fear of being among people may be excessive or unreasonable, but they are unable to overcome it. They often worry for days or weeks in advance of a dreaded situation. It is important, however, to note that simple performance anxiety, stage fright or shyness in social situations should not be diagnosed in DSM-IV as social phobia unless the anxiety or avoidance are so marked and persistent that they lead to clinically significant impairment or subjective suffering. For a diagnosis, DSM-IV further demands that the fear is not due to the effect of a substance or a medical condition, and is not better accounted for by another mental disorder.

1.5
Obsessive–Compulsive Disorders

Obsessive–compulsive disorder (OCD) is characterized by recurring and extremely time-consuming obsessions or compulsions that cause marked distress or significant impairment in daily functioning. Obsessions can be described as recurrent, intrusive or inappropriate thoughts, images or impulses that cause feelings of anxiety. Obsessions often involve preoccupations with contamination, symmetry, pathological doubting or uncertainty or harm to self or others, as well as preoccupations with sexual or violent thoughts. The persons often have the feeling that omitting the ritual will lead to disastrous results. Compulsions can be characterized as repetitive behaviours that the person feels driven to perform in an attempt to avoid feelings of tension or anxiety. They can include both repetitive physical behaviours (such as stereotypic counting or arranging, checking or cleaning behaviours) and also mental rituals (such as repeating specific verbal rituals). Obsessions are unpleasant and provoke anxiety, whereas carrying out a compulsion may reduce anxiety. Although compulsions are defined by DSM-IV as repetitive behaviours that a person feels driven to perform in response to an obsession, the diagnostic criteria allow the diagnosis of OCD also to be made when a person reports only compulsions. In order to meet DSM-IV criteria for OCD, obsessions and the performance of the compulsions must be of significant intensity and/or frequency to cause significant distress or marked impairment. The person has further to recognize, at least in part, the irrational nature of the obsessive–compulsive symptoms, yet he or she is not able to stop them. The differential diagnosis of OCD includes other mental disorders in the context of which repetitive behaviours and thoughts can occur. For a diagnosis of OCD, the content of the obsessions/compulsions cannot be completely explained by another disorder. The obsessions of OCD must further be distinguished from the ruminations of major depression, racing thoughts of mania, and psychotic features of schizophrenia. The compulsions of OCD must be distinguished from the stereotypic movements found in individuals with mental retardation or autism, the tics of Tourette syndrome, the stereotypies of complex partial seizures and the ritualized self-injurious behaviours of borderline personality disorder.

1.6 Posttraumatic Stress Disorder

Posttraumatic stress disorder (PTSD) is an anxiety disorder that can develop after exposure to a terrifying event in which severe physical harm occurred or in which the person was threatened. According to DSM-IV the diagnosis of PTSD is based on an extreme response to an extremely threatening and stressful event (e.g. natural disasters, wars, actual or threatened injuries, violent personal assaults). As a reaction to the traumatic event, the person develops an intense feeling of fear, horror and helplessness. The person may suffer from repeated involuntary re-experiences of some aspects of the situation in the form of flashback episodes or dreams. These re-traumatizations may occur spontaneously or subtle cues linked to the event may trigger them. In addition, the person shows physical or emotional avoidance of disturbing memories and a numbing of general responsiveness. This may imply feelings of detachment or estrangement from others, a restricted range of affect and a sense of a foreshortened future. Finally, there are symptoms of increased arousal including sleep disturbance, difficulties with memory and concentration, hypervigilance, irritability or angry outbursts as well as an exaggerated startle response. DSM-IV allows for the specification of symptoms of PTSD to be acute (less than 3 months duration of symptoms), chronic (more than 3 months duration of symptoms) or delayed (onset of the symptoms at least 6 month after the trauma). For a diagnosis, symptoms of re-experience, avoidance and increased arousal must be present for more than 1 month and must be associated with remarkable distress or impairment in everyday life.

1.7 Generalized Anxiety Disorder

The essential clinical feature of generalized anxiety disorder (GAD) is long-lasting, excessive and unrealistic anxiety or worry about a number of life circumstances occurring. The worry and tension is causeless and more severe than the degree of anxiety most people experience. The anxiety further is associated with increased cognitive and physiological arousal. Usually, people with GAD expect the worst: they worry excessively about money, health, family or work, even when there are no signs of trouble at all. For a diagnosis to be made according to DSM-IV, the person must experience the worries over a period of 6 months or more, must find it difficult to control the worries and the symptoms must cause marked distress or significant impairment in daily life. At least three of the following six symptoms also need to be persistent: restlessness, being easily fatigued, difficulty in concentrating, irritability, muscle tension and sleep disturbance. This diagnosis needs to be distinguished from anxiety arising as part of a mood disorder, or anxiety related to another DSM-IV axis I disorder.

2 Part II: Epidemiology

2.1 Prevalence

In the last few decades several large epidemiological studies have estimated the prevalence of anxiety disorders in the community; most of them were carried out in industrialized countries.

Table 2 summarizes lifetime, 12-month, 6-month and point prevalence findings for anxiety disorders across major community studies that have been conducted since the introduction of the DSM-III in 1980. Table 2 in addition provides information about diagnostic criteria, instruments used, and sample sizes.

Before trying to come to a conclusion about the prevalence of anxiety disorders in the general population, some comments should be made about the reasons why one must be careful when interpreting the results. As we see in Table 2, we are facing different studies conducted in different countries in various settings. Across the cited studies, different diagnostic instruments, different sampling procedures, inclusion of different age groups, different criteria used to generate diagnoses, different time frames for the diagnoses (e.g. lifetime, six-month prevalence or point estimates) and different severity ratings for diagnostic decisions have been used. All these methodological differences may at least partially explain the, at the first sight, remarkable degree of heterogeneity and inconsistency of findings across the reviewed studies.

Table 2 Prevalence rates of anxiety disorders in the general population according to DSM-III, DSM-III-R and DSM-IV

Study (country); reference	Assess-ment in-strument[2]	n Age	Time frame	Anxiety disorder (any)	Subtypes						
					Panic disorder[1]	Agora-phobia[2]	Specific phobia	Social phobia	GAD	OCD	PTSD
DSM-III											
ECA (five districts in the USA)	DIS	n=20,291	Lifetime	14.6[3]	1.6 (0.5)	5.2	10.0	2.8	4.1–6.6[4]	2.5	-
Regier et al. 1990		Age: >18	12-month	-	-	-	-	-	2.0–3.6[4]	1.7	-
			6-month	8.9[3]	0.8 (0.2)	3.4	6.4	1.5	-	1.5	-
			Point (1-month)	7.3[3]	0.5 (0.2)	2.9	5.1	1.3	-	1.3	-
Christchurch (New Zealand)	DIS	n=1,498	Lifetime	10.5[5]	2.2	8.1[6]	-	3.0	31.1	2.2	-
Wells et al. 1989		Age: 18–64	12-month	-	-	-	-	-	-	-	-
			6-month	-	-	-	-	-	-	-	-
			Point	-	-	-	-	-	-	-	-
Seoul (Korea)	DIS	n=3,134	Lifetime	9.2[7]	1.1	2.1 (0.7)	5.4	0.5	3.6	2.3	-
Lee et al. 1990		Age: 18–65	12-month	-	-	-	-	-	-	-	-
			6-month	-	-	-	-	-	-	-	-
			Point	-	-	-	-	-	-	-	-
MFS, Munich (Germany)	DIS	n=483	Lifetime	13.9[3]	2.4	5.7	8.0[8]/5.2[9]	2.2[9]	-	2.0	-
Wittchen et al. 1992		Age: 18–55	12-month	-	-	-	-	-	-	-	-
Wittchen and Perkonigg 1996			6-month	8.1[3]	1.1	3.6	4.1[8]		-	1.8	-
			Point (1-month)	7.2[3]	1.0	2.9	3.7[8]	-	-	1.4	-
Zurich (Switzerland)[10]	SPIKE	n=546	Lifetime	15.5[11]	2.7	4.5	11.8	5.3	2.9	1.0	-
Angst and Dobler-Mikula 1985		Age: 19–20	12-month	-	0.2	1.6 (0.7)	1.2	-	-	-	-
Angst 1993			3-month	-	0.1	1.5 (0.7)	1.2	-	-	-	-
			Point (n.r.)	-	0.1	1.5 (0.7)	1.2	-	-	-	-
Florence (Italy)	FPI	n=1,110	Lifetime	-	1.4	0.4 (0.9)	0.6	1.0	5.4	0.7	0.2
Faravelli et al. 1989		Age >14	12-month	-	-	-	-	-	-	-	-
			6-month	-	-	-	-	-	-	-	-
			Point (n.r.)	-	0.3	0.3 (0.7)	0.5	0.5	2.8	0.6	0.1
Puerto Rico	DIS	n=1,551	Lifetime	13.6[3]	1.7	6.9	8.6	1.6	-	3.2	-
Canino et al. 1987		Age: 17–64	12-month	-	-	-	-	-	-	-	-
			6-month	7.5[3]	1.1	3.9	4.4	1.1	-	1.8	-
			Point	-	-	-	-	-	-	-	-

Table 2 (continued)

Study (country); reference	Assess-ment in-strument[2]	*n* Age	Time frame	Anxiety disorder (any)	Subtypes						
					Panic disorder[1]	Agora-phobia[2]	Specific phobia	Social phobia	GAD	OCD	PTSD
Edmonton (Canada)	DIS	*n*=3,258	Lifetime	11.2[3]	1.2	2.9	7.2	1.7	-	3.0	-
Bland et al. 1988		Age >18	12-month	-	-	-	-	-	-	-	-
			6-month	-	-	-	-	-	-	-	-
			Point	-	-	-	-	-	-	-	-
DSM-III-R											
NCS (USA)	CIDI	*n*=8,098	Lifetime	24.9[11]	3.5	5.3	11.3	13.3	5.1	-	7.8[12]
Kessler et al. 1994		Age: 15–54	12-month	17.2[11]	2.3	2.8	8.8	7.9	3.1	-	-
			6-month	-	-	-	-	-	-	-	-
			Point	-	-	-	-	-	-	-	-
NEMESIS (Netherlands)	CIDI	*n*=7,076	Lifetime	19.3[7]	3.8	3.4	10.1	7.8	2.3	0.9	-
Bijl et al. 1998		Age: 18–64	12-month	12.4[7]	2.2	1.6	7.1	4.8	1.2	0.5	-
			6-month	-	-	-	-	-	-	-	-
			Point (1-month)	9.7[7]	1.5	1.0	5.5	3.7	0.8	0.3	-
ECA-SP (Brazil)	CIDI	*n*=1,464	Lifetime	17.4[13]	-	-	-	-	-	-	-
Andrade et al. 1996[14]		Age>18	12-month	10.9[13]	-	-	-	-	-	-	-
			Point (1-month)	8.7[13]	-	-	-	-	-	-	-
MHS-OHS (Canada)	UM-CIDI	*n*=6,261	Lifetime	21.3[13]	-	-	-	-	-	-	-
Offord et al. 1994[14]		Age: 18–54	12-month	12.4[13]	1.1	1.6	6.4	6.7	1.1	-	-
Offord et al. 1996			6-month	-	-	-	-	-	-	-	
			Point (1-month)	6.2[13]	-	-	-	-	-	-	
EPM (Mexico)	UM-CIDI	*n*=1,734	Lifetime	5.6[13]	-	-	-	-	-	-	-
Caraveo et al. 1998[14]		Age: 18–54	12-month	4.0[13]	-	-	-	-	-	-	-
			Point (1-month)	2.3[13]	-	-	-	-	-	-	-
Mental Health Program Turkey	CIDI	*n*=6,095	Lifetime	7.4[13]	-	-	-	-	-	-	-
Kýlýc 1998[14]		Age: 18–54	12-month	5.8[13]	-	-	-	-	-	-	-
			Point (1-month)	5.0[13]	-	-	-	-	-	-	-
MAPSS (USA)	CIDI	*n*=3,012	Lifetime	16.8[15]	1.7	7.8	7.4	7.4	-	-	-
Vega et al. 1998		Age: 18–59	12-month	-	-	-	-	-	-	-	-
			6-month	-	-	-	-	-	-	-	-
			Point	-	-	-	-	-	-	-	-

Table 2 (continued)

Study (country); reference	Assess-ment in-strument[2]	*n* Age	Time frame	Anxiety disorder (any)	Subtypes						
					Panic disorder[1]	Agora-phobia[2]	Specific phobia	Social phobia	GAD	OCD	PTSD
Oslo (Norway)	CIDI	*n*=2,066	Lifetime	-	4.5	6.1	14.4	13.7	4.5	1.6	-
Kringlen et al. 2001		Age: 18–65	12-month	-	2.6	3.1	11.1	7.9	1.9	0.7	-
			6-month	-	-	-	-	-	-	-	-
			Point	-	-	-	-	-	-	-	-
Basel (Switzerland)	CIDI	*n*=470	Lifetime	28.7[11]	1.3 (2.1)	10.8	4.5	16.0	1.9	-	-
Wacker et al. 1992		Age: 18–65	12-month	-	-	-	-	-	-	-	-
			6-month	-	-	-	-	-	-	-	-
			Point	-	-	-	-	-	-	-	-
DSM-IV											
EDSP, Munch	M-CIDI	*n*=3,021	Lifetime	14.4[16]	1.6	2.6	2.3	3.5	0.8	0.7	1.3
Wittchen et al. 1998a		Age: 14–24	12-month	9.3[16]	1.2	1.6	1.8	2.6	0.5	0.6	0.7
			6-month	-	-	-	-	-	-	-	-
			Point	-	-	-	-	-	-	-	-
GHS-MHS (Germany)	CIDI	*n*=4,181	Lifetime	-	3.9	-	-	-	-	-	-
Wittchen and Jacobi 2001		Age: 18–65	12-month	14.5[17]	2.3	2.0	7.6	2.0	1.5	0.7	-
Jacobi et al. 2004			6-month	-	-	-	-	-	-	-	-
			Point (4 weeks)	9.0	1.1	-	-	-	1.2	0.4	-
Dresden-Study	F-DIPS	*n*=1,538	Lifetime	27.2[18]	2.1 (0.8)	2.3	12.3	12.0	2.4	1.3	3.0
Becker et al. 2000		Age: 18–25 (w)	12-month	-	-	-	-	-	-	-	-
			6-month	-	-	-	-	-	-	-	-
			Point (7 days)	17.9[18]	0.3 (0.5)	1.8	9.8	6.7	1.4	0.8	0.4
TACOS	M-CIDI	*n*=4,075	Lifetime	15.1[19]	0.9 (1.3)	1.1	10.6	1.9	0.8	0.5	1.4
Meyer et al. 2000		Age: 18–64	12-month	-	-	-	-	-	-	-	-
			6-month	-	-	-	-	-	-	-	-
			Point	-	-	-	-	-	-	-	-

Table 2 (continued)

Study (country); reference	Assessment instrument[2]	*n* Age	Time frame	Anxiety disorder (any)	Subtypes						
					Panic disorder[1]	Agoraphobia[2]	Specific phobia	Social phobia	GAD	OCD	PTSD
NSMHW (Australia)	CIDI	*n*=10,641	Lifetime	-							
Andrews et al. 2001		Age: 18+	12-month	5.6[20]	1.1	0.5	-	1.3	2.6	0.7	1.3
			6-month	-	-	-	-	-	-	-	-
			Point (1-month)	3.8[20]	0.5	0.2	-	1.0	2.0	0.5	0.9
South Florida-Study (USA)	CIDI	*n*=1,803	Lifetime	15.2[21]	2.1	-	-	2.5	1.4	-	11.7
Turner and Gil 2002		Age: 19–21	12-month	-	1.6	-	-	-	-	-	8.4
			6-month	-	-	-	-	-	-	-	-
			Point	-	-	-	-	-	-	-	-
ESEMED	CIDI	*n*=21,425	Lifetime	13.6[22]	2.1	0.9	7.7	2.4	2.8	-	1.9
Alonso et al. 2004		Age: 18–65+	12-month	6.4	0.8	0.4	3.5	1.2	1.0	-	0.9
			6-month	-	-	-	-	-	-	-	-
			Point	-	-	-	-	-	-	-	-

CIDI, Composite International Diagnostic Interview; DIS, Diagnostic Interview Schedule; ECA, Epidemiologic Catchment Area program (if not annotated differently, all data refer to all five districts); ECA-SP, Epidemiologic Catchment Area study in the city of Sao Paolo; EDSP, Early Developmental Stages of Psychopathology study, in reference to the base analysis of ages 14 to 24; EPM, Epidemiology of Psychiatric Comorbidity Project; ESEMED, European Study of the Epidemiology of Mental Disorders; F-DIPS, Diagnostisches Interview bei psychischen Störungen—Forschungsversion.; FPI, Florence Psychiatric Interview; GAD, generalized anxiety disorder; GHS-MHS, General Health Survey—Mental Health Supplement; MAPSS, The Mexican American Prevalence and Services Survey; M-CIDI, Munich-Composite International Diagnostic Interview; MFS, Munich Follow-up Study; MHS-OHS, The Mental Health Supplement to the Ontario Health Survey; n.r., not reported; NCS, National Comorbidity Survey; NEMESIS, Netherlands Mental Health Survey and Incidence Study; NSMHW, National Survey of Mental Health and Wellbeing; OCD, obsessive–compulsive disorder; PTSD, post-traumatic stress disorder; SPIKE, Structured psychopathology interview and rating of the social consequences of psychic disturbances for epidemiology; TACOS, Transitions in Alcohol Consumptions and Smoking; UM-CIDI, University of Michigan modified version of the Composite International Diagnostic Interview. [1]DSM-III: panic disorder without agoraphobia (with agoraphobia); DSM-III-R/DSM-IV: with/without agoraphobia. [2]DSM-III: agoraphobia with/without panic attacks (with panic attack); DSM-III-R/DSM-IV: Agoraphobia without panic attack. [3]Phobias (agora-, social, specific), panic disorder, obsessive–compulsive disorder. [4]Quoted from Carter et al. 2001; data refer to the districts Durham, St. Louis und Los Angeles. [5]Phobias (agora-, specific), panic disorder, obsessive–compulsive disorder, somatoform disorder. [6]Agoraphobia or specific phobia. [7]Phobias (agora-, social, specific), panic disorder, obsessive–compulsive disorder, GAD. [8]Including social phobia. [9]Wittchen 1993. [10]Lifetime: cumulative lifetime prevalence with 30 years; other data refer to ages 22–23. [11]Phobias (agora-, social, specific), panic disorder, GAS. [12]Quoted from Kessler et al. 1995. [13]Phobias (agora-, social, specific), panic disorder and/or GAS. [14]Quoted and cited from ICPE (Andrade et al. 2000). [15]Phobias (agora-, social, specific), panic disorder. [16]Phobias (agora-, social, specific, NBB), panic disorder, GAD, obsessive–compulsive disorder, PTSD. [17]Phobias (agora-, social, specific, NOS), panic disorder, GAD, obsessive–compulsive disorder. [18]Phobias (agora-, social, specific, NOS), panic disorder, GAS, obsessive–compulsive disorder, PTSD, acute stress disorder. [19]Phobias (agora-, social, specific, NOS), panic disorder, GAD, obsessive–compulsive disorder, PTSD, anxiety disorder based on a medical condition. [20]Phobias (agora-, specific), panic disorder, obsessive–compulsive disorder, PTSD. [21]Social phobia, panic disorder, GAD, PTSD. [22]GAD, social phobia, specific phobia, agoraphobia, PTSD.

Looking now into the prevalence rates reported from the several studies, lifetime prevalence rates for all anxiety disorders lumped together in the different studies range between 5.6% and 28.7%. The estimated median of the included studies is 15.1%. Lifetime prevalence estimates describe the proportion of persons in the population who have developed the disorder under consideration at least once in their life. Based on this estimated median, anxiety disorders occur in approximately 1 in 7 persons in the general population at some point in their life. Looking closer at specific anxiety disorders, it becomes evident that the high lifetime prevalence of anxiety disorders is mostly due to the high frequency of simple phobia and social phobia.

Across studies, lifetime prevalence estimates for specific phobia range from 0.6% to 14.4%. Table 3 presents, additionally, lifetime prevalence rates for specific fears and phobias, based on the findings from the U.S. National Comorbidity Survey (NCS; Kessler et al. 1994; Curtis et al. 1998). As can be seen, the most prevalent specific phobias were animal phobia (5.7%) and height phobia (5.3%), confirming previous research findings from the Epidemiological Catchment Area Survey (Bourdon et al. 1988). Across the studies, lifetime prevalence of social phobia was estimated to range between 0.5% and 16%. Community surveys assessing the lifetime prevalence of social phobia according to the DSM-III criteria—by using the Diagnostic Interview Schedule (DIS, Robins et al. 1981)—found lifetime prevalence rates of DSM-III social phobia to range from 0.5% to 3.0%, while more recently conducted surveys assessing social phobia according to DSM-IV by using the Composite International Diagnostic Interview (CIDI; WHO 1990) have found considerably higher lifetime prevalence rates. For example, in the U.S. NCS (Kessler et al. 1994) DSM-III-R social phobia was found to have lifetime prevalence rate of 13.3%. As part of the Munich Early Developmental Stages of Psychopathology Study (EDSP;

Table 3 Lifetime prevalence of specific phobias according to the specific stimuli and situations in the NCS (adapted from Curtis et al. 1998)

Stimulus/situation	Lifetime prevalence (%)
Height	5.3
Flying	3.5
Close spaces	4.2
Being alone	3.1
Storms	2.9
Animals	5.7
Blood	4.5
Water	3.4
Any	11.3

Wittchen et al. 1998a; Lieb et al. 2000a), lifetime prevalence of DSM-IV social phobia was investigated in a community sample of 14- to 24-year-olds. Using a computerized version of the CIDI (DIAX/M-CIDI; Wittchen and Pfister 1997), the lifetime prevalence of social phobia was found to be 3.5% among adolescents and young adults. Table 2 shows that for GAD among adults, the estimated lifetime prevalence rates range from a low of 1.9% in Switzerland (Wacker et al. 1992) to a high of 31.1% in New Zealand (Wells et al. 1989). Most estimates range between 2.3 and 4.5%. For agoraphobia, lifetime prevalence rates between 0.4% and 10.9% have been reported across the studies. OCD and panic disorder seem to be less frequent in the general population, with lifetime prevalence rates between 0.5% and 3.2% and 0.5% and 4.5%, respectively. Compared to the other anxiety disorder, less is known about the frequency of PTSD in the general population, since only a few studies before DSM-IV included the assessment of PTSD. With the exception of the relatively high estimates of 7.8% and 11.7% for U.S. samples (Kessler et al. 1995; Turner and Gil 2002), studies outside the U.S. found consistently low lifetime prevalence rates, ranging from 0.2% to 3.0%.

Table 2 also indicates that the prevalence estimates for 12-month, 6-month and point prevalences are lower when compared to the lifetime estimates. This can be seen as one indicator of the fluctuating character of anxiety disorders. As discussed above, variation across studies is probably mainly due to differences in study characteristics. Overall, the 12-month prevalence rates for any anxiety disorder result in an estimated median of 11%, indicating that 1 in 10 people were affected by an anxiety disorder in the year preceding the assessment.

2.2 Age of Onset

Epidemiological studies in which age of first manifestation of anxiety was investigated consistently show that, with the exception of panic disorder and generalized anxiety disorder, anxiety disorders seem to start early in life, in the first and second decade of life.

Data from the WHO International Consortium in Psychiatric Epidemiology (ICPE), which carried out cross-national comparative studies on the prevalence and correlates of mental disorders, investigated the distribution of age of onset for the overall diagnostic group of anxiety disorder including panic disorder, agoraphobia, simple phobia, social phobia and generalized anxiety disorder. Fig. 1 shows graphically the findings that this group obtained across six countries.

In these analyses, the Kaplan-Meier method was used to generate age-of-onset-curves. Figure 1 shows not only that the onset distributions were similar across countries, but also that more than 50% of the cases had their first onset before age 20. Even the proportion of individuals with a first manifestation of anxiety disorder before age 10 is remarkably high. After age 40, the risk for

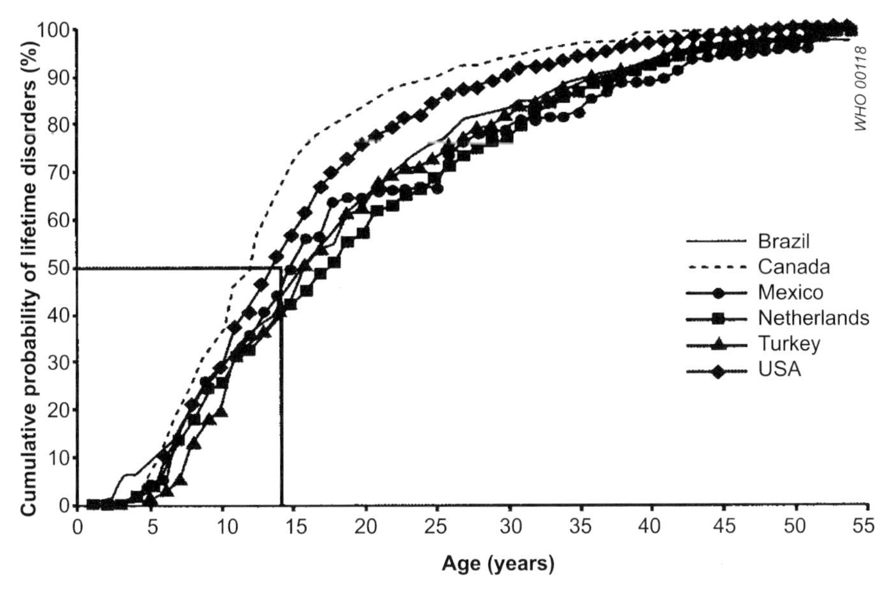

Fig. 1 Age of onset distribution of any anxiety disorders in the ICPE surveys (WHO-ICPE-Analyses, Andrade et al. 2000; reprinted with permission)

first onset of an anxiety disorder becomes lower. In the ICPE analyses, anxiety disorders were estimated to have a median age of onset of 15 years, ranging from 12 years in Canada to 18 years in the Netherlands (Andrade et al. 2000). Similar results have also been obtained in the German Mental Health Survey as part of the German National Health Interview and Examination Survey (GHS-MHS; Jacobi et al. 2004). Data of this population survey suggest the median of retrospectively reported first onset of anxiety disorder to be at age 18.

Regarding the specific types of anxiety disorders, specific and social phobia show the earliest manifestations (see Wittchen et al. 1999a). The main risk period for these anxiety disorders lies in childhood or adolescence, and after age 20 the probability for first onset considerably decreases. In contrast, generalized anxiety disorder, panic disorder and OCD manifest somewhat later, during late adolescence until middle adulthood (Burke et al. 1990; Magee et al. 1996; Wittchen et al. 1999a). Thus, for generalized anxiety disorder, remarkable risks of onset begin in the teens but then cumulate through the 50s (Bijl et al. 1998; Kessler et al. 2002). For panic disorder, ECA results suggest the average age of onset to be at the end of the third decade. The ECA data further suggest different risk periods for males and females; the probability for first onset was highest between ages 30 and 44 years for females and between ages 30 and 44 years for males. Recent results, however, from the Munich EDSP study suggest that first onset of panic disorder can already be observed during early adolescence (Wittchen et al. 1998b).

Data on retrospectively reported age of first onset of OCD suggest that the second and third decade in life seem to be a critical period for first manifestation. Thus, partially dependent on the age range of the study sample, mean ages of first onset for OCD have been reported to range between 12.8 (Flament et al. 1988) and 35.5 years (Weissman et al. 1994; Degonda et al. 1992; Grabe et al. 2001). Also using retrospective collected age-of-onset information but applying the more sophisticated life table methods, the ECA study found highest hazard rates between age 15 and 39 years, suggesting this age frame as most important for the first onset of OCD (Burke et al. 1990).

2.3 Correlates

2.3.1 Gender

Data from epidemiological studies consistently have shown that anxiety disorders are more common in women than in men. On average, anxiety disorders are about twice as frequent in women (Kessler et al. 1994; Alonso et al. 2004; Jacobi et al. 2004). Although there are variations across the specific forms of anxiety disorders (female:male ratio ranging between 1.5 and 2.5), the overall higher risk for women remains stable. The lifetime and 12-month prevalences of agoraphobia, specific phobia, generalized anxiety disorder, panic disorder and posttraumatic stress disorders are approximately twice as prevalent among women as men (Eaton et al. 1991; Kessler et al. 1994; Magee et al. 1996; Bijl et al. 1998; Alonso et al. 2004; Jacobi et al. 2004). Across the surveys, smaller sex differences were found for social phobia and OCD (Magee et al. 1996; Bijl et al. 1998; Alonso et al. 2004; Jacobi et al. 2004).

2.3.2 Sociodemographic Factors

Apart from gender, other reported sociodemographic correlates for anxiety disorders include education (Eaton et al. 1991; Magee et al. 1996; Bijl et al. 1998; Wittchen et al. 1998a; Andrews et al. 2001), marital status (Bland et al. 1988a,b; Regier et al. 1993; Magee et al. 1996; Andrews et al. 2001; Alonso et al. 2004; Jacobi et al. 2004), urbanicity (Magee et al. 1996; Bijl et al. 1998; Wittchen et al. 1998a; Andrews et al. 2001; Alonso et al. 2004), employment status (Magee et al. 1996; Bijl et al. 1998; Andrews et al. 2004; Alonso et al. 2004; Jacobi et al. 2004) and financial situation (Eaton et al. 1991; Kessler et al. 1994; Bijl et al. 1998; Wittchen et al. 1998a).

2.4 Risk Factors

2.4.1 Family–Genetic Factors

One of the major risk factors for the development of an anxiety disorder is a family history of psychopathology. In several epidemiological studies the familial aggregation of anxiety disorders was demonstrated (Angst 1998; Kendler et al. 1997; Kessler et al. 1997; Bromet et al. 1998; Lieb et al. 2000b; Wittchen et al. 2000a; Chartier et al. 2001; Bijl et al. 2002). On the basis of the Munich EDSP-study, Lieb et al. (2000b) could demonstrate that offspring of parents with social phobia have an increased risk for social phobia vs offspring mentally healthy of parents (see Fig. 2). This study found, in addition, that risk for social phobia seems also to be elevated among offspring of parents with other psychopathology, e.g. other anxiety disorders, depression or alcohol use disorders, suggesting a familial cross-transmission of these disorders.

In another study, Lieb et al. (2002) confirmed the cross-aggregation of anxiety disorders and depression by showing that offspring of parents with depressive disorders have not only an elevated risk for depressive disorders but also for anxiety disorders. Considering parental comorbidity, the cross-aggregation between parental depression and general anxiety disorder in off-

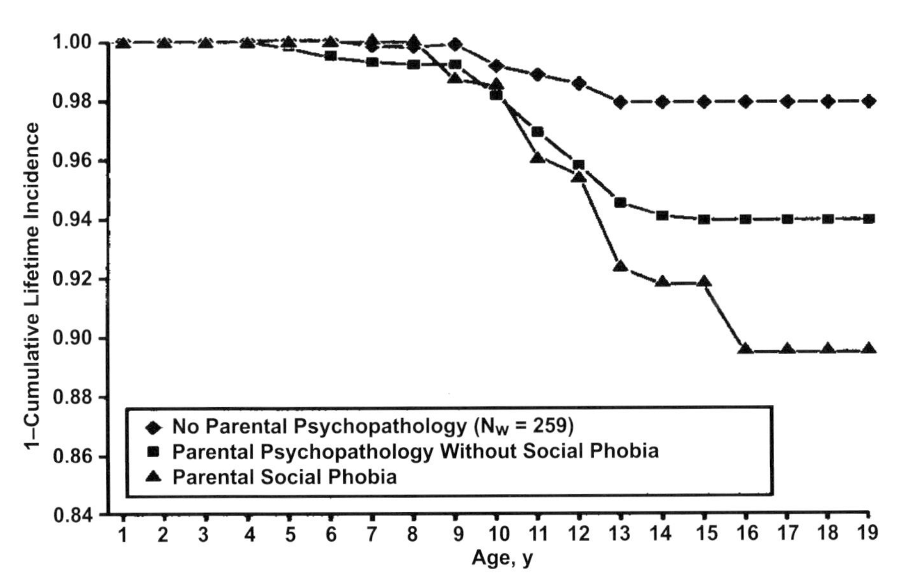

Fig. 2 Onset of social phobia among respondents with parents with social phobia, parents with psychopathology, excluding social phobia, and those whose parents had no psychopathology (Lieb et al. 2000; from *Arch Gen Psychiatry*, 57:859–66, Fig. 563 therein Copyright (2000) by permission of American Medical Association. All rights reserved)

spring remained significant. These results are similar to the results of Kendler et al. (1997) who investigated the familial aggregation of mental disorders by using the family history data of the NCS. Although these researchers found some specificity in the familial transmission of generalized anxiety disorder, the cross-aggregation between major depression and generalized anxiety disorder remained stable after controlling for comorbidity. The influence of a family history of anxiety disorders on the risk to develop anxiety disorders was also demonstrated in several studies by using the family- and high-risk design (Hettema et al. 2001). Since the genetic epidemiology of anxiety disorder is intensively discussed in another chapter (see Merikangas and Low, this volume), further results will not be discussed in more detail here.

2.4.2 Behavioural Inhibition and Parenting Style

Behavioural inhibition describes the tendency to be shy, timid and constrained in unfamiliar situations (Kagan et al. 1984).This disposition is very stable, can be observed early in life and is assumed to be genetically determined. Offspring of parents with anxiety disorders show higher behavioural inhibition than offspring of parents without anxiety disorders (Rosenbaum et al. 1991). Behavioural inhibition has seldom been evaluated in epidemiological studies. However, those that included it consistently found associations between behavioural inhibition during childhood and subsequent development of anxiety disorders. Inconsistent findings, however, have been reported regarding the specificity of behavioural inhibition, i.e. whether behavioural inhibition increases specifically the risk for specific forms of anxiety disorders (social anxiety), all anxiety disorders or other mental disorders (Wittchen et al. 2000a; Biederman et al. 2001).

Considering the influence of parenting style in the development of anxiety disorders, the available epidemiological studies present a rather inconsistent picture (Ernst and Angst 1997; DeWit et al. 1999; Magee 1999; Wittchen et al. 2000a). For social phobia, Lieb et al. (2000a) demonstrated on the basis of the prospective-longitudinal EDSP study, that specifically parental overprotection and rejection seem to increase children's risk for developing this anxiety disorder. In contrast, aspects of family climate could not be shown to be associated with an increased risk for anxiety disorders (Chartier et al. 2001; Merikangas et al. 2002).

2.4.3 Life Events

The impact of life events as potential risk factor in the development of anxiety disorders has been evaluated in several epidemiological studies. In almost all studies evidence was found for an association between childhood adversities and the subsequent development of anxiety disorders (Ernst et al. 1993; Fergusson et al. 1996; Kessler et al. 1997; Bijl et al. 1998; Chartier et al. 2001; DeGraaf et al. 2002). Molnar et al. (2001) evaluated on the basis of the NCS data the relationship between child sexual abuse and subsequent mental disorders. They found that, among women, child sexual abuse increases the risk for agoraphobia, panic disorder, posttraumatic stress disorder and social phobia. Among men, only posttraumatic stress disorder was associated with child sexual abuse. Similar specific findings were recently reported by MacMillan et al. (2001) who found significant associations between child sexual abuse and subsequent anxiety disorders only among women. Other life events that have been investigated include parental divorce, death of parents and early separation from parents, but to date these factors have not consistently been proved as risk factors for the development of anxiety disorders.

2.5 Comorbidity

The term comorbidity was developed in the context of chronic diseases and was defined by Feinstein (1970) as "any distinct additional clinical entity that has existed or that may occur during the clinical course of a patient who has the index disease under study" (pp. 456–457). In psychiatric epidemiology, comorbidity describes the co-occurrence of at least two mental disorders within the same person within a defined time period (e.g. lifetime, 12 months; Maser and Cloninger 1990; Wittchen 1996). The lifetime perspective of comorbidity is specifically important for the identification of the temporal patterns of comorbidity, which is especially important for the development of models about the pathogenetic relationship between different forms of mental disorders (see Merikangas and Stevens 1998). Studies of diagnostic patterns in general population samples (e.g. Kessler et al. 1996; Merikangas et al. 1998; Kessler 2001; Jacobi et al. 2004) and in clinical samples (e.g. Mezzich et al. 1990; Sartorius et al. 1996) have shown that comorbidity among anxiety disorders is highly prevalent. The ECA investigators were the first to demonstrate that comorbidity is widespread among subjects with a lifetime history of an anxiety disorder and that comorbidity seems more the rule than the exception; more than two third of subjects with lifetime DSM-III anxiety disorders meet additional lifetime criteria for at least one other mental disorder (Robins et al. 1991). Similar results were found in the NCS: over 70% of the respondents with a lifetime history of at least one anxiety disorder also had another mental disorder (Kessler et al. 1994), with highest comorbidity rates for panic disorder (92%) and generalized anxiety disorder (Kessler 1997). Also findings of the more recently conducted German GHS-MHS are consistent with this pattern: 62% of subjects with a 12-month diagnosis of DSM-IV anxiety disorder were found to have a second 12-month diagnosis as well (Jacobi et al. 2004). Comparable to the NCS results, highest comorbidity rates were found for panic disorder (88%)

and generalized anxiety disorder (94%). In the GHS-MHS population sample, more than half of the subjects with panic disorder or generalized anxiety disorder, more than 60% with OCD, and more than 20% of the subjects with a phobic disorder fulfilled diagnostic criteria for three or more other mental disorders. All these findings suggest that the co-occurrence of anxiety disorders with other mental disorders is a pervasive feature of anxiety disorders. However, although epidemiological studies have shown that comorbidity in anxiety disorders is a valid phenomenon, the meaning of comorbidity is still poorly understood. Although several models have been proposed to explain comorbidity (Kessler and Price 1993), we are just beginning to understand its pathogenetic and treatment implications.

2.5.1 Comorbidity with Affective Disorders

Several population-based studies have consistently found a remarkable association between anxiety disorders and affective disorders, particularly with major depression (Weissman et al. 1994; Angst 1993; Lewinsohn et al. 1997; Merikangas et al. 1998; Regier et al. 1998; Kessler 2001). According to NCS data, subjects with an anxiety disorder have an almost five times greater chance of developing major depression, compared to subjects without any anxiety disorder (Kessler 2001). Cross-sectional studies that have investigated the temporal pattern of onset of anxiety and depression in comorbid cases have demonstrated that major depression generally develops secondary to anxiety, suggesting that anxiety disorders increase the risk for subsequent depression (Kessler et al. 1996; Regier et al. 1998). Recently, several prospective analyses of the EDSP study have shown that indeed almost all forms of anxiety disorders increase the risk for first onset of major depression (Wittchen et al. 2000a; Stein et al. 2001; Bittner et al. 2004). Together, on the basis of the available empirical evidence, it can be concluded that primary anxiety disorders increase the risk of developing a secondary depressive disorder.

2.5.2 Comorbidity with Substance Use Disorders

Epidemiological studies have also documented the comorbidity between anxiety disorders and substance use disorders. In the U.S. NCS, one in five people with a lifetime history of a DSM-III-R anxiety disorder fulfilled, in addition, diagnostic criteria for alcohol dependence (range across specific forms of anxiety disorders: 21%–30%) and about 15% fulfilled criteria for drug dependence (range 15%–23%; Kessler et al. 1996). In a more recent cross-sectional investigation of patterns of co-morbidity between substance use and anxiety disorders in six studies participating in the ICPE (Merikangas et al. 1998), the investigators found across all sites strong associations between anxiety disorders and alcohol and drug dependence. The odds ratios (ORs) as a measure of association were ranging between 1.8 and 2.7 for alcohol dependence and between 3.3 and 5.2 for drug dependence. Considering the observed ranges of associations within the specific substance dependencies, the available epidemiological findings do not argue for single combinations of disorders having consistently stronger associations than others. Risks rather seem to be of similar magnitude. The ICPE analyses also investigated the temporal ordering of the onset of the comorbid disorders and found that in general, anxiety disorders precede the onset of substance problems/disorders. Applying more sophisticated prospective analyses, Zimmermann et al. (2003) could demonstrate based on the 4-year follow-up data of the EDSP that specifically panic disorder and social phobia are predictors of subsequent alcohol problems among adolescents and young adults.

Concerning the comorbidity between anxiety disorders and nicotine dependence, less epidemiological research findings are available. Those studies that investigated this issue, however, argue for an association between nicotine dependence and anxiety disorders (Breslau et al. 1994; Johnson et al. 2000; Sonntag et al. 2000). Recently published EDSP findings have shown a prospective association specifically between prior nicotine dependence panic and the development of subsequent (Isensee et al. 2003).

2.5.3 Comorbidity Within the Anxiety Disorders

Although less studied, epidemiological investigation has also shown that there is a considerable degree of overlap within the anxiety disorders. In the NCS, associations (in terms of ORs) within different forms of anxiety disorders were found to range between 3.8 and 12.3 for generalized anxiety disorder, 5.8 and 11.9 for agoraphobia, 4.9 and 8.5 for specific phobia, and 3.8 and 7.8 for social phobia (Wittchen et al. 1994; Magee et al. 1996). The strongest comorbidity was found between panic disorder and agoraphobia, due to the fact that agoraphobia with panic disorder and agoraphobia without panic were not distinguished in the diagnostic criteria of agoraphobia. Interestingly, only about one third of the respondents who meet criteria for DSM-III-R agoraphobia additionally reported panic attacks. This result confirms earlier results found in the ECA and Zurich study (Angst and Dobler-Mikola 1985; Weissman et al. 1986) that panic seems to be involved only in a minority of people with agoraphobia. Similar findings have recently been obtained in the general population sample of the EDSP study. Wittchen et al. (1998b) demonstrated that among adolescents and young adults, most subjects with agoraphobia reported nether full nor limited panic attacks or panic experiences. Other analyses from this study suggest that panic attacks seem to be rather unspecific for the development of subsequent disorder, since pre-existing panic attacks prospectively increased the risk of onset of any secondary anxiety disorder, social phobia, specific phobia or generalized anxiety disorder, as well as of any secondary alcohol use disorder (Goodwin et al. 2004).

3
Summary

The studies reviewed here have shown than anxiety disorders are common mental disorders in the general population. Overall, anxiety disorders typically start early in life and have a high degree of comorbidity with other anxiety, affective and substance use disorders. Comorbid anxiety disorders, specifically specific and social phobia, are often the temporally primary disorders and are associated with an elevated risk for the subsequent onset of psychopathology. The observation that temporally primary anxiety disorders increase the risk for secondary psychopathology raises the question of whether early prevention would prevent the onset of subsequent psychopathology. More epidemiological analyses and findings are needed to fully understand the mechanisms of comorbidity: Why do people with primary anxiety develop secondary affective or substance use disorders? Specifically longitudinal surveys that include such early manifestations of anxiety disorders are needed in order to understand the developmental pathway from anxiety to subsequent psychopathology. Not discussed in this reviews are findings concerning impairments associated with anxiety disorders and patterns of treatment seeking. Briefly, several studies have demonstrated that people with anxiety disorders experience reduced quality of life as well as remarkable impairment in work productivity and role functioning (Magee et al. 1996; Wittchen et al. 2000b). Other studies have demonstrated that anxiety disorders are associated with substantial costs to the health care systems (Greenberg et al. 1999). To conclude, anxiety disorders are prevalent and serious health concerns that should be taken in both practice and science as seriously as mood or substance use disorders. However, further epidemiological research is warranted in order to elucidate the various components involved into the aetiology of anxiety disorders as well as to learn more about the mechanisms by which anxiety disorders lead to other psychopathology.

References

Alonso J, Angermeyer MC, Bernert S, et al (2004) Prevalence of mental disorders in Europe: results from the European Study of the Epidemiology of Mental Disorders (ESEMeD) project. Acta Psychiatr Scand Suppl 420:21–27

Andrade L, Caraveo-Anduaga JJ, Berglund P, Bijl R, Kessler RC, Demler O, Walters E, Kylyc C, Offord D, Ustun TB, Wittchen H-U (2000) Cross-national comparisons of the prevalences and correlates of mental disorders. Bull World Health Organ 78:413–426

Andrews G, Henderson S, Hall W (2001) Prevalence, comorbidity, disability and service utilisation. Overview of the Australian Mental Health Survey. Br J Psychiatry 178:145–153

Angst J (1993) Comorbidity of anxiety, phobia, compulsion and depression. Int Clin Psychoneuropharm 8(Suppl 1):21–25

Angst J (1998) Panic disorder: history and epidemiology. Eur Psychiatry 13(Suppl 2):51–55

Angst J, Dobler-Mikola A (1985) The Zurich Study. V. Anxiety and phobia in young adults. Eur Arch Psychiatry Neurol Sci 235:171–178

Becker ES, Türke V, Neurmer S, Soeder U, Krause P, Margraf J (2000) Incidence and prevalence rates of mental disorders in a community sample of young women: results of the "Dresden Study". In: Manz R, Kirch W (eds) Public health research and practice: report for the Public Health Research Association, Saxony. Roderer, Regensburg, pp 259–291

Biederman J, Hirshfield-Becker DR, Rosenbaum JF, Hérot C, Friedman D, Snidman N, Kagan J, Faraone SV (2001) Further evidence of association between behavioural inhibition and social anxiety in children. Am J Psychiatry 158:1673–1679

Bijl RV, Ravelli A, van Zessen G (1998) Prevalence of psychiatric disorder in the general population: results of The Netherlands Mental Health Survey and incidence Study (NEMESIS). Soc Psychiatry Psychiatr Epidemiol 33:587–595

Bijl RV, Cuijpers P, Smit F (2002) Psychiatric disorders in adult children of parents with a history of psychopathology. Soc Psychiatry Psychiatr Epidemiol 37:7–12

Bittner A, Goodwin RD, Wittchen HU, Beesdo K, Höfler M, Lieb R (2004) What characteristics of primary anxiety disorders predict subsequent major depression? J Clin Psychiatry 65:618–626

Bland RC, Newman SC, Orn H (1988a) Age of onset of psychiatric disorders. Acta Psychiatr Scand 77(Suppl 338):43–49

Bland RC, Orn H, Newman SC (1988b) Lifetime prevalence of psychiatric disorders in Edmonton. Acta Psychiatr Scand 77(Suppl 338):24–32

Bourdon KH, Boyd JH, Rae DS, Burns BJ, Thompson JW, Locke BZ (1988) Gender differences in phobias: results of the ECA community study. J Anxiety Disord 2:227–241

Breslau N, Kilbey MM, Andreski P (1994) DSM-III-R nicotine dependence in young adults: prevalence, correlates and associated psychiatric disorders. Addiction 89:743–754

Bromet E, Sonnega A, Kessler RC (1998) Risk factors for DSM-III-R posttraumatic stress disorder: findings from the National Comorbidity Survey. Am J Epidemiol 147:353–361

Brown TA, Barlow DH (2002) Classification of anxiety and mood disorders. In: Barlow D (ed) Anxiety and its disorders: the nature and treatment of anxiety and panic, 2nd edn. Guillford Press, New York, pp 292–327

Burke KC, Burke JD, Regie DA, Rae DS (1990) Age at onset of selected mental disorders in five community populations. Arch Gen Psychiatry 47:511–518

Canino GS, Bird HR, Shrout PE, Rubio-Stipec M, Bravo M, Martinez R, Sesman M, Guevara LM (1987) The prevalence of specific psychiatric disorders in Puerto Rico. Arch Gen Psychiatry 44:727–735

Carter RM, Wittchen H-U, Pfister H, Kessler R (2001) One-year prevalence of subthreshold and threshold DSM-IV generalized anxiety disorder in a nationally representative sample. Depress Anxiety 13:78–88

Chartier MJ, Walker JR, Stein MB (2001) Social phobia and potential childhood risk factors in a community sample. Psychol Med 31:307–315

Curtis GC, Magee WJ, Eaton WW, Wittchen HU, Kessler RC (1998) Specific fears and phobias. Epidemilogy and classification. Br J Psychiatry 173:212–217 de Graaf R, Bijl R, SmitFilipVollebergh WAM, Spijker J (2002) Risk factors for 12-month comorbidity of mood, anxiety, and substance use disorders: findings from the Netherlands Mental Health Survey and Incidence Study. Am J Psychiatry 159:620–629

Degonda M, Wyss M, Angst J (1992) The Zurich Study. XVIII. Obsessive-compulsive disorders and syndromes in the general population. Eur Arch Psychiatry Clin neurosci 243:16–22

DeWit DJ, Ogborne A, Offord DR, MacDonald K (1999) Antecedents of the risk of recovery from DSM-III-R social phobia. Psychol Med 29:569–582

DSM-III (1980) American Psychiatric Association diagnostic and statistical manual of mental disorders, 3rd edn. American Psychiatric Press, Washington

DSM-III-R (1987) Diagnostic and statistical manual of mental disorders, 3rd edn, revised. American Psychiatric Press, Washington

DSM-IV (1994) Diagnostic and statistical manual of mental disorders, 4th edn. American Psychiatric Press, Washington

DSM-IV-R (2000) Diagnostic and statistical manual of mental disorders, 4th edn, revised. American Psychiatric Press, Washington

Eaton WW, Dryman A, Weissman MM (1991) Panic and phobia. In: Robins LN, Regier DA (eds) Psychiatric disorders in America. The Epidemiologic Catchment Area Study. The Free Press, New York, pp 155–179

Ernst C, Angst J (1997) The Zurich Study XXIV. Structural and emotional aspects of childhood and later psychopathology. Eur Arch Psychiatry Clin Neurosci 247:81–86

Ernst C, Angst J, Földényi M (1993) The Zurich Study. XVII Sexual abuse in childhood. Frequency and relevance for adult morbidity data of a longitudinal epidemiological study. Eur Arch Psychiatry Clin Neurosci 242:293–300

Faravelli C, Guerinni Degl'Innocenti B, Giardinelli L (1989) Epidemiology of anxiety disorders in Florence. Acta Psychiatr Scand 79:308–312

Feinstein A (1970) The pre-therapeutic classification of comorbidity in chronic disease. J Chronic Dis 23:455–468

Fergusson DM, Horwood J, Lynskey MT (1996) Childhood sexual abuse and psychiatric disorder in young adulthood: II. Psychiatric outcomes of childhood sexual abuse. J Am Acad Child Adolesc Psychiatry 34:1365–1374

Flament MF, Whitaker A, Rapaport JL, Davies M, Berg CZ, Kalikow K, Sceery W, Shaffer D (1988) Obsessive compulsive disorder in adolescence: An epidemiological study. J Am Acad Child Adolesc Psychiatry 27:764–771

Goodwin RD, Lieb R, Höfler M, Pfister H, Bittner A, Beesdo K, Wittchen HU (2004) Panic attack as a risk factor for severe psychopathology among young adults. Am J Psychiatry 161:2207–2214

Grabe HJ, Meyer C, Hapke U, Rumpf HJ, Freyberger HJ, Dilling H, John U (2001) Lifetime-comorbidity of obsessive-compulsive disorder and subclinical obsessive-compulsive disorder in northern Germany. Eur Arch Psychiatry Clin Neurosci 251:130–135

Hettema JM, Neale MC, Kendler KS (2001) A review and meta-analysis of the genetic epidemiology of anxiety disorders. Am J Psychiatry 158:1568–1578

Isensee B, Wittchen HU, Stein MB, Höfler M, Lieb R (2003) Smoking increases the risk of panic: findings from a prospective community study. Arch Gen Psychiatry 60:692–700

Jacobi F, Wittchen HU, Hölting M, Höfler M, Pfister H, Müller N, Lieb R (2004) Prevalence, co-morbidity and correlates of mental disorders in the general population: results from the German Health Interview and Examination Survey (GHS). Psychol Med 34:1–15

Johnson JG, Cohen P, Pine DS, Klein DF, Kasen S, Brook JS (2000) Association between cigarette smoking and anxiety disorders during adolescence and early adulthood. JAMA 284:2348–2351

Kagan J, Reznick JS, Clarke C, Snidman N, Garcia-Coll C (1984) Behavioral inhibition to the unfamiliar. Child Dev 55:2212–2225

Kendler KS, Davis CG, Kessler RC (1997) The familial aggregation of common psychiatric and substance use disorders in the National Comorbidity Survey: a family history study. Br J Psychiatry 170:541–548

Kessler RC (1997) The prevalence of psychiatric comorbidity. In: Wetzler S, Sanderson WC (eds) Treatment strategies for patients with psychiatric comorbidity. Wiley, New York, pp 23–48

Kessler RC (2001) Comorbidity of depression and anxiety disorders. In: Montgomery SA, den Boer JA (eds) SSRIs in depression and anxiety. Wiley, Chichester, pp 87–106

Kessler RC, Price RH (1993) Primary prevention of secondary disorders: a proposal and agenda. Am J Community Psychol 21:607–633

Kessler RC, McGonagle KA, Zhao S (1994) Lifetime and 12-month prevalence of DSM-III-R psychiatric disorders in the United States. Results from the National Comorbidity Survey. Arch Gen Psychiatry 51 (1):8–19

Kessler RC, Sonnega A, Bromet E, Hughes M, Nelson CB (1995) Posttraumatic stress disorder in the National Comorbidity Survey. Arch Gen Psychiatry 52:1048–1060

Kessler RC, Nelson CB, McGonagle KA, Edlund MJ, Frank RG, Leaf PJ (1996) The epidemiology of co-occurring addictive and mental disorders: implications for prevention and service utilization. Am J Orthopsychiatry 66:17–31

Kessler RC, Crum RM, Warner LA, Nelson CB, Schulenberg J, Anthony JC (1997a) Lifetime co-occurrence of DSM-III-R alcohol abuse and dependence with other psychiatric disorders in the National Comorbidity Survey. Arch Gen Psychiatry 54:313–321

Kessler RC, Davis CG, Kendler KS (1997b) Childhood adversity and adult psychiatric disorder in the US National Comorbidity Survey. Psychol Med 27:1001–1119

Kessler RC, Andrade LH, Bijl RV, Offord DR, Demler OV, Stein DJ (2002) The effects of co-morbidity on the onset and persistence of generalized anxiety disorder in the ICPE surveys. Psychol Med 32:1213–1225

Klein DF (1964) Delineation of two drug responsive anxiety syndromes. Psychopharmacologia 5:397–408

Kringlen E, Torgerson S, Cramer V (2001) A Norwegian psychiatric epidemiological study. Am J Psychiatry 158:1091–1098

Lee CK, Kwak YS, Yamamoto J, et al (1990) Psychiatric epidemiology in Korea. Part I: Gender and age differences in Seoul. J Nerv Ment Dis 178:242–246

Lewinsohn PM, Zinbarg R, Seeley JR, Lewinsohn M, Sack WH (1997) Lifetime comorbidity among anxiety disorders and between anxiety disorders and other mental disorders in adolescents. J Anxiety Disord 11:377–394

Lieb R, Isensee B, von Sydow K, Wittchen HU (2000a) The early stages of psychopathology study (EDSP): a methodological update. Eur Addict Res 6:170–182

Lieb R, Wittchen HU, Höfler M, Fuetsch M, Stein M, Merikangas KR (2000b) Parental psychopathology, parenting styles, and the risk of social phobia in offspring. A prospective-longitudinal community study. Arch Gen Psychiatry 57:859–866

Lieb R, Isensee B, Höfler M, Pfister H, Wittchen HU (2002) Parental major depression and the risk of depressive and other mental disorders in offspring: a prospective-longitudinal community study. Arch Gen Psychiatry 59:365–374

MacMilian HL, Fleming JE, Streiner DL, Lin E, Boyle MH, Jamieson E, Duku EK, Walsh CA, Wong MYY, Beardslee WR (2001) Childhood abuse and lifetime psychopathology in a community sample. Am J Psychiatry 158:1878–1883

Magee WJ (1999) Effects of negative life experiences on phobia onset. Soc Psychiatry Psychiatr Epidemiol 34:343–351

Magee WJ, Eaton WW, Wittchen HU, McGonagle KA, Kessler RC (1996) Agoraphobia, simple phobia and social phobia in the National Comorbidity Survey. Arch Gen Psychiatry 53:159–168

Marshall RD, Klein DF (2003) Conceptual antecedents of the anxiety disorders. In: Nutt D, Ballenger J (eds) Anxiety disorders. Blackwell, Oxford, pp 3–22

Maser JD, Cloninger CR (1990) Comorbidity of anxiety and mood disorders: introduction and overview. In: JD Maser, CR Clonninger (eds) Comorbidity of mood and anxiety disorders. American Psychiatric Press, Washington, pp 3–12

Merikangas KR, Mehta RL, Molnar BE, Walters EE, Swendsen JD, Aguilar-Gaziola S, Bijl R, Borges G, Caraveo-Anduaga JJ, Dewit DJ, Kolody B, Vega WA, Wittchen HU, Kessler RC (1998) Comorbidity of substance use disorders with mood and anxiety disorders: results of the international consortium in psychiatric epidemiology. Addict Behav 23:893–907

Merikangas KR, Stevens DE (1998) Models of transmission of substance use and comorbid psychiatric disorders. In: Kranzler HR, Rounsaville BJ (eds) Dual diagnosis and treatment. New York: Marcel Dekker, pp 31–53

Merikangas KR, Avenevoli S, Acharyya S, Zhang H, Angst J (2002) The spectrum of social phobia in the Zurich Cohort Study of young adults. Biol Psychiatry 51:81–91

Meyer C, Rumpf H-J, Hapke U, Dilling H, John U (2000) Lebenszeitprävalenz psychischer Störungen in der erwachsenen Allgemeinbevölkerung. Nervenarzt 71:535–542

Mezzich JE, Ahn CW, Fabrega H, Pilkonis PA (1990) Patterns of psychiatric comorbidity in a large population presenting for care. In: Maser JD, Cloninger CR (eds) Comorbidity of mood and anxiety disorders. American Psychiatric Press, Washington, pp 189–204

Molnar BE, Buka SL, Kessler RC (2001) Child sexual abuse and subsequent psychopathology: results from the National Comorbidity Survey. Am J Public Health 91:753–760

Offord DR, Boyle MH, Campbell D, Goering P, Lin E, Wong M, Racine YA (1996) One-year prevalence of psychiatric disorder in Ontarians 15 to 64 years of age. Can J Psychiatry 41:559–563

Regier DA, Narrow WE, Rae DS (1990) The epidemiology of anxiety disorders. The Epidemiologic Catchment Area (ECA) experience. J Psychiatr Res 24(Suppl 2):3–14

Regier DA, Farmer ME, Rae DS, Mayers JK, Kramer M, Robins LN, George LK, Karno M, Locke BZ (1993) One-month-prevalence of mental disorders in the United States and sociodemographic characteristics: The Epidemiologic Catchment Area Study. Acta Psychiatr Scand 88:35–47

Regier DA, Rae DS, Narrow WE, Kaelber CT, Schatzberg AF (1998) Prevalence of anxiety disorders and their comorbidity with mood and addictive disorders. Br J Psychiatry 34:24–28

Robins LN, Helzer JE, Croughan J, Ratcliff KS (1981) National Institute of Mental Health Diagnostic Interview Schedule: its history, characteristics, and validity. Arch Gen Psychiatry 38:381–389

Robins LN, Locke BZ, Regier DA (1991) An overview of psychiatric disorders in America. In: Robins LN, Regier DA (eds) Psychiatric disorders in America. The Epidemiologic Catchment Area Study. The Free Press, New York, pp 328–366

Rosenbaum JF, Biederman J, Hirshfeld DR, Bolduc EA, Faraone SV, Kagan J, Snidman N, Reznick JS (1991) Further evidence of an association between behavioral inhibition and anxiety disorders: results from a family study of children from a non-clinical sample. J Psychiatr Res 25:49–65

Sartorius N, Üstün TB, Lecrubier Y, Wittchen HU (1996) Depression comorbid with anxiety: results from the WHO study on psychological disorders in primary health care. Br J Psychiatry 168(Suppl 30):29–34

Sonntag H, Wittchen HU, Höfler M, Kessler RC, Stein MB (2000) Are social fears and DSM-IV social anxiety disorder associated with smoking and nicotine dependence in adolescents and young adults. Eur Psychiatry 15:67–74

Stein MB, Fuetsch M, Müller N, Höfler M, Lieb R, Wittchen HU (2001) Social anxiety and the risk of depression. A prospective community study of adolescents and young adults. Arch Gen Psychiatry 58:251–256

Turner RJ, Gil AG (2002) Psychiatric and substance use disorders in South Florida. Racial/ethnic and gender contrasts in a young adult cohort. Arch Gen Psychiatry 59:43–50

Vega WA, Kolody B, Aguilar-Gaxiola S, Alderete E, Catalano R, Caraveo-Anduaga J (1998) Lifetime prevalence of DSM-III-R psychiatric disorders among urban and rural Mexican Americans in California. Arch Gen Psychiatry 55:771–778

Wacker HR, Müllejans R, Klein KH, Battegay R (1992) Identification of cases of anxiety disorders and affective disorders in the community according to ICD-10 and DSM-III-R by using the Composite International Diagnostic Interview (CIDI). Int J Methods Psychiatr Res 2:91–100

Weissman MM, Leaf PJ, Blazer DG, Boyd JH, Florio L (1986) The relationship between panic disorder and agoraphobia: an epidemiologic perspective. Psychopharmacol Bull 22:787–791

Weissman MM, Bland RC, Canino GJ, Greenwald S, Hwu HG, Lee CK, Newman SC, Oakley-Browne MA, Rubio-Stipec M, Wickramaratne PJ, et al (1994) The cross national epidemiology of obsessive-compulsive disorder: The Cross National Collaborative Group. J Clin Psychiatry 55 (Suppl):5–10

Wells JE, Bushnell JA, Hornblow AR, Joyce PR, Oakley-Browne MA (1989) Christchurch psychiatric epidemiology study. I. Methodology and lifetime prevalence for specific psychiatric disorders. Aust N Z J Psychiatry 23:315–326

Wittchen H-U, Jacobi F (2001) Die Versorgungssituation psychischer Störungen in Deutschland. Eine klinisch-epidemiologische Abschätzung des Bundesgesundheitssurveys 1998. BGG 44:993–1000

Wittchen H-U, Perkonigg A (1996) Epidemiologie psychischer Störungen. Grundlagen, Häufigkeit, Risikofaktoren und Konsequenzen. In: Ehlers A, Hahlweg K (eds) Enzyklopädie der Psychologie, Themenbereich D Praxisgebiete. Band 1, Grundlagen der Klinischen Psychologie. Hogrefe, Göttingen, pp 69–144

Wittchen HU (1993) Epidemiologie und Komorbidität. In: Holsboer F, Phillipp M (eds) Angststörungen. Pathogenese—Diagnostik—Therapie. SM Verlagsgesellschaft, Gräfelfing, pp 40–58

Wittchen HU (1996) What is comorbidity. Fact or artefact? (editorial). Br J Psychiatry 168(Suppl 30):7–8

Wittchen HU, Pfister H (1997) DIA-X-Interviews: Manuals für Screening Verfahren und Interview; Interviewheft Längsschnittuntersuchung (DIA-X-Lifetime); Ergänzungsheft Querschnittsuntersuchung (DIA-X-12 Monate); PC-Programm zur Durchführung des Interviews (Längs- und Querschnittuntersuchung), Auswertungsprogramm. Swets and Zeitlinger, Frankfurt

Wittchen HU, Essau CA, vn Zerssen D, Krieg JC, Zaudig M (1992) Lifetime and six-month prevalence of mental disorders in the Munich Follow-Up Study. Eur Arch Psychiatry Clin Neurosci 241:247–258

Wittchen HU, Zhao S, Kessler RC, Eaton WW (1994) DSM-III-R generalized anxiety disorder in the National Comorbidity Survey. Arch Gen Psychiatry 51:355–364

Wittchen HU, Nelson CB, Lachner G (1998a) Prevalence of mental disorders and psychosocial impairments in adolescents and young adults. Psychol Med 28:109–126

Wittchen HU, Reed V, Kessler RC (1998b) The relationship of agoraphobia and panic in a community sample of adolescents and young adults. Arch Gen Psychiatry 55:1017–1024

Wittchen HU, Lieb R, Schuster P, Oldehinkel AJ (1999a) When is onset? Investigations into early developmental stages of anxiety and depressive disorders. In: JL Rapaport (ed) Childhood onset of "adult" psychopathology, clinical and research advances. American Psychiatric Press, Washington, pp 259–302

Wittchen HU, Kessler RC, Pfister H, Höfler M, Lieb R (2000a) Why do people with anxiety disorders become depressed? A prospective-longitudinal community study. Acta Psychiatr Scand 102(Suppl 406):14–23

Wittchen HU, Carter RM, Pfister H, Montgomery SA, Kessler RC (2000b) Disabilities and quality of life in pure and comorbid generalized anxiety disorder and major depression in a national survey. Int Clin Psychopharmacol 15:319–328

World Health Organization (1990) Composite international diagnostic interview (CIDI, Version 1.0). World Health Organization, Geneva

Zimmermann P, Wittchen HU, Höfler M, Lieb R (2003) Primary anxiety disorders and the development of subsequent alcohol use disorders: a four-year community study of adolescents and young adults. Psychol Med 33:1211–1222

HEP (2005) 169:433–447

Transcultural Issues

M. T. Lin (✉) · K.-M. Lin

1124 W. Carson St. B4-South, Torrance CA, 90502, USA
margarettlin@yahoo.com

Abstract Pharmacogenetics as a field of research is increasing the basis of knowledge on the use of psychotropics in different ethnic patient populations. This chapter summarizes current knowledge on the metabolism of anxiolytic agents with emphasis on pharmacogenetics and ethnic variations in drug responses.

Keywords Pharmacogenetics · Ethnicity · Drug responses

1 Introduction

A growing body of literature suggests the importance of culture and ethnicity in the psychopharmacologic management of psychiatric disorders including anxiety disorders. It is well known that ethnic differences exist in terms of response to various non-psychotropic medications (Kalow 1992). For psychotropic medications, however, data suggesting inter-ethnic variations have emerged more gradually. Early studies examining dosage and adverse events found that Asian patients experience adverse side effects at lower dosages of psychotropic medication than Caucasian patients (Lin 1983). Along with results from more recent studies, there is a move toward using pharmacogenetics to determine whether pharmacokinetic and pharmacodynamic differences between different ethnic groups can be predicted and whether these in turn can explain the differences in clinical response and adverse event profile. This chapter will briefly outline current knowledge about the pharmacologic treatment for anxiety as it relates to the pharmacogenetic studies examining the pharmacokinetic and pharmacodynamic variations in different ethnic groups.

2 Pharmacogenetics

2.1 Cytochrome P450

The focus of pharmacogenetic studies has largely been on those genes that encode enzymes responsible for the metabolism of medications. However, ethnic differences may also be affected by genes controlling the function and response of therapeutic targets. A well-established example of the difference that exists between different ethnic groups is the metabolism of alcohol. One of the enzymes responsible for the metabolism of alcohol is acetaldehyde dehydrogenase (ALDH), and 40%–50% of Asian subjects have a mutation that renders this enzyme inactive. The result is an uncomfortable "flushing" response, well known among many Asians even with a very small amount of alcohol (Agarwal and Goedde 1992; Yoshida 1993). It is now known that a mutation in a single nucleotide is responsible for the production of the inactive form of ALDH (Novoradovsky et al. 1995; Goedde et al. 1986).

Psychotropics, including medications for the management of anxiety disorders, are mainly metabolized by the cytochrome P450 enzymes. There are approximately 20 of these enzymes and they are often responsible for the rate-limiting step of drug metabolism. Of these, the three that are most commonly involved in psychotropic drug metabolism are CYP2D6, CYP2C19, and CYP3A4.

CYP2D6 is involved in the metabolism of more than half of the psychotropic drugs (Table 1). In terms of the pharmacologic treatment for anxiety, it is important to point out that many of the selective serotonin reuptake inhibitors (SSRIs) and the tricyclic antidepressants (TCAs) are metabolized largely by CYP2D6. This cytochrome enzyme has very complicated mutation patterns. The number of functional genes varies from 0 to 13 copies. The functional risk of mutations encoding this enzyme is important. For example, nortriptyline is metabolized very slowly in patients who lack a functional gene. However, in individuals who have more than two copies of functional genes (due to gene duplication and multiplication) the drug is rapidly metabolized (Dalen

Table 1 Psychotropic drugs metabolized by CYP2D6

Antidepressants	
	Amitriptyline, clomipramine, imipramine, desipramine, nortriptyline, trimipramine, *N*-desmethylclomipramine, fluvoxamine, norfluoxetine, paroxetine, venlafaxine, sertraline
Neuroleptics	
	Chlorpromazine, thioridazine, perphenazine, haloperidol, reduced haloperidol, risperidone, clozapine, sertindole
Others	
	Codeine, opiates, propranolol, dextromethorphan, etc.

et al. 1998). A similar effect is also seen with venlafaxine, its metabolism being slower in those individuals lacking a functional gene (Fukuda et al, 2000; Veefkind et al. 2000). CYP2D6 is also involved in many drug–drug interaction problems (DeVane 1994; Ereshefsky and Dugan 2000). For example, quinidine, a medication commonly used for the control of cardiac arrhythmia, is an inhibitor of CYP2D6. If the enzyme is inhibited, the individual becomes a slow metabolizer regardless of how many copies of the functional gene are present.

Ethnic differences that exist in terms of the activity of CYP2D6 are largely determined by the genetic polymorphisms that exist across different ethnic groups. Some groups rapidly metabolize psychotropic medication and so need a higher dose compared with poor metabolizer groups who are very sensitive to even low doses of medication. A genetic polymorphism, known as *2XN appears to increase the activity of this enzyme and exists in 19%–29% of Arabs and Ethiopians and 1%–5% in others (Aklillu et al. 1996; Masimirembwa and Hasler 1997). On the other hand, *17 appears to result in a reduced CYP2D6 activity and is found in 25%–40% of sub-Saharan African blacks and 0% in others. Similarly, *10 also reduces CYP2D6 activity and is found in 47%–70% of Asians and 5% in others. Another specific mutation, *4, leads to a complete loss of enzyme activity, and is responsible for the bimodal distribution of CYP2D6 phenotype, with 5%–9% of Caucasians classified as poor metabolizers. This allele is rare in other populations. Thus, the pharmacogenetics of CYP2D6 serves as a dramatic example demonstrating the importance of genotypic profiles in determining the phenotypic expression of the enzyme as well as the pharmacokinetics of medications metabolized by this enzyme. The advances made in molecular analysis may explain in part what others have observed clinically for many years; that is, Asians and African Americans often are more prone to developing side effects from psychotropics than other ethnic groups.

CYP2C19 is another of the cytochrome P450 enzymes that are involved in the metabolism of a number of psychotropic drugs. Major ethnic differences exist in terms of the activity of this enzyme. Across various groups the percentage of poor metabolizers ranges from 3% to 20%. The reduction in activity of this enzyme is caused by two specific mutations, one of which, *3, is specific to Asian individuals and is not found outside of Asian populations. The percentage of poor metabolizers has been reported to be approximately 15%–20% in Asians, 5% in Hispanics, and 3% in Caucasians (de Morais et al. 1994; Goldstein et al. 1997). The rate in African American populations is unclear, with incidences of between 4% and 19% being reported (Masimirembwa and Hasler 1997). As with CYP2D6, drug–drug interaction is also an important consideration in the prescription of medications metabolized by CYP2C19. Fluvoxamine is a potent inhibitor of 2C19 along with other SSRIs, such as paroxetine and fluoxetine. Medications for the management of anxiety that are metabolized by CYP2C19 include diazepam, imipramine, amitriptyline, clomipramine and citalopram. However, only a few drugs are selectively dependent on 2C19 for metabolism, such as the barbiturates, hexobarbital, and mephobarbital, which are substrates of 2C19.

CYP3A4 is involved in the metabolism of 80%–90% of all currently available drugs (Table 2). Its presence accounts for the majority of the cytochrome enzymes in the liver. It has attracted attention from both physicians and pharmaceutical companies, as there have been incidences of fatal drug interactions, which have resulted in the withdrawal of these drugs from the market (terfenadine in 1997, astemizole in 1999, and cisapride in 2000). Combinations of potent CYP3A4 inhibitors and substrates can drive drug levels to the toxic range. The

Table 2 Drugs metabolized by CYP3A4

Typical antipsychotics
Thioridazine, haloperidol
Atypical antipsychotics
Clozapine, quetiapine, risperidone, sertindole, ziprasidone
Antidepressants
Nefazodone, sertraline, mirtazapine, tricyclic antidepressants
Mood stabilizers
Carbamazepine, gabapentin, lamotrigine
Benzodiazepines
Alprazolam, clonazepam, diazepam, midazolam, triazolam, zolpidem
Calcium channel blockers
Diltiazem, nifedipine, nimodipine, verapamil
Others
Androgens, estrogens, erythromycins, terfenadine, cyclosporine, dapsone, ketoconazole, lovastatin, lidocaine, alfentanil, amiodarone, astemizole, codeine, sildenafil

use of these drug combinations needs to be carefully monitored or avoided to prevent untoward consequences. Important inhibitors of CYP3A4 include: SSRIs, nefazodone, azole antifungals, macrolide antibiotics, and antiretrovirals. Interestingly, CYP3A4 is also potently inhibited by grapefruit juice and is significantly induced by St. John's Wort. Studies suggest ethnic differences in the activity of CYP3A4. An example is a study examining the metabolism of the calcium channel blocker nifedipine (Rashid et al. 1995; Sowunmi et al. 1995). Asian Indians were found to metabolize nifedipine at a slower rate than British Caucasians, as determined by AUC values. Another study reported similar differences between Caucasian and Asian volunteers in the rate of metabolism of alprazolam (Lin et al. 1988). In this study, Asian volunteers had higher plasma levels of alprazolam than Caucasian subjects following intravenous and oral administration of the same dose. Recent studies continue to find several single nucleotide polymorphisms with racial variability in their frequency. However, functional significance of these polymorphisms has not been clearly established (Dai et al. 2001).

These examples demonstrate that both genetic and environmental factors are involved in determining the activity of this and other cytochrome enzymes. Distinct patterns of genetic polymorphisms exist across ethnic groups and these can be tested and investigated alongside possible environmental and dietary factors that may cause differential expression of these genes.

2.2 Genes Encoding Therapeutic Targets

Polymorphisms in genes controlling the function of neurotransmitter systems (e.g., transporters, receptors, etc.) are thought to be related to the pathogenesis of many psychiatric disorders including anxiety disorders as well as temper-

Table 3 Genes possibly associated with increased susceptibility for psychiatric disorders

Genes encoding for the biosynthesis and catabolism of neurotransmitters
Tryptophan hydroxylase (*TPH*)
Tyrosine hydroxylase (*TH*)
Catechol-O-methyltransferase (*COMT*)
Monoamine oxidase (*MAO*)
Receptor genes
5-HT2A, 5-HT1A, DRD2
Transporter genes
Serotonin transporter (*5-HTT*)
Norepinephrine transporter (*NET*)
Dopamine transporter (*DAT*)

ament, personality disorders or personality traits (Table 3). Variations across ethnic groups also have been observed in terms of the rate of genetic polymorphisms for these genes. For example, the rate of the serotonin transporter gene (*SLC6A4*) polymorphism long allele ranges from approximately 20% in eastern Asians (e.g., Japanese and Chinese) to approximately 70% in African Americans (Gelernter et al. 1997) (Fig. 1). To the extent that this allele may be regarded as a risk factor for depression, suicide, or other psychiatric conditions (Greenberg et al. 1998; Jonsson et al. 1998), such findings pose intriguing questions regarding whether genetic polymorphisms might lead to differential vulnerabilities in psychopathology across ethnic groups.

The dopamine D2 receptor gene (*DRD2*) A1 polymorphism remains controversial in terms of whether or not it is functional (Baron 1993). Again there is a large variation in its prevalence across ethnic groups, ranging from 9% in Yemenite Jews to 79% in some of the American Indians (Barr and Kidd 1993; Wu et al. 2000).

Catechol-*O*-methyltransferase (COMT) catalyses the *O*-methylation of neurotransmitters, catechol hormones, and drugs such as levodopa and methyldopa. COMT activity is caused by a single mutation. This means that it is possible to have homozygous low-activity allele subjects, i.e., those having the lowest enzyme activity, homozygous high-activity subjects, and also those who are heterozygous and have intermediate activity. Ethnic differences in COMT activity have been observed in several populations with major differences occurring between Asian and Caucasian populations in terms of the percentage incidence of low-activity COMT (18% and 50%, respectively) (Palmatier et al. 1999; McLeod et al. 1998). This is a functional polymorphism that may be clinically important in terms of the risk of psychopathology (either schizophrenia or mood disorders) and the treatment of many neuropsychiatric disorders (Davidson et al. 1979; Henderson et al. 2000; Murphy et al. 1999; Horowitz et al. 2000; Kotler et al. 1999).

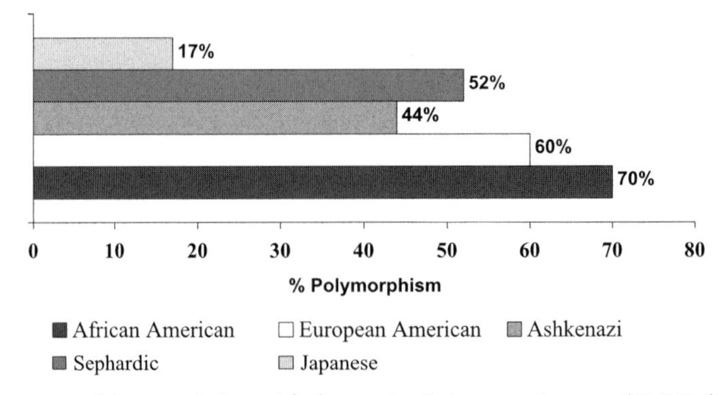

Fig. 1 Percentage of the population with the serotonin transporter gene (*SLC6A4*) polymorphism among different ethnic groups. (Data from Gelernter et al. 1997)

A recent Italian study of the *5-HTT* promoter region polymorphism and response to fluvoxamine found that individuals with the short allele were less likely to respond to treatment than those with the long allele (Smeraldi et al. 1998). Such a relationship was recently confirmed by Pollock et al. (2000); four other studies conducted in the U.S. and in Europe, as well as one report from Taiwan, involving the use of fluoxetine and paroxetine. However, the exact opposite was found in two studies conducted in Korea and Japan, where the short allele was predictive of significantly better response (Kim et al. 2000). In one of these latter studies, the promoter region polymorphism was in linkage disequilibrium with another polymorphism in intron 2, which might in part explain the discrepancy. Nevertheless, these observations do highlight the importance of ethnicity in studying genetic and other biological variables in relation to psychotropic responses, as well as the vulnerability of psychiatric conditions. Since ethnic differences are known to exist in the rate of the long allele, this may explain some of the ethnic differences in treatment response. However, further studies are needed to investigate whether an individual's likelihood of responding can be predicted.

3 Anxiolytic Agents and Ethnic Variations

3.1 Benzodiazepines

Benzodiazepines comprise the most frequently prescribed subclass of antianxiety agents. These agents, first introduced in the early 1960s, quickly replaced the use of barbiturates as the pharmacological approach to anxiety. The popularity of these agents can be attributed to their generally quick onset of action and wider safety margin in overdose compared to the barbiturates. However, the potential of these agents to elicit physical dependence also quickly became apparent. In addition to the frequent use of these agents as anxiolytics, benzodiazepines are also commonly used for muscle tension, insomnia, status epilepticus (diazepam), myoclonic epilepsy (clonazepam), preoperative anesthesia, and alcohol withdrawal. Importantly, Ativan (lorazepam) is often used in the emergency room and inpatient setting to manage acute agitation in patients. Controlled studies involving Asians and Caucasians demonstrated significant pharmacokinetic differences involving use of the benzodiazepines (Ghoneim et al. 1981; Kumana et al. 1987). The volume of distribution of diazepam in these studies was found to be lower, and both serum diazepam and desmethyldiazepam levels were higher in Asian than in white physically and psychiatrically healthy volunteers. These differences became statistically insignificant, however, after controlling for ethnic differences in skinfold thickness and the ratio of actual to ideal body weight, suggesting that ethnic differences may be secondary to differences in the percentage of body fat.

Lin et al. (1988) studied plasma alprazolam concentrations in 14 American-born Asian, 14 foreign-born Asian, and 14 Caucasian healthy male volunteers. Both Asian groups had greater AUCs and peak plasma concentrations and lower total plasma clearance than did the Caucasian group, after both oral and intravenous administration of alprazolam. Pharmacodynamically, the only significant difference was that foreign-born Asians experienced more sedation compared with both Caucasian and American-born Asian subjects. In a more recent study, Ajir et al. (1997) also reported that Asians had higher maximum serum concentrations, larger AUCs, and lower clearance of both adinazolam and its major active metabolite than did their Caucasian and African American counterparts. Together, the findings support the concept that Asian patients require smaller doses of adinazolam than do Caucasian patients to achieve similar levels of the parent drug and its metabolite.

Similar to Asians, African Americans have been found to have slower clearance of the benzodiazepines. Furthermore, many studies have reported that African Americans have greater cognitive effects and more anxiety reduction from the benzodiazepines when compared to Caucasians on the same dose of medications. In a study by Ajir et al. (1997), African Americans were found to have increased clearance of adinazolam. However, the AUC of its metabolite *N*-desmethyladinazolam was found to be higher in African Americans and may be responsible for the larger drug effects on African Americans in terms of adverse effects such as slower psychomotor performance, despite the higher metabolic capacity for adinazolam in this ethnic group.

In terms of the ethnic response to benzodiazepines in Hispanics, studies have implicated the important role of dietary effect. When Mexican subjects were fed a corn-rich diet, the metabolism of nifedipine by 3A4 appears to be reduced (Palma-Aguirre et al. 1994). Many studies suggest that the flavonoid quercetin is effective at inhibiting the intestinal agglomeration of CYP3A4. Corn is rich in quercetin and is a dietary staple for Hispanics. Importantly, many benzodiazepines such as alprazolam, midazolam, and triazolam are all metabolized by CYP3A4, and their use in Hispanics should be downwardly adjusted with caution to take into account the dietary effect on CYP3A4 activity. Similar considerations may apply for citrus-loving Hispanics, since flavonoid naringin found in grapefruit juice is a powerful inhibitor of CYP3A4.

3.2 Tricyclic Antidepressants

Imipramine, a TCA, was the first pharmacological agent noted to treat panic disorder (Klein 1964). Other TCAs, notably clomipramine, have also been found to have significant anxiolytic properties (den Boer et al. 1990; Modigh 1992). Studies of ethnic differences in the pharmacokinetics of the TCAs in

Asian Americans have led to inconclusive results. Of six studies comparing Asians with Caucasians, three revealed that Asians metabolize TCAs significantly slower than their Caucasian counterparts; however, the differences observed in the other three studies did not reach statistical significance, particularly after controlling for body weight (Kishimoto 1984; Rudorfer 1984; Schneider 1991). Other studies also did not find significant differences in the pharmacokinetics of nortriptyline between Mexican Americans and Caucasians. Pharmacodynamically, results from two clinical studies in Asia indicated that severely depressed hospitalized Asian patients responded clinically to lower combined concentrations of imipramine and desipramine (130 ng/ml) than studies previously reported on North American and European patients (180–200 ng/ml). These results suggest that differential brain receptor responsivity might also play a role in determining ethnic differences in tricyclic dosage requirement (Hu 1983; Yang 1985).

Systematic investigations examining the psychopharmacologic responses to TCAs in African Americans suggest significant differences in the pharmacokinetics between African Americans and Euro-Americans. The differential effects of chlorpromazine hydrochloride and imipramine hydrochloride were studied in 159 African American and 555 Euro-American inpatients. Standard psychometric scales were used to assess the symptoms and the responses to the medications administered. The results suggested that African American men showed more rapid improvement and that they were therapeutically more responsive to imipramine (Raskin et al. 1975). The high rate of CYP2D6*17 in African Americans might be responsible for the higher plasma levels of TCAs and the more rapid improvement and the overall greater clinical efficacy. In a study of outpatients from Tanzania, the patients responded to clomipramine hydrochloride in much lower doses than those recommended in Western textbooks. At the low-to-moderate dose of 125 mg of clomipramine hydrochloride, drowsiness and tremulousness were notable (Kilonzo 1994). Overall, most studies indicate that African Americans treated with TCAs have higher plasma levels per dose, more adverse effects with equivalent plasma levels, and earlier onset of action (Strickland et al. 1997). Obtaining plasma TCA levels may be a helpful adjunct to clinical interview in identifying poor metabolizers and in determining when an adequate dose is being used, as well as in ascertaining whether there has been an adequate trial of the medication. Monitoring the electrocardiograms in African American patients receiving TCAs may be of particular importance because of their tendency to have higher plasma levels and therefore, they may be more sensitive to the adverse cardiotoxic effects.

3.3 Selective Serotonin Reuptake Inhibitors

Several SSRIs have been FDA approved for one or more specific anxiety diagnoses [e.g., paroxetine for social anxiety, generalized anxiety disorder (GAD) and posttraumatic stress disorder (PTSD); sertraline for obsessive–compulsive disorder (OCD) and PTSD]. In clinical practice, the SSRIs have become the "first-line" treatment for panic disorder because of their overall safety and tolerability, their safety in overdose, and their low potential for addiction and withdrawal.

A review on the metabolism of SSRIs by Preskorn revealed a substantial inhibitory effect on the P450 isoenzyme system CYP2D6, with Prozac also inhibiting CYP3A4 (Preskorn 1994). Furthermore, as discussed previously, functionally significant genetic polymorphisms exist in most of the CYPs, especially CYP2D6, that metabolize most of the SSRIs including fluoxetine, paroxetine, and sertraline. CYP2D6*4 (CYP2D6B) leads to the production of defective proteins and is found in 25% of Caucasians but is rarely identified in other ethnic groups. This mutation is mainly responsible for the high percentage of poor metabolizers among Caucasians (5%–9%) who are sensitive to drugs metabolized by CYP2D6 (Ruiz 2000). A high frequency of CYP2D6*17 in African Americans and of CYP2D6*10 in Asian Americans was found to be associated with lower enzyme activities and slower metabolism of CYP2D6 substrates and may be responsible in part for the slower pharmacokinetic profiles and lower dose ranges observed in Asians with regard to the SSRIs (Lin and Poland 1995). CYP2D6 also is unique in that the gene is often duplicated, resulting in faster enzyme activity for individuals possessing these duplicated genes. It is estimated that 29% of Ethiopians may have this genotype and are superextensive metabolizers of substrates of CYP2D6. All in all, information on the ethnic variations in the metabolism of the SSRIs should be considered when clinicians use SSRIs to treat anxiety disorders in patients of different ethnic origin to account for the potential differential rates of metabolism and adjust the dose of medications to minimize side effects or adverse drug–drug interactions.

3.4
Monoamine Oxidase Inhibitors

Monoamine oxidase inhibitors (MAOIs) have been shown to be effective in the treatment of anxiety disorders such as social anxiety and panic disorder. In a 12-week, placebo-controlled trial of patients with panic disorder, Sheehan and colleagues reported that phenelzine (45 mg/day) was better than placebo; however, higher doses of the MAOI (e.g., 60–90 mg/day) may be more effective (Sheehan 1980; Buiges1987). Because of the potential risk of a hypertensive crisis with a tyramine-containing diet, the MAOIs have grown less in favor with clinicians, especially since the advent of the SSRIs. MAOIs, however, remain clinically effective agents in the treatment of atypical depression and anxiety disorders, and therefore their metabolism by patients of different ethnic origin remains an important topic of research. MAOIs are predominantly metabolized by CYP2C19 (Bezchlibnzyk-Buler and Jeffries 1999). Its pharmacokinetic properties are significant for auto-inhibition or metabolite-induced inhibition. Thus, clearance at higher dosages is decreased and caution should be exercised when prescribing these agents at higher dosages. Pharmacogenetic findings suggest a higher percentage of Asian Americans and African Americans are poor metabolizers of CYP2C19 substrates. Pharmacokinetics and pharmacodynamics studies with MAOIs are scarce with respect to potential variations in ethnic groups and may offer further information on the use of these agents in safer and clinically more effective fashions.

3.5 β-Adrenergic Blocking Agents

Investigations into the use of noradrenergic agents as anxiolytics were first directed toward their use in anxious musical performers. β-Blockers such as propranolol were found to be useful in alleviating symptoms of anxiety (e.g., palpitations, sweating). Years later, clonidine was shown by Gold et al. (1978) to be effective in blocking physiological symptoms associated with opioid withdrawal. Although not found to be effective in blocking panic, agents such as propranolol, atenolol, and nadolol have been found to be useful when used adjunctively with other agents in reducing symptoms of autonomic arousal associated with panic and social anxiety (Rosenbaum et al. 1998). Importantly, propranolol is metabolized primarily by CYP2D6 and should probably be used in lower dosages in Asians who are slower metabolizers of CYP2D6 substrates.

In addition to pharmacokinetics, differences in receptor sensitivity between Asians and Caucasians also have been demonstrated, rendering the former even more susceptible to the side effects of propranolol (Zhou et al. 1992).

3.6 Buspirone

Buspirone, marketed as Buspar, a nonbenzodiazepine and generally nonsedating anxiolytic, was the first prominent anxiolytic introduced after the benzodiazepines. Its antianxiety effects are believed to be secondary to its acting as a partial agonist of the 5-HT_{1A} receptor. Buspirone is as effective as diazepam and superior to placebo in double-blind trials involving anxious outpatients (Schatzberg et al. 2003). The parent drug is metabolized by CYP3A4 while its metabolite is metabolized by CYP2D6. Important drug–drug interactions with buspirone include agents that may inhibit CYP3A4 and 2D6. Agents such as the antifungal Itraconazole, calcium channel blockers, verapamil, and diltiazem can all potentially increase plasma buspirone level several fold by inhibiting CYP3A4 activity and may result in increased incidents of undesirable side effects with buspirone. Reports of serotonin syndrome, euphoria, seizures, or dystonia have also been found in cases where buspirone was combined with fluoxetine, fluvoxamine, and MAOIs (Bezchlibnyk-Buler and Jeffries 1999).

Furthermore, as already discussed, important ethnic variations exist in the metabolic activity of both CYP3A4 and 2D6 and caution is needed when administering buspirone to Asians who are more likely to be slow metabolizers of CYP2D6 substrates and may also be on other medications that further inhibit CYP2D6 activity.

4 Conclusion

This chapter provides a brief discussion on pharmacogenetics as it relates to the ethnic variations in drug responses in the treatment of anxiety disorders. Anxiety disorders are common and can be debilitating. Fortunately, many medications are available to help manage the symptoms of anxiety. The field of pharmacogenetics and the study of ethnic variations in drug response are important. Research in these areas is helping to build a foundation of knowledge. However, it still behooves the clinicians prescribing these medications to actually utilize what is known about the pharmacokinetics and pharmacodynamics of these agents in order to maximize treatment response while minimizing side effects.

References

Agarwal DP, Goedde HW (1992) Pharmacogenetics of alcohol metabolism and alcoholism. Pharmacogenetics 2:48–62

Ajir K, Smith M, Lin KM, Poland RE, Fleishaker JC, Chambers JH, Anderson D, Nuccio C, Zheng YP (1997) The pharmacokinetics and pharmacodynamics of adinazolam: multiethnic comparisons. Psychopharmacology (Berl) 129:265–270

Aklillu E, Persson I, Bertilsson L, Johansson I, Rodrigues F, Ingelman-Sundberg M (1996) Frequent distribution of ultrarapid metabolizers of debrisoquine in an Ethiopian population carrying duplicated and multiduplicated functional CYP2D6 alleles. J Pharmacol Exp Ther 278:441–446

Baron M (1993) The D2 dopamine receptor gene and alcoholism: a tempest in a wine cup? J Soc Biol Struct 34:204–209

Barr CL, Kidd KK (1993) Population frequencies of the A1 allele at the dopamine D2 receptor locus. Biol Psychiatry 34:204–209

Bezchlibnzyk-Buler KZ, Jeffries JJ (1999) Clinical handbook of psychotropic drugs. Hogrefe and Huber Publishers, Seattle, pp 103–110

Buiges J, Vallego J (1987) Therapeutic response to phenelzine in patients with panic disorder and agoraphobia with panic attacks. J Clin Psychiatry 48:55–59

Dai D, Tang J, Rose R, Hodgson E, Bienstock RJ, Mohrenweiser HW, Goldstein JA (2001) Identification of variants of CYP3A4 and characterization of their abilities to metabolize testosterone and chlorpyrifos. J Pharmacol Exp Ther 299:825–831

Davidson JR, McLeod MN, Turnbull CD, et al (1979) Catechol-O-methyltransferase activity and classification of depression. Biol Psychiatry 46:557–567

de Morais SM, Wilkinson GR, Blaisdell J, et al (1994) The major genetic defect responsible for the polymorphism of S-mephenytoin metabolism in humans. J Biol Chem 269:15419–15422
den Boer JA, Westenberg HGM, Karmerbeek WDJ, Verhoeven WM, Kahn RS (1990) Serotonin function in panic disorder: a double-blind placebo controlled study with fluvoxamine and ritanserin. Psychopharmacology (Berl) 102:85–94
DeVane CL (1994) Pharmacogenetics and drug metabolism of newer antidepressant agents. J Clin Psychiatry 55(Suppl):39–45
Ereshefsky L, Dugan D (2000) Review of the pharmacokinetics, pharmacogenetics, and drug interactions potential of antidepressants: focus on venlafaxine. Depress Anxiety 12(Suppl 1):30–44
Fukuda T, Nishida Y, Imaoka S, et al (2000) The decreased in vivo clearance of CYP2D6 substrates by CYP 2D6*10 might be caused not only by the low-expression but also by low affinity of CYP2D6. Arch Biochem Biophys 380:303–308
Gelernter J, Kranzler H, Cubells JF, et al (1997) Serotonin transporter protein (SLC6A4) allele and haplotype frequencies and linkage disequilibria in African-and European-American and Japanese populations in alcohol-dependent subjects. Hum Genet 101:243–246
Ghoneim MM, Korttila K, Chiang CK, Jacobs L, Schoenwald RD, Newaldt SP, Lauaba KO (1981) Diazepam effects and kinetics in Caucasians and Orientals. Clin Pharmacol Ther 29:749–756
Goedde HW, Agarwal DP, Harada S, et al (1986) Aldehyde dehydrogenase polymorphism in North American, South American, and Mexican Indian populations. Am J Hum Genet 38:395–399
Gold MS, Redmond DE Jr, Kleber HD (1978) Clonidine in opiate withdrawal. Lancet 1:929–930
Goldstein JA, Ishizaki T, Chiba K, et al (1997) Frequencies of the defective CYP2C19 alleles responsible for the mephenytoin poor metabolizer phenotype in various Oriental, Caucasian, Saudi Arabian and American black populations. Pharmacogenetics 7:59–64
Greenberg BD, McMahon FJ, Murphy DL (1998) Serotonin transporter candidate gene studies in affective disorders and personality: promises and potential pitfalls [guest editorial]. Mol Psychiatry 3:186–189
Henderson AS, Korten AE, Jorm Af, et al (2000) COMT and DRD3 polymorphisms, environmental exposures, and personality traits related to common mental disorders. Am J Med Genet 96:102–107
Horowitz R, Kotler M, Shufman E, et al (2000) Confirmation of an excess of the high enzyme activity COMT val allele in heroin addicts in a family-based haplotype relative risk study. Am J Med Genet 96:599–603
Hu WH, Lee CF, Yang YY, Tseng YT (1983) Imipramine plasma levels and clinical response. Bull Chinese Soc Neurol Psychiatry 9:40–49
Jonsson EG, Goldman D, Spurlock G, et al (1997) Tryptophan hydroxylase and catechol-O-methyltransferase gene polymorphisms: relationships to monoamine metabolite concentrations in CSF of healthy volunteers. Eur Arch Psychiatry Clin Neurosci 247:297–302
Kalow W (1992) Pharmacogenetics of drug metabolism. Pergamon Press, New York
Kilonzo GP, Kaaya SF, Rweikiza JK, Kassam M, Moshi G (1994) Determination of appropriate clomipramine dosage among depressed African outpatients in Dares Salaam, Tanzania. Cent Afr J Med 40:178–182
Kim DL, Lim SW, Lee S, et al (2000) Serotonin transporter gene polymorphism and antidepressant response. Neuroreport 11:215–219Kalow W
Kishimoto A, Hollister LE (1984) Nortriptyline kinetics in Japanese and Americans. J Clin Psychopharmacol 4:171–172

Klein DF (1964) Delineation of two-drug responsive anxiety syndromes. Psychopharmacologia 5:397–408

Kotler M, Barak P, Cohen H, et al (1999) Homicidal behavior in schizophrenia associated with a genetic polymorphism determining low catechol O-methyltransferase (COMT) activity. Am J Med Genet 88:628–633

Kumana CR, Lauder IJ, Chan M, Ko W, Lin HF (1987) Differences in diazepam pharmacokinetics in Chinese and white Caucasians—relation to body lipid stores. Eur J Clin Pharmacol 32:211–215

Lin KM, Finder E (1983) Neuroleptic dosage in Asians. Am J Psychiatry 140:490–491

Lin KM, Smith MW (2000) Psychopharmacotherapy in the context of culture and ethnicity. In: Ruiz P (ed) Review of psychiatry: ethnicity and psychopharmacology. American Psychiatric Press, Washington, pp 1–36

Lin KM, Lau JK, Smith R, Phillips P, Antal E, Poland RE (1988) Comparison of alprazolam plasma levels and behavioral effects in normal Asian and Caucasian male volunteers. Psychopharmacology (Berl) 96:365–369

Lin KM, Poland RE, Anderson D (1995) Psychopharmacology, ethnicity, and culture. Transcult Psychiatr Res Rev 32:1–40

Masimirembwa CM, Hasler JA (1997) Genetic polymorphism of drug metabolizing enzymes in African populations: implications for the use of neuroleptics and antidepressants. Brain Res Bull 44:561–571

McLeod HL, Syvanen AC, Githang'a J, et al (1998) Ethnic differences in catechol-O-methyltransferase pharmacogenetics: frequency of the codon 108/158 low activity allele is lower in Kenyan than Caucasian or Southwest Asian individuals. Pharmacogenetics 8:195–199

Modigh K, Westberg P, Eriksson E (1992) Superiority of clomipramine over imipramine in the treatment of panic disorder: a placebo-controlled trial. J Clin Psychopharmacol 12:251–261

Murphy DC, Jones LA, Owen MJ (1999) High rates of schizophrenia in adults with velo-cardio-facial syndrome. Arch Gen Psychiatry 56:940–945

Novoradovsky AG, Kidd J, Kidd K, Goldman D (1995) Apparent monomorphism of ALDH2 in seven American Indian populations. Alcohol 12:163–167

Palmatier MA, Kang AM, Kidd KK (1999) Global variation in the frequencies of functionally different catechol-O-methyltransferase alleles. Biol Psychiatry 46:557–567

Pollock BG, Ferrell RE, Mulsant BH, et al (2000) Allelic variation in the serotonin transporter promoter affects onset of paroxetine treatment response in late-life depression. Neuropsychopharmacology 23:587–590

Preskorn S (1994) Targeted pharmacotherapy in depression management: comparative pharmacokinetics of fluoxetine, paroxetine and sertraline. Int Clin Psychopharmacol 9(Suppl 3):13–19

Rashid TJ, Martin U, Clarke H, et al (1995) Factors affecting the absolute bioavailability of nifedipine. Br J Clin Pharmacol 40:51–58

Raskin A, Thomas H, Crook MA (1975) Antidepressants in black and white inpatients. Arch Gen Psychiatry 32:643–649

Rosenbaum JF, Pollack RA, Jordan SK, Pollack MH (1998) The pharmacotherapy of panic disorder. Bull Menninger Clin 60(2 Suppl A):A54–A75

Rudorfer MV, Lan EA, Chang WH, et al (1984) Desipramine pharmacokinetics in Chinese and Caucasian volunteers. Br J Clin Pharmacol 17:433–440

Schneider L, Pawluczyk S, Dopheide J, Lyness SA, Suckow RF, Copper TB (1991) Ethnic differences in protriptyline metabolism. New Research Program and Abstracts, American Psychiatric Association 144th Annual Meeting, New Orleans. American Psychiatric Press

Sheehan DV, Ballenger JC, Jacobsen G (1980) Treatment of endogenous anxiety with phobic, hysterical, and hypochondriacal symptoms. Arch Gen Psychiatry 37:511–519

Smeraldi E, Zanardi R, Benedetti F, et al (1998) Polymorphism within the promoter of the serotonin transporter gene and antidepressant efficacy of fluvoxamine. Mol Psychiatry 3:508–511

Sowunmi A, Rashid TJ, Akinyinka OO, et al (1995) Ethnic differences in nifedipine kinetics: comparisons between Nigerians, Caucasians and South Asians. Br J Clin Pharmacol 40:489–493

Strickland TL, Stein R, Lin KM, Risby Emile, Fong R (1997) The pharmacologic treatment of anxiety and depression in African Americans. Arch Fam Med 6:371–374

Veefkind AH, Haffmans PM, Hoencamp E (2000) Venlafaxine serum levels and CYP2D6 genotype. Ther Drug Monit 22:202–208

Wu X, Hudmon KS, Detry MA, et al (2000) D2 dopamine receptor gene polymorphisms among African-Americans and Mexican-Americans: a lung cancer case-control study. Cancer Epidemiol Biomarkers Prev 9:1021–1026

Yang YY (1985) Prophylactic efficacy of lithium and its effective plasma levels in Chinese bipolar patients. Acta Psychiatr Scand 71:171–175

Yoshida A (1993) Genetic polymorphisms of alcohol-metabolizing enzymes related to alcohol sensitivity and alcoholic diseases. In: Lin KM, Poland RE, Nakasaki G (eds) Psychopharmacology and psychobiology of ethnicity. American Psychiatric Press, Washington, pp 169–186

Zhou HH, Koshakji RP, Siolberstein D, Wilkinson G, Wood A (1989) Altered sensitivity to and clearance of propranolol in men of Chinese descent as compared with American white. N Engl J Med 320:565–570

HEP (2005) 169:449–468

Challenge Studies in Anxiety Disorders

M. E. Keck (✉) · A. Ströhle

Max Planck Institute of Psychiatry, Kraepelinstrasse 2-10, 80804 Munich, Germany
keck@mpipsykl.mpg.de

Abstract In psychiatry, the use of pharmacological challenges in panic disorder is unique in that the clinical phenomenon of central interest (i.e., the panic attack) can be provoked readily and assessed in the clinical laboratory setting. During the past 20 years pharmacological challenge studies have increased our knowledge concerning the neurobiology of panic disorder remarkably and may ultimately result in novel and more causal treatment strategies. Moreover, the differences in sensitivity to certain panicogens such as serotonergic agents, lactate, carbon dioxide and cholecystokinin tetrapeptide are likely to be fruitful in serving as biological markers of subtypes of panic disorders and should be a major focus of research, as the identification of reliable endophenotypes is currently one of the major rate-limiting steps in psychiatric genetic studies.

Keywords Anxiety · Panic disorder · Challenge · Lactate · CCK · CCK-4 · Yohimbine · Carbon dioxide · HPA

1 Introduction

In medicine, diagnostic symptom provocation has a long-lasting tradition; in diagnosing gestational diabetes, for example, physicians primarily depend upon the results of the oral glucose tolerance test. Similarly, in cardiovascular medicine, exercise ECG testing is used for the diagnosis of ischemic heart disease. One of the main stimuli of biological research in the etiology of panic disorder has been a series of observations that a number of agents can provoke panic attacks in predisposed subjects (Table 1).

The pharmacological challenge strategy involves administering a test agent under carefully controlled conditions to elucidate some aspect of biological or behavioral function in the organism studied. In psychiatry, the use of pharmacological challenges in panic disorder is unique in that the clinical phenomenon of central interest (i.e., the panic attack) can be provoked readily and assessed in the clinical laboratory setting. Uses of this approach allow for (1) the generation and testing of hypotheses regarding the underlying neurobiology of the disorder, (2) the identification of pathophysiologically distinct diagnostic subtypes, (3) the delineation of the effects and mechanisms of action of various treatments, (4) the study of new treatment approaches, and (5) clinical applications as a diagnostic test, as a means of assessing treatment adequacy and as part of cognitive-behavioral therapy (Table 2). For a variety of reasons, in the case of panic attacks, clinically relevant neurobiological processes are otherwise difficult to study. Despite the fact that several experimental animal models of panic attacks have been developed, these vary markedly in the extent to which they meet criteria for face, predictive, and construct validity (Blanchard et al. 2001; Griebel et al. 1996). The promise of the human challenge paradigm for elucidating pathophysiology, therefore, is still most exciting (Coplan and Klein 1996; Wiedemann et al. 2001; Ströhle et al. 2003).

Table 1 Experimentally induced panic attacks

Provocation	Mechanism	Respiration	HPA
Lactate	pH, pCO_2	⇑	=
CO_2	pCO_2	⇑	=
CCK-4	CCK_B receptor	⇑	⇑ =
mCPP, fenfluramine	5-HT receptor	=	⇑
Yohimbine	α_2 receptor	=	⇑
Inverse BZD agonist	$GABA_A$ receptor	=	⇑

BZD, benzodiazepine; CCK, cholecystokinin; mCPP, m-chlorophenylpiperazine.

Table 2 Sensitivity to treatment of experimentally induced panic attacks

	TCA	SSRI	BZD	CBT
Lactate	+	+/−	+	+
CO_2	+	+	+	?
CCK-4	+	+	+	?
Yohimbine	−	+	+	?

+/−, Conflicting results (i.e., evidence is equivocal); ?, not investigated sufficiently; BZD, benzodiazepines; CBT, cognitive-behavioral therapy; SSRI, selective serotonin reuptake inhibitors; TCA, tricyclic antidepressants.

1.1 Signaling Pathways

The acute pharmacological effects of psychotropic medications support the hypothesis that specific neurotransmitter systems may be directly involved in the onset, maintenance, and progression of the illness. Several major transmitter systems have been implicated in the pathophysiology of panic disorder, including noradrenergic, serotonergic, and γ-aminobutyric acid (GABA)-ergic neuronal circuitries. The ultimate common mechanism of action for this heterogeneous group of agents may, at least in part, be the attenuation of activity of the anxiety-modulating neuropeptides corticotropin releasing hormone (CRH) and vasopressin (AVP) at both hypothalamic and extrahypothalamic sites, where they serve an important function as neurotransmitters and neuromodulators (Keck and Holsboer 2001).

Much of the investigation into the function of these systems in panic has utilized physiological and pharmacological challenges to elicit panic reactions. Among the substances successfully used to evoke panic attacks in patients suffering from panic disorder or in healthy controls are sodium lactate, inhaled carbon dioxide (CO_2), isoproterenol, yohimbine, *m*-chlorophenylpiperazine (mCPP), fenfluramine, and cholecystokinin (CCK) agonists (Table 1; Coplan and Klein 1996). The success of such a diverse group of agents in eliciting panic attacks suggests a low threshold for the triggering of panic pathways by a variety of mechanisms that ultimately may lead to a final common pathway.

2 Noradrenergic/Adrenergic Challenges

In animal experiments, stimulation of the locus coeruleus, a collection of noradrenalin-producing neurons located bilaterally in the pons area with extensive projection sites, produces fear responses resembling those when animals are threatened by a predator (e.g., Redmond 1981). Studies of noradrenergic neurotransmission in panic disorder have principally relied on pharmacological challenge paradigms using the imidazoline derivative clonidine, an α_2-agonist, and yohimbine, an indole alkaloid with α_2-antagonist properties. The α_2-autoreceptor mechanism, which regulates firing of the noradrenergic neuron by mediating negative feedback inhibition, is interrupted by α_2-antagonists, whereas α_2-agonists, such as the centrally acting antihypertensive drug clonidine, enhance negative feedback. The latter, therefore, reduces noradrenergic firing and diminishes anxiety, at least on a temporary basis. Yohimbine administration, in turn, was repeatedly shown to produce panic attacks in patients with panic disorder (e.g., Charney and Heninger 1986; Abelson et al. 1992; Uhde et al. 1992). The relative prefrontal cortical blood flow was found to be decreased in panic disorder patients relative to control subjects following yohimbine challenge as measured with single photon emission tomography (SPECT) (Table 1; Woods et al. 1988b). Phenomenologically, however, yohimbine does not produce a panic attack that is identical to naturally occurring episodes or attacks provoked by other agents such as, e.g., sodium lactate. Intriguingly, long-term treatment with the antidepressant and antipanic agent imipramine, which significantly affects noradrenergic neurotransmission, does not attenuate yohimbine-induced panic attacks, as do alprazolam and the selective serotonin reuptake inhibitor (SSRI) fluvoxamine (Charney and Heninger 1985; Goddard et al. 1993).

Clonidine normally stimulates the hypothalamic release of growth hormone through α_2-agonism. The growth hormone release in panic patients, however, was found to be diminished, which could reflect downregulation of postsynaptic α_2-receptors in response to chronic noradrenergic discharge (e.g., Nutt et al. 1992). In addition, several studies indicated that the degree of blood pressure decrease, cortisol secretion, and noradrenergic turnover, as measured by plasma 3-methoxy-4-hydroxyphenylglycol (MHPG) concentrations, regularly induced by clonidine administration, was exaggerated in panic disorder (e.g., Charney et al. 1992; Uhde et al. 1986; Coplan et al. 1997a,b). Altered α_2-adrenoceptor sensitivity is also evidenced by findings that yohimbine produces exaggerated cardiovascular responses, enhanced plasma MHPG, and cortisol increases in panic disorder patients relative to control subjects (Cameron et al. 2000; Yeragani et al. 1992).

Taken together, panic patients consistently display altered responses to both α_2-adrenoceptor agonists and antagonists, suggesting that the noradrenergic system may be dysregulated. In this context it is of interest to note that the hypothalamic–pituitary–adrenocortical (HPA) axis and the noradrenergic system are closely interrelated and appear to work in synchrony. In the hypothalamic paraventricular nucleus, noradrenergic terminals originating from the locus coeruleus not only synapse onto growth hormone-containing neurons but also on CRH neurons, leading to the release of these neuropeptides. On the other hand, the locus coeruleus is known to be richly innervated with CRH immunoreactive fibers (Valentino et al. 1993). Stress- or CRH-induced increases in locus coeruleus neuronal firing are blocked by CRH receptor antagonists and chronic antidepressant treatment (e.g., Valentino et al. 1990; Schulz et al. 1996), suggesting that anxiogenic effects of CRH are mediated through its actions on the locus coeruleus noradrenergic system (Butler et al. 1990).

Adrenaline (or epinephrine) as an endogenous amine does not penetrate the blood–brain barrier and produces peripheral symptoms of arousal. Anxiety during adrenaline provocation is therefore unlikely to result from a central neurochemical mechanism. Consequently, anxiogenic effects of peripherally acting sympathomimetics are assumed to be mediated by secondary cognitive interpretation (Nutt and Lawson 1992). Adrenaline, therefore, has been suggested as a suitable agent to test psychological panic models, i.e., fear of stress-related bodily symptoms and anticipatory anxiety (e.g., Reiss 1991). The available data, however, are sparse and controversial, and it has been concluded that cognitive factors are not of major importance in explaining adrenaline's panicogenic properties (Veltman et al. 1996).

Isoproterenol, a synthetic sympathomimetic amine acting selectively at both β_1- and β_2-adrenoceptors, is also able to induce panic attacks in a subset of patients suffering from panic disorder. There is, however, a discrepancy in the findings, and the reliability and mechanisms of isoproterenol-induced panic remain to be clarified. It should also be emphasized that isoproterenol is not able to cross the blood–brain barrier.

3 Serotonergic Challenges

Despite the fact that the serotonergic system is a major area of focus for panic disorder, evidence for an etiological role of serotonergic dysfunction in panic disorder still remains to be clarified. This particular area of focus, however, has been directly stimulated by the uniform and superior antipanic effects of the SSRIs (e.g., Boyer 1995). Serotonin (5-hydroxytryptamin, 5-HT), however, has been implicated in almost every conceivable physiologic or behavioral function, and most drugs currently used for the treatment of psychiatric disorders are thought to act, at least partially, through serotonergic mechanisms. From the approximately 15 serotonin receptors discovered so far, inadequate 5-HT_{1A} receptor function is considered to further anxiety and avoidant behavior (Deakin 1996; Lucki et al. 1996; Parks et al. 1998). The somatodendritic autoreceptors of serotonergic neurons in both the dorsal raphé nucleus and the nucleus raphé magnus appear to be predominantly of the 5-HT_{1A} subtype. A variety of drugs with 5-HT_{1A} selectivity such as the anxiolytics buspirone and ipsapirone share the ability to inhibit raphé cell firing (e.g., Pan et al. 1993). Dense concentrations of 5-HT_{1A} binding sites are also found in the hippocampal pyramidal cell layer and the cerebral cortex (Miquel et al. 1991). Activation of the HPA system during panic and anxiety is also likely to contribute to an altered serotonergic neurotransmission, as it is known that glucocorticoids disrupt hippocampal 5-HT_{1A} neurotransmission by uncoupling the 5-HT_{1A} receptor from its intracellular G protein messenger system (Lesch and Lerer 1991). Moreover, corticosteroids are well known to play an inhibitory role in 5-HT_{1A} mRNA and protein expression (for review, Chaouloff 1995). Other serotonin receptors potentially involved in anxiety include the 5-HT_{2A}, 5-HT_{2C}, and 5-HT_3 receptors, and it has been suggested that 5-HT_{1A} and 5-HT_{2A} receptors may play reciprocal roles in mediating anxiety (Lucki et al. 1996). It has been hypothesized that the serotonergic innervation of both the amygdala and the hippocampus mediates anxiogenic effects by 5-HT_{2A} receptor stimulation, whereas serotonergic innervation of hippocampal 5-HT_{1A} receptors suppresses the association of a conditioned stimulus with an unconditioned stimulus and provides resilience to aversive events (Graeff et al. 1993).

Studies of 5-HT neurotransmission in panic disorder have principally employed the mixed serotonin agonist–antagonist mCPP and the indirect serotonin agonist fenfluramine (Table 1). mCPP has complex effects on brain 5-HT systems: it appears to act as an agonist at the 5-HT_{1C} and 5-HT_{1A} receptors, whereas effects at the 5-HT_3 receptor seem primarily antagonistic. Mixed agonist and antagonist activity has been found at the 5-HT_2 site (e.g., Kahn et al. 1991). 5-HT-releasing properties and binding to α_2-adrenergic sites have also been reported. Fenfluramine, a phenylethylamine derivative, potently releases presynaptic 5-HT and inhibits 5-HT reuptake, with weaker action as a postsynaptic 5-HT agonist. In neuroimaging studies, cerebral blood flow significantly increased in the anterior cingulate cortex in healthy subjects but not in subjects with panic disorder during fenfluramine challenge (Meyer et al. 2000). Patients with panic disorder have demonstrated increased rates of anxiety, but not necessarily overt panic attacks, in response to these agents (e.g., Hollander et al. 1990; Targum et al. 1992; Wetzler et al. 1996). The enhanced responsivity to serotonin-releasing agents raises the possibility of postsynaptic 5-HT hypersensitivity in panic disorder due to chronically decreased serotonergic neurotransmission. The complex pharmacology of the agents used, however, makes it difficult to draw conclusions about specific abnormalities within the 5-HT system. Ipsapirone, an azapirone derivative, acts selectively as full agonist at presynaptic 5-HT_{1A} autoreceptors and as a partial agonist at postsynaptic 5-HT_{1A} sites. In panic disorder patients, corticotropin (ACTH), cortisol, and hypothermic responses were blunted but anxiety responses did not differ from controls (Lesch et al. 1991). These findings support some role of 5-HT_{1A} receptors in the pathogenesis of panic disorder (Lesch et al. 1992). It has to be noted, however, that treatment with the 5-HT_{1A} partial agonist buspirone has proved effective in generalized anxiety disorder but not in panic disorder (DeMartinis et al. 2000; Sheehan et al. 1993).

Serotonin-precursor loading, i.e., administration of L-tryptophan or L-hydroxytryptophan, is not anxiogenic either in panic patients nor in healthy volunteers (Charney and Heninger 1986; Westenberg and Den Boer 1989; van Vliet et al. 1996). The discrepancy of these studies with the mCPP and fenfluramine findings may be due to differential receptor subtype activation or to presynaptic effects of the precursors. Moreover, although the effects of intravenous L-tryptophan are believed to result from the central synthesis and release of 5-HT, other mechanisms, such as, e.g., decreased availability of tyrosine for dopamine synthesis, might also be involved (Price et al. 1995). Tryptophan depletion by giving a tryptophan-free amino acid mixture preceded by a 24-h low tryptophan diet did not prove anxiogenic in unmedicated panic disorder study subjects (Goddard et al. 1994; but see also Sect. 6 for a combination with CO_2 challenge).

Taken together, so far, pharmacologic challenge studies involving 5-HT have been similarly unable to establish a primary role for 5-HT in the pathophysiology in panic disorder (Charney and Drevets 2002).

4 Lactate Infusion Challenge

Sodium lactate infusion studies constitute the largest single body of pharmacological challenge research with robust and reproducible findings (Table 1). Due to a paucity of relevant preclinical data, however, the mechanisms underlying lactate-induced panic are still unclear. The limited information provided on neuronal pathways influenced by this provocation test therefore limits its value in elucidating the pathophysiology of anxiety disorders. It is important to note that, as with all challenge paradigms, only a certain percentage of patients suffering from panic disorder are lactate sensitive (approx. 60%–80%).

An analysis of data from sodium lactate studies revealed that self-reported fear, high cortisol levels, and low partial pressure of CO_2, due to hyperventilation prior to lactate infusion, were the strongest predictors of panic (Coplan et al. 1998a). These findings raise the possibility that a particular biological and emotional state sets the stage for the occurrence of a panic attack in predisposed subjects. Moreover, the individual components of the triad (fear, cortisol, CO_2) were found to be correlated, suggesting the activation of a putative common neural substrate (Coplan et al. 1998a). This is consistent with central amygdalar activation affecting changes in cortical areas (cognitive misappraisal), the hypothalamic paraventricular nucleus (HPA system activation), and the pontine parabrachial nucleus, implicated in fear-driven hyperventilation (e.g., Davis et al. 1986).

Intriguingly, although living through a panic attack represents an intense stressor, most spontaneous clinical panic attacks as well as those induced by lactate, CO_2, and bicarbonate show no HPA activation (Levin et al. 1987;

Hollander et al. 1989; Woods et al. 1988a; Kellner et al. 1998). In general, in most studies cortisol elevation in patients with panic disorder is reliably observed only during the anticipation of panic attacks (Coplan et al. 1998a), not during the attacks themselves. As the anxiolytic neuropeptide atrial natriuretic peptide (ANP) has been demonstrated to inhibit the stimulated HPA system (Kellner et al. 1992; Ströhle et al. 1998) and as this peptide is released in response to infusion of sodium lactate (Kellner et al. 1995), this might in part explain the observed quiescence of the peripheral HPA system. In line with this, central CRH neuropeptidergic circuits other than those driving the peripherally accessible HPA system may well be overactive and could be therapeutic targets of antagonist actions in panic disorder (for review, Holsboer 2003). Sympathetic responses during the panicogenic challenge with sodium lactate are also absent, which again can be explained by an inhibitory action of elevated plasma ANP concentrations (Seier et al. 1997).

Lactate is metabolized to bicarbonate, resulting in a peripheral metabolic alkalosis, and bicarbonate in turn is metabolized to CO_2, which stimulates both medullary chemoreceptors and the locus coeruleus, causing panic in vulnerable individuals. It has been suggested therefore that panic patients have enhanced sensitivity of ventral medullary chemoreceptors to fluctuations in pH and that panic attacks would result from the chemoreceptors' misperception of life-threatening central hypoxia and acidosis secondary to cerebrovascular vasoconstriction due to lactate-induced peripheral metabolic alkalosis (Carr and Sheehan 1984). Consistent with the assumption of locus coeruleus stimulation is the finding that clonidine partially attenuates lactate-induced panic (Charney et al. 1992). A comprehensive theory suggests that both CO_2 (see Sect. 6 and lactate induce panic by triggering a suffocation false alarm in susceptible individuals with a hypersensitive suffocation detector (Klein 1993). According to this hypothesis, a physiological misinterpretation by a suffocation monitor misfires an evolved suffocation alarm system. This produces sudden respiratory distress followed swiftly by a brief hyperventilation, panic, and the urge to flee. Lactate administered during sleep provokes greater fluctuations in cardiac and respiratory activity in panic-prone subjects than normal controls (Koenigsberg et al. 1994). The finding, therefore, cannot be attributed to anticipatory anxiety.

By use of both the sodium lactate and the CCK-4 (see Sect. 5) challenge paradigms it could be shown that panic attacks elicited significant decreases in plasma concentrations of the neuroactive steroids allopregnanolone (3α,5α-tetrahydroprogesterone) and pregnanolone (3α,5β-tetrahydroprogesterone) (Ströhle et al. 2003). These findings are comparable with a decreased GABA-ergic tone, as neuroactive steroids, i.e., derivatives of progesterone, are potent positive allosteric $GABA_A$-receptor modulators, which alter neuronal excitability through rapid nongenomic effects at the cell surface (for review, Rupprecht and Holsboer 1999). The association between changes in plasma neuroactive steroid concentrations with experimentally induced panic attacks and the well-documented pharmacological properties of these compounds as $GABA_A$ receptor modulators suggest that neuroactive steroids may play an important role in the pathophysiology of panic attacks in patients with panic disorder. Because long-term antidepressant drug treatment has been shown to influence the composition of neuroactive steroids contrary to panic-induced changes (Romeo et al. 1998; Uzunova et al. 1998; Ströhle et al. 1999, 2000), the antipanic efficacy of antidepressants in the treatment of panic disorder may in part be mediated by stabilizing the equilibrium of endogenous neuroactive steroid concentrations.

Positron emission tomography (PET) studies with a ^{15}O-labeled tracer demonstrated that patients with panic disorder, who were vulnerable to lactate at baseline, had abnormally greater right than left parahippocampal blood flow, blood volume, and oxygen metabolism, as well as an abnormally high whole-brain metabolism (Reiman et al. 1986). In a follow-up study using ^{18}F-fluorodeoxyglucose, significant differences in glucose metabolism were found. However, the left parahippocampal areas had higher rates of glucose metabolism. In addition, low glucose metabolism was found in the right inferior parietal and right superior temporal brain regions (Bisaga et al. 1998).

It is of interest to note that in a small open trial, six patients with panic disorder who had panicked during sodium lactate infusion were given cognitive-behavioral treatment for 12–24 weeks. After treatment they underwent another lactate infusion, and four patients were rated as having no panic. As similar results can be obtained also after treatment with antidepressant drugs and benzodiazepines, these findings suggest that reduced vulnerability to lactate accompanies remission of panic (Shear et al. 1991).

5
Cholecystokinin Tetrapeptide Challenge

The neuropeptide CCK is an octapeptide found regionally in the gastrointestinal tract and brain (brain-gut peptide), where it acts as a neurotransmitter and neuromodulator. Its most abundant form in the brain is the C-terminal sulfated octapeptide fragment CCK-8, which interacts with the same affinity with both CCK receptor subtypes CCK-A and CCK-B. Extensive pharmacological studies have been carried out during the past few years suggesting that CCK may participate in the neuroendocrine responses to stress (e.g., Harro et al. 1993; Daugé and Léna 1998).

Together with CRH, CCK belongs to the most extensively studied neuropeptides in anxiety models. Generally, CCK is thought to induce anxiogenic-like effects, although the results of those animal studies have been highly variable and sometimes contradictory (Griebel 1999). The heterogeneity of behavioral responses produced by CCK can further be explained by the fact that the neuropeptide has been infused in different brain areas in order to delineate the anatomical substrate of CCK-induced anxiogenic-like effects. Local administration of CCK-8 directly into the amygdala, for example, produced hyper-anxiety (Frankland et al. 1997). This was not the case when CCK-8 was infused into the anterior nucleus accumbens (Daugé et al. 1990), suggesting that regional differences in CCK receptors are probably responsible for this discrepancy. As has recently been shown in CCK receptor gene knockout mice, however, the role of the receptor subtypes in anxiety-related behavior is still controversial (Miyasaka et al. 2002).

CCK-8 concentrations were found to be lower in panic patients than in normal control subjects (Brambilla et al. 1993), and the CCK-B receptors were hypersensitive in panic disorders (Akiyoshi et al. 1996). A significant association between panic disorder and a single nucleotide polymorphism found in the coding region of the CCK-B receptor gene has been reported (Kennedy et al. 1999). If confirmed by replication, these data would suggest that a CCK-B receptor gene variation might be involved in the pathogenesis of panic disorder. Clinical trials, however, have provided rather disappointing and inconclusive data about the anxiolytic potential of the CCK-B antagonists available so far (Kramer et al. 1995; Shlik et al. 1997; Pande et al. 1999).

CCK-B receptor agonists such as pentagastrin or CCK-4 (25–50 µg i.v.) have panic-like anxiogenic effects in humans (Table 1). Panic patients and patients suffering from posttraumatic stress disorder, however, are more sensitive than healthy controls to the anxiogenic effects of CCK-4 (e.g., DeMontigny et al. 1989; Radu et al. 2002). Pretreatment with a CCK-B receptor antagonist is able to reverse both the autonomic and anxiogenic effects of pentagastrin (Lines et al. 1995). The safety, reliability, and dose-dependence of the anxiogenic effects of CCK-4 as well as the similarity of effect to naturalistic panic are strong, rendering CCK-4 an attractive probe of anxiety. CCK-4-induced panic has a characteristic physiological activation curve that only lasts about 2–3 min (Bradwejn et al. 1995). Like sodium lactate-induced panic, CCK-4-induced attacks are accompanied by hyperventilation. Increases in regional cerebral blood flow as measured by PET scans have been described in the claustrum-insular, hypothalamic, amygdala, cerebellar vermis, and the anterior cingulate regions during CCK-4-induced panic attacks in healthy subjects (Benkelfat et al. 1995; Javanmard et al. 1999). Interestingly, as outlined above, CCK-4-induced panic attacks induce significant decreases in plasma concentrations of the neuroactive steroids allopregnanolone (3α, 5α-tetrahydroprogesterone) and pregnanolone (3α,5β-tetrahydroprogesterone) (Ströhle et al. 2003). In contrast to sodium lactate and CO_2, CCK-4 induced panic is accompanied by an increase in ACTH and it has been suggested therefore that CRH mediates the panicogenic effects of CCK-4 (DeMontigny et al. 1989; Kellner et al. 1997; Shlik et al. 1997; Koszycki et al. 1998; Ströhle et al. 2000).

Similar to sodium lactate- and CO_2-induced panic, CCK-4-induced attacks can be blocked by antipanic treatment with antidepressants (e.g., Bradwejn and Koszycki 1994; Shlik et al. 1997) or benzodiazepines (DeMontigny et al.

1989; Zwanzger et al. 2003). Administration of the neuropeptide ANP was effective in reducing the CCK-4-elicited panic reaction in patients with panic disorder and to a lesser extent in healthy controls. Moreover, ANP inhibited the CCK-4-induced rise of ACTH in both patients and controls, which may be attributed to a reduced hypothalamic CRH release (Jessop 1999; Wiedemann et al. 2001). Interestingly, ACTH release in response to CCK-4 was found to be blunted in patients when compared to healthy controls. This finding is possibly due to a chronic hypersecretion of CRH with a subsequent downregulation of CRH receptors in such patients (Wiedemann et al. 2001).

Among all panicogens, CCK-4 is the only one that fulfils the criteria for a neurotransmitter/neuromodulator and that, unlike sodium lactate and CO_2, spares issues such as volume overload and acid base-mediated alterations. Hence, it is likely that CCK-4 may be among the most suitable panicogenic challenges.

6 Carbon Dioxide

Voluntary hyperventilation leading to hypocapnia [i.e., a low partial pressure (p)CO_2] can precipitate panic in panic disorder patients. Training habitual hyperventilators to breathe appropriately has therapeutic benefits (Salkovskis et al. 1986) but it has to be kept in mind that not all hyperventilators are anxious (Bass and Gardner 1989). Hyperventilation, therefore, may rather be the consequence, and not the cause, of panic. Although reducing pCO_2 can trigger panic, the role of CO_2 is further complicated because increased pCO_2 can evoke panic as well: Hypercapnia via CO_2 inhalation can induce panic attacks that benzodiazepines are able to prevent (Woods et al. 1986; Gorman et al. 1988). Thus, although hypocapnia and hypercapnia have different effects on, for example, the cerebral circulation, they both cause anxiety-related symptoms. Hypersensitivity to CO_2 inhalation is one of the most widely studied laboratory markers of panic disorders, and many studies have clearly demonstrated the ability of single- or double-vital capacity inhalations of gas mixtures with varying concentrations of CO_2 (5%–35%) (Table 1). Inhaling CO_2 acts like lactate to produce respiratory stimulation and provoke panic, perhaps related to an altered and hypersensitive central suffocation detector (Klein 1993).

Anxiety reactivity to 35% CO_2 inhalations has been reported not to be significantly influenced by clinical characteristics of the disorder such as baseline anxiety, frequency of panic attacks, severity of agoraphobia, duration of illness, and age (Perna et al. 1994). On the other hand, several studies suggest a relevant role of genetic factors in 35% CO_2-induced panic attacks, and it has been concluded that CO_2-induced panic might be considered a phenotypic expression of a genetic vulnerability to panic disorder even before the clinical onset of panic disorder (e.g., Perna et al. 1995; Bellodi et al. 1999).

CO_2 hypersensitivity, therefore, might be the expression of impairment, under genetic control, at some level of the respiratory system and may constitute a promising trait marker (Coryell 1997). Treatment with antipanic agents significantly modulates CO_2 hypersensitivity in panic patients (e.g., Bertani et al. 1997; Perna et al. 1997) and the main neurotransmitters modulated by these antipanic medications (i.e., serotonin, noradrenalin) have been reported to influence respiration (Bonham 1995). Serotonergic system activation and α-adrenoceptor or cholinergic-receptor blockade reduce CO_2 sensitivity (Mueller et al. 1982). It has been claimed therefore that the modulation of these neurotransmitters plays a role in the pathogenetic mechanisms of panic disorder (e.g., Coplan et al. 1992). It is of note that the inhibition of the panicogenic effect of a 35% CO_2 challenge may be used as a predictor for the antipanic properties of a compound, as it has been shown that the reduction of reaction to a 35% CO_2 challenge after 1 week of treatment predicts a therapeutic effect later on in treatment (Perna et al. 1997). Exaggerated responses to CO_2 inhalation in subjects with panic disorder could be also modulated through blockade of central muscarinic cholinergic receptors by pretreatment with biperiden (Battaglia et al. 2001). Central cholinergic receptors are therefore likely to contribute to an increased sensitivity to hypercapnia in predisposed subjects.

Studies in which the availability of serotonin is manipulated in combination with a laboratory panic challenge has considerably increased insight into the relationship between serotonin and panic disorder. Tryptophan depletion caused an increased panic response to a 5% or 35% CO_2 challenge in panic disorder patients (Miller et al. 2000; Schruers et al. 2000). Conversely, panic anxiety and symptoms, as well as the number of panic attacks following 35% CO_2 inhalation, were significantly reduced by pretreatment with the serotonin precursor L-hydroxytryptophan, suggesting that under certain circumstances serotonin may act to inhibit panic (Schruers et al. 2002).

Taken together, abundant research work describes associations between respiratory perturbation and acute anxiety. This association has been demonstrated most convincingly in panic disorder, where various forms of respiratory stimulation, including lactate infusion and CO_2 inhalation, consistently produce high degrees of anxiety and more pronounced perturbations in respiratory physiological parameters (Merikangas and Pine 2002). It is important to note, however, that these associations may be extended beyond the specific diagnosis of panic disorder, because enhanced sensitivity to respiratory challenges is also found in conditions that exhibit strong familial or phenomenological associations with panic disorder, including limited symptom panic attacks, certain forms of situational phobias, and high ratings on anxiety sensitivity scales (Schmidt et al. 1997). The mechanisms that contribute to such enhanced sensitivity remain poorly specified. At a cognitive level, such hypersensitivity might result from an overall sensitivity to somatic sensations, consistent with data linking high degrees of anxiety to future panic attacks (Schmidt et al. 1999). On the other hand, enhanced sensitivity to respiratory sensations appears more closely tied to panic attacks than sensitivity to other somatic factors (Merikangas and Pine 2002). Therefore, as outlined above, at the physiological level such hypersensitivity is likely to result from perturbations in brain systems involved in respiratory regulation (Klein 1993).

7
GABA–Benzodiazepine Receptor Complex

Excessive or inappropriate anxiety can be controlled by enhancing inhibitory synaptic neurotransmission mediated by GABA using clinically effective benzodiazepines. Beyond that long-standing clinical experience, the significance of the GABA–benzodiazepine receptor complex in the mediation of anxiety has been firmly established in preclinical literature (e.g., Löw et al. 2000). Inverse agonists such as the β-carboline FG 7142 are anxiogenic and activate the HPA system (Table 1; Dorow et al. 1983). These findings have made the GABA–benzodiazepine receptor complex a subject of significant research interest. In panic disorder patients, decreases in benzodiazepine binding were found particularly in the orbitofrontal cortex and insula by use of ^{11}C-flumazenil PET (Malizia et al. 1998). Challenge studies aimed at characterizing putative alterations in benzodiazepine receptor sensitivity, however, have so far led to conflicting results. Because flumazenil (a clinically well-known benzodiazepine antagonist with additional inverse agonistic and partial agonistic effects) produced panic attacks in panic disorder patients but not in healthy controls in one study, it has been concluded that a shift in the benzodiazepine receptor set-point towards a pronounced inverse agonistic action of flumazenil exists in panic patients or in a specific subgroup (Nutt et al. 1990). It has been generalized therefore that benzodiazepine receptor functioning would be shifted in panic patients so that antagonists are recognized as partial inverse agonists. Other investigations, in contrast, failed to demonstrate anxiogenic or panic-provoking effects of flumazenil in panic disorder patients. Moreover, neither ACTH release nor cardiovascular parameters indicated that flumazenil exerted an inverse agonistic activity in the patients studied (Ströhle et al. 1998, 1999). In conclusion, the hypothesized shift in the benzodiazepine receptor set point cannot be generalized to all patients with panic disorder. Further studies are needed to clarify whether or not there is a subgroup of patients with panic disorder characterized by a different response pattern to flumazenil and an altered $GABA_A$-benzodiazepine receptor complex function. Moreover, the possibility exists that more subtle changes in $GABA_A$ receptors such as, e.g., α2-subunit functioning, might be of clinical relevance (Löw et al. 2000).

8 Conclusion and Outlook

In summary, the pharmacological challenge studies have increased our knowledge concerning the neurobiology of panic disorder remarkably and may finally result in novel and more causal treatment strategies such as, for example, the use of neurosteroids or ANP (e.g., Wiedemann et al. 2001; Ströhle et al. 2003). With respect to the pathophysiological changes underlying laboratory-induced panic, however, most findings—the vast majority, in fact—are far from being satisfactorily explained. This most probably reflects the complexity of the etiologic factors. The differences in sensitivity to certain panicogens, therefore, might be fruitful in serving as biological markers of subtypes of panic disorders and should be a major focus of research, as the identification of reliable endophenotypes is currently one of the major rate-limiting steps in psychiatric genetic studies (e.g., Smoller and Tsuang 1998).

The heterogeneity of agents capable of producing panic attacks in susceptible patients and the inconsistency of autonomic responses during a panic attack have led to the assumption that panic originates in an abnormally sensitive fear network, which includes the prefrontal cortex, insula, thalamus, amygdala, and amygdalar projections to the brainstem and hypothalamus (Gorman et al. 2000). Substances that cause panic attacks act to provoke a sensitized brain network that has been conditioned to respond to noxious stimuli. In individual patients, or even subgroups of patients, various projections from the central nucleus of the amygdala (i.e., the center of the hypersensitive network) to brainstem sites, such as the locus coeruleus (blood pressure, heart rate), periaqueductal gray region (defensive behavior), lateral nucleus of the hypothalamus (sympathetic nervous system activation), and parabrachial nucleus (respiratory rate), may be stronger or weaker resulting in differences in the pattern of autonomic and neuroendocrine responses during panic (Gorman et al. 2000).

References

Abelson JL, Glitz D, Cameron OG (1992) Endocrine, cardiovascular and behavioral responses to clonidine in patients with panic disorder. Biol Psychiatry 32:18–25

Abelson JL, Le Mélledo JM, Bichet DG (2000) Dose response of arginine vasopressin to the CCK-B agonist pentagastrin. Neuropsychopharmacology 24:161–169

Akiyoshi J, Moriyama T, Isogawa K, Miyamoto M, Sasaki I, Kuga K, Yamamoto H, Yamada K, Fugii I (1996) CCk-4-induced calcium mobilization in T cells is enhanced in panic disorder. J Neurochem 66:1610–1615

Bass C, Gardner W (1989) Hyperventilation in clinical practice. Br J Hosp Med 41:73–81

Battaglia M, Bertella S, Ogliari A, Bellodi L, Smeraldi E (2001) Modulation by muscarinic antagonists of the response to carbon dioxide challenge in panic disorder. Arch Gen Psychiatry 58:114–119

Bellodi L, Perna G, Caldirola D, Arancio C, Bertani A, Di Bella D (1998) CO2-induced panic attacks: a twin study. Am J Psychiatry 155:1184–1188

Benkelfat C, Bradwejn J, Meyer E (1995) Functional neuroanatomy of CCk-4-induced anxiety in normal healthy volunteers. Am J Psychiatry 152:1180–1184

Bertani A, Perna G, Caldirola D, Arancio C, Bellodi L (1997) Pharmacologic effect of paroxetine, sertraline and imipramine on reactivity to the 35% CO2 challenge: a double blind, random, placebo controlled study. J Clin Psychopharmacol 17:97–101

Bisaga A, Katz JL, Antonini A, Wright CE, Margouleff C, Gorman JM, Eidelberg D (1998) Cerebral glucose metabolism in women with panic disorder. Am J Psychiatry 155:1178–1183

Blanchard C, Hynd AL, Minke KA, Minemoto T, Blanchard RJ (2001) Human defensive behaviors to threat scenarios show parallels to fear- and anxiety-related defense patterns of non-human mammals. Neurosci Biobehav Rev 25:761–770

Bonham AC (1995) Neurotransmitters in the CNS control of breathing. Respir Physiol 101:219–230

Boyer W (1995) Serotonin uptake inhibitors are superior to imipramine and alprazolam in alleviating panic attacks: a metaanalysis. Int Clin Psychopharmacol 10:45–49

Bradwejn J, Koszycki D (1994) Imipramine antagonism of the paniogenic effects of cholecystokinin tetrapeptide in panic disorder patients. Am J Psychiatry 151:261–263

Bradwejn J, Koszycki D, Paradis M, Reece P, Hinton J, Sedman A (1995) Effect of CI-988 on cholecystokinin tetrapeptide-induced panic symptoms in healthy volunteers. Biol Psychiatry 38:742–746

Brambilla F, Bellodi L, Perna G, Garberi A, Panerai A, Sacerdote P (1993) T cell cholecystokinin concentrations in panic disorder. Am J Psychiatry 150:1111–1113

Briggs AC, Stretch DD, Brandon S (1993) Subtyping of panic disorder by symptom profile. Br J Psychiatry 163:201–209

Butler PD, Weiss JM, Stout JC, Nemeroff CB (1990) CRF produces fear-enhancing and behavioral activating effects following infusion into the locus coeruleus. J Neurosci 10:176–183

Cameron OG, Zubieta JK, Grunhaus L (2000) Effects of yohimbine on cerebral blood flow, symptoms, and physiological function in humans. Psychosom Med 62:549–559

Carr DB, Sheehan DV (1984) Panic anxiety: a new biological model. J Clin Psychiatry 45:323–330

Chaouloff F (1996) Regulation of 5-HAT receptors by corticosteroids: where do we stand? Fundam Clin Pharmacol 9:219–233

Charney DS, Drevets WC (2002) Neurobiological basis of anxiety disorders. In: Davis KL, Charney D, Coyle JT, Nemeroff C (eds) Neuropsychopharmacology—the fifth generation of progress. Lippincott Williams and Wilkins, Philadelphia, pp 901–930

Charney DS, Heninger GR (1986) Noradrenergic function and the mechanism of action of antianxiety treatment. I. The effect of alprazolam treatment. Arch Gen Psychiatry 42:458–467

Charney DS, Woods SW, Krystal JH, Nagy LM, Heninger GR (1992) Noradrenergicneuronal dysregulation in panic disorder: the effects of intravenous yohimbine and clonidine in panic disorder patients. Acta Psychiatr Scand 86:273–282

Coplan JD, Klein DF (1996) Pharmacologic probes in panic disorder. In: Westenberg HGM, den Boer JA, Murphy DL (eds) Advances in the neurobiology of panic disorder. Wiley, New York, pp 179–204

Coplan JD, Goetz R, Klein DF (1998a) Plasma cortisol concentrations preceding lactate-induced panic. Psychological, biochemical, and physiological correlates. Arch Gen Psychiatry 55:130–136

Coplan JD, Trost R, Owens MJ (1998b) Cerebrospinal fluid concentrations of somatostatin and biogenic amines in grown primates reared by mothers exposed to manipulated foraging conditions. Arch Gen Psychiatry 55:473–477

Coryell W (1997) Hypersensitivity to carbon dioxide as a disease-specific trait marker. Biol Psychiatry 41:259–263

Daugé V, Léna I (1998) CCK in anxiety and cognitive processes. Neurosci Biobehav Rev 22:815–825

Daugé V, Bohme GA, Crawley JN, Durieux C, Stutzmann JM (1990) Investigation of behavioral and electrophysiological responses induced by selective stimulation of CCKB receptors by using a new highly potent CCK analog, BC 264. Synapse 6:73–80

Davis M (1986) Pharmacological and anatomical analysis of fear conditioning using the fear-potentiated startle paradigm. Behav Neurosci 100:814–824

De Martinidis N, Rynn M, Rickels K (2000) Prior benzodiazepine use and buspirone response in the treatment of generalized anxiety disorder. J Clin Psychiatry 61:91–94

De Montigny C (1989) Cholecystokinin tetrapeptide induces panic like attacks in healthy volunteers. Arch Gen Psychiatry 46:511–517

Deakin JFW (1996) 5-HAT antidepressant drugs and the psychosocial origins of depression. J Psychopharmacol 10:31–38

Dorow R, Horowski R, Paschelke G, Amin M (1983) Severe anxiety induced by FG 7142, a beta-carboline ligand for benzodiazepine receptors. Lancet 2:98–99

Frankland PW, Josselyn SA, Bradwejn J, Vaccarino FJ, Yeomans JS (1997) Activation of amygdala cholecystokinin B receptors potentiates the acoustic startle response in the rat. J Neurosci 17:1838–1847

Goddard AW, Sholomskas DE, Augeri FM (1994) Effects of tryptophan depletion in panic disorders. Biol Psychiatry 36:775–777

Goldstein R, Wickramaratne P, Horwath E, Weissmann M (1997) Familial aggregation and phenomenology of "early"-onset (at or before age 20 years) in panic disorder. Arch Gen Psychiatry 54:271–278

Gorman JM, Fyer MR, Goetz R (1988) Ventilatory physiology of patients with panic disorder. Arch Gen Psychiatry 45:31–39

Gorman JM, Kent JM, Sullivan GM, Coplan JD (2000) Neuroanatomical hypothesis of panic disorder, revised. Am J Psychiatry 157:493–505

Graef FG, Silveira MC, Nogueira RL (1993) Role of the amygdala and periaqueductal gray in anxiety and panic. Behav Brain Res 58:123–131

Griebel G, Blanchard DC, Blanchard RJ (1996) Predator-elicited flight responses in Swiss-Webster mice: an experimental model of panic attacks. Prog Neuropsychopharmacol Biol Psychiatry 20:185–205

Harro J, Vasar E, Bradwejn J (1993) Cholecystokinin in animal and human research of anxiety. Trends Pharmacol Sci 14:244–249

Hollander E, Liebowitz MR, Gorman JM, Cohen B, Fyer AJ, Klein DF (1989) Cortisol and sodium lactate-induced panic. Arch Gen Psychiatry 46:135–140

Holsboer F (2003) The role of peptides in treatment of psychiatric disorders. J Neural Transm 64:17–34

Javanmard M, Shlik J, Kennedy SH, Vaccarino FJ, Houle S, Bradwejn J (1999) Neuroanatomic correlates of CCK-4-induced panic attacks in healthy humans: a comparison of two time points. Biol Psychiatry 45:872–882

Jessop DS (1999) Central non-glucocorticoid inhibitors of the hypothalamo–pituitary–adrenal axis. J Endocrinol 160:169–180

Kahn RS, Wetzler S (1991) m-Chlorophenylpiperazine as a probe of serotonin function. Biol Psychiatry 30:1139–1166

Keck ME, Holsboer F (2001) Hyperactivity of CRH neuronal circuits as a target for therapeutic interventions in affective disorders. Peptides 22:835–844

Kellner M, Wiedemann K, Holsboer F (1992) ANF inhibits the CRH-stimulated secretion of ACTH and cortisol in man. Life Sci 50:1835–1842

Kellner M, Herzog L, Yassouridis A, Holsboer F, Wiedemann K (1995) A possible role of atrial natriuretic hormone in pituitary-adrenocortical unresponsiveness in lactate-induced panic disorder. Am J Psychiatry 152:1365–1367

Kennedy JL, Bradwein J, Koszycki D (1999) Investigation of cholecystokinin system genes in panic disorder. Mol Psychiatry 4:284–285

Klein DF (1993) False suffocation alarms, spontaneous panics, and related conditions. Arch Gen Psychiatry 50:306–317

Klein DF (1998) Panic and phobic anxiety: phenotypes, endophenotypes, and genotypes. Am J Psychiatry 155:1147–1149

Koenigsberg HW, Pollak CP, Fine J, Kakuma T (1994) Cardiac and respiratory activity in panic disorder: effects of sleep and sleep lactate infusions. Am J Psychiatry 151:1148–1152

Kramer MS, Cutler NR, Ballenger JC, Patterson WM, Mendels J (1995) A placebo-controlled trial of L-365,260, a CCK antagonist, in panic disorder. Biol Psychiatry 37:462–466

Le Mellédo JM, Bradwein J, Koszycki D, Bichet DG, Bellavance F (1998) The role of the β-noradrenergic system in cholecystokinin-tetrapeptide-induced panic symptoms. Biol Psychiatry 44:364–366

Lesch KP, Hoh A, Schulte HM (1991) Long-term fluoxetine treatment decreases 5-HT1A receptor responsivity in obsessive compulsive disorder. Psychopharmacology (Berl) 105:415–420

Lesch KP, Wiesmann M, Hoh A (1992) 5-HT1A receptor-effector system responsivity in panic disorder. Psychopharmacology (Berl) 106:111–117

Levin AP, Doran AR, Liebowitz MR, Fyer AJ, Klein DF, Paul SM (1987) Pituitary adrenocortical unresponsiveness in lactate-induced panic. Psychiatry Res 21:23–32

Lines C, Challenor J, Traub M (1995) Cholecystokinin and anxiety in normal volunteers—an investigation of the anxiogenic properties of pentagastrin and reversal by the cholecystokinin receptor subtype-b-antagonist L-365,260. Br J Clin Pharmacol 39:235–242

Löw K, Crestani F, Keist R, Benke D, Brünig I (2000) Molecular and neuronal substrate for the selective attenuation of anxiety. Science 290:131–134

Lucki I (1996) Serotonin receptor specificity in anxiety disorders. J Clin Psychiatry 57(Suppl 6):5–10

Malizia AL, Cunningham VJ, Bell CJ, Liddle PF, Jones T, Nutt DJ (1998) Decreased brain GABAA-benzodiazepine receptor binding in panic disorder. Arch Gen Psychiatry 55:715–720

Mc Nally RJ (1990) Psychological approaches to panic disorder: a review. Psychol Bull 108:403–419

Merikangas KR (2002) Genetic and other vulnerability factors for anxiety and stress disorders. In: Davis KL, Charney D, Coyle JT, Nemeroff C (eds) Neuropsychopharmacology—the fifth generation of progress. Lippincott Williams and Wilkins, Philadelphia, pp 867–882

Meyer JH, Swinson R, Kennedy SH (2000) Increased left posterior parietal-temporal cortex activation after D-fenfluramine in women with panic disorder. Psychiatry Res 98:133–143

Mezey E, Reisine TD, Skirboll L, Beinfeld M, Kiss JZ (1985) CCK in the medial parvocellular subdivision of the paraventricular nucleus: coexistence with CRH. Ann NY Acad Sci 448:152–156

Miller HE, Deakin JF, Anderson IM (2000) Effect of acute tryptophan depletion on CO2-induced anxiety in patients with panic disorder and normal volunteers. Br J Psychiatry 176:182–188

Miquel MC, Doucet E, Boni C (1991) Central serotonin 1A receptor: respective distributions of encoding mRNA, receptor protein and binding sites by in situ hybridization histochemistry, radioimmunohistochemistry and autoradiographic mapping in the rat brain. Neurochem Int 19:453–465

Miyasaka K, Kobayashi S, Ohta M, Kanai S, Yoshida Y, Nagata A, Matsui T, Noda T, Takuguchi S, Takata Y, Kawanami T, Funakoshi A (2002) Anxiety-related behaviors in cholecystokinin-A,B, and AB receptor gene knockout mice in the plus-maze. Neurosci Lett 335:115–118

Mueller AR, Lundberg DBA, Breese GR, Jonason Y (1982) The neuropharmacology of respiration control. Pharmacol Rev 34:255–279

Nutt DJ, Lawson CW (1992) Panic attacks: a neurochemical overview of models and mechanisms. Br J Psychiatry 160:165–178

Nutt DJ, Glue P, Lawson CW, Wilson S (1990) Flumazenil provocation of panic attacks: evidence for altered benzodiazepine receptor sensitivity in panic disorder. Arch Gen Psychiatry 47:917–925

Pan ZZ, Wessendorf MW, Williams JT (1993) Modulation by serotonin of the neurons in rat nucleus raphe magnus in vitro. Neuroscience 54:421–429

Pande AC, Greiner M, Adams JB (1999) Placebo-controlled trial of the CCK-B antagonist, CI-988 in panic disorder. Mol Psychiatry 46:860–862

Parks CL, Robinson PS, Sibille E (1998) Increased anxiety of mice lacking the serotonin1A receptor. Proc Natl Acad Sci USA 95:10734–10739

Perna G, Battaglia M, Garberi A, Arancio C, Bellodi L (1994) Carbon dioxide/oxygen challenge test in panic disorder. Psychiatry Res 52:159–171

Perna G, Gabriele A, Caldirola D, Bellodi L (1995) Hypersensitivity to inhalation of carbon dioxide and panic attacks. Psychiatry Res 57:267–273

Perna G, Bertani A, Gabriele A, Politi E, Bellodi L (1997) Modification of 35% carbon dioxide hypersensitivity across one week of treatment with clomipramine and fluvoxamine: a double-blind, randomised, placebo-controlled study. J Clin Psychopharmacol 17:173–178

Price LH, Goddard AW, Barr LC, Goodman WK (1995) In: Bloom, FE, Kupfer, DJ (eds) Psychopharmacology—the fourth generation of progress. Raven Press, New York, pp 1311–1323

Radu D, Ahlin A, Svanborg P, Lindefors N (2002) Anxiogenic effects of the CCKB agonist pentagastrin in humans and dose-dependent increase in plasma C-peptide levels. Psychopharmacology (Berl) 161:396–403

Redmond DE Jr (1981) Clonidine and the primate locus coeruleus: evidence suggesting anxiolytic and anti-withdrawal effects. Prog Clin Biol Res 71:147–163

Reiman EM, Raichle ME, Robins E, Butler FK, Herscovitch P, Fox P, Perlmutter P (1986) The applications of positron emission tomography to the study of panic disorder. Am J Psychiatry 143:469–477

Reiss S (1991) Expectancy model of fear, anxiety and panic. Clin Psychol Rev 11:141–153

Romeo E, Ströhle A, Di Michele F, Holsboer F, Pasini A, Rupprecht R (1998) Effects of antidepressant treatment on neuroactive steroids in major depression. Am J Psychiatry 155:910–913

Rupprecht R, Holsboer F (1999) Neuroactive steroids: mechanisms of action and neuropsychopharmacological perspectives. Trends Neurosci 22:410–416

Salkovskis P, Jones D, Clark D (1986) Respiratory control in the treatment of panic attacks: replication and extension with concurrent measurement of behaviour and pCO2 . Br J Psychiatry 148:526–532

Schmidt N, Lerew D, Jackson R (1997) The role of anxiety sensitivity in the pathogenesis of panic: prospective evaluation of spontaneous panic attacks during acute stress. J Abnorm Psychol 106:355–364

Schruers K, Klaassen T, Pols H, Overbeek T, Deutz NE, Griez E (2000) Effects of tryptophan depletion on carbon dioxide provoked panic in panic disorder patients. Psychiatry Res 93:179–187

Schruers K, van Diest R, Overbeek T, Griez E (2002) Acute L-5-hydroxytryptophan administration inhibits carbon-dioxide-induced panic in panic disorder patients. Psychiatry Res 113:237–243

Schulz DW, Mansbach RS, Sprouse J, Braselton JP, Collins J (1996) CP-154,526: a potent and selective nonpeptide antagonist of CRF receptors. Proc Natl Acad Sci USA 93:10477–10482

Seier FE, Kellner M, Yassouridis A, Heese R, Strian F, Wiedemann K (1997) Autonomic reactivity and hormonal secretion in lactate-induced panic attacks. Am J Physiol 272:H2630–H2638

Shear MK, Fyer AJ, Josephson S, Fitzpatrick M, Klein DF (1991) Vulnerability to sodium lactate in panic disorder patients given cognitive therapy. Am J Psychiatry 148:795–797

Shlik J, Aluoja A, Vasar V, Vasar E, Podar T, Bradwein J (1997) Effects of citalopram on behavioral, cardiovascular, and neuroendocrine response to cholecystokinin tetrapeptide challenge in patients with panic disorder. J Psychiatry Neurosci 22:332–340

Smoller JW, Tsuang MT (1998) Panic and phobic anxiety: defining phenotypes for genetic studies. Am J Psychiatry 155:1152–1162

Ströhle A, Romeo E, Hermann B, Di Michele F, Holsboer F (1999) Concentrations of 3α-reduced neuroactive steroids and their precursors in plasma of patients with major depression and after clinical recovery. Biol Psychiatry 45:274–277

Ströhle A, Holsboer F, Rupprecht R (2000) Increased ACTH concentrations associated with cholecystokinin tetrapeptide-induced panic attacks in patients with panic disorders. Neuropsychopharmacology 22:251–256

Ströhle A, Romeo E, di Michele F, Pasini A, Hermann B (2003) Induced panic attacks shift γ-aminobutyric acid type A receptor modulatory neuroactive steroid composition in patients with panic disorder. Arch Gen Psychiatry 60:161–168

Targum SD (1992) Cortisol response during different anxiogenic challenges in panic disorder patients. Psychoneuroendocrinology 17:453–458

Uhde TW, Tancer ME, Rubinow DR (1992) Evidence for hypothalamo-growth hormone dysfunction in panic disorder: profile of growth hormone responses to clonidine, yohimbine, caffeine, glucose, GRF, and TRH in panic disorder patients vs. healthy volunteers. Neuropsychopharmacology 6:101–118

Uzunova V, Sheline Y, Davis JM, Rasmusson A, Uzunov DP, Costa E, Guidotti A (1998) Increase in the cerebrospinal fluid content of neurosteroids in patients with unipolar major depression who are receiving fluoxetine or fluvoxamine. Proc Natl Acad Sci USA 95:3239–3244

Valentino RJ, Curtis AL, Parris DG, Wehby RG (1990) Antidepressant actions on brain noradrenergic neurons. J Pharmacol Exp Ther 253:833–840

Valentino RJ, Foote SL, Page ME (1993) The locus coeruleus as a site for integrating CRF and noradrenergic mediation of stress response. Ann NY Acad Sci 697:173–188

van Vliet IM, Slaap BR, Westenberg HG, Den Boer JA (1996) Behavioral, neuroendocrine and biochemical effects of different doses of 5-HTP in panic disorder. Eur Neuropsychopharmacol 6:103–110

Veltman DJ, van Zijderveld GA, van Dyck R (1996) Epinephrine infusions in panic disorder: a double-blind placebo-controlled study. J Affect Disord 39:133–140

Westenberg HGM, Den Boer JA (1989) Serotonin function in panic disorder: effect of L-5-hydroxytryptophan in patients and controls. Psychopharmacology (Berl) 98:283–285

Wetzler S, Asnis GM, DeLecuona JM (1996) Serotonin function in panic disorder: intravenous administration of meta-chlorophenylpiperazine. Psychiatry Res 62:77–82

Wiedemann K, Jahn H, Yassouridis A, Kellner M (2001) Anxiolyticlike effects of atrial natriuretic peptide on cholecystokinin tetrapeptide-induced panic attacks. Arch Gen Psychiatry 58:371–377

Woods SW, Charney DS, Lake J, Goodman WK, Redmond DE, Heninger DR (1986) Carbon dioxide sensitivity in panic anxiety: ventilatory and anxiogenic response to carbon dioxide in healthy subjects and panic and anxiety patients before and after alprazolam treatment. Arch Gen Psychiatry 43:900–909

Woods SW, Charney DS, Goodman WK, Heninger GR (1988a) Carbon dioxide-induced anxiety: behavioral, physiologic, and biochemical effects of carbon dioxide in patients with panic disorders and healthy subjects. Arch Gen Psychiatry 45:43–52

Woods SW, Kosten K, Krystal JH (1988b) Yohimbine alters regional cerebral blood flow in panic disorder [letter]. Lancet 2:678

Yeragani VK, Berger R, Pohl R (1992) Effects of yohimbine on heart rate variability in panic disorder patients and normal controls: a study of power spectral analysis of heart rate. J Cardiovasc Pharmacol 20:609–618

Zwanzger P, Eser D, Aicher S, Schule C, Baghai TC (2003) Effects of alprazolam on cholecystokinin-tetrapeptide-induced panic and hypothalamic-pituitary-adrenal-axis activity: a placebo-controlled study. Neuropsychopharmacology 28:979–984

HEP (2005) 169:469–501

Pharmacotherapy of Anxiety

J.R. Nash (✉) · D.J. Nutt

Psychopharmacology Unit, School of Medical Sciences, University of Bristol,
Bristol BS8 1TD, UK
Jon.Nash@bristol.ac.uk

Abstract The pharmacological treatment of anxiety has a long and chequered history, and recent years have seen a rich development in the options available to prescribers. Most of the currently used anxiolytic agents act via monoaminergic (chiefly serotonin) or amino acid (GABA or glutamate) neurotransmitters, and this chapter describes the pharmacology of the major drug groups. Clinical applications are discussed with respect to the five major anxiety disorders, as well as simple phobia and depression with concomitant anxiety. Prospective future developments in the field are considered.

Keywords Pharmacology · Anxiety · GABA · Serotonin · Benzodiazepines · Antidepressants

1 Introduction

These are exciting times for physicians involved in the treatment of anxiety disorders. Therapeutic options are increasing whilst the level of public interest in the field has never been greater. Patients can equip themselves to be active partners in the therapeutic process using the various available sources of medical information. Longstanding controversies, such as the relative merits of psychological therapies versus medication and the safety of long-term medical treatments of anxiety, are debated in the national media. Perversely, at a time when psychiatrists have more to offer their anxious patients than ever before, the validity of their role is challenged from some quarters. Nevertheless, medical practice is now based on a substantial volume of clinical experience and evidence from controlled trials, and we can justify with confidence many of the treatment options we put before our patients.

Since the publication of the work of Donald Klein (1964) that described the discrimination of panic disorder from other neuroses, the diagnostic classification of anxiety disorders has undergone a progressive evolution, with new diagnostic categories emerging. Discrepancies remain between different classifications, with the accepted gold standard being the Diagnostic and Statistical Manual of the American Psychiatric Association (DSM-IV 1994), but general agreement has been reached for the diagnostic validity of the main categories. This has allowed the quantitative measurement of anxiety disorders in epidemiological surveys (Kessler et al. 1994; Wittchen H-U et al. 1998), and the demonstration of the prevalence and economic burden of anxiety has been a key factor driving research into anxiolytic therapies.

The rapid development of the psychiatry of anxiety over the past 15 years has been accelerated by developments in a diversity of other disciplines, including cognitive and experimental psychology, preclinical and clinical pharmacology, and neuroscience (particularly neuroimaging). Research from areas such as genetics and molecular biology is only just beginning to have an impact. With the current pace of scientific discovery, physicians can anticipate an increasing range of effective and acceptable treatment options for anxiety disorders. This chapter describes the state of clinical psychopharmacology for anxiety in 2005, but may soon become out of date.

2 Clinical Management of Anxiety

Optimal treatment of an anxious patient involves far more than the prescription of medication (Fig. 1). The skill of the psychiatrist in establishing the

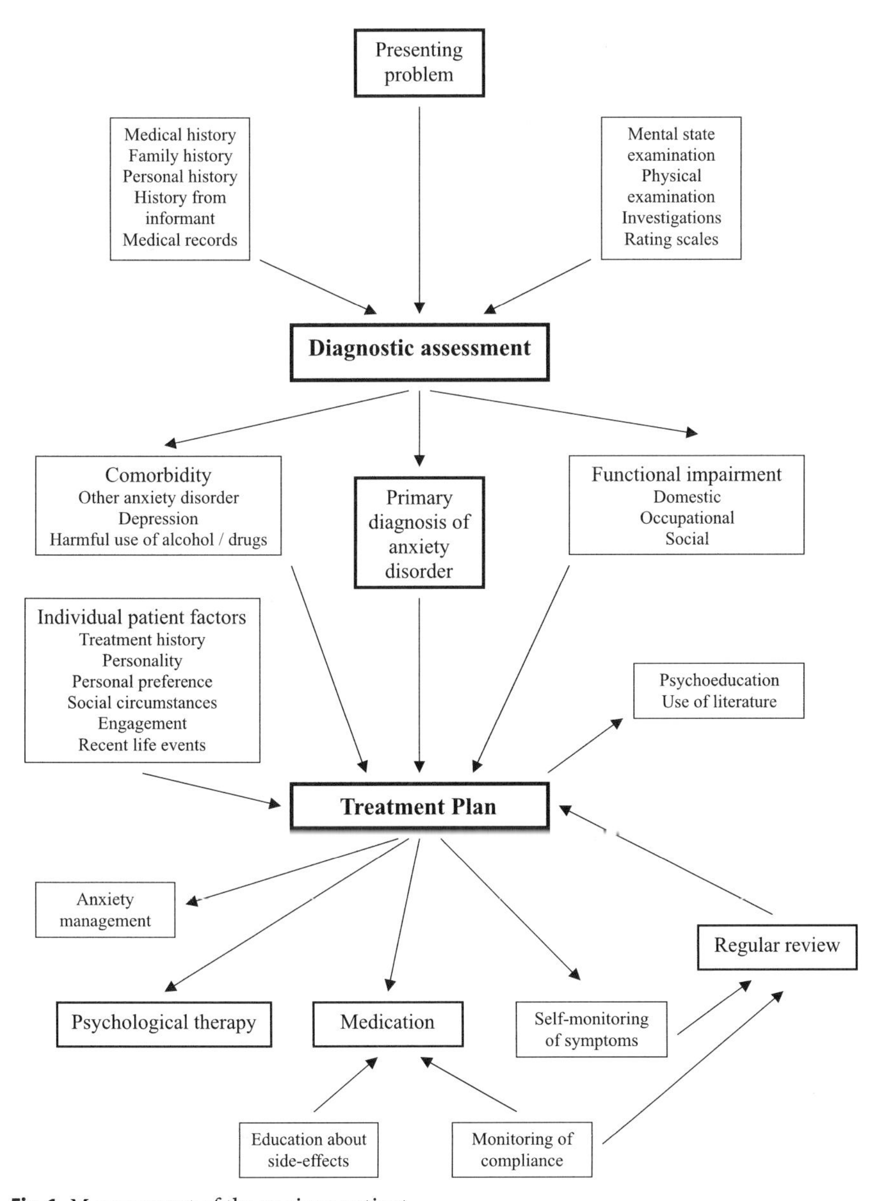

Fig. 1. Management of the anxious patient

diagnosis of a specific anxiety disorder is an invaluable part of the assessment process, as a correct diagnosis has a great influence on the treatment offered. Anxiety disorders frequently present with comorbid conditions, particularly depression, alcohol or substance use problems, and other anxiety disorders. These must be detected and managed appropriately.

Patients tend to present when their anxiety impairs their occupational, social or domestic functioning, and identification of the key complaints and motivations for seeking treatment is critical in drawing up an effective management plan. For example, a patient with generalized anxiety disorder (GAD) may present because her resulting insomnia is impairing her ability to work; management should include strategies to improve sleep efficiency as well as treatment of the anxiety.

Education about the nature of anxiety helps engagement and improves recovery. Models have been described for most anxiety disorders in both the biological and psychological dimensions, and patients benefit from an explanation, tailored to their level of understanding, in each dimension. This can be reinforced by the use of educative literature (Dannon et al. 2002). Patients should be encouraged to record and monitor their symptoms, as this can improve treatment efficacy (Febbraro and Clum 1998).

Once a diagnosis has been made and explained to the patient, a treatment plan should be negotiated. A range of biological and psychological treatments may be suitable and should be put to the patient, who is offered an element of choice alongside the recommendations of the physician. A combination of drug and psychological therapies can be more effective than either alone (Barlow et al. 2000). The patient may have preconceptions about specific therapies, often as a result of their anxiety, e.g. a patient with panic disorder fears the effects of drugs and a patient with social anxiety baulks at the suggestion of group therapy. An open discussion of benefits and adverse effects, including long-term side-effects, is likely to improve compliance. Although medications are generally well-tolerated, some side-effects commonly occur, and anxious patients experience more than others (Davies et al. 2003). Progress with treatment should be encouraged by regular review, particularly in the early stages.

3
Anxiolytic Drugs

The use of substances for their anxiolytic properties dates to the beginning of recorded human history. The twentieth century saw a substantial development of their use for medical purposes, and major progress was made in the 1990s, "the decade of anxiety", as advances in neuroscience provided a basis for the targeted design of new treatments. There is now a greater range of drugs available that are better tolerated, although not necessarily more effective, than their predecessors. However, despite increased knowledge of the complex physiology of the brain, the actions of current pharmacotherapeutic agents are moderated via a relatively small number of neurotransmitter systems, with the most important being the amino acid neurotransmitters (chiefly γ-aminobutyric acid, GABA, but also glutamate) and the monoaminergic neurotransmitters (serotonin, noradrenaline and to a lesser extent dopamine).

3.1 Drugs Acting via Amino Acid Neurotransmission

Glutamate is the major excitatory amino acid in the brain. It has a key role in learning and memory and is involved in the mediation of the response to stress. Glutamate receptors are present throughout the central nervous system but differ widely according to their localisation and function (Kent et al. 2002), and as a result have not been easy to identify as targets for pharmacological manipulation.

GABA is formed by the decarboxylation of glutamate, and is the major inhibitory neurotransmitter. In recent years the $GABA_A$ receptor has been identified as the mediator of the anxiolytic and sedative effects of drugs such as alcohol and the benzodiazepines. Abnormalities of this receptor have been identified in humans with anxiety disorders (Nutt and Malizia 2001).

For much of the second half of the twentieth century the benzodiazepines were the mainstay of the treatment of anxiety. Despite well-publicised concerns about their long-term safety, they remain an important therapeutic option. The anticonvulsants contain a number of drugs that act via GABA or glutamate neurotransmission and have a limited but interesting role in the treatment of particular anxiety disorders.

3.1.1 Benzodiazepines

The efficacy of benzodiazepines in most anxiety disorders has been proved through extensive clinical experience and controlled trials (Faravelli et al. 2003), although it is important to note that they are not effective at treating post-traumatic stress disorder or comorbid depression, and there is less evidence to support their use in obsessive–compulsive disorder (OCD). Their anxiolytic effects have an immediate onset and in contrast to many other drugs, they do not cause a worsening of anxiety when therapy is initiated.

3.1.1.1 Tolerability and Safety

Benzodiazepines are generally well-tolerated (Table 1), although side-effects such as sedation, loss of balance and impaired psychomotor performance may be problematic for some patients. There are reported associations with road

Table 1 Benzodiazepines in anxiety

Efficacy	Panic disorder Generalised anxiety disorder Social anxiety disorder Specific phobias	
Side-effects	**Common**	**Uncommon**
	Depressed CNS functioning (sedation, muscle weakness, light-headedness, confusion, ataxia, impaired psychomotor performance)	Paradoxical aggression Headache Hypotension Weight gain Sexual dysfunction
	Discontinuation effects (see Table 2)	Respiratory depression (in respiratory disease)
Toxic effects	Coma Aspiration of gastric contents Respiratory depression	

traffic accidents (Barbone et al. 1998) and with falls and fractures in the elderly (Wang et al. 2001). They are relatively safe in overdose (Buckley et al. 1995), although the risk is increased if taken in combination with alcohol or other sedative drugs.

3.1.1.2
Discontinuation Problems

The major controversy surrounding the use of benzodiazepines has concerned the risks of long-term treatment, specifically tolerance, abuse, dependence and withdrawal effects. From being the most widely prescribed psychotropic drug they suffered a major backlash, but a more balanced view of their place in treatment is emerging (Williams and McBride 1998). After 40 years of clinical experience there is little evidence of tolerance to the anxiolytic effects of benzodiazepines (Rickels and Schweizer 1998). Abuse (taking in excess of the prescribed dose) is uncommon except in individuals with a history of abuse of other drugs, who may not be suitable for benzodiazepine therapy (Task Force Report of the American Psychiatric Association 1990). There is, however, a consensus that adverse effects on discontinuation are more common than with other anxiolytics (Schweizer and Rickels 1998). A careful clinical assessment is indicated in this situation, as these effects may be caused by recurrence or rebound (recurrence with increased intensity) of the original anxiety symptoms.

Table 2 Benzodiazepine withdrawal syndrome

Symptoms	Hyperarousal: anxiety, irritability, insomnia, restlessness	Autonomic lability: sweating, tachycardia, hypertension, tremor, dizziness
	Neuropsychological effects: dysphoria, perceptual sensitisation, tinnitus, confusion, psychosis	Seizures
Risk factors	**Treatment factors:** treatment duration > 6 months; high dose; short-acting drug; abrupt cessation	**Patient factors:** severe premorbid anxiety; alcohol/substance use disorder; female; dysfunctional personality; panic disorder
Therapeutic strategies	Gradual tapering: Switch to long-acting drug, e.g. diazepam; Cover with secondary agent (anticonvulsant, antidepressant); Cognitive behavioural therapy	

A benzodiazepine withdrawal syndrome has been described in some patients discontinuing therapy (Table 2). Although potentially serious, it is generally mild and self-limiting (up to 6 weeks), but may accompany or provoke a recurrence of anxiety symptoms and cause great concern to the patient. As with any other treatment, the risks and benefits of benzodiazepine therapy should be carefully assessed and discussed with the patient. Monotherapy will not be first-line treatment for the majority of patients, but benzodiazepines offer a valuable option that should not be discounted.

3.1.1.3 Drug Interactions

The potential for interaction with other medications comes largely from two sources: (1) the exacerbation of sedation and impaired psychomotor performance by other drugs also causing these effects, and (2) alterations in the hepatic metabolism of benzodiazepines by drugs that are either inducers or inhibitors of cytochrome P450 (CYP450) enzymes. The increased toxicity in combination with alcohol is mostly pharmacodynamic but may partly be due to the inhibition of metabolism of some benzodiazepines by high alcohol concentrations. Other drugs that may have additive effects on sedation include tricyclic antidepressants, antihistamines, opioid analgesics and the α_2-adrenoceptor agonists clonidine and lofexidine.

Most benzodiazepines undergo oxidative metabolism in the liver that may be enhanced by enzyme inducers (e.g. carbamazepine, phenytoin) or slowed by inhibitors (sodium valproate, fluoxetine, fluvoxamine). Oxazepam, lorazepam and temazepam are directly conjugated and are not subject to these interactions.

3.1.1.4 Clinical Usage

The specific clinical use of the numerous available benzodiazepines depends on their individual pharmacokinetic and pharmacodynamic properties. Drugs with a high affinity for the $GABA_A$ receptor (alprazolam, clonazepam, lorazepam) have high anxiolytic efficacy; drugs with a short duration of action (temazepam) are used as hypnotics to minimise daytime sedative effects. Diazepam has a long half-life and duration of action and may be favoured for long-term use or when there is a history of withdrawal problems; oxazepam has a slow onset of action and may be less susceptible to abuse.

Guidance on the clinical indications for benzodiazepine therapy is available from various sources (Task Force Report of the American Psychiatric Association 1990; Ballenger et al. 1998a; Bandelow et al. 2002). Long-term therapy is most likely to present problems with discontinuation and is usually reserved for cases that have proved resistant to treatment with antidepressants alone. Patients may benefit from a 2–4 week course of a benzodiazepine whilst antidepressant therapy is initiated, as this counteracts the increased anxiety caused by some drugs (Goddard et al. 2001). A benzodiazepine may be useful as a hypnotic in some cases of anxiety disorder, and can be used by phobic patients on an occasional basis before exposure to a feared situation.

3.1.2 Anticonvulsants

There is some overlap between the clinical syndromes of anxiety and epilepsy: panic disorder and post-traumatic stress disorder can present with symptoms similar to temporal lobe seizures; alcohol and drug withdrawal states can cause both anxiety and seizures; and some drugs (e.g. barbiturates and benzodiazepines) act as both anticonvulsants and anxiolytics. Most anticonvulsant drugs act via the neurotransmission of GABA or glutamate, and in recent years have offered a promising field for the development of novel anxiolytic therapies (Kent et al. 2002). Although there is solid preclinical research demonstrating their anxiolytic properties, the evidence base in humans is less impressive and they tend to be reserved for second-line or adjunctive therapy. Drug interactions mediated via hepatic enzymes are a significant feature of these drugs.

3.1.2.1 Carbamazepine

No satisfactory randomised controlled trials have been published demonstrating the efficacy of carbamazepine in anxiety disorders, although it has a history of use as an anxiolytic in panic disorder and PTSD. It has an unfavourable side-effect profile (nausea, dizziness, ataxia) and multiple drug interactions due to induction of liver enzymes.

3.1.2.2 Gabapentin and Pregabalin

Gabapentin acts by increasing GABA activity, although its exact mechanism of action is unclear. It causes dose-related sedation and dizziness. It has been shown in randomised controlled trials to be effective in social anxiety disorder (Pande et al. 1999) and to benefit some patients with panic disorder (Pande et al. 2000). Pregabalin is a related compound that has recently demonstrated efficacy in GAD in a phase III study (Pande et al. 2003).

3.1.2.3 Lamotrigine

This anticonvulsant drug blocks voltage-gated sodium channels and inhibits release of glutamate. A controlled study found efficacy in PTSD (Hertzberg et al. 1999). Important side-effects include fever and skin reactions.

3.1.2.4 Sodium Valproate

As with carbamazepine, the historical use of valproate for anxiety is not supported by robust clinical trials. A randomised study showed efficacy in panic disorder (Lum et al. 1991) and benefit has been reported in open studies in OCD and PTSD. The major side-effects are tremor, nausea, ataxia and weight gain and there is the potential for drug interactions via inhibition of hepatic enzymes.

3.1.2.5 Other Drugs

Tiagabine blocks neuronal uptake of GABA and has reported benefits in panic disorder and PTSD (Lydiard 2003). Topiramate has complex actions on GABA and glutamate and was found to be helpful for some symptoms of PTSD (Berlant and van Kammen 2002). Vigabatrin inhibits GABA metabolism and has been shown to block induced panic attacks in healthy volunteers (Zwanzger et al. 2001).

3.2 Drugs Acting via Monoaminergic Neurotransmission

Aside from the $GABA_A$ receptor, most research into the neurochemistry of anxiety has explored the role of the monoamine transmitters serotonin (5-HT), noradrenaline and dopamine. This interest originated with the serendipitous discovery of drugs that were later found to exert their anxiolytic effects by actions on monoamine function. Advances in neuroscience research techniques have, rather than clarifying the role of these neurotransmitters, tended to present an increasingly complex picture (Argyropoulos and Nutt 2003). Nevertheless these advances have led to the development of "designer drugs" with selective effects on neurotransmitter function that have been successfully tested as anxiolytics. These drugs do not exceed their predecessors in terms of efficacy, but better tolerability has led to their adoption as first-line treatments for anxiety disorders.

The biology of the monoamines is described in detail elsewhere. In simple terms, they facilitate transmission in neural pathways that originate in nuclei of the brainstem and have descending projections to the autonomic nervous system and widespread ascending projections to sites in the limbic system and cortex. These pathways modulate many aspects of behavioural function as well as anxiety responses. Of the three monoamines, the role of serotonin in anxiety is best understood, but the picture is complex as increased serotonergic activity may be anxiogenic or anxiolytic depending on the site of action (Bell and Nutt 1998).

Anxiolytic drugs alter monoaminergic neurotransmission by increasing synaptic availability or by direct action on postsynaptic receptors. Mechanisms for increasing monoamine availability include increasing release by blocking inhibitory autoreceptors, decreasing reuptake by blocking transporters, and decreasing metabolism by inhibiting oxidative enzymes. Monoamines are also implicated in the pathophysiology of depression, and drugs that increase their synaptic availability tend to have antidepressant effects. These drugs have been traditionally classified as antidepressants, although they have a primary role as anxiolytics. Other anxiolytic drugs acting via monoamine neurotransmission are the postsynaptic serotonin receptor partial agonist buspirone, the β-adrenoceptor blockers and drugs classed as antipsychotics.

3.2.1 Antidepressants

The growth during the 1990s in the use of antidepressants, particularly selective serotonin reuptake inhibitors (SSRIs), for the treatment of anxiety disorders represented a major advance in the pharmacotherapy of anxiety. The efficacy of tricyclic antidepressants (TCAs) and monoamine oxidase inhibitors (MAOIs) had been established alongside their antidepressantw actions several decades

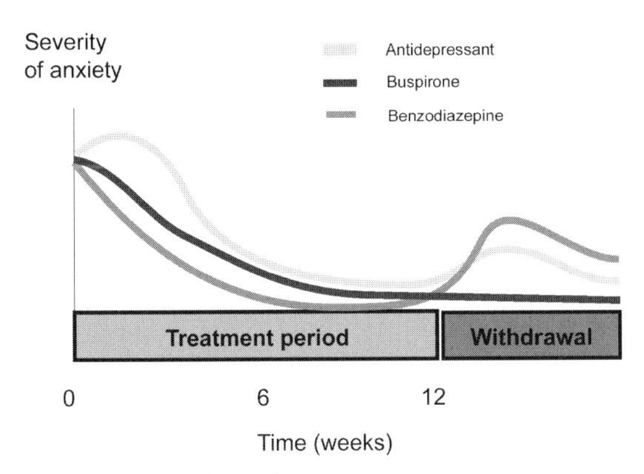

Fig. 2. Treatment response to anxiolytic drugs

previously, but the launch of new, better-tolerated medications coincided with the backlash against benzodiazepines and an increase in the profile of anxiety disorders.

Taken together, the efficacy of antidepressants covers the spectrum of anxiety disorders, although there are important differences between drugs in the group (Table 3). Several new antidepressants have been marketed since the SSRIs: venlafaxine and mirtazapine are discussed later (Sects. 3.2.1.2 and 3.2.1.4); nefazodone, a serotonin reuptake inhibitor and postsynaptic 5-HT_2 blocker showed promise in early studies but was recently withdrawn by its manufacturers; reboxetine, a noradrenaline reuptake inhibitor (NARI) showed benefits in panic disorder in one published study (Versiani et al. 2002) and further evidence of its anxiolytic efficacy is awaited.

Antidepressants differ from benzodiazepines in the onset and course of their actions (Fig. 2). Most cause an increase in anxiety on initiation of therapy, and anxiolytic effects occur later. In comparative studies, improvement matches that on benzodiazepines after 4 weeks (Rocca et al. 1997). Withdrawal effects, particularly rebound, are less problematic with antidepressants, although stopping treatment is associated with a significant rate of relapse, and a withdrawal syndrome has been described for most of the shorter-acting drugs.

3.2.1.1 SSRIs

These drugs increase synaptic serotonin by selectively blocking the serotonin reuptake transporter. In preclinical and human studies acute doses tend to be anxiogenic (Bell and Nutt 1998) but chronic administration has anxiolytic effects, possibly due to downregulation of presynaptic autoreceptors (Blier et al. 1990). There are five SSRIs widely available: citalopram, fluoxetine, fluvoxamine, paroxetine and sertraline. Escitalopram, the *S*-enantiomer of citalopram,

Table 3 Antidepressants in anxiety disorders

Antidepressant	Efficacy[a]	Tolerability	Safety	Discontinuation syndrome
MAOI/RIMA	Panic disorder	Significant short- and long-term side-effects; special dietary requirements; moclobemide better tolerated	Significant overdose toxicity (less with moclobemide)	Reported
	Social anxiety disorder PTSD			
Mirtazapine	(Panic disorder, PTSD)	Few side-effects on initiation; few long-term side-effects	Relatively safe in overdose	Not reported
TCA	Panic disorder	Onset worsening; side-effects on initiation; some long-term effects	Significant overdose toxicity	Well-described
	OCD (GAD, PTSD)			
SSRI/SNRI	GAD	Onset worsening; side-effects on initiation; few long-term effects	Relatively safe in overdose (venlafaxine possibly less safe)	Well-described; more common with paroxetine; uncommon with fluoxetine
	OCD			
	Panic disorder			
	PTSD			
	Social anxiety disorder			

[a]Parentheses indicate where evidence is less strong.

was recently licensed in the UK for the treatment of panic disorder and is likely to have the same spectrum of efficacy as citalopram (Waugh and Goa 2003).

The SSRIs as a class are now widely considered to be appropriate first-line anxiolytic drugs; in particular paroxetine, the most potent 5-HT reuptake blocker, has been licensed in the UK for the treatment of each of the major anxiety disorders. Short-term efficacy has been clearly demonstrated in randomised controlled trials, but in common with other antidepressants, research evidence is lacking for long-term efficacy and necessary duration of treatment.

Tolerability and Safety An advantage of the SSRIs has been their improved tolerability relative to their predecessors, the tricyclic antidepressants and benzodiazepines. This has been demonstrated in comparative studies of drugs from these classes (e.g. Zohar and Judge 1996). Nevertheless they are not without side-effects: on initiation nausea, anxiety, jitteriness and insomnia are related to the starting dose; later sedation, asthenia, headache, sweating and sexual dysfunction may occur. Hyponatraemia occurs mostly in the elderly. Some effects are particular to individual drugs within the class; for example paroxetine has anticholinergic properties and can cause dry mouth, constipation and urinary hesitancy; sertraline is more likely to cause dyspepsia and diarrhoea; fluoxetine has agonist activity at 5-HT_{2C} receptors causing headache, agitation and loss of appetite (Goodnick and Goldstein 1998).

Although SSRI overdose can cause seizures, coma and cardiac abnormalities (Barbey and Roose 1998), these toxic effects occur only in large overdoses or in combination with other drugs. Fatality rates are substantially lower than with TCA overdose (Mason et al. 2000). Public attention has been drawn to reports of suicidal and aggressive thoughts and behaviour associated with initiating SSRIs (Healy 2003). The scientific basis for this assertion is disputed and continues to be debated, but it does not appear that SSRI treatment is associated with increased suicidality on a population level (Carlsten et al. 2001; Khan et al. 2003).

Discontinuation Problems Further controversy has surrounded misleading claims in the lay media that SSRIs have "addictive" properties. These centre around reports of patients suffering symptoms when trying to discontinue medication. As with the benzodiazepines, these symptoms may be a recurrence of the premorbid anxiety, although rebound anxiety has not been clearly demonstrated. Self-limiting symptoms associated with SSRI withdrawal have been widely reported (Haddad 1998), and are generally described as the "SSRI discontinuation syndrome" (Table 4). The most frequently occurring symptoms are dizziness, nausea and headache.

The syndrome is more common with paroxetine, possibly due to its anticholinergic activity, and is very uncommon with fluoxetine due to the long half-life of its metabolites (Michelson et al. 2000). It can start 48 h after the final dose, and although most cases resolve within 2–3 weeks, symptoms may rarely last longer than this.

Drug Interactions SSRIs interact with other drugs that have effects on 5-HT neurotransmission, including TCAs, buspirone, sumatriptan and tryptophan, but particularly important is the interaction with MAOIs that can lead to a synergistic increase in synaptic serotonin. This can result in the serotonin syndrome, comprising restlessness, irritability, tremor, sweating and hyperreflexia. The syndrome can be lethal (Sternbach 1991). In general clinical practice, there should be a washout of 2 weeks between discontinuing MAOI therapy and starting SSRI; a washout of 1–2 weeks should follow SSRI discontinuation (5 weeks for fluoxetine).

The drugs have variable potential for drug interactions via hepatic CYP450 enzymes (Table 5). Escitalopram has the lowest potential for interactions.

Clinical Usage Expert sources recommend SSRIs as first-line treatments of anxiety disorders (American Psychiatric Association 1998; Ballenger et al. 1998a; Bandelow et al. 2002). In preparation for treatment, a full discussion of potential benefits and anticipated side-effects (including discontinuation effects) should be held with the patient (Bull et al. 2002). Some patients have difficulty initiating treatment because of anxiety about side-effects. In these cases the drug may be increased slowly from a low starting dose, if necessary using the

Table 4 SSRI discontinuation syndrome

Symptoms	Neurological symptoms: dizziness, tremor, vertigo, paraesthesia/shooting pains	Somatic distress: nausea, headache, lethargy
	Psychological symptoms: anxiety, confusion, memory problems	Hyperarousal: agitation, restlessness, insomnia, irritability
Risk factors	**Treatment factors:** longer duration of treatment; rapid discontinuation; short half-life drug; possibly increased dose	**Patient factors:** possibly younger age; any psychiatric diagnosis
Therapeutic strategies	Careful assessment: Reassurance Reinstitute therapy if necessary Taper slowly (over 1 month) Switch to fluoxetine	

Table 5 SSRIs and hepatic cytochrome P450 enzymes

SSRI	Inhibitor of enzymes
Citalopram	–
Fluoxetine	2D6 (potent) 3A4 (potent)
Fluvoxamine	1A2 (potent) 3A4 (potent) 2D6 (moderate)
Paroxetine	2D6 (potent) 3A4 (moderate)
Sertraline	2D6 (moderate)

–, No significant enzyme inhibition.

syrup form of fluoxetine or paroxetine, or a benzodiazepine may be used to cover the initiation period.

There is little research evidence to guide a decision on duration of treatment. Some studies have shown continued improvement for up to 12 months, and for most disorders there is a significant relapse rate when treatment is stopped (Lecrubier and Judge 1997; Michelson et al. 1999). Guidelines suggest a duration of 12–24 months if treatment is successful, but if there are risk factors for relapse treatment may be required for much longer. Treatment discontinuation should be carefully planned and medication tapered.

3.2.1.2 Venlafaxine

Venlafaxine is a serotonin and noradrenaline reuptake inhibitor (SNRI). It shares these properties with the TCAs amitriptyline, clomipramine and imipramine, but it is the first selective SNRI, with low affinity for muscarinic, histaminic and α-adrenergic receptors. At low doses serotonergic effects predominate, but at higher doses the reuptake of noradrenaline is significantly blocked (Melichar et al. 2001). It is available as immediate and extended release (XR) preparations.

There is a large evidence base for the antidepressant efficacy of venlafaxine, but fewer studies have been carried out in anxiety disorders. The best evidence is for GAD (Allgulander et al. 2001) and anxiety symptoms associated with depression (Silverstone and Ravindran 1999). Side-effects on initiation of therapy are similar to those of SSRIs, with nausea being the most common. Higher doses can cause raised blood pressure. A discontinuation syndrome similar to that seen with SSRIs has been reported. Toxicity causes cardiac conduction problems, seizures and coma, and venlafaxine overdose is associated with a higher mortality than that of the SSRIs (Buckley and McManus 2002). Although metabolised by CYP2D6, venlafaxine does not inhibit this enzyme and has a low potential for drug interactions.

3.2.1.3 Tricyclic Antidepressants

This group includes compounds with actions on a range of neurotransmitter systems. Their antidepressant efficacy is mediated by reuptake inhibition of serotonin and noradrenaline, although side-effects such as sedation may also be useful. Their use in anxiety disorders is supported by a long history of clinical experience and a reasonable evidence base from controlled trials. Studies support the use of clomipramine (a potent serotonin reuptake inhibitor) in panic disorder and OCD (Lecrubier et al. 1997; Clomipramine Collaborative Study Group 1991), of imipramine in panic disorder and GAD (Cross-National Collaborative Panic Study 1992; Rickels et al. 1993), and of amitriptyline in PTSD (Davidson et al. 1993a). No controlled studies support the use of TCAs in social anxiety disorder.

A meta-analysis of controlled studies suggested superior efficacy of clomipramine over SSRIs in OCD (Kobak et al. 1998), but this has not been demonstrated in direct comparisons and the use of SSRIs has superseded that of TCAs because of advantages in safety and tolerability (Zohar and Judge 1996). Side-effects of TCAs include anticholinergic effects (drowsiness, dry mouth, blurred vision and constipation), antihistaminergic effects (drowsiness and weight gain) and postural hypotension caused by α_1-adrenoceptor blockade, as well as the side-effects common to SSRIs. Some effects are dose-related, and usual practice is to titrate the dose slowly upwards. A discontinuation syndrome similar to that with SSRIs is well-described, and withdrawal should be tapered. Overdose causes hypotension, cardiac arrhythmias, metabolic acidosis, seizures and coma and is associated with a significant mortality. Interactions can occur with other drugs with CNS effects (particularly MAOIs), and with drugs that affect hepatic metabolism.

3.2.1.4 Mirtazapine

Mirtazapine has a novel mechanism of action that in theory should promote anxiolytic effects, although evidence from studies of anxiety disorders is awaited. It increases synaptic release of serotonin and noradrenaline via blockade of presynaptic inhibitory α_2-adrenoceptors, as well as blocking postsynaptic 5-HT_2 and 5-HT_3 serotonin receptors and H_1 histamine receptors. Mirtazapine has good efficacy for anxiety symptoms associated with depression (Fawcett and Barkin 1998), and in controlled studies was superior to placebo in PTSD (Davidson et al. 2003) and equivalent to fluoxetine in panic disorder (Ribeiro et al. 2001).

The actions of mirtazapine lead to a unique side-effect profile. Important effects are sedation, drowsiness, dry mouth, increased appetite and weight gain. It does not cause initial worsening of anxiety. Tolerance to the sedative properties occurs after a few weeks and paradoxically higher doses tend to be less sedating. The main effect of overdose is sedation. It has the potential for interaction with drugs that inhibit the CYP450 2D6 and 3A4 isoenzymes, although reports of interactions are rare. Discontinuation symptoms have not yet been reported.

3.2.1.5 Inhibitors of Monoamine Oxidase

MAOIs increase synaptic availability of serotonin, noradrenaline and dopamine by inhibiting their intracellular metabolism. The classical MAOIs phenelzine and tranylcypromine bind irreversibly to monoamine oxidase, whilst the newer drug moclobemide is a reversible inhibitor of monoamine oxidase A (RIMA). The long history of use of MAOIs in panic disorder, PTSD and social anxiety disorder is supported by controlled trials (Sheehan et al. 1980; Frank et al. 1988; Versiani et al. 1992). The evidence for moclobemide is less conclusive, with both positive and negative studies in panic disorder and social anxiety disorder, and meta-analysis suggests a lower response rate in social anxiety disorder than with SSRIs (van der Linden et al. 2000). Brofaromine, a RIMA, was effective in a controlled trial in social anxiety disorder but is no longer marketed.

MAOIs have a significant side-effect profile, including dizziness, drowsiness, insomnia, headache, postural hypotension and anticholinergic effects. Asthenia, weight gain and sexual dysfunction can occur during long-term use. A hypertensive reaction (cheese reaction) may follow the ingestion of foods containing tyramine, which must therefore be removed from the diet. Overdose can be fatal due to seizures, cardiac arrhythmias and hypotension. Interactions can occur with sympathomimetics, antihypertensives and most psychoactive drugs, and a washout of 2 weeks is advised when switching from an MAOI to another antidepressant. Moclobemide is better tolerated than MAOIs, although at high doses (>900 mg daily) dietary restrictions should be observed. The main side-effects are dizziness and insomnia. Overdose toxicity is less, although fatalities have been reported.

3.2.2 Buspirone

Buspirone differs from the antidepressants in that its effects are mediated solely via 5-HT$_{1A}$ receptors. It is a partial agonist at postsynaptic 5-HT$_{1A}$ receptors in the limbic system but a full agonist at autoreceptors in the raphé (Blier and Ward 2003). Acute dosage inhibits serotonin release, but this recovers with continued administration. Anxiolytic effects take several weeks to emerge (Fig. 2). Buspirone is effective in the treatment of GAD (Enkelmann 1991) and for anxiety symptoms in depression (Rickels et al. 1991), either as monotherapy or combined with an SSRI. Response is less favourable if the patient has recently taken a benzodiazepine (DeMartinis et al. 2000). In comparison with benzodiazepines for the treatment of GAD, the onset of the anxiolytic effects are slower but equate to benzodiazepines at 4–6 weeks (Enkelmann 1991). Evidence is lacking to support the use of buspirone in other anxiety disorders.

Buspirone is well-tolerated, with the main side-effects being dizziness, anxiety, nausea and headache. It is tolerated by the elderly (Bohm et al. 1990). It does not cause sexual dysfunction and does not appear to be associated with a discontinuation syndrome. Overdose causes drowsiness but there are no reports of serious toxic effects. A potential for interaction with drugs that inhibit the CYP450 3A4 isoenzyme is not a significant problem in clinical practice. GAD is usually a chronic condition and buspirone is suitable for long-term treatment. Patients should be advised to expect a slow onset of benefits and be reviewed regularly in the early stages of treatment.

3.2.3
β-Blockers

The rationale for using β-adrenoceptor blockers for the treatment of anxiety is twofold: first for the control of symptoms caused by autonomic arousal (e.g. palpitations, tremor) and second because there is a postulated but poorly understood involvement of central noradrenergic activity in anxiety pathways. There is a history of clinical use of these drugs in each of the five major anxiety disorders, but evidence is lacking from controlled clinical trials, and positive findings have often been superseded by later negative studies. Early trials were carried out with propranolol and the more cardioselective atenolol, which has mainly peripheral effects. The efficacy of atenolol in performance anxiety suggests that not all of the effects are centrally mediated (Gorman et al. 1985). Recently there has been interest in pindolol, a β-blocker that also blocks 5-HT_{1A} autoreceptors and may promote serotonergic neurotransmission. Studies using pindolol to augment SSRI treatment of anxiety disorders have had mixed results (Hirschmann et al. 2000; Dannon et al. 2000).

β-Blockers commonly cause side-effects including bradycardia, hypotension, fatigue and bronchospasm. Overdose can cause fatal cardiogenic shock. Because of the doubtful evidence for efficacy and poor tolerability and safety, their use in anxiety disorders is limited. They may have a circumscribed role in the prevention of performance anxiety (Elman et al. 1998).

3.2.4
Antipsychotics

This category contains various drugs that are licensed for the treatment of psychotic disorders. Their effects are mediated via antagonism of D_2 dopamine receptors in the limbic system and cortex. They are loosely divided into two groups: older "classical" drugs such as haloperidol and chlorpromazine that are potent D_2 blockers; and "atypical" antipsychotics that have a lower affinity for D_2 receptors but also block 5-HT_2 receptors. The history of clinical use of classical antipsychotics as "major tranquillisers" has little support from controlled trials (El-Khayat and Baldwin 1998). Evidence is greatest in OCD for the augmentation of SSRI treatment with haloperidol (McDougle et al. 1994) and the atypical drugs risperidone (McDougle et al. 2000) and quetiapine (Atmaca et al. 2002). Recent controlled trials have reported benefits for the atypical drug olanzapine in social anxiety disorder (Barnett et al. 2002) and in addition to SSRIs in PTSD (Stein et al. 2002). Open studies are reporting efficacy for atypical antipsychotics in anxiety disorders and it may be that their clinical use expands in the future.

Atypical antipsychotics have advantages in tolerability and safety over the older drugs. They have a lower incidence of extrapyramidal movement disorders, but may cause sedation and weight gain. Their metabolism by CYP450 enzymes leads to a potential for interaction with many co-prescribed drugs.

3.3
Drugs with Other Mechanisms of Action

3.3.1
Antihistamines

The longstanding use in some countries of hydroxyzine, a centrally-acting H_1-histamine receptor antagonist, is supported by positive findings in controlled trials in GAD (Ferreri and Hantouche 1998; Lader and Scotto 1998). Hydroxyzine promotes sleep and its anxiolytic effects have an early onset. Although it causes sedation, tolerance to this effect often occurs and effects on psychomotor performance are smaller than with benzodiazepines (de Brabander and Deberdt 1990). It is well-tolerated and withdrawal effects have not been reported. Although the evidence for its efficacy is not large, hydroxyzine provides an option for some patients with GAD for whom standard treatments are unsuitable.

3.3.2
Lithium

Lithium is effective in the treatment of mood disorders. Its mechanism of action is unclear but is likely to be via modification of intracellular second messenger systems. There are no controlled trials demonstrating the efficacy of lithium in anxiety disorders, but there have been case reports of its use as an augmenting agent in panic disorder and OCD. The high toxicity and poor tolerability of lithium limit its use in anxiety in the absence of a stronger evidence base.

4 Diagnostic Aspects

Although every method for categorising anxiety disorders has its shortcomings, in current clinical practice the diagnostic criteria of the American Psychiatric Association (DSM-IV 1994) is most commonly used. "Anxiety disorder" is broken down into sub-syndromes with clear operational criteria. In particular, the criteria are clearly stated for a symptomatic individual to become a case, and this diagnostic threshold is usually defined in terms of impairment of occupational, social or domestic functioning. Although many patients will have symptoms from more than one diagnostic category, it is important to elicit the primary diagnosis, as this will influence the recommended treatment. Comorbid disorders, usually a second anxiety disorder, mood disorder or substance use disorder are common and should be detected. The key diagnostic criteria for the major anxiety disorders are given in Table 6.

5 Pharmacotherapy of Anxiety Disorders

5.1 Generalised Anxiety Disorder

GAD is a prevalent, chronic, disabling anxiety disorder. It is comorbid with other anxiety or mood disorders in the majority of cases (Ballenger et al. 2001). Whilst it is a relatively new diagnostic concept, longitudinal studies have reinforced its validity (Kessler et al. 1999). The core symptoms are chronic worry and tension, and GAD frequently presents with somatic complaints such as headache, myalgia or insomnia (Lydiard 2000). The diagnosis requires symptoms to be present for at least 6 months, although the duration of illness at presentation is usually much longer than this. The presence of comorbidity leads to a worse prognosis (Yonkers et al. 1996). Cognitive behavioural therapy (CBT) has been shown to be effective in GAD and should be considered if available (Durham et al. 1994).

Recommended drugs for GAD are antidepressants, benzodiazepines, buspirone and hydroxyzine (Ballenger et al. 2001). The use of antipsychotics is not supported by controlled trials and is discouraged due to their poor long-term

Table 6 DSM-IV classification of anxiety disorders

Generalised anxiety disorder	Excessive worry/anxiety about various matters for at least 6 months Difficulty in controlling worry Accompanying somatic symptoms (effects of chronic tension) Clinically important distress or impairment of functioning
Obsessive–compulsive disorder	Presence of obsessions (thoughts) or compulsions (behaviours) Symptoms are felt by patient to be unreasonable or excessive Clinically important distress or impairment of functioning
Panic disorder (± agoraphobia)	Severe fear or discomfort peaking within 10 minutes Characteristic physical/psychological symptoms Episodes are recurrent and some are unexpected Anxiety about further attacks or consequences of attacks (Agoraphobia: anxiety about place/situation where panic attack is distressing or escape difficult; situation is avoided, endured with distress or companion is required)
Post-traumatic stress disorder	Severe traumatic event that threatened death or serious harm Felt intense fear, horror or helplessness Repeated reliving experiences Phobic avoidance of trauma-related stimuli Hyperarousal Symptoms last >1 month and cause clinically important distress or impairment of functioning
Social anxiety disorder	Recurrent fears of social or performance situations Situations avoided or endured with distress Clinically important distress or impairment of functioning
Specific phobia	Persistent fear/avoidance of specific object or situation Phobic stimulus immediately provokes anxiety response Clinically important distress or impairment of functioning

tolerability. Pregabalin (related to the anticonvulsant gabapentin) was effective in preliminary trials and may be a future treatment option (Pande et al. 2003).

Recent evidence has brought about a shift in prescribing in GAD and now the usual choice for first-line treatment will be an antidepressant. These are effective, well-tolerated, suitable for long-term use and will treat comorbid mood and anxiety disorders. Suitable drugs include venlafaxine (Allgulander et al. 2001) and the SSRI paroxetine (Stocchi et al. 2003). A non-sedating TCA such as imipramine could also be used if tolerated and where the risk of suicide is deemed to be low (Rickels et al. 1993). Little research is available to guide a decision on treatment duration. The recommendation for panic disorder is to continue therapy for at least 12 months following clinical improvement and this seems a reasonable practice to follow in other anxiety disorders (American Psychiatric Association 1998). Buspirone is also appropriate for long-term therapy in the absence of comorbid depression (Rakel 1990).

Benzodiazepines are effective as monotherapy (Rickels et al. 1993) but are rarely used as first-line in this context because of their side-effect profile. They have a useful short-term role for the rapid control of anxiety symptoms or for the control of somatic symptoms such as muscle tension and insomnia, particularly in the early stages of antidepressant therapy. Hydroxyzine has a limited role but can be considered if other treatments are unsuitable (Lader and Scotto 1998).

5.2 Obsessive–Compulsive Disorder

OCD is a disabling disorder that tends to run a chronic or recurrent course (Sasson et al. 1997). It is diagnosed by the presence of obsessions (recurrent, intrusive thoughts, images or impulses that are experienced as irrational and unpleasant) or compulsions (repetitive behaviours that are performed to reduce a feeling of unease). The symptoms are present for at least 1 h every day and cause impairment of important functions. Prevalence has been measured in various populations and is generally 1%–2%. Symptoms start as early as the first decade and have often been present in excess of 10 years at presentation (Hollander et al. 1996). Depression occurs in more than 50% of cases and there is significant comorbidity with other anxiety disorders, eating disorders and tic disorders. Although classified with the anxiety disorders, OCD is distinct from the rest of this group in its epidemiological profile and neurobiology. In clinical terms, OCD symptoms respond to drugs that enhance serotonergic neurotransmission but not to noradrenergic drugs, and they respond poorly to benzodiazepines.

The recommended first-line drugs for OCD are SSRIs and the TCA clomipramine (Pigott and Seay 1999). The required dose is generally higher than that required for other disorders (e.g. clomipramine 150–250 mg, paroxetine 40–60 mg) and SSRIs have advantages in safety and tolerability. Long-term treatment may be required. There is a good evidence base for the efficacy of CBT, and there may be added benefits from combining psychological and pharmacological therapies (Hohagen et al. 1998). In cases poorly responsive to SSRI treatment, augmentation with the antipsychotics haloperidol, risperidone or quetiapine has support from clinical trials, and addition of buspirone, lithium and the serotonin precursor L-tryptophan have also been tried. In severe treatment-resistant cases the neurosurgical procedure stereotactic cingulotomy should be considered (Jenike et al. 1991).

5.3 Panic Disorder and Agoraphobia

Panic disorder is also a common, chronic and disabling disorder with its peak incidence in young adulthood (Ballenger et al. 1998a). A panic attack is defined as the sudden onset of anxiety symptoms, rising to a peak within 10 min. DSM-IV requires 4 of 13 defined symptoms to be present. The symptoms are physical symptoms corresponding to those caused by autonomic arousal and psychological symptoms (fear and depersonalisation/derealisation, an altered perception of oneself or the world around). Panic disorder occurs when there are recurrent panic attacks, some of which are uncued or unexpected, and there is fear of having further attacks. Agoraphobia is present in around half of cases (Wittchen et al. 1998) and is a poor prognostic indicator. For some patients the anticipatory anxiety or agoraphobia may be considerably more disabling than the panic attacks themselves.

Panic disorder is comorbid with episodes of depression at some stage in the majority of cases (Stein et al. 1990), with social anxiety disorder and to a lesser extent GAD and PTSD, and with alcohol dependence and personality disorder. Comorbidity results in increased severity and poor response to treatment. Panic disorder is associated with a significantly increased risk of suicide, and this is increased further by the presence of comorbid depression (Lepine et al. 1993).

There is solid evidence for pharmacotherapy of panic disorder with SSRIs (Boyer 1995), the TCAs clomipramine and imipramine (Lecrubier et al. 1997; Cross-National Collaborative Panic Study 1992) and the benzodiazepines alprazolam, clonazepam, lorazepam and diazepam (Ballenger et al. 1988; Beauclair et al. 1994; Charney and Woods 1989; Noyes et al. 1996). Therapy is likely to be required for a minimum of 12 months, and the favourable tolerability of SSRIs will usually lead to their choice as first-line therapy. Patients with panic disorder are sensitive to drug side-effects, so a low initial dose should be used and titrated up to the recommended treatment dose (e.g. paroxetine 10 mg titrated up to 40 mg). Coadministration of a benzodiazepine with an SSRI for the first 2–4 weeks may reduce initial agitation and hasten clinical improvement (Goddard et al. 2001). Once improvement has been achieved, the dose may be slowly reduced to a lower maintenance level. Stopping treatment is associated with discontinuation effects and an increased risk of relapse and should be approached with caution. CBT is an effective treatment for panic disorder and additional benefits may be gained from combination therapy (Oehrberg et al. 1995). Other drugs effective in controlled studies include the antidepressants phenelzine (Sheehan et al. 1980), moclobemide (Tiller et al. 1999), venlafaxine (Pollack et al. 1996), mirtazapine (Ribeiro et al. 2001) and reboxetine (Versiani et al. 2002), and the anticonvulsants sodium valproate (Lum et al. 1991) and gabapentin (Pande et al. 2000).

5.4
Post-traumatic Stress Disorder

This is another anxiety disorder that is common although underdiagnosed, frequently chronic and usually severely disabling (Ballenger et al. 2000). The diagnosis is given when specific psychological and physical symptoms follow exposure to a traumatising event that invokes fear, horror and helplessness. Symptoms fall into three categories: re-experiencing phenomena (flashbacks, nightmares, distress when memories of trauma are triggered); persistent avoidance of triggers to memory of the trauma and general numbing; hyperarousal (insomnia, irritability, poor concentration, hypervigilance, increased startle response). Symptoms must persist for more than 1 month after the trauma.

PTSD is highly comorbid with depression (Kessler et al. 1995) and substance use disorders, and is associated with a previous exposure to trauma and a previous history of anxiety disorders. PTSD probably carries the highest risk of suicide among the anxiety disorders (Davidson et al. 1991). Without effective treatment the disorder generally runs a chronic, unremitting course.

The evidence base for pharmacotherapy is shallow although improving. Efficacy is established for the SSRIs, particularly paroxetine (Tucker et al. 2001), fluoxetine (Connor et al. 1999) and sertraline (Brady et al. 2000) and the TCA amitriptyline (Davidson et al. 1993a). Treatment is started at standard dose but may be required to be titrated upwards (e.g. paroxetine 20–50 mg). Results from long-term studies are awaited but treatment should be continued for a minimum of 12 months. Medication is given alongside psychotherapy, usually cognitive and exposure therapies (Foa 2000). Other treatments include the antidepressants phenelzine and mirtazapine, the anticonvulsants lamotrigine, sodium valproate, carbamazepine and tiagabine, and augmentation with the atypical antipsychotic olanzapine. The use of benzodiazepines is not advised, as their efficacy is not established and withdrawal symptoms may be particularly distressing. If insomnia is problematic then a non-benzodiazepine hypnotic may be prescribed.

5.5
Social Anxiety Disorder

This disorder is characterised by anxiety symptoms in social or performance situations, accompanied by a fear of embarrassment or humiliation. Situations are avoided or endured with distress. There may be a specific fear of one or two situations (most commonly public speaking), or of three or more situations in the generalized subtype. Epidemiological studies find this to be the most prevalent anxiety disorder among the general population (Magee et al. 1996). Its peak onset is around the time of adolescence, and the resulting impairments can have a profound effect on social and occupational development. If untreated it tends to follow a chronic, unremitting course. Social anxiety disorder is frequently comorbid with depression, other anxiety disorders, alcohol problems and eating disorders. It is associated with an increased rate of suicide that is significantly higher in the presence of comorbidity (Schneier et al. 1992).

Drug studies have focussed on the generalized subtype (Ballenger et al. 1998b). The largest evidence base is for the SSRIs, which are accepted to be the drug treatment of choice. Treatment is started at standard dose and increased as necessary (e.g. paroxetine 20–50 mg). Duration of treatment is usually for at least 12 months, and there is benefit from combination with CBT (Blomhoff et al. 2001). The other class of antidepressant to be considered is the MAOIs, as phenelzine and moclobemide have controlled trial data to support their use (Versiani et al. 1992). TCAs have no proven efficacy and evidence for venlafaxine and mirtazapine is awaited. Among the benzodiazepines only clonazepam has been shown to be effective as monotherapy, possibly due to its effects on $5HT_{1A}$ receptors (Davidson et al. 1993b). Benzodiazepines may also be used to augment SSRI treatment. Other drugs to consider are the anticonvulsant gabapentin and the antipsychotic olanzapine. β-Blockers are not effective in generalized social anxiety disorder but have a role in symptomatic control in specific performance anxiety.

5.6 Specific Phobia

In specific phobia disorder the patient has an inappropriate or excessive fear of a particular stimulus or situation, such as animals, heights or thunder. An anxiety reaction is consistently and rapidly evoked on exposure to the stimulus, and there is anticipatory anxiety. Population studies have found a surprisingly high prevalence and associated disability, for example a lifetime prevalence of 12% in the National Comorbidity Survey (Magee et al. 1996). The standard treatment for specific phobia is behavioural therapy, and patients rarely present for pharmacological treatment. Nevertheless, there are clinical and pharmacological similarities between patients with specific phobias and those with other anxiety disorders (Verburg et al. 1994), and it might be predicted that anxiolytic medications would have beneficial effects. A small controlled study found an improvement in measures of fear and avoidance after a 4-week trial of the SSRI paroxetine (Benjamin et al. 2000), and there is also a role for the use of a short-acting benzodiazepine to control anxiety prior to exposure to the feared stimulus.

5.7 Depression with Concomitant Anxiety

The prevalence of depression in patients with anxiety disorders is high, as is the prevalence of anxiety in patients with depression (Tylee et al. 1999; Kessler et al. 1998). Among patients presenting for treatment of anxiety symptoms, a large proportion will have a primary diagnosis of depression. In these situations it is critical to offer a treatment plan that will prove effective against both anxiety and depression (Nutt 2000). The presence of both disorders together causes an increase in disability, increased severity of symptoms, a higher likelihood of suicidal thoughts and a poor response to treatment (Lepine et al. 1997).

Antidepressants would be the obvious drug class to select in this patient group, and a number of controlled studies have demonstrated their efficacy. Both SSRIs and TCAs are effective, with the most evidence being for the SSRI paroxetine and the TCAs clomipramine and amitriptyline (Feighner et al. 1993; Ravindran et al. 1997; Stott et al. 1993). Comparative studies favour the SSRIs because of their better tolerability, and safety is also a factor in a group at high risk of suicide. Recent studies have demonstrated the efficacy of the new antidepressants venlafaxine (Silverstone and Ravindran 1999) and mirtazapine (Fawcett and Barkin 1998) in this group, and as their tolerability matches that of the SSRIs they should also be considered as first-line treatment. Benzodiazepines produce a rapid improvement in anxiety but are ineffective at treating depression (Lenox et al. 1984) and are not suitable for long-term treatment in this context. They have a short-term role on initiation of antidepressant therapy in selected patients.

6 Conclusions and Future Directions

It has been shown that the recent shift in clinical practice towards the use of antidepressants, particularly SSRIs, for the first-line treatment of anxiety disorders is supported by research evidence from randomised controlled trials. The use of these drugs is likely to be refined in future years as important gaps in the current knowledge base are filled. These include the optimal duration of treatment, the identification of patients at particular risk of relapse, the benefits of combining drugs with psychotherapy and suitable options for patients resistant to first-line treatments. New drugs available for the treatment of depression may also prove to be effective for anxiety disorders. The prime position of the SSRIs has been reinforced by evidence for the role of serotonin in anxiety; the newer antidepressants tend to have a dual action on serotonergic and noradrenergic neurotransmission, and clarification of the role of noradrenaline in anxiety is likely to occur.

It is only in recent years that drugs acting via GABA neurotransmission have been supplanted as first-line treatments, and new drugs in this class with improved tolerability compared to the benzodiazepines are likely to be marketed in the near future (Ashton and Young 2003). Further down the line, agonists that are selective for specific subunits of the $GABA_A$ receptor offer the prospect of drugs that are anxiolytic but with fewer sedative properties (Nutt and Malizia 2001). Overall it is remarkable that current pharmacological strategies are centred around such a small number of brain mechanisms. Future strategies may involve glutamate neurotransmission (Kent et al. 2002) and neuropeptides such as corticotrophin releasing factor antagonists (Gutman et al. 2001) and substance P antagonists (Argyropoulos and Nutt 2000), and a continued expansion in the range of anxiolytic therapies should be anticipated.

References

Allgulander C, Hackett D, Salinas E (2001) Venlafaxine extended release (ER) in the treatment of generalized anxiety disorder: twenty-four-week placebo-controlled dose-ranging study. Br J Psychiatry 179:15–22

American Psychiatric Association (1998) Practice guidelines for the treatment of patients with panic disorder. American Psychiatric Press, Washington

Argyropoulos SV, Nutt DJ (2000) Substance P antagonists: novel agents in the treatment of depression. Expert Opin Investig Drugs 9:1871–1875

Argyropoulos SV, Nutt DJ (2003) Neurochemical aspects of anxiety. In: Nutt DJ, Ballenger JC (eds) Anxiety disorders. Blackwell Science, Oxford, pp 183–199

Ashton CH, Young AH (2003) GABA-ergic drugs: exit stage left, enter stage right. J Psychopharmacol 17:174–178

Atmaca M, Kuloglu M, Tezcan E, Gecici O (2002) Quetiapine augmentation in patients with treatment resistant obsessive-compulsive disorder: a single-blind, placebo-controlled study. Int Clin Psychopharmacol 17:115–119

Ballenger JC, Burrows GD, DuPont RL Jr, Lesser IM, Noyes R Jr, Pecknold JC, Rifkin A, Swinson RP (1988) Alprazolam in panic disorder and agoraphobia: results from a multicenter trial. I. Efficacy in short-term treatment. Arch Gen Psychiatry 45:413–422

Ballenger JC, Davidson JR, Lecrubier Y, Nutt DJ, Baldwin DS, den Boer JA, Kasper S, Shear MK (1998a) Consensus statement on panic disorder from the International Consensus Group on Depression and Anxiety. J Clin Psychiatry 59 Suppl 8:47–54

Ballenger JC, Davidson JR, Lecrubier Y, Nutt DJ, Bobes J, Beidel DC, Ono Y, Westenberg HG (1998b) Consensus statement on social anxiety disorder from the International Consensus Group on Depression and Anxiety. J Clin Psychiatry 59 Suppl 17:54–60

Ballenger JC, Davidson JR, Lecrubier Y, Nutt DJ, Foa EB, Kessler RC, McFarlane AC, Shalev AY (2000) Consensus statement on posttraumatic stress disorder from the International Consensus Group on Depression and Anxiety. J Clin Psychiatry 61 Suppl 5:60–66

Ballenger JC, Davidson JR, Lecrubier Y, Nutt DJ, Borkovec TD, Rickels K, Stein DJ, Wittchen HU (2001) Consensus statement on generalized anxiety disorder from the International Consensus Group on Depression and Anxiety. J Clin Psychiatry 62 Suppl 11:53–58

Bandelow B, Zohar J, Hollander E, Kasper S, Moller HJ (2002) World Federation of Societies of Biological Psychiatry (WFSBP) guidelines for the pharmacological treatment of anxiety, obsessive-compulsive and posttraumatic stress disorders. World J Biol Psychiatry 3:171–199

Barbey JT, Roose SP (1998) SSRI safety in overdose. J Clin Psychiatry 59 Suppl 15:42–48

Barbone F, McMahon AD, Davey PG, Morris AD, Reid IC, McDevitt DG, MacDonald TM (1998) Association of road-traffic accidents with benzodiazepine use. Lancet 352:1331–1336

Barlow DH, Gorman JM, Shear MK, Woods SW (2000) Cognitive-behavioral therapy, imipramine, or their combination for panic disorder: a randomized controlled trial. JAMA 283:2529–2536

Barnett SD, Kramer ML, Casat CD, Connor KM, Davidson JR (2002) Efficacy of olanzapine in social anxiety disorder: a pilot study. J Psychopharmacol 16:365–368

Beauclair L, Fontaine R, Annable L, Holobow N, Chouinard G (1994) Clonazepam in the treatment of panic disorder: a double-blind, placebo-controlled trial investigating the correlation between clonazepam concentrations in plasma and clinical response. J Clin Psychopharmacol 14:111–118

Bell CJ, Nutt DJ (1998) Serotonin and panic. Br J Psychiatry 172:465–471

Benjamin J, Ben-Zion IZ, Karbofsky E, Dannon P (2000) Double-blind placebo-controlled pilot study of paroxetine for specific phobia. Psychopharmacology (Berl) 149:194–196
Berlant J, van Kammen DP (2002) Open-label topiramate as primary or adjunctive therapy in chronic civilian posttraumatic stress disorder: a preliminary report. J Clin Psychiatry 63:15–20
Blier P, Ward NM (2003) Is there a role for 5-HT(1A) agonists in the treatment of depression? Biol Psychiatry 53:193–203
Blier P, de Montigny C, Chaput Y (1990) A role for the serotonin system in the mechanism of action of antidepressant treatments: preclinical evidence. J Clin Psychiatry 51:S14–S20
Blomhoff S, Haug TT, Hellstrom K, Holme I, Humble M, Madsbu HP, Wold JE (2001) Randomised controlled general practice trial of sertraline, exposure therapy and combined treatment in generalized social phobia. Br J Psychiatry 179:23–30
Bohm C, Robinson DS, Gammans RE, Shrotriya RC, Alms DR, Leroy A, Placchi M (1990) Buspirone therapy in anxious elderly patients: a controlled clinical trial. J Clin Psychopharmacol 10(Suppl 3):47S–51S
Boyer W (1995) Serotonin uptake inhibitors are superior to imipramine and alprazolam in alleviating panic attacks: a meta-analysis. Int Clin Psychopharmacol 10:45–49
Brady K, Pearlstein T, Asnis GM, Baker D, Rothbaum B, Sikes CR, Farfel GM (2000) Efficacy and safety of sertraline treatment of posttraumatic stress disorder: a randomized controlled trial. JAMA 283:1837–1844
Buckley NA, McManus PR (2002) Fatal toxicity of serotoninergic and other antidepressant drugs: analysis of United Kingdom mortality data. BMJ 325:1332–1333
Buckley NA, Dawson AH, Whyte IM, O'Connell DL (1995) Relative toxicity of benzodiazepines in overdose. BMJ 310:219–221
Bull SA, Hu XH, Hunkeler EM, Lee JY, Ming EE, Markson LE, Fireman B (2002) Discontinuation of use and switching of antidepressants: influence of patient-physician communication. JAMA 288:1403–1409
Carlsten A, Waern M, Ekedahl A, Ranstam J (2001) Antidepressant medication and suicide in Sweden. Pharmacoepidemiol Drug Saf 10:525–530
Charney DS, Woods SW (1989) Benzodiazepine treatment of panic disorder: a comparison of alprazolam and lorazepam. J Clin Psychiatry 50:418–423
Clomipramine Collaborative Study Group (1991) Clomipramine in the treatment of patients with obsessive-compulsive disorder. Arch Gen Psychiatry 48:730–738
Connor KM, Sutherland SM, Tupler LA, Malik ML, Davidson JR (1999) Fluoxetine in posttraumatic stress disorder. Randomised, double-blind study. Br J Psychiatry 175:17–22
Cross-National Collaborative Panic Study (1992) Drug treatment of panic disorder. Comparative efficacy of alprazolam, imipramine, and placebo. Br J Psychiatry 160:191–202
Dannon PN, Sasson Y, Hirschmann S, Iancu I, Grunhaus LJ, Zohar J (2000) Pindolol augmentation in treatment-resistant obsessive compulsive disorder: a double-blind placebo controlled trial. Eur Neuropsychopharmacol 10:165–169
Dannon PN, Iancu I, Grunhaus L (2002) Psychoeducation in panic disorder patients: effect of a self-information booklet in a randomized, masked-rater study. Depress Anxiety 16:71–76
Davidson JR, Hughes D, Blazer DG, George LK (1991) Post-traumatic stress disorder in the community: an epidemiological study. Psychol Med 21:713–721
Davidson JR, Kudler HS, Saunders WB, Erickson L, Smith RD, Stein RM, Lipper S, Hammett EB, Mahorney SL, Cavenar JO (1993a) Predicting response to amitriptyline in posttraumatic stress disorder. Am J Psychiatry 150:1024–1029

Davidson JR, Potts N, Richichi E, Krishnan R, Ford SM, Smith R, Wilson WH (1993b) Treatment of social phobia with clonazepam and placebo. J Clin Psychopharmacol 13:423–428
Davidson JR, Weisler RH, Butterfield MI, Casat CD, Connor KM, Barnett S, van Meter S (2003) Mirtazapine vs. placebo in posttraumatic stress disorder: a pilot trial. Biol Psychiatry 53:188–191
Davies SJ, Jackson PR, Ramsay LE, Ghahramani P (2003) Drug intolerance due to nonspecific adverse effects related to psychiatric morbidity in hypertensive patients. Arch Intern Med 163:592–600
de Brabander A, Deberdt W (1990) Effects of hydroxyzine on attention and memory. Hum Psychopharmacol 5:357–362
DeMartinis N, Rynn M, Rickels K, Mandos L (2000) Prior benzodiazepine use and buspirone response in the treatment of generalized anxiety disorder. J Clin Psychiatry 61:91–94
DSM-IV (1994) American Psychiatric Association diagnostic and statistical manual of mental disorders, 4th edn. American Psychiatric Press, Washington
Durham RC, Murphy T, Allan T, Richard K, Treliving LR, Fenton GW (1994) Cognitive therapy, analytic psychotherapy and anxiety management training for generalized anxiety disorder. Br J Psychiatry 165:315–323
El-Khayat R, Baldwin DS (1998) Antipsychotic drugs for non-psychotic patients: assessment of the benefit/risk ratio in generalized anxiety disorder. J Psychopharmacol 12:323–329
Elman MJ, Sugar J, Fiscella R, Deutsch TA, Noth J, Nyberg M, Packo K, Anderson RJ (1998) The effect of propranolol versus placebo on resident surgical performance. Trans Am Ophthalmol Soc 96:283–291
Enkelmann R (1991) Alprazolam versus buspirone in the treatment of outpatients with generalized anxiety disorder. Psychopharmacology (Berl) 105:428–432
Faravelli C, Rosi S, Truglia E (2003) Treatments: benzodiazepines. In: Nutt DJ, Ballenger JC (eds) Anxiety disorders. Blackwell Science, Oxford, pp 315–338
Fawcett J, Barkin RL (1998) A meta-analysis of eight randomized, double-blind, controlled clinical trials of mirtazapine for the treatment of patients with major depression and symptoms of anxiety. J Clin Psychiatry 59:123–127
Febbraro GA, Clum GA (1998) Meta-analytic investigation of the effectiveness of self-regulatory components in the treatment of adult problem behaviors. Clin Psychol Rev 18:143–161
Feighner JP, Cohn JB, Fabre LF Jr, Fieve RR, Mendels J, Shrivastava RK, Dunbar GC (1993) A study comparing paroxetine placebo and imipramine in depressed patients. J Affect Disord 28:71–79
Ferreri M, Hantouche EG (1998) Recent clinical trials of hydroxyzine in generalized anxiety disorder. Acta Psychiatr Scand 393 Suppl:102–108
Foa EB (2000) Psychosocial treatment of posttraumatic stress disorder. J Clin Psychiatry 61 Suppl 5:43–48
Frank JB, Kosten TR, Giller EL Jr, Dan E (1988) A randomized clinical trial of phenelzine and imipramine for posttraumatic stress disorder. Am J Psychiatry 145:1289–1291
Goddard AW, Brouette T, Almai A, Jetty P, Woods SW, Charney D (2001) Early coadministration of clonazepam with sertraline for panic disorder. Arch Gen Psychiatry 58:681–686
Goodnick PJ, Goldstein BJ (1998) SSRIs in affective disorders. I. Basic pharmacology. J Psychopharmacol 12(Suppl B):S5–S20
Gorman JM, Liebowitz MR, Fyer AJ, Campeas R, Klein DF (1985) Treatment of social phobia with atenolol. J Clin Psychopharmacol 5:298–301
Gutman DA, Owens MJ, Nemeroff CB (2001) CRF receptor antagonists: a new approach to the treatment of depression. Pharm News 8:18–25

Haddad P (1998) The SSRI discontinuation syndrome. J Psychopharmacol 12:305–313
Healy D (2003) Lines of evidence on the risks of suicide with selective serotonin reuptake inhibitors. Psychother Psychosom 72:71–79
Hertzberg MA, Butterfield MI, Feldman ME, Beckham JC, Sutherland SM, Connor KM, Davidson JR (1999) A preliminary study of lamotrigine for the treatment of posttraumatic stress disorder. Biol Psychiatry 45:1226–1229
Hirschmann S, Dannon PN, Iancu I, Dolberg OT, Zohar J, Grunhaus L (2000) Pindolol augmentation in patients with treatment-resistant panic disorder: a double-blind, placebo-controlled trial. J Clin Psychopharmacol 20:556–559
Hohagen F, Winkelmann G, Rasche-Ruchle H, Hand I, Konig A, Munchau N, Hiss H, Geiger-Kabisch C, Kappler C, Schramm P, Rey E, Aldenhoff J, Berger M (1998) Combination of behaviour therapy with fluvoxamine in comparison with behaviour therapy and placebo. Results of a multicentre study. Br J Psychiatry Suppl 35:71–78
Hollander E, Greenwald S, Neville D, Johnson J, Hornig CD, Weissman MM (1996) Uncomplicated and comorbid obsessive-compulsive disorder in an epidemiologic sample. Depress Anxiety 4:111–119
Jenike MA, Baer L, Ballantine T, Martuza RL, Tynes S, Giriunas I, Buttolph ML, Cassem NH (1991) Cingulotomy for refractory obsessive-compulsive disorder. A long-term follow-up of 33 patients. Arch Gen Psychiatry 48:548–555
Kent JM, Mathew SJ, Gorman JM (2002) Molecular targets in the treatment of anxiety. Biol Psychiatry 52:1008–1030
Kessler RC, McGonagle KA, Zhao S, Nelson CB, Hughes M, Eshleman S, Wittchen H-U, Kendler KS (1994) Lifetime and 12-month prevalence of DSM-III-R psychiatric disorders in the United States: results from the National Comorbidity Survey. Arch Gen Psychiatry 51:8–19
Kessler RC, Sonnega A, Bromet E, Hughes M, Nelson CB (1995) Posttraumatic stress disorder in the National Comorbidity Survey. Arch Gen Psychiatry 52:1048–1060
Kessler RC, Stang PE, Wittchen HU, Ustun TB, Roy-Burne PP, Walters EE (1998) Lifetime panic-depression comorbidity in the National Comorbidity Survey. Arch Gen Psychiatry 55:801–808
Kessler RC, DuPont RL, Berglund P, Wittchen HU (1999) Impairment in pure and comorbid generalized anxiety disorder and major depression at 12 months in two national surveys. Am J Psychiatry 156:1915–1923
Khan A, Khan S, Kolts R, Brown WA (2003) Suicide rates in clinical trials of SSRIs, other antidepressants, and placebo: analysis of FDA reports. Am J Psychiatry 160:790–792
Klein DF (1964) Delineation of two drug-responsive anxiety syndromes. Psychopharmacologia 5:397–408
Kobak KA, Greist JH, Jefferson JW, Katzelnick DJ, Henk HJ (1998) Behavioral versus pharmacological treatments of obsessive compulsive disorder: a meta-analysis. Psychopharmacology (Berl) 136:205–216
Lader M, Scotto JC (1998) A multicentre double-blind comparison of hydroxyzine, buspirone and placebo in patients with generalized anxiety disorder. Psychopharmacology (Berl) 139:402–406
Lecrubier Y, Judge R (1997) Long-term evaluation of paroxetine, clomipramine and placebo in panic disorder. Collaborative Paroxetine Panic Study Investigators. Acta Psychiatr Scand 95:153–160
Lenox RH, Shipley JE, Peyser JM, Williams JM, Weaver LA (1984) Double-blind comparison of alprazolam versus imipramine in the inpatient treatment of major depressive illness. Psychopharmacol Bull 20:79–82

Lepine JP, Chignon JM, Teherani M (1993) Suicide attempts in patients with panic disorder. Arch Gen Psychiatry 50:144–149
Lepine JP, Gastpar M, Mendlewicz J, Tylee A (1997) Depression in the community: the first pan-European study DEPRES (Depression Research in European Society). Int Clin Psychopharmacol 12:19–29
Lum M, Fontaine R, Elie R, Ontiveros A (1991) Probable interaction of sodium divalproex with benzodiazepines. Prog Neuropsychopharmacol Biol Psychiatry 15:269–273
Lydiard RB (2000) An overview of generalized anxiety disorder: disease state—appropriate therapy. Clin Ther 22(Suppl A):3–19
Lydiard RB (2003) The role of GABA in anxiety disorders. J Clin Psychiatry 64 suppl 3:21–27
Magee WJ, Eaton WW, Wittchen HU, McGonagle KA, Kessler RC (1996) Agoraphobia, simple phobia, and social phobia in the National Comorbidity Survey. Arch Gen Psychiatry 53:159–168
Mason J, Freemantle N, Eccles M (2000) Fatal toxicity associated with antidepressant use in primary care. Br J Gen Pract 50:366–370
McDougle CJ, Goodman WK, Leckman JF, Lee NC, Heninger GR, Price LH (1994) Haloperidol addition in fluvoxamine-refractory obsessive-compulsive disorder. A double-blind, placebo-controlled study in patients with and without tics. Arch Gen Psychiatry 51:302–308
McDougle CJ, Epperson CN, Pelton GH, Wasylink S, Price LH (2000) A double-blind, placebo-controlled study of risperidone addition in serotonin reuptake inhibitor-refractory obsessive-compulsive disorder. Arch Gen Psychiatry 57:794–801
Melichar JK, Haida A, Rhodes C, Reynolds AH, Nutt DJ, Malizia AL (2001) Venlafaxine occupation at the noradrenaline reuptake site: in-vivo determination in healthy volunteers. J Psychopharmacol 15:9–12
Michelson D, Pollack M, Lydiard RB, Tamura R, Tepner R, Tollefson G (1999) Continuing treatment of panic disorder after acute response: randomised, placebo-controlled trial with fluoxetine. The Fluoxetine Panic Disorder Study Group. Br J Psychiatry 174:213–218
Michelson D, Fava M, Amsterdam J, Apter J, Londborg P, Tamura R, Tepner RG (2000) Interruption of selective serotonin reuptake inhibitor treatment. Double-blind, placebo-controlled trial. Br J Psychiatry 176:363–368
Noyes R, Burrows GD, Reich JH, Judd FK, Garvey MJ, Norman TR, Cook BL, Marriott P (1996) Diazepam versus alprazolam for the treatment of panic disorder. J Clin Psychiatry 57:349–355
Nutt D (2000) Treatment of depression and concomitant anxiety. Eur Neuropsychopharmacol 10 Suppl 4:433–437
Nutt DJ, Malizia AL (2001) New insights into the role of the GABA(A)-benzodiazepine receptor in psychiatric disorder. Br J Psychiatry 179:390–396
Oehrberg S, Christiansen PE, Behnke K, Borup AL, Severin B, Soegaard J, Calberg H, Judge R, Ohrstrom JK, Manniche PM (1995) Paroxetine in the treatment of panic disorder. A randomised, double-blind, placebo-controlled study. Br J Psychiatry 167:374–379
Pande AC, Davidson JR, Jefferson JW, Janney CA, Katzelnick DJ, Weisler RH, Greist JH, Sutherland SM (1999) Treatment of social phobia with gabapentin: a placebo-controlled study. J Clin Psychopharmacol 19:341–348
Pande AC, Pollack MH, Crockatt J, Greiner M, Chouinard G, Lydiard RB, Taylor CB, Dager SR, Shiovitz T (2000) Placebo-controlled study of gabapentin treatment of panic disorder. J Clin Psychopharmacol 20:467–471

Pande AC, Crockatt JG, Feltner DE, Janney CA, Smith WT, Weisler R, Londborg PD, Bielski RJ, Zimbroff DL, Davidson JR, Liu-Dumaw M (2003) Pregabalin in generalized anxiety disorder: a placebo-controlled trial. Am J Psychiatry 160:533–540

Pigott TA, Seay SM (1999) A review of the efficacy of selective serotonin reuptake inhibitors in obsessive-compulsive disorder. J Clin Psychiatry 60:101–106

Pollack MH, Worthington JJ 3rd, Otto MW, Maki KM, Smoller JW, Manfro GG, Rudolph R, Rosenbaum JF (1996) Venlafaxine for panic disorder: results from a double-blind, placebo-controlled study. Psychopharmacol Bull 32:667–670

Rakel RE (1990) Long-term buspirone therapy for chronic anxiety: a multicenter international study to determine safety. South Med J 83:194–198

Ravindran AV, Judge R, Hunter BN, Bray J, Morton NH (1997) A double-blind, multicenter study in primary care comparing paroxetine and clomipramine in patients with depression and associated anxiety. J Clin Psychiatry 58:112–118

Ribeiro L, Busnello JV, Kauer-Sant'Anna M, Madruga M, Quevedo J, Busnello EA, Kapczinski F (2001) Mirtazapine versus fluoxetine in the treatment of panic disorder. Braz J Med Biol Res 34:1303–1337

Rickels K, Schweizer E (1998) Panic disorder: long-term pharmacotherapy and discontinuation. J Clin Psychopharmacol 18:12S–18S

Rickels K, Amsterdam JD, Clary C, Puzzuoli G, Schweizer E (1991) Buspirone in major depression: a controlled study. J Clin Psychiatry 52:34–38

Rickels K, Downing R, Schweizer E, Hassman H (1993) Antidepressants for the treatment of generalized anxiety disorder. A placebo-controlled comparison of imipramine, trazodone, and diazepam. Arch Gen Psychiatry 50:884–895

Rocca P, Fonzo V, Scotta M, Zanalda E, Ravizza L (1997) Paroxetine efficacy in the treatment of generalized anxiety disorder. Acta Psychiatr Scand 95:444–450

Sasson Y, Zohar J, Chopra M, Lustig M, Iancu I, Hendler T (1997) Epidemiology of obsessive-compulsive disorder: a world view. J Clin Psychiatry 58 Suppl 12:7–10

Schneier FR, Johnson J, Hornig CD, Liebowitz MR, Weissman MM (1992) Social phobia. Comorbidity and morbidity in an epidemiologic sample. Arch Gen Psychiatry 49:282–288

Schweizer E, Rickels K (1998) Benzodiazepine dependence and withdrawal: a review of the syndrome and its clinical management. Acta Psychiatr Scand 393:S95–S101

Sheehan DV, Ballenger J, Jacobsen G (1980) Treatment of endogenous anxiety with phobic, hysterical, and hypochondriacal symptoms. Arch Gen Psychiatry 37:51–59

Silverstone PH, Ravindran A (1999) Once-daily venlafaxine extended release (XR) compared with fluoxetine in outpatients with depression and anxiety. J Clin Psychiatry 60:22–28

Stein MB, Tancer ME, Uhde TW (1990) Major depression in patients with panic disorder: factors associated with course and recurrence. J Affect Disord 19:287–296

Stein MB, Kline NA, Matloff JL (2002) Adjunctive olanzapine for SSRI-resistant combat-related PTSD: a double-blind, placebo-controlled study. Am J Psychiatry 159:1777–1779

Sternbach H (1991) The serotonin syndrome. Am J Psychiatry 148:705–713

Stocchi F, Nordera G, Jokinen RH, Lepola UM, Hewett K, Bryson H, et al. (2003) Efficacy and tolerability of paroxetine for the long-term treatment of generalized anxiety disorder. J Clin Psychiatry 64:250–258

Stott PC, Blagden MD, Aitken CA (1993) Depression and associated anxiety in primary care: a double-blind comparison of paroxetine and amitriptyline. Eur Neuropsychopharmacol 3:324–325

Task Force Report of the American Psychiatric Association (1990) Benzodiazepine dependence, toxicity and abuse. American Psychiatric Press, Washington

Tiller JW, Bouwer C, Behnke K (1999) Moclobemide and fluoxetine for panic disorder. International Panic Disorder Study Group. Eur Arch Psychiatry Clin Neurosci 249 Suppl 1:7–10

Tucker P, Zaninelli R, Yehuda R, Ruggiero L, Dillingham K, Pitts CD (2001) Paroxetine in the treatment of chronic posttraumatic stress disorder: results of a placebo-controlled, flexible-dosage trial. J Clin Psychiatry 62:860–868

Tylee A, Gastpar M, Lepine JP, Mendlewicz J (1999) Identification of depressed patient types in the community and their treatment needs: findings from the DEPRES II (Depression Research in European Society II) survey. Int Clin Psychopharmacol 14:153–165

van der Linden GJ, Stein DJ, van Balkom AJ (2000) The efficacy of the selective serotonin reuptake inhibitors for social anxiety disorder (social phobia): a meta-analysis of randomized controlled trials. Int Clin Psychopharmacol 15 Suppl 2:S15–S23

Verburg C, Griez E, Meijer J (1994) A 35% carbon dioxide challenge in simple phobias. Acta Psychiatr Scand 90:420–423

Versiani M, Nardi AE, Mundim FD, Alves AB, Liebowitz MR, Amrein R (1992) Pharmacotherapy of social phobia. A controlled study with moclobemide and phenelzine. Br J Psychiatry 161:353–360

Versiani M, Cassano G, Perugi G, Benedetti A, Mastalli L, Nardi A, Savino M (2002) Reboxetine, a selective norepinephrine reuptake inhibitor, is an effective and well-tolerated treatment for panic disorder. J Clin Psychiatry 63:31–37

Wang PS, Bohn RL, Glynn RJ, Mogun H, Avorn J (2001) Hazardous benzodiazepine regimens in the elderly: effects of half-life, dosage, and duration on risk of hip fracture. Am J Psychiatry 158:892–898

Waugh J, Goa KL (2003) Escitalopram: a review of its use in the management of major depressive and anxiety disorders. CNS Drugs 17:343–362

Williams DD, McBride A (1998) Benzodiazepines: time for reassessment. Br J Psychiatry 173:361–362

Wittchen HU, Reed V, Kessler RC (1998) The relationship of agoraphobia and panic in a community sample of adolescents and young adults. Arch Gen Psychiatry 55:1017–1024

Yonkers KA, Warshaw MG, Massion AO, Keller MB (1996) Phenomenology and course of generalized anxiety disorder. Br J Psychiatry 168:308–313

Zohar J, Judge R (1996) Paroxetine versus clomipramine in the treatment of obsessive-compulsive disorder. Br J Psychiatry 169:468–474

Zwanzger P, Baghai TC, Schuele C, Strohle A, Padberg F, Kathmann N, Schwarz M, Moller HJ, Rupprecht R (2001) Vigabatrin decreases cholecystokinin-tetrapeptide (CCK-4) induced panic in healthy volunteers. Neuropsychopharmacology 25:699–703

HEP (2005) 169:503–526

New Pharmacological Treatment Approaches for Anxiety Disorders

A. Ströhle

Klinik für Psychiatrie und Psychotherapie, Charité Campus Mitte,
Charité - Universitätsmedizin Berlin, Schumannstr. 20/21, 10117 Berlin, Germany
andreas.stroehle@charite.de

Abstract New developments in the pharmacological treatment of anxiety disorders will have distinct backgrounds: characterization of pathophysiological processes including evolving techniques of genomics and proteomics will generate new drug targets. Drug development design will generate new pharmacological substances with specific action at specific neurotransmitter and neuropeptide receptors or affecting their reuptake and metabolism. New anxiolytic drugs may target receptor systems that only recently have been linked to anxiety-related behavior. This includes the *N*-methyl-D-aspartate (NMDA), *S*-α-amino-3-hydroxy-5-methyl-4-isoxazolepropionic acid (AMPA), and the cannabinoid receptors. In addition, signal transduction pathways, neurotrophic factors, and gases such as nitric oxide or carbon monoxide may be new drug targets. Combining psychopharmacological and psychotherapeutical interventions is a further field where benefits for the treatment of anxiety disorders could be achieved. Although the road of drug development is arduous, improvements in the pharmacological treatment of anxiety disorders are expected for the near future.

Keywords Anxiety · Anxiety disorders · Anxiolytic drugs · New drugs · Pharmacotherapy of anxiety disorders

1 Introduction

Pharmacology has provided powerful tools to characterize the neurochemical pathways of stress and anxiety in the brain, and how these pathways are involved in the pathophysiology and treatment of anxiety disorders. In the past, this work has largely focused on "classical" neurotransmitter systems, including the synthesis, release, and metabolism of monoamines and receptor subtypes that control presynaptic release of neurotransmitters and their postsynaptic effects. Increasing the specificity of drugs but also the combination of mechanisms has been pursued to improve anxiolytic drugs.

New drug targets have been generated by characterizing the importance of hormones and second messenger systems in the pathophysiology and treatment of anxiety disorders. Neuropeptides and neuroactive steroids are at least in part synthesized and released in the brain independent from their peripheral activity.

Combining pharmacological and psychological treatment is common in clinical practice. However, the possible interaction of these two treatment approaches has been the subject of a long debate. A new strategy is the specific administration of drugs to accelerate behavioral treatment. Exposure therapies are central in the treatment of most anxiety disorders. Data from animal experiments suggest that exposure therapies could become more efficient if they would be combined with pharmacotherapy (Myers and Davis 2002; Davis and Myers 2002) that targets, for instance, the glycine recognition site of the *N*-methyl-D-aspartate (NMDA) receptor (Walker et al. 2002), the endocannabinoid system of the brain (Marsicano et al. 2002), or protein kinases (Lu et al. 2001; Cohen 2002). And indeed, the administration of D-cycloserine seems to improve exposure therapy in patients with simple phobias (Ressler et al. 2004).

Reduced side effects, increased response rate, and accelerated onset of action, as well as decreased toxicity, increased remission rates, and shortening of treatment duration are some of the demands for new anxiolytic drugs. With the implementation of genomics, proteomics, and pharmacogenomics, many novel drug targets will be generated in the future. Current research strategies for the development of new anxiolytic drugs will be highlighted.

2
Classical Neurotransmitters

The classical neurotransmitter systems targeted by anxiolytic drugs are the serotonergic, the γ-aminobutyric acid (GABA)-ergic, and to a lesser extent the noradrenergic system. While the tricyclic antidepressants target a plethora of neurotransmitter systems, the selective serotonin reuptake inhibitors (SSRIs) more or less selectively target the reuptake of serotonin (5-HT). Evidence from depression studies suggests that combining serotonergic and noradrenergic reuptake may improve clinical effectiveness. However, for anxiety disorders no such studies are available at the moment

2.1
Serotonin

Modulation of the serotonergic system is a successful strategy for the pharmacological treatment of anxiety disorders. It is now suggested that altering serotonergic neurotransmission by pharmacological manipulation is a complex process involving presynaptic autoreceptors ($5\text{-HT}_{1A/1D}$), the 5-HT reuptake transporter site, and at least 14 different postsynaptic receptor subtypes, of which several are suggested to be important (5-HT_{1A}, 5-HT_{2A}, 5-HT_{2C}, 5-HT_3) for anxiety-related behavior (Hoyer et al. 2002).

SSRIs have been approved for the treatment of the majority of anxiety disorders, except agoraphobia and specific phobia. The mechanisms of action responsible for SSRIs' anxiolytic activity remain to be fully delineated. Understanding of pre- and postsynaptic receptor regulation with chronic treatment and cross-system effects are critical in furthering our understanding of these drugs. Increasing specificity may improve clinical efficacy.

Recently developed specific 5-HT receptor subtype agonists and antagonists are now being studied in preclinical models to further elucidate their role in anxiety modulation. Additionally, knockout strategies have improved our knowledge of 5-HT receptors and their role in anxiety-related behavior. However, we are only at the beginning of understanding the complex role of the 5-HT system in anxiety disorders and their treatment. Additionally, many of these agents fail to show significant clinical efficacy in patients with specific anxiety disorders. Improving animal models of anxiety disorders is therefore paramount.

The 5-HT_{2A} receptors are strategically located on GABA-containing interneurons in the deep layers of the cerebral cortex, allowing them a modulatory role and link between the noradrenergic and serotonergic systems. These modulatory interneurons are a potential site of the anxiolytic activity of 5-HT_{2A} antagonists. The $5\text{-HT}_{2A/2C}$ receptor antagonists ritanserin and mianserin are anxiolytic in patients (Ceulemens et al. 1985; Conti and Pinde 1979) and can block the effects of *m*-chlorophenylpiperazine (m-CPP) as shown in both preclinical (Kennett et al. 1989) and clinical (Pigott et al. 1991) studies. The withdrawn antidepressant nefazodone possesses 5-HT_{2A} receptor antagonistic activity along with weak 5-HT and norepinephrine reuptake inhibition properties and has been shown to be more effective than imipramine in reducing depression-associated anxiety and to be effective in panic disorder (Bystritsky et al. 1999). Mirtazapine is an antidepressant with potential anxiolytic activity (Ribeiro et al. 2001); among its many effects is its ability to block 5-HT_{2A} receptors. Specific 5-HT_{2A} receptor antagonists and drugs that reduce 5-HT_{2A} receptor density may be anxiolytic. In line with this, atypical antipsychotics with prominent 5-HT_{2A} receptor blockade are being studied for their potential effects in anxiety disorder patients.

Although promising in preclinical models of anxiety, the 5-HT_3 receptor antagonist ondansetron has limited efficacy in panic disorder (Schneier et al. 1996); however, at higher doses (1 mg), ondansetron was superior to placebo in a study of patients with generalized anxiety disorder (Freeman et al. 1997).

2.2 Noradrenaline

The noradrenergic system, originating in the locus coeruleus (LC) and other medullary and pontine nuclei, has extensive connections with fear and anxiety circuits and, in addition to the hypothalamus–pituitary–adrenocortical (HPA) system, represents the physiological response to stress. The LC projects to the prefrontal and entorhinal cortices, the amygdala, the bed nucleus of the stria terminalis, the hippocampus, the periaqueductal gray, the thalamus, the hypothalamus, and the nucleus of the solitary tract (Chouros and Gold 1992). Efferents from the LC activate the sympathetic adrenomedullary and the parasympathetic branch. Preclinical studies demonstrate that increases in norepinephrine release in the LC, hypothalamus, and amygdala are associated with anxiety, fear, and uncontrollable stress.

Catecholamine depletion or α_2-adrenergic receptor antagonists before stress exposure have been shown to affect frontal cortex function, as manifested by impaired working memory (Li and Mei 1994). Preclinical studies have further demonstrated that α_1-adrenergic receptor agonists impair cognitive performance, whereas α_1-adrenergic receptor antagonists such as urapidil and prazosin are capable of reversing these cognitive deficits under stress but not under nonstressed conditions (Li and Mei 1994; Birnbaum et al. 1999). New drugs targeting the α_1-adrenergic receptor may have a use in the treatment of specific components of anxiety-related behavior.

The effects of β-adrenergic blockade on the consolidation of traumatic memories has been an area of special interest for the treatment of posttraumatic stress disorder (PTSD), and recently the first randomized controlled study on the effects of propranolol in the prevention of PTSD was published. Pittmann and coworkers (2002) could demonstrate that propranolol may reduce PTSD symptoms if treatment is started within several hours after the traumatic event.

2.3 γ-Aminobutyric Acid

GABA is the predominant inhibitory neurotransmitter in the CNS. It is formed by decarboxylation of glutamate, the major central excitatory amino acid, utilizing the enzyme l-glutamic acid decarboxylase (GAD). GABA receptors consists of two different superfamilies: $GABA_A$ and $GABA_B$. Traditional anxiolytics and sedative drugs such as barbiturates and benzodiazepines modulate the $GABA_A$ receptors.

The pharmacological relevance of the multitude of structurally diverse $GABA_A$ receptor subtypes has only recently been characterized. Based on point mutation strategy, α_1-$GABA_A$ receptors were found to mediate sedation, anterograde amnesia, and part of the anticonvulsant activity, whereas α_2-$GABA_A$ receptors, but not α_3-$GABA_A$ receptors mediate anxiolysis (Löw et al. 2000). "A new benzodiazepine pharmacology" therefore has been suggested to evolve with subtype-specific ligands (Möhler et al. 2002). Targeting the α_2-$GABA_A$ receptors, which represent only 15% of all diazepam-sensitive $GABA_A$ receptor-selective ligands, is expected to be devoid of the major side effects that accompany the classical benzodiazepine anxiolytics.

Consequently, the first subtype-specific $GABA_A$ receptor-modulating drugs have been developed: L-838417 failed to modulate the GABA response at α_1-receptors but enhanced the GABA response at α_2-, α_3-, and α_5-receptors and displayed anxiolytic and anticonvulsant activity without impairing motor performance (McKernan et al. 2000). The pyrido-indole-4-carboxamide derivative SL651498 showed higher affinity for α_1-, α_2-, and α_3-$GABA_A$ receptors compared to α_5-receptors. In addition, it acted as full agonist at α_2- and α_3-receptors but as partial agonist at α_1-$GABA_A$ receptors. In line with its selectivity for the activation of α_2- and α_3-receptors, the compound showed potent anxiolytic action in animal models but did not impair motor coordination or working memory (Scatton et al. 2000). Further improvements may be achieved by focusing the ligand affinity or efficacy more specifically on α_2-receptors.

Recently, an additional GABA-ergic control component of spatial and temporal memory became apparent involving α_5GABA_A receptors. The α_5GABA_A receptor subtype has a privileged site of expression on hippocampal pyramidal cells being located extrasynaptially at the base of the spines that receive the excitatory input and on the adjacent shaft of the dendrite (Fritschy et al. 1998b; Crestani et al. 2002). The α_5GABA_A receptors were therefore considered able to modulate the transduction of the signal arising at excitatory synapses and, by doing so, would operate as a control element of learning and memory in their own right. Pharmacological modulation of the α_5GABA_A receptors may promote hippocampus-dependent learning during, for example, behavioral treatment of anxiety disorders.

The GABA receptors have also been the target of drug development strategies for the treatment of epilepsy. For this indication, several drugs have been developed with the potential to have anxiolytic activity as well. Manipulation of GABA concentrations by interfering with the reuptake or metabolization of GABA may be another anxiolytic treatment approach. The anticonvulsant tiagabine increases GABA levels by selectively inhibiting the GABA transporter-1 (GAT-1) responsible for the reuptake of GABA in the CNS. Although placebo-controlled studies have to be performed, preclinical and clinical studies suggest that tiagabine may be useful as an anxiolytic drug. In line, vigabatrin, which selectively and irreversibly inhibits GABA transaminase, seemed to be a promising antipanic drug (Zwansger et al. 2001). However, potential side effects hinder the usage and further study of this antiepileptic drug.

Besides effects on calcium channels and other possible mechanisms, pregabalin and gabapentin regulate the GABA transporter GAT-1. Although gabapentin has been studied in placebo-controlled trials in several anxiety disorders, including social phobia (Pande et al. 1999b) and panic disorder (Pande et al. 2000), the results suggest no improved efficacy over SSRIs. In generalized anxiety disorder, pregabalin was found to have significant efficacy versus placebo and efficacy at the end of the first week of treatment, comparable to alprazolam (Rickels et al. 2002)

2.4
Glutamate

As the major excitatory amino acid, L-glutamic acid (glutamate) is integral to the functioning of up to 40% of all brain synapses (Coyle et al. 2002). Glutamate is primarily derived from intermediary glucose metabolism and can be formed directly from glial cell-synthesized glutamine stores. Several glutamate transporters regulate synaptic transmission and concentrations of glutamate, with astroglial transporters providing the primary mode of inactivation of glutamate in the forebrain (Maragakis and Rothstein 2001). There is a high functional and regional diversity of the glutamate receptor subtype combinations. Postsynaptic ionotropic glutamate receptors, such as NMDA, S-α-amino-3-hydroxy-5-methyl-4-isoxazolepropionic acid (AMPA), and kainate, mediate fast excitations and synaptic plasticity associated with sodium and calcium ligand-gated ion channels. Pre- and postsynaptic metabotropic glutamate receptors modulate postsynaptic excitability and provide feedback for the further release of neurotransmitters. The diversity in localization and function within the glutamatergic system calls for the further identification of specific drug targets involved in the development and treatment of anxiety disorders.

Although the neuronal basis of fear acquisition is well characterized, we are just beginning to understand the mechanisms involved in fear inhibition, suppression, and unlearning. While GABA seems to be primarily involved in the expression of fear that has already been acquired, glutamate seems to be critically involved in plasticity underlying the development of inhibitory learning. Preclinical studies demonstrate that D-cycloserine, a partial agonist at the strychnine-insensitive glycine binding site on the NMDA receptor complex, facilitates extinction (Walker et al. 2002). As the first preliminary data show, modulation of glutamatergic systems during exposure therapy may increase the effectiveness of treatment (Ressler et al. 2004).

3 Neuropeptides

Neuropeptides are among the most promising new drug targets for anxiolytics. They are short amino acid neuromodulators central for emotional behavior, stress response, and anxiety-related behavior. The list of neuropeptides involved in the modulation of this behavior is ever increasing. Rodents with mutations in genes encoding neuropeptides and their receptors have been developed and behaviorally characterized. Further progress will be achieved with time- and region-specific knockout strategies. Additionally, there is progress in the development of specific and highly potent small-molecule neuropeptide receptor ligands that can readily cross the blood–brain barrier, and first clinical studies have been started.

Because neuropeptides have a more discrete neuroanatomical localization than classical neurotransmitters, it is expected that they produce less disturbances of physiological processes if modulated by drugs; and antagonists are suggested to be less likely to produce tolerance or dependency. Additionally, these drugs are not expected to disrupt normal physiology in the absence of neuropeptide release, i.e., activation of the system.

3.1 Corticotropin Releasing Hormone

Most studies exploring behavioral effects of CRH in animals have used intracerebroventricular or site-specific effects of corticotropin-releasing hormone (CRH), and all agree that CRH mediates numerous anxiogenic and fear-related aspects of stress. These include the CRH-induced potentiation of acoustic startle, suppression of social interaction, and an increase in stress-induced freezing behavior (Dunn and Berridge 1990). This is further supported by transgenic mice overexpressing CRH: These mice have deficits in emotionality and were used as a genetic model of anxiogenic behavior (Stenzel-Poore et al. 1994).

Decreased CRH neurotransmission has been studied by administering antisense oligodeoxynucleotides corresponding to the start-coding region of CRH mRNA. The application of this kind of gene therapy to stressed rats produced a decrease in CRH biosynthesis and led to the reduction of anxiety-related behavior (Skutella et al. 1994). Comparison of the behavioral effects of antisense probes that were either directed against CRH-R1 or against -R2 receptor mRNA suggested that CRH-R1 is more likely to convey anxiety-related, and possibly depression-related, signaling (Liebsch et al. 1999; Skutella et al. 1998). Complementary evidence was provided by the generation of CRH-R1 receptor-deficient mouse mutants, which proved to be less anxious than normal mice (Smith et al. 1998; Timpl et al. 1998). Conflicting results with respect to anxiety-like behavior were described for CRH-R2 receptor knockout mice (Kishimoto et al. 2000; Bale et al. 2000).

The most straightforward strategy to restrain the anxiogenic and depressogenic effect of excessive CRH production and release is the administration of CRH-R1 antagonists. One of these compounds (R121919), a pyrazolopyrimidine, has been tested: In the first open label trial with this substance, depressed patients had significantly reduced depression and anxiety scores as was seen in both clinician and patient ratings, suggesting that this type of compound may have considerable therapeutic potential (Zobel et al. 2000). The further development of this drug was stopped due to potential side effects. However, other nonpeptidergic CRH-R1 antagonists have been developed and are being studied now. Although especially promising, to date there are no data on the therapeutic potential of CRH-R1 antagonists in the treatment of anxiety disorders.

3.2 Arginine Vasopressin

The nonapeptide vasopressin (AVP) is synthesized in the paraventricular nucleus of the hypothalamus (PVN) and the nucleus supraopticus. Besides its role in fluid regulation, AVP is also a key modulator of the HPA system, where it potentiates the effects of CRH on adrenocorticotropic hormone (ACTH) release. Extrahypothalamic AVP-containing neurons are localized in the medial amygdala and the bed nucleus of the stria terminalis. AVP applied intracerebroventricularly or to the lateral septum has been shown to affect cognition, social behavior, and anxiety-like behavior in rodents (Insel et al. 2001).

In the brain, the effects of AVP are mediated through G protein-coupled receptors (V_{1A} and V_{1B}). The V_{1A} is expressed in the amygdala, septum, and hypothalamus (Ostrowski et al. 1992; Tribollet et al. 1999). While the V_{1B} receptor is primarily localized in the anterior pituitary, it has also been detected in various brain areas, including the amygdala, the hypothalamus, and the hippocampus (Lolait et al. 1995; Hernando et al. 2001) and has recently been shown—in addition to the V_{1A} receptor (Landgraf et al. 1995)—to be involved in the regulation of anxiety-related behavior (Griebel et al. 2002). V_{1B} receptor knockout mice, however, did not differ in elevated plus maze-related parameters from their wildtype littermates (Wersinger et al.

2002), raising concerns about possible developmental confounds. The V_2 receptor is found in the kidney and is responsible for the antidiuretic effects of AVP.

The first nonpeptide V_{1A} or V_{1B} receptor antagonists have been developed and are now being studied for their potential anxiolytic and antidepressant activity. SSR149415 is such a promising compound (Griebel et al. 2002; Serrandeil et al. 2002), which needs an activated stress response for its activity and which has yet to be studied in patients.

3.3
Atrial Natriuretic Peptide

Whereas several peptides besides AVP are known to act synergistically with CRH, the only peptide candidate in humans that inhibits the HPA system at all regulatory levels of the system seems to be atrial natriuretic peptide (ANP). ANP has been shown to inhibit the stimulated release of CRH and ACTH in vitro and in vivo. This could be observed in humans as well, where ANP inhibits the CRH-induced ACTH (Keller et al. 1992), prolactin (Wiedemann et al. 1995), and cortisol secretion (Ströhle et al. 1998). ANP is not only synthesized by atrial myocytes (deBold et al. 1985) and released into the circulation, but is also found in neurons of different brain regions (Tanala et al. 1984) where specific receptors have been found. ANP receptors and immunoreactivity have been found in periventricular and paraventricular hypothalamic nuclei, the LC, and the central nucleus of the amygdala.

Intracerebroventricular administration of ANP elicited anxiolytic activity in the open field, the social interaction, and the elevated plus maze tests (Biro et al. 1999; Bhattacharya et al. 1996). The effects of central and peripheral administration of atriopeptin II, a 23 amino acid residue peptide of ANP, was furthermore investigated in the elevated plus maze test in rats previously exposed to a social defeat stress. Results show that the intracerebroventricular, intra-amygdala, and intraperitoneal administration of atriopeptin II produced anxiolytic effects without affecting spontaneous locomotor activity (Ströhle et al. 1997).

In patients with panic disorder, basal ANP concentrations are lower when compared to healthy control subjects, but ANP concentrations are faster and more pronounced during experimentally induced panic attacks (Kellner et al. 1995). In line with these findings, there is evidence for an anxiolytic activity of ANP in humans: ANP decreases CCK-4-induced panic anxiety in patients with panic disorder (Ströhle et al. 2001) and healthy control subjects, and attenuates HPA system activity by decreasing ACTH and cortisol stimulation (Wiedemann et al. 2001). Modulation of ANP concentrations or nonpeptidergic ANP receptor ligands may be ultimately used in the pharmacological treatment of anxiety disorders, such as panic disorder.

3.4 Cholecystokinin

The neuropeptide cholecystokinin (CCK) is an octapeptide found regionally in the gastrointestinal tract and brain (brain–gut peptide), where it acts as a neurotransmitter and neuromodulator. Its most abundant form in the brain is the C-terminal sulfated octapeptide fragment CCK-8, which interacts with the same affinity with both CCK receptor subtypes CCK-A and CCK-B. With a distinct distribution, CCK has been found to be colocalized with a number of classical neurotransmitters such as GABA, dopamine, and 5-HT. Extensive pharmacological studies during the past few years suggest that CCK may participate in the neuroendocrine responses to stress (e.g., Harro et al. 1993; Daugé and Léna 1998).

Together with CRH, CCK belongs to the most extensively studied neuropeptides in anxiety models. Generally, CCK is thought to induce anxiogenic-like effects, although the results of those animal studies have been highly variable and sometimes contradictory (Griebel 1999). As has recently been shown in CCK receptor gene knockout mice, however, the role of the receptor subtypes in anxiety-related behavior is still controversial (Miyasaka et al. 2002).

CCK-8 concentrations were found to be lower in panic patients than in normal control subjects (Brambilla et al. 1993), and the CCK-B receptors were hypersensitive in panic disorders (Akiyoshi et al. 1996). A significant association between panic disorder and a single nucleotide polymorphism found in the coding region of the CCK-B receptor gene has been reported (Kennedy et al. 1999). If confirmed by replication, these data would suggest that a CCK-B receptor gene variation may be involved in the pathogenesis of panic disorder.

CCK-B receptor agonists such as pentagastrin or CCK-4 have panic-like anxiogenic effects in humans. Panic disorder patients are more sensitive than healthy controls to the anxiogenic effects of CCK-4 (e.g., DeMontigny et al. 1989). Pretreatment with a CCK-B receptor antagonist is able to reverse both the autonomic and anxiogenic effects of pentagastrin (Lines et al. 1995). Clinical trials, however, have provided rather disappointing and inconclusive data about the anxiolytic potential of the CCK-B antagonists available so far (Kramer et al. 1995; Shlik et al. 1997; Pande et al. 1999a). Bioavailability has been discussed as a problem with the currently available compounds.

3.5 Neuropeptide Y

Neuropeptide Y (NPY) is one of the most common neuropeptides and it has at least three identified receptors (Y_1, Y_2, and Y_5); Y_1 and Y_2 are G protein-coupled receptors. NPY receptors are located in a wide range of brain regions, including the cortex and several subcortical structures involved in fear neurocircuitry such as the amygdala, the hypothalamus, and brain stem nuclei

(Dumont et al. 1995). Activation of Y_1 and Y_5 receptors in the basolateral amygdala produces dose-dependent anxiolytic-like effects in rodents (Heilig et al. 1993). In contrast, presumably through presynaptic inhibition of NPY release, Y_2 receptor activation is anxiogenic (Sajdyk et al. 2002). The anxiolytic effects of NPY are reversed by α_2-adrenergic receptor antagonists, but not by $GABA_A$ receptor ligands, implicating the noradrenergic system in NPY's anxiolytic effects.

Mutant mice lacking NPY show increased anxiety-related behavior (Palmiter et al. 1998); a full description of the behavioral phenotype of NPY receptor-null mutant mice is not available yet. However, data from Y_2 receptor-null mutants support an anti-stress activity of NPY (Tschenett et al. 2003). In addition to the anxiolytic and anti-stress effects of NPY, a relationship to alcohol intake has been described. Voluntary ethanol consumption is increased in NPY and Y_1 receptor-null mutant mice, whereas either NPY overexpression or potentiation of NPY signaling through blockade of Y_2 receptors suppresses rodent alcohol intake (Thiel et al. 1998, 2002; Thorsell et al. 2002). Thus, in addition to anxiety and depression, alcohol dependency may be a promising clinical field for newly developed NPY receptor ligands.

3.6
Tachykinins and Substance P

The peptide tachykinins are widely distributed throughout the brain, spinal cord, and peripheral nervous system. Although research has primarily focused on pain and inflammation, it was well known that tachykinins are located in brain areas implicated in the pathophysiology of mood and anxiety disorders. Since its discovery in the 1930s, the 11 amino acid peptide substance P has been one of the most extensively studied neuropeptides. Its effects are mediated through G protein-coupled tachykinin (NK_1) receptors, while neurokinin shows greatest affinity for the NK_3 receptor. Substance P is frequently co-localized within neurons containing other neurokinins or neurotransmitters, such as GABA, dopamine, glutamate, 5-HT, and acetylcholine, often influencing their synaptic release (Otsuka and Yoshioka 1993).

A great number of preclinical studies describes anxiogenic effects of substance P (Commons and Valentino 2002), and anxiolytic effects of NK_1 antagonists (Vassout et al. 2000). Disruption of the NK_1 receptor by knockout techniques results in mice with reduced anxiety in response to stress (Santarelli et al. 2001), and it has been suggested that the effects of substance P on 5-HT and opioid transmission within the periaqueductal gray-dorsal raphé regions may be indirectly mediated through glutamatergic neurons (Commons and Valentino 2002). Additionally, it has been hypothesized that the anxiolytic effects of NK_1 receptor antagonists are a result of a reduced autoinhibition of the LC.

With the development of highly specific NK_1 receptor antagonists, clinical studies could be started and despite negative results in pain studies, two trials show an antidepressant activity of the NK_1 receptor antagonists MK0869 and L759274 (Kramer et al. 1998). However, to date there are no clinical studies published of NK_{1-3} receptor antagonists in patients with anxiety disorders.

3.7 Galanin

The 29–30 amino acid neuropeptide galanin coexists with noradrenaline in the LC, and with 5-HT in the nucleus dorsalis raphé where it has been shown to act as an inhibitory neuromodulator. Galanin and its receptors are found in limbic regions including the amygdala, BNST, hippocampus, and hypothalamus. The three known G protein-coupled receptor subtypes (GAL1–3) have a differential localization in the brain and the periphery, suggesting that the subtypes mediate differential functional effects of galanin.

Galanin mediates the neuronal, neuroendocrine, and sympathetic response to stress, and it has been shown that stress upregulates the galanin gene expression in the hypothalamus, amygdala, and LC (Holmes et al. 2003). Moreover, exogenous galanin and peptidergic galanin receptor antagonists modulate anxiety in a region- and task-specific manner (Bing et al. 1993; Moller et al. 1999). Under stress and high noradrenergic activity, endogenous galanin in the amygdala has been associated with anxiogenic effects (Morilak et al. 2003). Mutant mice with a conditional overexpression of galanin in noradrenaline-containing neurons are relatively insensitive to the anxiogenic effects of a noradrenaline challenge (Holmes et al. 2002), while GAL1 receptor-null mutant mice selectively show increased anxiety-like behavior under stressful conditions (Wrenn et al. 2001).

Although our understanding of the role of galanin in stress and anxiety is at an early stage, preclinical studies suggest that targeting the galanin system might be of therapeutic benefit in disorders where noradrenergic overactivity is pathophysiologically relevant.

4 Further Anxiolytic Drug Targets

4.1 Neuroactive Steroids

In the last decade, considerable evidence has emerged that certain steroids may alter neuronal excitability via their action at the cell surface through interaction with certain neurotransmitter receptors. For steroids with these particular properties, the term "neuroactive steroids" has been used (Majewska et al. 1986; Paul and Purdy 1992; Rupprecht and Holsboer 1999).

The first behavioral observations related to these steroids date back to Selye, who over 50 years ago reported that progesterone and deoxycorticosterone

(DOC) have a strong sedative action through their A ring-reduced metabolites. These two steroids, 3α,5α-tetrahydroprogesterone (3α,5α-THP) and 3α,5α-tetrahydrodeoxycorticosterone (3α,5α-THDOC), bind at $GABA_A$ receptors to enhance GABA-induced chloride currents. In rats, 3α,5α-THDOC and 3α,5α-THP are elevated in cortical and hypothalamic tissue after stress (Paul and Purdy 1992), and they have been shown to be anxiolytic and hypnotic, respectively, as predicted by electrophysiology, where a benzodiazepine-like action was demonstrated (Rupprecht et al. 1993). Interestingly, 3α,5α-THP dampens the activity of the HPA system and counteracts CRH-induced anxiety. In addition, neonatal treatment of rats with 3α,5α-THDOC abolishes the behavioral and neuroendocrine consequences of adverse early life events (Patchev et al. 1997). Several other neuroactive steroids have opposite effects. For example, the sulfated form of pregnenolone has been observed to antagonize $GABA_A$ receptor-mediated chloride currents by reducing the channel-open frequency (Majewska et al. 1986), being therefore proconvulsant.

A new line of research in regard to the mechanisms of antidepressant drugs was stimulated by the observation that in animal studies the SSRI fluoxetine, which is widely used for the treatment of depression and anxiety, may enhance the concentrations of 3α,5α-THP in the rat brain (Uzunov et al. 1996). At the molecular level it has been demonstrated that SSRIs shift the activity of the 3α-hydroxysteroid oxidoreductase, which catalyzes the conversion of 5α-DHP into 3α,5α-THP, towards the reductive direction, thereby enhancing the formation of 3α,5α-THP (Griffin et al. 1999). Additionally, 3α,5α-THP has been suggested to possess antidepressant-like effects in mice using the Porsolt forced swim test (Khisti et al. 2000). These preclinical findings suggest that 3α-reduced neuroactive steroids such as 3α,5α-THP may play a role in treatment with antidepressant drugs. Indeed, the concentrations of GABA agonistic neuroactive steroids 3α,5α-THP and 3α,5β-THP were reduced in plasma of depressed patients, while there was an increase in 3β,5α-THP, an antagonistic isomer of 3α,5α-THP (Romeo et al. 1998; Uzunova et al. 1998). In contrast to preclinical data, tri- and tetracyclic antidepressants also interfered with the composition of neuroactive steroids, in a similar way to SSRIs (Romeo et al. 1998). In addition, antidepressants do not generally shift the activity of the 3α-hydroxysteroid oxidoreductase towards the reductive direction. The concentrations of 3α,5α-THDOC were elevated during depression, probably as a consequence of hypercortisolemia, and reduced by fluoxetine (Ströhle et al. 2000), but not by tri- or tetracyclic antidepressants (Ströhle et al. 1999). Thus, the effects of antidepressants on neuroactive steroids also appear to be substrate specific.

While no data on the role of 3α-reduced neuroactive steroids in PTSD or its treatment in panic disorder patients have been published to date, opposite changes to those seen in major depression have emerged. At baseline, patients with panic disorder had significantly increased concentrations of the positive allosteric modulators 3α,5α-THP and 3α,5β-THP, together with significantly decreased concentrations of 3β,5α-THP, a functional antagonist for $GABA_A$ agonistic steroids, which might result in an increased GABA-ergic tone. SSRI treatment did not influence these changes of neuroactive steroid concentrations (Ströhle et al. 2002). Most strikingly, during experimentally induced panic attacks, drastic changes of neuroactive steroid concentrations occurred, paralleling psychopathological changes and resulting in a dramatically reduced GABA-ergic tone (Ströhle et al. 2003), supporting the assertion that the increased baseline concentrations of $GABA_A$ agonistic neuroactive steroids may serve as a counterregulatory mechanism against the occurrence of spontaneous panic attacks. When attempting to pharmacologically modify the equilibrium of neuroactive steroids as a treatment for psychiatric disorders (Guidotti and Costa 1998), consideration must also be given to baseline concentrations and possible counterregulatory mechanisms.

4.2 Neurotrophic Factors and Second Messenger Systems

The stress dependency and the dynamic—in most cases chronic—nature of anxiety disorders along with the anxiolytic activity of antidepressant drugs suggest that neuronal plasticity plays a role in the pathophysiology and treatment of anxiety disorders. Antidepressants regulate intracellular signaling pathways and induce molecular, cellular, and structural changes. In addition, signal proteins such as cyclic AMP response element binding protein (CREB) have been implicated in fear conditioning and extinction processes.

Over the past decade, new drug targets have been identified through an increased understanding of the signal transduction pathways involved in monoamine-based therapies. For example, most antidepressants stimulate adenylyl cyclase through G protein-coupled receptors, such as β-adrenergic and $5\text{-}HT_{4,6,7}$. Elevated levels of cyclic adenosine monophosphate (cAMP) are known to stimulate the cAMP-dependent protein kinase A (PKA), which regulates the phosphorylation of specific proteins such as CREB. Duman and coworkers (1997) hypothesized that activated CREB, resulting from chronic antidepressant treatment, increases concentrations of mRNA encoding brain-derived neurotrophic factor (BDNF) in the hippocampus. BDNF has been shown to protect 5-HT and dopamine neurons against insult and the damaging effects of stress. Although the transcriptional control of BDNF mRNA in the CNS is complex (Timmusk et al. 1993), AMPA receptor activation is known to increase BDNF mRNA (Zafra et al. 1990). Novel AMPA potentiators such as LY392098 increase the expression of BDNF and have, at least in animal models, an antidepressant-like effect (Li et al. 2001). In addition, the selective phosphodiesterase type IV inhibitor rolipram had anxiolytic (Griebel et al. 2001) and antidepressant activity (Duman et al. 2000). Because of nausea as a side effect, this drug was not further developed.

The cellular mechanisms underlying fear conditioning and extinction involve additional signaling pathways which we are only beginning to understand. LeDoux and coworkers characterized the cellular events during fear conditioning including calcium entry through both NMDA receptors and L-type voltage-gated calcium channels in the lateral amygdala as a result of the associative pairing of condition stimuli (CS) and unconditioned stimuli (US) (Bauer et al. 2002). They suggest a role of Ca^{2+} entry through NMDA receptors in short-term memory and Ca^{2+} entry through L-type voltage-gated calcium channels in long-term memory formation. Subsequent activation of protein kinases and extracellular signal-regulated kinase (ERK)/mitogen-activated protein kinase (MAPK) are thought to contribute to CREB phosphorylation and CREB-dependent gene expression (Dolmetsch et al. 2001). The resulting changes in RNA and protein synthesis are thought to represent the synaptic and behavioral plasticity that is experimentally observed as long-term potentiation (LTP) and conditioned fear memory. The extinction of fear memory is an active process, presumably utilizing similar mechanisms as those used in the acquisition of conditioned fear. Extinction is facilitated by enhanced signaling via NMDA receptors, PKA, MAPK, and calcium-calmodulin kinase II (CaMKII) (Szapiro et al. 2003). Phosphatase activity (Mansuy et al. 1998) and calcineurin (Lin et al. 2003) seem to be also crucially involved in memory extinction. As we further understand the intracellular and molecular mechanisms of the fear response and its extinction and habituation, new pharmacological drug targets can be developed for patients with anxiety disorders.

4.3 Nitric Oxide

Nitric oxide (NO) and carbon monoxide are atypical neurotransmitters. They are not stored in synaptic vesicles, are not released in by exocytosis, and do not act at postsynaptic membrane receptor proteins. NO is generated in a single step from the amino acid arginine through the action of the NO synthase (NOS). The form of NOS initially purified was designated nNOS (neuronal NOS), the macrophage form is termed inducible NOS (iNOS), and the endothelial from is called eNOS.

NOS-containing neurons have a very discrete localization in the CNS, representing only 1% of neuronal cells. However, their axons ramify so extensively that virtually every cell in the brain may encounter a NOS nerve terminal. As a diatomic gas, NO is freely diffusible and thus can readily enter adjacent neuronal cells. Once inside the target cell, NO binds the iron in heme contained within the active site of soluble guanylyl cyclase, activating the enzyme to form cyclic guanosine monophosphate (GMP). The activity of NO is therefore mediated by an "enzyme receptor." In neurons, NO is formed in response to calcium influx reminiscent of calcium-dependent exocytotic release of neurotransmitters.

Insight into a physiological role for NO in the brain comes from behavioral studies of nNOS knockout mice. Depending on testosterone, these mice were extremely aggressive (Nelson et al. 1995; Kriegsfeld et al. 1997). Kandel and coworkers could demonstrate that NO plays also a role in learning and memory: Mice with a deletion of both eNOS and nNOS show a clear decrease of long-term potentiation (Son et al. 1996). Stroke damage is markedly reduced after treatment with NO inhibitors and nNOS knockout mice (Huang et al. 1994). Additionally, glutamate neurotoxicity is diminished in cultures from nNOS knockout mice or after treatment with NO inhibitors (Dawson et al. 1996), giving evidence for a role of NO in stroke.

NO has also been implicated in the anxiolytic effects of $GABA_A$ receptor activation and the behavioral effects of the anesthetic gas nitrous oxide: Pharmacological inhibition of NOS can decrease the anxiolytic effects of a benzodiazepine, a $GABA_A$ receptor agonist, or nitrous oxide (Caton et al. 1994) and inhibition of NO function has a similar effect (Li et al. 2003), with the nNOS being implicated in this effect (Li et al. 2001). In the absence of other anxiety-modulating drugs, anxiolytic (Vale et al. 1998) and anxiogenic (Dunn et al. 1998) effects have been described in studies of pharmacological inhibition of NOS.

5
Pharmacogenomics and Pharmacoproteomics

The progress made in genome research raises the question of whether the new knowledge will eventually lead to better anxiolytics. Besides hypothesis-driven research to identify new drug targets, the future will bring an upsurge of systematic research in biotechnology-driven drug discovery efforts. A further specification of the clinical phenotype including neuroendocrine, neuropsychological, neurophysiological, neuroimaging, and treatment response data will have to be correlated with data from genotyping studies. To achieve the goal of genotype-/phenotype-based differential therapy, large-scale efforts, including large sample sizes and genotyping capacities, are needed.

Despite all the shortcomings of the currently available pharmacogenetic studies, this field holds promise for the future of anxiolytic drug therapy. The pharmacokinetics of anxiolytic drugs involves cytochrome P-related metabolization and P-glycoprotein activity. Pharmacogenomics may allow us to develop sets of single nucleotide polymorphisms (SNPs) that could be combined into easily used assays that will rapidly classify patients according to their likely response to pharmacotherapy. Psychiatrists could then base the treatment decision on more objective parameters then the ones used today. This will limit unwanted side effects, adverse drug reactions, and could reduce response times. A more individualized pharmacotherapy will then be possible.

Pharmacogenomics might also lead to a better understanding of the mechanism of action of antidepressant and anxiolytic drugs. The identification of novel candidate genes would allow the development of novel drug targets and therapeutic compounds. One might also identify subgroups of patients in which different pathophysiological changes lead to the development of specific anxiety disorders. In this way, an individual targeting of the pathological pathway may be realizable, again shortening time to response and reducing side effects.

6 Conclusions

As we are beginning to understand the pathophysiology and treatment of anxiety disorders, new drug targets are being identified. They include classical neurotransmitters as well as neuropeptides and neuroactive steroids. Novel neurotransmitters such as NO, neurotrophic factors, and second messenger systems may eventually also be targeted by new anxiolytic drugs. In addition, pharmacogenomics and pharmacoproteomics will further improve drug development and treatment.

References

Akiyoshi J, Moriyama T, Isogawa K, Miyamoto M, Sasaki I, Kuga K, Yamamoto H, Yamada K, Fugii I (1996) CCk-4-induced calcium mobilization in T cells is enhanced in panic disorder. J Neurochem 66:1610–1615

Bale TL, A Contarino, Smith GW, Chan R, LH Gold, Sawchenko PE, GF Koob, WW Vale, KF Lee (2000) Mice deficient for corticotropin-releasing hormone receptor-2 display anxiety-like behavior and are hypersensitive to stress. Nat Genet 24:410–414

Bauer E, LeDoux JE, Nader K (2001) Fear conditioning and LTP in the lateral amygdala are sensitive to the same stimulus contingencies. Nat Neurosci 4:687–688

Bhattacharya SK, Chakrabarti A, Sandler M, Glover V (1996) Anxiolytic activity of intracerebroventricularly administered atrial natriuretic peptide in the rat. Neuropsychopharmacology 15:199–206

Bing O, Moller C, Engel JA, Soderpalm B, Heilig M (1993) Anxiolytic-like action of centrally administered galanin. Neurosci Lett 164:17–20

Birnbaum S, Gobeske K, Auerbach J, Taylor J, Arnsten A (1999) A role for norepinephrine in stress-induced cognitive deficits: alpha-1-adrenoceptor mediation in the prefrontal cortex. Biol Psychiatry 46:1266–1274

Biro E, Sarnyai Z, Penke B, Szabo G, Telegdy G (1999) Role of endogenous corticotropin-releasing factor in mediation of neuroendocrine and behavioral response to cholecystokinin octapeptide sulfate ester in rats. Neuroendocrinology 57:340–345

Brambilla F, Bellodi L, Perna G, Garberi A, Panerai A, Sacerdote P (1993) T cell cholecystokinin concentrations in panic disorder. Am J Psychiatry 150:1111–1113

Bystritsky A, Rosen P, Suri R, Vapnik T (1999) Pilot open-label study of nefazodone in panic disorder. Depress Anxiety 10:137–139

Caton P, Tousman SA, Quock RM (1994) Involvement of nitric oxide in nitrous oxide anxiolysis in the elevated plus-maze. Pharmacol Biochem Behav 48:689–692
Ceulemens D, Hoppenbrouers M, Gelders Y, Reyntjens A (1985) The influence of ritanserin, a serotonin antagonist, in anxiety disorders: A double-blind placebo controlled study versus lorazepam. Pharmacopsychiatry 18:303–305
Chrousos G, Gold P (1992) The concept of stress and stress system disorders: overview of physical and behavioral homeostasis. JAMA 267:1244–1252
Cohen P (2002) Protein kinases—the major drug targets of the twenty-first century? Nat Rev Drug Discov 1:309–315
Commons K, Valentio R (2002) Cellular basis of substance P in the periaqueductal gray and dorsal raphe nucleus. J Comp Neurol 447:82–97
Conti L, Pinder R (1979) A controlled comparative trial of mianserin and diazepam in the treatment of anxiety states in psychiatric outpatients. J Int Med Res 7:185–189
Coyle J, Leski M, Morrison J (2002) The diverse roles of L-glutamic acid in brain signal transduction. In: Davis K, Charney D, Coyle J, Nemeroff C (eds) Neuropsychopharmacology: and the fifth generation of progress. Lippincott Williams and Wilkins, Philadelphia, pp 71–90
Crestani F, Keist R, Fritschy JM, Benke D, Vogt K, Prut L, Bluethmann H, Möhler H, Rudolph U (2002) Trace fear conditioning involves hippocampal α5 GABAA receptors. Proc Natl Acad Sci U S A 99:8980–8985
Daugé V, Léna I (1998) CCK in anxiety and cognitive processes. Neurosci Biobehav Rev 22:815–825
Davis M, Myers KM (2002) The role of glutamate and gamma-aminobutyric acid in fear extinction: clinical implications for exposure therapy. Biol Psychiatry 52:998–1007
Dawson VL, Kizushi VM, Huang PL, Snyder SH, Dawson TM (1996) Resistance to neurotoxicity in cortical cultures from neuronal nitric oxide synthase-deficient mice. J Neurosci 16:2479–2487
de Bold AJ (1985) Atrial natriuretic factor a hormone produced by the heart. Science 230:767–770
De Montigny C (1989) Cholecystokinin tetrapeptide induces panic like attacks in healthy volunteers. Arch Gen Psychiatry 46:511–517
Dolmetsch R, Pajvani U, Fife K, Spotts JM, Greenberg ME (2001) Signaling to the nucleus by an L-type calcium channel-calmodulin complex through the MAP kinase pathway. Science 294:333–339
Duman R, Malberg J, Nakagawa S, D'Sa C (2000) Neuronal plasticity and survival in mood disorders. Biol Psychiatry 48:732–739
Duman RS, Heninger GR, Nestler EJ (1997) A molecular and cellular theory of depression. Arch Gen Psychiatry 54:597–606
Dumont Y, Fournier A, St-Pierre S, Quirion R (1995) Characterization of neuropeptide Y binding sites in rat brain membrane preparations using [125I]Leu31, [Pro34]peptide YY and [125I]peptide YY3-36 as selective Y1 and Y2 radioligands. J Pharmacol Exp Ther 272:673–680
Dunn A, Berridge CW (1990) Physiological and behavioral responses to corticotropin releasing factor administration: Is CRF a mediator of anxiety or stress response. Brain Res Brain Res Rev 15:71–100
Dunn R, Reed TA, Copeland PD, Frye CA (1998) The nitric oxide synthase inhibitor 7-nitroindazole displays enhanced anxiolytic efficacy without tolerance in rats following subchronic administration. Neuropharmacology 37:899–904
Freeman AR, Westphal J, Norris G, Roggero B, Webb P, Freeman K (1997) Efficacy of ondansetron in the treatment of generalized anxiety disorder. Depress Anxiety 5:140–141

Fritschy JM, Johnson DK, Mohler H, Rudolph U (1998) Independent assembly and subcellular targeting of GABAA receptor subtypes demonstrated in mouse hippocampal and olfactory neurons in vivo. Neurosci Lett 249:99–102

Griebel G (1999) Is there a future for neuropeptide receptor ligands in the treatment of anxiety disorders. Pharmacol Ther 82:1–61

Griebel G, Perrault G, Soubrie P (2001) Effects of SR48968, a selective non-peptide NK2 receptor antagonist on emotional processes in rodents. Psychopharmacology (Berl) 158:241–251

Griebel G, Simiand J, Serradeil-LeGal C, Wagnon J, Pascal M, Scatton B, Maffrand JP, Soubrie P (2002) Anxiolytic- and antidepressant-like effects of the non-peptide vasopressin V1b receptor antagonist, SSR 149415, suggest an innovative approach for the treatment of stress-related disorders. Proc Natl Acad Sci U S A 99:6370–6375

Griffin LD, Mellon SH (1999) Selective serotonin reuptake inhibitors directly alter activity of neurosteroidogenic enzymes. Proc Natl Acad Sci U S A 96:13512–13517

Guidotti A, Costa E (1998) Can the antidysphoric and anxiolytic profiles of selective serotonin inhibitors be related to their ability to increase brain 3α,5α-tetrahydroprogesterone (allopregnanolone) availability? Biol Psychiatry 44:865–873

Harro J, Vasar E, Bradwejn J (1993) Cholecystokinin in animal and human research of anxiety. Trends Pharmacol Sci 14:244–249

Heilig M, McLeod S, Brot M, Heinrichs SC, Menzaghi F, Koob GF, Britton KT (1993) Anxiolytic-like action of neuropeptide Y mediation by Y1 receptors in amygdala, and dissociation from food intake effects. Neuropsychopharmacology 8:357–363

Hernando F, Schoots O, Lolait SJ, Burbach JPH (2001) Immunohistochemical localization of the vasopressin V1b receptor in the rat brain and pituitary gland: anatomical support for its involvement in the central effects of vasopressin. Endocrinology 142:1659–1668

Holmes A, Yang RJ, Murphy DL, Crawley JN (2002) Evaluation of antidepressant-related behavioral responses in mice lacking the serotonin transporter. Neuropsychopharmacology 27:914–923

Holmes A, Kinney JW, Wrenn CC, Li Q, Yang RJ, Ma L, Vishwanath J, Saavedra MC, Innerfield CE, Jacoby AS, Shine J, Iismaa TP, Crawley JN (2003) Galanin GAL-R1 receptor null mutant mice display increased anxiety-like behavior specific to the elevated plus-maze. Neuropsychopharmacology 28:1031–1044

Holsboer F (1999) The rationale for corticotropin-releasing hormone receptor (CRH-R) antagonists to treat depression and anxiety. J Psychiatr Res 33:181–214

Hoyer D, Hannon J, Martin G (2002) Molecular, pharmacological and functional diversity of 5-HT receptors. Pharmacol Biochem Behav 71:533–554

Kellner M, Wiedemann K, Holsboer F (1992) ANF inhibits the CRH-stimulated secretion of ACTH and cortisol in man. Life Sci 50:1835–1842

Kellner M, Herzog L, Yassouridis A, Holsboer F, Wiedemann K (1995) A possible role of atrial natriuretic hormone in pituitary-adrenocortical unresponsiveness in lactate-induced panic disorder. Am J Psychiatry 152:1365–1367

Kennedy JL, Bradwein J, Koszycki D (1999) Investigation of cholecystokinin system genes in panic disorder. Mol Psychiatry 4:284–285

Kennett G, Whitton P, Shah K, Curzon G (1989) Anxiogenic like effects of mCPP and TFMPP in animal models are opposed by 5-HT1C receptor antagonists. Eur J Pharmacol 164:445–454

Khisti RT, Chopde CT, Jain SP (2000) Antidepressant-like effect of the neurosteroid 3α-hydroxy-5α-pregnan-20-one in mice forced swim test. Pharmacol Biochem Behav 67:137–143

Kishimoto T, Radulovic M, Lin CR, Hooshmand F, Hermanson O, Rosenfeld MG, Spiess J (2000) Deletion of the CRH2 reveals an anxiolytic role for corticotropin-releasing hormone receptor-2. Nat Genet 24:415–419

Kramer MS, Cutler NR, Ballenger JC, Patterson WM, Mendels J (1995) A placebo-controlled trial of L-365,260, a CCK antagonist, in panic disorder. Biol Psychiatry 37:462–466

Kramer MS, Cutler N, Feighner J, Shrivastava R, Carman J, Sramek IJ, Reines SA, Snavely D, Wyatt-Knowles E, Hayle EJ (1998) Distinct mechanism for antidepressant activity by blockade of central substance P receptors. Science 281:1640–1645

Kriegsfeld LJ, Dawson TM, Dawson VL, Nelson RJ, Snyder SH (1997) Aggressive behavior in male mice lacking the gene for neuronal nitric oxide synthase requires testosterone. Brain Res 769:66–70

Landgraf R, Gerstberger R, Montkowski A, Probst JC, Wotjak CT, Holsboer F, Engelmann M (1995) V1 vasopressin receptor antisense oligodeoxynucleotide into septum reduces vasopressin binding, social discrimination abilities, and anxiety-related behavior in rats. J Neurosci 15:4250–4258

Li B-M, Mei ZT (1994) Delayed response deficit induced by local injection of the alpha-2 adrenergic antagonist yohimbine into the dorsolateral prefrontal cortex in young adult monkeys. Behav Neural Biol 62:134–139

Li S, Yusuke O, Yang D, Quock RM (2003b) Antagonism of nitrous oxide-induced anxiolytic-like behavior in the mouse light/dark exploration procedure by pharmacologic disruption of endogenous nitric oxide function. Psychopharmacology (Berl) 166:366–372

Li X, Tizzano JP, Griffey K (2001) Antidepressant-like action of an AMPA receptor potentiator (LY392098). Neuropsychopharmacology 40:1028–1033

Liebsch G, Landgraf R, Engelmann M, Lörscher P, Holsboer F (1999) Differential behavioural effects of chronic infusion of CRH1 and CRH2 receptor antisense oligonucleotides into the rat brain. J Psychiatr Res 33:153–163

Lin C-H, Yeh S-H, Leu T-H, Chang W-C, Wang S-T, Gean P-W (2003) Identification of calcineurin as a key signal in the extinction of fear memory. J Neurosci 23:1574–1579

Lin D, Parsons L (2002) Anxiogenic-like effect of serotonin1B receptor stimulation in the rat elevated plus-maze. Pharmacol Biochem Behav 71:581–587

Lines C, Challenor J, Traub M (1995) Cholecystokinin and anxiety in normal volunteers—an investigation of the anxiogenic properties of pentagastrin and reversal by the cholecystokinin receptor subtype-b-antagonist L-365,260. Br J Clin Pharmacol 39:235–242

Lolait SJ, O'Carroll AM, Mahan LC, Felder CC, Button D, Young III WS (1995) Extrapituitary expression of the rat V1b vasopressin receptor gene. Proc Natl Acad Sci USA 92:6783–6787

Low K, Crestani F, Kleist R, Benke D, Brunig I, Benson J, Möhler H (2000) Molecular and neuronal substrate for the selective attenuation of anxiety. Science 290:131–140

Lu KT, Walker DL, Davis M (2001) Mitogen-activated protein kinase cascade in the basolateral nucleus of amygdala is involved in extinction of fear-potentiated startle. J Neurosci 21:RC162

Majewska MD, Harrison NL, Schwartz RD, Barker JL, Paul SM (1986) Steroid hormone metabolites are barbiturate-like modulators of the GABA receptor. Science 232:1004–1007

Mansuy I, Mayford M, Jacob B, Kandel ER, Bach ME (1998) Restricted and regulated overexpression reveals calcineurin as a key component in the transition from short-term to long-term memory. Cell 92:39–49

Maragakis N, Rothstein J (2001) Glutamate transports in neurologic disease. Arch Neurol 58:365–370

Marsicano G, Wotjak CT, Azad SC, Bisogno T, Rammes G, Cascio MG, Hermann H, Tang J, Hofmann C, Zieglgansberger W, Di Marzo V, Lutz B (2002) The endogenous cannabinoid system controls extinction of aversive memories. Nature 418:530–534
McKeman R, Rosdahl T, Reynolds D, Sur C, Wafford K, Atack J (2000) Sedative but not anxiolytic properties of benzodiazepines are mediated by the GABA-A receptor alpha-1 subtype receptors. Nat Neurosci 3:587–592
Miyasaka K, Kobayashi S, Ohta M, Kanai S, Yoshida Y, Nagata A, Matsui T, Noda T, Takuguchi S, Takata Y, Kawanami T, Funakoshi A (2002) Anxiety-related behaviors in cholecystokinin-A,B, and AB receptor gene knockout mice in the plus-maze. Neurosci Lett 335:115–118
Möhler H, Fritschy JM, Rudolph U (2002) A new benzodiazepine pharmacology. J Pharmacol Exp Ther 300:2–8
Moller C, Sommer W, Thorsell A, Heilig M (1999) Anxiogenic-like action of galanin after intra-amygdala administration in the rat. Neuropsychopharmacology 21:507–512
Morilak DA, Cecchi M, Khoshbouei H (2003) Interactions of norepinephrine and galanin in the central amygdala and lateral bed nucleus of the stria terminalis modulate the behavioral response to acute stress. Life Sci 73:715–726
Myers KM, Davis M (2002) Behavioral and neural analysis of extinction. Neuron 36:567–584
Nelson RJ, Demas GE, Huang PL, Fishman MC, Dawson VL, Dawson TM, Snyder SH (1995) Behavioral abnormalities in mice lacking neuronal nitric oxide synthase. Nature 378:383–386
Ostrowski NL, Lolait SJ, Bradley DJ, O'Carroll AM, Brownstein MJ, Young WS (1992) Distribution of V1a and V2 vasopressin receptor messenger ribonucleic acids in liver, kidney, pituitary and brain. Endocrinology 131:533–535
Otsuka M, Yoshioka K (1993) Neurotransmitter functions of mammalian tachykinins. Physiol Rev 73:229–308
Palmiter RD, Erickson JC, Hollopeter G, Baraban SC, Schwartz MW (1998) Life without neuropeptide Y. Recent Prog Horm Res 53:163–199
Pande A, Davidson J, Jefferson J, Janney C, Katzelnick D, Weisler R (1999b) Treatment of social phobia with gabapentin: A placebo-controlled study. J Clin Psychopharmacol 19:341–348
Pande A, Davidson J, Jefferson J, Janney C, Chouinard G, Lydiard R (2000) Placebo-controlled study of gabapentin treatment of panic disorder. J Clin Psychopharmacol 20:467–471
Pande AC, Greiner M, Adams JB (1999a) Placebo-controlled trial of the CCK-B antagonist, CI-988 in panic disorder. Mol Psychiatry 46:860–862
Patchev VK, Montkowski A, Rouskova D, Koranyi L, Holsboer F, Almeida O (1997) Neonatal treatment of rats with the neuroactive steroid tetrahydrodeoxycorticosterone (THDOC) abolishes the behavioral and neuroendocrine consequences of adverse early life events. J Clin Invest 99:962–966
Paul SM, RH Purdy (1992) Neuroactive steroids. FASEB J 6:2311–2322
Pigott T, Zohar J, Hill J, Bernstein S, Grover G, Zohar-Kadouch R (1991) Metergoline blocks the behavioral and neuroendocrine effects of orally administered m-chlorophenylpiperazine in patients with obsessive compulsive disorder. Biol Psychiatry 29:418–426S
Pitman R, Sanders K, Zusman R, Healy A, Cheema F, Lasko N (2002) Pilot study of secondary prevention of posttraumatic stress disorder with propranolol. Biol Psychiatry 51:189–192
Ressler KJ, Rothbaum BO, Tannenbaum L, Anderson P, Graap K, Zimand E, Hodges L, Davis M (2004) Cognitive enhancers as adjuncts to psychotherapy. Arch Gen Psychiatry 61:1136–1144

Ribeiro L, Busnello J, Kauer-Sant'Anna M, Madruga M, Quevedo J, Busnello E (2001) Mirtazapine versus fluoxetine in the treatment of panic disorder. Braz J Med Biol Res 34:1303–1307

Rickels K, Pollack M, Lydiard R (2002) Efficacy and safety of pregabalin and alprazolam in generalized anxiety disorder. American Psychiatric Association Annual Meeting, vol. NR 162. American Psychiatric Press, Philadelphia

Romeo E, A Ströhle, F di Michele, G Spaletta, B Hermann, F Holsboer, A Pasini, R Rupprecht (1998) Effects of antidepressant treatment on neuroactive steroids in major depression. Am J Psychiatry 155:910–913

Rupprecht R, Holsboer F (1999) Neuroactive steroids: mechanisms of action and neuropsychopharmacological perspectives. Trends Neurosci 22:410–416

Sajdyk TJ, Schober DA, Gehlert DR (2002) Neuropeptide Y receptor subtypes in the basolateral nucleus of the amygdala modulate anxiogenic responses in rats. Neuropharmacology 43:1165–1172

Santarelli L, Gobbi G, Debs P, Sibille E, Blier P, Hen R (2001) Genetic and pharmacological disruption of neurokinin 1 receptor function decreases anxiety-related behaviors and increases serotonergic function. Proc Natl Acad Sci USA 98:1912–1917

Scatton B, Deportere H, George P, Servin M, Benavides J, Schoemaker H, Perrault G (2000) Selectivity for GABAA receptor α subunits as a strategy for developing hypnoselective and anxioselective drugs. Int J Neuropsychopharmacol 3:S41–S43

Schneier F, Garfinkel R, Kennedy B, Campeas R, Fallon B, Marshall R (1996) Ondansetron in the treatment of panic disorder. Anxiety 2:199–202

Shlik J, Aluoja A, Vasar V, Vasar E, Podar T, Bradwein J (1997) Effects of citalopram on behavioral, cardiovascular, and neuroendocrine response to cholecystokinin tetrapeptide challenge in patients with panic disorder. J Psychiatry Neurosci 22:332–340

Skutella T, Montkowski A, Stöhr A, Probst JR, Landgraf R, Holsboer F, Jirikowski GF (1994) Corticotropin-releasing hormone (CRH) antisense oligodeoxynucleotide treatment attenuates social defeat-induced anxiety in rats. Cell Mol Neurobiol 14:579–588

Skutella T, Probst JC, Renner U, Holsboer F, Behl C (1998) Corticotropin-releasing hormone receptor (type I) antisense targeting reduces anxiety. Neuroscience 85:795–805

Smith GW, Aubry J-M, Dellu F, Contarino A, Bilezjian LM, Gold LH, Hause C, Bentley CA, Sawchenko PE, Koob GF, Vale W, Lee K-F (1998) Corticotropin-releasing factor receptor 1-deficient mice display decreased anxiety, impaired stress response, and aberrant neuroendocrine development. Neuron 20:1093–1102

Son H, Hawkins RD, Martin K, Kiebler M, Huang PL, Fishman MC, Kandel ER (1996) Long term potentiation is reduced in mice that are doubly mutant in endothelial and neuronal nitric oxide synthase. Cell 87:1015–1023

Stenzel-Poore MP, Heinrichs SC, Rivest S, Koob GF, Vale WW (1994) Overproduction of corticotropin-releasing factor in transgenic mice: a genetic model of anxiogenic behavior. J Neurosci 14:2579–2584

Ströhle A, Jahn H, Montkowski A, Liebsch G, Boll E, Landgraf R, Holsboer F, Wiedemann K (1997) Central and peripheral administration of atriopeptin is anxiolytic in rats. Neuroendocrinology 65:210–215

Ströhle A, Kellner M, Holsboer F, Wiedemann K (1998) Atrial natriuretic hormone decreases endocrine response to a combined dexamethasone corticotropin-releasing hormone test. Biol Psychiatry 43:371–375

Ströhle A, Romeo E, Hermann B, di Micelle F, Spaletta G, Pasini A, Holsboer F, Rupprecht R (1999) Concentrations of 3α-reduced neuroactive steroids and their precursors in plasma of patients with major depression and after clinical recovery. Biol Psychiatry 45:274–277

Ströhle A, Pasini A, Romeo E, Hermann B, Spalletta G, di Michele F, Holsboer F, Rupprecht R (2000) Fluoxetine decreases concentrations of 3α,5α-tetrahydrodeoxycorticosterone (3α,5α-THDOC) in major depression. J Psychiatr Res 34:183–186

Ströhle A, Kellner M, Holsboer F, Wiedemann K (2001) Anxiolytic activity of atrial natriuretic peptide in patients with panic disorder. Am J Psychiatry 158:1514–1516

Ströhle A, Romeo E, di Michele F, Pasini A, Yassouridis A, Holsboer F, Rupprecht R (2002) GABAA receptor modulatory neuroactive steroid composition in panic disorder and during paroxetine treatment. Am J Psychiatry 159:145–147

Ströhle A, Romeo E, di Michele F, Pasini A, Hermann B, Gajewsky G, Holsboer F, Rupprecht F (2003) Induced panic attacks shift GABAA receptor modulatory neuroactive steroid composition. Arch Gen Psychiatry 60:161–168

Szapiro G, Vianna MRM, McGaugh JL, Medina JH, Izquierdo I (2003) The role of NMDA glutamate receptors, PKA, MAPK, and CAMKII in the hippocampus in extinction of conditioned fear. Hippocampus 13:53–58

Tanaka IS, Misono KS, Inagami T (1984) Atrial natriuretic factor in rat hypothalamus, atria and plasma: determinations by specific radioimmunoassay. Biochem Biophys Res Commun 124:663–668

Thiele TE, Marsh DJ, Ste Marie L, Bernstein IL, Palmiter RD (1998) Ethanol consumption and resistance are inversely related to neuropeptide Y levels. Nature 396:366–369

Thiele TE, Koh MT, Pedrazzini T (2002) Voluntary alcohol consumption is controlled via the neuropeptide Y Y1 receptor. J Neurosci 22:RC208

Thorsell A, Rimondini R, Heilig M (2002) Blockade of central neuropeptide Y (NPY) Y2 receptors reduces ethanol self-administration in rats. Neurosci Lett 332:1–4

Timmusk T, Palm K, Metsis M (1993) Multiple promoter direct tissue-specific expression of rat BDNF gene. Neuron 10:475–489

Timpl P, Spanagel R, Sillaber I, Kresse A, Reul JHMH, Stalla GK, Blanquet V, Steckler T, Holsboer F, Wurst W (1998) Impaired stress response and reduced anxiety in mice lacking a functional corticotropin-releasing hormone receptor 1. Nat Genet 19:162–166

Tribollet E, Raufatse D, Maffrand JP, Serradeil-le Gal C (1999) Binding of the non-peptide vasopressin V1a receptor antagonist SR-49059 in the rat brain: an in vitro and in vivo autoradiographic study. Neuroendocrinology 69:113–120

Tschenett A, Singewald N, Carli M, Balducci C, Salchner P, Vezzani A, Herzog H, Sperk G (2003) Reduced anxiety and improved stress coping ability in mice lacking NPY-Y2 receptors. Eur J Neurosci 18:143–148

Uzunov DP, Cooper TB, Costa E, Guidotti A (1996) Fluoxetine-elicited changes in brain neurosteroid content measured by negative ion mass fragmentography. Proc Natl Acad Sci USA 93:12599–12604

Uzunova V, Sheline Y, Davis JM, Rasmusson A, Uzunov DP, Costa E, Guidotti A (1998) Increase in the cerebrospinal fluid content of neurosteroids in patients with unipolar major depression who are receiving fluoxetine or fluvoxamine. Proc Natl Acad Sci USA 95:3239–3244

Vale A, Green S, Montgomery AM, Shafi S (1998) The nitric oxide synthesis inhibitor L-NAME produces anxiogenic-like effects in the rat elevated plus-maze test, but not in the social interaction test. J Psychopharmacol 12:268–272

Vassout A, Veenstra S, Hauser K, Ofner S, Brugger F, Schilling W (2000) NKP608: A selective NK-1 receptor antagonist with anxiety-like effects in the social interaction and social exploration test in rats. Regul Pept 96:7–16

Walker DL, Ressler KJ, Lu KT, Davis M (2002) Facilitation of conditioned fear extinction by systemic administration or intra-amygdala infusions of D-cycloserine as assessed with fear potentiated startle. J Neurosci 22:2343–2351

Wersinger SR, Ginns EI, O'Carroll AM, Lolait SJ, Young III WS (2002) Vasopressin V1b receptor knockout reduces aggressive behavior in male mice. Mol Psychiatry 7:975–984

Wiedemann K, Herzog L, Kellner M (1995) Atrial natriuretic hormone inhibits corticotropin-releasing hormone induced prolactin release. J Psychiatr Res 29:51–58

Wiedemann K, Jahn H, Yassouridis A, Kellner M (2001) Anxiolytic activity of atrial natriuretic peptide on cholecystokinin tetrapeptide-induced panic attacks. Arch Gen Psychiatry 58:371–377

Wrenn CC, Crawley JN (2001) Pharmacological evidence supporting a role for galanin in cognition and affect. Prog Neuropsychopharmacol Biol Psychiatry 25:283–299

Zafra F, Hengerer B, Leibrock J (1990) Activity dependent regulation of BDNF and NGF mRNAs in the rat hippocampus is mediated by non-NMDA glutamate receptors. J Neurosci 9:3545–3550

Zobel AW, Nickel T, Künzel HE, Ackl N, Sonntag A, Ising M, F Holsboer (2000) Effects of the high-affinity corticotropin-releasing hormone receptor 1 antagonist R121919 in major depression: The first 20 patients treated. J Psychiatr Res 34:171–181

Zwansger P, Baghai TC, Schuele C, Ströhle A, Padberg F, Kathmann N, Schwarz M, Möller H-J, Rupprecht R (2001) Vigabatrin decreases cholecystokinin-tetrapeptide (CCK-4) induced panic in healthy volunteers. Neuropsychopharmacology 25:699–703

HEP (2005) 169:527–546

Pharmacogenomics

E.B. Binder · F. Holsboer (✉)

Max-Planck Institute of Psychiatry, Kraepelinstr. 10, 80804 Munich, Germany
holsboer@mpipsykl.mpg.de

Abstract So far no pharmacogenetic/genomic study has been conducted specifically for anxiety disorders. Some of the presented results, however, do pertain to such disorders. For example, pharmacokinetic aspects of antidepressant drug therapy likely also apply to patients with anxiety disorders, and several genetic polymorphisms in the cytochrome P450 (CYP) gene family and drug transporter molecules, such as the multidrug resistance (MDR) gene type 1, have been reported to influence the pharmacokinetics of antidepressant drugs. At this stage of pharmacogenomics research, it is difficult to interpret the relevance of pharmacodynamic–genetic association studies conducted in depressed patients for anxiety disorders. A number of studies have reported an influence of polymorphisms of genes mostly in the serotonergic pathway on the response to antidepressant drugs in patients suffering from depression. In order to know whether they can be extrapolated to patients with anxiety disorders, clinical studies are warranted. Despite all the shortcomings of the currently available pharmacogenetic studies, this field holds great promise for the treatment of anxiety disorders. In the future, psychiatrists may be able to base treatment decisions (i.e., the type and dose of prescribed drug) on more objective parameters than only the diagnostic algorithms used now. This will limit unwanted side effects and adverse drug reactions, and could reduce time to response, resulting in a more individualized pharmacotherapy.

Keywords Anxiety · Pharmacogenetic · Pharmacogenomic · Pharmacokinetic · Pharmacodynamic

1 Introduction

Pharmacotherapy is an effective therapeutic approach for anxiety disorders, and benzodiazepines and antidepressant drugs, including tricyclic antidepressants (TCA), selective serotonin reuptake inhibitors (SSRIs), monoamine oxidase (MAO) inhibitors, and others, are routinely prescribed in adjunction to psychotherapeutic approaches. While benzodiazepines and antidepressants drugs have proven anxiolytic effects (see previous chapters), several drawbacks have to be considered for each class. Benzodiazepines are rapidly effective against the acute symptoms of anxiety or panic attacks; however, their prolonged use can lead to the development of dependence and tolerance, and exaggerated sedation is a bothersome side effect. Antidepressant drugs do not appear to induce dependence, but the onset of their anxiolytic action is often delayed by weeks. In addition, these drugs can be associated with a variety of side effects, ranging from dry mouth, to weight gain, to sexual dysfunction. Furthermore, not all patients respond favorably to antidepressant therapy. In the absence of biologically based treatment guidelines, the right antidepressant can only be determined by trial and error.

Drug response can be influenced by a variety of factors, including environmental (for example nutrition and co-administered drugs) and genetic factors. Since the 1950s, inherited differences in drug response have been described (Roden and George 2002; Weinshilboum 2003) for a variety of different compounds, establishing the field of pharmacogenetics and later pharmacogenomics. While pharmacogenetics refers to gauging the effect of single genes, pharmacogenomics describes the use of genome-wide approaches to elucidate individual differences in the outcome of drug therapy, including both adverse events and drug response. These terms, however, are often used interchangeably.

1.1 The Sequence of the Human Genome

The knowledge of the sequence of the human genome, which has been publicly available to all researchers since February 2001 (Lander et al. 2001; Venter et al. 2001), has dramatically changed the possibilities for genetic approaches. It is now possible to screen the whole genome in a hypothesis-free approach for any association with drug response. To be able to investigate the whole genome, frequent and evenly distributed genetic markers are needed. In the last few years, single nucleotide polymorphisms (SNPs) have become the most promoted genetic markers for complex or common phenotypes (Collins et al. 1999; Kwok and Chen 1998; Risch and Merikangas 1996). These polymorphisms consist of a single base exchange and occur on average every 500–1,000 bp. SNPs can either be functionally relevant themselves or serve as markers for other nearby mutations with which they are in linkage disequilibrium (Brookes 1999). In a combined effort of academic and corporate research, over 1.4 millions SNPs have been identified so far (Sachidanandam et al. 2001) and are publicly available [e.g., SNP database of the NIH, dbSNP or HGBASE (Brookes et al. 2000; Smigielski et al. 2000)]. Of them, 60,000 are located in the coding region of genes. A main advantage of using SNPs as genetic markers is that they can be genotyped using high throughput methods, allowing for the rapid and affordable investigation of many SNPs, which is indispensable for a genome-wide approach (Kwok 2000; Syvanen 2001). With a total of presumably at least 3 million SNPs, the realization of a genome screen covering all genes using these single markers could necessitate genotyping hundreds of thousand of SNPs/individual (Carlson et al. 2001; Kruglyak 1999). So far, this is not readily realizable in most laboratories due to limitations in throughput capacities and the exorbitant cost. Two main approaches have so far been put forward to overcome these restrictions.

The first is the establishment of SNP-haplotype maps of the whole genome. These haplotypes represent a sequence of SNP alleles on the same chromosome. Within a given haplotype, SNP alleles are derived from the same ancestral chromosome. Recent studies have shown that the genome is likely structured into distinct larger haplotype blocks (up to hundreds of base pairs), suggesting the existence of hotspots of recombination. Genotyping only a few of these SNPs within a haplotype could be sufficient to represent all SNPs contained in this block. A haplotype map would make finding candidate genes a manageable task. Instead of searching through millions of SNPs, genotyping 10,000–50,000 SNPs that are characteristic for all existing haplotype blocks would allow us to cover the whole genome (Cardon and Abecasis 2003; Judson et al. 2002) (see Fig. 1).

Another possible approach is to identify candidate genes using genome-wide analyses of changes in mRNA expression using so called "microarrays" (Lockhart and Winzeler 2000). Microarrays allow researchers to investigate changes in mRNA expression patterns following certain treatments or in different conditions in parallel from all genes. Results from these studies may be useful in the identification of novel candidate genes. Another important tool for the discovery and validation of novel candidate genes, proteomics, is discussed in detail in the following chapter (see Turck, this volume).

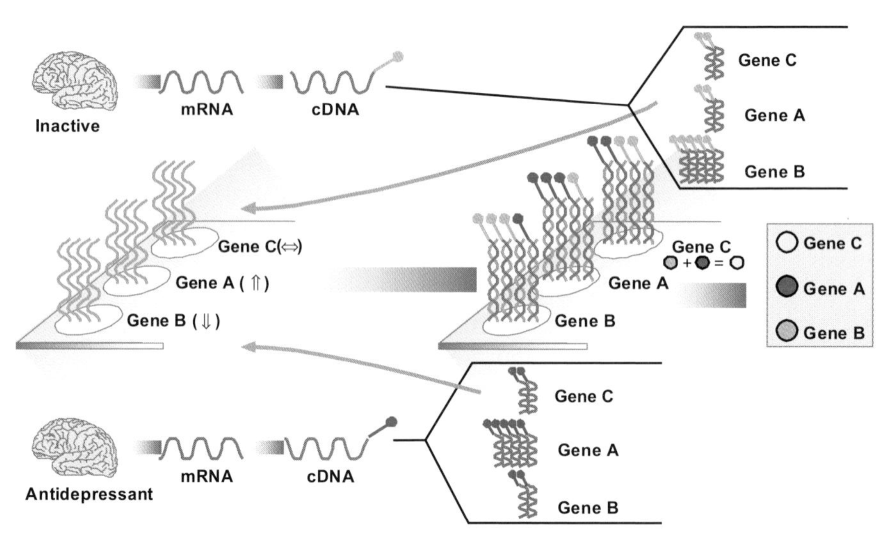

Fig. 1 Expression profiling of antidepressant-induced changes using complementary (c)DNA microarrays. Messenger (m)RNA is extracted from the brains of animals either treated with antidepressants or vehicle. mRNA is then transcribed into cDNA and labeled with a different fluorescent dye for each condition. The labeled cDNA from both conditions is then applied to the microarray in equal amounts. Genes in which expression is upregulated in the treatment condition will lead to more of the cDNA labeled with one fluorescent dye hybridizing to the spot where the complementary sequences have been deposited. The amount of each fluorescent dye on each spot are quantified, allowing investigators to assess whether gene expression has been upregulated (*gene A*), downregulated (*gene B*), or not been affected (*gene C*) by the treatment

1.2 General Issues on the Pharmacogenomics of Anxiety and Anxiolytic Drugs

As mentioned above, not all patients treated with anxiolytic or antidepressant drugs respond favorably to these treatments. There is some evidence from family studies that suggests an important contribution of genetic factors. Already in the early 1960s, studies on the effects TCAs were conducted in families (Angst 1961; Pare et al. 1962). O'Reilly et al. (1994) reported a familial aggregation of response to tranylcypromine, a MAO inhibitor in a family with eight members affected with major depression over two generations. These initial case reports were followed by only a few systematic studies. A study by Franchini et al. (1998) indicated a possible genetic basis of response to the SSRI fluvoxamine in 45 pairs of relatives. In light of these data, some groups have used response to a certain antidepressant drug or mood stabilizer as an additional phenotype in classical linkage analyses for mood disorders in the hope of identifying genetically more homogeneous families (Serretti et al. 1998; Turecki et al. 2001). Nonetheless, family studies supporting a genetic basis of response to antidepressants and even more so anxiolytic drugs are sparse, certainly due to difficulties in collecting such samples.

To date, no pharmacogenomic or pharmacogenetic studies have been explicitly performed in anxiety disorder. Nonetheless, general pharmacokinetic as well some pharmacodynamic genetic aspects of anxiolytic drugs and antidepressants may also pertain to anxiety disorders. Pharmacokinetics refers to processes influencing the delivery of a drug to the target, including absorption, distribution, metabolism, and elimination. Several genetic polymorphisms in key genes of this pathway, including the cytochrome P450 (CYP) gene family (oxidation of compounds), *N*-acetyl transferase (*N*-acetylation), thiopurine methyltransferase (conjugation), and drug transporter molecules, such as genes from the multidrug resistance (MDR) gene family, have been reported to influence the pharmacokinetics of drugs (see Roden and George 2001 for review). For anxiolytic drugs, especially antidepressants, pharmacokinetic influences of genetic polymorphisms in certain CYP genes have been reported, and preclinical data strongly suggest a regulation of intracerebral antidepressant drug concentrations by P-glycoprotein, the product of the MDR1 gene (Rasmussen and Brosen 2000; Steimer et al. 2001; Uhr et al. 2000). The term pharmacodynamics encompasses all processes influencing the relationship between the drug concentration and the resulting effect. Here direct genetic effects of target molecules of antidepressant drugs such as polymorphisms of the serotonin transporter (*SERT*) and indirect genetic effects of polymorphisms in molecules not primarily targeted by these drugs such the Gβ3 G protein subunit gene have been reported for anxiolytic drugs (Serretti et al. 2002; Staddon et al. 2002; Steimer et al. 2001). In the following sections, we will elaborate on previous findings in the genetics of the pharmacokinetics and pharmacodynamics of antidepressant drugs and their relevance to anxiety disorders. We will also expand on the possibilities of pharmacogenomic studies in the prediction of drug response in anxiety disorders, and finally we will briefly summarize the potential of pharmacogenomics in the discovery of novel anxiolytic agents.

2 Pharmacokinetic Aspects

2.1 The Cytochrome P450 Gene Family

Approximately 50 CYP enzymes, which are haem proteins, have been identified thus far. In humans there are about 10 important drug-metabolizing CYP genes. These are mainly expressed in the smooth endoplasmatic reticulum of the hepatocytes, but can also be found in gut mucosa, kidney, lung tissue, skin, and in the brain. Of these, *CYP2D6*, *CYP2C19*, *CYP3A4*, and *CYP1A2* are important in the metabolism of antidepressant drugs and benzodiazepines

(Staddon et al. 2002; Steimer et al. 2001) (see Table 1 for more details). Most studies so far have focused on the role of *CYP2D6* in the pharmacokinetics of antidepressant drugs. Over 70 functionally different alleles have been reported for *CYP2D6*, more than 15 of these encode an inactive or no enzyme at all, while others consist of gene duplications (Bertilsson et al. 2002). According to the inherited alleles, individuals can thus be grouped into poor (PM), intermediate (IM), extensive (EM), and ultra-rapid metabolizers (UM) (Nebert and Dieter 2000). An increased risk of toxic reactions has been reported in PM due to overdosing, while certain drugs may not reach therapeutic plasma concentrations in UM due to underdosing. The proportion of different types of metabolizers in a population varies with ethnicity, so that 7% of Caucasians but only 1% of Asians are PM. In addition, there are several population-specific alleles, only encountered in certain ethnicities (Bertilsson et al. 2002).

Dalen et al. (1998) reported a close correlation between the number of functional CYP2D6 gene copies and plasma levels of the TCA nortriptyline. From these single-dose experiments, Bertilsson et al. (2002) extrapolated that patients with no or only one functional copy of the gene would already reach therapeutic plasma levels with starting doses for nortriptyline and would easily reach potentially toxic concentrations with high-normal doses. Patients with 2–4 copies on the other hand would require high-normal doses to even reach therapeutic plasma levels. In the case of the one reported patient with 13 gene copies, even high-normal doses would not be sufficient for clinically relevant plasma concentrations. Similar polymorphism/plasma concentration correlations have been reported for the SSRI paroxetine (Ozdemir et al. 1999; Sindrup et al. 1992) and the combined serotonin norepinephrine reuptake

Table 1 Cytochrome P450 (CYP) isoenzymes and metabolism of antidepressant and anxiolytic drugs and important inhibitors and inductors

CYP isoenzymes	CYP1A4	CYP2C	CYP2D6	CYP3A4
Substrates	Imipramine; clomipramine; fluvoxamine	Amitriptyline; clomipramine; imipramine; moclobemide; diazepam	Fluoxetine; fluvoxamine; paroxetine; venlafaxine; nortriptyline; desipramine	Diazepam; imipramine; sertraline
Inhibitors	Fluvoxamine	Moclobemide; fluvoxamine; tranylcypromine	Paroxetine	Ketoconazole; omeprazole; grapefruit juice
Inductors	Tobacco smoke	Barbiturates; omeprazole; rifampicin		Carbamazepine; phenytoin; barbiturates

inhibitor (SNRI) venlafaxine (Fukuda et al. 2000; Veefkind et al. 2000). For the latter, a relationship between PM status and the increased occurrence of cardiovascular side effects or toxicity has been reported (Lessard et al. 1999). In summary, knowledge of the CYP2D6 metabolizer status could be helpful in individualizing dose escalation schemes for certain antidepressants. This could be especially helpful in the case of TCAs, were relatively small dose–response windows have been reported for their antidepressant effect (Burke and Preskorn 1999; Preskorn et al. 1988). For SSRIs, on the other hand, no clear dose–response relationship has been reported, at least for the treatment of depressive symptoms, and so far no threshold toxic concentrations have been defined (Corruble and Guelfi 2000; Preskorn and Lane 1995).

Specific dose recommendation based on CYP2D6 genotypes have already been put forward (Kirchheiner et al. 2001) with doses of TCA halved for PM. The proposed doses adjustments for SSRIs were significantly smaller than otherwise recommended, and some authors even question the relevance of genotype-adjusted dosing for SSRIs, given their flat dose–response curve (Brosen and Naranjo 2001). Nonetheless, an identification of PM may prevent overdosing and the occurrence of specific side effects with SSRIs or SNRIs. In addition, knowledge of the metabolizer status of a patient may also be helpful in predicting problems with drug interactions. Brosen et al. (1993) report that pharmacokinetic interactions of paroxetine (an inhibitor of CYP2D6) and the TCA desipramine (extensively metabolized by CYP2D6) are dependent on the metabolizer status. Co-administration of the two drugs in EM who have at least two functional copies of the CYP2D6 gene leads to a fivefold decrease in desipramine clearance. In PM who lack functional CYP2D6 genes, desipramine clearance was not influenced by paroxetine, suggesting alternate metabolic pathways in PM.

In summary, most data are available on the influence of CYP genes on the pharmacokinetics of antidepressants. Genotype-adjusted dose escalation schemes have already been put forward and should be especially useful in TCA treatment. It has to be noted, though, that all studies concerning dose escalation and response have been performed in depressed patients. Dose–response relationships may, however, be different in patients with anxiety disorders, so that further studies are warranted.

2.2 P-Glycoprotein

P-glycoprotein is a member of the highly conserved superfamily of ATP-binding cassette (ABC) transporter proteins (see, for example, Amdukar et al. 1999 for review). This 170-kDa glycoprotein is encoded by the MDR1 gene (now *ABCB1*) on chromosome 16. It is a plasma membrane protein with two transmembrane domains containing each six membrane-spanning helices and an ATP-binding site that actively transports its substrates against a concentration gradient. P-glycoprotein is expressed in the apical membrane of the intestinal epithelial cells, the biliary canalicular membrane of hepatocytes, and the luminal membrane of proximal tubular epithelial cells in the kidney. In addition, it is also found in high levels in the luminal membranes of the endothelial cells that line the small blood capillaries which form the blood–brain and blood–testis barrier (Cordon-Cardo et al. 1989; Thiebaut et al. 1987). The MDR1 gene was first discovered as one of the causes of resistance of tumor cells against chemotherapy. Subsequent studies have discovered that its function is not limited to tumor cells but that P-glycoprotein protects cells throughout the healthy organism against many drugs by acting as an efflux pump for xenobiotics. Substrates besides anti-neoplastic drugs include certain antibiotics, analgesics, cardiotropic drugs, and immunosuppressants. Because of its location at the blood–brain barrier, P-glycoprotein is in a unique position to also regulate the concentration of psychotropic drug in the brain and may limit the brain accumulation of many drugs (Schinkel et al. 1996). Experiments in transgenic mice lacking *mdr1a* or *mdr1a* and *mdr1b*, both homologs of the human MDR1 gene, show that also intracerebral concentrations of antidepressant drugs are regulated by this molecule (Uhr and Grauer 2003; Uhr et al. 2000). These studies conclude that the CNS bioavailability of the SSRI citalopram and the TCAs trimipramine and amitriptyline is regulated by these molecules, while this may not be true for the SSRI fluoxetine. Mouse mutants where MDR1 genes were deleted showed up to over five times higher intracerebral concentrations of the first named drugs than their wildtype littermates, while fluoxetine concentrations where equal (Uhr and Grauer 2003; Uhr et al. 2000). Since P-glycoprotein appears to regulate access to the brain for some antidepressants, it is perceivable that functional polymorphisms in this gene may influence intracerebral antidepressant concentration. Over one hundred SNPs are listed in public SNP databases for *P-glycoprotein*. The most studied polymorphism is a silent SNP in exon 26, often referred to as C3435T, that exchanges a C against a T (see Brinkmann et al. 2001 for review). This SNP has been associated with an altered intestinal digoxin uptake that correlates with intestinal *P-glycoprotein* mRNA expression levels (Hoffmeyer et al. 2000). Homozygotes for the T allele of this SNP, who show a low intestinal P-glycoprotein expression, represent 25% of all Caucasians (Cascorbi et al. 2001). As for the CYP genes, large inter-ethnic differences in allele frequencies have been reported for this SNP, with less than 5% of Africans carrying the TT genotype (Schaeffeler et al. 2001).

While effects of *P-glycoprotein* polymorphisms have been reported for intestinal uptake, no such studies exist for effects on blood–brain barrier penetration. If certain polymorphisms were to alter intracerebral concentrations of specific antidepressants, prior knowledge of the patients relevant P-glycoprotein genotypes could prevent the administration of a drug that might never reach therapeutic intracerebral levels despite a normal plasma concentration.

3
Pharmacodynamic Aspects

The genetics of pharmacodynamic aspects of antidepressant and anxiolytic drugs covers both genes that are direct drug targets, such as the serotonin reuptake transporter (*SERT*), serotonin (*5-HT*) receptors or γ-aminobutyric acid (GABA)-ergic receptor subunits, and genes that are indirectly involved in drug action, such a G proteins. Even though the primary drug targets of antidepressants are known, it is still unclear which neurotransmitter systems and subsequent signaling pathways are ultimately targeted to lead to clinical improvement. A concatenation of data indicates that altering monoaminergic neurotransmission alone is not sufficient to elicit an amelioration of depressive or anxiety symptoms. One obvious argument is that reuptake inhibition occurs within minutes while it takes several weeks until clinical effects are seen. This implies that the majority of candidate genes relevant for response to these drugs are still unknown. So far, mostly candidate genes from the monoaminergic system have been investigated. It is clear that future pharmacogenetic studies have to take a less hypothesis-driven approach and focus more on "unbiased" whole genome approaches. These approaches may not only give insight into the pharmacogenetics of antidepressants and anxiolytics but also lead to the discovery of novel drug targets.

The next paragraphs will first summarize previous pharmacogenetic studies (mostly on antidepressant drugs) and then expand on the techniques available for whole genome approaches.

3.1
Monoaminergic Candidate Genes

Most pharmacogenetic studies for antidepressants have been conducted on candidate genes from monoaminergic pathways. The relevance of monoaminergic systems for anxiety disorders and their treatment have been extensively covered in previous chapters.

The most thoroughly studied gene is *SERT*, located on chromosome *17q* (Lesch et al. 1993; Ramamoorthy et al. 1993). Several polymorphisms have been described for this gene. A 44-bp insertion/deletion polymorphism (called SERTPR or 5-HTTLPR in the literature) in the promoter region has been associated with different basal activity of the transporter, most likely related to differential transcriptional activity (Heils et al. 1996; Lesch et al. 1996). The long variant (l allele) of this polymorphism has been shown to lead to a higher serotonin reuptake by the transporter. Other polymorphisms include a variable tandem repeat (VNTR) polymorphism in intron 3 as well as several non-synonymous SNPs in the coding region (see Hahn and Blakely 2002 for review). The latter polymorphisms have, however, been less studied with regards to pharmacogenetic aspects than SERTPR. The effects of this polymorphism on response to treatment with SSRIs have now been investigated in nine publications (see Table 2). All studies conducted in Caucasian samples have so far shown an association of the short form of the SERTPR (s allele) with a slower, less favorable response to antidepressant drugs. The s allele being associated with a reduced basal activity of the SERT, one could imagine that the effects of SSRIs on synaptic serotonin concentration may be less pronounced than with the l allele. Studies in Asian patients, however, revealed a somewhat different picture. In a Korean sample, Kim et al. (2000) observed a better response to fluvoxamine and paroxetine in patients homozygous for the s allele. This finding was replicated in Japanese patients by Yoshida et al. (2002), while Ito et al. (2002) found no association between SERTPR variants and response to fluvoxamine. Finally, Yu et al. (2002) described a positive association of the l-l genotype with response to fluoxetine in Chinese patients, which is similar to the results observed in Caucasian patients. These contradictory findings in Asian and Caucasian samples may result from ethnically different allele frequencies, the s allele being present in 50% of Caucasians but 75% of Asians (Gelernter et al. 1999). The group of patients homozygous for the l allele is thus smaller in

Table 2 The influence of SERTPR genotype and response to antidepressant drugs

Type of antidepressant	Study	Positive association with response	Ethnicity
Fluvoxamine	$n = 99$ (BP + MP) Smeraldi et al. 1998	L-allele $p = 0.017$	Caucasian
Fluvoxamine	$n = 155$ (BP + MP) Zanardi et al. 2001	L-allele $p = 0.029$	Caucasian
Paroxetine	$n = 64$ (BP + MP) Zanardi et al. 2000	L-allele (s-allele slower) $p < 0.001$	Caucasian
Paroxetine	$n = 95$ (late-life depression) Pollock et al. 2000	L-allele (s-allele slower) $p = 0.028$	Caucasian
Citalopram	$n = 102$ (MP) Arias et al. 2001	L-allele (s-allele more with no remission) $p = 0.006$	Caucasian
Fluoxetine +paroxetine	$n = 120$ (MP/Korean) Kim et al. 2000	S-allele $p = 0.007$	Asian
Fluvoxamine	$n = 66$ (MP/Japanese) Yoshida et al. 2002	S-allele	Asian
Fluoxetine	$n = 121$ (MP/Chinese) Yu et al. 2002	L-allele $p = 0.013$	Asian
Fluvoxamine	$n = 66$ (MP/Japanese) Ito et al. 2002	No association	Asian

BP, bipolar disorder; MP, mono/unipolar depression.

Asian samples, possible hampering the detection of a positive association of this genotype with response. It is also possible that different polymorphisms in the SERT gene are relevant for response in different ethnic groups. Further studies addressing this issue are certainly warranted.

SERTPR variants may also not be specifically associated with response to SSRIs but confer a more general predisposition to respond to any kind of antidepressant treatment targeting the serotonin system. The l allele of SERTPR has not only associated been associated with a better response to SSRIs but also a good response to total sleep deprivation (Benedetti et al. 1999). An enhancement of serotonergic transmission has been proposed as one possible mechanism of action of sleep deprivation (Gardner et al. 1997).

Of the other *SERT* polymorphisms, only the VNTR polymorphism has been investigated for its association with response to antidepressant drugs (Kim et al. 2000).

Pharmacogenetic studies also exist for several other genes of the monoaminergic systems, including *tryptophan hydroxylase* (*TPH*), *monoamine oxidase A* (*MAOA*), *5-HT receptors* (*2a* and *6*), *dopamine receptors*, and the *G protein* β*3 subunit*. Smeraldi's group detected an association of an intronic SNP in *TPH* with response to fluvoxamine and paroxetine in two separate samples (Serretti et al. 2001b; Serretti et al. 2001c). This association was not replicated in a Japanese sample (Yoshida et al. 2002b). Three separate studies found no association of a VNTR polymorphism in *MAOA*, affecting gene transcription, with response to MAO inhibitors and SSRIs (Cusin et al. 2002; Muller et al. 2002; Yoshida et al. 2002b). Three different SNPs in *5HT2A* have been investigated in three different studies, with two of the studies reporting an association with response to antidepressant treatment (Cusin et al. 2002; Minov et al. 2001; Sato et al. 2002). No association was shown for a silent SNP in exon 1 of *5HT6* (Wu et al. 2001). Also no association was found between two SNPs causing amino acid exchanges in the *dopamine receptors type 2* and *4* and response to fluvoxamine and paroxetine (Serretti et al. 2001a). As most monoaminergic receptors belong to the class of G protein-coupled receptors, G protein subunits, such as the β3 subunits, are candidate genes for the pharmacogenetics of antidepressant drugs. A SNP leading to altered signal transduction, most likely via alternative splicing (Siffert 2003), was found to be associated with response to antidepressant treatment in two independent studies (Serretti et al. 2003; Zill et al. 2000).

The association results with these genes are so far less convincing than those with *SERT*. For some studies the number of investigated patients are small [the smallest sample size being 34 (Wu et al. 2001)]. For others, different polymorphisms have been investigated for the same genes, rendering it more difficult to compare the results across studies (e.g., *5HT2A*). Replications of these results in different ethnic groups with large sample sizes are needed for a conclusive evaluation of the importance of these genes in the pharmacogenetics of antidepressant drugs.

3.2 Other Candidate Systems

So far only one study has investigated pharmacogenetic aspects of benzodiazepine treatment. This study was conducted in children of alcoholics and measured the effects on eye movement measures (Iwata et al. 1999). A SNP leading to an amino acid exchange in the *$GABA_A$ receptor* α*6 subunit* was investigated. The authors observed that this SNP was associated with less diazepam-induced impairment of saccadic velocity, concluding that this polymorphism may play a role in sensitivity to benzodiazepines. It is difficult to judge the relevance of these data for anxiety disorders. On the one hand the sample size is small ($n = 51$) and on the other hand the subjects all had a family history of alcoholism. It has been previously shown that a family history of alcoholism may be associated with a diminished sensitivity to benzodiazepines (Cowley et al. 1992; Cowley et al. 1994). Studies on the effects of GABA receptor subunit genotypes on the anxiolytic effects and addiction potential of benzodiazepines are needed.

Several studies suggest that a normalization in the hypothalamic–pituitary–adrenal axis hyperactivity, including the sensitivity of the glucocorticoid receptor may be required for a response to antidepressive treatment, at least in depressed patients (see Holsboer 2000 for review). Our group has investigated the influence of polymorphism in genes regulating this axis in response to antidepressant drugs in the Munich Antidepressant Response Signature (MARS) sample (Holsboer 2001). We found a strong association between polymorphisms in a co-chaperone of the GR and response to antidepressant drugs (Binder et al. 2004). This co-chaperone has been found to regulate GR sensitivity. This study is the first to consider genes that are likely involved in the final common pathway of response to antidepressant drug for pharmacogenetics. Further studies investigating these candidate gene pathways in patients with anxiety disorder are currently being conducted.

4 The Search for Novel Candidate Genes

As stated above, the ultimate mechanism of action of the antidepressant and anxiolytic effects of antidepressant drugs have not been fully elucidated yet. It is therefore necessary to identify novel candidate genes using unbiased genome-wide strategies.

4.1 Gene Expression Profiling Experiments

High-density DNA arrays or microarrays allow investigators to examine mRNA levels of all known genes in one experiment (see Lockhart and Winzeler 2000 for review). DNA sequences either complementary to longer [complementary DNA (cDNA) sequences transcribed from mRNA] or shorter stretches (synthesized oligonucleotides up to 60 bp long) of the expressed mRNA sequence are attached at a precise location on the surface of a glass slide or any other kind of appropriate surface (chips). DNA sequences are either deposited by spotting minute amounts on the surface (cDNA or oligonucleotide arrays) or synthesized in situ (the company Affymetrix uses photolithography). Then a soluble mixture of the mRNA, transcribed into cDNA and labeled is hybridized to the chip. The amount of signal is measured for each spot and quantified. If two groups are to be compared on cDNA arrays, the respective mRNAs are often labeled with two different fluorochromes (usually Cy3 and Cy5). To quantify differences between the groups, the relative amounts of the signal of the two fluorochromes are measured. In the case of a gene unaffected by the group difference, equal signal from each fluorochrome is expected (see Fig. 1). Data on ten thousands of genes are then gathered and analyzed using a variety of bioinformatic strategies.

Using microarrays, differences in gene expression can be detected on a whole genome level for different behavioral responses, treatment modalities, or brain regions. Several groups have already tried to use this technique to identify novel candidate genes for depression and anxiety and their treatment. In the search of the common final pathway of antidepressant action, gene expression changes following chronic antidepressant treatment have been analyzed (Yamada et al. 2000, 2001). Our group has searched for genes commonly regulated by two antidepressants with different receptor binding profiles, paroxetine and mirtazapine (Landgrebe et al. 2002). Another approach is to identify genes differentially regulated in brain regions that have been associated with different cognitive functions, including anxiety behavior (see, for example, Dent et al. 2001). So far, however, no strong candidate that has also been validated in human genetic studies has been brought forward. A series of technical and methodological problems have to be considered when interpreting gene expression profiling for neurobiological issues (Luo and Geschwind 2001; Mirnics 2001; Watson et al. 2000). Neurons even within a given brain region can show very heterogeneous expression of neurotransmitter, receptors, and connections to other brain regions, leading to possibly opposite changes in mRNA expression in neighboring neurons following the same stimulus so that single neurons may have to be dissected out of the tissue. Several researchers have already used laser capture microdissections to address this issue. In addition, methods allowing investigators to amplify minute amounts of mRNA from single cells are being developed and tested. Finally, access to human brain tissue for pharmacogenetic-oriented studies is difficult, and a multitude of unknown confounding factors can still invalidate the results.

4.2 Genome-Wide Association Studies

Truly genome-wide association studies using SNP markers have still not been published, possibly due to limitations of genotyping throughput and cost, and the necessity for large patient samples. Nonetheless, technological advances in this field are being accomplished at an enormous speed. Together with the promises of a reduced genotyping demand with the establishment of haplotypes maps, the feasibility of genome-wide screens is moving into the nearer future.

It is very likely that pharmacogenetic properties are as complex as the genetics for the treated diseases. One therefore has to expect multiple response-modifying genes each only contributing a small effect. To detect these effects, large patients samples (over 1,000 individuals) might be needed (McCarthy and Hilfiker 2000), especially when having to account for multiple testing in whole genome studies. Screening in samples of diverse ethnicity will also be necessary. Already on a single gene basis it becomes evident that pharmacogenetic associations may be specific for a certain ethnic group. Ethnic differences in allele frequencies and pharmacogenetic effects, for example, have been reported for the CYP genes, *P-glycoprotein* and *SERT* (Ameyaw et al. 2001; Bertilsson et al. 2002; Gelernter et al. 1999).

5 Conclusions

Up to this point, no pharmacogenetic/genomic study has been conducted specifically for anxiety disorders. Some of the presented results, however, also pertain to those disorders. Pharmacokinetic aspects of antidepressant drug therapy likely also apply to patients with anxiety disorders. Even tough no clear dose–response relationship has been established for SSRIs, knowledge of the CYP-related metabolizer status may help to reduce the incidence of side effects that may interfere with the compliance of the patient. Genetic studies with P-glycoprotein may reveal genotype-specific intracerebral concentrations of antidepressant drugs, which should pertain to all psychiatric disorders responsive to these compounds. A recent study has already published an association of a certain P-glycoprotein genotype with response to antiepileptic treatment (Siddiqui et al. 2003). It is more difficult to interpret the relevance of pharmacodynamic genetic associations in depression for anxiety disorders. Here studies in patients with anxiety disorders are warranted. Separate sets of novel candidate genes may be responsible for the antidepressant and anxiolytic effects of these drugs. In addition to studies in different patient samples, this may also necessitate the development of innovative animal models able to separate these effects.

Despite all the shortcomings of the currently available pharmacogenetic studies, this field holds great promise for the treatment of anxiety disorders.

Pharmacogenomics may allow us to develop sets of SNPs that could be combined into easily used assays that will rapidly classify patients according to their likely response to pharmacotherapy. The psychiatrist will then be able to base the treatment decision concerning the type and dose of a prescribed drug on more objective parameters than the ones used currently. This will limit unwanted side effects and adverse drug reactions, and could reduce time to response. A more individualized pharmacotherapy will then be possible. Pharmacogenomics might also lead to a better understanding of the mechanism of action of antidepressant and anxiolytic drugs. The identification of novel candidate genes would allow for the development of novel drug targets and therapeutic compounds. One might also identify subgroups of patients in which different pathophysiological changes lead to the development of anxiety disorders. In this way, an individual targeting of the pathological pathway may be realizable, again shortening time to response and reducing side effects. It is of note, however, that a drug individually tailored to the pathology of a specific patient may work optimally in this patient but potentially will be ineffective in other patients. Thus, the more specific the drugs get, the more we need to know about the specific pathophysiological and genetic background of each individual patient. This will lead to a fragmentation of a market that so far has been dominated by a few "blockbuster" drugs. It is to be hoped that market-oriented considerations will not put a hold on the exploitation of genotype-based medicine. The fact that anxiolytic treatments work in too few people, take too long before clinical improvement is seen, and have too many adverse effects, calls for a drug discovery initiative as outlined in this article.

References

Ameyaw MM, Regateiro F, Li T, et al (2001) MDR1 pharmacogenetics: frequency of the C3435T mutation in exon 26 is significantly influenced by ethnicity. Pharmacogenetics 11:217–221

Angst J (1961) A clinical analysis of the effects of tofranil in depression: longitudinal and follow-up studies. Treatment of blood-relations. Psychopharmacologia 2:381–407

Arias B, Catalan R, Gasto C, et al (2001) Genetic variability in the promotor region of the serotonin transporter gene is associated with clinical remission of major depression after long-term treatment with citalopram. World J Biol Psychiatry 2:9S

Benedetti F, Serretti A, Colombo C, et al (1999) Influence of a functional polymorphism within the promoter of the serotonin transporter gene on the effects of total sleep deprivation in bipolar depression. Am J Psychiatry 156:1450–1452

Binder EB, Salyakina D, Lichtner P, Wochnik GM, Ising M, Putz B, Papitol S, Seaman S, Lucae S, Kohli MA, Nickel T, Kunzel HE, Fuchs B, Majer M, Pfennig A, Kern N, Brunner J Modell S, Baghai T, Deiml T, Zill P, Bondy B, Rupprecht R, Messer T, Kohnlein O, Dabitz H, Bruckl T, Muller N, Pfister H, Lieb R, Mueller JC, Lohmussaar E, Strom TM, Betttecken T, Meitinger T, Uhr M, Rein T, Holsboer F, Muller-Myhsok B (2004) Polymorphisms in FKBP5 are associated with increased recurrence of depressive episodes and rapid response to antidepressant treatment. Nat Genet 36(12):1319–1325

Bertilsson L, Dahl ML, Dalen P, Al-Shurbaji A (2002) Molecular genetics of CYP2D6: clinical relevance with focus on psychotropic drugs. Br J Clin Pharmacol 53:111–122

Brinkmann U, Roots I, Eichelbaum M (2001) Pharmacogenetics of the human drug-transporter gene MDR1: impact of polymorphisms on pharmacotherapy. Drug Discov Today 6:835–839

Brookes AJ (1999) The essence of SNPs. Gene 234:177–186

Brookes AJ, Lehvaslaiho H, Siegfried M, et al (2000) HGBASE : a database of SNPs and other variations in and around human genes. Nucleic Acids Res 28:356–360

Brosen K, Naranjo CA (2001) Review of pharmacokinetic and pharmacodynamic interaction studies with citalopram. Eur Neuropsychopharmacol 11:275–283

Brosen K, Hansen JG, Nielsen KK, Sindrup SH, Gram LF (1993) Inhibition by paroxetine of desipramine metabolism in extensive but not in poor metabolizers of sparteine. Eur J Clin Pharmacol 44:349–355

Burke MJ, Preskorn SH (1999) Therapeutic drug monitoring of antidepressants: cost implications and relevance to clinical practice. Clin Pharmacokinet 37:147–165

Cardon LR, Abecasis GR (2003) Using haplotype blocks to map human complex trait loci. Trends Genet 19:135–140

Carlson CS, Newman TL, Nickerson DA (2001) SNPing in the human genome. Curr Opin Chem Biol 5:78–85

Cascorbi I, Gerloff T, Johne A, et al (2001) Frequency of single nucleotide polymorphisms in the P-glycoprotein drug transporter MDR1 gene in white subjects. Clin Pharmacol Ther 69:169–174

Collins A, Lonjou C, Morton NE (1999) Genetic epidemiology of single-nucleotide polymorphisms [see comments]. Proc Natl Acad Sci U S A 96:15173–15177

Cordon-Cardo C, O'Brien JP, Casals D, et al (1989) Multidrug-resistance gene (P-glycoprotein) is expressed by endothelial cells at blood-brain barrier sites. Proc Natl Acad Sci U S A 86:695–698

Corruble E, Guelfi JD (2000) Does increasing dose improve efficacy in patients with poor antidepressant response: a review. Acta Psychiatr Scand 101:343–348

Cowley DS, Roy-Byrne PP, Godon C, et al (1992) Response to diazepam in sons of alcoholics. Alcohol Clin Exp Res 16:1057–1063

Cowley DS, Roy-Byrne PP, Radant A, et al (1994) Eye movement effects of diazepam in sons of alcoholic fathers and male control subjects. Alcohol Clin Exp Res 18:324–332

Cusin C, Serretti A, Zanardi R, et al (2002) Influence of monoamine oxidase A and serotonin receptor 2A polymorphisms in SSRI antidepressant activity. Int J Neuropsychopharmacol 5:27–35

Dalen P, Dahl ML, Ruiz ML, Nordin J, Bertilsson L (1998) 10-Hydroxylation of nortriptyline in white persons with 0, 1, 2, 3, and 13 functional CYP2D6 genes. Clin Pharmacol Ther 63:444–452

Dent GW, O'Dell DM, Eberwine JH (2001) Gene expression profiling in the amygdala: an approach to examine the molecular substrates of mammalian behavior. Physiol Behav 73:841–847

Franchini L, Serretti A, Gasperini M, Smeraldi E (1998) Familial concordance of fluvoxamine response as a tool for differentiating mood disorder pedigrees. J Psychiatr Res 32:255–259

Fukuda T, Nishida Y, Zhou Q, Yamamoto I, Kondo S, Azuma J (2000) The impact of the CYP2D6 and CYP2C19 genotypes on venlafaxine pharmacokinetics in a Japanese population. Eur J Clin Pharmacol 56:175–180

Gardner JP, Fornal CA, Jacobs BL (1997) Effects of sleep deprivation on serotonergic neuronal activity in the dorsal raphe nucleus of the freely moving cat. Neuropsychopharmacology 17:72–81

Gelernter J, Cubells JF, Kidd JR, Pakstis AJ, Kidd KK (1999) Population studies of polymorphisms of the serotonin transporter protein gene. Am J Med Genet 88:61–66

Hahn MK, Blakely RD (2002) Monoamine transporter gene structure and polymorphisms in relation to psychiatric and other complex disorders. Pharmacogenomics J 2:217–235

Heils A, Teufel A, Petri S, et al (1996) Allelic variation of human serotonin transporter gene expression. J Neurochem 66:2621–2624

Hoffmeyer S, Burk O, von Richter O, et al (2000) Functional polymorphisms of the human multidrug-resistance gene: multiple sequence variations and correlation of one allele with P-glycoprotein expression and activity in vivo. Proc Natl Acad Sci U S A 97:3473–3478

Holsboer F (2000) The corticosteroid receptor hypothesis of depression. Neuropsychopharmacology 23:477–501

Holsboer F (2001) Stress, hypercortisolism and corticosteroid receptors in depression: implications for therapy. J Affect Disord 62:77–91

Ito K, Yoshida K, Sato K, et al (2002) A variable number of tandem repeats in the serotonin transporter gene does not affect the antidepressant response to fluvoxamine. Psychiatry Res 111:235–239

Iwata N, Cowley DS, Radel M, Roy-Byrne PP, Goldman D (1999) Relationship between a GABAA alpha 6 Pro385Ser substitution and benzodiazepine sensitivity. Am J Psychiatry 156:1447–1449

Judson R, Salisbury B, Schneider J, Windemuth A, Stephens JC (2002) How many SNPs does a genome-wide haplotype map require? Pharmacogenomics 3:379–391

Kim DK, Lim SW, Lee S, et al (2000) Serotonin transporter gene polymorphism and antidepressant response. Neuroreport 11:215–219

Kirchheiner J, Brosen K, Dahl ML, et al (2001) CYP2D6 and CYP2C19 genotype-base d dose recommendations for antidepressants: a first step towards subpopulation-specific dosages. Acta Psychiatr Scand 104:173–192

Kruglyak L (1999) Prospects for whole-genome linkage disequilibrium mapping of common disease genes. Nat Genet 22:139–144

Kwok PY (2000) High-throughput genotyping assay approaches. Pharmacogenomics 1:95–100

Kwok PY, Chen X (1998) Detection of single nucleotide variations. Genet Eng (N Y) 20:125–134

Lander ES, Linton LM, Birren B, et al (2001) Initial sequencing and analysis of the human genome. Nature 409:860–921

Landgrebe J, Welzl G, Metz T, et al (2002) Molecular characterisation of antidepressant effects in the mouse brain using gene expression profiling. J Psychiatr Res 36:119–129

Lesch KP, Wolozin BL, Estler HC, Murphy DL, Riederer P (1993) Isolation of a cDNA encoding the human brain serotonin transporter. J Neural Transm Gen Sect 91:67–72

Lesch KP, Bengel D, Heils A, et al (1996) Association of anxiety-related traits with a polymorphism in the serotonin transporter gene regulatory region. Science 274:1527–1531

Lessard E, Yessine MA, Hamelin BA, O'Hara G, LeBlanc J, Turgeon J (1999) Influence of CYP2D6 activity on the disposition and cardiovascular toxicity of the antidepressant agent venlafaxine in humans. Pharmacogenetics 9:435–443

Lockhart DJ, Winzeler EA (2000) Genomics, gene expression and DNA arrays. Nature 405:827–836

Luo Z, Geschwind DH (2001) Microarray applications in neuroscience. Neurobiol Dis 8:183–193

McCarthy JJ, Hilfiker R (2000) The use of single-nucleotide polymorphism maps in pharmacogenomics. Nat Biotechnol 18:505–508

Minov C, Baghai TC, Schule C, et al (2001) Serotonin-2A-receptor and -transporter polymorphisms: lack of association in patients with major depression. Neurosci Lett 303:119–122

Mirnics K (2001) Microarrays in brain research: the good, the bad and the ugly. Nat Rev Neurosci 2:444–447

Muller DJ, Schulze TG, Macciardi F, et al (2002) Moclobemide response in depressed patients: association study with a functional polymorphism in the monoamine oxidase A promoter. Pharmacopsychiatry 35:157–158

Nebert DW, Dieter MZ (2000) The evolution of drug metabolism. Pharmacology 61:124–135

O'Reilly RL, Bogue L, Singh SM (1994) Pharmacogenetic response to antidepressants in a multicase family with affective disorder. Biol Psychiatry 36:467–471

Ozdemir V, Tyndale RF, Reed K, et al (1999) Paroxetine steady-state plasma concentration in relation to CYP2D6 genotype in extensive metabolizers. J Clin Psychopharmacol 19:472–475

Pare C, Rees L, Saisbury M (1962) Differentiation of two genetically specific types of depression by the response to antidepressants. Lancet 29:1340–1343

Pollock BG, Ferrell RE, Mulsant BH, et al (2000) Allelic variation in the serotonin transporter promoter affects onset of paroxetine treatment response in late-life depression. Neuropsychopharmacology 23:587–590

Preskorn SH, Lane RM (1995) Sertraline 50 mg daily: the optimal dose in the treatment of depression. Int Clin Psychopharmacol 10:129–141

Preskorn SH, Dorey RC, Jerkovich GS (1988) Therapeutic drug monitoring of tricyclic antidepressants. Clin Chem 34:822–828

Ramamoorthy S, Leibach FH, Mahesh VB, Ganapathy V (1993) Partial purification and characterization of the human placental serotonin transporter. Placenta 14:449–461

Rasmussen BB, Brosen K (2000) Is therapeutic drug monitoring a case for optimizing clinical outcome and avoiding interactions of the selective serotonin reuptake inhibitors? Ther Drug Monit 22:143–154

Risch N, Merikangas K (1996) The future of genetic studies of complex human diseases [see comments]. Science 273:1516–1517

Roden DM, George AL Jr (2002) The genetic basis of variability in drug responses. Nat Rev Drug Discov 1:37–44

Sachidanandam R, Weissman D, Schmidt SC, et al (2001) A map of human genome sequence variation containing 1.42 million single nucleotide polymorphisms. Nature 409:928–933

Sato K, Yoshida K, Takahashi H, et al (2002) Association between −1438G/A promoter polymorphism in the 5-HT(2A) receptor gene and fluvoxamine response in Japanese patients with major depressive disorder. Neuropsychobiology 46:136–140

Schaeffeler E, Eichelbaum M, Brinkmann U, et al (2001) Frequency of C3435T polymorphism of MDR1 gene in African people. Lancet 358:383–384

Schinkel AH, Wagenaar E, Mol CA, van Deemter L (1996) P-glycoprotein in the blood-brain barrier of mice influences the brain penetration and pharmacological activity of many drugs. J Clin Invest 97:2517–2524

Serretti A, Franchini L, Gasperini M, Rampoldi R, Smeraldi E (1998) Mode of inheritance in mood disorder families according to fluvoxamine response. Acta Psychiatr Scand 98:443–450

Serretti A, Zanardi R, Cusin C, et al (2001a) No association between dopamine D(2) and D(4) receptor gene variants and antidepressant activity of two selective serotonin reuptake inhibitors. Psychiatry Res 104:195–203

Serretti A, Zanardi R, Cusin C, Rossini D, Lorenzi C, Smeraldi E (2001b) Tryptophan hydroxylase gene associated with paroxetine antidepressant activity. Eur Neuropsychopharmacol 11:375–380

Serretti A, Zanardi R, Rossini D, Cusin C, Lilli R, Smeraldi E (2001c) Influence of tryptophan hydroxylase and serotonin transporter genes on fluvoxamine antidepressant activity. Mol Psychiatry 6:586–592

Serretti A, Lilli R, Smeraldi E (2002) Pharmacogenetics in affective disorders. Eur J Pharmacol 438:117–128

Serretti A, Lorenzi C, Cusin C, et al (2003) SSRIs antidepressant activity is influenced by G beta 3 variants. Eur Neuropsychopharmacol 13:117–122

Siddiqui A, Kerb R, Weale ME, et al (2003) Association of multidrug resistance in epilepsy with a polymorphism in the drug-transporter gene ABCB1. N Engl J Med 348:1442–1448

Siffert W (2003) Effects of the G protein beta 3-subunit gene C825T polymorphism: should hypotheses regarding the molecular mechanisms underlying enhanced G protein activation be revised? Focus on "A splice variant of the G protein beta 3-subunit implicated in disease states does not modulate ion channels". Physiol Genomics 13:81–84

Sindrup SH, Brosen K, Gram LF, et al (1992) The relationship between paroxetine and the sparteine oxidation polymorphism. Clin Pharmacol Ther 51:278–287

Smeraldi E, Zanardi R, Benedetti F, Di Bella D, Perez J, Catalano M (1998) Polymorphism within the promoter of the serotonin transporter gene and antidepressant efficacy of fluvoxamine. Mol Psychiatry 3:508–511

Smigielski EM, Sirotkin K, Ward M, Sherry ST (2000) dbSNP: a database of single nucleotide polymorphisms. Nucleic Acids Res 28:352–355

Staddon S, Arranz J, Mancama D, Mata I, Kerwin R (2002) Clinical application of pharmacogenetics in psychiatry. Psychopharmacology (Berl) 162:18–23

Steimer W, Muller B, Leucht S, Kissling W (2001) Pharmacogenetics: a new diagnostic tool in the management of antidepressive drug therapy. Clin Chim Acta 308:33–41

Syvanen AC (2001) Accessing genetic variation: genotyping single nucleotide polymorphisms. Nat Rev Genet 2:930–942

Thiebaut F, Tsuruo T, Hamada H, Gottesman MM, Pastan I, Willingham MC (1987) Cellular localization of the multidrug-resistance gene product P-glycoprotein in normal human tissues. Proc Natl Acad Sci U S A 84:7735–7738

Turecki G, Grof P, Grof E, et al (2001) Mapping susceptibility genes for bipolar disorder: a pharmacogenetic approach base d on excellent response to lithium. Mol Psychiatry 6:570–578

Uhr M, Grauer MT (2003) abcb1ab P-glycoprotein is involved in the uptake of citalopram and trimipramine into the brain of mice. J Psychiatr Res 37:179–185

Uhr M, Steckler T, Yassouridis A, Holsboer F (2000) Penetration of amitriptyline, but not of fluoxetine, into brain is enhanced in mice with blood-brain barrier deficiency due to mdr1a P-glycoprotein gene disruption. Neuropsychopharmacology 22:380–387

Veefkind AH, Haffmans PM, Hoencamp E (2000) Venlafaxine serum levels and CYP2D6 genotype. Ther Drug Monit 22:202–208
Venter JC, Adams MD, Myers EW, et al (2001) The sequence of the human genome. Science 291:1304–1351
Watson SJ, Meng F, Thompson RC, Akil H (2000) The "chip" as a specific genetic tool. Biol Psychiatry 48:1147–1156
Weinshilboum R (2003) Inheritance and drug response. N Engl J Med 348:529–537
Wu WH, Huo SJ, Cheng CY, Hong CJ, Tsai SJ (2001) Association study of the 5-HT(6) receptor polymorphism (C267T) and symptomatology and antidepressant response in major depressive disorders. Neuropsychobiology 44:172–175
Yamada M, Yamazaki S, Takahashi K, et al (2000) Identification of a novel gene with RING-H2 finger motif induced after chronic antidepressant treatment in rat brain. Biochem Biophys Res Commun 278:150–157
Yamada M, Yamazaki S, Takahashi K, et al (2001) Induction of cysteine string protein after chronic antidepressant treatment in rat frontal cortex. Neurosci Lett 301:183–186
Yoshida K, Ito K, Sato K, et al (2002a) Influence of the serotonin transporter gene-linked polymorphic region on the antidepressant response to fluvoxamine in Japanese depressed patients. Prog Neuropsychopharmacol Biol Psychiatry 26:383–386
Yoshida K, Naito S, Takahashi H, et al (2002b) Monoamine oxidase: a gene polymorphism, tryptophan hydroxylase gene polymorphism and antidepressant response to fluvoxamine in Japanese patients with major depressive disorder. Prog Neuropsychopharmacol Biol Psychiatry 26:1279–1283
Yu YW, Tsai SJ, Chen TJ, Lin CH, Hong CJ (2002) Association study of the serotonin transporter promoter polymorphism and symptomatology and antidepressant response in major depressive disorders. Mol Psychiatry 7:1115–1119
Zanardi R, Benedetti F, Di Bella D, Catalano M, Smeraldi E (2000) Efficacy of paroxetine in depression is influenced by a functional polymorphism within the promoter of the serotonin transporter gene. J Clin Psychopharmacol 20:105–107
Zanardi R, Serretti A, Rossini D, et al (2001) Factors affecting fluvoxamine antidepressant activity: influence of pindolol and 5-HTTLPR in delusional and nondelusional depression. Biol Psychiatry 50:323–330
Zill P, Baghai TC, Zwanzger P, et al (2000) Evidence for an association between a G-protein beta3-gene variant with depression and response to antidepressant treatment. Neuroreport 11:1893–1897

HEP (2005) 169:547–560

Pharmacoproteomics

C. W. Turck

Max Planck Institute of Psychiatry, Kraepelinstr. 2, 80804 Munich, Germany
turck@mpipsykl.mpg.de

Abstract Proteomics, the comprehensive analysis of the protein complement of the genome of an organism, is becoming an increasingly important discipline for the identification of disease targets. In addition, the effects of drug treatment and metabolism can now be studied on the protein level in a comprehensive manner.

Keywords Proteomics · Disease targets · Protein analysis technologies

1 Background

In order to understand complex physiological pathways and the pathogenesis of diseases, it is often not sufficient to elucidate the genome of a cell or organism. Although there are diseases caused by an exchange of single base pairs within the genome, most diseases are multi-factorial and are caused by several genetic as well as environmental factors. Consequently, the discovery of markers of multi-factorial diseases requires a global approach. As the functionally active macromolecules in a cell, proteins are prime candidates as disease targets. The term proteome that was created in 1995 indicates the PROTEins expressed by a genOME (Wilkins et al. 1996). Unlike the genome and similar to the transcriptome (all mRNAs expressed by a genome) the proteome varies from tissue to tissue within the same organism. After the completion of the sequence analysis for several species, it is now obvious that humans do not have many more genes than lower organisms (Lander et al. 2001; Venter et al. 2001). The question that presents itself is how does *Homo sapiens* manage to be so complex? It is an interesting proposal that proteins, not genes, are responsible for an organism's complexity and the interactions of proteins in networks determine how an organism functions. It is therefore only logical that it is now widely accepted that the key to understanding health and disease is to study the organism's proteins. This includes where each protein is located in a cell, when the protein is present and for how long, and which other proteins it is interacting with. Comprehensive proteome characterization therefore needs to provide not only protein sequence but various types of other information including the protein's abundance, localization, and state of post-translational modifications.

Unlike the genome the proteome is not a static but a dynamic and constantly changing entity that is cell- and tissue-specific and dependent on the environment. Because of the dynamic nature of protein expression and function, these properties need to be determined quantitatively in a time-dependent manner. Proteomics, the study of the proteome, involves the analysis of the complete pattern of the expressed proteins and their post-translational modifications in a cell, tissue, or body fluid. An integrated view of any living system hence requires an analysis that takes into account the spatial as well as temporal distribution of all the proteins in a cell or tissue. The analytical effort that is necessary to deliver such an integrated view is by several orders of magnitude more complicated than that of the recently finished human genome (Lander et al. 2001; Venter et al. 2001).

The traditional paradigm of one gene equaling one protein is not valid anymore, and although there may be "only" 30,000 genes in the human genome (Lander et al. 2001; Venter et al. 2001) there are probably millions of proteins derived form these genes that are due to splice variants and post-translational modifications (Fig. 1). In this regard it is estimated that under normal physiological conditions most cellular proteins are post-translationally modified. Currently, there are approximately 200 such post-translational modifications known (Krishna and Wold 1993). These modifications can affect protein conformation, stability, localization, binding interactions, and function. Consequently, the differences between disease and normal tissue are not restricted to the quantity of particular proteins, but may also be reflected by different processing or the degree of post-translational modifications. The comparison of the proteome of diseased and healthy tissues and the subsequent identification of the proteins that are different from normal in disease therefore represents a critical technology to unravel the pathogenesis of disease, to identify therapeutic targets, and to develop diagnostic tests.

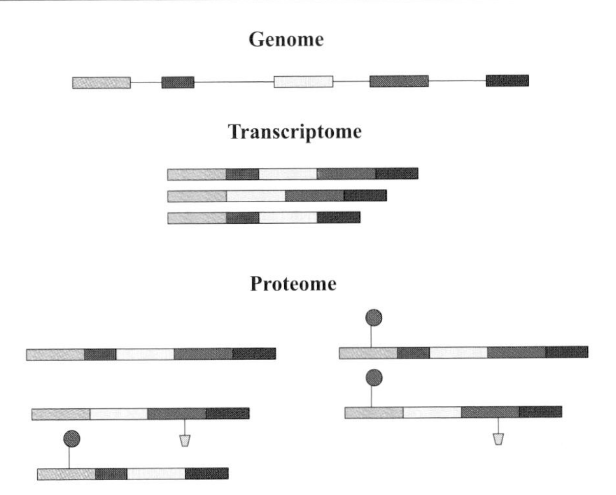

Fig. 1 From genome to transcriptome to proteome. Through alternate splicing a single gene can be transcribed into several mRNAs. Every mRNA that is subsequently translated can give rise to more than one protein caused by differential post-translational modifications as indicated by *purple symbols*

2 Proteomics: A New Term for an Old Science

Attempts of a global analysis of proteins in tissues and cells (now called proteomics) has been around for almost 30 years. These studies were often carried out with a technique called two-dimensional polyacrylamide gel electrophoresis (2D PAGE) (O'Farrel 1975; Klose 1975), which will be explained in more detail in Sect. 5.1. Although these studies allowed for the visualization of protein patterns from different tissues or cells, the determination of the identity of the differences was only possible for a limited number of samples. This limitation in one's ability to identify proteins at low levels changed dramatically in the early 1990s when powerful new technologies for protein identification and analysis became available (Hillenkamp and Karas 1990; Fenn et al. 1989). In addition, the completion of the sequence analysis of several genomes, including the human genome (Lander et al. 2001; Venter et al. 2001), now makes available in public databases the sequence of every protein of an organism. All this has led to what is now referred to as the post-genomic era, which is set to play a major role in the study of biological systems and mechanisms of disease at the molecular level.

Proteomics can be divided into two major areas of research—global and targeted proteomics (Pandey and Mann 2000). Global proteomics compares protein expression in a cell, tissue, or body fluid in a comprehensive and global fashion. In this approach, proteins derived from a total cellular or tissue extract or from a body fluid are typically displayed on two-dimensional gels and

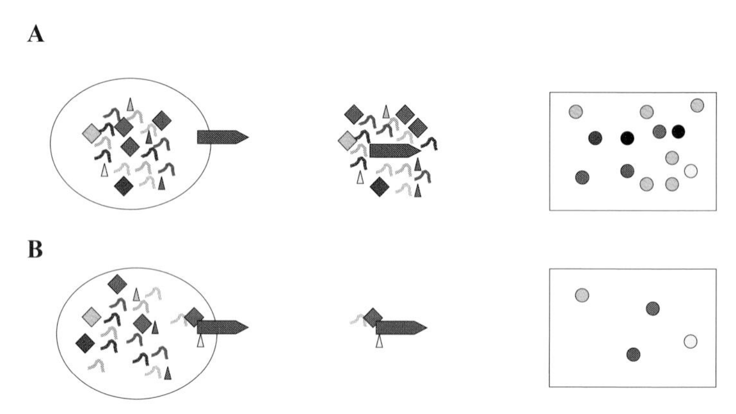

Fig. 2A, B Global (A) vs targeted (B) proteomics. In a global proteomics approach, all the protein constituents of a cell are isolated and subjected to 2D PAGE. In a targeted proteomics experiment, only a subset of proteins is analyzed. An example is the complex that is formed between a transmembrane receptor and cytoplasmic signaling proteins upon ligand activation of the receptor

every protein spot is identified by mass spectrometry (MS) (Fig. 2A). This approach, which is also referred to as expression and comparative proteomics, is the mainstay in medical applications when proteins from normal and diseased cells or tissues are compared and the differences identified as potential molecular disease markers (Pandey and Mann 2000). Other applications of this approach are in the pharmacoproteomics arena and are concerned with the analysis of drug influence, toxicology, and drug-target validation studies.

Targeted proteomics approaches, on the other hand, are employed for specific problems and are often concerned with the analysis of the components that are part of protein complexes (Fig. 2B). An example is the analysis of cellular signaling complexes that are formed upon receptor activation after ligand binding (Pandey and Mann 2000). Next to protein identification, these studies also include the determination of post-translational modifications of the receptor and the proteins involved in the signaling pathway. Likewise, cell-map proteomics is classified as a targeted proteomics approach that deals with the determination of the subcellular location of proteins and protein networks (Pandey and Mann 2000). Since the most proteins in the cell are not found as free entities but in protein complexes, the elucidation of these complexes is critical for the understanding of biochemical pathways and protein function in general. For the isolation of the individual protein components of a complex, so-called bait proteins are created that represent one component of the complex and that will bind to the other partners upon exposure to cellular lysates. The individual protein components are then separated by gel electrophoresis and identified by MS (Pandey and Mann 2000).

3
Proteomics Methods

Proteomics includes a variety of technologies that include differential protein display on gels, protein chips, quantitation of protein amounts, analysis of post-translational modifications, characterization of protein complexes and networks and bioinformatics. All this information in combination with genome and phenotype studies will ultimately yield a comprehensive picture of a cellular or tissue proteome (Wasinger and Corthals 2002).

The reason for the dramatic expansion in recent years of the field of proteomics in many areas of the life and health sciences was caused by the development of methods that make possible the rapid and high-throughput analysis of small amounts of proteins. Due to the limited amounts of protein in cells, tissues, and body fluids, the methods employed for protein identification and analysis of post-translational modifications need to be very sensitive. In this regard, the required ability to identify ultralow levels of proteins has made a great leap forward. This is in great part due to advances that have been made in the MS analysis of peptides and proteins (Hillenkamp and Karas 1990; Fenn et al. 1989). The development of ultrasensitive mass spectrometers in combination with protein sample preparation methods that avoid losses of small amounts of material now allow for the analysis of low femtomole or even attomole amounts of protein. The other aspect that is making MS the method of choice in today's protein identification efforts is the rapidly increasing number of entries in protein and DNA databases. The entire human genome has been sequenced and allows for the use of MS data from enzymatic protein digests to rapidly identify human proteins with high confidence with the help of specially developed search algorithms (Eng et al. 1994). Electrospray (ES) tandem MS in combination with on-line liquid chromatography (LC/MS/MS) and matrix-assisted laser desorption ionization (MALDI) MS have become the methods of choice to identify proteins at low levels and can now be applied to proteomics projects that are geared towards the identification of differences in protein expression levels between healthy and diseased cells and tissues. The great sensitivities of these MS techniques combined with the completion of the human genome sequencing endeavor make this methodology ideally suited for the identification of disease-associated proteins in tissues and body fluids.

4
Proteomics in Medicine

Proteomic technologies promise to be of great value in molecular medicine, particularly in the detection and discovery of disease markers. Since it was demonstrated that there is a poor correlation between mRNA and protein abundance (Gygi et al. 1999a), protein profiling will ultimately result in a better understanding of disease mechanisms and the molecular effects of drugs.

However, applying the new technologies to analyze the entire proteome of a clinical specimen is not always straightforward (Anderson and Anderson 1998). The major hurdle when it comes to patient samples is the limiting amount of starting material that is available to carry out the analysis. Clinical samples such as needle biopsies and body fluids frequently cannot be obtained in large enough amounts. Furthermore, unlike RNA, protein samples cannot be amplified by a process such as the polymerase chain reaction. Another reason for the difficult analysis of protein samples is due to the fact that, unlike genomics, with only two to three orders of magnitude differences between the highest and lowest copy number transcripts, protein samples from tissues and body fluids often have up to ten log orders of magnitude differences between the highest and lowest expressed proteins (Anderson and Anderson 1998).

In pharmacoproteomics, protein amounts and differential modifications are potentially relevant markers in tissues and body fluids for monitoring perturbations introduced by drug-treatment effects. In this regard the comparison of protein patterns before and after treatment with a drug makes possible the identification of changes in biochemical pathways that may be related to the drug's efficacy or toxicity. The ultimate goal of proteomics in medicine is the capability to provide qualitative and quantitative data of patient sample proteins that reflect a disease state or a response of disease treatment and to eventually complement or even replace current diagnostic laboratory technologies (Anderson and Anderson 1998). Proteomic studies may also provide prognostic data and lead to the identification of common pathways of drug action, drug resistance, and drug efficacy. Such investigations will ultimately lay the groundwork for the establishment of comprehensive databases of the mechanisms that are involved in the pathogenesis of disease and will allow for the development of therapies that are more direct than the ones offered by conventional genomics approaches. Currently, the pharmaceutical industry has available only 400–500 drug targets, of which many represent transmembrane receptors. With the new developments in the proteomics field, it is anticipated that the number of drug targets will expand tremendously in the very near future (Anderson and Anderson 1998).

5 Analytical Techniques

Proteome analysis in general involves two stages: protein separation and subsequent identification and analysis. Multidimensional separations are required in order to result in an adequate resolution of complex protein or peptide

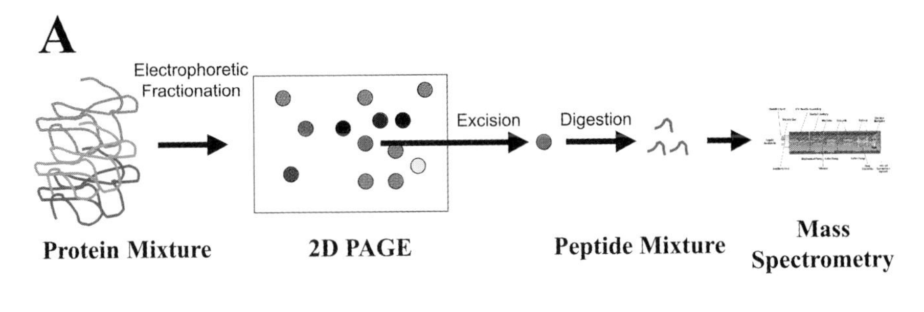

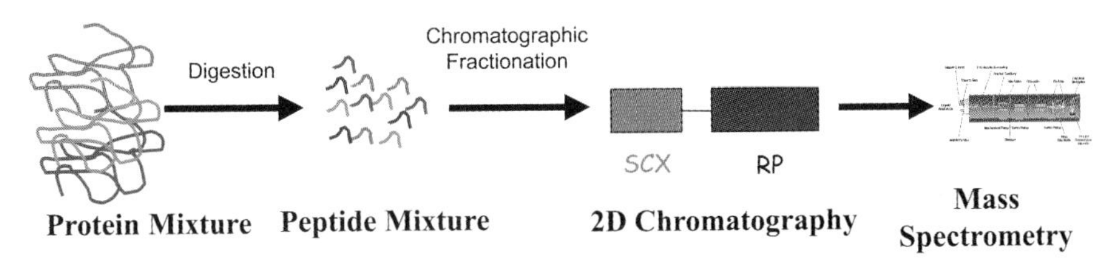

Fig. 3A, B Strategies for the analysis of complex protein mixtures. **A** Proteins are separated by 2D PAGE and the relevant protein spots excised from the gel, digested, and analyzed by mass spectrometry. **B** Proteins are first digested into small peptides and then fractionated by two rounds of chromatography before each peptide is analyzed by mass spectrometry

mixtures. Fractionation on the protein level is in general carried out by 2D PAGE (O'Farrell 1975; Klose 1975). This technique produces high-resolution protein separations resulting in the display of potentially thousands of protein spots (Fig. 3A). Alternatively, in the "shotgun" approach, proteins are digested by specific enzymes into small peptides and subsequently separated by multi-dimensional chromatography techniques followed by analysis of the peptides by MS (Fig. 3B; Link 2002). Both of these methods in combination with improvements in MS instrumentation, the implementation of protein arrays, and the development of robust informatics software are providing sensitive and high-throughput technologies for the large-scale identification and quantitation of protein expression, analysis of protein modifications, subcellular localization, protein–protein interactions, and protein function.

5.1 Protein Fractionation

Although almost 30 years old, the most widely used technology in comparative proteomics is still 2D PAGE. This applies to the global analysis of proteins from tissues, cells, and body fluids. Proteins are separated according to their charge in one dimension (isoelectric focusing) and according to their size in the second dimension [sodium dodecyl sulfate (SDS) electrophoresis]. The combination of these two separation techniques allows for the resolution of thousands of proteins in a single gel. At the same time, the relative abundance of proteins between two or more samples can be assessed after staining the proteins in the gel. This is achieved by software programs that match proteins according to their position in the gel, which is a reflection of their isoelectric point and size and determining their relative abundance by comparing the staining intensities.

However, analysis of protein mixtures derived from cells, tissues, and body fluids by 2D PAGE by no means represents a comprehensive picture of the proteins in the mixture. Proteins with extreme isoelectric points, large proteins, small proteins, and hydrophobic proteins are commonly not amenable to 2D PAGE and hence can be easily missed. Furthermore, low abundant proteins are often not detected in 2D gels when proteins of high abundance are present. This limitation is particularly relevant when analyzing serum or other body fluids, where protein amounts vary by ten orders of magnitude (Anderson and Anderson 1998).

Due to these limitations of 2D PAGE, alternative fractionation methods for complex protein mixtures have been developed. One method relies on the separation of tryptic peptide mixtures that are derived from the protein mixture by two rounds of chromatography (Link 2002). Peptides resulting from a protein digest of the sample are loaded onto a cation exchange column and eluted by applying a salt gradient in a stepwise fashion. Peptides eluting from the cation exchange column in each salt step are then loaded onto a reversed phase column and eluted by applying a shallow gradient of the organic modifier. Subsequently, the fractionated peptides are either collected and analyzed by MALDI MS or directly infused into an ES mass spectrometer. This method allows for the identification of thousands of proteins via their peptide constituents and is therefore referred to as "shotgun" proteomics. A major advantage of the "shotgun" approach is that low abundant proteins can be identified in the presence of large abundant proteins, a scenario that is often encountered when analyzing protein mixtures from body fluids such as serum or cerebrospinal fluid (CSF).

Alternative, non-gel based quantitative methods for assessing the relative amounts of proteins in different samples have been developed as well. Instead of relying on the staining intensities of protein spots in 2D gels, chemical procedures for tagging proteins are now available (Gygi et al. 1999b). These procedures rely on the same reactivity of two chemical tags that slightly differ in mass due to the introduction of heavy isotopes into one of the tags. As a result, proteins from two sample populations are tagged and subsequently differ in mass. This small difference in mass can be analyzed by MS and allows relative quantitation of the two protein populations.

5.2
Protein Identification and Analysis

Protein identification and analysis of their post-translational modifications is nowadays achieved by MS. Several types of mass spectrometers have been developed for this purpose. Most commonly used are MALDI MS (Hillenkamp and Karas 1990) and ES MS (Fenn et al. 1989). Both techniques are so-called soft ionization methods that allow for the ionization of proteins and peptides without destroying them. Typically, proteins that are to be identified are first digested into small peptides with the help of specific enzymes. The reason for digesting the proteins into smaller peptides is that the latter are more amenable to MS analysis. Both methods can be automated and furthermore allow for the fragmentation of peptides, thus allowing for the determination of the amino acid sequence of the peptide. The fragmentation data generated by the mass spectrometer are compared to theoretical fragmentation data from all the known proteins in a database and this allows for the identification of the protein that the peptide is derived from (Eng et al. 1994).

5.3
Protein Chips

Due to the great success of DNA microarrays there are intense efforts underway to develop protein chips for rapid and high-throughput screening projects on the protein level. As with any chip technology the goal is to get a comprehensive look at the proteins expressed in a cell, tissue, or body fluid. The chip technology allows for the simultaneous analysis of thousands of proteins in a single experiment. In the case of short peptide chips the individual peptides can be made by on-chip synthesis analogous to oligonucleotide chips (Lee and Mrksich 2002). In the case of proteins the molecules are spotted onto the chip by contact printing or ink jet technology. In the protein chip approach a variety of "bait" proteins such as antibodies can be immobilized in an array format. The surface is then probed with the sample of interest and only the proteins that bind to the relevant antibodies remain bound to the chip (Fig. 4). An example of this approach is the use of antibodies against a variety of tissue-derived proteins on a chip that is probed with proteins from a body fluid of a patient. The outcome is a comprehensive knowledge of what proteins are present to what extent in the body fluid. A major challenge for the protein chip technology is the above-mentioned complexity and size of the proteome. In order to generate a comprehensive chip one would need antibodies against all the constituents of a proteome, including all the post-translationally modified proteins. This is necessary since an alteration in protein modification could be a critical determinant in the disease. An example is the activation of the all-important tyrosine kinase family of receptors that undergo multiple phosphorylation events upon ligand binding to the receptor (Ullrich and

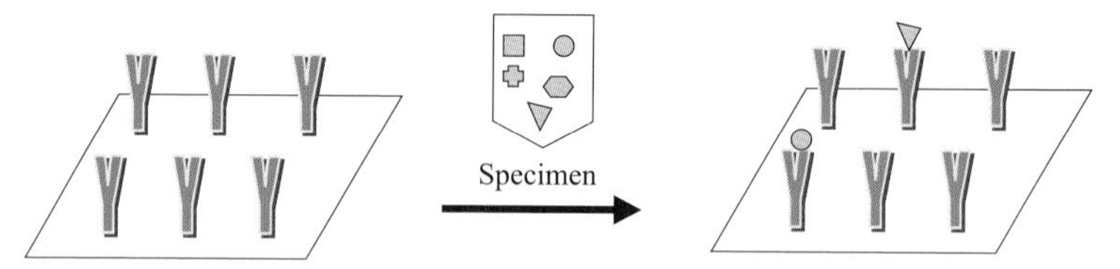

Fig. 4 Antibody chip technology. Chips with immobilized antibodies are probed with a specimen and assessed for binding of specific disease marker proteins

Schlessinger 1990). These phosphorylation events stand at the beginning of specific signaling cascades that ultimately lead to transcription initiation and cell mitosis or differentiation. Because of the complexity of the proteome, it is at this stage most reasonable to think about protein chips that have a defined number of antibodies against known disease markers.

In a related approach, arrays with different types of surface chemistries such as hydrophobic, hydrophilic, anionic, and affinity are used to absorb certain protein groups from biological or patient samples. The chip-absorbed proteins are then directly detected by surface-enhanced laser desorption/ionization time-of-flight MS (SELDI-TOF MS) (Issaq et al. 2002). The resulting protein masses can be used in pattern analysis and thereby provide a useful diagnostic tool.

The protein chip technology may turn out to be of great value for what is now being referred to as individualized medicine. In pharmacoproteomics, body fluids from patients that undergo drug treatments can rapidly be assessed using the chip technology and the treatment tailored to each patient's unique background.

6 Applications

Since disease processes lead to protein changes, it is of paramount importance to consider the relationship between disease and therapy at the protein level (Anderson and Anderson 1998). The objective of modern pharmacoproteomics is the identification of phenotypic differences in drug metabolism or response and the subsequent examination of candidate proteins for variations that underlie the observed phenotypes. Instead of focusing on a few protein targets, pharmacoproteomics takes a more global approach in order to capture complicated patterns of protein expression (Anderson and Anderson 1998; Petricoin III et al. 2002).

One important issue in medical proteomics is that of individual variations in protein expression and structure (Anderson and Anderson 2002). The protein "noise" that is generated by individual polymorphic variations can be significant and make the identification of specific marker proteins difficult.

Since human populations are outbred and therefore characterized by genetic variation, these differences are also observed at the protein level. In addition, factors such as the environment and lifestyle have presumably a great influence on genetic and protein backgrounds. In summary, it is critical to filter out the disease-specific changes from changes that are the result of polymorphic differences (Anderson and Anderson 2002).

The obvious targets for medical proteomics research are body fluids, including human plasma and CSF. These body fluid proteomes are ideally suited for disease diagnosis and therapeutic monitoring. Next to the abundant classical proteins, plasma also contains many tissue proteins as markers. The plasma proteome has a large dynamic range (up to 10 orders of magnitude), which makes it a difficult human proteome to analyze. This is because very abundant proteins, such as albumin (55%), and many proteins present in minute quantities are present in the same sample (Anderson and Anderson 2002). Similarly, the CSF proteome is made up of predominantly serum proteins such as albumin and, immunoglobulin. Only approximately 20% of the proteins in CSF are actually derived from the brain. These are very likely the ones that are of interest to the researcher studying psychiatric and neurological diseases. In order to improve the quality of 2D protein analysis of CSF, we are using a procedure that depletes the specimen from two major proteins, namely albumin and immunoglobulin (Thompson et al. 1998). Figure 5 shows the 2D images before and after depletion, illustrating the improved quality of the gel in the latter case.

Another limiting factor is the relatively small amount of CSF that is usually available from lumbar puncture. Typically, from 3 ml of CSF, after depletion of

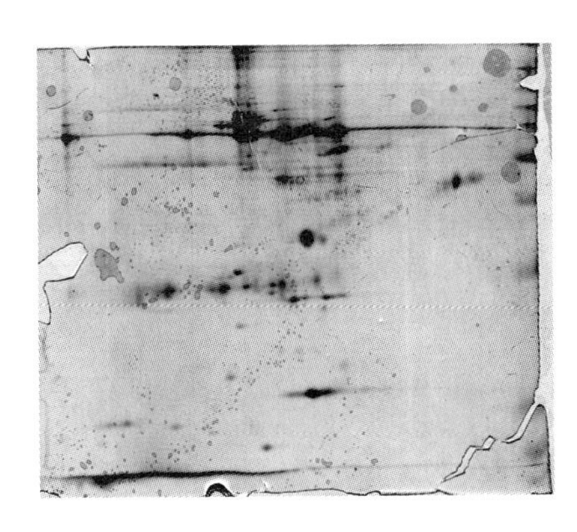

Fig. 5 2D PAGE of CSF before and after depletion with Cibacron blue/protein G affinity resins

Table 1 "Shotgun" analysis of CSF. The top 30 proteins that were identified by mass spectrometry (MS) are shown in the left column under "Reference". The score generated by the MS data-identification program SEQUEST (Eng et al. 1994) is shown in the right column

Reference	Xcorr Score
gi|4507725|ref|NP_000362.1| (NM_000371) transthyretin	2,150.4
gi|4502027|ref|NP_000468.1| (NM_000477) albumin precursor; PRO0883 protein	1,030.5
gi|4557871|ref|NP_001054.1| (NM_001063) transferrin precursor; PRO1557 protein [Homo sapiens]	746.5
gi|14577919|ref|NP_009224.1| (NM_007293) complement component 4A	260.3
gi|139641|sp|P02774|VTDB_HUMAN VITAMIN D-BINDING PROTEIN PRECURSOR (DBP)	190.3
gi|9257232|ref|NP_000598.1| (NM_000607) orosomucoid 1 precursor; Orosomucoid-1 (alpha-1-acid glycoprotein)	170.4
gi|113585|sp|P01877|ALC2_HUMAN IG ALPHA-2 CHAIN C REGION	160.4
gi|112892|sp|P04217|A1BG_HUMAN ALPHA-1B-GLYCOPROTEIN	130.3
gi|14748212|ref|XP_028322.1| (XM_028322) hypothetical protein XP_028322 [Homo sapiens]	130.3
gi|14724978|ref|XP_011125.3| (XM_011125) hypothetical protein XP_011125 [Homo sapiens]	130.3
gi|4507473|ref|NP_003235.1| (NM_003244) TGFB-induced factor	122.4
gi|410564|gb|AAB27961.1| beta-trace N-terminal [human, cerebrospinal fluid]	120.4
gi|15149461|ref|NP_149129.1| (NM_033138) caldesmon 1, isoform 1	114.5
gi|4557225|ref|NP_000005.1| (NM_000014) alpha 2 macroglobulin precursor	110.4
gi|10120703|pdb|1E3F|A Chain A, Structure Of Human Transthyretin Complexed	108.3
gi|4502005|ref|NP_001613.1| (NM_001622) alpha-2-HS-glycoprotein	100.3
gi|4758978|ref|NP_004639.1| (NM_004648) protein tyrosine phosphatase, non-receptor type substrate 1	90.4
gi|4557287|ref|NP_000020.1| (NM_000029) angiotensinogen precursor	90.3
gi|494652|pdb|1TLM|A Chain A, Transthyretin (also called Prealbumin)	90.3
gi|4502807|ref|NP_001810.1| (NM_001819) chromogranin B precursor	80.3
gi|4557485|ref|NP_000087.1| (NM_000096) ceruloplasmin (ferroxidase)	70.5
gi|4503107|ref|NP_000090.1| (NM_000099) cystatin C	70.3
gi|112908|sp|P02750|A2GL_HUMAN LEUCINE-RICH ALPHA-2-GLYCOPROTEIN (LRG)	60.5
gi|4826762|ref|NP_005134.1| (NM_005143) haptoglobin [Homo sapiens]	60.4
gi|11321561|ref|NP_000604.1| (NM_000613) hemopexin [Homo sapiens]	60.4
gi|7706781|ref|NP_057528.1| (NM_016444) zinc finger protein 226	60.3
gi|70058|pir||A2HU Ig alpha-2 chain C region—human	60.3
gi|4826908|ref|NP_005018.1| (NM_005027) phosphoinositide-3-kinase	50.4
gi|4557843|ref|NP_000440.1| (NM_000449) regulatory factor X, 5 [Homo sapiens]	42.2
gi|4757826|ref|NP_004039.1| (NM_004048) beta-2-microglobulin [Homo sapiens]	40.4
gi|14757106|ref|XP_029171.1| (XM_029171) hypothetical protein XP_029171 [Homo sapiens]	40.4

Proteins from albumin/immunoglobulin-depleted CSF (see Fig. 4) were digested with trypsin and the resultant peptides fractionated by two rounds of chromatography using cation exchange and reversed phase columns.

the major proteins, there is only enough protein material for one analysis by 2D PAGE. Alternative methods for a more sensitive analysis of this precious body fluid are therefore needed. To this end we are exploring the above-described "shotgun" technique (Link 2002) to gain a better understanding of the protein constituents that are part of CSF (Table 1; Maccarrone et al. 2004). To date only very few proteins are used in routine clinical diagnosis from body fluids such as serum and CSF. With the continued development of methods for body fluid proteome analysis, it is anticipated that many more protein diagnostic markers will be at hand in the near future.

References

Anderson NL, Anderson NG (1998) Proteome and proteomics: new technologies, new concepts, and new words. Electrophoresis 19:1853–1861

Anderson NL, Anderson NG (2002) The human plasma proteome. Mol Cell Proteomics 1:845–867

Eng J, McCormack JR, Yates J (1994) An approach to correlate tandem mass spectral data of peptides with amino acid sequences in a protein database. J Am Soc Mass Spectrom 5:976

Fenn JB, Mann M, Meng CK, Wong SF, Whitehouse CM (1989) Electrospray ionization for mass spectrometry of large biomolecules. Science 246:64–71

Gygi SP, Rochon Y, Franza BR, Aebersold R (1999a) Correlation between protein and mRNA abundance in yeast. Mol Cell Biol 19:1720–1730

Gygi SP, Rist B, Gerber SA, Turecek F, Gelb MH, Aebersold R (1999b) Quantitative analysis of complex protein mixtures using isotope-coded affinity tags. Nat Biotechnol 17:994–999

Hillenkamp F, Karas M (1990) Mass spectrometry of peptides and proteins by matrix-assisted ultraviolet laser desorption/ionization. Methods Enzymol 193:280–295

Issaq HJ, Veenstra TD, Conrads TP, Felschow D (2002) The SELDI-TOF MS approach to proteomics: protein profiling and biomarker identification. Biochem Biophys Res Commun 292:587–592

Klose J (1975) Protein mapping by combined isoelectric focusing and electrophoresis of mouse tissues. A novel approach to testing for induced point mutations in mammals. Humangenetik 26:231–243

Krishna RG, Wold F (1993) Post-translational modification of proteins. Adv Enzymol Relat Areas Mol Biol 67:265–298

Lander ES, Linton LM, Birren B, Nusbaum C, Zody MC, Baldwin J, Devon K, Dewar K, Doyle M, FitzHugh W, Funke R, Gage D, Harris K, Heaford A, Howland J, Kann L, et al (2001) Initial sequencing and analysis of the human genome. Nature 409:860:921

Lee Y-S, Mrksich M (2002) Protein chips: from concept to practice. Trends Biotechnol Suppl 20:S14S18

Link AJ (2002) Multidimensional peptide separations in proteomics. Trends Biotechnol Suppl 20:S8–S13

Maccarrone G, Milfay D, Birg I, Rosenhagen M, Grimm R, Bailey J, Zolotarjova N, Turck CW (2004) Mining the human CSF proteome by immunodepletion and shotgun mass spectrometry. Electrophoresis 25:2402–2412

O'Farrel PH (1975) High resolution two-dimensional electrophoresis of proteins. J Biol Chem 250:4007–4021

Pandey A, Mann M (2000) Proteomics to study genes and genomes. Nature 405:837–846

Petricoin EF, Ardekani AM, Hitt BA, Levine PJ, Fusaro VA, Steinberg SM, Mills GB, Simone C, Fishman DA, Kohn EC, Liotta LA (2002) Use of proteomic patterns in serum to identify ovarian cancer. Lancet 359:572–577

Thompson PM, Rosenberger C, Holt S, Perrone-Bizzozero NI (1998) Measuring synaptosomal associated protein-25 kDa in human cerebral spinal fluid. J Psychiatr Res 32:297–300

Ullrich A, Schlessinger J (1990) Signal transduction by receptors with tyrosine kinase activity. Cell 61:203–212

Venter JC, Adams MD, Myers EW, Li PW, Mural RJ, Sutton GG, Smith HO, Yandell M, Evans CA, Holt RA, Gocayne JD, Amanatides P, Ballew RM, Huson DH, et al (2001) The sequence of the human genome. Science 291:1304

Wasinger VC, Corthlas GL (2002) Proteomic tools for biomedicine. J Chromatogr B Analyt Technol Biomed Life Sci 771:33–48

Wilkins MR, Sanchez JC, Gooley AA, Appel RD, Humphery-Smith I, Hochstrasser DF, Williams KL (1996) Progress with proteome projects: why all proteins expressed by a genome should be identified and how to do it. Biotechnol Genet Eng Rev 13:19–50

Subject Index